MASSIVE STARS NEAR AND FAR

IAU SYMPOSIUM 361

COVER ILLUSTRATION:

Hubble Space Telescope image of the cluster of massive stars, Pismis 24, and part of the emission nebula NGC 6357 (Credit:NASA/ESA/Hubble and Jesús Maíz Apellániz) overlaid on an image of the Poulnabrone Dolmen in the West of Ireland

Credit: Steve Ford Elliott. Poster credit: Morgan Fraser, Sophie Murray.

IAU SYMPOSIUM PROCEEDINGS SERIES

INTERNATIONAL ASTRONOMICAL UNION

UNION ASTRONOMIQUE INTERNATIONALE

International Astronomical Union

MASSIVE STARS NEAR AND FAR

PROCEEDINGS OF THE 361st SYMPOSIUM OF THE INTERNATIONAL ASTRONOMICAL UNION BALLYCONNELL, REPUBLIC OF IRELAND 8–13 MAY 2022

Edited by

JONATHAN MACKEY

Dublin Institute for Advanced Studies, Ireland

JORICK S. VINK

Armagh Observatory and Planetarium, Northern Ireland

and

NICOLE ST-LOUIS

Université de Montréal, Canada

CAMBRIDGE
UNIVERSITY PRESS

CAMBRIDGE UNIVERSITY PRESS
University Printing House, Cambridge CB2 8BS, United Kingdom
1 Liberty Plaza, Floor 20, New York, NY 10006, USA
10 Stamford Road, Oakleigh, Melbourne 3166, Australia

First published 2024

Printed in Great Britain by Henry Ling Limited, The Dorset Press, Dorchester, DT1 1HD

Typeset in System LaTeX 2ε

A catalogue record for this book is available from the British Library of Congress Cataloguing in Publication data

This journal issue has been printed on FSC^{TM}-certified paper and cover board. FSC is an independent, non-governmental, not-for-profit organization established to promote the responsible management of the world's forests. Please see www.fsc.org for information.

ISBN 9781108490658 hardback
ISSN 1743-9213

Table of Contents

Preface to the Proceedings xvi

Editors of the Proceedings xx

List of Participants xxi

Open questions in massive star research 1
N. Langer

Chapter I: Observations of massive stars near and far

ULLYSES and Complementary Surveys of Massive Stars in the Magellanic Clouds 15
Paul A. Crowther

Massive Stars in Low-metallicity Galaxies in the CLASSY Survey 26
Danielle A. Berg, Beth L. James, Teagan King, Meaghan McDonald, Zuyi Chen, John Chisholm, Timothy Heckman, Crystal L. Martin, Dan P. Stark and The CLASSY Team

Observations of Very Low-Metallicity Massive Stars 36
Miriam Garcia

Global Properties of Star-Forming Galaxies from Ultraviolet Spectroscopy 46
Claus Leitherer, Ilyse Clark, Karla Arellano-Córdova and Danielle A. Berg

One step closer to the First Stars: +150 OB stars in the metal-poor galaxy Sextans A 52
Marta Lorenzo, Miriam Garcia and Francisco Najarro

NGC 3125-A1 revisited at higher spectral resolution with COS G160M 58
A. Wofford, A. Sixtos, L. Smith, S. Charlot and G. Bruzual

Massive stars in metal-poor dwarf galaxies are often extreme rotators 63
Abel Schootemeijer, Danny J. Lennon, Miriam Garcia, Norbert Langer, Ben Hastings and Christoph Schürmann

Constraining the behaviour of the young massive stars through interferometry 70
Emma Bordier, Abigail J. Frost, Hugues Sana and Antoine Mérand

The Gaia view on massive stars: EDR3 and what to expect from DR3 76
Jesús Maíz Apellániz, Rodolfo H. Barbá, Michelangelo Pantaleoni González, Michael Weiler, B. Cameron Reed, Román Fernández Aranda, Pablo Crespo Bellido, Alfredo Sota, Emilio J. Alfaro and J. Alejo Molina Lera

Dynamical vs Supernova Acceleration of OB Stars in the Small Magellanic Cloud 82
M. S. Oey, J. Dorigo Jones, G. D. Phillips, N. Castro, M. M. Dallas and M. Moe

Dissecting the very metal-poor galaxy SBS0335-052E with MUSE: massive stars vs. nebular HeII emission 88
C. Kehrig, J. M. Vílchez, M. A. Guerrero, J. Iglesias-Páramo, L. K. Hunt, S. Duarte Puertas and G. Ramos-Larios

Hot stars in the Gaia-ESO Survey 95
R. Blomme, S. Daflon, M. Gebran, A. Herrero, A. Lobel, L. Mahy, F. Martins, T. Morel, S. R. Berlanas, E. Gosset, J. Maíz Apellániz and GES Working Group 13

An infrared view of NGC3603 97
Raúl Castellanos, Francisco Najarro, Miriam Garcia, Lee R. Patrick, Christopher J. Evans, Elizabeth S. Bartlett and J. S. Clark

The largest and most ambitious quantitative spectroscopic analysis of B-type supergiants in the Milky Way 100
Abel de Burgos, Sergio Simon-Díaz and Miguel A. Urbaneja

Apparently isolated massive stars within Galactic HII regions 103
Diego de la Fuente, Amparo Marco, Ignacio Negueruela, Lee R. Patrick, Artemio Herrero and Stephen S. Eikenberry

MAGIC Intensity Interferometer as a powerful tool to understand massive stars 105
Irene Jiménez Martínez, V. A. Acciari, E. Colombo, J. Cortina, C. Delgado, C. Díaz, D. Fink, M. Fiori, D. Guberman, T. Hassan, E. Lyard, S. Mangano, M. Mariotti, G. Martínez, R. Mirzoyan, G. Naletto, T. Njoh Ekoume, N. Produit, M. Polo, J. J. Rodríguez, T. Schweizer, R. Walter, C. Wunderlich and L. Zampieri for the MAGIC collaboration

Fundamental parameters for 222 B-type stars for the X-shooter Spectral Library 108
C. J. K. Larkin, Kristiina Verro, Anaïs Gonneau and S. C. Trager

Alicante Survey of Massive Stars in H II regions (A-SMASHER) 111
A. Marco, I. Negueruela, D. de la Fuente, L. R. Patrick, A. Herrero and S. S. Eikenberry

Infrared studies of the massive stellar population at the Galactic Center and the inner disk 114
Francisco Najarro, J. S. Clark, Marcus Lohr, Lee R. Patrick, Miriam García, Raúl Castellanos, Gonzalo Muñoz-Sánchez and Diego de la Fuente

Unraveling massive star and binary physics in the nearby low-Z galaxy SMC as a proxy for high-redshift galaxies 116
Daniel Pauli

Revisiting the Cygnus OB associations 119
Alexis L. Quintana and Nicholas J. Wright

An exploration of Cygnus OB2 and perspectives for the upcoming WEAVE High-Resolution Cygnus Survey 122
Sara R. Berlanas and Artemio Herrero

Chapter II: Stellar atmospheres and winds

Stellar Atmospheres and Supernovae: Systematic Errors 127
D. John Hillier

Spectroscopic analyses of massive stars at different metallicities 139
Wolf-Rainer Hamann

Next generation spectroscopic analysis for large samples of massive stars 145
Joachim M. Bestenlehner

MUSE 3D spectroscopy of BA-type supergiants in NGC 300 151
Gemma González-Torà, Miguel A. Urbaneja, Norbert Przybilla, Stefan Dreizler, Martin M. Roth, Sebastian Kamann and Norberto Castro

SPAMMS: applications and use cases for the 3D spectroscopic analysis technique to study deformed massive stars 157
Michael Abdul-Masih

Mass loss and the Eddington parameter 163
Joachim M. Bestenlehner

New empirical mass-loss rates and clumping properties of massive stars 168
Calum Hawcroft, Hugues Sana, Laurent Mahy, J. O. Sundqvist, M. Abdul-Masih, J. C. Bouret, S. A. Brands, A. de Koter, F. A. Driessen and J. Puls

ISOSCELES: Grid of stellar atmosphere and hydrodynamic models of massive stars 174
Ignacio Araya, Natalia Machuca, Michel Curé and Catalina Arcos

Access platform to our database of CMFGEN spectra for atmospheric studies of massive stars 177
A. Payen-Sandoval, L. Arias, J. Zargó, A. Arrieta, C. R. Fierro-Santillán and J. Klapp

CMD for atmospheres of massive stars created by using CMFGEN synthetic spectra 179
Anabel Arrieta, Agustín Payen-Sandoval, Lorena Arias, Janos Zsargo, Celia Fierro-Santillán and Jaime Klapp

Clumping and X-rays in cool B-Supergiants 182
Matheus Bernini-Peron, W. L. F. Marcolino and A. A. C. Sander

The clumped winds of the most massive stars 184
Sarah A. Brands, Alex de Koter, Joachim M. Bestenlehner, Paul A. Crowther, Jon O. Sundqvist, Joachim Puls, Saida M. Caballero-Nieves, Michael Abdul-Masih, Florian A. Driessen, Miriam García, Sam Geen, Götz Gräfener, Calum Hawcroft, Lex Kaper, Zsolt Keszthelyi, Norbert Langer, Hugues Sana, Fabian R. N. Schneider, Tomer Shenar and Jorick S. Vink

New self-consistent wind parameters to fit optical spectra of O-type stars 190
Alex Camilo Gormaz-Matamala

Structure in stellar winds of early B type supergiants 192
Timothy N. Parsons, Raman K. Prinja, Lidia M. Oskinova and Derck L. Massa

Upper Mass-Loss Limits and Clumping in the Intermediate and Outer Wind Regions of OB stars 194
M. M. Rubio-Díez, J. O. Sundqvist, F. Najarro, A. Traficante, J. Puls, L. Calzoletti and D. Figer

Astro+: Design, construction, and scientific exploitation of a large-scale massive star spectroscopic database 197
Klaus Rübke, Amparo Marco, Ignacio Negueruela, Artemio Herrero, Sergio Simon-Diaz, Hugo Tabernero and Lee Patrick

Quantitative spectroscopy of mid B-type supergiants 200
David Weßmayer, Norbert Przybilla and Keith Butler

Chapter III: Physical processes in massive stars

Wind line variability and intrinsic errors in observational mass loss rates 205
Derck Massa and Raman Prinja

Magnetic massive stars from stellar mergers 212
Fabian R. N. Schneider, Sebastian T. Ohlmann, Philipp Podsiadlowski, Friedrich K. Röpke, Steven A. Balbus, Rüdiger Pakmor and Volker Springel

Asteroseismology reveals the near-core magnetic field strength in the early-B star HD 43317 218
D. M. Bowman, D. Lecoanet and T. Van Reeth

Constraining physical processes in pre-supernovae massive star evolution 224
Erin R. Higgins, Jorick S. Vink, Andreas Sander and Raphael Hirschi

3D effects of rotation on spectroscopic observables 230
Michael Abdul-Masih

First results of a magnetic survey of classical Cepheids 233
James A. Barron, Gregg A. Wade, Colin P. Folsom and Oleg Kochukhov

A comprehensive empirical investigation of Galactic fast-rotating O-type stars . 236
N. Britavskiy, G. Holgado, S. Simón-Díaz, J. Maíz Apellániz, M. Pantaleoni González, Y. Nazé, S. Burssens and J. J. Eldridge

Can we detect deep axisymmetric toroidal magnetic fields in stars? 238
Hachem Dhouib, Stéphane Mathis, Lisa Bugnet, Timothy Van Reeth and Conny Aerts

Apsidal motion in NGC 6231: Sounding the internal structure of massive stars . . 241
S. Rosu, G. Rauw, M.-A. Dupret, A. Noels and M. Farnir

Apsidal Motion in Massive Binaries: CPD-41° 7742, an Extreme Case? 248
S. Rosu, G. Rauw, Y. Nazé, E. Gosset and C. Sterken

Stability analysis of radial oscillation modes in massive stars 251
J. P. Sánchez Arias, S. Guha and M. Kraus

Spins of stripped B stars support magnetic internal angular momentum transport . . . 253
Christoph Schürmann, Norbert Langer, Xiao-Tian Xu and Chen Wang

Slowly-rotating nitrogen-rich O stars in 30 Doradus . . . 256
F. Tramper

Nitrogen production in population III stars . . . 259
S. Tsiatsiou, C. Georgy, S. Ekström and G. Meynet

Obliquely Rotating Massive Star Magnetospheres . . . 261
Asif ud-Doula and Stanley P. Owocki

Chapter IV: Stellar multiplicity

The Nature of Unseen Companions in Massive Single-Line Spectroscopic Binaries . . . 267
Hugues Sana, Michael Abdul-Masih, Gareth Banyard, Julia Bodensteiner, Dominic M. Bowman, Karan Dsilva, C. Eldridge, Matthias Fabry, Abigail J. Frost, Calum Hawcroft, Soetkin Janssens, Laurent Mahy, Pablo Marchant, Norbert Langer, Timothy Van Reeth, Koushik Sen and Tomer Shenar

Multiplicity of Galactic Luminous Blue Variable stars . . . 273
Laurent Mahy, Cyprien Lanthermann, Eric Gosset, Damien Hutsemékers, Jacques Kluska, Alex Lobel, Maddalena Reggiani and Hugues Sana

Hunting for red supergiant binaries: UVIT photometry of the SMC . . . 279
Lee R. Patrick, David Thilker, Danny Lennon, Luciana Bianchi, Abel Schootemeijer, Ricardo Dorda, Norbert Langer and Ignacio Negueruela

The impact of binary interaction on the main-sequence morphology of young star clusters . . . 286
Chen Wang, Norbert Langer, Abel Schootemeijer, Antonino Milone, Ben Hastings, Xiao-Tian Xu, Julia Bodensteiner, Hugues Sana, Norberto Castro, D. J. Lennon, Pablo Marchant, A. de Koter and Selma E. de Mink

The spin rate properties of Galactic massive O-type stars . . . 292
G. Holgado, S. Simón-Díaz, A. Herrero and R. H. Barbá

Near and far: a hunt for binary-interaction products . . . 299
J. Bodensteiner, H. Sana, L. Mahy, T. Shenar, G. Banyard and C. Wang

Formation of black holes in the pair-instability mass gap: Hydrodynamical simulation of a massive star collision & evolution of a post-collision star . . . 302
Guglielmo Costa, Alessandro Ballone, Michela Mapelli, Alessandro Bressan and Morgan MacLeod

Constraining the dynamical mass of the massive binary 9 Sagittarii . . . 305
Matthias Fabry, Calum Hawcroft, Abigail J. Frost, Laurent Mahy, Pablo Marchant, Jean-Baptiste Le Bouquin and Hugues Sana

HR 6819 – a post-interaction binary system originally thought to be a triple system containing a black hole 307
Abigail J. Frost, Julia Bodensteiner, Thomas Rivinius, Dietrich Baade, Antoine Merand, Fernando Selman, Michael Abdul-Masih, Gareth Banyard, Emma Bordier, Karan Dsilva, Calum Hawcroft, Laurent Mahy, Maddalena Reggiani, Tomer Shenar, Mauricio Cabezas, Petr Hadrava, Marianne Heida, Robert Klement and Hugues Sana

Contact Tracing of Massive Binary Stars 311
Jan Henneco, Fabian R. N. Schneider, Saskia Hekker and Eva Laplace

Hunting for massive binaries with a black-hole component using Gaia data . . . 313
Soetkin Janssens, T. Shenar, H. Sana, S. Faigler, N. Langer, P. Marchant, T. Mazeh, C. Schürmann and S. Shahaf

Multiplicity of Northern bright O-type stars with optical long baseline interferometry. Results of the pilot survey 316
Cyprien Lanthermann

Deciphering hard disks 319
Yaël Nazé, Gregor Rauw, Myron A. Smith and Christian Motch

Study of the Be/X-ray binary IGR J21343+4738 physical parameters based on the 15-years monitoring 321
Evgeniia Nikolaeva and Ilfan Bikmaev

X-ray emission characteristics of an O+O binary HD 93205 323
Bharti Arora and Jeewan C. Pandey

Massive stars and their large-separation low-mass companions in Sco OB1 326
T. Pauwels, M. Reggiani, A. Rainot and H. Sana

Phase-resolved spectroscopic analysis of the eclipsing black hole X-ray binary M33 X-7 329
Varsha Ramachandran

MWC 656: A Be+BH or a Be+sdO? 332
Th. Rivinius, R. Klement, S. D. Chojnowski, D. Baade, K. Shepard and P. Hadrava

Observations south of HD 5980 334
Andrés Sixtos, Aida Wofford and Antonio Peimbert

Plaskett's Star: a fundamental revision of the architecture of the system 336
E. Stacey, G. A. Wade, J. H. Grunhut, C. P. Folsom and O. Kochukhov

Evolution and final fate of massive post-common-envelope binaries 339
Dandan Wei, Fabian R. N. Schneider, Philipp Podsiadlowski, Eva Laplace, Jan Henneco, Friedrich K. Röpke and Marco Vetter

Chapter V: Stellar structure and evolution of single stars

Stellar Structure and Evolution of Massive Rotating Single Stars 343
R. Hirschi, E. Kaiser, P. Eggenberger, S. Ekström, C. Georgy, A. Maeder and G. Meynet

Entrainment in 3D hydrodynamics simulations of neon burning 353
F. Rizzuti, R. Hirschi, C. Georgy, W. D. Arnett, C. Meakin and A. StJ. Murphy

Two-dimensional models of fast rotating stars and mixing processes 359
Michel Rieutord and Damien Gagnier

Very Massive Stars: Near and Far 369
Sébastien Martinet, Georges Meynet, Sylvia Ekström, Cyril Georgy, Lionel Haemmerlé, Devesh Nandal and Raphael Hirschi

Massive star interiors revealed by gravity wave asteroseismology and high-resolution spectroscopy 376
Dominic M. Bowman

Fundamental Properties of O and B Stars with Optical Interferometry 382
Kathryn D. Gordon, Douglas R. Gies and Gail H. Schaefer

On the elusive detection of O-type stars close to the ZAMS: the empirical birthline of massive stars 388
G. Holgado, S. Simón-Díaz and L. Haemmerlé

How to make an 85 Solar Mass Black Hole 391
Ethan Winch, Jorick S. Vink, Erin Higgins and Gautham Sabhahit

Chapter VI: Cool supergiants

Variable mass loss on 100 yr timescales from massive yellow hypergiants 397
René D. Oudmaijer and Evgenia Koumpia

Analytic, Turbulent Pressure Driven Mass Loss from Red Supergiants 404
N. Dylan Kee and the MAESTRO Project

Strong lithium lines in red supergiants at different metallicities 410
Ignacio Negueruela, Javier Alonso-Santiago, Ricardo Dorda and Lee R. Patrick

On the Z-(in)dependence of the Humphreys-Davidson Limit 416
Gautham N. Sabhahit, Jorick S. Vink, Erin R. Higgins and Andreas A. C. Sander

Red Supergiants in M31: The Humphreys-Davidson limit at high metallicity . . . 422
Sarah L. E. McDonald, Ben Davies and Emma R. Beasor

Luminous red supergiants in the Magellanic Clouds 429
S. De Wit, A. Z. Bonanos, F. Tramper, M. Yang, G. Maravelias, K. Boutsia, N. Britavskiy and M. Zapartas

ASSESSing evolved massive stars in NGC 6822 and IC 10 432
G. Munoz-Sanchez, G. Maravelias, A. Z. Bonanos, F. Tramper, S. de Wit and M. Yang

Chapter VII: Unsteady mass loss

Luminous blue variables and B[e] supergiants 437
Andrea Mehner

Introducing the ASSESS project: Episodic Mass Loss in Evolved Massive Stars - Key to Understanding the Explosive Early Universe 447
A. Z. Bonanos, G. Maravelias, M. Yang, F. Tramper, S. de Wit, M. Zapartas, K. Antoniadis, E. Christodoulou and G. Munoz-Sanchez

Using machine learning to investigate the populations of dusty evolved stars in various metallicities 454
Grigoris Maravelias, Alceste Z. Bonanos, Frank Tramper, Stephan de Wit, Ming Yang, Paolo Bonfini, Emmanuel Zapartas, Konstantinos Antoniadis, Evangelia Christodoulou and Gonzalo Munoz-Sanchez

Wray 15-906 low mass Luminous Blue Variable in a pre-supernova stage 460
O. Maryeva, V. Gvaramadze, A. Kniazev and L. Berdnikov

Chapter VIII: WR stars and Stellar Feedback

Wolf-Rayet stars: recent advances and persisting problems 465
Tomer Shenar

The enigmatic winds of Wolf-Rayet stars: Results from dynamically consistent atmosphere modelling 473
Andreas A. C. Sander

X-raying massive stars and their feedback near and far 479
Lidia Oskinova

Properties of Primitive Galaxies 485
Sara R. Heap, I. Hubeny, J.-C. Bouret, T. Lanz and J. Brinchmann

Intrinsic polarization of Wolf-Rayet stars due to the rotational modulation of the stellar wind 490
Slah Abdellaoui, Jiří Krtička and Petr Kurfürst

Dependence of Wolf-Rayet wind clumping on the surface temperature 493
André-Nicolas Chené, Nicole St-Louis, Guillaume Lenoir-Craig, Anthony F. J. Moffat and Kenneth Gayley

The multiplicity of Galactic Wolf-Rayet stars 496
Karan Dsilva, Tomer Shenar, Hugues Sana and Pablo Marchant

M 1-67 and RCW 58: nebulae around WN8h stars formed through CE evolution 499
P. Jiménez-Hernández, S. J. Arthur, J. A. Toalá and A. P. Marston

Double Vision: Combining X-ray and Spectropolarimetric Observations of WR Binaries 502
R. A. Johnson, Y. Nazé, T. Panzera, A. G. Fullard, J. R. Lomax, J. L. Hoffman and K. H. Nordsieck

Investigating the impact of different velocity fields on the spectral appearance of Wolf-Rayet stars 505
R. R. Lefever, T. Shenar, A. A. C. Sander, L. Poniatowski, K. Dsilva and H. Todt

A search for correlations between Wolf-Rayet stochastic wind variability and stellar parameters 507
Guillaume Lenoir-Craig, Nicole St-Louis, Anthony F. J. Moffat and Herbert Pablo

PION: Simulating bow shocks and circumstellar nebulae around massive stars . . 510
Jonathan Mackey, Samuel Green, Maria Moutzouri, Thomas J. Haworth, Robert D. Kavanagh, Maggie Celeste, Robert Brose and Davit Zargaryan

Metal-poor Wolf-Rayet star enriching the ISM by nitrogen in the galaxy NGC 4068 513
Anastasiya Yarovova, Oleg Egorov, Alexei Moiseev and Olga Maryeva

A SITELLE view of the Galactic Wolf-Rayet Nebulae NGC6888 and NGC2359 515
N. St-Louis, C. Dumontier, M. Ruest and L. Drissen

Chapter IX: Massive star formation near and far

Formation of Very Low-metallicity Stars 521
Kazuyuki Omukai

The origin of massive stellar systems via disk fragmentation 528
G. André Oliva and Rolf Kuiper

Metallicity Dependences of Massive Star Formation from Theoretical and Observational Perspectives 534
K. E. I. Tanaka, Y. Zhang, T. Shimonishi and R. Matsukoba

Formation of Supermassive Stars and the Direct Collapse to Black Holes 539
John A. Regan

Constraints on the stellar upper mass limit from simulations of UV disk ablation 550
N. Dylan Kee and Rolf Kuiper

Binarity in Massive Young Stellar Objects 556
Robert G. Shenton, René D. Oudmaijer and Stuart L. Lumsden

In-Situ Field Massive Star Formation in the Small Magellanic Cloud 559
I. Vargas-Salazar, M. S. Oey, J. R. Barnes, X. Chen, N. Castro, K. M. Kratter and T. A. Faerber

Chapter X: Stellar endpoints and gravitational waves

The Diversity of Massive Stellar Transients Found in Sky-surveys 565
Joe Lyman

Weak Mass Loss from the Red Supergiant Progenitor of SN 2021yja 575
Griffin Hosseinzadeh

Constraining massive star mass loss through supernova radio properties 580
Takashi Moriya

Exploring a direct observational method to measure high-redshift cloud collapse timescales and GRB progenitor lifetimes 584
Jeff Cooke, Nandita Khetan, Sandra Savaglio, Jielai Zhang and Mark Suhr

Metallicity distributions of core-collapse supernovae within 30Mpc: Evidence for a lack of single massive Ib progenitors at low metallicities 590
Joanne L. Pledger, Rudi Ganss, Anne E. Sansom, Philip A. James, Joachim Puls and Stacey M. Habergham-Mawson

The evolution of binary-stripped stars: consequences for supernovae and black hole formation 596
Eva Laplace, Fabian Schneider, Philipp Podsiadlowski, Selma de Mink, Stephen Justham, Mathieu Renzo, Ylva Götberg, Rob Farmer and David Vartanyan

Low-frequency radio view of a fast-blue optical transient - AT 2018cow 602
A. J. Nayana and Poonam Chandra

How Much Hydrogen is in Type Ib and IIb SN Progenitors? 605
Avishai Gilkis and Iair Arcavi

Supplying angular momentum to the jittering jets explosion mechanism using convection in inner star layers 607
Dmitry Shishkin

Observations and modeling of two Type IIP supernovae in M61: similar yet so different 610
Rishabh Singh Teja, G. C. Anupama and D. K. Sahu

Progenitors and companions of stripped-envelope supernovae 612
E. Zapartas, O. Fox, T. Fragos, M. Renzo, J. Andrews, S. S. Bavera, A. Dotter, M. Drout, S. van Dyk, D. Milisavljevic, K. Kovlakas, D. Misra, K. Rocha, N. Smith, B. Williams and Z. Xing

Numerical modeling of IR SEDs of dusty CCSN within a Bayesian framework . . 615
Szanna Zsíros, Ilse De Looze and Tamás Szalai

Chapter XI: Future instrumentation and facilities

Ultraviolet observations of massive stars and the instrumentation to come 621
Ana I. Gómez de Castro

The CubeSpec space mission: Asteroseismology of massive stars from time-series optical spectroscopy 630
D. M. Bowman, B. Vandenbussche, H. Sana, A. Tkachenko, G. Raskin, T. Delabie, B. Vandoren, P. Royer, S. Garcia, T. Van Reeth and the CubeSpec Collaboration

Polstar : a FUV Spectropolarimetry Mission 633
Nicole St-Louis and the Polstar consortium

Author Index 637

Preface to the Proceedings

The organisers of IAU Symposium 361, "Massive Stars Near and Far", were delighted to finally welcome participants to Ireland from 8-13th May 2022. This was the first IAU Symposium hosted in Ireland since IAUS 230: "Populations of High-Energy Sources in Galaxies" in 2005, and followed a milestone event for the development of Astronomy research in Ireland: the accession of Ireland to the European Southern Observatory (ESO) in 2018. The conference was originally scheduled for May 2020 but was among the first wave of meetings that had to be postponed indefinitely following the outbreak and global spread of the COVID-19 pandemic in February and March 2020. At the time we couldn't imagine that all of our lives would be so fundamentally changed in the intervening period, or that it would take so long just to reschedule an international conference! During this time the LOC had two PhD graduations, one retirement, babies born and many career paths changed.

The Symposium was the latest in the long-standing series of Massive Star conferences, taking place roughly every four years since the first one in 1971 in Buenos Aires, Argentina (IAU Symposium 49: "Wolf-Rayet and High-Temperature Stars"). The last meeting in this series was the very successful IAU Symposium 329 in Auckland, NZ in 2016, "The lives and death-throes of massive stars". The Ireland meeting was originally to take place in Puerto Rico but, following the devastating hurricane that hit the island, the organisers realised it would be impossible to host the conference, and so a call went out in January 2018 from IAU Commission G2 for expressions of interest proposing a new venue. Once Ireland was selected as the host, we set up an all-island LOC including researchers from the Republic and Northern Ireland, and chose the Slieve Russell Hotel and Country Club, near the border in Ballyconnell, Co. Cavan, as the meeting hotel. The SOC was drawn from a diverse group of researchers from all over the world. Little did the LOC and SOC members expect that they would be serving for 4 years, through planning for two in-person meetings (the first one postponed) and one virtual preview meeting, including 3 calls for abstracts that had to be reviewed by the SOC. We are very grateful to the enthusiastic and committed members of the LOC and SOC, who remained engaged and contributed hugely to making the meetings a success.

The short-notice postponement of the conference in May 2020 was difficult for both organisers and participants as many costs had already been paid for or committed. We are grateful to the vast majority of participants for their support and patience as we refunded registration fees. As it became clear that the pandemic would continue for some time, we decided to host a Virtual Preview Meeting in the week of 3-7 May 2021, so that the early career researchers with invited and contributed talks could finally present their research to the community. This virtual meeting was a great success, regularly having 180-200 participants logged-on for the talks, with an archive of talks on a private youtube channel for those in timezones that prevented live participation. A standout memory is Dylan Kee giving his talk live from Hawaii in the small hours of the morning, and still being awake enough to ask questions to the other speakers in the session.

The organisers are grateful to the IAU for allowing us to schedule the in-person meeting for 8-13 May 2022 and renewing the travel grant offers, enabling 35 researchers from 19 countries to come to the meeting in-person who otherwise would not have had the resources to participate. In total we had 211 participants from 32 countries representing all continents except Antarctica. The meeting venue, the Slieve Russell Hotel in Ballyconnell, Co. Cavan, Ireland, is located in a rural setting with almost all participants staying at the meeting hotel, ensuring plenty of informal interaction during the week over meals and walking on the extensive grounds. The excellent conference facilities helped with social distancing and ventilation - high ceilings and a room that holds over 500 allowed us to seat everyone classroom style with generous spacing between tables, and with a large space for posters at the back of the hall.

The Symposium brought together observational and theoretical astrophysicists to discuss all aspects of massive stars: their formation, evolution, demise as supernovae and GRBs, and (for the first time in this series of Massive Stars Symposia!) gravitational waves from mergers of stellar-remnant neutron stars and black holes. The special focus for this meeting was on massive stars in the Early Universe, how they were born, lived and died, how they compare with massive stars in our Galaxy, how we can observe them and their imprint on the cosmos, and what we can learn from very low-metallicity galaxies in the local Universe. The scientific sessions were:

1.	Observations of massive stars near and far	(Chair: Jorick Vink)
2.	Stellar Atmospheres and Winds	(Chair: Nicole St Louis)
3.	Physical Processes in Massive Stars	(Chair: Dany Vanbeveren)
4.	Stellar Multiplicity	(Chair: Alceste Bonanos)
5.	Stellar Structure and Evolution of Single Stars	(Chair: Paco Najarro)
6.	Cool Supergiants	(Chair: Lidia Oskinova)
7.	Unsteady Mass Loss	(Chair: Morgan Fraser)
8.	Wolf-Rayet Stars and Stellar Feedback	(Chair: Raman Prinja)
9.	Massive Star Formation Near and Far	(Chair: Asif Ud-Doula)
10.	Stellar End-Points and Gravitational Waves	(Chair: Heloise Stevance)
11.	Future Instrumentation and Facilities	(Chair: Jonathan Mackey)

"Massive stars: Near & Far" offered the opportunity for researchers working on massive stars in the Local Universe to interact with those working on massive stars at cosmological distances. After an opening by Prof Langer on the key outstanding questions from the birth to death of massive stars, Paul Crowther discussed the HST UV Legacy Project ULLYSES that targets local massive stars in nearby low metallicity galaxies, such as the LMC and the SMC.

After this introduction to massive stars in the local Universe ("Near"), we moved up in redshift, where the rest-frame UV is shifted towards higher wavelengths. Allison Strom started her talk explaining the relevance of "Cosmic Noon" where star formation in the Universe peaked, and massive stars look very different from those nearby. The speakers on the remainder of the first day intermittently discussed observations Near & Far. The final discussion was very lively, as it became clear that the combination of projects such as ULLYSES and CLASSY will be critical for understanding the data from the first generations of massive stars with JWST (first results had not yet come out at the time of the meeting, although there was great excitement and anticipation among the community).

The second & third day focused on the physics and evolution of single & binary massive stars nearby. One highlight was the discovery of a huge Mega-Gauss interior magnetic field at the core-envelope boundary from asteroseismology, by Dominic Bowman and collaborators. It is clear that multi-dimensional simulations are helping to progress our understanding of stellar interiors and surface layers, and promise significant advances in this decade. For studying populations of massive stars and identifying unusual objects, new surveys producing huge quantities of data, together with machine learning and other analysis tools, are rapidly expanding our knowledge of stellar diversity.

On Thursday the focus shifted to the formation of massive stars, both locally and in the early Universe, including a very informative overview talk by Ralf Klessen on "The First Stars". As if massive stars and very massive stars (VMS), with masses over 100 solar masses, are not sufficiently spectacular, the discussion moved on to the formation and existence of supermassive stars of order 100,000 Solar masses, which seem to be required to explain the formation of supermassive black holes already at very early times in the Universe.

It is an exciting time for transient science, with new discoveries from ongoing surveys and new facilities coming online in the next few years. Strategies for dealing with the deluge of data so that they can be dealt with by humans was a recurring theme of the session on stellar end-points and gravitational waves on Thursday. On Friday morning Ana Gomez de Castro told us about the planned future UV observatories, a key waveband for observations of massive stars and currently covered only by HST.

The week ended, like the life of a massive star, but this time not as a supernova or direct collapse to a Black Hole but with a golf analogy by Alex de Koter. The Summary was both informative and entertaining and participants departed in good spirits with huge amounts of energy and new ideas, after talking to old and new colleagues and friends after a 2 year delay.

Despite the limitations imposed on us by COVID-19, we still were able to run some of our planned public engagement and education activities related to massive stars. Deirdre Kelleghan (artist, amateur astronomer and educator) visited 4 primary schools in the Cavan area on the 9th and 10th of May, talked to them about astronomy, stars, telescopes including JWST, and the beauty of our Universe through painting and drawings. The Armagh Observatory and Planetarium generously donated their mobile planetarium and staff time for school visits on the 11th of May, a fabulous experience for the students.

There were two conference excursions, one to the Marble Arch Caves and the Cuilcagh Lakelands UNESCO Global Geopark, and the other to the seaside surfing town of Bundoran on the Wild Atlantic Way. The organisers got great feedback on both tours. The weather had not been great for the first two days, and so everyone appreciated a fresh and windy outdoor afternoon with a good walk in the hills or along the coastal cliffs, and even (for the brave few) sea swimming and surfing lessons.

During and after the meeting we received a lot of mostly positive feedback, mainly about how great it was for people to finally meet up again in-person with colleagues and friends from around the world. The conference facilities of the Slieve Russell Hotel were excellent and many people expressed surprise and delight that the B&B room rates were so reasonable for such facilities. Despite the Covid-precautions we took, we could not escape it completely with so many participants from many different countries at different stages of the pandemic. Quite a few of the participants reported to have tested positive for COVID after the end of the conference; thankfully most people had mild symptoms and as far as we know all made a complete recovery. There is no doubt that the meeting benefited from the feel-good factor of people being at a face-to-face meeting for the first time in more than 2 years, and this was reflected in the overwhelmingly positive spirit pervading the discussions, questions and answers, and social interaction throughout the week. The secluded location with a large residential conference hotel also contributed to this.

It bears repeating that we are especially grateful to the long-serving SOC and LOC who in the end planned two in-person meetings and one virtual meeting, and to the Organising Committee of Commission G2 and divisions G, B and C who supported us. The SOC went through nearly 500 abstracts over the 3 years of their service for the original meeting, the virtual preview, and the rescheduled meeting. Significant in-kind support for administration and finances from the Dublin Institute for Advanced Studies (DIAS) is also greatly appreciated. Local organisation was a marathon team effort from researchers and students at DIAS, Armagh Observatory and Planetarium, University College Dublin, Maynooth University and Trinity College Dublin.

In the months since the meeting took place we have seen the first spectacular results from JWST, fully justifying the investment of time and resources that this observatory required, and surpassing the hopes and expectations of the community with new discoveries about massive stars in our Galaxy and at the highest redshifts. Similarly, the first results from ULLYSES and related surveys are out and are enabling new discoveries about the low-metallicity population of massive stars in the local Universe. With the new Gravitational Wave observing runs starting in mid-2023, it is an exciting time to be working in our field. We look forward with great anticipation to the next Massive Stars Symposium in a few years time, and wish the organisers a successful meeting.

The Editors: Jonathan Mackey, Jorick S. Vink, Nicole St-Louis

Editors of the Proceedings

- Jonathan Mackey (Dublin Institute for Advanced Studies, Ireland)
- Jorick S. Vink (Armagh Observatory and Planetarium, Northern Ireland)
- Nicole St-Louis (Université de Montréal, Canada)

Scientific Organising Committee

• Paula Benaglia	Instituto Argentino de Radioastronomía, Argentina
• Alceste Bonanos	National Observatory of Athens, Greece
• Alex Carciofi	Universidade de São Paulo, Brazil
• Sylvia Ekström	University of Geneva, Switzerland
• Ana Gómez de Castro	Universidad Complutense de Madrid, Spain
• Jose Groh	Trinity College Dublin, Ireland
• Claus Leitherer	Space Telescope Science Institute, USA
• Emily Levesque	University of Washington, USA
• Jonathan Mackey (co-chair)	Dublin Institute for Advanced Studies, Ireland
• Fabrice Martins	CNRS, France
• Elena Rossi	University of Leiden, Netherlands
• Nicole St-Louis	Université de Montréal, Canada
• Jorick S. Vink (co-chair)	Armagh Observatory and Planetarium, Northern Ireland
• Daniel Whalen	University of Portsmouth, UK
• Naoki Yoshida	University of Tokyo, Japan

Local Organising Committee

• Ioana Boian	Trinity College Dublin, Ireland
• Seán Brennan	University College Dublin, Ireland
• Morgan Fraser	University College Dublin, Ireland
• Anne Grace	Dublin Institute for Advanced Studies, Ireland
• Samuel Green	Dublin Institute for Advanced Studies, Ireland
• Erin Higgins	Armagh Observatory and Planetarium, Northern Ireland
• Jonathan Mackey	Dublin Institute for Advanced Studies, Ireland
• Kate Maguire	Trinity College Dublin, Ireland
• Maria Moutzouri	Dublin Institute for Advanced Studies, Ireland
• Sophie Murray	Dublin Institute for Advanced Studies, Ireland
• John Regan	Maynooth University, Ireland
• Gautham Sabhahit	Armagh Observatory and Planetarium, Northern Ireland
• Jorick S. Vink	Armagh Observatory and Planetarium, Northern Ireland

List of Participants

Abdellaoui, Slah
Abdul-Masih, Michael
Aghakhanloo, Mojgan
Agliozzo, Claudia
Aguilera-Dena, David R.
Allan, Andrew
Andrassy, Robert
Antoniadis, Konstantinos
Araya, Ignacio
Arias, Lorena
Arora, Bharti
Arrieta, Anabel
Backs, Frank
Banyard, Gareth
Barron, James
Beasor, Emma
Berg, Danielle
Bernini Peron, Matheus
Bestenlehner, Joachim
Björklund, Robin
Blanchard, Peter
Blomme, Ronny
Bodensteiner, Julia
Bonanos, Alceste
Bordier, Emma
Bostroem, Azalee
Bowman, Dominic
Brands, Sarah
Brennan, Sean
Britavskiy, Nikolay
Broekgaarden, Floor
Brose, Robert
Burssens, Siemen
Burton, Michael
Byrne, Robert
Cannon, Emily
Castellanos Sánchez, Raúl
Chene, Andre-Nicolas
Christodoulou, Evangelia
Cooke, Jeff
Costa, Guglielmo
Crowther, Paul
Davies, Ben
de Burgos Sierra, Abel
de Koter, Alex
de la Fuente Guillen, Diego
de Mink, Selma
de Wit, Stephan
de Wit, Willem-Jan
Derkink, Annelotte
Dhouib, Hachem
Dorda, Ricardo
Drout, Maria
Dsilva, Karan
Ekström, Sylvia
Fabry, Matthias
Farrell, Eoin
Fitzpatrick, Beth
Fraser, Morgan
Frost, Abigail
Gagnier, Damien
Garcia, Miriam
Gilkis, Avishai
Goetberg, Ylva
Goldberg, Jared
Gomez de Castro, Ana
González-Torà, Gemma
Gordon, Kathryn
Gormaz Matamala, Alex
Graefener, Goetz
Grasha, Kathryn
Grassitelli, Luca
Green, Samuel
Gull, Maude
Haemmerle, Lionel
Hamann, Wolf-Rainer
Hawcroft, Calum
Heap, Sara
Henneco, Jan
Herrington, Nicholas
Higgins, Erin
Hillier, John
Hirschi, Raphael
Ho, Anna
Holgado, Gonzalo
Hosseinzadeh, Griffin
Hussaini, Maryam
J. Santos, Wilton
Jacobson-Galan, Wynn
Janssens, Soetkin
Jiménez Martínez, Irene
Jiménez-Hernández, Palmira
Johnson, Rachel
Johnston, Cole

Jung, Moo-Keon
Kamp, Keefe
Karamehmetoglu, Emir
Kee, N. Dylan
Kehrig, Carolina
Keszthelyi, Zsolt
Klement, Robert
Klencki, Jakub
Klessen, Ralf
Kolaczek-Szymanski, Piotr
Kotysz, Krzysztof
Kravtsov, Ihor
Kremer, Kyle
Kummer, Floris
Labadie-Bartz, Jonathan
Langer, Norbert
Lanthermann, Cyprien
Laplace, Eva
Larkin, Cormac
Lazzarini, Margaret
Lefever, Roel
Leitherer, Claus
Lenoir-Craig, Guillaume
Lorenzo, Marta
Lyman, Joe
Mackey, Jonathan
Madura, Thomas
Mahy, Laurent
Maíz Apellániz, Jesús
Maravelias, Grigoris
Marchant, Pablo
Marco, Amparo
Martinet, Sébastien
Maryeva, Olga
Massa, Derck
McDonald, Sarah
McQuinn, Kristen
Mehner, Andrea
Moens, Nicolas
Moriya, Takashi
Morris, Patrick
Moutzouri, Maria
Munoz Sanchez, Gonzalo
Murphy, Laura
Murray, Sophie
Najarro, Francisco
Nandal, Devesh
Nazarenar Rodriguez, Cinthya
Naze, Yael
Negueruela, Ignacio
Nikolaeva, Evgeniia
O'Brennan, Hannah
O'Grady, Anna
O'Neill, David
Oey, Sally
Oliva, G. Andre
Olivier, Grace
Omand, Conor
Omukai, Kazuyuki
Oskinova, Lidia
Oudmaijer, Rene
Owocki, Stanley
Pantaleoni, Michelangelo
Parsons, Timothy
Patrick, Lee
Pauli, Daniel
Pauwels, Tinne
Penny, Laura
Pistinner, Shlomi
Pledger, Joanne
Poniatowski, Luka
Prentice, Simon
Prinja, Raman
Quintana Isasi, Alexis
R. Aguilera-Dena, David
Ramachandran, Varsha
Regan, John
Ren, Yi
Renzo, Mathieu
Reyero Serantes, Sabela
Rickard, Matthew
Ricker, Paul
Rieutord, Michel
Rivinius, Thomas
Rizzuti, Federico
Roberti, Lorenzo
Rodriguez Berlanas, Sara
Røed Ødegaard, Knut Jørgen
Rosen, Anna
Rosu, Sophie
Rubio Díez, Maria del Mar
Rübke, Klaus
Sabhahit, Gautham
Sanchez, Julieta
Sander, Andreas
Schneider, Fabian
Schootemeijer, Abel
Schultz, William
Schürmann, Christoph
Sen, Koushik
Senchyna, Peter
Shenar, Tomer

Shenton, Robert
Shishkin, Dmitry
Simón-Díaz, Sergio
Sixtos, Andres
Smartt, Stephen
Smith, Keith
Smith, Nathan
St-Louis, Nicole
Stacey, Erik
Stanway, Elizabeth
Stevance, Heloise
Stoop, Mitchel
Strom, Allison
Sushch, Iurii
Tanaka, Kei
Teja, Rishabh Singh
Telford, Grace
Tinyanont, Samaporn
Toonen, Silvia
Tramper, Frank
Tsiatsiou, Sophie
Tuquet, Selin
ud-Doula, Asif
van Son, Lieke
Vanbeveren, Dany
Vargas-Salazar, Irene
Villasenor, Jaime
Vink, Jorick S.
Wade, Gregg
Wang, Chen
Weßmayer, David
Williamson, Marc
Winch, Ethan
Wofford, Aida
Woods, Paul
Woods, Tyrone
Yarovova, Anastasiya
Yoon, Sung-Chul
Zapartas, Manos
Zsíros, Szanna
Sana, Hugues
Humphreys, Roberta
A.J., Nayana
Dorn-Wallenstein, Trevor
Wei, Dandan
Massey, Philip
Neugent, Kathryn
Cantiello, Matteo

Massive Stars Near and Far
Proceedings IAU Symposium No. 361, 2024
J. Mackey, J. S. Vink, & N. St-Louis, eds.
doi:10.1017/S1743921322003088

Open questions in massive star research

N. Langer

Universität Bonn and Max Planck Institut für Radioastronomie Bonn
email: nlanger@astro.uni-bonn.de

Abstract. In discussing open question in the field of massive stars, I consider their evolution from birth to death. After touching upon massive star formation, which may be bi-modal and not lead to a zero-age main sequence at the highest masses, I consider the consequences of massive stars being close to their Eddington limit. Then, when discussing the effects of a binary companion, I highlight the importance of massive Algols and contact binaries for understanding the consequences of mass transfer, and the role of binaries in forming Wolf-Rayet stars. Finally, a discussion on pair instability supernovae and of superluminous supernovae is provided.

Keywords. Massive stars, Binary stars, Supernovae, Compact objects

1. Introduction

For the task to get us started with this scientific conference on Massive Stars, which includes many different aspects of research, I decided to highlight what can be considered as important open questions in the field. I quickly realised that it is impossible to do this in a complete, or even in a comprehensive way. This contribution does therefore not attempt to review the state-of-the art or the recent progress in any of the topics discussed below. Specialists on the mentioned topics may be missing depth and detail. This paper is rather written for the non-specialists — which perhaps we all are in most aspects of massive star research.

The intention here is to identify and discuss unsolved problems which are significant not only for the listed topic but also for a wider scope. Clearly, there are many more of them than the ones brought forward here. While I tried to remain within the topical boundaries set by the conference program, it appears again impossible to cover all the contributing science fields, and it was also hard to fight my personal bias. I must therefore happen that readers feel important questions are missing here, and my hope is that they enjoy reading this paper nevertheless.

2. How do massive stars form?

The question of massive star formation is important because the answer sets the initial state for all of the further evolution. Any mistake in the assumptions on the initial state of massive stars may propagate. In stellar evolution calculations, mostly one of two different approaches are followed. In one of them, chemically homogeneous stellar models are initially put to hydrostatic and thermal equilibrium, on the so called zero age main sequence (e.g., Brott et al. 2011, Ekström et al. 2012). Alternatively, the models start with a fully convective structure in hydrostatic equilibrium on their Hayashi line, and further-on contract to eventually ignite hydrogen burning (e.g., Choi et al. 2016). Both approaches lead to nearly identical internal structures. However, they are pursued because they are simple, but they ignore at least two longstanding problems.

One problem is that young massive stars may still be accreting at the time when hydrogen burning ignites (cf., fig. 1 in Yorke 1986). This issue may be related to the question why we see essentially no young massive stars (see Sect. 2.3). The second problem is that the vast majority of massive stars are born in binary or higher order multiple systems. When stars are born near the Hayashi line, they are big and do not allow for short orbital periods. We discuss this issue further in Sect. 2.1. For an in-depth review on massive star formation see Zinnecker & Yorke (2007).

2.1. *Which fraction of binaries merge as pre-main sequence stars?*

One may wonder why any pre-main sequence binaries should merge, because while after core hydrogen ignition the stars become bigger in size, the radii of pre-main sequence stars generally decrease (Siess et al. 2000). However, star formation occurs in a dense environment, and residual accretion as well as tides induced by circum-binary material may lead to shrinking orbits in pre-main sequence binaries (Korntreff et al. 2012). Based on a simple toy model which includes these processes, Tokovinin & Moe (2020) predict 33% of all early B type binaries to merge during their pre-main sequence evolution. Their model explains the observed B star binary fraction as well as the prevalence of short orbital periods in B type pre-interaction binaries (Villaseñor et al. 2021).

A high fraction of pre-main sequence binary mergers implies a bi-modality in massive stars. One type would form via disk accretion, the other would gain a major fraction of their mass through a binary merger. The question is whether we can recognise this dichotomy in the long lived core-hydrogen burning state. One stellar quantity which may contain a memory of the formation process is the stellar spin. While merger products could have a large spin due to the surplus of orbital angular momentum (de Mink et al. 2013), Schneider et al. (2019, 2020), via detailed numerical simulations, find that outflows in the puffed-up post-merger state drain the merger product of angular momentum. Merger products could therefore be exceptionally slow rotators.

Indeed, in field stars, a bi-modal rotational velocity distribution has been found in early B (Dufton et al. 2013) and A type (Zorec & Royer 2012) stars. The same is true in rich, young open star clusters, where this phenomenon gives rise to a split main sequence band in high precision colour-magnitude diagrams (Milone et al. (2013, 2018); Marino et al. 2018). Star counts identify 10...30% of the main sequence stars as slow rotators. Here, the idea of different formation mechanisms of stars in the two main sequence components is supported by the fact that their mass functions are distinctly different (Wang et al. 2022).

2.2. *What causes large scale magnetic fields in stars?*

Between 5% and 10% of the early type main sequence stars has been found to host large scale magnetic fields (Grunhut et al. 2017, Schöller et al. 2017). While several formation pathways for these magnetic fields have been suggested (cf., Langer 2014), the idea that large scale, stable fields can be created in stellar mergers has recently gained momentum (Ferrario et al. 2009, Tutukov & Fedorova 2010, Schneider et al. 2016, 2019, Pelisoli et al. 2022), and is supported by the low binary fraction of magnetic stars (Kochukhov et al. 2018).

However, further work is needed to verify this idea. The number of slow rotators amongst early B type stars, for example, seems clearly larger than the number of magnetic early B stars. Sticking to the merger idea, this could imply that not each merger generates a long-term stable B-field, perhaps because stability requires the intertwined toroidal-poloidal large scale topology found by Braithwaite & Spruit (2004), which may or may not form as remnant of the decay of an unstable turbulent or toroidal field generated during the merger. Alternatively, Fossati et al. (2016) suggested that large scale magnetic fields in massive stars may decay on a timescale comparable to their main sequence lifetime. Both scenarios have different consequences for the late time and post-core collapse evolution of massive stars, and magnetic stellar evolution models to explore them are highly demanded (Fuller et al. 2019, Keszthelyi et al. 2019, Takahashi & Langer 2021).

2.3. *Are young very massive stars embedded?*

During the last decade or so, large spectroscopic surveys of massive stars have shown a dearth of young stars above $\sim 20\,M_\odot$. This has been found in the Galaxy (Castro et al. 2014, Holgado *et al.* 2020), the LMC (Schneider et al. 2018), and the SMC (Schootemeijer et al. 2021) to comparable extent. Perhaps, the missing stars are still embedded in their birth clouds or even accreting (cf., Sect. 2.1). However, to solve the problem, they need to remain hidden at optical wavelengths for roughly half their total lifetime, but a corresponding population of IR bright stars seems also not observed.

Ongoing accretion during a fraction of their core hydrogen burning evolution would severely alter the internal composition profiles of massive main sequence stars. They would encounter the problem of rejuvenation, as the mass gainers in mass transferring close binaries (Braun & Langer 1995; cf., Sect. 4.2), introducing a large uncertainty on their post-main sequence evolution.

3. Which effects has the Eddington limit?

Massive main sequence stars are known to follow a steeper than linear mass-luminosity relation, which means that above a certain mass limit they must approach or even exceed the Eddington limit. Usually, the Eddington limit is considered at the stellar surface, where it may have strong consequences for the stellar wind (Owocki et al. 2017). When evaluating the Eddington limit in the stellar envelope, Langer et al. (2012) showed that solar metallicity main sequence stars above $\sim 30\ M_\odot$ are expected to exceed it. However, instead of posing an upper mass limit, perhaps stars above the Eddington limit are inflated and variable.

3.1. *Do stellar envelopes inflate near the Eddington limit?*

A super-Eddington radiative luminosity in a stellar envelope does not imply the loss of hydrostatic equilibrium. Instead, in corresponding stellar structure models, a density inversion produces an inwards directed force compensating for the excess acceleration produced by the radiation. Ishii et al. (1999) investigated zero-age main sequence and helium main sequence models and found this effect to lead to inflated envelopes.

Petrovic et al. (2006) found inflated envelopes in models of massive Wolf-Rayet stars, and showed that the radius increase depends on the stellar wind mass loss rate. Gräfener et al. (2012), and Sanyal et al. (2015, 2017) showed the extensive degree of envelope inflation in models of evolved massive stars, where a radius increase of up to a factor of 40 has been found. Notably, the effects of density inversion and envelope inflation have also been retrieved in 3D-radiation-hydro models of Jiang et al. (2015), and are thus not an artefact of 1D models. We also need 2D-models of rotating stars near the Eddington limit in order to understand how the centrifugal force affects the envelope inflation.

3.2. *Are Luminous Blue Variable stars near their Eddington limit?*

The empirical verification of envelope inflation is difficult. The most direct way may relate to stability analyses of inflated models. Many inflated stellar models turn out to be pulsationally unstable (Grassitelli et al. 2016). However, the corresponding brightness fluctuations may be small and difficult to observe. On the other hand, Sanyal et al. (2015) found that models of evolved massive stars with a radius increase due to inflation of more than a factor of three where all found beyond the hot edge of the empirical LBV instability strip identified by Smith et al. (2004).

Indeed, Grassitelli et al. (2021) find strong radius and brightness variations resembling those of S Doradus type LBVs in time-dependant stellar evolution models which exhibit strong inflation. The variations occurred on a decade-long timescale, which is much larger than the dynamical timescale of those models, and relates to the thermal timescale of the inflated envelopes of these models. Grassitelli et al. showed that for plausible stellar wind mass loss recipes, these models can never reach a thermal equilibrium configuration, but keep undergoing cyclic large-amplitude variations. These models support the long-standing idea of a connection between LBVs and the Eddington limit (e.g., Lamers & Fitzpatrick 1988).

Whether the great LBV eruptions à la η Carina also relate to the Eddington limit is not clear yet. In a recent model, Hirai et al. (2021) explain η Car's great eruption as the result of a stellar merger in a very massive triple system. More 3D models of binary mergers (Schneider et al. 2019) are needed to systematically study their mass outflows, and to investigate their features when they occur near the Eddington limit.

4. Which effects has a binary companion?

The majority of massive stars is born with a binary companion in such a close orbit that Roche lobe overflow and mass transfer is unavoidable (Sana et al. 2012, Moe & Di Stefano 2017). As mass is the single most important parameter of stellar evolution, mass transfer affects the evolution of both components drastically. Whereas the evolution of the mass donor can be predicted with large certainty (Sect. 4.3), the evolution of the mass gainer depends strongly on several uncertain physical processes, in particular on the mass transfer efficiency (Sect. 4.1) and on rejuvenation (Sect. 4.2).

4.1. *How efficient is mass transfer?*

When mass is transferred from the donor to the mass gainer, either by direct impact of the accretion stream (Dessart et al. 2003) or via the formation of an accretion disk (Lin & Pringle 1976), either not all of the transferred matter may end up on the mass gainer or the mass gainer may immediately eject some of it. The mass transfer efficiency is the ratio between the amount of mass which remains on the mass gainer to the amount of mass transferred by the mass donor. While it is naturally a time dependant quantity (Langer 2012), often it is used as the time average over a mass transfer episode. Conservative evolution implies a mass transfer efficiency of one, and for non-conservative evolution, one needs to specify the angular momentum carried by any material which leaves the binary system in addition to the mass transfer efficiency.

As theoretical models for the mass transfer efficiency are sparse (e.g., Oka et al. 2002), in most evolutionary models of massive binaries a constant value for the mass transfer efficiency is adopted, which is then independent of time and of the binary parameters. This is problematic because we know that some binary systems evolved through mass transfer rather conservatively while other did not. E.g., some massive Algol binaries with accretor-to-donor mass ratios above three require near-conservative mass transfer (e.g., Sen et al. 2022), while we know WR+O star binaries which require highly non-conservative evolution (Petrovic et al. 2005). In recent large grids of detailed binary evolution models, we explored a mass transfer scheme in which the accretor stops to take in mass once it is spun-up to critical rotation (Langer 2012). This leads to very

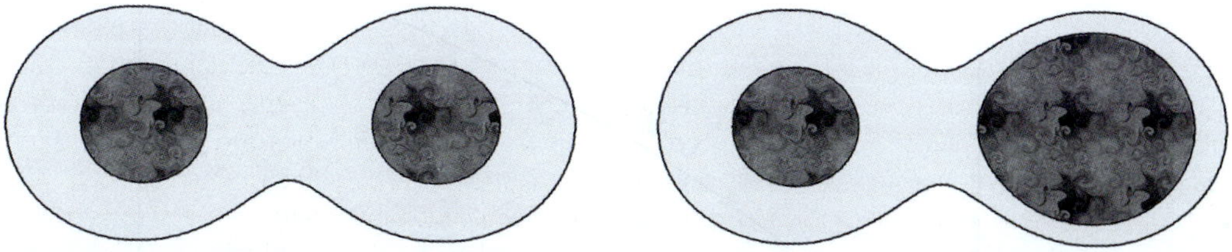

Figure 1. Schematic view of contact binaries in a thermal equilibrium state where the masses of both component stars, and thus their Roche volumes, are equal. In the left figure, complete rejuvenation is assumed, implying the same convective core masses and radii. The orbital period and the mass ratio (q=1) will not change further-on. In the right figure, it is assumed that rejuvenation did not occur, and therefore the core masses and radii are unequal, implying different timescales of envelope expansion for both components.

inefficient mass transfer for initially wide binaries in which the mass gainer is spun-up quickly (Packet 1981), but to rather conservative evolution in short period binaries where tides can prevent or slow down the spin-up. Detailed comparisons with the observed populations of massive Algols, Be stars, Be/X-ray binaries and Wolf-Rayet binaries imply that our mass transfer scheme is dissatisfactory for binaries with initial primary masses near $10\,M_\odot$, which produce most of the Be/X-ray binaries (see also Vinciguerra et al. 2020). On the other hand, it leads to results which agree well with observations for initial primary masses above $\sim 20\,M_\odot$, which produce most of the WR stars (Pauli et al. 2022) and black holes (Langer et al. 2020). A refined mass transfer scheme which accounts for the strong mass dependence of the mass transfer efficiency appears therefore warranted.

4.2. *Do mass gainers rejuvenate?*

As mentioned above, there is strong observational evidence for a high mass transfer efficiency in many binary systems. In other words, some mass gainers — for which we consider here only core hydrogen burning stars which is the most common case — increase their mass substantially. As more massive stars have more massive convective cores, accretion generally leads to an increase of the core mass. This implies an increase of the hydrogen mass fraction in the core, also called rejuvenation. To what extent mass gainers are rejuvenated is one of them most important and most longstanding unsolved questions in binary evolution. Often, it is assumed that the chemical composition profile of the mass gainer will adjust to that of a single star of the same mass and core hydrogen abundance (Hellings 1983). The problem is that any mixing of hydrogen into the core has to overcome the strongly stabilising mean molecular weight gradient. This process is likely controlled by semiconvection (Braun & Langer 1995), a thermal timescale mixing process of uncertain efficiency (e.g., Schootemeijer et al. 2019). It turns out that contact binaries have the potential to elucidate the issue.

Menon et al. (2021) have computed a large grid of short period binary evolution models with MESA, which covers the initial parameter space of binaries which encounter a nuclear timescale contact phase, during which both stars overfill their Roche volumes (Fig. 1). These models assumed conservative mass transfer, and efficient semiconvection and thus complete rejuvenation. When comparing to observations of massive contact binaries, Menon et al. found that many more contact binaries with a mass ration close to one were predicted than observed. For complete rejuvenation, when contact binaries encounter a mass ratio of one, there will be no further mass transfer, as both stars have the same internal structure, and thus expand at the same rate. Also the orbital period P will not change any more, leading to very large predicted values of $P/\dot{P}$, which is again not observed (Abdul-Masih et al. 2022). While the models of Menon et al. are the

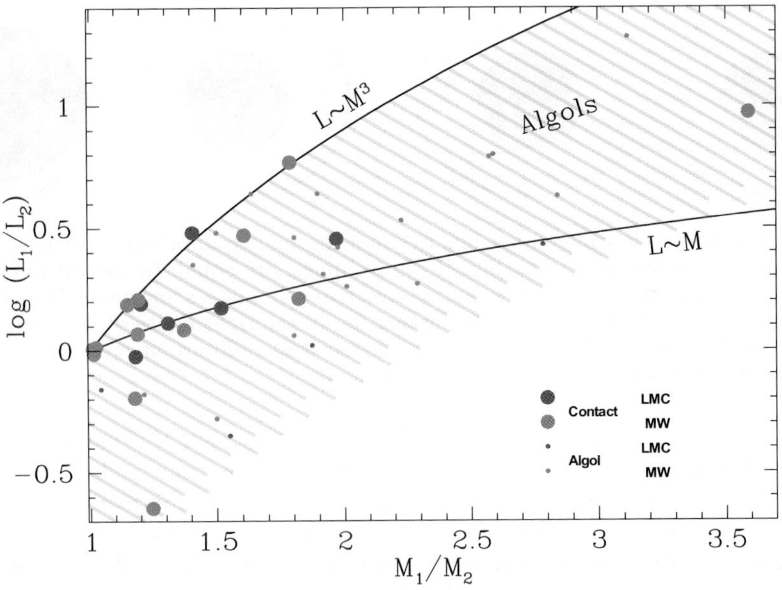

Figure 2. Mass and logarithmic luminosity ratios of the components of observed short period binaries. Subscript 1 refers to the more massive components, which are in the mass range $\sim 8\,M_\odot \ldots 40\,M_\odot$. The plot shows LMC (red) and MW (blue) systems, observationally classified as contact (large symbols) or semi-detached binaries (small symbols), as listed in tables 1 and 2 of Menon et al. (2021), and in tables 1 and 2 of Sen et al. (2022), respectively. The upper solid line indicates the locus of binaries in which the stars follow the single star mass luminosity relation, here assumed as $L \sim M^3$, es expected for pre-interaction binaries. The lower solid line indicates the relation $L \sim M$, which has been proposed to hold for contact binaries. The hatched area marks the mass-luminosity ratio range for Algol binaries (semi-detached) in the considered primary mass range as predicted by Sen et al. (2022; see their figure 13).

first of their kind, the contact scheme in MESA does not account for a possible energy transfer inside the common envelope. I.e., both components of the MESA models can have different effective temperatures, whereas the observed temperatures of the components of contact binaries are similar Abdul-Masih et al. 2021, as theoretically expected (Lucy 1986, Tassoul 2000). In any case, while this implies that both components of equal mass contact binaries are expected to have nearly identical luminosities and radii, they would only maintain a mass ratio of one if their internal structures were also identical, i.e., if complete rejuvenation would apply. On the other hand, a single $P/\dot{P}$-measurement well above the nuclear timescale would imply that complete rejuvenation works in nature, unless the observed binary was born with a mass ratio of one (Fig. 1).

In contact binaries, the von Zeipel theorem and Roche geometry also demand the luminosity ratio to equal the mass ratio of both components. This may be used as a consistency check in contact binary systems. In Fig. 2, we show the luminosity ratios in binaries which were observationally classified as contact binaries (large dots). While some are indeed close to the $L \sim M$-relation, others are very close to the $L \sim M^3$-relation, indicating they might be pre-interaction systems. Some systems are also found in between both relations, or even at luminosity ratios below one. This figure suggests that either the simple theoretical expectation is wrong, or that sometimes pre-interaction binaries with nearly Roche-lobe filling components or semi-detached Algol binaries are mistakenly classified as contact binaries. The accurate determination of the system parameters of a larger sample of such binaries will allow to settle this important question in the near future.

4.3. *Which fraction of the Wolf-Rayet stars were stripped by a companion?*

The question raised above is discussed controversially since long time (cf., T. Shenar, this volume). Single star evolution calculations have been explaining the properties of the observed WR populations with some success, in particular when strong rotationally induced mixing as well as strong stellar wind mass loss have been adopted (Maeder 1983, Langer et al. 1994, Meynet & Maeder 2005). However, it has been realised during the last decade that the majority of massive stars is born in close binary systems. As such, it appears unavoidable that binary stripping produces a significant fraction of the observed Wolf-Rayet stars.

In fact, in this respect, the predictions from binary evolution models are more robust than those from single star models. The mass donors are found to simply loose most of their H-rich envelope due to mass transfer, while the mass loss of single stars, in particular in the LBV and RSG stages, is still very uncertain. In a recent large binary model grid for LMC metallicity produced with MESA, Pauli et al. (2022) found the evolution of the mass donors to agree very well with comparable models produced with BPASS (Eldridge et al. 2017). The models of Pauli et al. reproduce the number and luminosity distributions of the different types of Wolf-Rayet stars — which presumably from a complete population (Breysacher et al. 1999, Hainich et al. 2014) — to significant detail, without the need to include Wolf-Rayet stars produced by single stars.

A fraction of the observed WR stars may also originate from the initially less massive star in binaries. Once the initially more massive star has evolved via its Wolf-Rayet phase into a compact object, it may strip off the envelope of its companion either in a stable mass transfer phase or in a common envelope evolution. The fraction of Wolf-Rayet stars produced in this way is uncertain, mostly due to uncertainties in the formation kicks of compact objects, and in the common envelope physics. However, they may have black hole companions, and are good candidates to produce merging compact objects (van den Heuvel et al. 2017).

5. What is the origin of peculiar supernovae?

A large fraction of peculiar supernovae, and supernova subtypes, is likely originating from close binary evolution (e.g., Yoon et al. 2010, Zapartas et al. 2019, Schneider et al. 2021). Sorting out the connections between observed supernovae and their progenitor evolution will keep refining our understanding of binary evolution. Here, I want to point out two questions relating to non-standard explosion mechanisms.

5.1. *Where are the pair-instability supernovae?*

In contrast to iron core collapse supernovae, where only about 1% of the available energy is transformed into the kinetic energy of the supernova, the predictions of pair-instability supernova (PISN) models have been extremely robust since the discovery of the instability (Rakavy & Shaviv 1967; Fraley 1968). There is no doubt that helium cores with masses above $\sim 60\,M_\odot$ will collapse before oxygen ignition, and produce a supernova due to explosive oxygen burning, as long as their mass is smaller than $\sim 135\,M_\odot$ (Heger & Woosley 2002, Heger et al. 2003), However, as yet, no observed supernova has been doubtlessly identified as PISN.

Obviously, the high required mass would make them rare events. In addition, at solar metallicity, stellar wind mass loss would likely reduce the final helium core mass well below 60 $M_\odot$, whatever the initial mass might have been (Yusof et al. 2022). On the other hand, stellar evolution models including the then most recent stellar mass loss prescriptions lead Langer et al. (2007) to conclude that PISNe could occur at metallicies of up to $Z_\odot/3$, which would result in one PCSN per 1000 core-collapse supernovae.

Notably, based on more recent mass loss rate prescriptions, Higgins et al. (2021) find a PCSN metallicity threshold of only half of solar metallicity. There may be two ways to resolve this apparent contradiction.

Firstly, it is generally believed that PISNe are very bright and easy to spot (cf., Sect. 5.2). However, they may result from compact WR/He-stars, and up to He-core masses of $80\,M_\odot$ they produce essentially no radioactive ^{56}Ni (Heger & Woosley 2002). E.g., Herzig et al. (1990) find a peak magnitude of only $M_V \simeq 14\,$m in a PISN model from a $61\,M_\odot$ Wolf-Rayet star. Secondly, it is possible that a star undergoing a PISN is still covered by a massive and extended hydrogen envelope. This would drastically increase the peak brightness of the explosion. E.g., for such a case, with a helium core mass of $70\,M_\odot$, Kozyreva et al. (2014) find a peak magnitude of $M_V \simeq 19\,$m. However, the resulting light curve strongly resembled that of a bright Type II-plateau supernova. Therefore, there might be PISNe not recognized as such amongst the observed Type II supernovae, which is are known to show a large diversity.

For the upper mass range of PISNe, i.e., for helium core masses above $\sim 100\,M_\odot$, the nickel masses produced in the explosion models exceed several solar masses. The corresponding explosions are extremely bright and fall into the regime of superluminous supernovae. Perhaps, those are restricted to the Early Universe (Whalen et al. 2013). Near the lower mass limit, the collapse induced by $e^\pm$-production becomes too weak to lead to explosive burning which can disrupt the star. Instead, in so called pulsational pair instability supernovae (which end up forming black holes) only parts of the envelope of the star may be ejected (Woosley 2017, Marchant et al. 2019). Observational evidence for these events is accumulating (Woosley et al. 2007, Mauerhan et al. 2013, Pastorello et al. 2013, Woosley & Smith 2022).

New supernova surveys may elucidate this question in the near future. They will also test the predictions of chemically homogeneous evolution in very massive binaries (de Mink et al. 2009), which leads to a viable channel to produce pairs of merging massive black holes (Marchant et al. 2016, Mandel & de Mink 2016). However, it is also producing pair instability supernovae, in the helium core mass range mentioned above. The cosmic population synthesis of these events by du du Buisson et al. (2020) shows that even dim PISN should be detected with rates above one per year in surveys with magnitude limits of $\sim 22\,m$.

5.2. *What causes superluminous supernovae?*

During the last two decades, supernovae have been discovered which were more than 100 times more luminous that previously known ones, i.e., displaying $\sim 10^{51}$ erg in optical photons (cf., Quimby et al. 2011), called superluminous supernovae (SLSNe). Three different types of explanations for their high luminosity have been put forward. The first is a high amount of ^{56}Ni, i.e., several solar masses, which seems possible only in the context of PISNe (Sect. 5.1). The second explanation assumes the supernova ejecta to ram into a dense wall of gas, i.e., solar masses of nearby circumstellar matter (as perhaps produced by pulsational pair instability supernovae). This breaks the supernova ejecta and converts their kinetic energy ($\sim 10^{51}$ erg for a typical supernova) into heat and photons. The third explanation assumes that the core collapse leading to the supernova produces an extremely rapidly rotating and extremely magnetised neutron star, a millisecond magnetar, which injects part of its rotational energy ($\sim 10^{52}$ erg) into the expanding supernova ejecta (Sukhbold & Woosley 2016).

For the most common type, the Type Ic SLSNe, the magnetar scenario appears to be favoured. Observationally, they appear to be related to the long-duration gamma-ray bursts, as both types of explosions have a preference to occur in low metallicity dwarf galaxies (Lunnan et al. 2014, Schulze et al. 2018), and possibly also to fast radio bursts (FRBs; Metzger et al. 2017). In an analysis of 38 Ic SLSNe, Nicholl et al. (2017) derive the required neutron star spins and magnetic field strengths within the magnetar model, as well as the ejecta masses and kinetic energies of the supernovae, which were found to correspond remarkably well to the parameters obtained from models of low-metallicity, chemically homogeneously evolving massive stars (Aguilera-Dena et al. 2018, 2020), which have also been put forward as progenitors of long-duration gamma ray bursts (Yoon & Langer 2005, Woosley & Heger 2006, Yoon et al. 2006).

However, so far chemically homogeneously evolving stars have not yet been unambiguously identified, which leaves a big open question for this scenario. Furthermore, there are other types of SLSNe, most remarkably the hydrogen-rich SLSNe, for which the origin is not yet clarified (e.g., Kangas et al. 2022). With several large upcoming supernovae surveys in the UV, in the optical/IR, at radio wavelengths, and the resumption of the gravitational wave searches in the near future, it is likely that transient universe will keep to amaze us.

References

Abdul-Masih, M., Sana, H., Hawcroft, C., et al., 2021, *A&A*, 651, A96
Abdul-Masih, M., Escorza, A., Menon, A, et al., 2022, *A&A*, in press, arXiv:2208.01671
Aguilera-Dena, D. R., Langer, N., Moriya, T. J., et al., 2018, *ApJ*, 858, 115
Aguilera-Dena, D. R., Langer, N., Antoniadis, J., et al. 2020, *ApJ*, 901, 114
Braithwaite, J., Spruit, H.C., 2004, *Nature*, 431, 891
Braun, H., Langer, N., 1995, *A&A*, 297, 483
Breysacher, J., Azzopardi, M., Testor, G., 1999, *A&AS*, 137, 117
Brott, I., de Mink, S. E., Cantiello, M., et al., 2011, *A&A*, 530, A115
Castro, N.; Fossati, L.; Langer, N., et al., 2014, *A&A*, 570, L13
Choi, J., Dotter A., Conroy, C., *et al.* 2016, *ApJ*, 823, 102
de Mink, S. E., Cantiello, M., Langer, N., et al. 2009, *A&A*, 497, 243
de Mink, S. E., Langer, N., Izzard, R. G., et al., 2013, *ApJ*, 764, 166
Dessart, L., Langer, N., Petrovic, J., 2003, *A&A*, 404, 991
du Buisson, L., Marchant, P., Podsiadlowski, P., et al., 2020, *MNRAS*, 499, 5941
Dufton, P. L., Langer, N., Dunstall, P. R., et al., 2013, *A&A*, 550, A109
Ekström, S., Georgy, C., Eggenberger, P., *et al.* 2012, *A&A*, 537, A146
Eldridge, J. J., Stanway, E. R., Xiao, L., et al., 2017, *PASA*, 34, 58
Ferrario, L., Pringle, J. E., Tout, C. A., Wickramasinghe, D. T., 2009, *MNRAS*, 400, L71
Fraley, G. S., 1968, *Ap&SS*, 2, 96
Fossati, L., Schneider, F. R. N., Castro, N., 2016, *A&A*, 592, A84
Fuller, J., Piro, A. L.. Jermyn, A. S., 2019, *MNRAS*, 485, 3661
Gräfener, G., Owocki, S. P., Vink, J. S., 2012, *A&A*, 538, A40
Grassitelli, L., Chene, A.-N., Sanyal, D., et al., 2016, *A&A*, 590, A12
Grassitelli, L., Langer, N., Mackey, J, et al., 2021, *A&A*, 647, A99
Grunhut, J. H., Wade, G. A., Neiner, C., 2017, *MNRAS*, 465, 2432
Hainich, R., Rühling, U., Todt, H., et al., 2014, *A&A*, 565, A27
Heger, A., Woosley, S. E., 2002, *ApJ*, 567, 532
Heger, A., Fryer, C. L., Woosley, S. E., et al., 2003, *ApJ*, 591, 288
Herzig, K., El Eid, M. F., Fricke, K. J., et al., 1990, *A&A*, 233, 462
Hellings, P., 1983, *Ap& SS*, 96, 37
Higgins, E. R., Sander, A. A. C., Vink, J. S., Hirschi, R., 2021, *MNRAS*, 505, 4874
Hilditch, R. W., Howarth, I. D., Harries, T. J., 2005, *MNRAS*, 357, 304
Hirai, R, Podsiadlowski, P., Owocki, S P., et al., 2021, *MNRAS*, 503, 4276
Holgado, G.; Simon-Diaz, S.; Haemmerle, L., et al., 2020, *A&A*, 638, A157
Ishii, M., Ueno, M, Kato, M, 1999, *PASJ*, 51, 417
Jiang, Y.-F., Cantiello, M., Bildsten, L., et al., 2015, *ApJ*, 813, 74

Kangas, T., Yan, Lin, Schulze, S., et al., arXiv:2207.12059
Keszthelyi, Z., Meynet, G., Georgy, C., et al., 2019, *MNRAS*, 485, 5843
Kochukhov, O., Johnston, C., Alecian, E., 2018, *MNRAS*, 478, 1749
Korntreff, C., Kaczmarek, T., Pfalzner, S., 2012, *A&A*, 543, A126
Lamers, H. J. G. L. M., Fitzpatrick, E. L., 1988, ApJ, 324, 279
Langer, N., 2012, *ARAA*, 50, 107
Langer, N., 2014, *IAUS*, 302, 1
Langer, N., Hamann, W.-R.; Lennon, M., et al., 1994, *A&A*, 290, 819
Langer, N., Norman, C. A., de Koter, A., et al., 2007, *A&A*, 475, L19
Langer, N., et al., 2012, in: *Wolf-Rayet Stars*, W.-R. Hamann et al., eds., Univ.-Verlag Potsdam, p.241
Langer, N., Schürmann, C., Stoll, K., et al., 2020, *A&A*, 638, A39
Lin, D. N. C., Pringle, J. E., 1976, *IAUS*, 73, 237
Lucy, L.B., 1986, *ApJ*, 151, 1123
Lunnan, R., Chornock, R., Berger, E., et al., 2014, *ApJ*, 787, 138
Maeder, A., 1983, *A&A*, 120, 113
Mandel, I., de Mink, S. E., 2016, *MNRAS*, 458, 2634
Marino, A. F., Przybilla, N., Milone, A. P., et al., 2018, *AJ*, 156, 116
Marchant, P., Langer, N., Podsiadlowski, P., et al., 2016, *A&A*, 588, A50
Marchant, P., Renzo, M., Farmer, R., et al., 2019, *ApJ*, 882, 36
Mauerhan, J. C.; Smith, N.; Filippenko, A. V., et al., 2013, *MNRAS*, 430, 1801
Menon, A., Langer, N., de Mink, S. E., *et al.* 2021, *MNRAS*, 507, 5013
Metzger, B. D., Berger, E., Margalit, B., 2017, *ApJ*, 841, 14
Meynet G., Maeder, A., 2005, *A&A*, 429, 581
Milone, A. P., Bedin, L. R., Cassisi, S., et al., 2013, *A&A*, 555, A143
Milone, A. P., Marino, A. F., Di Criscienzo, M., 2018, *MNRAS*, 477, 2640
Moe, M., Di Stefano, R., 2017, **ApJS**, 230, 15
Nicholl, M., Guillochon, J., Berger, E, 2017, *ApJ*, 850, 55
Oka, K., Nagae, T., Matsuda, T., 2002, *A&A*, 394, 115
Owocki, S. P., Townsend, R. H. D., Quataert, E, 2017, *MNRAS*, 472, 3749
Packet, W., 1981, *A&A*, 102, 17
Pastorello, A., Cappellaro, E., Inserra, C., et al. 2013, *ApJ*, 767, 1
Pauli, D., Langer, N., Aguilera-Dena, D. R., et al., 2022, *A&A*, in press, 2022arXiv220810194P
Pelisoli, I., Dorsch, M., Heber, U., et al., 2022, *MNRAS*, 515, 2496
Petrovic, J., Pols, O., Langer, N, 2006, *A&A*, 450, 219
Petrovic, J., Langer, N., van der Hucht, K. A., 2005, *A&A*, 435, 1013
Quimby R.M., Kulkarni S.R., Kasliwal M.M., et al., 2011, *Nature*, 474, 487
Rakavy, G., Shaviv, G., 1967, *ApJ*, 148, 803
Sana, H., de Mink, S. E., de Koter, A., et al., 2012, *Science*, 337, 444
Sanyal, D., Grassitelli, L., Langer, N., et al., 2015, *A&A*, 580, A20
Sanyal, D., Langer, N., Szecsi, D., et al., 2017, *A&A*, 597, A71
Schneider, F. R. N., Podsiadlowski, P., Müller, B., 2021, *A&A*, 645, A5
Schneider, F. R. N., Podsiadlowski, P, Langer, N., et al., 2016, *MNRAS*, 457, 2355
Schneider, F. R. N., Sana, H., Evans, C. J., et al. 2018, *Science*, 359, 69
Schneider, F. R. N., Ohlmann, S. T., Podsiadlowski, P, et al., 2019, *Nature*, 574, 211
Schneider, F. R. N., Ohlmann, S. T., Podsiadlowski, P, et al., 2020, *MNRAS*, 495, 2796
Schöller, M., Hubrig, S., Fossati, L., et al., 2017, *A&A*, 599, A66
Schootemeijer, A., Langer, N., Grin, N. J., et al., 2019, *A&A*, 625, A132
Schootemeijer, A., Langer, N., Lennon, D., et al., 2021, *A&A*, 646, A106
Schulze, S., Krühler, T., Leloudas, G., et al., 2018, *MNRAS*, 473, 1258
Sen, K., Langer, N., Marchant, P. *et al.* 2022, *A&A*, 659, A98
Siess, L., Dufour, E., Forestini, M 2000, *A&A*, 358, 593
Smith, N., Vink, J. S., de Koter, A. 2004, *ApJ*, 615, 475
Sukhbold, T., Woosley, S. E., 2016, *ApJL*, 820, L38

Takahashi, K., Langer, N., 2021, *A&A*, 646, A19
Tassoul, J.-L., 2000, *Stellar Rotation*, Cambridge Univ. Press, p. 232
Tokovinin, A., Moe, M. 2020, *MNRAS*, 491, 5158
Tutukov, A. V., Fedorova, A. V., 2010, A.Rep, 54, 156
van den Heuvel, E. P. J., Portegies Zwart, S. F., de Mink, S. E., 2017, *MNRAS*, 471, 4256
Villasenor, J. I., Taylor, W. D., Evans, C. J., et al., 2021, *MNRAS*, 507, 5348
Vinciguerra, S., Neijssel, C. J., Vigna-Gomez, A, et al., 2020, *MNRAS*, 498, 4705
Wang, C., Langer, N., Schootemeijer, A., et al., 2022, *NatAs*, 6, 480
Whalen, D. J., Even, W., Frey, L. H., et al., 2013, *ApJ*, 777, 110
Woosley, S. E., 2017, *ApJ*, 836, 244
Woosley, S. E., Heger, A., 2006, *ApJ*, 637, 914
Woosley, S. E., Smith, N., 2022, arXiv:2205.06386
Woosley, S. E., Blinnikov, S., Heger, A., 2007, *Nature*, 450, 390
Yoon, S.-C., Langer, N, 2005, *A&A*, 443, 643
Yoon, S.-C., Langer, N., Norman, C., 2006, *A&A*, 460, 199
Yoon, S.-C., Woosley, S. E., Langer, N, 2010, *ApJ*, 725, 940
Yorke, H.W. 1986, *ARAA*, 24, 49
Yusof, N., Hirschi, R., Eggenberger, P., et al., 2022, *MNRAS*, 511, 2814
Zapartas, E., de Mink, S. E., Justham, S, et al., 2019, *A&A*, 631, A5
Zinnecker, H., Yorke, H. W. 2007, *ARAA*, 45, 481
Zorec, J., Royer, F., 2012, *A&A*, 537, A120

Chapter I:
Observations of massive stars near and far

Massive Stars Near and Far
Proceedings IAU Symposium No. 361, 2024
J. Mackey, J. S. Vink, & N. St-Louis, eds.
doi:10.1017/S1743921322002435

ULLYSES and Complementary Surveys of Massive Stars in the Magellanic Clouds

Paul A. Crowther

Department of Physics & Astronomy, University of Sheffield
Hounsfield Road, Sheffield, S3 7RH, UK
email: paul.crowther@sheffield.ac.uk

Abstract. An overview is provided of the scientific goals of the Magellanic Cloud component of the STScI Directors Discretionary UV initiative ULLYSES, together with the complementary spectroscopic survey XShootU (VLT/Xshooter) and other ancillary datasets. Together, ULLYSES and XShootU permit the first comprehensive, homogeneous study of wind densities and velocities in metal-poor massive stars, plus UV/optical spectroscopic libraries for population synthesis models and a large number of interstellar sight-lines towards the Magellanic Clouds.

Keywords. Stars: early-type - Stars: massive - Stars: evolution - Stars: winds, outflows - Stars: abundances - Stars: fundamental parameters

1. Introduction

The evolution of single and binary massive stars is inherently linked to a broad range of astrophysical transients involving their ultimate fate, including core collapse supernovae and gravitational wave mergers such as kilonovae. However, many details of their evolution remain unclear, including mass-loss, internal mixing, and close binary evolution.

In the Milky Way, initiatives such as the Galactic O Star Spectroscopic Survey (GOSSS Maiz Apellaniz et al. 2019) together with extensive high-resolution UV spectroscopy from IUE/SWP (Howarth & Prinja 1989) provide reference datasets for the study of mass-loss at solar composition, and UV templates for metal-rich star-forming galaxies.

However, some exotic astrophysical phenomenon are preferentially observed in metal-deficient host galaxies (superluminous supernovae, long gamma-ray bursts) while typical high-redshift star-forming galaxies are metal-poor. Initiatives such as the VLT FLAMES Tarantula Survey (VFTS Evans et al. 2011) have provided high quality optical spectroscopy of large numbers of metal-poor massive stars, but extensive UV spectroscopy of individual stars has remained elusive until now.

In this review I will provide an overview of a new UV initiative, ULLYSES, administered under the STScI Director's Discretionary scheme, which seeks to address this deficiency, focusing on Magellanic Clouds targets. I will also summarise complementary surveys such as XShootU which seek to exploit the full potential of the programme.

2. UV Legacy stellar physics programme

In 2018 the STScI Director Ken Sembach convened a science definition Working Group, chaired by Sally Oey, to identify a UV Legacy programme on star formation and associated stellar physics, recognising the finite lifetime of HST's large aperture UV spectroscopic capability.

The Hubble UV Legacy Library of Young Stars as Essential Standards (ULLYSES) programme is based on the Working Group's recommendations, and is devoting ∼1000

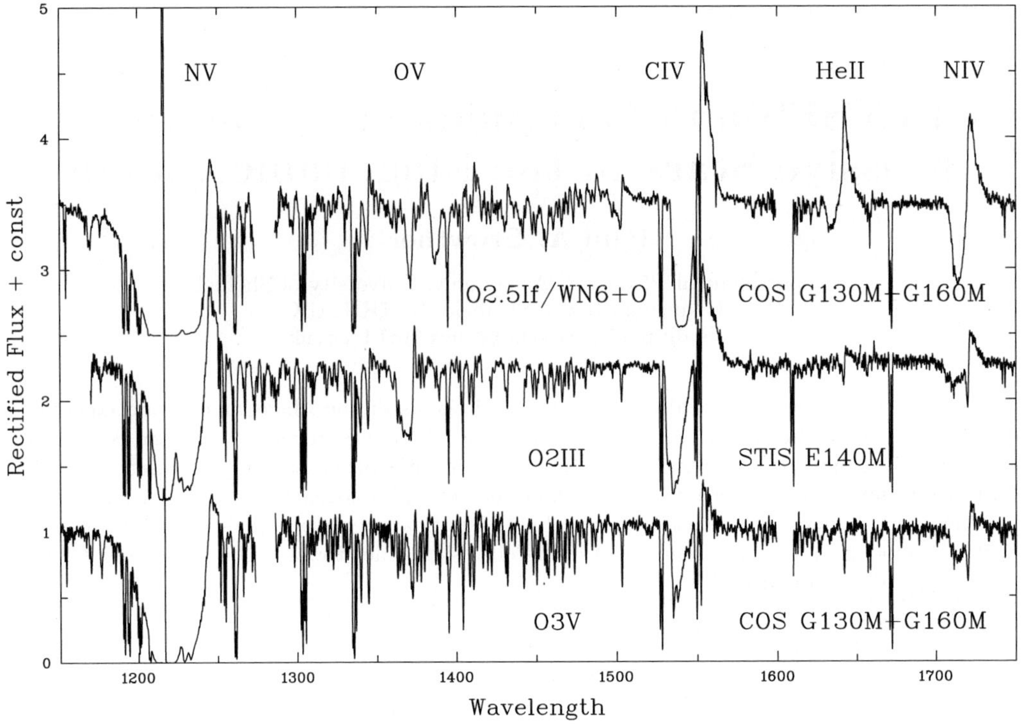

Figure 1. Representative HST UV spectra of LMC early O stars (VFTS 482 = Mk 39; Sk−67° 211 = HDE 269810; PGMW 3058 from top to bottom) from ULLYSES. These spectra also highlight the forest of Fe V-VI lines in the λλ1250–1500 region for such stars.

HST orbits to an ultraviolet spectroscopic library of young stars, with Julia Roman-Duval responsible for leading the core implementation team, supported by a scientific advisory committee. This overview focuses on half of the programme involving the acquisition of high quality, high resolution COS and STIS observations of massive stars in metal-poor environments, predominantly in the Large and Small Magellanic Clouds, with present-day metallicities of 1/2 $Z_{\odot}$ and 1/5 $Z_{\odot}$, respectively.

Amongst nearby metal-poor star-forming galaxies, only the Large and Small Magellanic Clouds with distances of 50 kpc (Pietrzyński et al. 2013) and 62 kpc (Graczyk et al. 2014), respectively, are sufficiently close for medium to high resolution UV spectroscopy with HST.

2.1. *ULLYSES science*

The total ULLYSES sample comprises ∼250 massive stars in the Magellanic Clouds, primarily with O and B spectral types, plus a few Wolf-Rayet stars and related objects Roman-Duval et al. (2020). A small subset of targets are known binary systems, (e.g. Sk−67°105, Ostrov & Lapasset 2003), while others have been identified as binaries since the sample was finalised (e.g. VFTS 66, Mahy et al. 2020). All targets include the 'standard' far ultraviolet region (Lyα to N IV λ1718), for which representative spectra are presented in Fig. 1 highlighting stellar wind (N V λλ1238–42, C IV λλ1548–51), photospheric iron forest and interstellar features. The diagnostic-rich λλ900–1200 window is included for a subset of the sample (FUSE or COS/G130M/1096) while the STIS/E230M

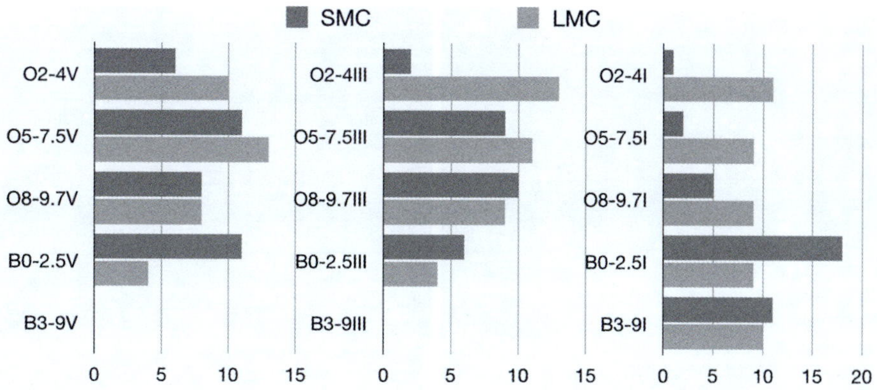

Figure 2. Subtype distribution of SMC (blue/dark grey) and LMC (yellow/light grey) OB stars from ULLYSES. A few Wolf-Rayet stars, Of/WN stars and related objects are also included in the survey. ULLYSES+ targets are excluded from these charts, some of which have uncertain luminosity class.

grating is added for late O and B supergiants (1978 setting, plus 2707 setting for late B supergiants).

As of June 2022, Data Release 5 (DR5) of ULLYSES includes HST spectroscopy of 233 massive stars in the Magellanic Clouds, plus FUSE spectroscopy of 122 targets. The spectral type distribution of SMC and LMC OB stars from the complete ULLYSES survey are presented in Fig. 2. DR5 also includes archival UV datasets for stars which were not included in the original target list (S/N or spectral coverage or uncertain luminosity class). I shall refer to these as ULLYSES+ targets throughout this overview.

UV spectroscopy provides unique information on the wind properties of early-type stars, specifically wind clumping (e.g. Brands et al. 2022). Theoretical studies (Vink et al. 2001) point to reduced wind densities in metal-poor environments, since early-type winds are radiatively-driven (Vink 2021). However, empirical studies spanning the Milky Way, LMC and SMC (e.g. Mokiem et al. 2007) have been largely based on $H\alpha$ which is difficult to determine for stars with relatively weak winds and is insensitive to clumping.

ULLYSES offers the prospect of a definitive determination of stellar wind properties of high mass stars at low metallicities, with the far-UV region especially important since it includes unsaturated P Cygni profiles such as P V $\lambda\lambda$1118-1128 (Crowther et al. 2002).

The primary diagnostic for wind velocities in early-type stars involves the blueward trough of UV P Cygni profiles (e.g. C IV $\lambda\lambda$1548–51) which ULLYSES/ULLYSES+ permit the extension of Galactic calibrations (Prinja et al. 1990) to metal poor environments (C. Hawcroft et al. in prep), which have historically involved theoretical predictions (Leitherer et al. 1992). Comparisons of wind properties between the Milky Way, LMC and SMC (e.g. Mokiem et al. 2007) often include (reduced) wind momenta, $\dot{M} v_{\infty} R^{0.5}$ (Kudritzki et al. 1999), yet commonly adopt historical Milky Way OB scalings (Lamers et al. 1995) which may not apply to metal-poor counterparts (Prinja & Crowther 1998).

Together ULLYSES/ULLYSES+ comprise ~15% of the total O star population in the SMC (Kalari et al. 2018), versus 4% of the LMC O star population, whose star-formation rate is a factor of five times higher (Kennicutt et al. 2008). The scarcity of SMC early O-type (super)giants in Fig. 2 arises from its modest star formation rate (see also Schootemeijer et al. 2021).

ULLYSES promises to provide important diagnostics beyond mass-loss through abundance determinations of CNO elements. To date, optical studies of O stars to date have

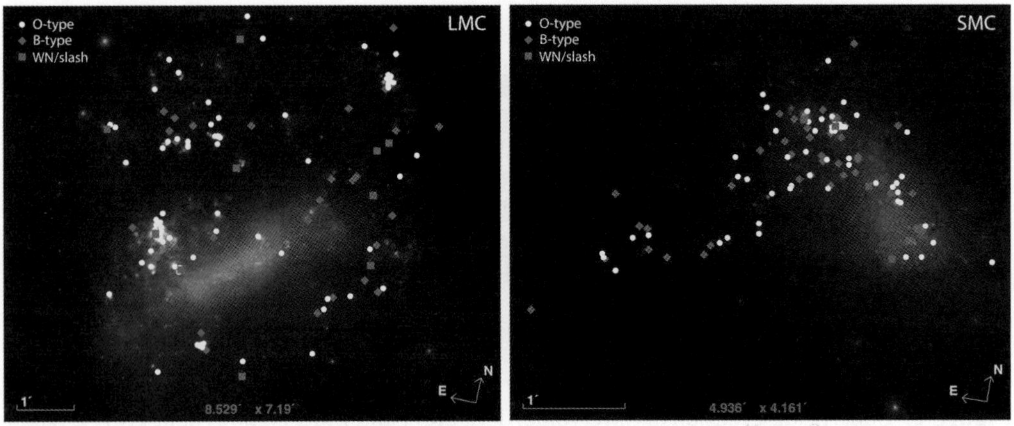

Figure 3. Locations of the ULLYSES targets in the LMC (left) and SMC (right). Circles are O-type stars, diamonds are B-type stars, and squares are Wolf-Rayet and Of/WN stars. The background consists of DSS2 colour images. Scales of 8.5° × 7.2° and 4.9° × 4.2° for the LMC and SMC correspond to physical scales of 7.4×6.5 kpc and 5.3×4.6 kpc, given their distances of 50 kpc and 62 kpc, respectively.

permitted the determination of He and N abundances (Hunter et al. 2008), with the addition of far-UV spectroscopy critical to the determination of C/N and O/N abundances, which will help to constrain mixing in metal-poor massive stars.

Beyond stellar astrophysics, ULLYSES will also significantly extend the number of interstellar Magellanic Cloud H I and metal sight-lines beyond those from previous surveys (e.g. METAL, Roman-Duval et al. 2019), plus H_2 column densities (Tumlinson et al. 2002; Welty et al. 2012). The spatial location of the ULLYSES Magellanic Cloud targets is presented in Figure 3. The latter Lyman-Werner bands contaminate far-UV spectral diagnostics to varying degrees (Walborn et al. 2002).

2.2. *UV population synthesis at low metallicity*

Historically, UV population synthesis tools (e.g. Starburst99, Leitherer et al. 2014) have incorporated templates of Milky Way OB stars from extensive samples from IUE/SWP (Howarth & Prinja 1989) but Magellanic Cloud OB samples from HST have been so scarce (Walborn et al. 2000) that combined LMC/SMC templates have been adopted, or theoretical spectra have been used (Rix et al. 2004). Large numbers of ULLYSES targets permits the construction of templates at LMC and SMC metallicities, of application to metal-poor galaxies (e.g. CLASSY, Berg et al. 2022).

In general, scaled Solar abundances are adopted for stars in the Magellanic Clouds, but there is considerable element-to-element variation from H II regions, stars and supernova remnants (for details see J. Vink et al, in prep), as shown in Figure 4. Adopting the latest Magg et al. (2022) solar values, the average present day LMC metallicity is 0.4 $Z_\odot$, with [O/H]=−0.4 dex and [Fe/H]=−0.3 dex, while the average SMC metallicity is 0.15 $Z_\odot$ with [O/H]=[Fe/H]=−0.7 dex.

3. Complementary datasets

Extensive optical and IR photometry exists for bright stars in the Magellanic Clouds, including Spitzer IRAC+MIPS (e.g. Bonanos et al. 2009), and the VISTA VMS near-IR public survey (Cioni et al. 2011). However, complementary optical spectroscopic datasets are required to exploit the full potential of ULLYSES.

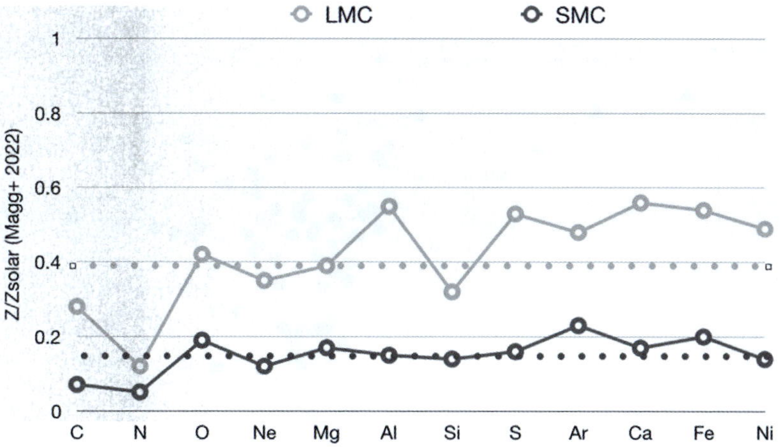

Figure 4. Present day abundances of the LMC (pale green/grey) and SMC (dark green/grey) from H II regions, stars and supernova remnants (see J. Vink et al. in prep for details) with respect to the solar value Magg et al. (2022).

Optical spectroscopic studies for the majority of ULLYSES targets have been undertaken (e.g Massey et al. 2009; Evans et al. 2011), albeit involving heterogeneous tools and datasets, while properties for the remainder can be obtained from contemporary spectral type calibrations (Doran et al. 2013). In Figure 5 we present Hertzsprung-Russell diagrams for ULLYSES(+) targets in each galaxy, together with evolutionary tracks and isochrones from Brott et al. (2011) and Köhler et al. (2015). The typical mass range for SMC OB stars is 15–60 $M_{\odot}$ versus 25–150 $M_{\odot}$ for LMC OB stars.

3.1. *XShootU*

Consequently, the large Hubble ULLYSES award has motivated the acquisition of complementary homogeneous spectroscopy, in particular the XShootU initiative, coordinated by the IAU Commission G2 (chaired by Jorick Vink between 2018–2021). XShootU involves VLT/XShooter spectroscopy of all Magellanic Cloud ULLYSES targets. Further details of XShootU are provided in J. Vink et al (in prep).

To date, calibrated reductions for ∼70% of optical (UBV + VIS arms) XShooter datasets have been provided to the XShootU consortium by Working Group 2 (coordinator Hugues Sana) via the March 2022 eDR1 release. The remaining 30% of optical datasets and NIR arm will follow in due course.

A representative Xshooter UBV observation of Sk-67°111 (O6 Iaf) in the LMC is presented in Fig. 6, illustrating the high data quality of XShootU, which will permit the uniform determination of stellar temperatures, surface gravities, helium abundances and rotational velocities for the entire ULLYSES sample (ULLYSES+ targets were not acquired).

Together, ULLYSES and XShootU offer the potential to characterize the physical and wind properties of an unprecedented sample of Magellanic Cloud early-type stars. Preliminary CMFGEN (Hillier & Miller 1998) spectral fits to Sk-68° 16 (O7 III) are presented in Fig. 7, courtesy of JC Bouret, although FASTWIND (Puls et al. 2020) and PoWR (Gräfener et al. 2002) are also well suited to spectroscopic analysis of early-type stars with winds.

UV spectroscopic studies of high ionization P Cygni resonance lines (e.g. O VI $\lambda\lambda$1031, 1037) N V $\lambda\lambda$1238, 1242) are also sensitive to X-rays (e.g. Taresch et al. 1997), which are

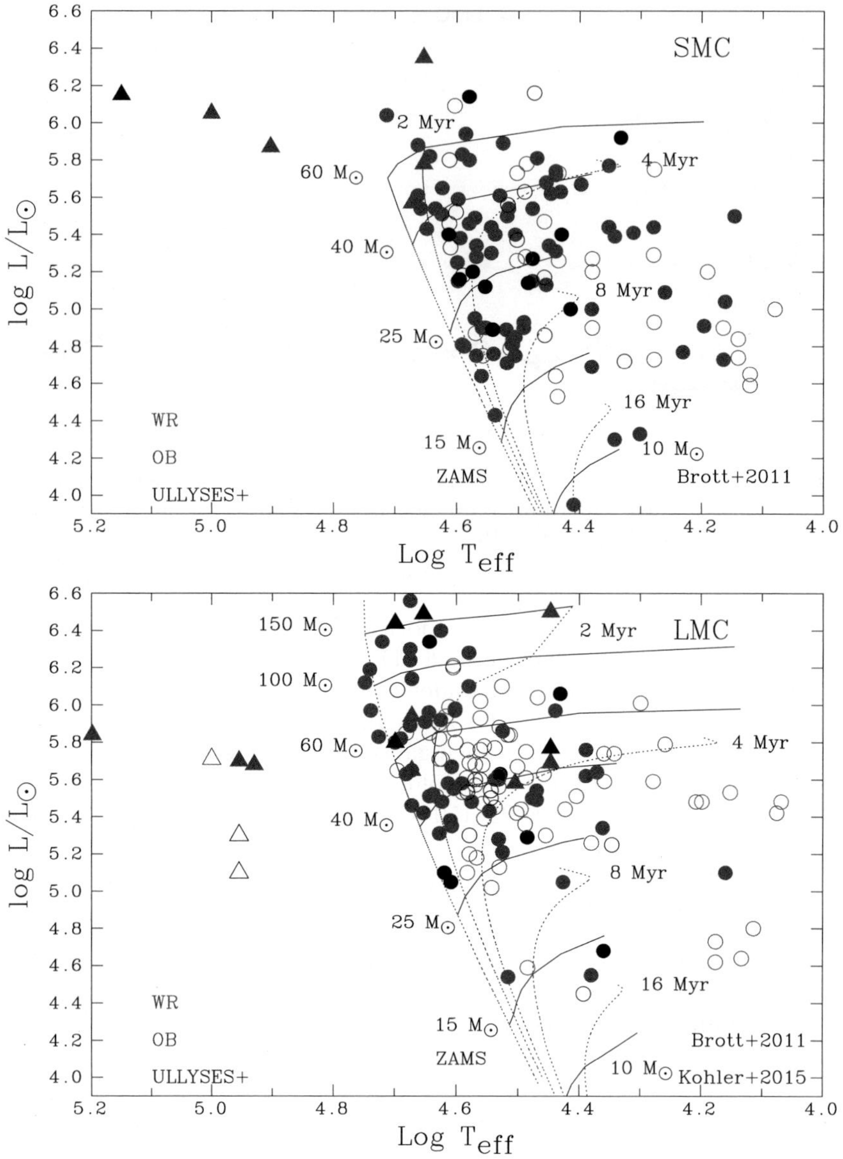

Figure 5. Hertzsprung-Russell diagram of SMC (top) and LMC (bottom) ULLYSES targets, OB stars in circles (red), WR stars in triangles, based on previous spectroscopic fitting (filled symbols) or calibrations (open symbols), plus ULLYSES+ stars in black. $0.2Z_{\odot}$ or $0.5Z_{\odot}$ evolutionary tracks and isochrones are from Brott et al. (2011) and Köhler et al. (2015) for very massive stars in LMC.

known to originate in the winds of early-type stars (Rauw 2022). To date the majority of X-ray studies of OB stars have involved Milky Way stars, although deep *Chandra* X-ray observations of a few star-forming regions in the Magellanic Clouds have been obtained.

In particular, T-ReX (PI Leisa Townsley), has involved a 2 Ms *Chandra* ACIS survey of the Tarantula Nebula, revealing over 100 detections of early-type sources. Fig. 8 compares

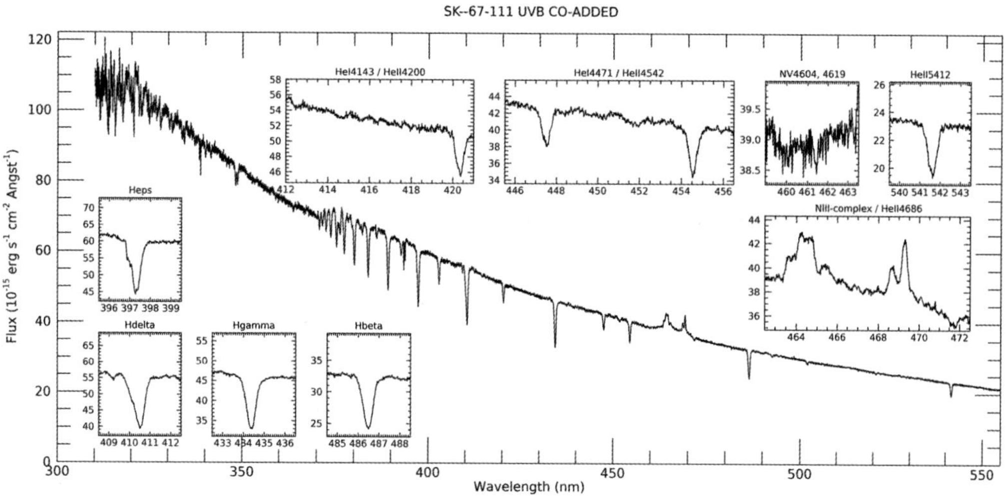

Figure 6. VLT/Xshooter UVB spectroscopy of Sk−67° 111 (O6 Iaf) from the XShootU initiative.

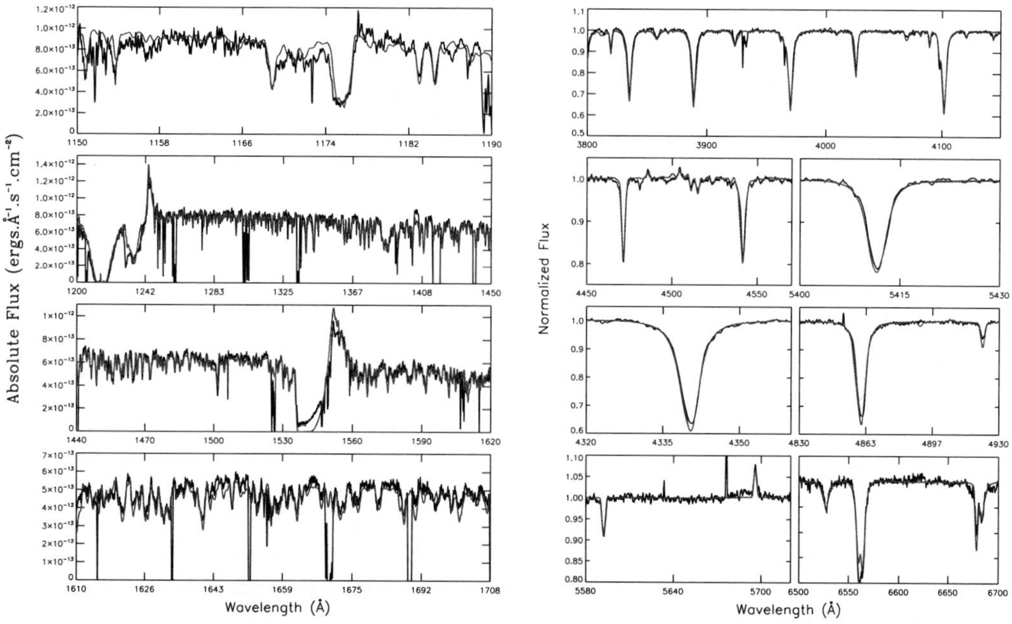

Figure 7. Preliminary CMFGEN fits (red/grey) to UV (HST/STIS/E140M from ULLYSES, left) and optical (VLT/Xshooter from XShootU, right) spectroscopy (black) of the LMC star Sk−68° 16 (O7 III), courtesy of JC Bouret.

bolometric to X-ray luminosities of O, Of/WN and WN stars in the Carina Nebula (Nazé et al. 2011) to the Tarantula Nebula (Crowther et al. 2022) revealing similar X-ray properties for luminous stars between both environments, such that similar X-ray assumptions are recommended for metal-poor massive stars.

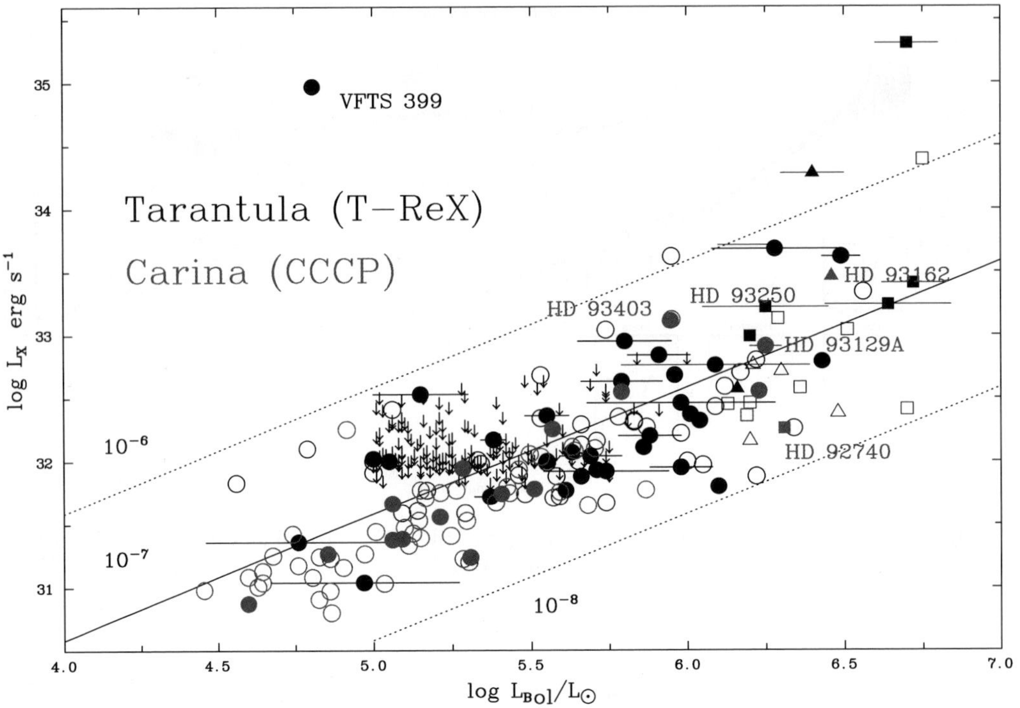

Figure 8. Comparison between X-ray and bolometric luminosities for single (open) and binary (filled) O stars (circles), Of/WN stars (triangles) and WN stars (squares) in the Tarantula Nebula from XSPEC fits to T-ReX observations (Crowther et al. 2022, black) with those in the Carina Nebula from CCCP (Nazé et al. 2011, red/grey).

3.2. *Magellan/MIKE*

The spectral resolution of Xshooter (R∼7,000) is sufficient for the majority of ULLYSES targets, although slowly rotating OB stars would benefit from higher spectral resolution in order to resolve spectral features and characterize their rotation rates. Nidia Morrell has obtained complementary Magellan/MIKE spectroscopy (R∼35,000–40,000) of 23 slowly rotating ULLYSES targets in Dec 2021, with an additional 3 night allocation for Dec 2022. Representative LMC and SMC examples are presented in Fig. 9.

3.3. *Population synthesis empirical templates at low metallicity*

The XShootU datasets have application beyond individual spectroscopic analysis. Population synthesis tools have historically included either theoretical spectra from model atmospheres, such as ATLAS (Castelli & Kurucz 2003) or MARCS (Gustafsson et al. 2008) or empirical spectra spanning a range of spectral coverage and resolution (e.g. Jacoby et al. 1984; Le Borgne et al. 2003).

In recent years, attempts have been made to extend templates to metal-poor populations across a range of temperatures and gravities (e.g. MaStar, Maraston et al. 2020). These have been successful for old, low mass populations within the Milky Way (Hill et al. 2022) but lack metal-poor templates amongst massive stars, which XShootU will address.

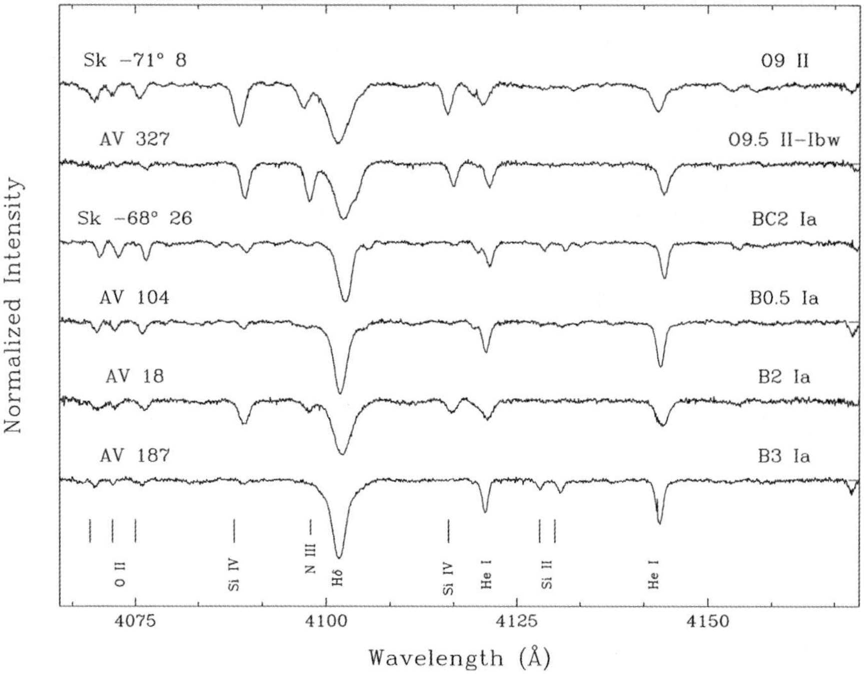

Figure 9. Representive Magellan/MIKE spectroscopy of slowly rotating LMC and SMC ULLYSES targets from Dec 2021, courtesy Nidia Morrell.

4. Summary and outlook

We provide a brief overview of the Magellanic Cloud component of the UV Legacy programme ULLYSES and complementary optical spectroscopic datasets, notably the VLT/Xshooter survey XShootU. Together, these offer the prospect of characterising the physical, wind and chemical properties of large samples of OB stars in the LMC and SMC, together with providing UV and optical spectroscopic templates for unresolved metal-poor populations and extensive interstellar sight-lines at intermediate to high spectral resolution.

In the near future, VISTA/4MOST will provide very large numbers of optical spectroscopic datasets for early-type stars in the Magellanic Clouds via the 1001MC survey (Cioni et al. 2019) although the prospect for high resolution UV spectroscopy of individual stars at metallicities below the Magellanic Clouds requires a much larger aperture space telescope such as the Large UV/Optical/IR Surveyor (LUVOIR).

Acknowledgements. Thanks to the following for their major contributions to ULLYSES and XshootU: Ken Sembach (STScI Director), Sally Oey (Working Group chair), Julia Roman-Duval (ULLYSES implementation team lead), Jorick Vink (IAU G2 Commission chair), plus JC Bouret for preliminary fits to Sk −68° 16 and Nidia Morrell for proposing and observing the Magellan MIKE datasets. PAC is supported by the Science and Technology Facilities Council research grant ST/V000853/1 (PI. V. Dhillon).

References

Berg, D. A., James, B. L., King, T., et al. 2022, arXiv e-prints, arXiv:2203.07357
Bonanos, A. Z., Massa, D. L., Sewilo, M., et al. 2009, AJ, 138, 1003
Brands, S. A., de Koter, A., Bestenlehner, J. M., et al. 2022, arXiv e-prints, arXiv:2202.11080

Brott, I., de Mink, S. E., Cantiello, M., et al. 2011, A&A, 530, A115
Castelli, F., & Kurucz, R. L. 2003, in Modelling of Stellar Atmospheres, ed. N. Piskunov, W. W. Weiss, & D. F. Gray, Vol. 210, A20
Cioni, M., R. L., Storm, J., Bell, C. P. M., et al. 2019, The Messenger, 175, 54
Cioni, M. R., Clementini, G., Girardi, L., et al. 2011, The Messenger, 144, 25
Crowther, P. A., Broos, P. S., Townsley, L. K., et al. 2022, arXiv e-prints, arXiv:2207.xxxxx
Crowther, P. A., Hillier, D. J., Evans, C. J., et al. 2002, ApJ, 579, 774
Doran, E. I., Crowther, P. A., de Koter, A., et al. 2013, A&A, 558, A134
Evans, C. J., Taylor, W. D., Hénault-Brunet, V., et al. 2011, A&A, 530, A108
Graczyk, D., Pietrzyński, G., Thompson, I. B., et al. 2014, ApJ, 780, 59
Gräfener, G., Koesterke, L., & Hamann, W. R. 2002, A&A, 387, 244
Gustafsson, B., Edvardsson, B., Eriksson, K., et al. 2008, A&A, 486, 951
Hill, L., Thomas, D., Maraston, C., et al. 2022, MNRAS, 509, 4308
Hillier, D. J., & Miller, D. L. 1998, ApJ, 496, 407
Howarth, I. D., & Prinja, R. K. 1989, ApJS, 69, 527
Hunter, I., Brott, I., Lennon, D. J., et al. 2008, ApJL, 676, L29
Jacoby, G. H., Hunter, D. A., & Christian, C. A. 1984, ApJS, 56, 257
Kalari, V. M., Vink, J. S., Dufton, P. L., & Fraser, M. 2018, A&A, 618, A17
Kennicutt, Robert C., J., Lee, J. C., Funes, J. G., et al. 2008, ApJS, 178, 247
Köhler, K., Langer, N., de Koter, A., et al. 2015, A&A, 573, A71
Kudritzki, R. P., Puls, J., Lennon, D. J., et al. 1999, A&A, 350, 970
Lamers, H. J. G. L. M., Snow, T. P., & Lindholm, D. M. 1995, ApJ, 455, 269
Le Borgne, J. F., Bruzual, G., Pelló, R., et al. 2003, A&A, 402, 433
Leitherer, C., Ekström, S., Meynet, G., et al. 2014, ApJS, 212, 14
Leitherer, C., Robert, C., & Drissen, L. 1992, ApJ, 401, 596
Magg, E., Bergemann, M., Serenelli, A., et al. 2022, A&A, 661, A140
Mahy, L., Sana, H., Abdul-Masih, M., et al. 2020, A&A, 634, A118
Maiz Apellaniz, J., Sota, A., Arias, J. I., et al. 2019, VizieR Online Data Catalog, J/ApJS/224/4
Maraston, C., Hill, L., Thomas, D., et al. 2020, MNRAS, 496, 2962
Massey, P., Zangari, A. M., Morrell, N. I., et al. 2009, ApJ, 692, 618
Mokiem, M. R., de Koter, A., Vink, J. S., et al. 2007, A&A, 473, 603
Nazé, Y., Broos, P. S., Oskinova, L., et al. 2011, ApJS, 194, 7
Ostrov, P. G., & Lapasset, E. 2003, MNRAS, 338, 141
Pietrzyński, G., Graczyk, D., Gieren, W., et al. 2013, Nature, 495, 76
Prinja, R. K., Barlow, M. J., & Howarth, I. D. 1990, ApJ, 361, 607
Prinja, R. K., & Crowther, P. A. 1998, MNRAS, 300, 828
Puls, J., Najarro, F., Sundqvist, J. O., & Sen, K. 2020, A&A, 642, A172
Rauw, G. 2022, arXiv e-prints, arXiv:2203.16842
Rix, S. A., Pettini, M., Leitherer, C., et al. 2004, ApJ, 615, 98
Roman-Duval, J., Jenkins, E. B., Williams, B., et al. 2019, ApJ, 871, 151
Roman-Duval, J., Proffitt, C. R., Taylor, J. M., et al. 2020, Research Notes of the American Astronomical Society, 4, 205
Schootemeijer, A., Langer, N., Lennon, D., et al. 2021, A&A, 646, A106
Taresch, G., Kudritzki, R. P., Hurwitz, M., et al. 1997, A&A, 321, 531
Tumlinson, J., Shull, J. M., Rachford, B. L., et al. 2002, ApJ, 566, 857
Vink, J. S. 2021, arXiv e-prints, arXiv:2109.08164
Vink, J. S., de Koter, A., & Lamers, H. J. G. L. M. 2001, A&A, 369, 574
Walborn, N. R., Fullerton, A. W., Crowther, P. A., et al. 2002, ApJS, 141, 443
Walborn, N. R., Lennon, D. J., Heap, S. R., et al. 2000, PASP, 112, 1243
Welty, D. E., Xue, R., & Wong, T. 2012, ApJ, 745, 173

Discussion

LEITHERER: Paul, you mentioned abundances (CNO) but how far away are we from determining iron abundances in the ultraviolet?

CROWTHER: Although fits to the UV forest in Magellanic Cloud stars for an adopted iron abundance are pretty good, accurate abundances from ULLYSES are very challenging given the complexity of atomic data involved, so alternative methods are preferred (e.g. optical Fe III lines in B stars, Thompson et al. 2008 MNRAS 383, 729).

XXX: Does the ULLYSES target list include stars which allow us to tackle the weak wind problem?

CROWTHER: Yes, both galaxies include late O and early B dwarfs (Fig. 2) which have been shown to exhibit significantly weaker winds than predicted by radiatively driven wind theory, based on UV/Hα diagnostics.

Massive Stars Near and Far
Proceedings IAU Symposium No. 361, 2024
J. Mackey, J. S. Vink, & N. St-Louis, eds.
doi:10.1017/S1743921323001771

Massive Stars in Low-metallicity Galaxies in the CLASSY Survey

Danielle A. Berg[1], Beth L. James[2], Teagan King[3], Meaghan McDonald[3], Zuyi Chen[4], John Chisholm[1], Timothy Heckman[5], Crystal L. Martin[6], Dan P. Stark[4] and The CLASSY Team[7]

[1]Department of Astronomy, The University of Texas at Austin, 2515 Speedway, Stop C1400, Austin, TX 78712, USA
email: daberg@austin.utexas.edu

[2]AURA for ESA, Space Telescope Science Institute, 3700 San Martin Drive, Baltimore, MD 21218, USA

[3]Space Telescope Science Institute, 3700 San Martin Drive, Baltimore, MD 21218, USA

[4]Steward Observatory, The University of Arizona, 933 N Cherry Ave, Tucson, AZ, 85721, USA

[5]Center for Astrophysical Sciences, Department of Physics & Astronomy, Johns Hopkins University, Baltimore, MD 21218, USA

[6]Department of Physics, University of California, Santa Barbara, Santa Barbara, CA 93106, USA

[7]39 collaborators worldwide

Abstract. Rest-frame far-ultraviolet spectra are fundamental to our understanding of star-forming galaxies, providing a unique window on massive star populations (MSs), chemical evolution, feedback processes, and reionization. JWST is ushering in a new era, pushing the FUV frontier beyond z=10. The success of such endeavors hinges on a comprehensive understanding of the MSs and gas conditions that power the observed spectra. The COS Legacy Archive Spectroscopic SurveY (CLASSY) is a powerful and promising solution providing high-quality, high-resolution FUV spectra of 45 nearby star-forming galaxies. The spectra contain a suite of features that simultaneously characterize the MSs that populate metal-poor galaxies, physical properties of large-scale outflows, and chemical abundance patterns. The CLASSY sample is consistent with the z 0 mass-metallicity relationship and spans 1.5 dex in metallicity. These unique properties make CLASSY the benchmark training set for studies of MSs in star-forming galaxies both across cosmic time and connecting metal-poor to metal-rich populations.

Keywords. atlases, galaxies: star clusters, stars: winds, outflows, stars: mass loss

1. Introduction

Massive stars are fundamental to our understanding of the baryon cycle and galaxy evolution in general. The size of the gas reservoir from which stars form is regulated by the balance between gas accretion from the cosmic web, the rate at which stars form from this gas, and gas loss from galaxies. Gas may be driven out of galaxies by feedback, including stellar winds, super novae explosions, and the energetic radiation from hot gas falling into super massive black holes found at the center of galaxies, called active galactic nuclei feedback. Additionally, the ionizing spectra of massive stars are not yet well-understood. The extreme-UV (EUV) radiation field they produce is reprocessed by the interstellar medium (ISM), powering the observed nebular emission lines. Uncertainties in the form of the ionizing spectrum thus have a significant effect on the interpretation of the nebular lines. While the implementation of new physics in stellar population synthesis (SPS) models (i.e., rotation, binaries) continues to improve predictions of the EUV radiation field (e.g., Levesque et al. 2012; Eldridge et al. 2017; Götberg et al. 2018), the shape of the ionizing spectrum remains very poorly constrained for the metal poor ($Z/Z_\odot < 0.2$) stellar populations that come to dominate at high redshift. Therefore, to truly understand cosmic galaxy evolution, we must study both the stars and the gas within the same systems.

1.1. *The Diagnostic Power of the FUV*

The far-ultraviolet (FUV), defined here as $\sim 1200 - 2000$ Å, is arguably the richest wavelength regime in diagnostic spectral features characterizing stellar and nebular processes and will provide an important window onto the first generation of galaxies in the forthcoming era of extremely large telescopes (ELTs) and the James Webb Space Telescope (JWST). In particular, several spectral features characterize the (1) hot, massive stars that dominate feedback processes and are observed as prominent wind features ($1240 < \lambda < 1720$) produced in the expanding atmospheres of O- and B-stars and are brightest in the FUV (e.g., Leitherer et al. 1999); (2) resonant interstellar absorption lines (see Veilleux et al. 2005) and associated non-resonant ("fine-structure") emission lines (e.g., Shapley et al. 2003) that trace both the static ISM and galactic outflows ($1240 < \lambda < 1854$); (3) emission lines ($1390 < \lambda < 1910$; e.g., Kinney et al. 1993) that are produced by collisional excitation or recombination in gas that is photoionized by the hot, massive stars (e.g., Schaerer 2003); and (4) Lyα profiles ($\lambda = 1215$ Å), which is typically nebular in origin, entrains details of the outflowing gas and predicts the escape of ionizing radiation (e.g., Wofford et al. 2013; Rivera-Thorsen et al. 2015; Verhamme et al. 2015; Gronke 2017). All of these features are shaped, in some way, by the radiation, energy, and/or momentum produced by massive stars. The range and richness of features in the FUV is demonstrated in Figure 1.

2. The CLASSY Survey

The new release of the COS Legacy Archive Spectroscopic SurveY (CLASSY; Berg et al. 2022) offers a powerful and promising opportunity to study massive star populations. The CLASSY atlas is the first high-quality, high-resolution FUV spectral catalog of star-forming galaxies at $z \sim 0$, enabling detections of the nebular lines together with faint stellar absorption lines. To optimize the scientific utility of the CLASSY FUV spectral atlas, it provides: (1) Broad wavelength coverage to simultaneously probe the continuum, absorption, and emission features that characterize the stellar populations and gas of the galaxy. (2) High spectral resolution. Theory predicts that the velocities of galactic outflows scale with the circular velocities of their host galaxies (e.g., Murray et al. 2005). Therefore, CLASSY has very high spectral resolution ($R > 10,000$ or velocity resolution less than 30 km s^{-1}) allowing measurements of the impact of stellar feedback in galaxies, event at low stellar-mass (see, e.g., McGaugh 2001). (3) High signal-to-noise (S/N) in the continuum, allowing faint stellar features to be significantly detected and measured.

2.1. *Sample Properties*

CLASSY combines 135 orbits of new Hubble Space Telescope (HST) / Cosmic Origins Spectrograph (COS) data with 177 orbits of archival HST data, for a total of 312

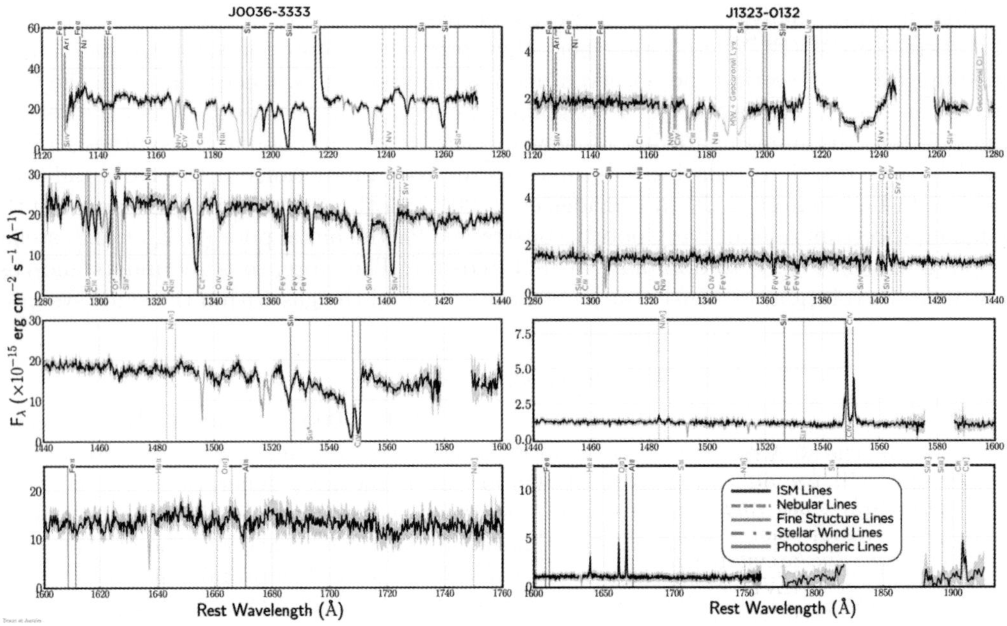

Figure 1. CLASSY spectra demonstrating the numerous diagnostic features present in the FUV. *Left:* The spectrum of the metal-rich galaxy J0036-3333 is rich with stellar wind features (dot-dashed green lines), such as the SiIV and CIV features, and interstellar medium absorption lines (solid blue lines), including the strong SiII and CII features. J0036-3333, however, contains no significant UV emission features besides Lyα λ1215. *Right:* The spectrum of the metal-poor galaxy J1323-0132 is strikingly different than that of J0036-3333. In particular, the spectrum has strong emission features from Lyα, CIV, HeII, OIII], and CIII] and additional weak detections from the high-ionization OIV, SIV, and NIV] lines, but minimal absorption features. Figures adopted from Berg et al. (2022).

orbits, to complete an atlas of 45 local star-forming galaxies. In order to achieve nearly-panchromatic FUV spectral coverage with the highest spectral resolution possible, CLASSY efficiently augments existing archival data with new HST/COS observations, uniting the high-resolution G130M, G160M, and G185M gratings for a sample of galaxies spanning broad parameter space. As a result, CLASSY provides a well-controlled local FUV sample with the requisite sensitivity and spectral resolution to enable synergistic studies of stars and gas within the *same* galaxies. The CLASSY sample properties are shown in Figure 2.

3. High Level Science Products

CLASSY is a treasury program that is providing a number of state-of-the-art data products to the astronomical community in digital form to ensure their enduring value and utility. The primary data products are the high-resolution, high-S/N coadded multi-grating FUV spectral templates of the CLASSY sample of 45 star-forming galaxies. Additional planned data products include stellar continuum fits, nebular abundance measurements, improved UV emission line diagnostics, feedback properties, radiative transfer models, and more. With these high-level science products (HLSPs), the CLASSY Treasury will complete the vital picture that is needed to diagnose the suite of star-forming galaxy properties that will be offered to us in the next decade. Given the uncertain future of observed- frame UV capabilities, the CLASSY templates and models embody an indispensable toolset.

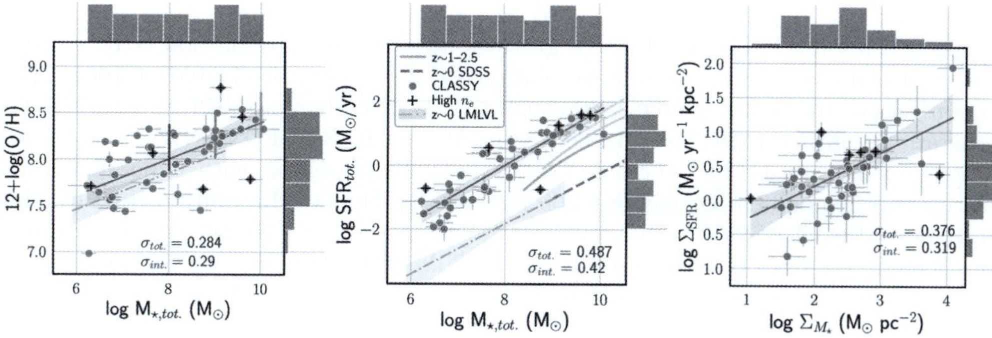

Figure 2. CLASSY sample properties, as measured from optical spectra and UV+optical photometry (see Berg et al. 2022), span a broad range of parameter space. *Left:* Gas-phase oxygen abundance is plotted relative to total stellar mass, showing the well-known mass-metallicity relationship. The CLASSY sample seems to follow the same trend as other normal star-forming galaxies (e.g., Berg et al. 2012; Curti et al. 2020), but with larger scatter. *Middle:* Star formation rate is plotted versus stellar mass for both the entire galaxy and in terms of surface density (*right*). The middle plot shows a trend that extends to lower masses and star formation rates, but lies above the previous galaxy sample studies at $z \sim 0$ (SDSS: Chang et al. (2015); blue dashed line and LMLVL: Berg et al. (2012) and Lee et al. (2009); cyan dot-dashed lines) and $z \sim 1 - 2.5$ (Whitaker et al. 2014, yellow lines). Figure adapted from Berg et al. (2022).

The first HLSP, the CLASSY FUV spectral atlas of star-forming galaxies, has been released on two main platforms. First, all CLASSY HLSPs will be available via the CLASSY HLSP home page: `https://archive.stsci.edu/hlsp/classy`, which also serves as the main website for the CLASSY Treasury and relevant information. Importantly, this website provides simple tarball downloads of individual CLASSY HLSPs, as well as the entire CLASSY HLSP data collection. In addition to full data access, the CLASSY HLSP home page provides machine-readable table downloads for the tables of properties in the CLASSY I paper (Berg et al. 2022), interactive spectra that allow quick-look examination of the coadded spectral features, user-friendly jupyter notebook guides to accessing and using the CLASSY HLSPs, and important links such as to the NASA Astrophysics Database System (ADS) library of CLASSY publications.

The second platform for accessing the CLASSY HLPSs is the Mikulski Archive for Space Telescopes (MAST) via the CLASSY MAST website: `https://mast.stsci.edu/search/ui/#/classy`. The MAST portal provides a unique search-and-selection method to access the HLSPs, where users can select a subsample of the data products based on HLSP type, filename, or target properties such as coordinates or stellar mass. This search tool will continue to be developed based on the specifics of each HLSP release in order to allow more efficient and straightforward access to data products that are of interest to the user. In this manner, appropriate subsets of the CLASSY HLSPs can easily be selected for a given scientific analysis, to match specific survey sample properties, and more.

3.1. *HLSP 1: The CLASSY Spectral Atlas*

The primary HLSPs of the CLASSY treasury survey are high S/N multi-grating coadded spectra. The coadded spectra were carefully constructed from multi-grating HST/COS datasets with different spectral resolution and varying degrees of vignetting and degradation in spectral resolution depending on the shape of their overall UV light profile entering the COS aperture. The complicated CLASSY dataset can also be used as a strength to produce four spectral coadds for each galaxy with different resolutions: (1) The Very High Resolution (VHR) Coadds: consisting of only G130M spectra, which have a nominal point source resolution of 9.97 mÅ/pixel or 0.060 Å/resel. This is the highest spectral resolution of the CLASSY gratings. (2) The High Resolution (HR) Coadds: consisting of the CLASSY medium-resolution FUV gratings, or G130M+G160M spectra, with a nominal point source resolution of 12.23 mÅ/pixel or 0.073 Å/resel. (3) The Moderate Resolution (MR) Coadds: consisting of the CLASSY FUV+NUV medium resolution gratings, or G130M+G160M+G185M+G225M spectra, with a nominal point source resolution of 33 mÅ/pixel or 0.200 Å/resel. (4) The Low Resolution (LR) Coadds: consisting of the CLASSY medium and low resolution gratings. Possible grating combinations include G130M+G160M+G140L and G130M+G160M+G185M+G225M+G140L, with a nominal point source resolution of 80.3 mÅ/pixel or 0.498 Å/resel.

Examples of high-quality moderate-resolution CLASSY coadded spectra is shown in Figure 1. The different spectral dispersion of the coadds allow CLASSY to be used for many scientific studies that require a range of spectral resolutions. These coadds provide both high resolution spectra over the $\sim 1200 - 1700$ Å range of each object that is important to studies of stellar and ISM absorption features and contiguous FUV $\sim 1200 - 2000$ Å wavelength coverage at low resolution that is necessary to simultaneously interpret the stars and gas.

Interestingly, few CLASSY galaxies are in previous FUV spectral atlases, making it difficult to directly compare them. Fortunately, International Ultraviolet Explorer (IUE) and CLASSY observations both exist for two galaxies: J0337-0502 (SBS 0335-052 E) and J0934+5514 (I Zw 18 NW). For a visual comparison, the HST and IUE spectra for J0337-0502 are plotted in Figure 3, showing substantial, but expected differences. The smaller aperture of COS (2.5") relative to the IUE (10"×20" aperture) records lower UV continuum fluxes (however, this would not be expected for most of the NUV-compact CLASSY sample) and some features are present in only the IUE spectrum, which also contains the W cluster region. For example, there appear to be weak emission features present just blueward of 1400 Å and at 1750 Å, likely N III], that are emitted by the E region but not the W region. Perhaps the most important difference, though, is the significant number of features and details revealed by the higher resolution of the CLASSY spectra (~ 0.1 Å resolution compared to the ~ 6 Å of the IUE spectrum). Specifically, a number of significant ISM absorption lines and nebular emission lines are recovered, such as the O I and Si II lines shown in the left-most inset window of Figure 3 and the C IV, He II, O III], and C III] lines shown in the other three inset windows.

Figure 3 clearly demonstrates that the CLASSY survey provides a dramatic leap forward in FUV spectral atlases and is an invaluable tool for the interpretation and modeling of star-forming galaxies across redshifts. The variety of CLASSY galaxies also results in a diversity of spectra. More metal-poor, low-mass galaxies are generally rich with nebular emission lines, but lacking in strong ISM absorption lines or stellar features. On the other hand, more metal-rich, massive galaxies tend to have weak to no emission lines, but many ISM absorption lines and stellar features.

3.2. *HLSP 2: The CLASSY Stellar Continuum Fits*

The UV stellar continuum is a powerful tool to characterize the ionizing stellar population, sensitive to the most recent ~ 40 Myr of star formation. It is also important to resolve the contribution of the stellar, ISM, and narrow nebular (intrinsic widths typically 10s of km s^{-1}) features to disentangle their contributions to the emission line profile. SPS model fits to the stellar continuum are largely constrained by two different stellar features (see Figure 1): (1) the broad wind features from young O-stars and (2) the weak photospheric absorption lines from both O- and B-stars. Even at low-metallicities, these two types of stellar continuum features constrain the stellar population age. The

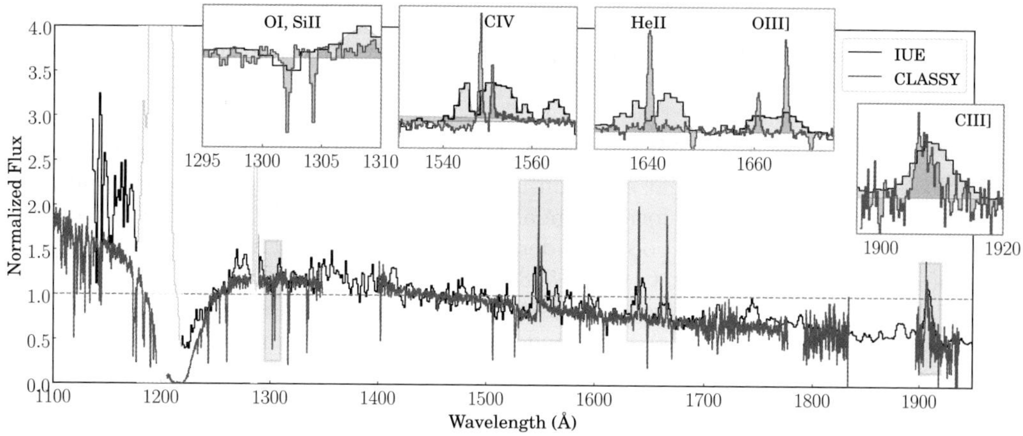

Figure 3. Comparison of rest-frame spectra for J0337-0502 (SBS 0335-052) to demonstrate the improvement of the CLASSY survey over previous FUV spectral atlases. The IUE spectrum is relatively low resolution ($\lambda/\Delta\lambda \sim 300$) and consists of the integrated light within its large 10" ×20" aperture. In contrast, the CLASSY coadded spectral templates are characterized by an unrivaled combination of contiguous 1200–2000 Å FUV wavelength coverage, high spectral resolution ($\lambda/\Delta\lambda \sim 15,000$), and high S/N (average S/N = 6.4 per resel). Additionally, the 2.5" COS aperture better targets compact regions such as massive star clusters. Both spectra are normalized at 1450 Å. Note that the breaks in the CLASSY spectrum are due to chip gaps and that the Geocoronal Lyα and OI emission have been whited-out. While few features are apparent in the IUE J0337-0502 spectra, the CLASSY spectrum reveals a significant number of absorption and emission features that enable simultaneous studies of the massive stellar population, ISM outflows, and nebular emission. Figure adopted from Berg et al. (2022).

spectra of young stars are dominated by the prominent P-Cygni features from the stellar winds which have deep and broad absorption extending to over ~ 5000 km s^{-1} from line center and broad emission redward of line center. P-Cygni features indicate that the stellar population is less than 5 Myr old and the P-Cygni features rapidly decrease at older ages. This is because stars older than 5 Myr do not drive significant stellar winds. These features provide robust age estimates of the young stellar population. The weak blue absorption portion of the P-Cygni features is sufficiently offset in velocity-space from the observed nebular emission (with observed emission only to ~ 1500 km s^{-1}), but the red P-Cygni emission will be blended with the strong high-ionization emission lines. It is, therefore, crucial to our science objectives to accurately model the P-Cygni profile and remove it from the nebular emission. Consequently, the narrow photospheric features provide a critical secondary feature to determine the age, and, in turn, the predicted continuum shape near the high-ionization lines.

CLASSY stellar continuum fits will soon be available on the CLASSY HLSP website and will be presented in Senchyna et al. (2022). The observed CLASSY stellar continua were fit using a linear combination of the single-aged fully-theoretical stellar continuum models. The fitting method follows the fully-tested fitting procedure of Chisholm et al. (2019), which accurately reproduces the shapes of high-ionization stellar features in local and high-redshift galaxies (e.g., Chisholm et al. 2016; Rigby et al. 2018). The CLASSY fits explore a suite of available stellar libraries, that include single and binary stars (i.e., STARBURST99, BPASS; Leitherer et al. 2010; Conroy & Gunn 2010; Stanway et al. 2016), updated models of Wolf-Rayet stars (the new Bruzual & Charlot models), and the full array of available stellar metallicities.

The resulting stellar population fits estimate mean stellar ages and continuum metallicities. Removing the fitted stellar continua isolates the true nebular emission, allowing

FUV emission lines to be used for nebular diagnostics diagnostic diagram. Additionally, the stellar continuum fits can be used as inputs to cloudy photoionization models to test the production of the observed emission lines.

4. Pilot Study of Two Metal-Poor Extreme Emission Line Galaxies

Bright high-ionization nebular emission-lines present our best tool to detect and characterize the most distant star-forming galaxies that JWST will observe, yet the physics governing their production is poorly understood. Galaxies containing prominent detections of these high-ionization C IV $\lambda\lambda$1548,50, He II λ1640, O III] $\lambda\lambda$1661,66, and C III] $\lambda\lambda$1907,09 are referred to as extreme emission line galaxies (EELGs). Such high-ionization emission requires especially hard ionizing-radiation, where current models of massive stars predict few photons. A number of alternative mechanisms have been proposed to produce the hard ionizing radiation in EELGs, including very massive metal poor stars, stripped stars (e.g., Götberg et al. 2017, 2019, 2020), ultra luminous X-ray sources (e.g., Schaerer et al. 2019), high-mass X-ray binaries (e.g., Eldridge et al. 2017), and accreting IMBHs (Thuan & Izotov 2005).

Here we investigate the massive star populations of two EELGs in CLASSY, J1044+0353 and J1418+2102, and how their ionizing spectra impact the observed FUV spectra. Recently, we fit the observed stellar continuum using a linear combination of single-aged fully-theoretical models in Olivier et al. (2022). The resulting fit, shown for J1044+0353 in Fig. 4, characterizes the ionizing spectral energy distribution (SED) of the current burst of star formation ($\lesssim$ 40 Myr), providing the luminosity-weighted age, metallicity, and reddening of the stellar population. Subtracting the stellar continuum fit reveals high-ionization UV emission lines that are well-fit by narrow, nebular Gaussian profiles. These lines (i.e., Si IV $\lambda\lambda$1393,1403, O IV $\lambda\lambda$1401,05,07, S IV $\lambda\lambda$1405,06,17, C IV $\lambda\lambda$1548,50, He II λ1640, and O III] $\lambda\lambda$1661,66) can be used as nebular diagnostics.

To test whether massive stars have drive the observed emission line fluxes, we ran a suite of `cloudy` photoionization models with the best-fit stellar models as the input ionizing source and the measured nebular properties as Gaussian priors. Comparing the resulting emission line fluxes produced by the massive star population to the observed nebular emission lines reveals that the stellar models are able to produce the entire suite of moderate- to low-ionization ($\lesssim$ 35 eV) emission lines, but failed to reproduce higher ionization lines. The difference between the predicted and observed high- and very-high ionization lines then provide tight constraints of the strength and shape of high-energy flux from non-stellar sources needed to match the observations.

Olivier et al. (2022) employed this method and demonstrated that employing a self-consistent model of the measured ionizing stellar population and full suite of emission lines spanning a wide range of energies does indeed highly constrain the full ionizing SED shape. They then added a blackbody to the stellar populations fit from the UV continuum to model the necessary high-energy photons to reproduce the very-high ionization lines (e.g., He II, [O IV]). As shown in Figure 5, they found that a contribution from a blackbody of 80,000 K with $\sim 60 - 70\%$ of the luminosity from the young stellar population is needed to reproduce the entire suite of observed emission lines spanning low-, intermediate-, high-, and very-high ionization potentials.

This self-consistent modeling of the ionizing spectra of two nearby EELGs indicates the presence of a previously unaccounted-for source of hard ionizing photons, but what could this extra source of very-high-energy ionizing photons be? The low-metallicity nature of these sources calls into question the accuracy of the low-metallicity stellar atmosphere models. Further, observations of individual O stars below the 20% solar metallicity ($Z_\odot$) of the Small Magellanic Cloud (SMC) are incredibly rare and, thus, hamper our ability to model the ionizing fluxes of metal-poor stellar populations.

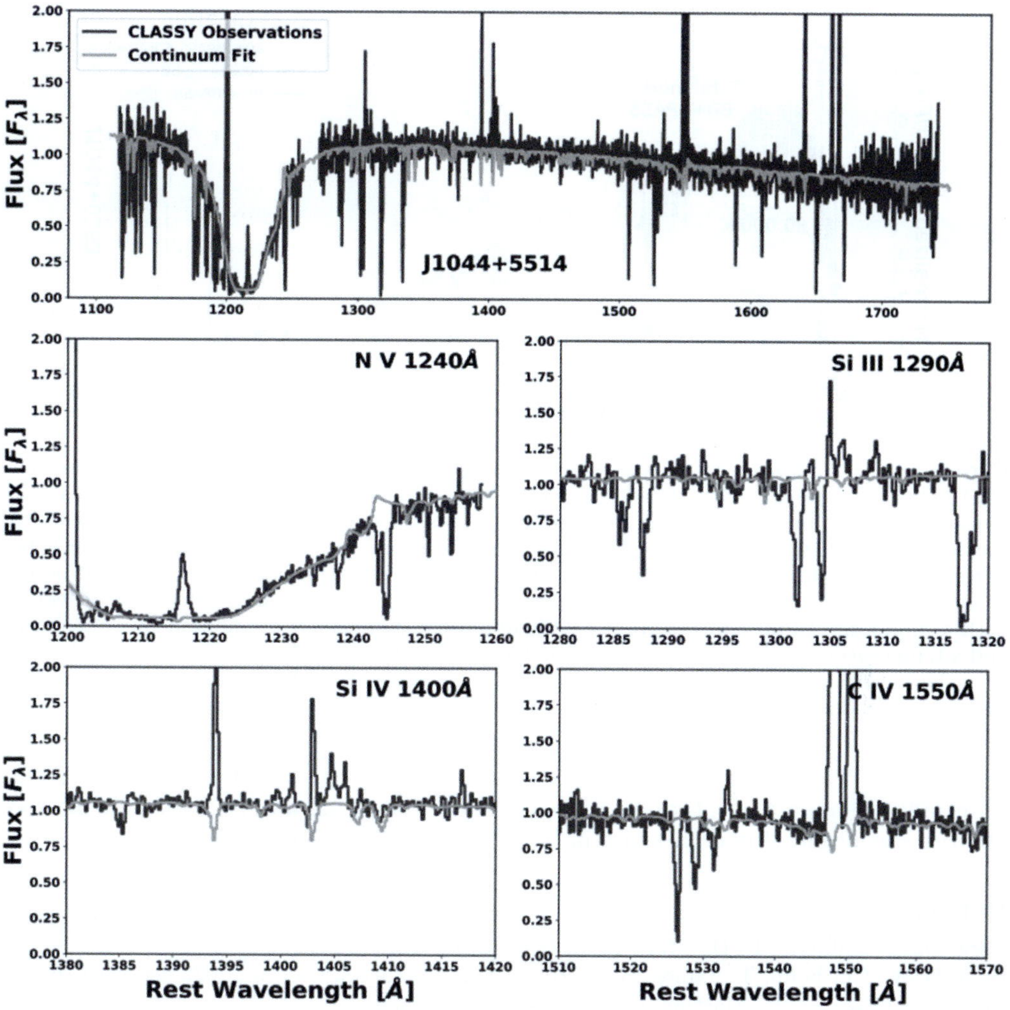

Figure 4. The CLASSY FUV spectrum of J1044+2102 and the best-fit massive star continuum model. The data are plotted in black and are normalized by their flux at 1495 Å. Over-plotted is the massive star population continuum fit from the STARBURST99 models (Leitherer et al. 1999) in gold.

To address this issue, Telford et al. (2021) presented the first-ever FUV spectra of three very metal-poor O-dwarf stars in the galaxies Leo P (3% $Z_\odot$), Sextans A (6% $Z_\odot$), and WLM (14% $Z_\odot$), shown in Fig. 6. Using this data, these authors demonstrated that the equivalent widths of photospheric metal lines and strengths of wind-sensitive features correlate with metallicity down to very low metallicities. They also used FUV through near-infrared SEDs to model the fundamental properties of the three stars. Comparing to FUV spectra of SMC stars with similar properties reveals two important insights about very metal poor stars: (1) the most metal-poor stars (represented by the star in Leo P) drive much weaker stellar winds than those seen at 20% $Z_\odot$ from the SMC counterparts and (2) very metal-poor stars, as determined from the stars in Leo P and Sextans A, are fast-rotators with projects rotation speeds of $v{\cdot}\sin(i) \geq 290$ km s^{-1} (Fig. 6). Such high rotation speeds were unexpected: The probability of finding two fast rotators is just a 3%–6% if the metal-poor stars are drawn from the same $v{\cdot}\sin(i)$ distribution observed for O dwarfs in the SMC. These observations suggest that model improvements are needed

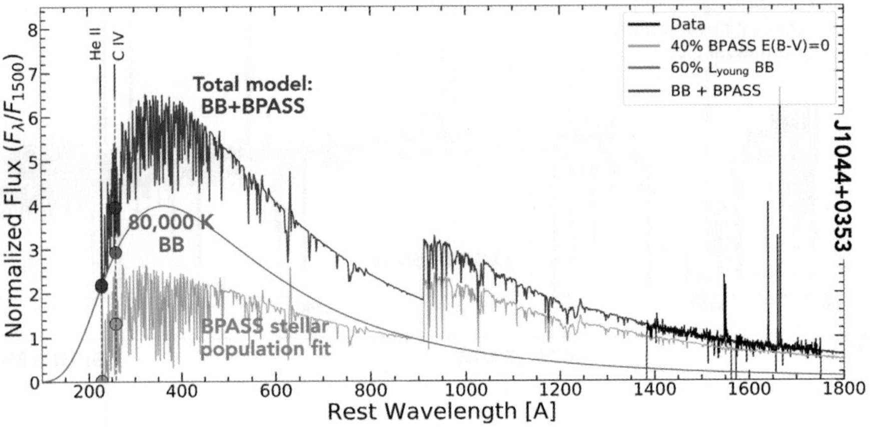

Figure 5. The ionizing X-ray+extreme-UV and non-ionizing FUV continua are shown for CLASSY galaxy J1044+0353 (labeled J1044. The overall self-consistent best-fit model is plotted as dark blue, and showing an excellent fit to both the observed massive star population continuum (black spectrum) and the very-high-ionization emission line strengths (dark blue circles). The best-fit model consists of (1) the binary-star (BPASS) massive star population continuum fit model to the observed FUV spectrum (lime green spectrum) plus (2) a blackbody with a luminosity of 60% of the total FUV luminosity (luminosity dominated by young stars) and a temperature of 80kK (teal spectrum). The teal circles show the intersection of each model with the wavelength required to produce HeII and CIV, where both the blackbody and the stellar population models alone fail to reproduce the observed stellar+nebular spectrum.

to include the impact of rotation and weak winds on ionizing flux to accurately interpret observations of metal-poor galaxies in both the near and distant universe.

References

Berg, D. A., Skillman, E. D., Marble, A., et al., 2012 *ApJ*, 754, 98

Berg, D. A., Chisholm, J., James, B. L., Heckman, T., Martin, C.et al., 2022 accepted to *ApJ*

Bouret, J. C., Lanz, T., Martins, F., et al., 2013 *A&A*, 555, A1

Chang, Yu-Yen, van der Wel, Arjen, da Cunha, Elisabete, Rix, & Hans-Walter, 2015 *ApJ*, 219, 8

Chisholm, J., Tremonti, C. A., Leitherer, C., Chen, Y., & Wofford, A., 2016 *MNRAS*, 457, 3133

Chisholm, J., Rigby, J. R., Berg, D. A., Bayliss, M., Dahle, H., Gladders, M., & Sharon, K., 2019 *ApJ*, 882, 182

Conroy, Charlie & Gunn, James E., 2010 *Astrophysics Source Code Library*, record ascl:1010.043

Curti, M., Mannucci, F., Cresci, G., & Maiolino, R., 2020 *MNRAS*, 491, 944

Eldridge, J. J., Stanway, E. R., Xiao, L., McClelland, L. A. S., et al., 2017 *PASA*, 34, 58

Eldridge, J. J. & Stanway, E. R., 2022 *arXiv*, arXiv:2202.01413

Götberg, Y., de Mink, S. E., & Groh, J. H., 2017 *A&A*, 615, 78

Götberg, Y., de Mink, S. E., Groh, J. H., Kupfer, T., Crowther, P. A., Zapartas, E., & Renzo, M., 2018 *A&A*, 608, 11

Götberg, Y., de Mink, S. E., Groh, J. H., Leitherer, C., & Norman, C., 2019 *A&A*, 634, 134

Götberg, Y., de Mink, S. E., McQuinn, M., Zapartas, E., Groh, J. H., & Norman, C., 2022 *A&A*, 634, 134

Gronke, Max, 2017 *A&A*, 608, 139

Kinney, A. L., Bohlin, R. C., Calzetti, D., Panagia, N., & Wyse, R. F. G., 1993 *ApJS*, 86, 5

Lee, J. C., Gil de Paz, A., Tremonti, C., Kennicutt, R. C., Jr., Salim, S., et al., 2009 *ApJS*, 706, 599

Leitherer, C., Schaerer, D., Goldader, J. D., et al., 1999 *ApJS*, 123, 3

Leitherer, C., Ortiz Otálvaro, P. A., Bresolin, F., et al., 2010 *ApJS*, 189, 309

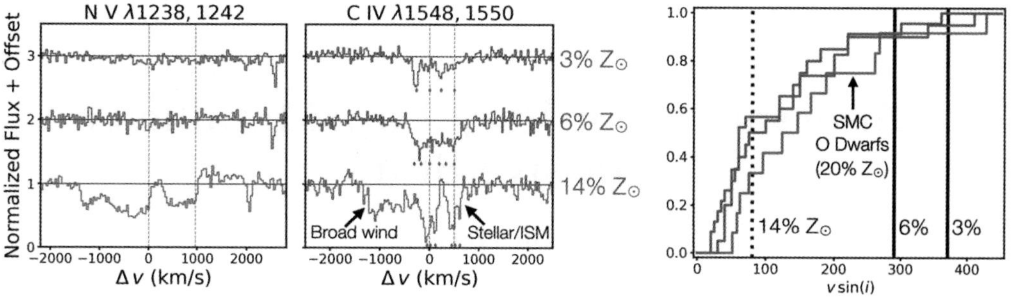

Figure 6. *Left:* Extremely low metallicity stars show weak stellar wind features, as demonstrated by the spectra of the LP26 star in Leo P (purple), the S3 star in Sextans A (blue), and the A15 star in WLM (red). The spectra are continuum-normalized, then offset, and plotted as a function NV 1238 Å, OV 1371 Å, SiIV 1393 Å, and CIV 1548 Å. Diamonds below each spectrum indicate the velocities of ISM SiII and CII absorption components detected. Broad, blueshifted CIV wind absorption is also present. A15 also shows wind absorption and emission in the NV and OV lines. All stars' SiIV lines are likely due to photospheric and ISM absorption rather than winds. *Right:* The cumulative distribution of $v{\cdot}\sin(i)$ measured for O dwarfs in the SMC are plotted as colored lines, where each is from a different sample/study (Penny & Gies 2009; Bouret et al. 2013; Ramachandran et al. 2019). In comparison, the $v{\cdot}\sin(i)$ measured for the three metal-poor O dwarfs from Telford et al. (2021) are indicated with black vertical lines. Both LP26 and S3 have $v{\cdot}\sin(i)$ higher than 90% of the measurements for SMC O dwarfs, The two most metal poor stars, LP26 and S3, have $v{\cdot}\sin(i)$ higher than 90% of the measurements for SMC O dwarfs. Both figures adapted from Telford et al. (2021).

Levesque, E. M., Leitherer, C., Ekstrom, S., Meynet, G., & Schaerer, D. 2012, *ApJ*, 751, 67
McGaugh, S. S., Rubin, V. C., & de Blok, W. J. G., 2001 *AJ*, 122, 2381
Murray, N., Quataert, E., & Thompson, T. A., 2005 *ApJ*, 618, 569
Olivier, G.M., Berg, D. A., Chisholm, J., Erb, D.K., Pogge, R.W., & Skillman, E.D., 2022 accepted to *ApJ*
Penny, L. R. & Gies, D. R., 2009 *ApJ*, 700, 844
Ramachandran, V., Hamann, W. R., Oskinova, L. M., et al., 2019 *A&A*, 625, A104
Rigby, J. R., Bayliss, M., Sharon, K., Gladders, M., Chisholm, J., et al., 2018 *AJ*, 155, 104
Rivera-Thorsen, T. E., Hayes, Matthew, Östlin, G., et al., 2015 *ApJ*, 805, 14
Schaerer, D., 2003 *A&A*, 397, 527
Schaerer, D., Fragos, T., & Izotov, Y. I., 2019 *A&A*, 622, 10
Senchyna, P., Chisholm, J., Stark, D., Berg, D. A., James, B. L., et al. *in preparation*
Shapley, A. E., Steidel, C. C., Pettini, M., & Adelberger, K. L., 2003 *ApJ*, 558, 65
Stanway, E. R., Eldridge, J. J., & Becker, G. D., 2016 *MNRAS*, 456, 485
Telford, O. G., Chisholm, J., McQuinn, K. B. W., & Berg, D. A., 2021 *ApJ*, 922, 191
Thuan, T. X. & Izotov, Y. I., 2005 *ApJ*, 161, 240
Veilleux, S., Cecil, G., & Bland-Hawthorn, J., 2005 *ARAA*, 43, 769
Verhamme, A., Orlitová, I., Schaerer, D., & Hayes, M., 2015 *A&A*, 578, A7
Whitaker, K. E., Franx, M., Leja, J., et al., 2014 *ApJ*, 795, 104
Wofford, A., Leitherer, C., & Salzer, J., 2013 *ApJ*, 765, 118

Massive Stars Near and Far
Proceedings IAU Symposium No. 361, 2024
J. Mackey, J. S. Vink, & N. St-Louis, eds.
doi:10.1017/S174392132200312X

Observations of Very Low-Metallicity Massive Stars

Miriam Garcia

Centro de Astrobiología (CAB), CSIC-INTA
Carretera de Ajalvir km4, 28850 Torrejón de Ardoz, Madrid, Spain
email: mgg@cab.inta-csic.es

Abstract. Very metal-poor massive stars hold the key to interpret high-redshift star-forming galaxies and the early reionization epoch, but also contemporary events such as gravitational waves. To study these objects in resolved environments, we need to resort to dwarf irregular galaxies far from the potential wells of M31 and the Milky Way, and therefore distant. While the archives, recently boosted by the ULLYSES and XSHOOTU programs, store a healthy dataset of massive stars in the Milky Way and the Magellanic Clouds, the number of observed targets with poorer metal content than the SMC (1/5 $Z_\odot$) is dramatically small. This paper reviews the state of observations of very metal-poor massive stars, assessing what can be realistically learned about their physics and evolution with current instrumentation, and arguing whether or not near-future facilities can remedy the gaps in the knowledge that remain.

Keywords. Low metallicity, Massive stars, Spectroscopic observations, Spectroscopic analyses

---

1. Introduction

Considering the monotonic (even if anisotropic) growth of metallicity (Z) of the Universe after the Big Bang, metal-poor massive stars are fundamental to model and interpret past cosmic epochs. Studies of high-redshift star forming galaxies, the use of γ-ray bursts (GRBs) and superluminous supernovae (SLSNe) as distance and star-formation rate diagnostics, the formation of the first stars, and the true identity of the sources that powered the reionization of the Universe, are some prominent open fronts that hold very metal-poor massive stars at their core. The interest in these objects was further rekindled by gravitational wave experiments reporting $\sim$30 $M_\odot$ double black holes, suggesting that the amount of mass lost during evolution is small, hence that low-metallicity may favor the formation of progenitors.

Up to now, massive stars in the Small Magellanic Cloud (SMC, 1/5 $Z_\odot$) make the current reference for the metal-poor Universe. Extensive samples and batteries of observations exist that enable studying the physics of these objects to high detail. However, we need to push further to lower metallicity values akin to the average metal content of the Universe at redshifts earlier than z=1 (e.g. Madau & Dickinson 2014) in order to tackle the problems described before.

The physical properties of sub-SMC metallicity massive stars during their evolution are still unknown. The weaker radiation-driven winds that can be powered by the decreased amount of iron, compared to solar analogs, induces strong changes in evolution. The smaller angular momentum loss favors fast rotating stars which may enter chemically homogeneous evolution (CHE). Pure CHE is a whole new channel of evolution that has never been observed to date, but which may have a dramatic impact on the feedback of metal-poor massive star populations (Szécsi et al. 2015). This is just an example of how important it is to characterize the evolutionary pathways of metal-poor massive stars. However, until we have well-populated (complete if possible) samples in very metal-poor galaxies we will not know how massive stars evolve, nor will we be able to link them as progenitors of GRBs, SNe and SLSNe. The role of binary evolution in these environments is also uncharted territory at the moment.

The past two decades have witnessed valiant efforts to improve our knowledge on these objects, by studying massive stars in resolved very metal-poor galaxies of the Local Group and vicinity with 8 and 10 m telescopes. They commenced with the exploratory works on WLM, IC 1613 and NGC 3109 framed in the Araucaria project (Bresolin et al. 2006, 2007; Evans et al. 2007). These galaxies were selected because of their low oxygen nebular abundances ($N(O/H) \sim 1/7\, N(O/H)_{\odot}$, hereafter denoted as $1/7\, O_{\odot}$), later-on confirmed by the analysis of early-B supergiants. Yet, independent studies of the optical spectrum of late-B/early-A supergiants (Hosek et al. 2014) and the ultraviolet spectrum of O-type stars (Garcia et al. 2014; Bouret et al. 2015), showed that the abundance of Fe-group elements was actually $1/5\, Fe_{\odot}$ and similar to the SMC's. Since radiation-driven winds are largely responsible for the expected changes in the evolution of very metal-poor massive stars, and mass loss is driven by Fe-group elements, these results implied that WLM, IC 1613 and NGC 3109 are not fit to study the very metal-poor regime.

Thus began the exploration of other galaxies at the very outskirts of the Local Group and beyond. The list includes systems with promising low nebular abundances of $1/10\, O_{\odot}$ (Sextans A), $1/20\, O_{\odot}$ (Leo A, SagDIG) and $1/30\, O_{\odot}$ (Leo P). Low iron abundances have already been confirmed for young stars in Sextans A and Leo P (Kaufer et al. 2004; Garcia et al. 2017; Telford et al. 2021). However, the galaxies are distant (Leo A – 0.8 Mpc, SagDIG – 1.1 Mpc, Sextans A – 1.3 Mpc, Leo P – 1.7 Mpc) or suffer from severe foreground extinction (SagDIG) and observations are challenging. In the following pages we will list the parameters needed to characterize the evolution of sub-SMC metallicity massive stars, we will advise an observational strategy to constrain them, and we will finally discuss what results can be expected according to the capabilities of current and near-future facilities.

2. Strategy

In order to characterize the evolution of massive stars at any metallicity, the following stellar parameters must be constrained for a large sample that covers different states of evolution: effective temperature $T_{\rm eff}$, gravity log g, luminosity $L_{\rm bol}$ (from these, stellar radii R_* and present-day masses M_* follow), and mass loss rate ($\dot{M}$). Rotational velocities (for which we measure the projection into the line of sight $v\sin i$) and surface abundances (He, C, N, O) of blue massive stars are needed to understand the mixing mechanisms. Understanding the role of binary evolution requires following-up a healthy sample of systems to obtain statistics for the period and mass fraction distributions, and identifying and studying post-binary interaction products (BIPs).

Since no massive stars were previously known in the sub-SMC metallicity galaxies, and bearing in mind that photometry does not suffice to assign spectral types (on the blue side O and B subtypes cannot be distinguished, and red supergiants may be confused with foreground red-clump stars), a four step approach is recommended:

1 Census: Optical-range spectroscopy of photometry-selected candidates can confirm massive stars and assign spectral sub-types. It may also provide the first list of binary candidate systems. Spectroscopic resolving power of $R=\lambda/\Delta\lambda \sim 700$–$1000$ and signal-to-noise ratio $S/N \sim 50$ suffice, although higher S/N would provide suitable data for step 2. The census must aim for completeness and contain of the order of 100–1000 stars.

2 Photospheric parameters: $T_{\rm eff}$ and log g can be constrained from R=1000, S/N~100 optical spectra (although see Sect. 3.1). Measuring projected rotational velocities and element abundances of blue-type massive stars, however, require a higher resolving power of R=5000. Ideally, photospheric parameters would be derived for all the stars in the census, although a sample of hundreds of stars should suffice.

3 Mass loss rates of all the stars studied in step 2 are needed in order to fulfill their evolutionary status, yet this is the most elusive parameter of all. The dust-driven $\dot{M}$ of red supergiants can be derived from their mid-infrared excess or by characterizing their circumstellar material (Mauron & Josselin 2011; Beasor et al. 2020). The winds of Wolf-Rayet (WR) stars and Luminous Blue Variables (LBV), should one such specimen be found, can be characterized from the data taken in step 2. As for OB-stars, however, only the ultraviolet range (UV) contains suitable $\dot{M}$ diagnostics because the Balmer series is not sensitive to the weak radiation-driven winds expected at sub-SMC metallicity. Considering the long required observing time induced by non-negligible extinction towards some of the candidates, and HST's oversubscription factor and remaining operation time, a sample of ten-some UV spectra of OB-stars can be realistically expected.

4 Multi-epoch observations: Finally, high spectral resolution follow-up of the candidate binaries detected in steps 1 and 2 is needed to produce the distribution of orbital periods and mass fractions of the systems. A dedicated monitoring program to detect additional systems would be necessary, but likely unfeasible considering the long observing times needed to reach massive stars at ∼1 Mpc with sufficient S/N. Specific techniques to pin down and characterize BIPs (e.g. stripped stars, Götberg et al. 2018) will also provide invaluable information on the role of binary evolution at this metallicity.

2.1. *Available instrumentation*

From this section on we will focus on the observations of OB-type stars. At a distance of ∼1 Mpc (25 distance modulus) the faintest O dwarfs will have V∼21. Obtaining S/N≥50 optical spectra of such faint V-magnitudes requires very long observing times and therefore multi-object spectroscopy (MOS) is the optimal choice. The lowest end of the initial mass function, the B dwarfs, will basically be out of reach.

Near-infrared (NIR) spectroscopy could theoretically enable the steps described in the previous section, however, stars are fainter in this spectral range ($(V-J)_0 \sim$ -0.65, $(V-K)_0 \sim$ -0.85, Martins & Plez 2006) and observations are further challenged by the strong background emission of the Earth's atmosphere. The short exposure times imposed by the sky coherence time in the K-band make observations of O-type stars at ∼1 Mpc unfeasible in this spectral range and challenging even for the Extremely Large Telescope (ELT). This may be a problem for observing very metal-poor massive stars with first-light ELT instrumentation, optimized for near- and mid-infrared wavelengths.

Figure 1 summarizes suitable instruments as a function of the number of objects they can observe simultaneously (multiplexing) and their resolving power. All 8 and 10m class telescopes have workhorse R∼500–2000 spectrographs operating in the optical range with long-slit and mask-based MOS modes. These can be used to build the census of massive stars within a galaxy and the spectra could also be used to determine $T_{\rm eff}$ and log g. Yet, the multiplexing is limited and the required observing campaigns would be of the order of ∼ 70 hours long (see e.g. Lorenzo et al. 2022a).

Large field, high spatial resolution integral field spectrographs (IFS) are much more efficient to register the population of massive stars within a galaxy. Since observations are not targeted, IFS have the added capability of unveiling reddened stars and other objects that cannot be easily identified from their photometric colors (e.g. WRs and LBVs). With $1' \times 1'$ field of view and $0.2'' \times 0.2''$ spatial sampling, MUSE at the Very Large Telescope (VLT) has an enormous discovery potential even though its $\lambda >$ 4650Å

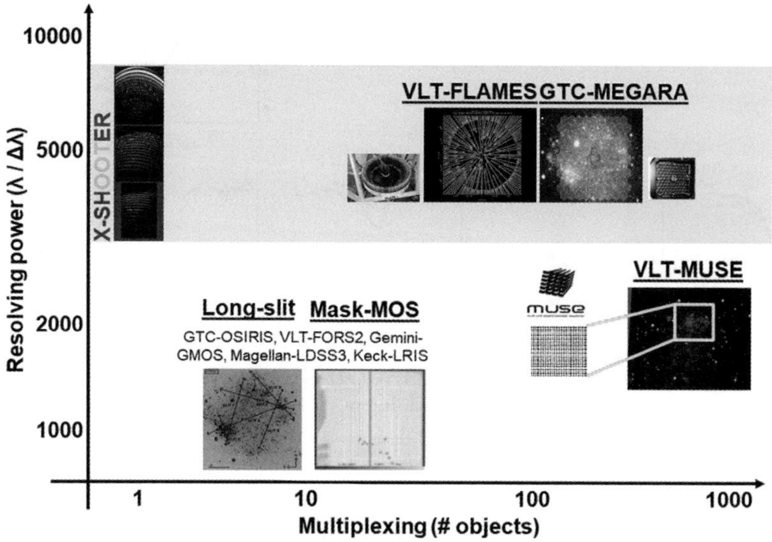

Figure 1. Summary of spectroscopic instrumentation suitable to study massive stars in Local Group galaxies. Instruments have been sorted according to spectral resolving power and multiplexing. Additional important parameters such as wavelength coverage, field of view, fiber aperture and fiber patrol area have not been considered.

wavelengh coverage is not optimal to study blue massive stars. Its successor, BlueMUSE (Richard et al. 2019), will unite MUSE's unique IFS capabilities with blue wavelength coverage, and will mean a breakthrough for the studies of massive stars in the Local Group. By targeting entire populations at once, both MUSE and BlueMUSE are also very well suited to spot spectroscopic binary systems.

Measuring element abundances and projected rotational velocities means a qualitative step up in spectral resolution to R$\geq$5000 and requires more sophisticated instrumentation. VLT-XSHOOTER has outstanding sensitivity and registers the spectra of objects from the optical range all the way to the NIR, with the required resolution. However, it only enables observations of one or two objects at once, hence observing a sample of $\sim$100 V$\sim$21 stars would be prohibitive in observing time. VLT-FLAMES and its 132 MEDUSA fibers meet the multiplexing requirement although its radial fiber positioning mechanism is not optimal for small galaxies, and the GIRAFFE spectrograph is rather insensitive in the blue spectral range. Furthermore, the fibers have $1.2''$ apertures on the sky, which hinders a good sky subtraction and will not resolve moderately close sources.

MEGARA mounted at the Gran Telescopio Canarias (GTC) makes for an interesting alternative (Gil de Paz et al. 2022). Similarly to FLAMES, it covers the whole 3600-5200Å range with R$\sim$5000 with two settings. However, its design is optimal to obtain spectroscopy of resolved objects within small fields. The instrument has 100 robot positioners for fibers, evenly distributed at the focal plane. Each of them can patrol an hexagon, and together they achieve coverage of the $3.5' \times 3.5'$ field of view. This fiber distribution is less prone to fiber obstruction as opposed to circle-based positioners such as FLAMES, and therefore it is better suited to study populations of massive stars that tend to be spatially concentrated. The instrument has the added advantage that each robot positioner handles a bundle of 7 fibers of $0.67''$ diameter each that keep information on the sky emission around each target. MEGARA will be crucial for advanced studies of very metal-poor massive stars once the robot positioners are commissioned and the multi-object mode is online.

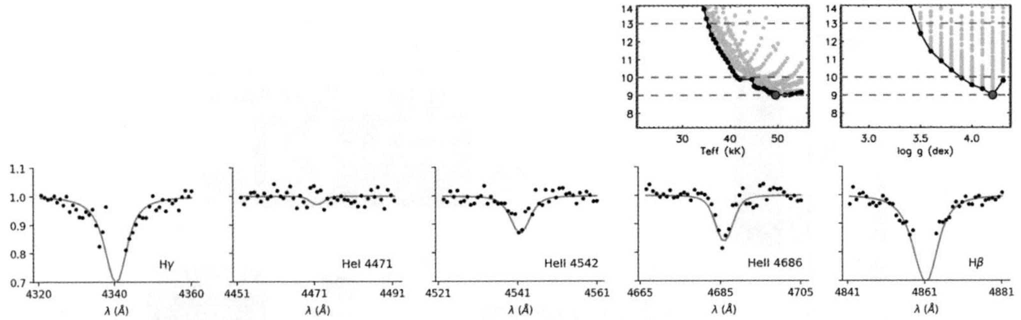

Figure 2. Preliminary fit to star LSS3.OB13206 (O5 Vz, V∼20.4) from Lorenzo et al. in prep. The spectrum (S/N=110) was analyzed with IACOB-GBAT. The top panels show well-defined, though broad, χ^2 distributions for $T_{\rm eff}$ and log g. The bottom panels show the best-fitting model for strategic Balmer, HeI and HeII lines.

3. Current status and a critical view of recent results

Even though previous work existed on massive stars in Local Group galaxies beyond the Magellanic Clouds, the work on very metal-poor dwarf irregulars has begun only recently. In this section we review progress according to the strategy outlined in Sect. 2, and discuss the precision of the derived parameters and the potential of different datasets.

3.1. *Census and first stellar parameters $T_{\rm eff}$, log g, vsin i*

Great advances have been made on registering massive stars in metal-poor galaxies from low-resolution spectroscopy. Ten years ago, only 6 stars had been observed in Sextans A (Kaufer et al. 2004), 10 in Leo A (Brown et al. 2007), SagDIG had not been explored and Leo P had not yet been discovered. Nowadays, the presence of massive stars has been confirmed in SagDIG (Garcia 2018) and Leo P (Evans et al. 2019), and Leo A is subject to close scrutiny (Gull 2022). Moreover, an extensive census of 150 OB-type stars in Sextans A is about to be published (Lorenzo et al. 2022a) and 7 RSGs have been reported in this galaxy (Britavskiy et al. 2019).

The low-resolution discovery spectra also enable the first spectroscopic analyses to constrain stellar parameters. Figure 2 shows the best fitting model obtained with the χ^2 algorithm IACOB-GBAT that runs on FASTWIND models (Simón-Díaz et al. 2011; Puls et al. 2005) for the R∼1000, S/N=110 spectra of an O-type star in Sextans A. While the algorithm returns a good fit and well-defined χ^2 distributions for $T_{\rm eff}$ and log g with clear minima, the distributions are wide and the error bars will be large. Poorer results are anticipated for most of the stars in the discovery census, observed with lower S/N, and for which log g will be largely undetermined (see also Camacho et al. 2016). In other words, R∼1000 spectra may not yield precise effective temperatures and gravities even if the S/N is high, and in the worse case scenario it may not constrain log g at all. Projected rotational velocities (therefore, macroturbulence) will neither be constrained unless $v\sin i$ is extremely high (see Fig. 3). At least R≥2000 and outstanding S/N spectra are needed to measure $v\sin i$.

VLT-MUSE has the potential to enhance the census of very metal-poor massive stars by one order of magnitude. It delivered the first list of OB-type stars in Leo P (Evans et al. 2019), and it also shows great promise out to 2 Mpc in the galaxy NGC 300 (McLeod et al. 2020). Recent MUSE observations in Sextans A (PI F. Tramper) will finally mine the population of OB-type stars in the large HII shells at the East of Sextans A.

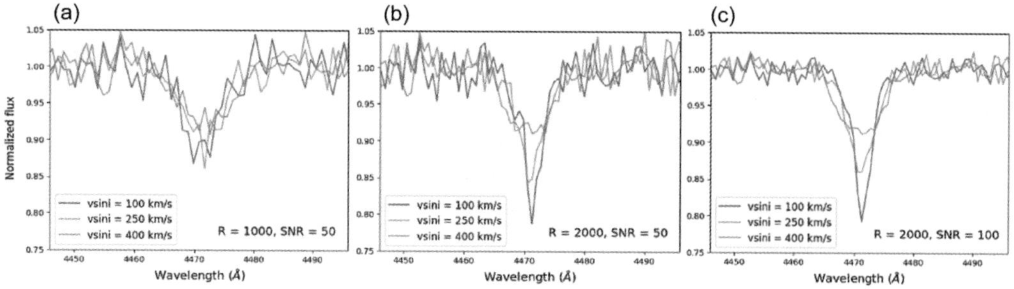

Figure 3. The HeI4471 line from a FASTWIND model has been broadened according to different *v*sin *i* values and simulating configurations with different spectral resolutions. Since metallic lines are very weak in the optical spectra of O-stars, *v*sin *i* will be determined from HeI lines. The rotational profiles cannot be distinguished at the S/N=50, R∼1000 typical spectra of discovery observations (panel a). An increase in resolution only (b) still cannot differentiate 100 and 250 km/s profiles, nor resolve the wings of the 400km/s model. Only at S/N= 100 and R = 2000 the rotational profiles can be discerned (c). *Courtesy: M. Lorenzo.*

MUSE's R=2000 spectral resolution is apt for quantitative spectroscopic analysis. However, the 4650-9300Å covered spectral range, modulated by the spectral energy distribution of O- and early-B type stars that make the observed fluxes to quickly drop at $\lambda > 7000$Å, offers only a limited number of diagnostic lines to study these objects (e.g. Martins & Palacios 2021). Effective temperatures and gravities of O-type stars can be determined from $H\beta$ and the He ionization equilibrium, yet S/N is again critical. McLeod et al. (2020) derived $T_{\rm eff}$ and log g of NGC 300 stars but with ± 3000 K and ±0.3 dex error bars respectively. In a MUSE study of stars in the Large Magellanic Cloud observed with better quality (S/N> 50), Castro et al. (2021) obtained more precise effective temperatures with typical errors of ± 1000 K, but gravity just as poorly constrained. Moreover, when comparing results for overlapping stars with the VLT-FLAMES Tarantula Survey (VFTS), Castro et al.'s MUSE log g estimates are systematically lower than those from the VFTS by ∼0.3 dex. Whether this offset is caused by MUSE's lower spectral resolution or by log g being determined mostly from $H\beta$, which is also sensitive to winds, remains to be further tested.

3.2. *Abundances and R∼5000 optical studies*

To assess whether constraining the surface abundances of very metal-poor massive stars is feasible, we resort to previous similar studies in IC 1613 (∼750 Kpc), NGC 3109 (∼1 Mpc) and NGC 55 (∼2 Mpc). Element abundances of He, C, N, O, Si and Mg can be derived from the R∼1000, 4000–5000Å spectra of early-B supergiants with a precision of ±0.2 dex as long as S/N≥100 (Bresolin et al. 2007; Castro et al. 2012). Likewise, the analysis of high quality spectra of late-B and early-A supergiants in the 4000–5500Å range can yield a global value for the abundances of Fe-group elements Fe, Cr and Ti (Hosek et al. 2014). In this case S/N≥100 warrants ±0.1 dex precision, while poorer S/N results in ±0.25 dex error bars or larger.

Deriving CNO surface abundances of O-type stars is even more challenging. Grin et al. (2017) provide some illustrative examples observed with VLT-FLAMES. The lines are weak and extremely high S/N, together with R≥5000, are required to ensure their detection and proper analysis (see also e.g. Martins et al. 2015; Carneiro et al. 2019). Alternatively, CNO abundances could be derived from photospheric ultraviolet lines (e.g. Bouret et al. 2013) although high R≥10000 resolution spectroscopy that allows disentangling the interstellar from the stellar component is again necessary.

What are the prospects for R≥5000 optical spectroscopy of very metal-poor massive stars? The highly sensitive VLT-XSHOOTER spectrograph would seem an ideal choice with its ample spectral coverage providing simultaneous observations of all the classical spectral diagnostics in the 4000-5500Å spectral range plus $H\alpha$. Tramper et al. (2011, 2014) obtained XSHOOTER observations of 10 O-type stars in IC 1613, WLM and NGC 3109 with R∼6000 in the UVB channel and S/N∼50-90 per resolution element. Their analysis obtained $T_{\rm eff}$ and log g with typical error bars of ±2000 K and ±0.2 dex or larger for log g, while the spectral quality prevented constraining CNO abundances. A bolder approach is needed, aiming for S/N well in excess of 100.

Considering that Tramper et al. devoted 3–4 hours for each NGC 3109 target, and the number of targets to be ideally observed in order to provide useful constraints to stellar evolution, a higher multiplexing spectrograph with similar spectral resolution is needed. The apparently obvious choice, FLAMES, lacks the required sensitivity in the UVB range. Moreover, the fiber positioning mechanism is not optimal to study resolved stars in very metal-poor galaxies, which have diameters smaller than 7′, as argued in Sect. 2.1. Because of its even distribution of fibers in the focal plane and its comparatively higher sensitivity in the 4000-5000Å spectral range, MEGARA is the optimal choice to obtain R∼5000 spectroscopy of very metal-poor massive stars and to determine surface abundances.

3.3. *The winds of very metal-poor OB-type stars from UV spectroscopy*

Diagnostic lines for the radiation-driven winds of blue-type stars exist across the electromagnetic spectrum, from X-ray to radio wavelengths (Kudritzki & Puls 2000). However, because very metal-poor massive stars are located at ∼1 Mpc, most of these stars are only accessible to space UV and optical facilities at the moment. Spectroscopy in the optical range would enable characterizing the strong winds of LBVs and WRs should one of these objects be found (Herrero et al. 2010; Tramper et al. 2013), although this is highly unlikely because radiation-driven mass loss decreases with metallicity (Mokiem et al. 2007; Björklund et al. 2021). For this same reason the classical diagnostic lines for the winds of OB-type stars in the optical range, the higher Balmer lines and HeII4686, will likely be insensitive to the expected low $\dot{M}$ values. This renders the strong resonance lines at 900–1900Å as the only means to characterize the winds of very metal-poor OB-type stars. The first UV observations of massive stars in Sextans A demonstrated that the spectra do show wind features (e.g. Fig. 4).

Ultraviolet spectroscopy comes with additional advantages. For moderate to weak winds, the terminal wind speed v_∞ can only be measured from the P Cygni profiles and blue troughs exhibited by the sensitive UV lines (Kudritzki & Puls 2000). On the other hand, the UV spectra of OB-type stars host a wealthier variety of metallic lines than the optical range. For instance, it is possible to measure CNO abundances of O-stars (Sect. 3.2) but, even more interestingly, Fe-group abundances can be constrained from multiple lines that deplete the continuum (Fig. 4). UV spectroscopy has been fundamental to demonstrate that the Leo P and Sextans A galaxies are not only more oxygen-poor but also more iron-poor than the SMC (Telford et al. 2021; Garcia et al. 2017).

Despite providing important information, UV spectroscopy exists only for a small number of very metal-poor OB-stars: 3 stars in Leo A, 9 stars in Sextans A and 1 star in Leo P. These numbers will improve slightly thanks to the *Hubble UV Legacy Library of Young Stars as Essential Standards* Director's discretionary program (ULLYSES†) which has among its goals to provide a library of very metal-poor massive stars. Unfortunately, the target list was severely limited by the number of objects that was known when the program was designed and only 3 additional stars in Sextans A will be included.

† `https://ullyses.stsci.edu/`

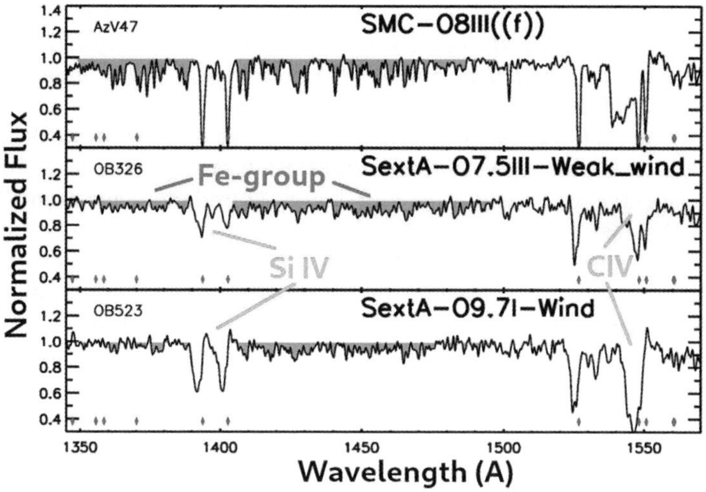

Figure 4. HST-COS observations of two O-type stars of Sextans A, compared to the HST-STIS spectrum of an SMC star of similar spectral type convolved to the equivalent spectral resolution. Diamond marks signal interstellar lines. The pink-shaded areas mark the location of multiple lines of iron-goup elements, roughly corresponding to FeVI and FeV. The stronger absorptions in the spectrum of the analog SMC star in these regions indicate that the Sextans A stars are more iron-poor. Both Sextans A stars exhibit SiIV and CIV wind profiles that will enable $\dot{M}$ and v_∞ determinations. The latter can be fully constrained in OB523, with a stronger wind and a well-developed CIV P Cygni profile.

Nonetheless, ULLYSES has provided another very important product: HST-WFC3 UV photometry of Sextans A that will enable more realistic exposure time estimates for COS spectroscopy, and an easier follow-up of interesting targets.

3.4. *Binary systems*

The number of binary systems in sub-SMC metallicity galaxies is scarce. Until now only one eclipsing binary has been studied (Bonanos 2013) and Lorenzo et al. (2022b) just delivered the first list of spectroscopic binary candidates. The number of known systems will likely increase thanks to the new MUSE observations of Sextans A, although the massive, multi-epoch, R∼5000 follow-up needed to produce period and mass fraction distributions is at the limit of current instrumentation (see Sect. 2.1).

The existing photometric databases may be more amenable to find post-binary interaction products. Lorenzo et al. (2022b) found a handful of O-type stars in Sextans A with bluer colors than the ZAMS that have been postulated as candidate stripped stars. Schootemeijer et al. (2022) identified a parallel feature to the main sequence in the $V - I$ *vs* V color-magnitude diagrams of very metal-poor galaxies, which is mostly populated by Be stars at absolute magnitudes fainter than $M_V > -5$ (see also Gull 2022). All these cases require confirmation by spectroscopic analysis at least for a subset of the candidates, but the possibility of identifying BIPs from photometry only greatly reduces the cost in observing time.

4. Summary and concluding remarks

Great advances are being made in the field of very-metal poor massive stars. Several groups are working in parallel to answer, in the long run, important questions such as *What is the evolutionary sequence of these objects?, Can we relate it to the relative frequency of GRBs, SNe and SLSNe?, What is the incidence of chemically-homogeneous evolution?, Does the metallicity-dependence of radiation driven winds hold at extremely low-Z?, What is the role of binary evolution and the contribution of BIPs to the integrated ionizing flux of massive star populations?, Do metal-poor environments really favor the formation of double black holes?*

Lacking a previous list of massive stars adequate for these studies, work began at unveiling the stars, which is already very demanding in observing time at 8–10 m telescopes. Some of these spectra will yield $T_{\rm eff}$ and log g, abundances for some B-supergiants, and $v\sin i$ for the fast rotators, but the bulk of the quantitative results will require S/N$\geq$100, R$\sim$5000 spectroscopy feasible with a very reduced number of spectrographs worldwide. UV spectroscopy exists for about a dozen objects but, because of extinction and the uncertain remaining lifetime of HST, the realistic expectation is that the sample will only be doubled at best. This is insufficient to constrain stellar winds at any given metallicity regime.

The advent of the ELT will supply the sensitivity needed to obtain spectra of very-metal poor massive stars with the required quality. However, its instrumentation is not optimally designed for the science of large samples of very metal-poor massive stars. The work carried out *today* by the research groups will be fundamental to craft the most optimal list of targets for follow-up with ELT-HARMONI or MOSAIC. As for sensitivity-enhanced UV observations, we can only enthusiastically support the construction of the next large IR/optical/UV observatory in space†.

Acknowledgements

M.G. gratefully acknowledges support by grants PID2019-105552RB-C41 and MDM-2017-0737 Unidad de Excelencia "María de Maeztu"-Centro de Astrobiología (CSIC-INTA), funded by the Spanish Ministry of Science and Innovation/State Agency of Research MCIN/AEI/10.13039/501100011033.

References

Beasor, E. R., Davies, B., Smith, N., *et al.* 2020, MNRAS, 492, 5994
Björklund, R., Sundqvist, J. O., Puls, J., et al. 2021, A&A, 648, A36.
Bonanos, A. Z. 2013, IAU Symposium, 289, 173
Bouret, J.-C., Lanz, T., Martins, F., *et al.* 2013, A&A, 555, A1
Bouret, J.-C., Lanz, T., Hillier, D. J., *et al.* 2015, MNRAS, 449, 1545
Bresolin, F., Pietrzyński, G., Urbaneja, M. A., *et al.* 2006, ApJ, 648, 1007
Bresolin, F., Urbaneja, M. A., Gieren, W., Pietrzyński, G., & Kudritzki, R.-P. 2007, ApJ, 671, 2028
Britavskiy, N. E., Bonanos, A. Z., Herrero, A., *et al.* 2019, A&A, 631, A95
Brown, W. R., Geller, M. J., Kenyon, S. J., *et al.* 2007, ApJ, 666, 231
Camacho, I., Garcia, M., Herrero, A., & Simón-Díaz, S. 2016, A&A, 585, A82
Carneiro, L. P., Puls, J., Hoffmann, T. L., *et al.* 2019, A&A, 623, A3
Castro, N., Urbaneja, M. A., Herrero, A., *et al.* 2012, A&A, 542, A79
Castro, N., Crowther, P. A., Evans, C. J., *et al.* 2021, A&A, 648, A65
Evans, C. J., Bresolin, F., Urbaneja, M. A., *et al.* 2007, ApJ, 659, 1198
Evans, C. J., Castro, N., Gonzalez, O. A., *et al.* 2019, A&A, 622, A129
Garcia, M., Herrero, A., Najarro, F., Lennon, D. J., & Alejandro Urbaneja, M. 2014, ApJ, 788, 64
Garcia, M., Herrero, A., Najarro, F., *et al.* 2017, The Lives and Death-Throes of Massive Stars, 329, 313
Garcia, M. 2018, MNRAS, 474, L66
Gil de Paz *et al.* 2022, A&A, submitted
Götberg, Y., de Mink, S. E., Groh, J. H., et al. 2018, A&A, 615, A78

† `https://nap.nationalacademies.org/resource/26141/interactive/`

Grin, N. J., Ramírez-Agudelo, O. H., de Koter, A., *et al.* 2017, A&A, 600, A82
Gull, M. 2022, these conference proceedings
Herrero, A., Garcia, M., Uytterhoeven, K., et al. 2010, A&A, 513, A70
Hosek, M. W., Jr., Kudritzki, R.-P., Bresolin, F., *et al.* 2014, ApJ, 785, 151
Kaufer, A., Venn, K. A., Tolstoy, E., Pinte, C., & Kudritzki, R.-P. 2004, AJ, 127, 2723
Kudritzki, R.-P., & Puls, J. 2000, ARA&A, 38, 613
Lorenzo, M., Garcia, M., & Najarro, F. 2022a, these conference proceedings
Lorenzo, M., Garcia, M., & Najarro, F., *et al.* 2022b, MNRAS, accepted
Madau, P., & Dickinson, M. 2014, ARA&A, 52, 415
Martins, F., & Plez, B. 2006, A&A, 457, 637
Martins, F., Simón-Díaz, S., Palacios, A., *et al.* 2015, A&A, 578, A109
Martins, F. & Palacios, A. 2021, A&A, 645, A67
Mauron, N. & Josselin, E. 2011, A&A, 526, A156
McLeod, A. F., Kruijssen, J. M. D., Weisz, D. R., *et al.* 2020, ApJ, 891, 25
Mokiem, M. R., de Koter, A., Vink, J. S., et al. 2007, A&A, 473, 603
Puls, J., Urbaneja, M. A., Venero, R., et al. 2005, A&A, 435, 669
Richard, J., Bacon, R., Blaizot, J., et al. 2019, arXiv:1906.01657
Schootemeijer, A., Lennon, D. J., Garcia, M., *et al.* 2022, these conference proceedings
Simón-Díaz, S., Castro, N., Herrero, A., et al. 2011, Journal of Physics Conference Series, 328, 012021.
Szécsi, D., Langer, N., Yoon, S.-C., et al. 2015, A&A, 581, A15
Telford, O. G., Chisholm, J., McQuinn, K. B. W., *et al.* 2021, ApJ, 922, 191
Tramper, F., Sana, H., de Koter, A., & Kaper, L. 2011, ApJ, 741, L8
Tramper, F., Gräfener, G., Hartoog, O. E., et al. 2013, A&A, 559, A72
Tramper, F., Sana, H., de Koter, A., Kaper, L., & Ramírez-Agudelo, O. H. 2014, A&A, 572, AA36

Discussion

No time was left for questions.

Massive Stars Near and Far
Proceedings IAU Symposium No. 361, 2024
J. Mackey, J. S. Vink, & N. St-Louis, eds.
doi:10.1017/S1743921322002137

Global Properties of Star-Forming Galaxies from Ultraviolet Spectroscopy

Claus Leitherer[1], Ilyse Clark[2], Karla Arellano-Córdova[2] and Danielle A. Berg[2]

[1]Space Telescope Science Institute,
3700 San Martin Dr, Baltimore, MD 21218, USA
email: leitherer@stsci.edu

[2]Department of Astronomy, The University of Texas at Austin,
2515 Speedway, Stop C1400, Austin, TX 78712, USA
email: ilyseclark@austin.utexas.edu, kzarellano@austin.utexas.edu,
daberg@austin.utexas.edu

Abstract. We analyzed archival HST and IUE ultraviolet spectra of 29 nearby star-forming galaxies. The range of aperture sizes permits studies of the galaxy properties over pc to kpc scales. We measured line strengths and spectral energy distributions over the 1200 - 3000 Åwavelength range and established trends with galaxy properties. Updated oxygen abundances were measured from ancillary optical data. Star-formation rates and internal dust attenuations were derived from comparison with synthesis models. The interstellar absorption lines are heavily saturated, yet scale with oxygen abundance. We interpret this as due to macroscopic velocities arising in a turbulent ISM and large-scale outflows. The stellar-wind lines also scale with oxygen abundance. As these lines are shaped by mass loss, which is driven by the Fe abundance, we can study the α-element/Fe ratio in these galaxies.

Keywords. stars: early-type, galaxies: abundances, galaxies: star clusters, galaxies: starburst, galaxies: stellar content

1. Motivation

The space-ultraviolet (UV) permits the most comprehensive view of the stellar populations in star-forming galaxies with low dust obscuration (Berg et al. 2022; Leitherer 2020). Over the past several decades, the Hubble Space Telescope (HST) has accumulated a substantial body of spectroscopic data in the HST Archive. While most of our knowledge of the UV properties of star-forming galaxies is based on HST, prior UV missions can still add value to a discussion of the UV properties of these galaxies. In this contribution we present data obtained with the International Ultraviolet Explorer (IUE) satellite, which operated between 1978 and 1996, and compare them to recent complementary HST observations. IUE employed a 45 cm telescope with a spectrograph covering the wavelength range 1150 – 3300 Å at a resolution of ∼6 Å. Notably, its large entrance aperture of 10" × 20" is unmatched by HST's UV spectrographs. Here we perform a comparison with the HST Cosmic Origins Spectrograph (COS) which uses a 2.5" circular entrance aperture and provides spectra at a resolving power of $R \approx 15{,}000$. Our goals are threefold: (i) investigate aperture-size effects; (ii) determine star-formation properties from pc to kpc scales; (iii) recognize the limitations and opportunities of low-resolution and low S/N data in view of upcoming missions, such as the James Webb Space Telescope (JWST).

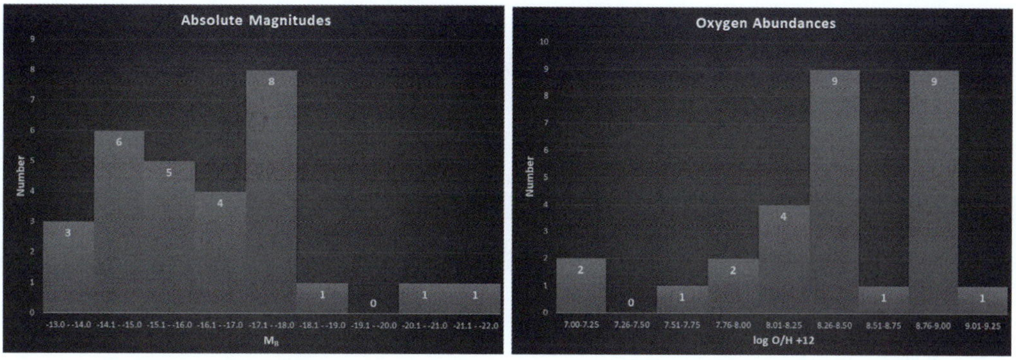

Figure 1. Histograms showing $M_{\rm B}$ (left) and log O/H + 12 (right) for the 29 galaxies.

2. Galaxy sample

We searched the IUE Archive and identified 29 galaxies with co-spatial IUE and HST spectra. We restricted our search to HST spectra obtained with COS and the Space Telescope Imaging Spectrograph (STIS). Here we discuss only the comparison with COS data. The 29 galaxies are 1: SBS 0335-052, 2: NGC 1705, 3: NGC 1741, 4: He 2-10, 5: IRAS 08339+6517, 6: I Zw 18, 7: NGC 3049, 8: NGC 3125, 9: NGC 3256, 10: Haro 2, 11: NGC 3310, 12: NGC 3351, 13: NGC 3353, 14: UGCA 219, 15: NGC 3690, 16: NGC 3991, 17: NGC 4194, 18: NGC 4214, 19: UGCA 281, 20: NGC 4449, 21: NGC 4670, 22: NGC 4861, 23: NGC 5236, 24: NGC 5253, 25: NGC 5457, 26: NGC 5996, 27: Tol 1924-416, 28: NGC 7552, 29: NGC 7714. Due to IUE sensitivity limitations, the galaxies generally are among the UV-brightest objects in the local universe at distances between a few to several tens of Mpc. In Fig. 1 we summarize the absolute blue magnitudes ($M_{\rm B}$) and the oxygen abundances (log O/H + 12) of the sample. The galaxies span a range of $-13 < M_{\rm B} < -22$, corresponding to very low-mass dwarfs and galaxies resembling Luminous Infrared Galaxies, with a typical luminosity similar to that of the Magellanic Clouds. The oxygen abundances mirror this trend. They fall into 7.0 < log O/H +12 < 9.2, i.e., between the lowest and highest abundances measured in local star-forming galaxies.

We processed both the IUE and HST spectra through the standard pipelines as well as with additional IDL and Python routines. In addition to standard processing, the spectra were corrected for Galactic foreground reddening and transformed into restframe wavelengths. Most IUE spectra cover the full wavelength range from 1200 to 3000 Å. Six galaxies lack near-UV spectra and terminate at 1950 Å. In Fig. 2 we compare the restframe spectra of NGC 7714 obtained with IUE and COS. This galaxy is a prototypical nuclear starburst (Weedman et al. 1981) with $M_{\rm B} = -20.0$ and log O/H + 12 = 8.3 at a distance of 39 Mpc. Note the overall agreement of the two spectra but at a much reduced spectral resolution of the IUE data.

3. Data analysis

The spectral energy distributions (SED) permit estimates of the intrinsic dust attenuation and current star-formation histories by comparing them to synthetic SEDs. The underlying physics is the independence of the spectral slope β around 1500 Å on the *stellar* properties, making it dependent only on the *dust* reddening. After reddening correction, the corrected SED is the dust-free UV luminosity, which can be converted into a UV star-formation rate (e.g., Leitherer et al. 2018). In our analysis we used theoretical SEDs generated with the Starburst99 code (Leitherer et al. 2014). The SEDs were

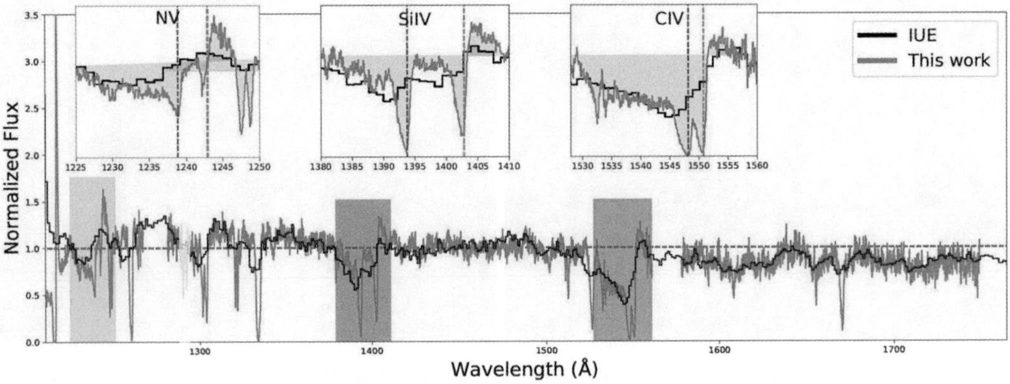

Figure 2. Comparison of IUE (black) and COS (green) spectra of NGC 7714. The lines in both spectra agree after accounting for the ∼50× lower IUE resolution.

obtained with stellar evolution models with rotation for solar ($Z_\odot$) and subsolar ($0.14Z_\odot$) chemical composition published by Ekström et al. (2012) and Georgy et al. (2013), respectively. Since these were the only chemical compositions available at the time this work was performed, we divided our sample into galaxies above and below log O/H + 12 = 8.3 and applied the solar and subsolar models accordingly. We assumed a Kroupa initial mass function (IMF; Kroupa 2008) and constant star formation rate over a time scale of 20 Myr. The IUE spectra indicate typical intrinsic dust reddening of $E(B-V) \approx 0.25$ and dust-corrected star-formation rates of ∼1 $M_\odot$ yr^{-1}.

We measured the equivalent widths (EW) of the strongest spectral lines in both the IUE and COS spectra. The discussion in this section refers to the measurements in the IUE data. The measured lines are Si II 1260, O I 1300, C II 1335, Si IV 1400, C IV 1550, Fe II 1607, and Al II 1670. See Leitherer et al. (2011) for detailed wavelengths and atomic data of these lines. Broad windows were used in order to encompass the full profile, including any emission component when P Cygni wind profiles were present. The spectral lines observed in the UV spectra of star-forming galaxies broadly fall into three groups: broad *stellar* lines, which sometimes show a P Cygni profile (e.g., C IV 1550), narrow *interstellar* lines (e.g., C II 1335), and features that are a blend of *both* (e.g., Si IV 1400).

Except for the broad stellar-wind lines, the IUE spectral resolution of ∼6 Å is too low to resolve the lines. This also precludes removing any contamination of EW by Galactic foreground components, which are present in many cases. In order to remove these components, we measured their equivalent widths on available COS spectra and subtracted them from the IUE measurements. This assumes that the Galactic foreground column density does not vary over arcsec scales, which is quite a reasonable assumption. We compared the derived EW of all lines to the available oxygen abundances and found significant correlations. Both the EW of the interstelllar and the stellar lines increase with increasing oxygen abundance. As the interstellar lines are strongly saturated, they are not expected to be abundance dependent. Rather, this behavior is caused by macroscopic velocities in the interstellar medium due to, e.g., outflows. This suggestion has been made before by Heckman et al. (1998).

The stellar-wind lines display an even tighter correlation with log O/H + 12 than the interstellar lines. In Fig. 3 we have plotted the relation of EW with oxygen abundance for C IV 1550 and Si IV 1400. Both lines increase with oxygen abundance over the range log O/H + 12 = 7.2 – 9.2. At the lowest abundances, C IV 1550 has strong contributions

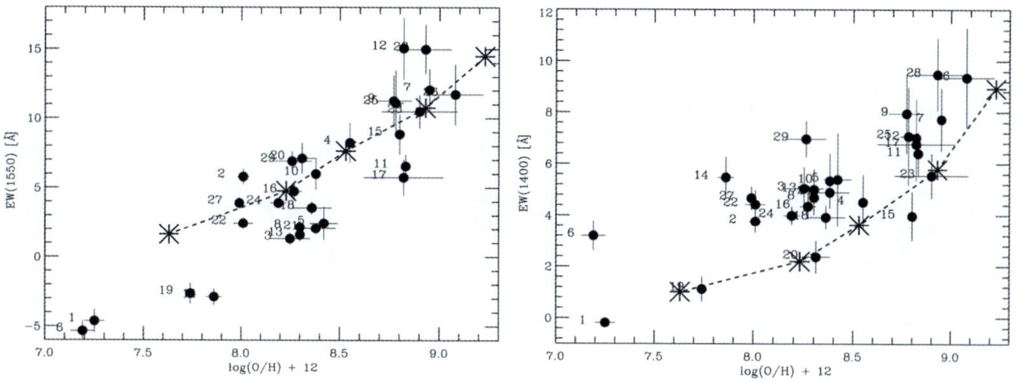

Figure 3. Relation between the measured C IV 1550 (left) and Si IV 1400 (right) equivalent widths in the IUE spectra (filled dots). The labels refer to the numbered galaxies in Section 2. The stars and the connecting dashed line are from Starburst99 population synthesis models. Positive EW values indicate absorption.

from nebular emission, leading to a net negative EW. Also shown is the theoretically predicted relation from Starburst99 population synthesis models. We used the previous generation of Geneva evolution models without rotation for broader abundance coverage (Schaller et al. 1992; Schaerer et al. 1993a,b; Charbonnel et al. 1993; Meynet et al. 1994). We verified that the relation in Fig. 3 is virtually unchanged when switching between the older and the more modern tracks. We used the theoretical stellar atmospheres based on WM-Basic as implemented in Starburst99. The model parameters are the same as those given above. The trend of the measured EW of C IV 1550 with oxygen abundance is in very good agreement with the model prediction. In contrast, the data points for Si IV 1400 are above the theoretical relation. We interpret this as due to an additional contribution from interstellar Si IV 1400, which is not included in the stellar models. We studied the influence of different values of the IMF on the theoretical relation. IMF changes have a very minor effect. This results from the compensating changes of the absorption and emission components of the wind lines, both of which increase for an IMF more heavily weighted towards massive stars. Obviously line profile studies at higher spectral resolution will be very sensitive to IMF variations.

The C IV 1550 stellar-wind line is deeply saturated. Therefore, its dependence on oxygen abundance is not primarily due to direct changes in the abundances. The carbon abundance scales with the oxygen abundance using solar abundance ratios in both the atmosphere and the evolution models. However, test calculations using PoWR models (A. Sander; private communication) demonstrate that changes in the carbon abundance over a range of $Z_\odot$ and 0.2 $Z_\odot$ have little effect on the C IV 1550 line of a representative individual star if all other element abundances are unchanged. In contrast, decreasing the Fe abundance by the same amount and leaving carbon unchanged leads to a trend expected to match that seen in Fig. 3. This can be understood in terms of the driving mechanism of the stellar winds. In the abundance regime covered in Fig. 3, Fe-group elements drive the winds (Vink 2022) and determine the wind velocities and mass-loss rates. Since these properties shape the line profiles, rather than abundances, the relation seen in Fig. 3 is partly, or even mostly, a relation between Fe and O. This may offer an opportunity to study the α-element/Fe relation in star-forming galaxies.

4. IUE versus COS

We performed corresponding measurements on available co-spatial COS data for comparison with the IUE data. The fluxes at 1500 Å are compared in Fig. 4 (left). As

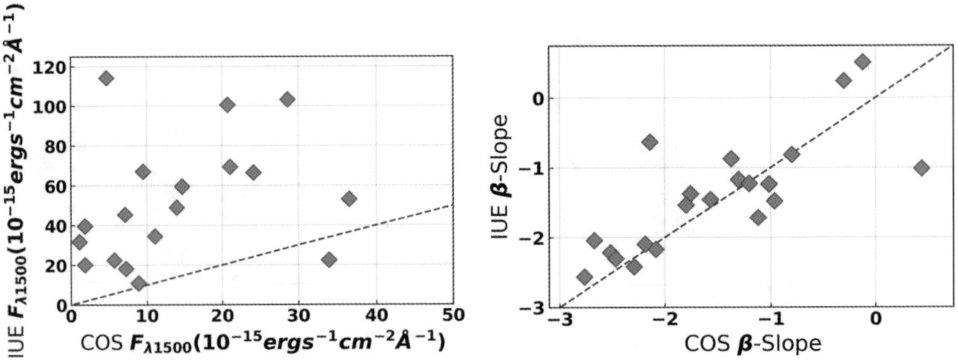

Figure 4. Left: continuum fluxes at 1500 Å in the IUE and COS data. Right: UV slope β determined for the IUE and COS data. The dashed line is the 1:1 relation.

expected, the larger IUE aperture encircles more UV light, leading to higher fluxes than observed through the COS aperture. Depending on the galaxy morphology and the presence of UV-bright star clusters, the ratio of the two measurements ranges between $\sim$1 and $\sim$100, with a typical value of $\sim$5.

We compared the UV slope β in both data sets. As discussed in Section 3, β is a proxy for the dust reddening and can be converted to $E(B-V)$ with the relation $\beta = -2.616 + 4.684E(B-V)$ of Reddy et al. (2018), which assumes the attenuation law derived by Calzetti et al. (2000). The fit to the data was performed with the windows of Calzetti et al. (1994) between 1268 and 1950 Å. Some galaxies observed with COS lack data longward of $\sim$1600 Å, and their β slopes refer to the continuum below this cut-off. The comparison of the two sets of measurements is reproduced in Fig. 4 (right). The figure suggests rather good agreement between the IUE and COS values, confirming no significant reddening variations over spatial scales of tens to hundreds of pc. There are two outliers in the figure. NGC 3125 at +0.44/–1.02 has a peculiar SED turning over at shorter wavelengths. As the COS data are limited to shorter wavelengths, β is more positive (redder). The COS pointing for NGC 3690 is slightly outside the IUE pointing, which may explain the offset of the data point at –2.14/–0.64.

Next, we discuss the equivalent width measurements. While the COS data permit detailed line-profile studies and measurements of absorption and emission components individually (when present), we used the same broad integration windows as in the IUE data for a meaningful comparison. The results are plotted in Fig. 5. The left panel shows the correlation for the low-ionization interstellar lines of Si II 1260, O I 1300, C II 1335, Fe II 1607, and Al II 1670. The high-ionization stellar (+interstellar) lines of Si IV 1400 and C IV 1550 are in the right panel. There is some scatter in both figures but no significant trends for any of the lines is detected, suggesting similar physical conditions of the interstellar medium and the stellar population over the probed spatial scales.

5. Take-away points

Our study confirms the value low-resolution, moderate S/N spectroscopic data such as those provided by IUE for studies of star-forming galaxies. The quality of upcoming data for galaxies close to the reionization age will resemble the one of the IUE data. We find that EW of several lines scales with O/H. In particular, C IV 1550 is an excellent proxy for O/H. This line is particularly useful as it is one of the strongest features in UV spectra of star-forming galaxies. This line may also hold the key to investigating

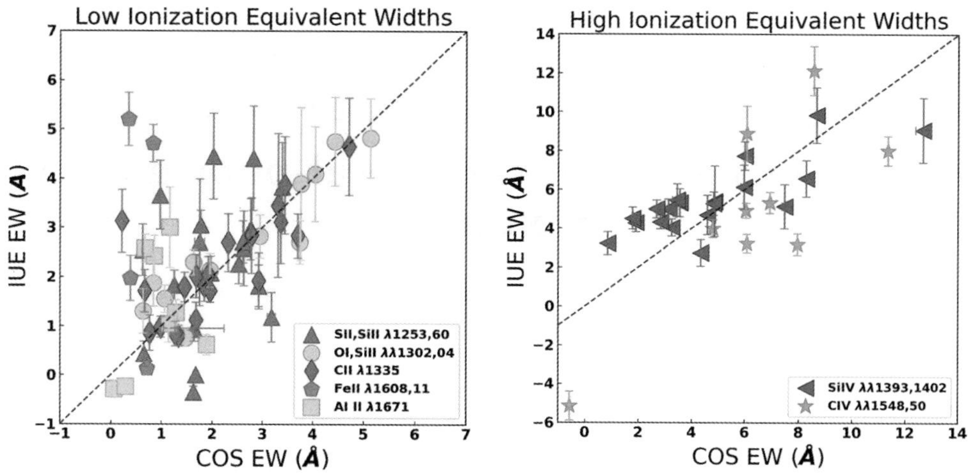

Figure 5. Comparison of the equivalent widths in the IUE and COS data for low- (left) and high-ionization lines (right). The dashed lines are the 1:1 relation.

α-element/Fe variations in galaxies. Comparison between IUE and COS data suggests no significant variations of the spectral properties over scales of tens to hundreds of pc.

Acknowledgements. Support for this work has been provided by NASA through grant number GO-15099 from the Space Telescope Science Institute, which is operated by AURA, Inc., under NASA contract NAS5-26555.

References

Berg, D. A., James, B. L., King, T., et al. 2022, *arXiv:2203.07357*
Calzetti, D., Kinney, A. L., & Storchi-Bergmann, T. 1994, *ApJ*, 429, 582
Calzetti, D., Armus, L., Bohlin, R. C., et al. 2000, *ApJ*, 533, 682
Charbonnel, C., Meynet, G., Maeder, A., Schaller, G., & Schaerer, D. 1993, *A&AS*, 101, 415
Ekström, S., Eggenberger, P., Meynet, G., et al. 2012, *A&A*, 537, 146
Georgy, C., Ekström, S., Eggenberger, P., et al. 2013, *A&A*, 558, 103
Heckman, T. M., Robert, C., Leitherer, C., Garnett, D., & van der Rydt, F. 1998, *ApJ*, 503, 646
Kroupa, P. 2008, in: J. H. Knapen, T. J. Mahoney, & A. Vazdekis (eds.), *Pathways Through an Eclectic Universe*, ASP Conf. Ser. 390, (San Francisco, CA: ASP), p. 3
Leitherer C. 2020, *Galaxies*, 8, 13
Leitherer, C., Byler, N., Lee, J. C., & Levesque, E. M. 2018, *ApJ*, 865, 55
Leitherer, C., Ekström, S., Meynet, G., et al. 2014, *ApJS*, 212, 14
Leitherer, C., Tremonti, C. A., Heckman, T. M., & Calzetti, D. 2011, *AJ*, 141, 37
Meynet, G., Schaller, G., Schaerer, D., & Charbonnel, C. 1994, *A&AS*, 103, 97
Reddy, N. A., Oesch, P. A., Bouwens, R. J., et al. 2018, *ApJ*, 853, 56
Schaerer, D., Charbonnel, C., Meynet, G., Maeder, A., & Schaller, G. 1993a, *A&AS*, 102, 339
Schaerer, D., Meynet, G., Maeder, A., & Schaller, G. 1993b, *A&AS*, 98, 523
Schaller, G., Schaerer, D., Meynet, G., & Maeder, A. 1992, *A&AS*, 96, 269
Vink, J. S. 2022, *arXiv:2109.08164*
Weedman, D. W., Feldman, F. R., Balzano, V. A., et al. 1981, *ApJ*, 248, 105

Massive Stars Near and Far
Proceedings IAU Symposium No. 361, 2024
J. Mackey, J. S. Vink, & N. St-Louis, eds.
doi:10.1017/S1743921322001922

One step closer to the First Stars: +150 OB stars in the metal-poor galaxy Sextans A

Marta Lorenzo*, **Miriam Garcia** and **Francisco Najarro**

Centro de Astrobiología, CSIC-INTA. Crtra. de Torrejón a Ajalvir km 4. 28850 Torrejón de Ardoz (Madrid), Spain
*email: mlorenzo@cab.inta-csic.es

Abstract. Local Group (LG) very metal-poor massive stars are the best proxy for the First Stars of the Universe, fundamental to model the early evolution of the first galaxies, and key to unravelling new evolutionary pathways restricted to low metallicities, such as chemically homogeneous evolution. Yet, due to the great leap in distance required to reach metal-poor galaxies of the LG and vicinity, no comprehensive spectroscopic studies have been carried out at sub-SMC metallicities so far.

We focus on the massive star population of the 1/10 $Z_\odot$ galaxy Sextans A. After five observing campaigns at the 10.4-m Gran Telescopio Canarias (GTC), we have assembled a spectroscopic catalogue of more than 150 OB stars. This catalogue will be fundamental to test stellar evolution at very low metallicity, to detect the first binary systems at 1/10 $Z_\odot$ and to unveil the most recent star formation history of this galaxy.

Keywords. Stars: early-type, stars: Population III, galaxies: stellar content, galaxies: individual (Sextans A), catalogs

1. Introduction

The study of massive stars in very low-metallicity environments is key to understanding the physics of analogue massive stars at the early epochs of the Universe. Yet, very little is known about their properties and evolution. A large sample of resolved massive stars with very low metal content is required to address the growing number of open questions in the field. These questions include the specific evolutionary pathways of massive stars at metallicities lower than the Small Magellanic Cloud (Groh et al. 2019; Szécsi et al. 2022), the incidence of chemically homogeneous evolution (CHE, (CHE, Szécsi et al. 2015), and the role of radiation driven winds and binary interactions in shaping the evolution of very metal-poor massive stars (Shenar et al. 2020).

Given its relative proximity, the Small Magellanic Cloud (SMC) constitutes the current standard of metal-poor regimes. However, its 1/5 $Z_\odot$ metallicity is too high to describe the Universe at redshift above $z = 1$. To extend the metallicity prescriptions of the physical properties of massive stars to the low-metallicity end, several groups have targeted massive stars in Local Group galaxies with lower metal content (1/10 - 1/20 $Z_\odot$). However, due to their long distances, the list of known massive stars with sub-SMC metallicities was scarce up to now.

In this work, we present the first major effort to mine the population of massive stars in the galaxy Sextans A. Located 1.34 Mpc away and presenting low foreground extinction ($E(B-V)_{fg} = 0.044$, Tammann et al. 2011), this dwarf irregular galaxy stands

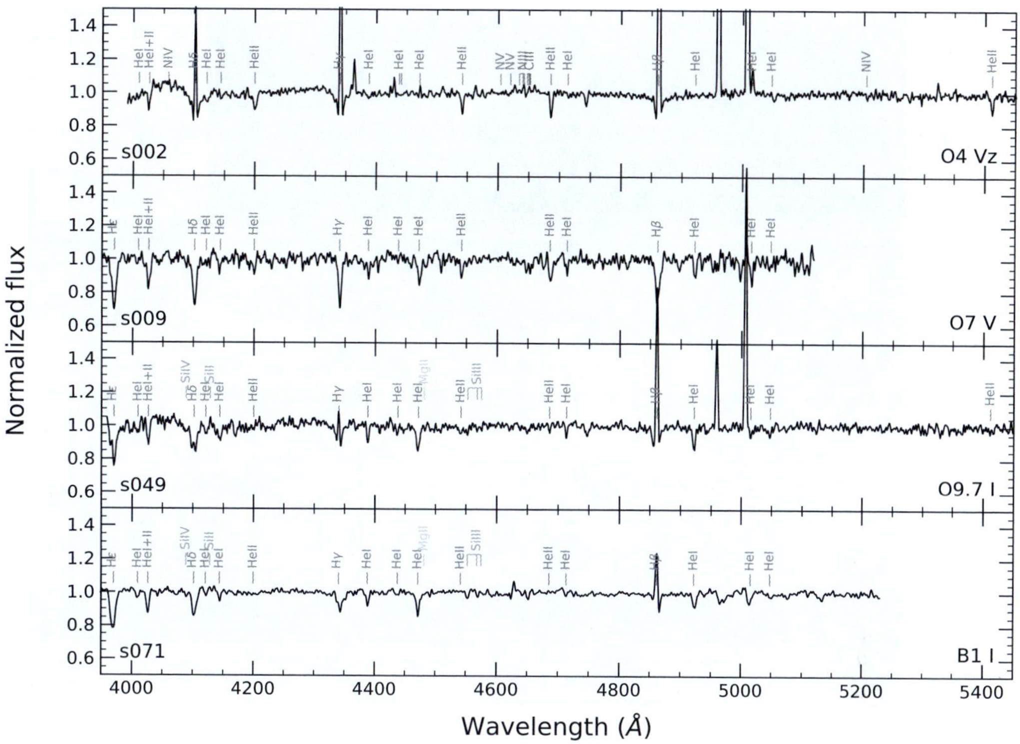

Figure 1. GTC-OSIRIS spectra of four OB stars in our catalogue. The data have been corrected by heliocentric and radial velocity, and smoothed for clarity based on the S/N of the spectrum.

out among other LG galaxies for its low metallicity values. Nebular oxygen abundance measurements range between 12+ log(O/H)= 7.49 – 7.71 (1/15 - 1/10 $O_{\odot}$, Skillman et al. 1989; Pilyugin 2001; Kniazev et al. 2005; Berg et al. 2012) and the iron abundance determined from blue stars is $\langle$[(Fe II, Cr II)/H]$\rangle$ = -0.99 ± 0.06 (1/10 $Fe_{\odot}$, Kaufer et al. 2004; Garcia et al. 2017), the lowest value confirmed by UV spectroscopy in the LG. The resulting sample will yield first-order information on the evolution of massive stars with sub-SMC metallicity and provide an extensive collection of spectra to characterise their physical properties at these very low-Z regimes.

2. The first extensive catalogue of OB stars in Sextans A

The catalogue was built through five observing campaigns, three runs based on long-slit spectroscopy (one of them already published in Camacho et al. (2016)) plus two multi-object runs. All of them were carried out with the Optical System for Imaging and low-Intermediate-Resolution Integrated Spectroscopy (OSIRIS) installed at the 10.4-m Gran Telescopio Canarias (GTC). The observations were designed to cover all the UV-bright sources of Sextans A and the massive stars located in the large bubbles of ionised gas of the galaxy. We obtained spectroscopy of 174 sources in total, with an average resolution of $R = \lambda/\Delta\lambda \sim 1000$ in the optical blue range, from 4000 to 5500 Å. Figure 1 shows four examples of the resulting spectra, covering different spectral types.

Of these 174 sources, 5 are foreground stars, 4 are red giants/supergiants members of Sextans A, 6 are A- and F-type stars, and 159 are OB-type stars. The OB sample includes 71 O stars, with types down to O3.5, and 38 BA supergiants. The latter collection will allow us to derive the abundances of α- and iron-elements and better constrain

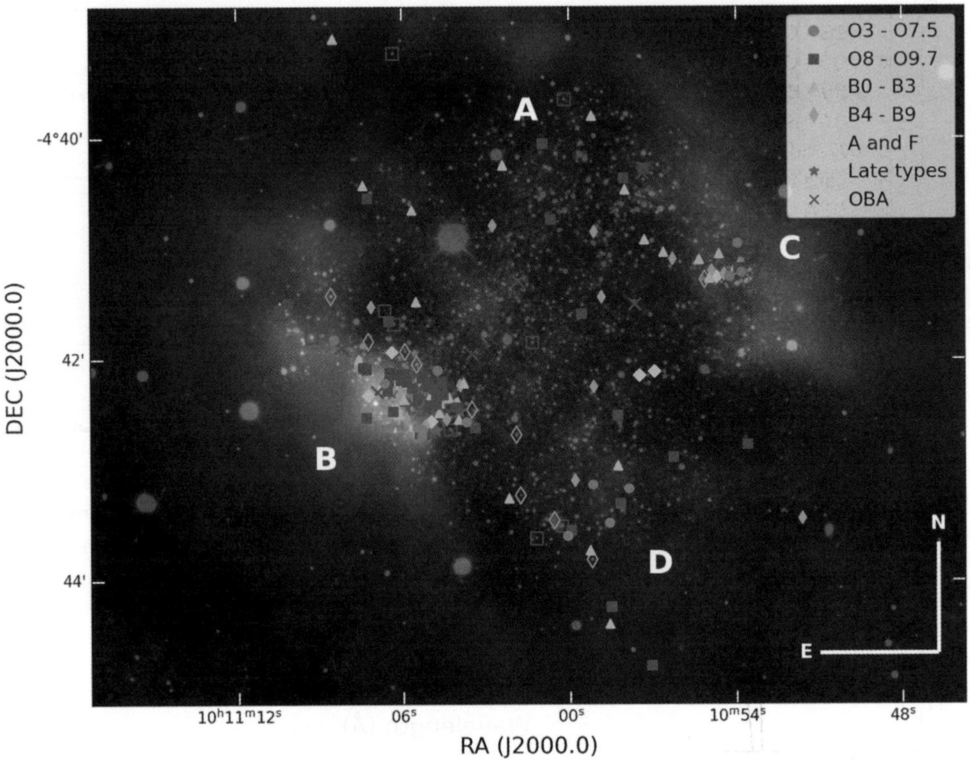

Figure 2. Sextans A. *LITTLE THINGS* neutral hydrogen map (Hunter et al. 2012) overlaid on a V–band image of Sextans A (Massey et al. 2007). Regions–A, –B, –C and –D with ongoing star formation are marked in the figure. We show the catalogue sources, excluding foreground stars, with spectral types coded as indicated in the legend. Stars with an undefined spectral subtype are drawn with empty markers.

the chemical composition of the galaxy. In addition, a total of 33 sources show some signatures of binarity in their spectra. These are the first binary candidates ever reported in Sextans A. However, their confirmation will need follow-up observations.

This makes this sample the largest spectroscopic catalogue of massive stars at metallicities lower than the Small Magellanic Cloud (SMC) to date and constitutes an increase of one magnitude with respect to previous lists in Sextans A. The catalogue will be published in Lorenzo et al., submitted.

3. The recent star forming history of Sextans A

In Fig. 2, we show the location of the sample stars in Sextans A and use the 159 OB sources to trace the star formation in the galaxy.

Most of the OB stars are located in regions–B and –C, overlapping with the high-column density of neutral gas and producing the large structures and bubbles of ionised hydrogen of the galaxy.

Region–B presents more dispersion in spectral types and hosts a large fraction of the earliest stars. The distribution of its massive population appears to be layered by spectral type and correlated with the column-density of the neutral gas. The early O-type stars overlap with the high gas density areas, and later types are located at the inner rim of the H I cloud. This apparent stratified pattern might suggest a temporal sequence of star formation in this region, starting from the inner edge of the cloud and moving outwards.

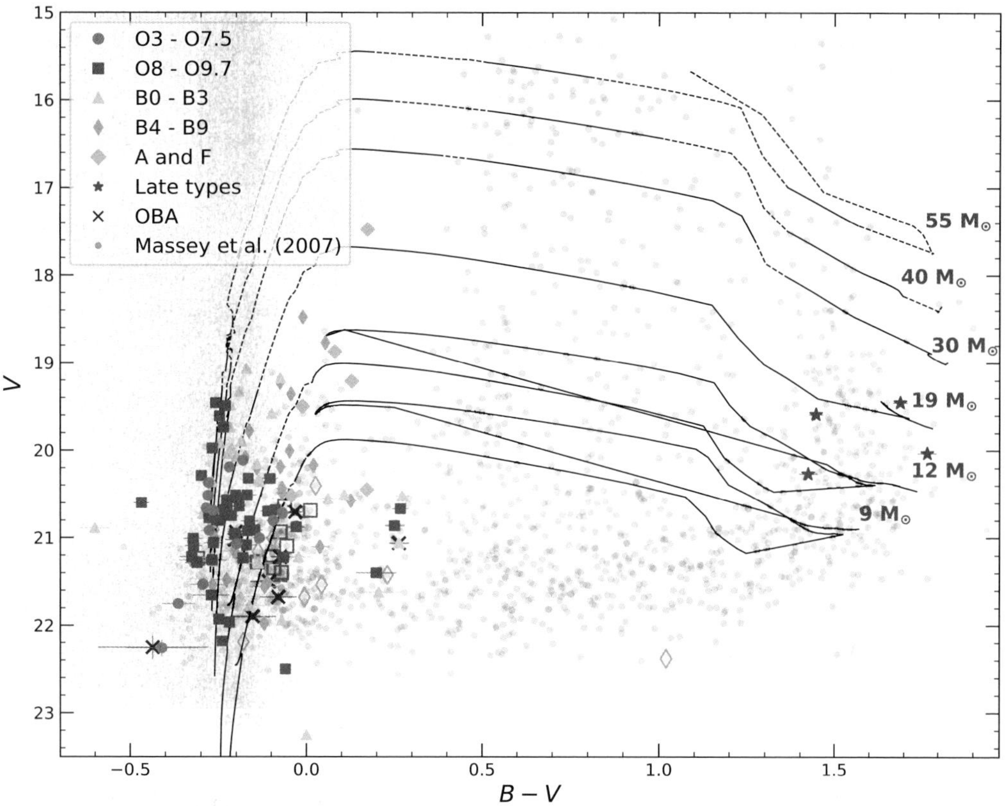

Figure 3. Colour magnitude diagram of Sextans A, showing Massey et al. (2007)'s catalogue in grey and our sample sources, excluding foreground stars, in different colours and symbols according to their spectral type, as listed in the legend. Those without a defined spectral subtype are drawn as empty symbols. $Z = 0.10\ Z_\odot$, $v \sin i = 100$ km s^{-1} evolutionary tracks by Szécsi et al. (2022) are also provided. The tracks have been shifted by $E(B-V)_{\rm fg} = 0.044$ (Skillman et al. 1989) to account for foreground extinction towards Sextans A and by the distance modulus of the galaxy ($\mu_0 = 25.63$, Tammann et al. 2011).

In addition, we also detect massive stars in unexpected areas where the column density of neutral gas is low. This was already pointed out by Garcia et al. (2019) in the South of Sextans A, but we detect some additional stars in the gas-void centre and other outer areas of the galaxy. Some reasons that could explain this odd star-forming locus are that these massive stars could be products of mergers of two or more lower-mass sources. On the other hand, star formation could occur in compact, CO-dark gas clouds undetected by H I maps.

4. A glimpse into the evolution of 1/10 $Z_\odot$ massive stars

Figure 2 shows Sextans A's colour magnitude diagram (CMD) built with Massey et al. (2007)'s catalogue and highlighting our sample stars with different colours and symbols.

The OB stars are spread across the blue plume. The observed dispersion can be explained mostly by the internal interstellar extinction in Sextans A. This stresses the need of untargeted spectroscopy to locate the massive star population in dwarf irregular galaxies with non-negligible redenning.

In addition, 18 OB-type stars present bluer colours than the evolutionary tracks. This number may increase after applying a detailed reddening correction towards individual sources, which we plan as future work. We proposed two scenarios to explain these blue outliers: chemically homogeneous evolution and systems hosting a stripped star. We have evaluated the different features that these two objects would exhibit under the constraints of the quality of our data and produced a list of candidates (see Lorenzo et al., submitted). However, follow-up observations with higher resolution and signal-to-noise ratio will be needed to confirm either scenario.

5. Conclusions and future work

Using the 10.4-m Gran Telescopio Canarias, we have assembled an extensive spectroscopic survey of more than 150 massive stars in Sextans A, a Local Group galaxy with 1/10 $Z_{\odot}$ metallicity. This is the largest census of massive stars at metallicities lower than the Small Magellanic Cloud and represents a fundamental first step to unveiling the properties, evolutionary pathways and fates of very metal-poor massive stars. We also provide the first list of candidates undergoing chemically homogeneous evolution or systems hosting a stripped star at 1/10 $Z_{\odot}$. In addition, we identify 33 sources with signs of multiplicity in their spectra.

We have already started the quantitative spectroscopic analysis of the sample O stars with higher spectral signal-to-noise ratio. These analyses will yield their stellar parameters which will help to constrain their evolution. We also plan to measure the current chemical composition of Sextans A and study the $[\alpha/\mathrm{Fe}]$ abundance ratio with our collection of BA supergiants. Finally, we have submitted a proposal for follow-up spectroscopy of the blue outliers with the instrument FORS2 of the Very Large Telescope. The observations request spectra with higher resolution and signal-to-noise that will enable discerning between the two proposed scenarios, CHE and systems hosting a stripped star. Finding even one single case of these objects will have critical implications for the stellar evolution theories, the interpretation of high-redshift galaxies and the progenitors of gravitational wave events.

6. Acknowledgements

Spanish grants PID2019-105552RB-C41 and Unidad de Excelencia "María de Maeztu" (MDM-2017-0737-19-3) supported this research.

References

Berg, D. A., et al. 2012, ApJ, 754, 98
Camacho, I., Garcia, M., Herrero, A., & Simón-Díaz, S. 2016, A&A, 585, A82
Garcia, M., Herrero, A., Najarro, F., Camacho, I., Lennon, D. J., Urbaneja, M. A., & Castro, N. 2017, in The Lives and Death-Throes of Massive Stars, ed. J. J. Eldridge, J. C. Bray, L. A. S. McClelland, & L. Xiao, Vol. 329, 313–321
Garcia, M., Herrero, A., Najarro, F., Camacho, I., & Lorenzo, M. 2019, MNRAS, 484, 422
Groh, J. H., et al. 2019, A&A, 627, A24
Hunter, D. A., et al. 2012, AJ, 144, 134
Kaufer, A., Venn, K. A., Tolstoy, E., Pinte, C., & Kudritzki, R.-P. 2004, AJ, 127, 2723
Kniazev, A. Y., Grebel, E. K., Pustilnik, S. A., Pramskij, A. G., & Zucker, D. B. 2005, AJ, 130, 1558
Massey, P., McNeill, R. T., Olsen, K. A. G., Hodge, P. W., Blaha, C., Jacoby, G. H., Smith, R. C., & Strong, S. B. 2007, AJ, 134, 2474
Pilyugin, L. S. 2001, A&A, 369, 594
Renzo, M., et al. 2019, A&A, 624, A66
Shenar, T., et al. 2020, A&A, 641, C2
Skillman, E. D., Terlevich, R., & Melnick, J. 1989, MNRAS, 240, 563
Szécsi, D., Agrawal, P., Wünsch, R., & Langer, N. 2022, A&A, 658, A125

Szécsi, D., Langer, N., Yoon, S.-C., Sanyal, D., de Mink, S., Evans, C. J., & Dermine, T. 2015, A&A, 581, A15
Tammann, G. A., Reindl, B., & Sandage, A. 2011, A&A, 531, A134

Discussion

RENZO: Could the isolated stars be runaways? What are the distances between the isolated stars and the confirmed star-forming regions? In Renzo et al. (2019), we measured that ejected stars present higher velocities at lower metallicities.

LORENZO: I do not have the numbers right now, but I can check them later and we can discuss that posibility.

MCQUINN: Do you know the column-density of the neutral gas where the isolated OB stars are located?

LORENZO: Our analysis has been only qualitative using the map of *LITTLE THINGS* survey. We plan to do a quantitative one in the future. However, I think that Garcia et al. (2019) listed a column-density of the order of 1 $M_{\odot}pc^{-2}$ in the southern region of Sextans A.

Massive Stars Near and Far
Proceedings IAU Symposium No. 361, 2024
J. Mackey, J. S. Vink, & N. St-Louis, eds.
doi:10.1017/S1743921322002502

NGC 3125-A1 revisited at higher spectral resolution with COS G160M

A. Wofford[1], A. Sixtos[1], L. Smith[2], S. Charlot[3] and G. Bruzual[4]

[1]Instituto de Astronomía, Universidad Nacional Autónoma de México, Unidad Académica en Ensenada, Km 103 Carr. Tijuana-Ensenada, Ensenada 22860, México
email: awofford@astro.unam.mx

[2]Space Telescope Science Institute, 3700 San Martin Drive, Baltimore, MD 21218, USA

[3]Sorbonne Université, UPMC-CNRS, UMR7095, Institut d'Astrophysique de Paris, F-75014 Paris, France

[4]Instituto de Radioastronomía y Astrofísica, UNAM, Campus Morelia, Michoacán, C.P. 58089, México

Abstract. Super star cluster (SSC) A1 in starburst galaxy NGC 3125 has the strongest broad He II λ1640 emission line ever observed in the nearby Universe and constitutes an important template for interpreting observations of galaxies that are located out to a redshift of $z \sim 3$. We use observations of SSC A1 obtained with the Cosmic Origins Spectrograph (COS) on board of the Hubble Space Telescope (HST) in order to check if there is a contribution of nebular emission to the He II line. In addition, we compare the COS G130M + G160M observations of A1 (1150 – 1750 Å) to the latest Charlot & Bruzual population synthesis models, which account for Very Massive Stars (VMS) of up to 300 $M_\odot$. A model with $Z = 0.008$ and age = 2.4 Myr provides a very reasonable fit to the C III λ1175, N V λ1240, C IV λ1550, He II λ1640, and N IV λ1718 stellar-wind features, although the O V λ1371 line is not well reproduced. Overall, our results show the great improvement of stellar evolution and population synthesis models over the past decade, and in particular, the improved formulation of stellar mass loss rates.

Keywords. techniques: spectroscopic, galaxies: starburst, galaxies: stellar content, ultraviolet: stars

1. Introduction

Starburst galaxies are characterized by massive violent bursts of star formation. They are powered by populations of hot stars with progenitor masses of 5 $M_\odot$ and above. Locally, they are responsible for 20% of the entire high-mass star formation (Brinchmann *et al.* 2004). Nearby starburst galaxies are the most obvious local counterparts of the normal star-forming galaxies discovered at high redshift, in particular galaxies located at redshifts of $z \sim 2 - 3$, i.e., at the epoch of 'Cosmic Noon'. The latter epoch is ideal to examine the mechanisms of star formation as it is when galaxies formed about half of their current stellar mass. Composite spectra of Cosmic-Noon galaxies (e.g., Shapley *et al.* 2003; Saxena *et al.* 2020) provide a glimpse of the dominant stellar populations present in the early Universe.

Nearby ($D = 11.5$ Mpc) starburst galaxy NGC 3125 (Tol 3) is of particular importance for studies of galaxies at Cosmic Noon. The galaxy has an ionized-gas oxygen abundance of 12+log(O/H)=8.3 (Hadfield & Crowther 2006), i.e., similar to that of the Large Magellanic Cloud (LMC). It is dominated by two emission regions, A and B. Region A can be subdivided into UV-bright and highly extinguished components

A1 and A2, respectively. In (Wofford *et al.* 2014, W14), we present HST observations of A1 obtained by PI Leitherer using the following two spectrographs and configurations: i) Space Telescope Imaging Spectrograph (STIS), 2"-wide long slit, and G140L grating; and ii) COS, 2.5"-wide circular aperture, and G130M grating. The analysis of these data by W14 shows that the spectroscopically-derived mass of A1 (1.7×10^5 M$_\odot$) is of the same order of magnitude as the binary-corrected virial mass of cluster NGC 2070 in the LMC (4.5×10^5 M$_\odot$, which is consistent with the mass determined from photometry; Bosch *et al.* 2009). The STIS data also show that A1 is characterized by broad ($FWHM \approx 1200\,\mathrm{km\,s^{-1}}$) He II λ1640 emission with a large equivalent width ($EW = 7.1 \pm 0.4$ Å; revised to ~ 5 Å based on the higher SNR data of this work); and stellar O V λ1371 absorption. Broad ($FWHM \approx 1000\,\mathrm{km\,s^{-1}}$) He II emission suggests stellar winds as the formation mechanism. In order to produce detectable stellar He II emission, one or more of the following conditions must be met: the stars must be hot to doubly ionize He, the mass-loss rates must be high to produce dense winds, and He must be overabundant. Since classical Wolf-Rayet (WR) stars meet all three conditions, the presence of He II is generally thought to indicate the presence of WR stars. Note however that H-rich Very Massive Stars (VMS), which have masses of $M > 150\,\mathrm{M_\odot}$, i.e., above the canonical upper mass limit of the stellar Initial Mass Function (IMF), have WR features, independent of He enrichment. In VMS, the reason for the increased EW of the He II λ4686 stellar-wind line is almost certainly the proximity to the Eddington limit. This is shown in figure 10 of Vink *et al.* (2011), where one sees that increasing the Eddington factor theoretically increases the mass loss rate, and also empirically, the He II EW. He II λ1640 is the UV counterpart to the optical He II line and a similar increase of the EW is expected for the UV line as well.

In W14, we rule out that the extraordinary He II emission and O V absorption of A1 are due to an extremely flat exponent of the upper IMF, and suggest that they originate in the winds of VMS. Such stars have been found in the core of NGC 2070, which is known as R136 and is $1 - 2.5$ Myr old (Brands *et al.* 2022). An example of a VMS is R136a3, which has an initial mass of 165 ± 30 and is classified as a hydrogen-rich WN5h star. An ultraviolet spectrum of this star, obtained with the Goddard High Resolution Spectrograph (GHRS, an instrument that is no longer on board of HST) is shown in figure 5 (top panel) of Crowther *et al.* (2010). Note the striking similarity between the GHRS spectrum of R136a3 and the COS spectrum of SSC A1, which we show in Figure 2.

NGC 3125 can be used as a template to interpret composite spectra of distant star-forming galaxies that show broad He II emission (Chandar *et al.* 2004). In composite spectra, the uncertainties in the redshifts of the galaxies that make up the spectrum contribute to the width of the composite He II line. There is however now strong evidence of the existence of broad He II emission at high-redshift is provided by the recent discovery, in a deep observation with the Very Large Telescope, of broad ($FWHM \approx 920\,\mathrm{km}^{-1}$) He II emission from a single non-lensed galaxy that is located at $z \sim 3$, and that is not an AGN (Cullen, private communication). In order to obtain the best possible interpretation of the latter and other distant systems, it is crucial to verify the hypothesis that there are VMS in NGC 3125-A1. This hypothesis can be tested by comparing the UV spectra of A1 with population synthesis models. Such models predict the photometric and spectroscopic properties of stellar systems.

2. Is the He II line purely stellar?

Prior to testing the VMS hypothesis, it is important to check if the integrated He II line of A1 has a contribution from nebular emission. For this purpose, COS G160M spectra of A1, with superior spectral resolution and SNR relative to the STIS data that we analyzed in W14, were obtained by PI Wofford. Two central wavelengths (1600 and

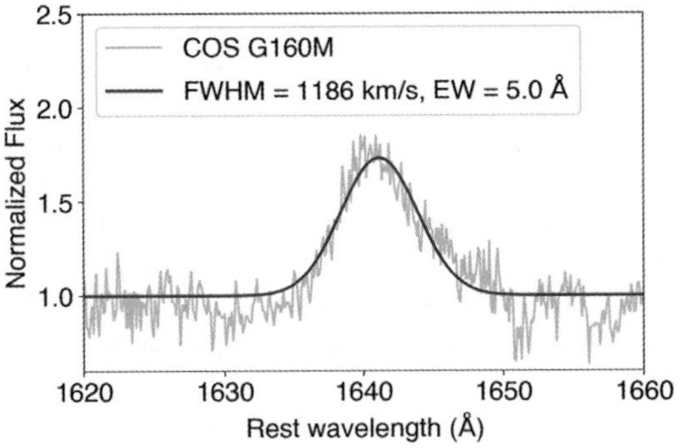

Figure 1. Portion of COS G160M spectrum of NGC 3125-A1 around He II λ1640 emission line (light blue curve) and Gaussian fit to the line (dark blue curve). The fit parameters are given in the legend.

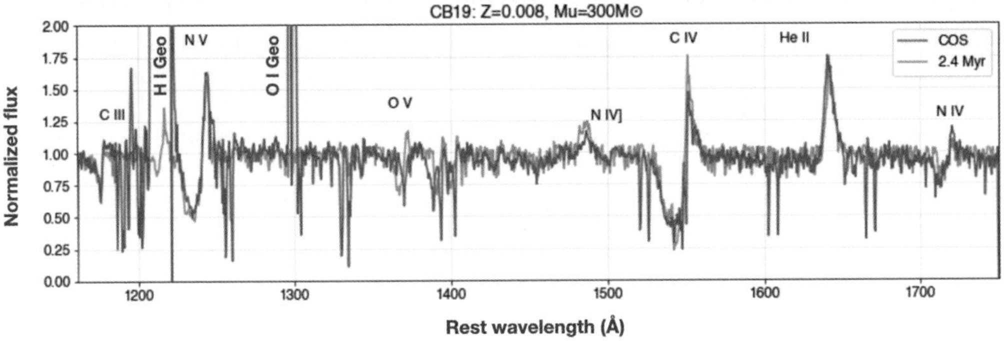

Figure 2. Comparison of COS G130M + COS G130M observation (blue curve) with best-fit model with VMS (orange curve). The strong emission lines at around 1210 and 1300 Å are due to Lyman-alpha and O I geocoronal emission.

1623 Å) were used in order to ensure continuous coverage in the wavelength range from ∼ 1400 to 1800 Å. A custom IDL program was used to co-add the exposures at the two central wavelengths. Figure 1 shows that the observed He II λ1640 emission line is purely stellar. The higher signal-to-noise in the continuum near the He II line of the G160M data relative to the STIS G140L observation enables a more accurate determination of the EW of the He II line. A Gaussian fit to the line in the G160M data yields $EW = 5\,Å$ and $FWHM = 1186$ km s^{-1}. The lower value of the EW obtained with the COS data relative to the STIS measurement is due to the better SNR in the nearby continuum of the COS observation, which enabled a more accurate fit to the line.

3. Preparation of spectrum for comparison with models

In order to prepare the COS observations for comparison with the models, we use custom programs that accomplish the following tasks: i) combine the G130M and G160M observations; normalize the spectrum by dividing it into sections that are normalized independently from each other (this is done by dividing each section by a straight-line fit to line-free portions within the section); iii) fit Voigt profiles to the Lyman-alpha aborption lines from the Milky Way and NGC 3125, after masking the nearby absorption or emission lines; iv) correct for the Lyman-alpha absorption lines; v) correct the spectrum for redshift; vi) degrade the observations to match the model spectral resolution. The observations cover the rest-frame UV wavelength range from 1150 to 1750 Å.

4. Models

It was not until recently that population synthesis models accounting for VMS became available. Martins & Palacios (2022) are able to reproduce A1's STIS He II line with a 2 Myr BPASS population synthesis model that accounts for close-binary evolution for stars of up to 100 $M_{\odot}$, and the added contribution of VMS. However, as shown in their figure 8, their solution results in a poor fit to the N V λ1240 and C IV λ1550 stellar wind features.

The COS observations that we use in this work have higher spectral resolution and SNR in the continuum compared to the STIS observation. They also cover a longer wavelength range. We use these higher quality observations to test the most recent Charlot & Bruzual population synthesis models (Plat *et al.* 2019), which account for the contribution of VMS of up to 300 $M_{\odot}$. The models use the PARSEC stellar evolution tracks of Chen *et al.* (2015), which are for single, non-rotating stars. The tracks take into consideration the formulation of mass loss rates by Vink *et al.* (2011). The model atmospheres and synthetic spectra of the Wolf-Rayet (WR) stars are those of the Potsdam group (PoWR, Gräfener, Hamann, Sander, Shenar, Hainich, Todt et al.). We assume a Simple Stellar Population (SSP), i.e., a single burst of star formation; and exclude the contribution of the ionized gas, as no nebular emission lines are present. We leave the metallicity and age of the stellar population as free parameters. We do not include the reddening due to dust in the model because the comparison with the observation is done in normalized space.

5. Main results

Figure 2 shows the comparison between the COS spectrum (blue curve) and our best-fit model with VMS (orange curve). The model has $Z = 0.008$ and age = 2.4 Myr. Note that interstellar absorption lines originating in the Milky Way and NGC 3125-A1 have not been removed from the observations. The model provides a very reasonable fit to the C III λ1175, N V λ1240, C IV λ1550, He II 1λ1640, and N IV λ1718 photospheric and stellar-wind features, although in the model, the emission component of the C IV P-Cygni like profile is a bit too strong. In addition, O V λ1371, which can be strong in VMS, is not very well reproduced. Note however that O V is known to be sensitive to clumping in the stellar wind Wofford *et al.* (2014). Overall, our results show the great improvement of stellar evolution and population synthesis models over the past decade, and confirm the importance of including VMS in population synthesis models.

6. Discussion

Several of the stellar-wind features that are observed in the COS spectrum of A1 are also observed in the spectra of classical WR stars. However, the presence of a strong O V λ1371 absorption line in A1's spectrum is an indication that this system is too young to already have classical WR stars. At a metallicity of $Z = 0.008$, classical WR stars start appearing at ~ 3 Myr and their contribution in a SSP peaks at ~ 4 Myr. At the age of our best-fit model, which is 2.4 Myr, VMS would still be present. This is why the presence of VMS is the preferred explanation for the observed spectrum of A1.

7. Acknowledgments

We thank J. Vink for pointing out that H-rich VMS have WR features, independent of He enrichment. We also thank S. Hernández for helping with the correction of A1's COS spectrum for Lyman-alpha absorption in the Milky Way and NGC 3125, which was crucial to test if the models fit well the nearby N V λ1240 stellar-wind doublet. Discussions with P. Crowther and T. Shenar during the conference helped clarify why VMS and not classical WR stars are the main contributors to the UV spectrum of A1.

References

Bosch, G., Terlevich, E., and Terlevich, R. 2009, *A J*, 137, 3437B
Brands S. A., de Koter, A., Bestenlehner, J. M., *et al.* 2022, *A & A*
Brinchmann, J., Charlot, S., White, S. D. M., *et al.* 2004, *MNRAS*, 351, 1151
Chandar, R., Leitherer, C., Tremonti, C. 2004, *ApJ*, 604, 153
Chen, Y., Bressan, Girardi, L., *et al.* 2015, *MNRAS*, 452, 1068
Crowther, P. A., Schnurr, O., Hirschi, R. *et al.* 2010, *MNRAS*, 408, 731
Eldridge, J. J., Stanway, E. R., Xiao, *et al.* 2017, *PASA*, 34, 58E
Hadfield, L. J., Crowther, P. A. 2006, *MNRAS*, 368, 1822
Martins, F., Palacios, A. 2022, *A & A*, 659, 163
Plat, A., Charlot, S., Bruzual, G., *et al.* 2019, *MNRAS*, 490, 978
Saxena, A., Pentericci, L., Mirabelli, M. *et al.* 2020, *A & A*, 636, 47
Shapley, A., Steidel, C. C., Pettini, M., *et al.* 2003, *ApJ*, 588, 65S
Vink, J. S., Muijres, L. E., Anthonisse, B., *et al.* 2011, *A & A*, 531A, 132V
Wofford, A., Leitherer, C., Chandar, R., *et al.* 2014, *ApJ*, 781, 122

Massive Stars Near and Far
Proceedings IAU Symposium No. 361, 2024
J. Mackey, J. S. Vink, & N. St-Louis, eds.
doi:10.1017/S1743921322002538

Massive stars in metal-poor dwarf galaxies are often extreme rotators

Abel Schootemeijer[1], Danny J. Lennon[2,3], Miriam Garcia[4], Norbert Langer[1,5], Ben Hastings[1,5] and Christoph Schürmann[1,5]

[1]Argelander-Institut für Astronomie, Universität Bonn, Auf dem Hügel 71, 53121 Bonn, Germany
email: aschoot@astro.uni-bonn.de

[2]Instituto de Astrofísica de Canarias, E-38200 La Laguna, Tenerife, Spain

[3]Departamento de Astrofísica, Universidad de La Laguna, E-38205 La Laguna, Tenerife, Spain

[4]Centro de Astrobiología, CSIC-INTA. Crtra. de Torrejón a Ajalvir km 4, E-28850 Torrejón de Ardoz (Madrid), Spain

[5]Max-Planck-Institut für Radioastronomie, Auf dem Hügel 69, 53121 Bonn, Germany

Abstract. We probe how common extremely rapid rotation is among massive stars in the early universe by measuring the OBe star fraction in nearby metal-poor dwarf galaxies. We apply a new method that uses broad-band photometry to measure the galaxy-wide OBe star fractions in the Magellanic Clouds and three more distant, more metal-poor dwarf galaxies. We find OBe star fractions of ~20% in the Large Magagellanic Cloud ($0.5Z_{\odot}$), and ~30% in the Small Magellanic Cloud ($0.2Z_{\odot}$) as well as in the so-far unexplored metallicity range $0.1 \lesssim Z/Z_{\odot} < 0.2$ occupied by the other three dwarf galaxies. Our results imply that extremely rapid rotation is common among massive stars in metal-poor environments such as the early universe.

Keywords. stars: early-type, stars: rotation, galaxies: Magellanic Clouds

1. Introduction

Massive stars in the early universe are thought to be metal-poor because after the Big Bang, the universe consisted almost exclusively of hydrogen and helium. Nearby dwarf galaxies are unique laboratories to study the early universe because they are both metal poor (Mateo 1998) and nearby enough to resolve individual massive stars. The evolution of these massive stars at low metallicity is thought to be strongly impacted by extremely rapid rotation (e.g., Langer 1992). In line with this is that superluminous supernovae, long-duration gamma-ray bursts, and ultra-luminous X-ray sources (all of which have been linked to rapid rotation - e.g., Aguilera-Dena et al. 2018; Marchant et al. 2016) are observed to occur mainly in low-metallicity dwarf galaxies (Lunnan et al. 2014; Kaaret et al. 2017). Diagnostics for extremely rapid rotation in stars are the presence of Hα emission and excess flux at long wavelengths, typically thought to originate from a decretion disk (see, e.g., Struve 1931; Poeckert & Marlborough 1976; Vink et al. 2009). Stars that display these features are classified as OBe stars. Here we investigate the prevalence of such rapidly rotating OBe stars in nearby metal-poor dwarf galaxies.

2. Methods

We employ archival broad-band photometry to separate OBe stars, main sequence (MS) stars without disks, and cooler helium-burning (HeB) stars. We use colors at short wavelengths to identify hot stars and colors at longer wavelengths to infer the presence of a disk. Below we describe the data sets used in this work.

2.1. *Magallanic Cloud data sets*

For the Small and Large Magellanic Cloud (SMC and LMC), we cross-correlate† the data from GAIA EDR3 (Gaia Collaboration et al. 2021), with classical *UB* photometry (Zaritsky et al. 2002, 2004) and infrared Spitzer-SAGE photometry (Meixner et al. 2006), adopting a 1" radius.

2.2. *Sextans A, Holmberg I, and Holmberg II data sets*

Sextans A, Holmberg I, and Holmberg II are about twenty to a hundred times further away than the Magellanic Clouds. For that reason, only data from the Hubble Space Telescope (HST) is deep enough for our method while at the same time having a high enough angular resolution. For Sextans A, we cross-correlate‡ $F336W$ and $F439W$ data from Bianchi et al. (2012) with $F555W$ and $F814W$ data from Holtzman et al. (2006). Before cross-correlating, we shift the data from Bianchi et al. (2012) by 0.6" in right ascension and 1" in declination. The cross-correlation radius is 0.35". For Holmberg I and Holmberg II we use the photometric data from the Legacy ExtraGalactic UV Survey (LEGUS; Sabbi et al. 2018).

3. Results

3.1. *The Magellanic Clouds*

The color-magnitude diagram (CMD) of the SMC (Fig. 1) shows several features. Here we focus on the two bluest bands, at $Gbp - Grp \approx -0.25$ and $Gbp - Grp \approx 0$. To investigate the nature of the sources in these bands, we plot the sources shown in our CMD also in a color-color diagram (Fig. 2, left). In this diagram, three groups form. At the right side of Fig. 2, we investigate their nature by plotting only sources for which spectral types have been determined (Martayan et al. 2007; Hunter et al. 2008; Bonanos et al. 2010; Dufton et al. 2019). Sources with $U - B > -0.5$ typically have spectral types of B6 and later, which is indicative of an effective temperature of $T_{\rm eff} < 15\,{\rm kK}$ (Schootemeijer et al. 2021). Such cool stars should be HeB stars, because the MS in the SMC does not extend down to temperatures this low (Schootemeijer et al. 2019). The sources with $U - B < -0.5$ and $J - [3.6] < 0.25$ tend to have spectral types of B5 and earlier, which implies that they are MS stars. The group of sources to the right of them consists almost exclusively of OBe stars (at a similar infrared color as found by Bonanos et al. 2010). Given how well these three groups of stars are separated in the color-color diagram, we conclude that this diagram can be used efficiently to tell apart MS, OBe, and HeB sources.

Turning back to Fig. 1, we see in the right panels that the bluest band with $Gbp - Grp \approx -0.25$ exclusively consists of MS stars without decretion disks. The band parallel to the MS band consists of a combination of HeB stars, and, perhaps more surprisingly, a significant number of OBe stars. We will quantify the fraction of OBe stars later on. The relatively high number of bright HeB stars – compared to MS stars – is explained by HeB stars being relatively bright in the visual. In the LMC, we use the same method to identify HeB, MS, and OBe stars, except that we place the threshold for HeB stars at $U - B > -0.3$ to account for redder intrinsic colors and higher extinction.

† Using `http://cdsxmatch.u-strasbg.fr/`
‡ Using the Aladin sky atlas

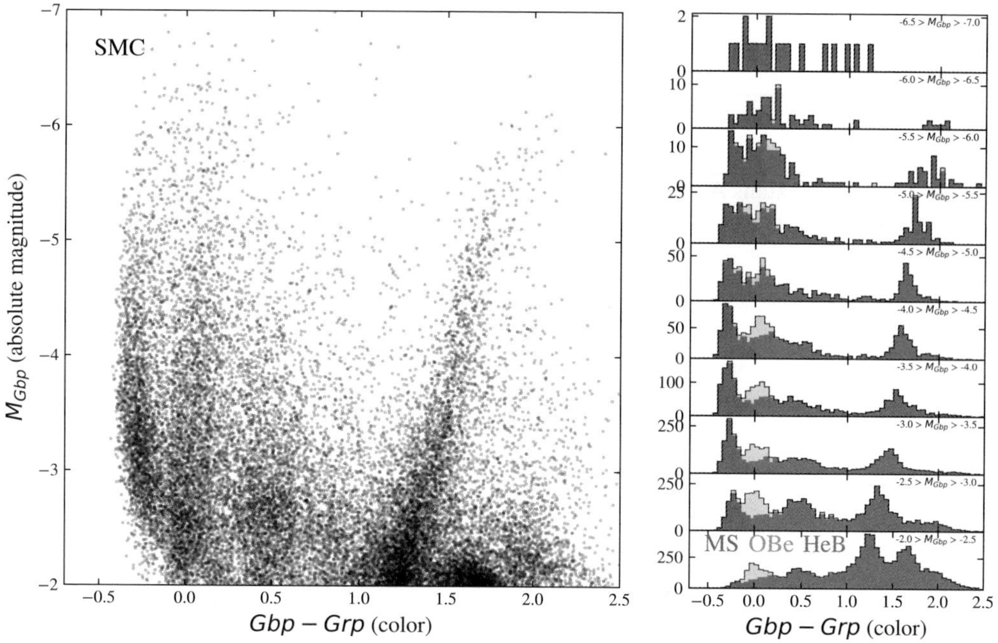

Figure 1. *Left:* color-magnitude diagram of bright sources in the Small Magellanic Cloud. *Right:* stacked color histograms showing the number of sources within the main sequence (MS, magenta), OBe (orange), and helium-burning (HeB, green) groups, as identified by the color-color cuts displayed in Fig. 2.

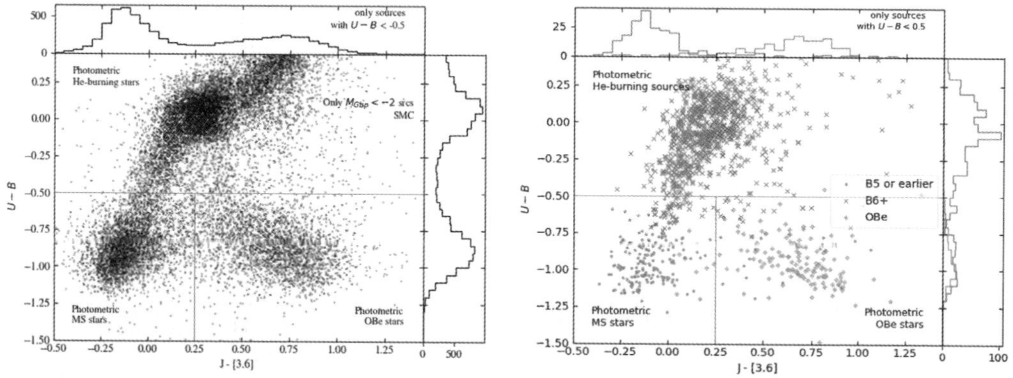

Figure 2. Color-color diagrams with histograms of both colors shown on the top and right side. *Left:* all sources brighter than $M_{Gbp} < -2$ in our data set. *Right:* same, but showing only sources for which spectral types have been determined (see text for more details).

3.2. *Sextans A, Holmberg I, and Holmberg II*

To identify MS, OBe, and HeB stars in Sextans A using HST data, we employ the color-color diagram shown in Fig. 3. As in Fig. 2, we see that three groups form, which we interpret as HeB, MS, and OBe stars. We have verified the OBe nature of the stars in the 'OBe' region of the color-color diagram in two ways: i) in the SMC, stars with known OBe spectral types form a separate group in a color-color diagram with colors similar to Fig. 3 ($Gbp - Grp$ versus $U - B$), and, ii), the sources in Sextans A identified by us

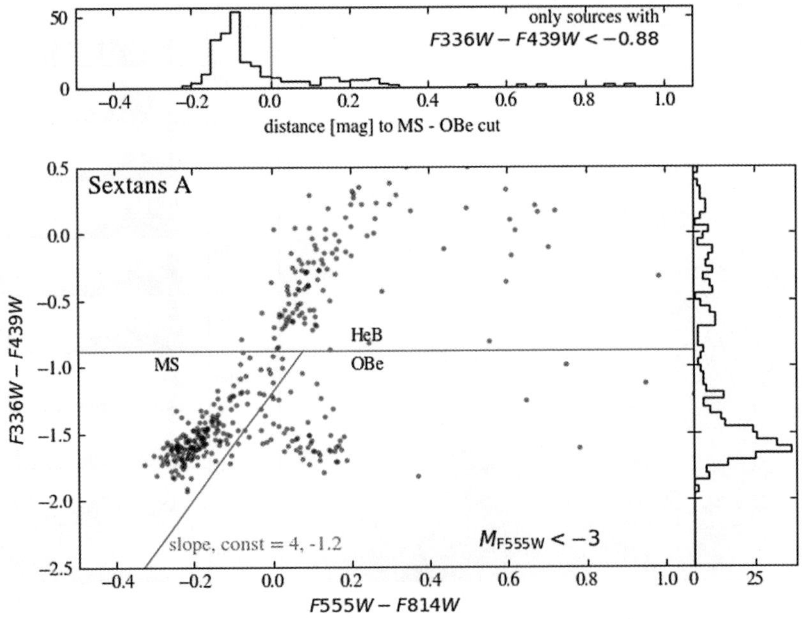

Figure 3. Color-color diagram of bright sources ($M_{\rm F555W} < -3$) in our Sextans A data set. On the right we show a histogram of their $F336W - F439W$ colors. At the top we show a histogram of the shortest distance to the main sequence (MS) - OBe cut of the sources that have $F336W - F439W < -0.88$. Sources with $F336W - F439W > -0.88$ are considered to be helium-burning (HeB) sources.

as OBe stars almost always have Hα in emission in according to narrow-band $F656N$ photometry of Dohm-Palmer et al. (2002).

For Holmberg I and Holmberg II, we apply the same strategy as for Sextans A, except that we use from the LEGUS data a $V - I$ color instead of $F555W - F814W$ and $U - B$ instead of $F336W - F439W$.

3.3. *OBe star fractions as a function of absolute magnitude*

In intervals of half a magnitude, based on our color-color diagrams we calculate the fraction of main sequence stars that are OBe stars: $f_{\rm OBe} = N_{\rm OBe}/(N_{\rm MS} + N_{\rm OBe})$. We do this for the SMC, LMC, Sextans A, Holmberg I, and Holmberg II. The result is shown in Fig. 4. We find high OBe star fractions of $f_{\rm OBe} \approx 0.2 - 0.3$, except at the bright end, where the OBe star fractions tend to decrease. This decrease is the strongest in the LMC, which could be related to stellar winds being relatively strong there (as the LMC is the highest-metallicity galaxy in our sample). As an indication, we write the average evolutionary masses (Schootemeijer et al. 2021) that have been determined for stars with spectral types of B5 and earlier in the SMC at the top of Fig. 4. Typical average masses in the considered absolute magnitude range vary from $10\,{\rm M}_\odot$ to $25\,{\rm M}_\odot$.

3.4. *OBe star fractions as a function of metallicity*

To investigate metallicity trends of the OBe star fraction, we consider one absolute magnitude range: $-3 > M_{\rm abs} > -5$. The top panel of Fig. 5 shows our result. At $0.1 \leq Z/Z_\odot \leq 0.2$ (i.e., in Sex A, Ho I, Ho II, and the SMC) as much as ∼30% of the hydrogen-burning sources is an OBe star. In the LMC, we find a value of ∼20%.

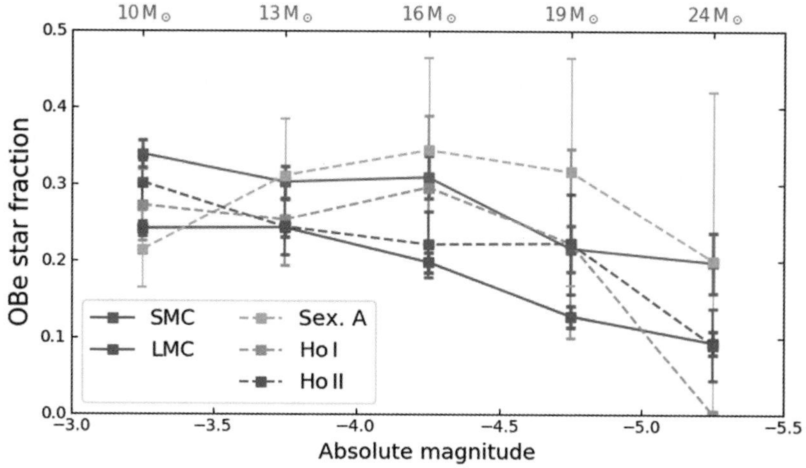

Figure 4. OBe star fractions as a function of absolute magnitude, measured in intervals of half a magnitude. We show these for the Small Magellanic Cloud (SMC), Large Magellanic Cloud (LMC), Sextans A (Sex. A), Holmberg I (Ho I), and Holmberg II (Ho II). At the top we show, as an indication, average evolutionary masses in the same absolute magnitude intervals (see text).

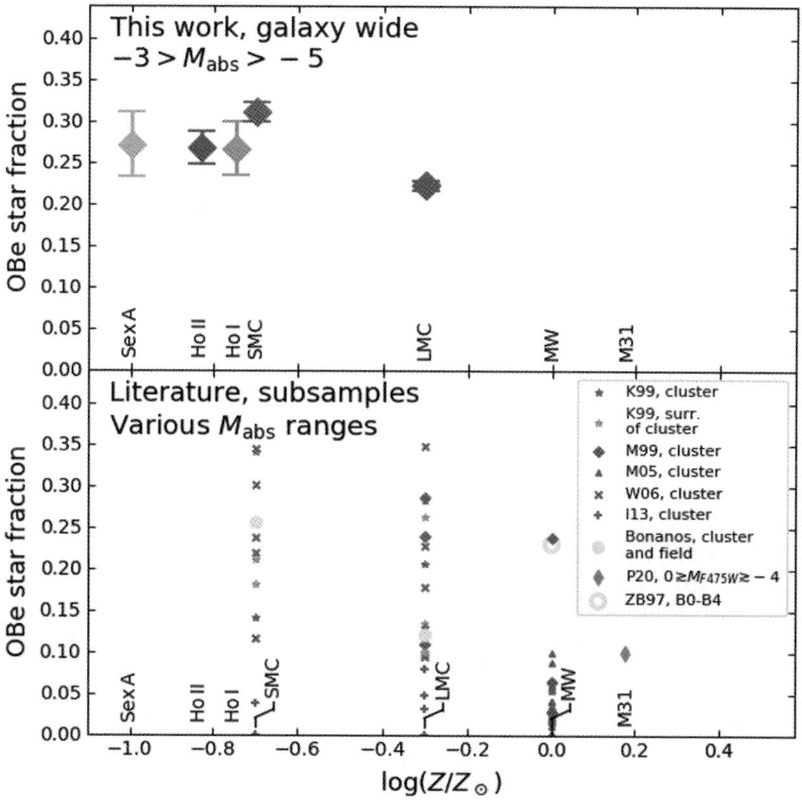

Figure 5. Metallicity trends of the OBe star fraction. *Top:* this work, in the absolute magnitude range $-3 > M_{\rm abs} > -5$. *Bottom:* literature OBe star fractions within stellar sub-samples of different galaxies, measured in various $M_{\rm abs}$ ranges. We also show Milky Way (MW) and M31 data. Included studies are K99 (Keller et al. 1999), M99 (Maeder et al. 1999), M05 (McSwain & Gies 2005), W06 (Wisniewski & Bjorkman 2006), I13 (Iqbal & Keller 2013), Bonanos (Bonanos et al. 2009, 2010), P20 (Peters et al. 2020), and ZB97 (Zorec & Briot 1997). The Sex A metallicity is based on Kaufer et al. (2004); for Ho I and Ho II and Ho II it is based on Croxall et al. (2009).

In the LMC and the SMC, we can compare our results with studies from the literature. The OBe star fraction has been studied before in different environments, albeit never in a galaxy-wide sample as we did here. Instead, typically a sub-sample of stars in a star cluster has been targeted. As the bottom panel of Fig. 5 shows, the result then depends on the cluster that is picked (e.g., because of cluster age), and also on which of its stars are included in the measurement. In the LMC and SMC, our OBe star fractions lie at the high end of the literature values.

4. Concluding remarks

We have devised a method that uses broad-band photometry to tell apart HeB, MS, and OBe stars. With data from the archive, we have determined galaxy-wide OBe star fractions. These include the first OBe star fractions at sub-SMC metallicty in Sextans A, Holmberg I, and Holmberg II. The latter two belong to the M81 group, meaning that OBe star fraction estimates have been extended beyond the Local Group. We found that in the range $0.1 - 0.2Z_{\odot}$, as much as thirty per cent of massive stars is an extremely rapid rotator. This result implies that a large fraction of massive stars in the early universe could be rapid rotators, and it sheds light on the preference for metal-poor dwarf galaxies that long-duration gamma-ray bursts, superluminous supernovae, and ultraluminous X-ray sources have.

References

Aguilera-Dena, D. R., Langer, N., Moriya, T. J., & Schootemeijer, A. 2018, ApJ, 858, 115

Bianchi, L., Efremova, B., Hodge, P., Massey, P., & Olsen, K. A. G. 2012, AJ, 143, 74

Bonanos, A. Z., et al. 2009, AJ, 138, 1003

—. 2010, AJ, 140, 416

Croxall, K. V., van Zee, L., Lee, H., Skillman, E. D., Lee, J. C., Côté, S., Kennicutt, Robert C., J., & Miller, B. W. 2009, ApJ, 705, 723

Dohm-Palmer, R. C., Skillman, E. D., Mateo, M., Saha, A., Dolphin, A., Tolstoy, E., Gallagher, J. S., & Cole, A. A. 2002, AJ, 123, 813

Dufton, P. L., Evans, C. J., Hunter, I., Lennon, D. J., & Schneider, F. R. N. 2019, A&A, 626, A50

Gaia Collaboration et al. 2021, A&A, 649, A1

Holtzman, J. A., Afonso, C., & Dolphin, A. 2006, ApJS, 166, 534

Hunter, I., Lennon, D. J., Dufton, P. L., Trundle, C., Simón-Díaz, S., Smartt, S. J., Ryans, R. S. I., & Evans, C. J. 2008, A&A, 479, 541

Iqbal, S., & Keller, S. C. 2013, MNRAS, 435, 3103

Kaaret, P., Feng, H., & Roberts, T. P. 2017, ARA&A, 55, 303

Kaufer, A., Venn, K. A., Tolstoy, E., Pinte, C., & Kudritzki, R.-P. 2004, AJ, 127, 2723

Keller, S. C., Wood, P. R., & Bessell, M. S. 1999, A&AS, 134, 489

Langer, N. 1992, A&A, 265, L17

Lunnan, R., et al. 2014, ApJ, 787, 138

Maeder, A., Grebel, E. K., & Mermilliod, J.-C. 1999, A&A, 346, 459

Marchant, P., Langer, N., Podsiadlowski, P., Tauris, T. M., & Moriya, T. J. 2016, A&A, 588, A50

Martayan, C., Frémat, Y., Hubert, A. M., Floquet, M., Zorec, J., & Neiner, C. 2007, A&A, 462, 683

Mateo, M. L. 1998, ARA&A, 36, 435

McSwain, M. V., & Gies, D. R. 2005, ApJS, 161, 118

Meixner, M., et al. 2006, AJ, 132, 2268

Peters, M., et al. 2020, AJ, 159, 119

Poeckert, R., & Marlborough, J. M. 1976, ApJ, 206, 182

Sabbi, E., et al. 2018, ApJS, 235, 23

Schootemeijer, A., Langer, N., Grin, N. J., & Wang, C. 2019, A&A, 625, A132

Schootemeijer, A., et al. 2021, A&A, 646, A106

Struve, O. 1931, ApJ, 73, 94

Vink, J. S., Davies, B., Harries, T. J., Oudmaijer, R. D., & Walborn, N. R. 2009, A&A, 505, 743
Wisniewski, J. P., & Bjorkman, K. S. 2006, ApJ, 652, 458
Zaritsky, D., Harris, J., Thompson, I. B., & Grebel, E. K. 2004, AJ, 128, 1606
Zaritsky, D., Harris, J., Thompson, I. B., Grebel, E. K., & Massey, P. 2002, AJ, 123, 855
Zorec, J., & Briot, D. 1997, A&A, 318, 443

Discussion

(COULDN'T HEAR A NAME SORRY): In Sextans A and the like where you do not have spectral types, how well can you separate early and late B stars?

SCHOOTEMEIJER: I agree that there is uncertainty in the exact location of the border between these stars. What I can say is that most of the O and early-B stars reside in a well-defined group, so the exact value that is chosen for the border should not affect our derived OBe star fraction too much.

Massive Stars Near and Far
Proceedings IAU Symposium No. 361, 2024
J. Mackey, J. S. Vink, & N. St-Louis, eds.
doi:10.1017/S1743921322002769

Constraining the behaviour of the young massive stars through interferometry

Emma Bordier[1,2], **Abigail J. Frost**[2], **Hugues Sana**[2] **and Antoine Mérand**[3]

[1]European Southern Observatory (ESO), Alonso de Cordova 3107, Vitacura, Santiago, Chile
email: Emma.Bordier@eso.org

[2]Institute of Astronomy, KU Leuven, Celestijnenlaan 200D, B-3001 Leuven, Belgium

[3]European Southern Observatory (ESO), Karl-Schwarzschild-Str. 2, Garching, Germany

Abstract. The formation of multiples has seen some significant progress over the past years mainly due to the advent and the expansion of high-angular resolution facilities. Star-forming regions are the laboratories where massive stars can be caught right after their formation phase. Still, the observational constraints and the properties of young multiple systems are poorly documented. These proceedings contain recent results about the multiplicity properties of six young O-type stars in the M17 star-forming region, observed by the means of near-IR interferometric observations, which have provided insight into the origin of massive close binaries in a cluster environment.

Keywords. stars: pre–main-sequence, (stars:) binaries (including multiple): close, stars: formation, instrumentation: high angular resolution, techniques: interferometric

---

1. Introduction

Our knowledge about high-mass star ($M_i > 8M_\odot$) formation is undoubtedly growing as a result of both observational and numerical efforts, even though unknowns are persisting. In addition to a rapid formation process ($\sim 10^5$ yrs, Tan et al. 2014), young massive stars are highly reddened with a spectral energy distribution peaking in the mid- to far-infrared (IR) and they form in dusty natal clouds that emit light at low frequency (Zinnecker & Yorke 2007). High-resolution, long-wavelength observations are thus necessary to detect forming stars. Also, physical processes involved in the formation of multiple stars span a large dynamic range, from ~1 au to dozens of pc. Consequently, the theoretical developments for multiple formation supersede the observational constraints.

Current multiplicity constraints: To date, only OB stars of a few Myr benefit from well characterised multiplicity properties. It is now unanimously acknowledged that the majority of massive stars experience interactions with (at least) a companion during their life, a fact that has strong repercussions on their final fate (Sana et al. 2012; Moe & Di Stefano 2017). The multiplicity distribution evolves during the lifetime of the system and might be altered through dynamical interactions. In itself, the multiplicity distribution measured among well formed massive stars do not necessarily reflect the primordial multiplicity (Offner et al. 2022). Constraining the true end product of massive multiple formation requires observations of very young (< 1Myr) massive star-forming regions.

Lack of short-period binaries in young clusters: It has been recently observed that 1D velocity dispersions of massive stars in young clusters increases as they get older

(Ramírez-Tannus et al. 2021). In other words, they observe an increase in the fraction of close binary systems as the clusters evolve. With its very young age (< 1Myr), the M17 high-mass star forming region is no exception to the rule. The analysis of 11 massive pre/near-main-sequence stars, revealed a low radial-velocity dispersion ($\sigma_{1D} = 5.6 \pm 0.2$ km s^{-1}, Sana et al. 2017), corresponding to a binary fraction of $f_{bin} = 0.12$, in stark contrast to that of fully formed main-sequence massive stars ($f_{bin} = 0.70$).

This intriguing lack of short-period binaries in one of the youngest populations of massive stars questions the formation processes leading to close binaries. Sana et al. (2017) and Ramírez-Tannus et al. (2021) put forward the migration scenario in which massive binaries are originally formed at large separations (100 $R_\odot$ or more) and harden on a time-scale of about 1.5 Myr. A strong evidence for this (almost) unexplored scenario is the presence of a significant number of relatively massive companions at separations corresponding to the expected size of the accretion disk (10-1000 au).

In this contribution, I make use of VLTI/GRAVITY observations of six young O stars (presumably single massive stars) in M17 and show that all the objects have at least one detectable companion in the 1-100 mas range, providing a new datapoint for the multiplicity of pre-main-sequence massive stars. I also discuss the separation range of the companions and explore the most likely channels for the formation of the host system.

2. Multiplicity among M17 massive objects

The VLTI/GRAVITY K-band instrument meets the optimal properties (near-IR wavelengths, high-angular resolution and sensitivity) for the search of companions around massive stars. The observational campaign targeted six pre-main-sequence O stars in the M17 star-forming region, known to be one of the youngest, closest and most luminous massive star nursery. The sample counts a modest amount of 6 objects as they were the brightest young massive stars in the region, with $K-$bands ranging from 7 to 9.3, reaching the limits of the GRAVITY+MACAO capabilities.

The main interferometric observables, the visibility amplitudes and closure phases were modelled using the python module PMOIRED (Mérand 2022). Binaries produce two fringe packets that will overlap to produce a sinusoidal signal in the visibility amplitudes. The separation between the peaks relates to the binary separation while the minimum visibility provides a measurement of the flux ratio between components. The closure phases are sensitive to the structure of the source and in particular to the asymmetries in the source distribution (different brightnesses, presence of an asymmetric circumstellar or -binary disk). A variety of models has been explored for each source, iteratively increasing the multiplicity order (single, binary, triple...) and complexity (fully resolved component, circumstellar disk). We infer the mass of the objects comparing the calibrated $K-$band spectra to a family of atmosphere models and evolutionary tracks (the full method can be found in Bordier et al. 2022).

Multiplicity and Companion fraction: The first direct derivable parameters from the analysis of the data are the multiplicity frequency (MF) and the companion fraction (CF). They respectively measure the ratio of multiple systems and the ratio of companions to the sample size. Fig. 1 shows the multiplicity and companion fraction of main sequence stars as a function of the primary mass, regardless of their environment (cluster, association, field). The total MF in our sample of six O stars in cluster (M17) exceeds 94% at the 68%-confidence interval and CF$= 2.3 \pm 0.6$. In other words, we found out that three O stars have a resolved companion (binaries) and the three others have two companions (triples) in the 1-100 mas separation range. We added 5 unresolved companions from the literature to the companions we detected with GRAVITY. These statistics are derived using a wide range of survey methods, still, the six systems observed in M17 attest to the accuracy of the MFs and CFs measured in other galactic regions for stars

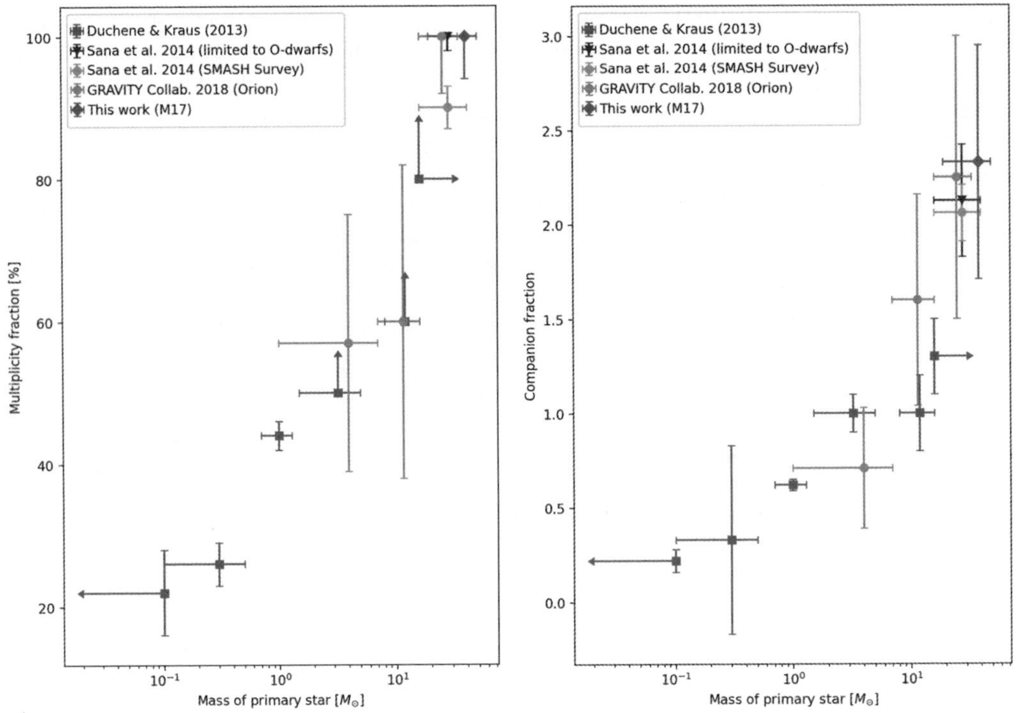

Figure 1. MF *(left)* and CF *(right)* measured with GRAVITY for the six O-stars in M17 as a function of the primary mass. For reference, values of Duchêne & Kraus (2013); Sana et al. (2014); Gravity Collaboration et al. (2018) are also presented for different mass ranges. The results from Sana et al. (2014) (SMASH Survey) presented here are those including companions up to 16000 au. Unresolved and resolved companions are taken into account in the final companion rate derived.

with $M > 20$ $M_\odot$. The young age of M17 and the cluster environment tends to show that massive star formation results in a multiple system with at least two companions on average.

We also display the separation distribution (a, assuming a distance of 1.7 kpc to M17) of all companions for different O-type star populations, all observed by the means of interferometry (see Fig. 2). This preliminary biased plot, shows a dearth of detections for companions within $a < 3$ mas and $\Delta mag > 2$, which are difficult sensitivities to reach with PIONIER and GRAVITY. We detect a handful of close binaries, with at least one companion below $a < 10$ au and wide binaries with $a > 10$ au.

3. Implications

Our results can be compared to the ones of Pomohaci et al. (2019) and Koumpia et al. (2021), two surveys targeting the multiplicity of massive young stellar objects (MYSOs) using K-band observations: high-contrast imaging for the former and interferometry for the latter. They highlight a multiplicity fraction of $31 \pm 8\%$ (>600 au) and $17 \pm 15\%$ (2-300 au) respectively. These results are interesting as they target the same class of objects but the statistics from Pomohaci et al. (2019) trace a different range of separations. The direct comparison might be biased due to the use of different techniques, non-uniform observations, sample size, mass ranges, physical distances and sensitivity limits. We highlight the fact that our targets are slightly older since they are suspected to have already reached the zero-age main-sequence. Still, it is difficult to attribute this

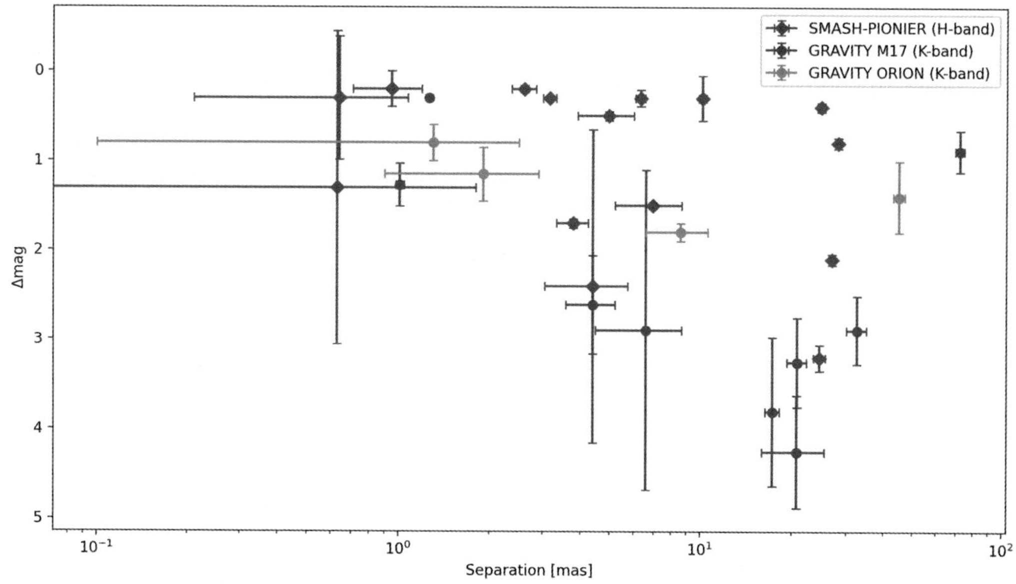

Figure 2. Magnitude difference (Δmag) per angular separation in mas. For reference, the 19 detected companions around O dwarfs as part of the SMaSH+ survey (blue; Sana et al. 2014) and the four resolved companions around high-mass stars in Orion (orange; Gravity Collaboration et al. 2018) have been plotted.

evident difference between the high MF derived in our study and the lower ones found by Pomohaci et al. (2019); Koumpia et al. (2021) to the above shortcomings. Taking all these limitations into consideration, our results (MF and CF) show more consistency with the results of Gravity Collaboration et al. (2018) and Sana et al. (2014) (limited to O-dwarfs), that are presented in Fig. 1.

Up to now, the most robust theories of multiple formation involve stellar migration, core accretion and fragmentation or disk fragmentation. The monolithic collapse involving core accretion and fragmentation would suggest that massive star formation is a scaled-up version of low-mass star formation. The best model would be the one reproducing the observed multiplicity statistics at birth and capable of forming very high-mass stars with tight companions when the systems reach the age of a few Myr. Meyer et al. (2018) and Oliva & Kuiper (2020) predict the formation of tight binaries with unequal components via disk fragmentation, in which early-stage companions have separations of tens of au to a couple of hundreds of au. To match the observed tight binaries (<1 au) in older populations, Ramírez-Tannus et al. (2021) show evidence for an inward migration of companions happening in $\sim$2 Myr, as a function of cluster age. The wide pairs may then harden and leave a tight binary behind. In the context of disk fragmentation theories, such an inward migration process may be driven by the interaction with the remnant of the accretion disk or with protostellar bodies within the cluster. Effects of migration on the binary period distribution will halt when the sinks of angular momentum disappear, i.e when disks and clusters disperse, or as other protostellar bodies are pushed to greater distances. Our findings fit with the predictions of disk fragmentation theories with regards to separations (<100 au) and the variety of systems (only 1 twin binary found). The presence of other companions within 100 au may support the migration channel where a third protostellar object pushes the first companion inward. Still, being about $\sim$1 Myr-old, the migration process inside M17 is likely underway: the wide range of separations and mass ratios might illustrate that the stellar population did not reach the expected time to form the final structures yet.

In order to confirm or discredit the striking high MF and CF found among this modest sample in M17, it is now important to conduct new observational campaigns of even younger protostars, targeting homogeneous samples and using similar observational techniques and sensitivities. A non-detection of companions within ~200 au would point out a different behaviour and invalidate the proposed scenario.

References

Bordier, E., Frost, A. J., Sana, H., Reggiani, M., Mérand, A., Rainot, A., Ramírez-Tannus, M. C., & de Wit, W. J. 2022, A&A, 663, A26

Duchêne, G., & Kraus, A. 2013, ARA&A, 51, 269

Gravity Collaboration et al. 2018, A&A, 620, A116

Koumpia, E., et al. 2021, A&A, 654, A109

Mérand, A. 2022, PMOIRED: Parametric Modeling of Optical Interferometric Data, Astrophysics Source Code Library, record ascl:2205.001

Meyer, D. M. A., Kuiper, R., Kley, W., Johnston, K. G., & Vorobyov, E. 2018, MNRAS, 473, 3615

Moe, M., & Di Stefano, R. 2017, ApJS, 230, 15

Offner, S. S. R., Moe, M., Kratter, K. M., Sadavoy, S. I., Jensen, E. L. N., & Tobin, J. J. 2022, arXiv e-prints, arXiv:2203.10066

Oliva, G. A., & Kuiper, R. 2020, A&A, 644, A41

Pomohaci, R., Oudmaijer, R. D., & Goodwin, S. P. 2019, MNRAS, 484, 226

Ramírez-Tannus, M. C., et al. 2021, A&A, 645, L10

Sana, H., Ramírez-Tannus, M. C., de Koter, A., Kaper, L., Tramper, F., & Bik, A. 2017, A&A, 599, L9

Sana, H., et al. 2012, Science, 337, 444

—. 2014, ApJS, 215, 15

Tan, J. C., Beltrán, M. T., Caselli, P., Fontani, F., Fuente, A., Krumholz, M. R., McKee, C. F., & Stolte, A. 2014, in Protostars and Planets VI, ed. H. Beuther, R. S. Klessen, C. P. Dullemond, & T. Henning, 149

Zinnecker, H., & Yorke, H. W. 2007, ARA&A, 45, 481

Discussion

?: Do you see any correlation between masses and separations?

EMMA BORDIER: We were not able to draw any conclusion on that respect. There is no clear trend for the companion mass according to the separation. We noticed that relatively massive companions are found around the primary stars regardless of their proximity to the host central star.

SILVIA TOONEN: It is very interesting to see this high fraction of triples directly in the observations. I was curious about their formation, are there any processes (regarding star formation) that are unique to triples/multiples, or can we look at it as forming a binary twice?

EMMA BORDIER: Indeed the formation process of such systems is still a subject of discussion. We discarded stellar mergers to form these high-mass stars as the M17 stellar density is too low. However, both the monolithic collapse and competitive accretion could work, even though they cannot straightforwardly result in what we observe, and there can be a mix of the 2 processes actually... For the companions themselves, disk fragmentation could be an explanation. Oliva et al. 2020 show that companions (2,3,4...) can be formed that way. Will they persist throughout the life of the central star, this is a mystery... We know that there is low chances for high-mass stars to form in isolation, so stellar-capture and dynamical interactions happening in the cluster could also be explored.

ANDRÉ OLIVA: I am wondering about the mass ratios of the companions. Are we talking about binaries/companions of similar masses or are the systems dominated by a primary surrounded by less-massive secondaries?

EMMA BORDIER: Constraining the mass of the objects was not an easy task. We used the medium resolution spectra from GRAVITY and applied the flux ratio to the spectra to get the contribution of each star (primary and companions). Then we compared with evolutionary tracks and atmosphere models. This resulted in wide mass ratio ranges. For companions of about the same brightness to half of the brightness, we talk about similar masses and pretty close companions (<10 au). For the triples, the second companion is mostly farther and the flux ratio drops so in that case the central binary dominates in terms of mass. However, we expect massive binaries to be surrounded by outer low-mass companions. This is the purpose of my upcoming project with NACO data in which we aim at finding far, low-mass companions (1000 fainter than the primary and >800 au).

KEI TANAKA:

- Does the similarity between your M17 results and previous studies suggest close binary star formation by gas disks? Or small body interaction?
- For $\sim$1 mas systems, could you see the orbital motions in yr timescale? If so, it would be fun!

EMMA BORDIER:

- From the fitting process, we could not model any circum-binary/stellar disk for example, meaning that there is no signature of a remnant disk. These stars have likely reached the ZAMS so they probably have blown out the material already. It would be interesting to look for the presence, dynamics and influence of the dust and gas surrounding the systems, for example by the means of ALMA. The fact that some systems are surrounded by an outer, smaller body would also be in favour of such interactions happening and supporting the migration of the first companion.
- We did not explore the orbital motions as there is only a single snapshot with GRAVITY but indeed it would be nice to monitor these objects and constrain the orbits (at least for the closest ones). We also had in mind to look for spectroscopic companions since for a couple of sources, there were no studies about the close environment.

Massive Stars Near and Far
Proceedings IAU Symposium No. 361, 2024
J. Mackey, J. S. Vink, & N. St-Louis, eds.
doi:10.1017/S1743921322003398

The Gaia view on massive stars: EDR3 and what to expect from DR3

Jesús Maíz Apellániz[1], Rodolfo H. Barbá[2]†, Michelangelo Pantaleoni González[1,3], Michael Weiler[4], B. Cameron Reed[5], Román Fernández Aranda[1,3,6], Pablo Crespo Bellido[1,3], Alfredo Sota[7], Emilio J. Alfaro[7] and J. Alejo Molina Lera[1,8]

[1]Centro de Astrobiología (CAB), CSIC-INTA. Campus ESAC. E-28 692 Madrid, Spain.
email: jmaiz@cab.inta-csic.es

[2]Dept. de Astronomía, Universidad de La Serena. Av. Cisternas 1200 Norte. La Serena, Chile.

[3]Dept. de Astrofísica y Física de la Atmósfera, U. Compl. de Madrid. E-28 040 Madrid, Spain.

[4]Dept. de Física Quàntica i Astrofísica, U. de Barcelona (IEEC-UB). 08 028. Barcelona, Spain.

[5]Department of Physics (Emeritus), Alma College. 48 801 Michigan. United States of America.

[6]Institute of Astrophysics, University of Crete. 70 013 Heraklion, Greece.

[7]Instituto de Astrofísica de Andalucía (IAA), CSIC. E-18 008 Granada, Spain.

[8]Instituto de Astrofísica de La Plata (CONICET, UNLP). 1900 La Plata, Argentina.

Abstract. At the time of this meeting, the latest *Gaia* data release is EDR3, published on 3 December 2020, but the next one, DR3, will appear soon, on 13 June 2022. This contribution describes, on the one hand, *Gaia* EDR3 results on massive stars and young stellar clusters, placing special emphasis on how a correct treatment of the astrometric and photometric calibration yields results that are simultaneously precise and accurate. On the other hand, it gives a brief description of the exciting results we can expect from *Gaia* DR3.

Keywords. astrometry — catalogs — Galaxy: structure — open clusters and associations: general — stars: kinematics and dynamics — stars: early-type

1. Introduction

The early third *Gaia* data release (EDR3, Brown et al. 2021) provides astrometry and G-band photometry for 1 467 744 819 and 1 806 254 432 sources, respectively. For the astrometry, 40% of the sources are "five-parameter solutions" in which the (pseudo-)color‡ of the source is known independently and is of higher quality than the remaining 60% of "six-parameter solutions" in which the (pseudo-)color has to be simultaneously calculated (Lindegren et al. 2021b). For the photometry, in addition to the G-band information, 86% of the sources also have $G_{\rm BP}$ and/or $G_{\rm RP}$ magnitudes extracted from the spectrophotometric detectors (Riello et al. 2021), whose wavelength-dependent information will not be available until the full third data release (DR3).

This contribution has four parts. First, we describe our calibration efforts to ensure the precision and accuracy of the astrometry and photometry for all sources, massive

† Deceased.
‡ See the *Gaia* documentation for its definition.

or not. Second, we present the integration between *Gaia* data and the Galactic OB stars of the Alma Luminous Star (ALS) catalog (Reed 2003), started with the previous *Gaia* data release (DR2, see Pantaleoni González et al. 2021) and which we have now updated with EDR3 data to produce the largest catalog of massive Galactic OB stars with accurate distances ever. Third, we discuss the Villafranca project of Galactic young stellar groups, which is cataloguing all stellar groups with OB stars in the solar neighborhood (Maíz Apellániz 2019; Maíz Apellániz et al. 2020, 2022a). Finally, we briefly comment on what to expect from the new data products that will become available in DR3.

2. Astrometric and photometric calibration

Lindegren et al. (2021a) describe that the catalog EDR3 parallaxes ϖ have a zero point $Z_{\rm EDR3}$ that depends on magnitude, color, and ecliptic latitude and that to minimize systematic effects one should use a corrected parallax $\varpi_{\rm c} = \varpi - Z_{\rm EDR3}$. In two papers (Maíz Apellániz et al. 2021a; Maíz Apellániz 2022) we have built on that work to produce a methodology that yields accurate and precise EDR3 parallaxes:

• Maíz Apellániz (2022) provides an alternative $Z_{\rm EDR3}$ with the same functional form as Lindegren et al. (2021a). The parameters are very similar for $G > 13$ but differ significantly for brighter stars.

• As described in Maíz Apellániz et al. (2021a) and elsewhere (e.g. Fabricius et al. 2021), EDR3 parallax uncertainties $\sigma_{\rm int}$ are too low and, if used directly, yield distances with underestimated uncertainties. Following the analysis of Maíz Apellániz (2022), calculating external uncertainties $\sigma_{\rm ext}$ as $\sigma_{\rm ext} = \sqrt{k^2\sigma_{\rm int}^2 + \sigma_{\rm s}^2}$, where k is a magnitude-dependent multiplicative constant and $\sigma_{\rm s}$ is described below, provides more realistic uncertainty estimates. k is close to 1 for the fainter stars, increases to values close to 2 around $G = 12$, decreases again to $\sim$1.3 around $G = 10$, and can become as large as 3.0 for the brightest stars, where the original $Z_{\rm EDR3}$ was based on a small number of sources.

• EDR3 parallaxes are correlated as a result of the presence of a residual angular covariance (Lindegren et al. 2021b). This has two consequences. [1] The existence of an uncorrected systematic uncertainty for individual parallaxes ($\sigma_{\rm s}$ in the previous formula) that is the square root of the angular covariance at zero separation. It has an estimated value of 10.3 μas and can be interpreted as the minimum uncertainty of any single EDR3 parallax (Maíz Apellániz et al. 2021a). [2] The need to add covariance terms when combining parallax uncertainties of objects assumed to be at the same distance (e.g. stellar clusters). As a corollary, stellar clusters that span a small angle in the sky have combined parallax uncertainties that approach $\sigma_{\rm s}$ asymptotically, leading to minimum distance uncertainties of $\sim$1% at 1 kpc and $\sim$3% at 3 kpc (Maíz Apellániz et al. 2022a).

• Parallaxes for objects with six-parameter solutions or large RUWE (Renormalized Unit Weight Error) are, in general, of worse quality but it is possible to use them correctly with an increased value of k (Maíz Apellániz 2022).

With respect to photometry, *Gaia* holds the promise of providing the first high dynamic range (17 mag), simultaneously accurate and precise (better that 10 mmag), whole-sky multiband optical photometric survey. However, as we discovered for *Gaia* DR1 (Maíz Apellániz 2017) and DR2 (Maíz Apellániz & Weiler 2018), eliminating the residual systematic offsets requires tweaking the filter passbands and applying small corrections. We have produced a similar analysis for EDR3 that will be submitted soon (Weiler et al. in preparation).

3. The ALS catalog and distances to OB stars

Finding OB stars via photometry alone, though possible (Maíz Apellániz & Sota 2008; Maíz Apellániz et al. 2014), is difficult due to the combined effects of extinction and photometric calibration of ground-based surveys. Two decades ago, Reed (2003) used a combination of photometric and (preferably) spectroscopic data to compile the ALS catalog of Galactic OB stars, which by 2005 contained 18 693 objects. However, the diverse quality of the data (which in many cases lacked even good-quality coordinates) hampered the use of the catalog without including a significant number of contaminants.

In Pantaleoni González et al. (2021), from now on ALS II, we undertook the task of cross-matching the ALS catalog with *Gaia* DR2, painstakingly comparing star by star between the existing data and the *Gaia* catalog. After eliminating duplicates, unmatched objects, and stars with bad astrometry or photometry, we were left with 15 662, of which 13 762 are suspected of truly being Galactic OB massive stars. Among the results from that paper, we point out the following:

- Neither Apsis (Bailer-Jones et al. 2013) nor StarHorse (Anders et al. 2019) provide good extinction measurements for OB stars based on *Gaia* DR2 data. Apsis apparently mistakes high-extinction OB stars for cooler, lower-extinction objects while StarHorse allows for some objects to have large negative extinctions and others to follow an anomalous relationship between color excess and amount of extinction.
- In order to convert parallaxes into distances one needs to use priors. Commonly used with *Gaia* DR2 data are those of Bailer-Jones et al. (2018) and Anders et al. (2019), which assume spatial distributions different from the one for OB stars. For that reason, we compared their distances with the ones derived from the prior of Maíz Apellániz (2001, 2005) and the updated parameters of Maíz Apellániz et al. (2008). We found good agreement among the distances derived from the three priors for values below 3 kpc but some differences beyond that.
- We produced a map of the location of the OB stars in the solar neighborhood. The most significant result is the discovery of a structure that extends from the Orion-Cygnus (or Local) and Perseus arms that we dub the Cepheus spur. It is located 50-100 pc above the Galactic plane and appears to be a 2-D extension of the 1-D Radcliffe wave (Dixon 1967; Alves et al. 2020).

After our work with ALS II, *Gaia* EDR3 yielded significantly improved astrometry and photometry. We continued our analysis of the data by attempting to cross-match the ALS stars that were not in the 15 662 sample above and reprocessing the whole sample to better differentiate between massive and non-massive stars. We have also expanded the sample by including stars with GOSSS Maíz Apellániz et al. (2011) and LiLiMaRlin (Maíz Apellániz et al. 2019) spectral classifications (a new paper of the GOSSS series, GOSSS IV, will be submitted soon) and from the Cygnus OB2 (Berlanas et al. 2020) and Carina OB1 (Berlanas et al. in preparation) associations. The new sample includes several thousand more stars and will be submitted as ALS III later this year.

The information that will be made available in DR3 (see below) and that from ground-based surveys such as GALANTE (Maíz Apellániz et al. 2021b) will allow us to significantly expand the ALS sample in the near future. Our estimate is that as many as 10^5 Galactic OB stars may be identified using those data.

4. The Villafranca project

The Villafranca catalog of Galactic OB stellar groups combines *Gaia* astrometry and photometry with spectroscopy from ground-based surveys to derive information about the young stellar clusters and associations in the solar neighborhood and beyond. A zeroth paper (Maíz Apellániz 2019) describes the method used to determine group membership, Villafranca I (Maíz Apellániz et al. 2020) analyzes 16 groups with O2-O3.5 stars using *Gaia* DR2 data, and Villafranca II (Maíz Apellániz et al. 2022a) revises those results with *Gaia* EDR3 data and adds ten new stellar groups with O stars.

With the Villafranca project we have been able to provide precise distances ($\sim$1% at 1 kpc, $\sim$3% at 3 kpc) whose accuracy has been verified for some examples using alternative geometric distances. We have also published detailed analyses of each cluster, describing the massive stars with accurate spectral types that belong to them and finding runaway candidates. For four nearby low-extinction clusters we have used the PMS seen in the cleaned *Gaia* CMDs to estimate their ages. Such CMD analyses will be extended in the future with the help of complementary ground-based surveys such as GALANTE (Maíz Apellániz et al. 2021b).

The most interesting result that has come out of the Villafranca project is the discovery of orphan clusters, very young stellar groups that have been disrupted as the consequence of multiple stellar interactions among their massive stars. The prototype object is the Bermuda cluster (Villafranca O-014 NW) at the North America nebula (Maíz Apellániz et al. 2022b), which through three ejection events expelled a total of at least nine stellar systems (twelve individual stars) between 1.9 and 1.5 Ma ago, including its three most massive stars (one of which, the primary of the Bajamar system, is the earliest-type O star within 1 kpc of the Sun). As a result, the cluster has lost over 200 $M_\odot$ in stars and has altered its mass function from one that was top-heavy to one that is compatible with a Kroupa-like function. Moreover, the cluster has lost so much mass that it is now expanding (Kuhn et al. 2020) and will likely dissolve several million years from now.

We are currently working on the third paper of the Villafranca series, which will concentrate on the region of the sky where Carina OB1† is located. Three stellar groups in Carina OB1 (Villafranca O-002, Villafranca O-003, and Villafranca O-025) and one in its background (Villafranca O-004) were already analyzed in previous papers and here we will add several more, some with O and early-B stars and some with just the latter. Future papers will expand the sample of Villafranca groups.

5. A peek into what the future holds in store

By the time this contribution is published, *Gaia* DR3 will be out in the wild but at the time of this writing access to the data is restricted to DPAC members (and their lips are sealed). What can we say about it without making fools of ourselves in the eyes of you, dear (future) reader, who may know more than us?

We can divide the DR3 results into two types: *raw* and *processed*. By *raw* we mean the direct output of the *Gaia* instruments without significant assumptions about the nature of the source and/or comparisons with models and by *processed* those results that imply an additional treatment such as measuring velocities, equivalent widths, or time series of the raw data. There will be three types of raw results: low-resolution BP+RP spectrophotometry for 219 197 643 objects, high-resolution z-band spectroscopy for 999 645 objects, and z-band photometry for 32 232 187 objects, in all cases averaged over time. The list of processed results is significantly longer and includes radial velocity and $v \sin i$ measurements, variable-star identifications and classifications, astrometric and spectroscopic orbit determinations, astrophysical parameters estimations, and Hα and DIB λ8621 equivalent widths, among others.

The information content resulting from *Gaia* DR3 will surpass that of previous data releases, as two dimensions will be added to a large number of objects: wavelength in the form of the two types of spectroscopy and time in the form of the analysis of variable stars. It is true that, as it was before, most of the information will not be for massive stars but this time there has been a concerted effort by DPAC to find hot stars (mostly of intermediate mass but some massive) in the sample through a combination of the

† Which, despite its name, should not be considered the Keel Kenobi association, something that was instead a result of Darth Vader's actions.

low-resolution spectrophotometry and the high-resolution spectroscopy. Furthermore, the added information will surely help finding new massive needles in the Galactic haystack.

Further down the road we will have *Gaia* DR4. While EDR3 and DR3 are based on 34 months of data, DR4 will use almost double that amount, 66 months. Such a longer baseline will undoubtedly improve the precision of most data products, especially proper motions. However, an even better improvement is likely to take place in the accuracy, that is, the reduction of systematic errors, as the experience from previous data releases will be used to characterize and correct them (for example, the comparison between DR2 and EDR3 was the base for the work of Lindegren et al. 2021a, see section 3 in that paper). And then there will be the data deluge in the form of the full astrometric, photometric, and radial-velocity catalogs: all data, all epochs, enough to maintain a generation of Galactic astronomers busy.

Acknowledgements

One of the authors, Rodolfo Barbá unexpectedly passed away after making significant contributions to most aspects of this work: we dedicate it to him. J.M.A., M.P.G., R.F.A., P.C.B., A.S., and J.A.M.L. acknowledge support from the Spanish Government Ministerio de Ciencia e Innovación through grant PGC2018-095 049-B-C22. M.W. acknowledges funding by the Spanish MICIN/AEI/10.130 39/501 100 011 033, by "ERDF A way of making Europe" by the "European Union" through grant RTI2018-095 076-B-C21, and by the Institute of Cosmos Sciences University of Barcelona (ICCUB, Unidad de Excelencia 'María de Maeztu') through grant CEX2019-000 918-M. E.J.A. acknowledges support from the State Agency for Research of the Spanish Government Ministerio de Ciencia e Innovación through the "Center of Excellence Severo Ochoa" award to the Instituto de Astrofísica de Andalucía (SEV-2017-0709) and through grant PGC2018-095 049-B-C21 This work has made use of data from the European Space Agency (ESA) mission *Gaia*†, processed by the *Gaia* Data Processing and Analysis Consortium (DPAC‡). Funding for the DPAC has been provided by national institutions, in particular the institutions participating in the *Gaia* Multilateral Agreement. The *Gaia* data is processed with the computer resources at Mare Nostrum and the technical support provided by BSC-CNS.

References

Alves, J., Zucker, C., Goodman, A. A., et al. 2020, Nature, 578, 237
Anders, F., Khalatyan, A., Chiappini, C., et al. 2019, A&A, 628, A94
Bailer-Jones, C. A. L., Rybizki, J., Fouesneau, M., Mantelet, G., & Andrae, R. 2018, AJ, 156, 58
Bailer-Jones, C. A. L., Andrae, R., Arcay, B., et al. 2013, A&A, 559, A74
Berlanas, S. R., Herrero, A., Comerón, F., et al. 2020, arXiv e-prints, arXiv:2008.09917
Brown, A. G. A., Vallenari, A., Prusti, T., et al. 2021, A&A, 649, A1
Dixon, M. E. 1967, MNRAS, 137, 337
Fabricius, C., Luri, X., Arenou, F., et al. 2021, A&A, 649, A5
Kuhn, M. A., Hillenbrand, L. A., Carpenter, J. M., & Avelar Menéndez, Á. R. 2020, ApJ, 899, 128
Lindegren, L., Bastian, U., Biermann, M., et al. 2021a, A&A, 649, A4
Lindegren, L., Klioner, S. A., Hernández, J., et al. 2021b, A&A, 649, A2
Maíz Apellániz, J. 2001, AJ, 121, 2737
Maíz Apellániz, J. 2005, in ESA Special Publication, Vol. 576, The Three-Dimensional Universe with Gaia, ed. C. Turon, K. S. O'Flaherty, & M. A. C. Perryman, 179
—. 2017, A&A, 608, L8
—. 2019, A&A, 630, A119 (Villafranca 0)
—. 2022, A&A, 657, A130

† https://www.cosmos.esa.int/gaia
‡ https://www.cosmos.esa.int/web/gaia/dpac/consortium

Maíz Apellániz, J., Alfaro, E. J., & Sota, A. 2008, arXiv:0804.2553
Maíz Apellániz, J., Barbá, R. H., Fernández Aranda, R., et al. 2022a, A&A, 657, A131 (Villafranca II)
Maíz Apellániz, J., Crespo Bellido, P., Barbá, R. H., Fernández Aranda, R., & Sota, A. 2020, A&A, 643, A138 (Villafranca I)
Maíz Apellániz, J., Pantaleoni González, M., & Barbá, R. H. 2021a, A&A, 649, A13
Maíz Apellániz, J., Pantaleoni González, M., Barbá, R. H., & Weiler, M. 2022b, A&A, 657, A72
Maíz Apellániz, J., & Sota, A. 2008, in RMxAC, Vol. 33, 44–46
Maíz Apellániz, J., Sota, A., Walborn, N. R., et al. 2011, in HSA 6, 467–472
Maíz Apellániz, J., Trigueros Páez, E., Jiménez Martínez, I., et al. 2019, in HSA 10, 420 (LiLiMaRlin)
Maíz Apellániz, J., & Weiler, M. 2018, A&A, 619, A180
Maíz Apellániz, J., Evans, C. J., Barbá, R. H., et al. 2014, A&A, 564, A63
Maíz Apellániz, J., Alfaro, E. J., Barbá, R. H., et al. 2021b, MNRAS, 506, 3138
Pantaleoni González, M., Maíz Apellániz, J., Barbá, R. H., & Reed, B. C. 2021, MNRAS, 504, 2968
Reed, B. C. 2003, AJ, 125, 2531
Riello, M., De Angeli, F., Evans, D. W., et al. 2021, A&A, 649, A3

Discussion

NATHAN SMITH: For the case of the North America cluster, some gas must have also been ejected in addition to the stars. Does that mean that the cluster was initially bound?

JESÚS MAÍZ APELLÁNIZ: Most likely. We estimate (making reasonable assumptions about the IMF) that the cluster formed 400-500 $M_\odot$ in stars and that it has ejected $\sim$200 $M_\odot$ of those. If you add the gas that must have been ejected, chances are that the cluster has retained significantly less than half of its initial total mass, which would be sufficient to unbound it.

ALEX DE KOTER: What is the origin of the ejections from the North America cluster? Dynamical interactions or supernova explosions?

JESÚS MAÍZ APELLÁNIZ: Dynamical interactions. The cluster is very young and there are no signs of supernovae at all. Indeed, the three ejections events can be dated quite accurately (to within a few tens of ka) to 1.5, 1.6, and 1.9 Ma ago and they serve as estimates of the age of the cluster ($\sim$2 Ma) if we assume that such events take place preferentially very early in the cluster's life. Those values are also consistent with the expansion rate observed in what remains of the cluster.

Massive Stars Near and Far
Proceedings IAU Symposium No. 361, 2024
J. Mackey, J. S. Vink, & N. St-Louis, eds.
doi:10.1017/S1743921322002757

Dynamical vs Supernova Acceleration of OB Stars in the Small Magellanic Cloud

M. S. Oey[1], J. Dorigo Jones[2], G. D. Phillips[1], N. Castro[3], M. M. Dallas[1,4] and M. Moe[5]

[1]University of Michigan, Astronomy Department, Ann Arbor, MI, 48109-1107, USA
email: msoey@umich.edu

[2]University of Colorado, Department of Astrophysical and Planetary Sciences, 2000 Colorado Ave., Boulder, CO 80309, USA

[3]Leibniz-Institut für Astrophysik, An der Sternwarte, 16 D-14482, Potsdam, Germany

[4]Current address: Space Telescope Science Institute, 3700 San Martin Drive, Baltimore, MD 21218, USA

[5]University of Arizona, Astronomy Department, Tucson, AZ, 85721, USA

Abstract. We use the RIOTS4 sample of SMC field OB stars to determine the origin of massive runaways in this low-metallicity galaxy using Gaia proper motions, together with stellar masses obtained from RIOTS4 data. These data allow us to estimate the relative contributions of stars accelerated by the dynamical ejection vs binary supernova mechanisms, since dynamical ejection favors faster, more massive runaways, while SN ejection favors the opposite trend. In addition, we use the frequencies of classical OBe stars, high-mass X-ray binaries, and non-compact binaries to discriminate between the mechanisms. Our results show that the dynamical mechanism dominates by a factor of 2 – 3. This also implies a significant contribution from two-step acceleration that occurs when dynamically ejected binaries are followed by SN kicks. We update our published quantitative results from Gaia DR2 proper motions with new data from DR3.

Keywords. massive stars — Be stars — runaway stars — interacting binary stars — field stars — Small Magellanic Cloud — open star clusters — multiple star evolution

1. Introduction

We know that at least 70% of most massive stars become interacting binaries (e.g., Sana et al. 2012; Moe & Di Stefano 2017), and that close binaries and multiples generate runaway stars. Therefore, field massive stars offer an important probe of the massive binary population. In particular, the field OB stars in the Small Magellanic Cloud (SMC) are beautifully accessible, since they are bright and easily identified in this Milky Way satellite galaxy. Since the SMC is a dwarf galaxy, the complete OB population can be observed. Moreover, the SMC is metal-poor, allowing us to quantitatively characterize conditions and processes at low metallicity.

Our group carried out the Runaways and Isolated O-Type Star Spectroscopic Survey of the SMC (RIOTS4; Lamb et al. 2016), which identified and spectroscopically confirmed a uniform sample of field OB stars in the SMC. The target stars are those at least 28 pc from other OB stars in this galaxy, photometrically identified as having $B \leq 15.21$ and reddening-free Q-parameter ≤ -0.84 (Oey et al. 2004). Vargas-Salazar et al. (2020, and these proceedings) find evidence that $\sim 5\%$ of the sample may have formed in situ (see also Oey et al. 2013; de Wit et al. 2004; Pflamm-Altenburg & Kroupa 2010). Therefore, $\sim 95\%$ of these field OB stars are ejected from clusters.

There are two ejection mechanisms. Following the nomenclature of Hoogerwerf et al. (2000), they are the dynamical ejection scenario (DES; e.g., Poveda et al. 1967) and the binary supernova scenario (BSS; e.g., Blaauw 1961). The DES mechanism relies heavily on the interaction of binaries with other stars and other binaries or multiples to accelerate stars above the escape velocities of their parent clusters. This includes also ejecting some tight binaries. Meanwhile, the BSS mechanism requires only a single close binary system, in which the SN of the primary unbinds the remaining star, which may also receive a kick from the explosion. On average, the DES produces faster, more massive runaways since it leverages the gravitational energy of multiple stars, while the BSS generates lower-velocity ejections of lower-mass stars. In binaries ejected by the DES, the explosion of the primary will re-accelerate the companion, resulting in a two-step ejection (Pflamm-Altenburg & Kroupa 2010).

2. OBe stars as BSS products

Classical OBe stars are known to be fast rotators that have generated decretion disks responsible for their characteristic Balmer emission (e.g., Rivinius et al. 2013; Dorigo Jones et al. 2020). These stars most likely acquired their high rotation velocities during binary mass exchange, which also transfers angular momentum (e.g., Kriz & Harmanec 1975; Pols et al. 1991). Therefore, many, if not most, of these stars are likely to be post-SN systems and thus BSS products. Following Smith & Tombleson (2015), we compare the spatial distributions of various populations using the distance of each star from the nearest O-star.

Figure 1 shows the cumulative distribution functions of nearest O-star distances for early and late O-stars, B-stars, Oe stars, Be stars, supergiants (luminosity class I and II), and high-mass X-ray binaries (HMXBs). As expected, early O-stars are closest to other O-stars, while late O-stars and B-stars have larger median distances since these longer-lived populations can disperse farther into the field. The Oe and Be stars have even farther median distances, supporting the scenario that they are dominated by BSS products ejected from clusters. In addition, the OBe distributions have a similar locus to the HMXBs, which are bona fide BSS products. We also see that the Oe and Be distributions are indistinguishable, within the uncertainties. This is again consistent with their mass-transfer origin, implying that their current masses are independent of their initial masses. Thus, several lines of evidence in our data point to OBe stars largely corresponding to BSS products. These results are discussed in more detail in Dallas & Oey (2022, in preparation).

Thus, to first order, we can distinguish the DES and BSS products by using the field OBe stars to represent BSS products and the remaining stars to represent DES products. We caution that these are unlikely to be exact representations of these populations, since some BSS systems may produce non-OBe stars, and the OBe phenomenon may also result from other, unrelated processes. We also noted above that some OB stars may have formed as in-situ field stars, which would not be ejected from clusters. However, we can use these observations to make a rough first estimate of the relative contributions of the two acceleration mechanisms.

3. Kinematics of DES versus BSS ejections

Adopting OBe stars as tracers of the BSS population, we use Gaia proper motions of the RIOTS4 sample to evaluate the relative contributions from the DES and BSS mechanisms, as well as the two-step mechanism. Figure 2 shows the proper motion velocity

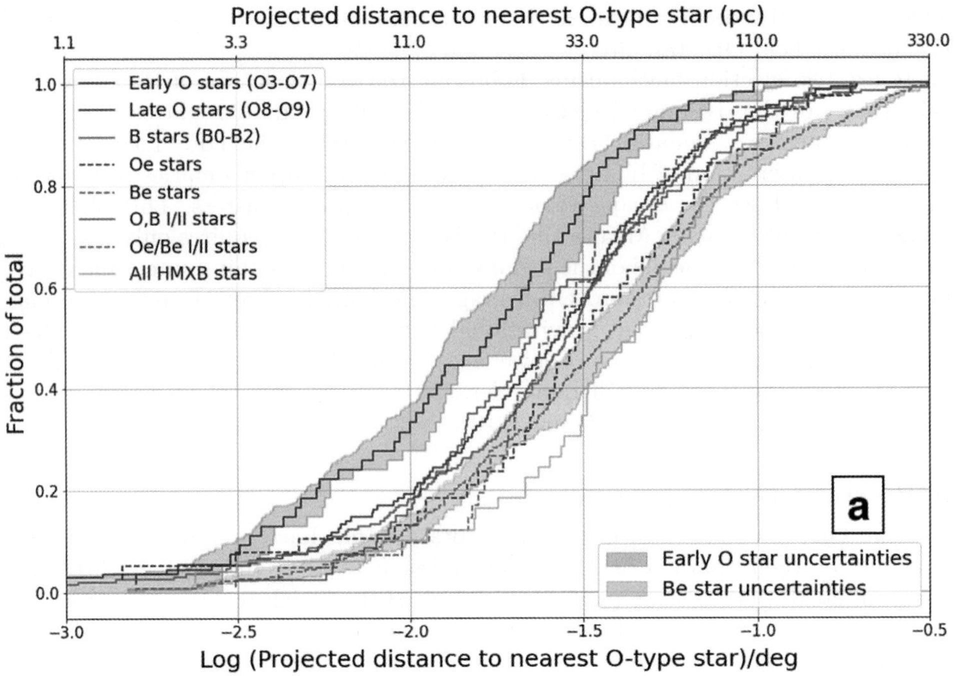

Figure 1. Cumulative distribution function of projected distances from nearest O-stars for different populations as shown. From Dallas, Oey & Castro (2022), in preparation. (A color version of this figure is available in the electronic version.)

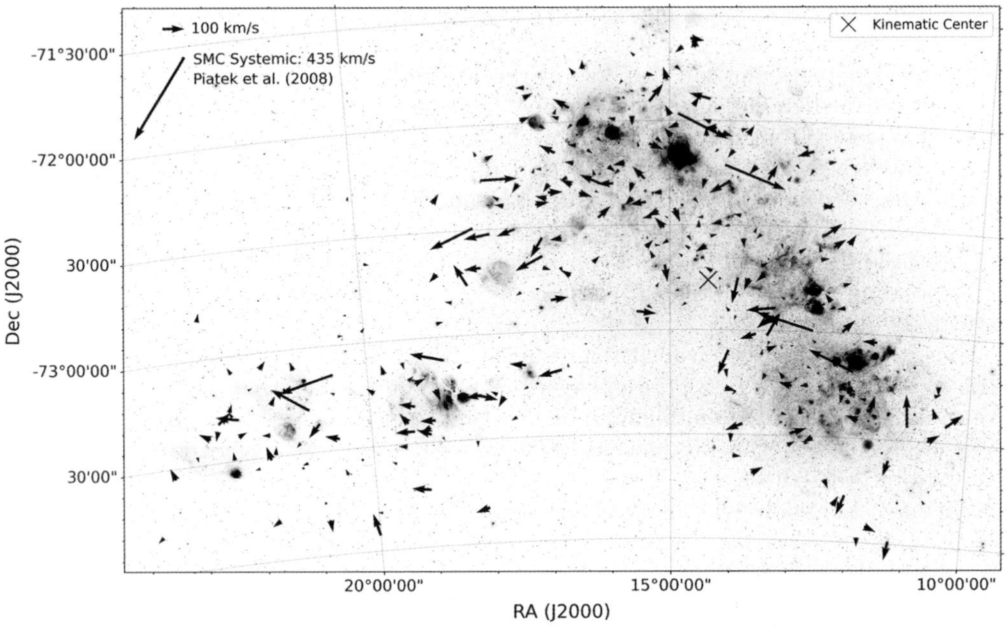

Figure 2. Map of the SMC with vectors showing residual transverse velocities Oey et al. (2018) relative to the local velocity fields within 5′ of each RIOTS4 target star, based on Gaia DR2 data for 304 RIOTS4 stars.

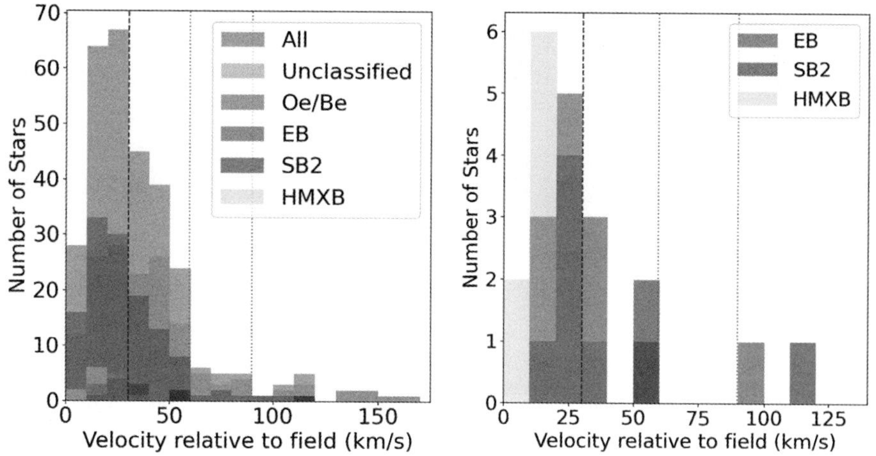

Figure 3. Velocity distribution for RIOTS4 field stars and various subsets, as shown. The right panel is a zoom showing the eclipsing binaries (EB), double-lined spectroscopic binaries (SB2), and high-mass X-ray binaries (HMXB). "Unclassified" objects are those that do not belong to any of the other subsets. The dashed vertical line shows a nominal 30 km s^{-1} value, and the dotted lines show 1 and 2 standard deviations above the median of 28.6 km s^{-1}. (A color version of this figure is available in the electronic version.)

vectors for the 304 RIOTS4 stars with available DR2 Gaia data (Dorigo Jones et al. 2020). These are calculated in the frame of the local velocity fields based on the blue stars having $(B-V) < 0.14$ and $M_{\rm bol} < -7.0$ from (Massey 2002) that are within $5'$ (90 pc) of each RIOTS4 target (Oey et al. 2018).

We show updated velocity distributions from Gaia EDR3 for 299 RIOTS4 stars in Figure 3. We see that the OBe stars, which are putative BSS products, are indeed at lower median velocities than the remaining stars, which conversely must be a population dominated by DES ejections. Eclipsing binaries (EBs) and double-lined spectroscopic binaries (SB2s) are non-compact, pre-SN binaries that are thus bona fide tracers of the DES mechanism. These can be compared to the HMXBs, which as mentioned above, are similarly tracers of BSS ejections. We likewise see that the HMXBs are restricted to lower velocities while the EBs and SB2s have distributions extending to much higher velocities, mirroring what we see for the OBe versus non-OBe stars.

Figure 3 shows that the OBe velocity distribution extends to relatively high velocities, well in excess of 30 km s^{-1}, whereas BSS products are largely expected to remain at "walkaway" velocities below this value. This may result from Gaia errors, which are asymmetric and tend to be positive, rather than negative. We find that the median velocity of the entire distribution decreases about 25%, from 39 km s^{-1} to 29 km s^{-1}, between DR2 and EDR3. However, another, real effect is that some OBe stars have been doubly accelerated, originating as non-compact DES ejections that have subsequently experienced a SN explosion. These are the "two-step" ejections (Pflamm-Altenburg & Kroupa 2010).

Figure 4 shows density plots of the non-OBe and OBe populations for stellar mass vs velocity. Their tracers, the EB and SB2 populations (DES), and the HMXBs (BSS), are overplotted. The data are again consistent with the expectation that DES products on average accelerate more massive runaways to faster velocities, while BSS products show opposite trends. The tracer populations are also consistent with these patterns.

Based on the DR2 proper motion data, Dorigo Jones et al. (2020) find that on the order of 26% of the SMC OB population corresponds to DES products, including $\sim 8\%$

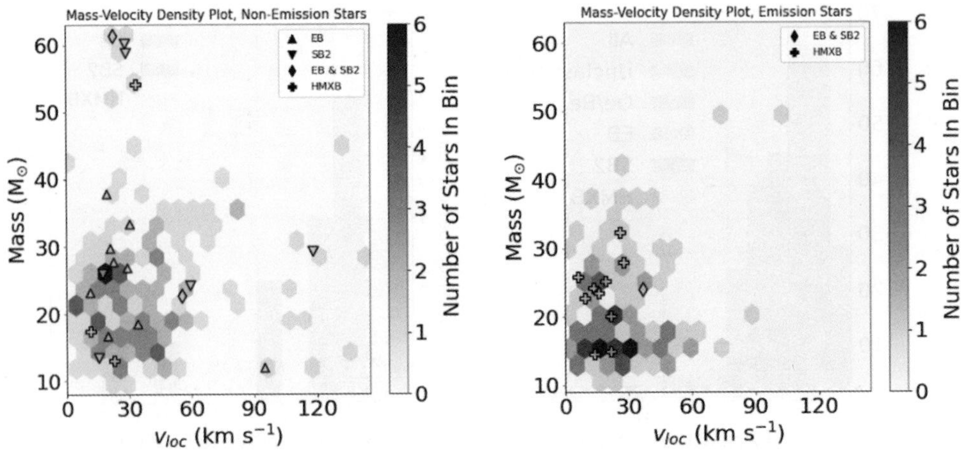

Figure 4. Distributions of mass vs transverse velocity for 283 stars in our RIOTS4 field star sample with mass determinations. The left panel shows non-OBe stars and the right panel shows OBe stars. Individual EBs, SB2s, and HMXBs are overplotted as shown.

non-compact binaries; while $\sim 9\%$ are BSS products and perhaps an additional $\sim 2\%$ are two-step ejections. These values are model-dependent and rely on parameters from N-body simulations for the DES mechanism by Oh & Kroupa (2016) and binary population synthesis models by Renzo et al. (2019) for the BSS. The total frequencies are larger than the SMC field population because they include ejections that do not meet the criteria for RIOTS4 field stars (see Section 5.1 of Dorigo Jones et al. 2020, for details). We caution that the measured velocities changed substantially between Gaia DR2 and EDR3, as noted above. Thus, our derived frequencies have substantial uncertainties and represent only a first crude estimate of the relative contributions of the different ejection mechanisms. We are currently in the process of updating these calculations (Phillips et al., in preparation). However, the ability to make such estimates based on empirical data promises major advances in our understanding of massive binary populations and cluster dynamics.

Acknowledgements

This work was supported in part by NSF grant AST-1514838 to M.S.O. N. C. acknowledges funding from the Deutsche Forschungsgemeinschaft (DFG), CA 2551/1-1.

References

Blaauw, A. 1961, Bull. Astron. Inst. Netherlands, 15, 265
de Wit, W. J., Testi, L., Palla, F., Vanzi, L., & Zinnecker, H. 2004, A&A, 425, 937
Dorigo Jones, J., Oey, M. S., Paggeot, K., Castro, N., & Moe, M. 2020, ApJ, 903, 43
Hoogerwerf, R., de Bruijne, J. H. J., & de Zeeuw, P. T. 2000, ApJL, 544, L133
Kriz, S., & Harmanec, P. 1975, Bulletin of the Astronomical Institutes of Czechoslovakia, 26, 65
Lamb, J. B., Oey, M. S., Segura-Cox, D. M., Graus, A. S., Kiminki, D. C., Golden-Marx, J. B., & Parker, J. W. 2016, ApJ, 817, 113
Massey, P. 2002, ApJS, 141, 81
Moe, M., & Di Stefano, R. 2017, ApJS, 230, 15
Oey, M. S., King, N. L., & Parker, J. W. 2004, AJ, 127, 1632
Oey, M. S., Lamb, J. B., Kushner, C. T., Pellegrini, E. W., & Graus, A. S. 2013, ApJ, 768, 66
Oey, M. S., et al. 2018, ApJL, 867, L8
Oh, S., & Kroupa, P. 2016, A&A, 590, A107
Pflamm-Altenburg, J., & Kroupa, P. 2010, MNRAS, 404, 1564

Pols, O. R., Cote, J., Waters, L. B. F. M., & Heise, J. 1991, A&A, 241, 419
Poveda, A., Ruiz, J., & Allen, C. 1967, Boletin de los Observatorios Tonantzintla y Tacubaya, 4, 86
Renzo, M., et al. 2019, A&A, 624, A66
Rivinius, T., Baade, D., Townsend, R. H. D., Carciofi, A. C., & Štefl, S. 2013, A&A, 559, L4
Sana, H., et al. 2012, Science, 337, 444
Smith, N., & Tombleson, R. 2015, MNRAS, 447, 598
Vargas-Salazar, I., Oey, M. S., Barnes, J. R., Chen, X., Castro, N., Kratter, K. M., & Faerber, T. A. 2020, ApJ, 903, 42

Discussion

SHENAR: Some later Be stars could originate from mass donors that are less massive and would not have exploded. Have you considered this in your velocity distributions?

OEY: No, we have not included this effect. Given that binaries tend to have high mass ratios, the contribution from such objects should be relatively small. There are also others that we ignored, for example, not all stars that undergo mass transfer will be seen as Be stars and similarly, there may also be a significant number of Be stars that originate from a completely unrelated mechanism. So we stress that our results are simply a first rough estimate of the relative contributions of the acceleration mechanisms.

Massive Stars Near and Far
Proceedings IAU Symposium No. 361, 2024
J. Mackey, J. S. Vink, & N. St-Louis, eds.
doi:10.1017/S1743921322002095

Dissecting the very metal-poor galaxy SBS0335-052E with MUSE: massive stars vs. nebular HeII emission

C. Kehrig[1], J. M. Vílchez[1], M. A. Guerrero[1], J. Iglesias-Páramo[1], L. K. Hunt[2], S. Duarte Puertas[1] and G. Ramos-Larios[3]

[1]Instituto de Astrofísica de Andalucía, CSIC,
Glorieta de la Astronomia s/n, 18008, Granada, Spain
email: kehrig@iaa.es

[2]INAF-Osservatorio Astrofisico di Arcetri, 50125 Firenze, Italy

[3]Instituto de Astronomía y Meteorología, Dpto. de Física, Universidad de Guadalajara
44130 Guadalajara, Jalisco, Mexico

Abstract. Nebular HeII emission implies the presence of energetic photons (E⩾54 eV). Despite the great deal of effort dedicated to understanding HeII ionization, its origin has remained mysterious, particularly in metal-deficient star-forming galaxies. Unfolding HeII-emitting, metal-poor starbursts at z∼0 can yield insight into the powerful ionization processes occurring in the primordial universe. Here we present a study on the origin of the extended nebular HeII emission in SBS 0335-052E, one of the most metal-poor ($Z \sim 3\%\ Z_{\odot}$) HeII-emitter starbursts known locally. Based on optical VLT/MUSE spectroscopic and Chandra X-ray observations, and current stellar models we found that the HeII-ionization budget of SBS 0335-052E can only be produced by peculiar, nearly metal-free ionizing stars (called here "PopIII-like" stars) with a top-heavy initial mass function. This result is in line with recent simulations for PopIII star formation down to z=0.

Keywords. galaxies: individual: SBS 0335-052E, galaxies: starburst, galaxies: dwarf, galaxies: ISM, galaxies: stellar content

1. Motivation

The presence of nebular HeII emission points to the existence of a hard radiation field as photon energies ⩾ 54 eV are necessary to ionize He^+. Highly ionized systems, like HeII emitters, are observed to be more frequent at high-z than locally (e.g., Kehrig et al. 2011; Cassata et al. 2013; Mainali et al. 2018). At the same time, observations of local, HeII-emitting star-forming (SF) galaxies suggest that the lower the metallicity (Z), the larger the nebular HeII line intensities (e.g., Senchyna et al. 2017). This is in line with the harder ionizing stellar spectra predicted at the lower metallicities typical in the early universe (e.g., Tumlinson & Shull 2000); the hard spectra from the first, hot Population III (PopIII) and nearly Z=0 stars can produce strong nebular HeII lines (e.g., Yoon et al. 2012; Nakajima & Maiolino 2022). Still, despite much observational and theoretical effort, including relevant advances in stellar modeling, the source of HeII ionization remains puzzling, especially in metal-poor SF galaxies (e.g., Kehrig et al. 2015a; Götberg et al. 2018; Stanway & Eldridge 2019; Pérez-Montero et al. 2020; Telford et al. 2021; Kehrig et al. 2021). Ultimately our understanding of metal-poor, hot massive

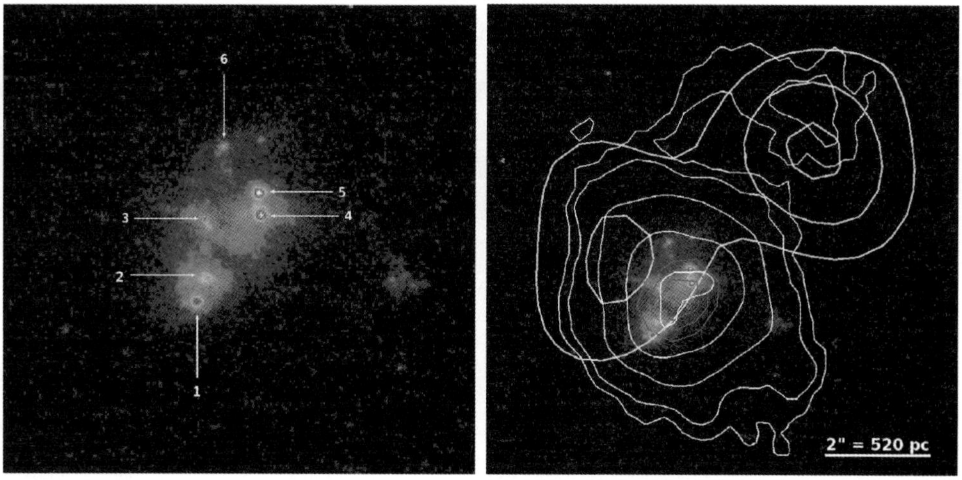

Figure 1. Left panel: HST ACS/F220W archival image of SBS 0335-052E (PI: G.Östlin). The brightest SSCs are labeled. The image is 6" × 6". Right panel: Soft (cyan lines) and hard (red lines) X-ray contours overlaid on the HST ACS/F220W image of SBS 0335-052E. The white curves represent the isocontours of the HeIIλ4686 emission line flux (see Fig. 2). In both panels, north is up and east is to the left (Figures taken from K18).

stars remains elusive which propagates to a lack of understanding of the formation of high-ionization lines as HeII.

The HeII line is also in comfortable reach of new and next generation telescopes (e.g., *JWST, ELT*) which will detect the rest-frame UV of many of galaxies during the epoch of reionization (e.g., Puech et al. 2018). Therefore detailed studies on the origin of nebular HeII at low redshifts are required to better interpret far-away high-ionizing emitters. Low metallicity, HeII-emitters at z $\sim$ 0 provide useful constraints on the little-known SEDs of metal-poor, hot massive stars, and are unique systems in which to test stellar population synthesis models at sub-SMC metallicities and gain a deeper understanding of the reionization era (see e.g., Kehrig et al. 2008, 2013, 2015a, 2015b; Senchyna et al. 2020; Telford et al. 2021).

In Kehrig et al. 2018 (hereafter K18) we combine *Multi-Unit Spectroscopic Explorer* (MUSE)/VLT optical integral field spectroscopy data (spectral resolving power is $R = \lambda/\delta\lambda = 2988$) with X-ray Chandra observations of SBS 0335-052E to obtain a new view on the origin of the nebular HeII excitation in this galaxy. SBS 0335-052E is among the most metal-poor (12+log(O/H)= 7.11 $\sim$ 3% $Z_{\odot}$; e.g., Papaderos et al. 2006) HeII-emitting SF galaxies known at z $\sim$ 0. It thus represents a nearby analog of distant metal-poor HeII emitters, providing a unique laboratory to study the high-ionization phenomenon expected to be common at z > 6. The left-panel of Fig. 1 shows the *HST/ACS* UV image of SBS 0335-052E where we can see its six brightest super-star clusters (SSCs) distributed over a region of $\sim$ 2 arcsec ($\sim$ 520 pc at the distance of 54 Mpc). Below we summarize the main results from K18.

2. What is ionizing nebular HeIIλ4686 in SBS 0335-052E ?

Based on the MUSE data we find a highly extended HeIIλ4686-emitting region which consists of a roundish oval shape component with a diameter of $\sim$ 5 arcsec ($\sim$ 1.3 kpc at the distance of 54 Mpc) over the galaxy core, and a shell-like structure (called here "HeII shell") which are connected to each other. The HeII-emitting zone extends out

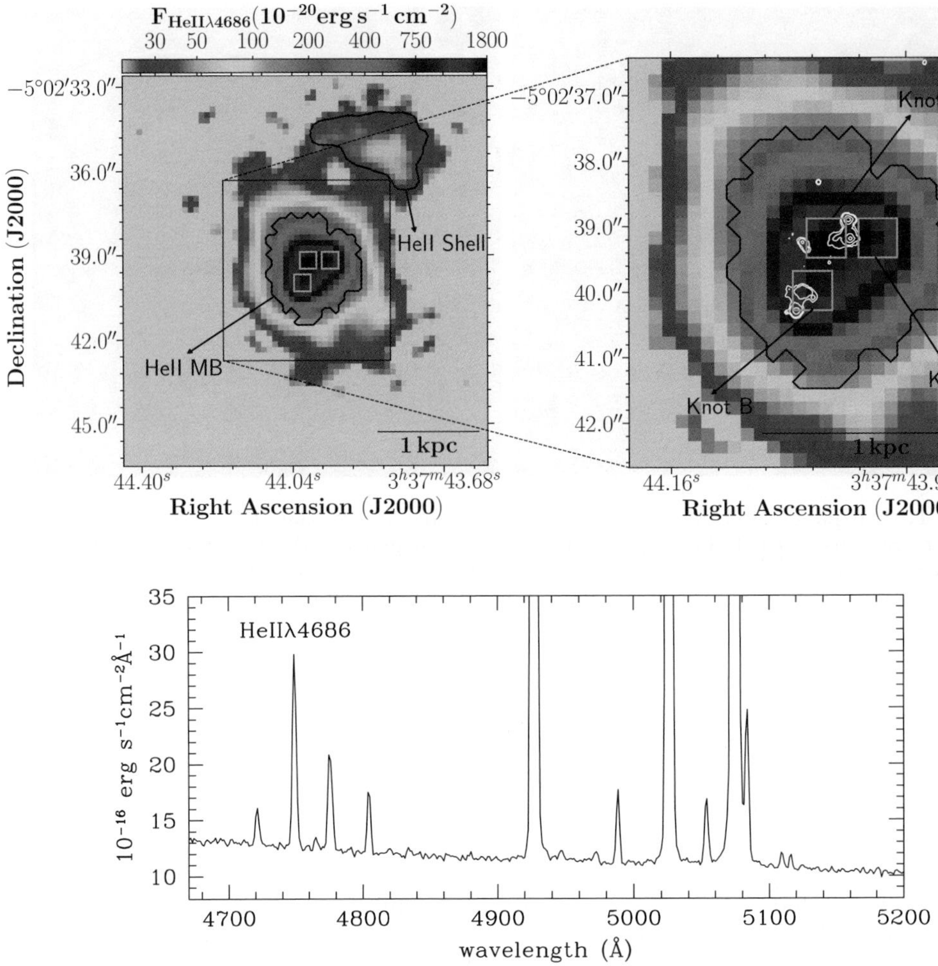

Figure 2. Top panel: The HeIIλ4686 flux map in two different scales showing the boundaries of the areas that we use to create the 1D spectra of several HeII-emitting regions, and the integrated spectrum of SBS 0335-052E (see K18 for details); the pink boxes are centered on the three HeII-emitting peaks; in the top-right panel the white lines mark the location of the brightest SSCs of SBS 0335-052E. Bottom panel: A zoom-in of the integrated spectrum of SBS 0335-052E over $\sim$ 4670-5200 Å(Figures taken from K18)

to distances $\gtrsim$ 6 arcsec ($\sim$ 1.5 kpc) from the youngest SSCs, and presents three peaks spatially displaced from the brightest stellar clusters (see top panels in Fig. 2)

We checked that the FWHM of the HeIIλ4686 line is comparable to that of the other nebular emission lines (e.g., Hβ, [OIII]λ5007). The values of the mean (μ) and standard deviation (σ) for the FWHM(HeII)/FWHM(Hβ) and FWHM(HeII)/ FWHM([OIII]λ5007) ratios are $\mu = 1.13$ ($\sigma = 0.09$) and $\mu = 1.10$ ($\sigma = 0.10$), respectively. The narrow line profile for the HeIIλ4686 emission and its spatial extent are evidence of its nebular nature.

We took benefit of our IFU data to create for the first time the integrated spectrum of SBS 0335-052E by adding the nebular emission across our MUSE FOV ($\sim$ 340 arcsec^2 $\sim$ 23 kpc^2; see bottom-panel in Fig. 2). Using this spectrum we recovered the entire nebular HeII emission and compute the corresponding total HeII-ionization budget, Q(HeII)$_{int}$ $\approx$ 3.17×10^{51} photon s^{-1}. As we showed in e.g., Kehrig et al. 2015a, 2020, this observational quantity is essential to perform a free-aperture investigation on the formation of the narrow HeII line.

In the righ-panel of Fig. 1 we show the Chandra X-ray emission from the different sources found in SBS 0335-052E. From our analysis of the combined MUSE and Chandra data, we infer that the Q(HeII) provided by the X-ray (soft and hard) emission from SBS 0335-052E is well below the needed value to ionize He^+ at the level measured in this galaxy.

Although we cannot discard some contribution to the HeII excitation from shocks, our MUSE IFU data favors hot massive stars as the main agent of the HeII ionization in SBS 0335-052E. The optical diagnostic diagrams shown in Fig. 3 indicate that for all positions in SBS 0335-052E our emission-line ratios fall in the general locus of SF objects according to the spectral classification scheme from Kewley et al. 2001. Accordingly, photoionization from hot massive stars seems to be the dominant excitation mechanism within SBS 0335-052E, regardless of the locus in the galaxy. Also, we find no sign of significant [SII]/Hα and/or [OI]λ6300/Hα enhancement (a frequent indication of shock-excited gas; e.g., Allen et al. 2008) associated with the HeII-emitting region (see top panels in Fig. 3). In particular, there is no [OI]λ6300 emission with $S/N \geq 5$ at the location of the HeII shell whose filamentary structure could suggest an important shock excitation; the HeII shell also shows low values of [SII]/Hα associated with (see Fig. 3). These observational facts reinforce that the nebular HeII in SBS 0335-052E is mainly due to stellar sources. Next, we explore the hot massive star scenario.

To investigate the stellar source scenario we compare observations with stellar models applying two approaches. A more simple one which makes use of models of single massive stars on a star-by-star basis gives us the following. The star-by-star study points out that Wolf-Rayet stars (WRs) solely cannot explain the HeII-ionizing energy budget derived for SBS 0335-052E. From models for fast-rotating Z=0 stars (Yoon et al. 2012), used here as an approximation for the HeII ionizing output of nearly metal-free stars, we found that $\sim$ 230 stars with mass $M_{ini} = 150\ M_\odot$ (with Q(HeII) $\approx 1.42 \times 10^{49}$ photon s^{-1} each) would be sufficient to account for the observed Q(HeII)$_{Int}$. Additionally, the ionizing spectra produced by these star models are hard enough to explain the highest He IIλ4686/Hβ values observed, assuming that ionization-bounded conditions are met.

Our second approach is based on the stellar population synthesis BPASSv2.1 (Eldridge et al. 2017). The outcomes from the BPASSv2.1 code indicate that while calling upon very massive stars ($M > 100\ M_\odot$) is not essential, a binary population with a 'top-heavy' IMF seem to be required to clear up the observed HeII flux budget. More interestingly, according to BPASSv2.1 models, stars with metallicity ($Z = 0.05$ % solar) much lower than that of the ionized gas ($Z \sim 3$ % solar) of SBS 0335-052E are to be invoked to explain the HeII ionization in this galaxy. In this hypothetical scenario, such nearly metal-free stars could not belong to the same SSCs which host more chemically evolved objects. Also, caution should be exercised when evaluating the $Z = 10^{-5}$ and 10^{-4} models; empirical data for stellar atmospheres at these low Z are not available.

Despite its nature as an entirely hypothetical scenario, both approaches suggest that (nearly) Z=0 stars (PopIII-like) are required to produce the total HeII ionizing photon flux. We highlight that similar results have been found to explain the HeII ionization in other nearby, metal-poor SF galaxies (see Kehrig et al. 2015a; Kehrig et al. 2015b; Kehrig et al. 2016; Senchyna et al. 2017).

3. Conclusions

Cosmic dawn marks a phase transition of the universe during which PoIII stars and their hosts put an end to the dark ages. The nature of these first sources is highly unconstrained (e.g., Schneider et al. 2002; Liu & Bromm 2020). The hard spectra of

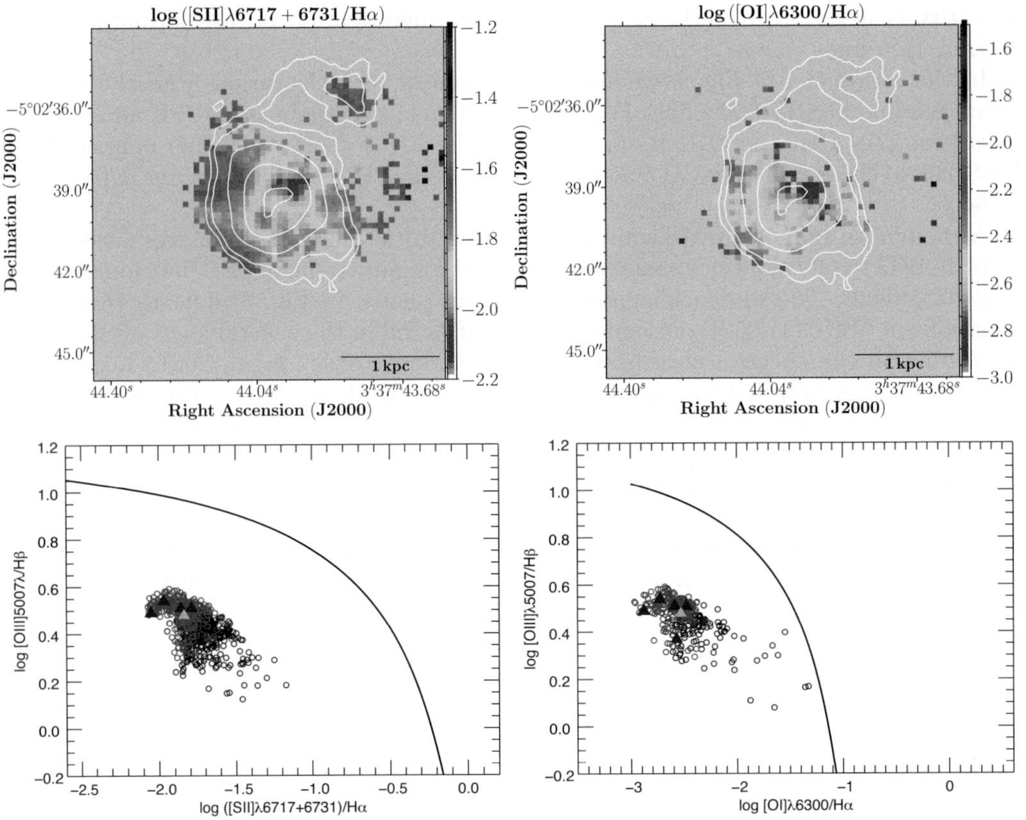

Figure 3. Top row: maps of the line ratios [SII]/Hα and [OI]/Hα. Only line fluxes with S/N (per pixel) $>$ 5 are plotted. All maps are presented in logarithmic scale. Isocontours of the HeIIλ4686 emission line flux are shown overplotted for reference. East is left and North is up. Bottom row: diagnostic diagrams: log ([OIII]/Hβ) vs. log ([SII]/Hα) and log ([OIII]/Hβ) vs. log ([OI]/Hα). Open circles correspond to individual spaxels from MUSE; red circles mark the individual HeII-emitting spaxels. Overlaid as triangles are the line ratios measured from the 1D spectra of selected galaxy regions; the solid curve is the maximum starburst model from Kewley et al. 2001 that isolate purely SF objects (below and to the left of the curve) from those where some other source of ionization rather than hot massive stars is required.

PopIII and nearly Z=0 stars can produce intense HeII emission. However, the origin of nebular HeII emission in metal-poor star-forming galaxies, near and far, is still unclear (e.g., Kehrig et al. 2015a; Saxena et al. 2017).

Metal-deficient HeII-emitting systems nearby are ideal to constrain the physics of hot, metal-poor (Z $<$ Z_{SMC}) massive stars and so the sources responsible for the Universe reionization, a major scientific objective of most of future facilities (e.g., ELT-MOSAIC, SKA). We have shown that IFU observations of such metal-poor HeII-emitters at z$\sim$0 provide extended empirical insight into the gas conditions and stars which support high-ionization emission (see Kehrig et al. 2008, 2013, 2015a, 2016, 2018, 2021)

In K18 we used MUSE IFU and Chandra observations to investigate the nebular HeII emission in the extremely metal-poor galaxy SBS 0335-052E found at $\sim$ 54 Mpc. We demonstrated that conventional HeII-ionizing sources (WRs, shocks, X-ray photons) are not sufficient to explain the observed HeII emission in this galaxy. From present models we found that nearly metal-free (Z$\sim$$10^{-5}$) hot stars ("PopIII-like" stars) and a top-heavy

IMF are required to reproduce the HeII-ionizing energy budget derived for SBS 0335-052E; Z$\sim$$10^{-5}$ is slightly larger than the predicted critical metallicity (Z$\sim$$10^{-4}$ solar) for IMF (and PopIII/PopII) transition (e.g., Schneider et al. 2002). We also want to note that recent theoretical studies predict that pockets of pristine gas may survive in weakly enriched ISM (Z$\sim$$10^{-5}$ - 10^{-4}) allowing PopIII star formation down to z=0 (see Liu & Bromm 2020).

The possibility remains that narrow HeII in metal-deficient galaxies is indeed powered by "PopIII-like stars", although HeII ionization keeps challenging current stellar models at the low Z regime (e.g., Kehrig et al. 2015a, 2015b, 2018).

References

Allen M.G. et al., 2008, *ApJS*, 178, 20
Cassata P. et al., 2013, *A&A*, 556, A68
Eldridge J. J. et al., 2017, *PASA*, 34, e058
Götberg Y. et al., 2018, *A&A*, 615, A78
Kehrig C., et al. 2008, *A&A*, 477, 813
Kehrig C., et al. 2011, *A&A*, 526, A128
Kehrig C., et al. 2013, *MNRAS*, 432, 2731
Kehrig C., et al. 2015a, *ApJ* (Letters), 801, L28
Kehrig C., et al. 2015b, in: W-R. Hamann, A. Sander & H. Todt (eds.), *Wolf-Rayet Stars: Proceedings of an International Workshop* (Universitätsverlag Potsdam), p.55-58
Kehrig C., et al. 2016, *MNRAS*, 459, 2992
Kehrig C., et al. 2018, *MNRAS*, 480, 1081
Kehrig C., et al. 2020, *MNRAS*, 498, 1638
Kehrig C., et al. 2021, *ApJ* (Letters), 908, L54
Kewley L., et al. 2001, *ApJ*, 556, 121
Liu B., Bromm V. 2020, *MNRAS*, 497, 2839
Mainali R., et al., 2018, *MNRAS*, 479, 1180
Nakajima K., Maiolino R., 2022, *MNRAS*, 513, 5134
Papaderos P., et al., 2006, *A&A*, 454, 119
Pérez-Montero E., et al., 2020, *A&A*, 643, A80
Puech M., et al., 2018, *Proceedings of the SPIE*, 10702, id. 107028R
Saxena A., et al., 2020, *MNRAS*, 496, 3796
Schneider, R., et al., 2002, *ApJ*, 571, 30
Senchyna P., et al., 2017, *MNRAS*, 472, 2608
Senchyna P., et al., 2020, *MNRAS*, 494, 941
Stanway E. R., Eldridge J. J., 2019, *A&A*, 621, A105
Telford O.G., et al., 2021, *ApJ*, 922, 191
Tumlinson J., Shull J. M., 2000, *ApJ*, 528, L65
Yoon S.-C., Dierks A., Langer N., 2012, *A&A*, 542, A113

Discussion

DE MINK: Do you think that X-ray binaries can play a role ?

KEHRIG: No, not really. For SBS 0335-052E we extracted its X-ray spectrum based on the spatial distribution of the hard X-ray emission and then we convolved the best fitting model with the energy dependent cross-section of He^+ to obtain the effective number of He^+-ionizing photons, and we got a number which is much lower than the observed one.

DE MINK: Is it possible that X-ray binaries deeply embedded from which we do not see the X-rays directly contribute to the HeII ionization?

KEHRIG: I do not think this is a possibility if you consider the large area over which nebular HeII emission is observed.

CROWTHER: We have a clear metallicity from oxygen from the HII regions, but how about iron ? Of course it is difficult to measure iron abundance but is there any contrain on the stellar metallicity based on iron abundance ?

KEHRIG: I am not sure about Fe abundance measurements for SBS 0335-052E but what I can tell you is that the metallicity measurements for the gas and stars are similar in this galaxy.

Massive Stars Near and Far
Proceedings IAU Symposium No. 361, 2024
J. Mackey, J. S. Vink, & N. St-Louis, eds.
doi:10.1017/S1743921322002253

Hot stars in the Gaia-ESO Survey

R. Blomme[1], S. Daflon[2], M. Gebran[3], A. Herrero[4,5], A. Lobel[1], L. Mahy[6,7,1], F. Martins[8], T. Morel[7], S. R. Berlanas[9], E. Gosset[7], J. Maíz Apellániz[10] and GES Working Group 13

[1]Royal Observatory of Belgium, Ringlaan 3, B-1180 Brussels, Belgium
email: ronny.blomme@oma.be

[2]Observatório Nacional/MCTIC, R. Gal. José Cristino 77, São Cristovão, 20921-400 Rio de Janeiro/RJ, Brazil

[3]Department of Chemistry and Physics, Saint Mary's College, Notre Dame, IN 46556, USA

[4]Instituto de Astrofísica de Canarias, E-38205 La Laguna, Tenerife, Spain

[5]Departamento de Astrofísica, Universidad de La Laguna, E-38206 La Laguna, Tenerife, Spain

[6]Instituut voor Sterrenkunde, KU Leuven, Celestijnlaan 200D, Bus 2401, 3001 Leuven, Belgium

[7]Space Sciences, Technologies, and Astrophysics Research (STAR) Institute, Université de Liège, Quartier Agora, Bât B5c, Allée du 6 août, 19c, 4000 Liège, Belgium

[8]LUPM-UMR 5299, CNRS & Université Montpellier, Place Eugène Bataillon, F-34095 Montpellier Cedex 05, France

[9]Departamento de Física Aplicada, Facultad de Ciencias, Universidad de Alicante, Carretera de San Vicente s/n, E-03690, San Vicente del Raspeig, Spain

[10]Centro de Astrobiología, CSIC-INTA, Campus ESAC, Camino bajo del castillo s/n, E-28 692 Villanueva de la Cañada, Spain

Abstract. The Gaia-ESO Survey (GES) is a large public spectroscopic survey that has collected spectra of about 100,000 stars. The survey provides not only the reduced spectra, but also the radial velocities, stellar parameters and surface abundances resulting from the analysis of the spectra. We present the work of the groups that analysed the spectra of the hottest stars in that Survey. The large temperature range that is covered ($T_{\rm eff}$ = 7,000 to 50,000 K) requires the use of different analysis codes by the different groups. Eight groups each analysed part of the data, with significant overlap that allowed cross-checks. In total 17,693 spectra of 6,462 stars were analysed, most of them in 37 open star clusters. The homogenisation of all this information led to stellar parameters for 5,584 stars. Abundances for at least one of the elements He, C, N, O, Ne, Mg, Al, Si and Sc were determined for 292 stars. The GES hot star data, as well as the Survey data in general, will be of considerable use in future studies of stellar evolution and open clusters.

Keywords. Surveys, catalogs, stars: fundamental parameters, stars: abundances, stars: early-type, techniques: spectroscopic

1. The data and their analysis

The Gaia-ESO Survey (GES; `https://www.gaia-eso.eu/`) used the multi-fibre FLAMES instrument on VLT-UT2. Both GIRAFFE and UVES spectra were obtained, with a number of different settings. Not all clusters and stars were observed with all the

Table 1. The different groups that analysed the data.

Group	Analysis technique	Parameters determined	Abundances determined
ROBGrid	χ^2 minimisation with grid of theoretical spectra	$T_{\rm eff}$, $\log g$, [M/H], $v_{\rm rad}$, $v \sin i$	
ROB	Fe - Fe^+ ionisation balance of diagnostic photospheric lines	$T_{\rm eff}$, $\log g$, [Fe/H], ξ, $v \sin i$	C, O, Mg, Al, Sc, Fe
MGNDU	Principal Component Analysis and Sliced Inverse Regression	$T_{\rm eff}$, $\log g$, [M/H], $v_{\rm rad}$, $v \sin i$	
Liège	χ^2 minimisation with grid	$T_{\rm eff}$, $\log g$, $v_{\rm rad}$, $v \sin i$	He, C, N, Ne, Mg, Si
ON	Non-LTE synthesis and Si ionisation balance	$T_{\rm eff}$, $\log g$, $v \sin i$	C, O, Si
IAC	χ^2 minimisation with grid of FASTWIND models	$T_{\rm eff}$, $\log g$, $v \sin i$, $v_{\rm macro}$	He
Mntp	χ^2 minimisation with grid of CMFGEN models	$T_{\rm eff}$, $\log g$, $v \sin i$, $v_{\rm macro}$	
LiègeO	CMFGEN	$T_{\rm eff}$, $\log g$, $v \sin i$, $v_{\rm macro}$	He, C, N

settings. HR03, HR04, HR05AB, HR06, HR14AB were used for the hotter stars, HR09B for the cooler stars. The UVES fibres were set on the brighter stars in the cluster.

One of the aims of the GES is to cover a large range of cluster parameters: age, metallicity, radial distance, and mass. We analysed stars from 37 open star clusters. Although we were mainly focussed on hot stars, the age range of the clusters is from a few Myr to a few Gyr.

The general policy of the GES is that a number of groups each analyse the spectra independently. At a later stage, the results are then compared and a homogenisation procedure is applied, giving a single set of parameters and abundances for each star (the recommended values).

For the hot stars, the temperature range covered requires the use of different analysis codes by the different groups. Table 1 lists the analysis technique used by each group, as well as the parameters and abundances that they determined.

2. Some results

The data analysed here allow a comparison between the FASTWIND (IAC) and CMFGEN codes (Mntp, LiègeO). $T_{\rm eff}$ differs on average by -400 ± 1300 K (CMFGEN minus FASTWIND), in acceptable agreement with the results of Massey et al. (2013) and Holgado et al. (2018). For $\log g$, the difference is $+0.04 \pm 0.12$ dex on average, which agrees with the Holgado et al. result, and is somewhat better than the Massey et al. one. The comparison between the ON results and that of the other groups shows the other groups to have higher $T_{\rm eff}$, with the difference increasing with $T_{\rm eff}$. While the agreement in $\log g$ is good, a small linear trend remains in the differences.

The details of this work are published in Blomme et al. (2022). The data analysed in that paper are then further homogenised in the GES consortium, to combine them with results from groups working on cooler stars. The results of that homogenisation are publicly available via ESO; see `https://www.eso.org/sci/publications/announcements/sciann17497.html` for details. Papers on scientific results are in preparation (Berlanas et al., Morel et al., Santos et al.).

References

Blomme, R., Daflon, S., Gebran, M., Herrero, A., Lobel, A., et al. 2022 *A&A*, 661, A120
Holgado, G., Simón-Díaz, S., Barbá, R.H., Puls, J., Herrero, A., et al. 2018 *A&A*, 613, A65
Massey, P., Neugent, K.F., Hillier, D.J., & Puls, J. 2013 *ApJ*, 768, 6

Massive Stars Near and Far
Proceedings IAU Symposium No. 361, 2024
J. Mackey, J. S. Vink, & N. St-Louis, eds.
doi:10.1017/S174392132200196X

An infrared view of NGC3603

Raúl Castellanos[1], Francisco Najarro[1], Miriam Garcia[1], Lee R. Patrick[1,2,3], Christopher J. Evans[4], Elizabeth S. Bartlett[5] and J. S. Clark†

[1]Centro de Astrobiología, CSIC-INTA, Carretera de Ajalvir km4, 28850 Torrejón de Ardoz, Madrid, Spain.
email: rcastellanos@cab.inta-csic.es

[2]Departamento de Física Aplicada, Universidad de Alicante, E-03690 San Vicente del Raspeig, Alicante, Spain.

[3]School of Physical Sciences, The Open University, Walton Hall, Milton Keynes MK76AA, UK.

[4]European Space Agency (ESA), ESA Office, Space Telescope Science Institute, 3700 San Martin Drive, Baltimore, MD 21218, USA.

[5]UK Astronomy Technology Centre, Royal Observatory Edinburgh, Blackford Hill, Edinburgh, UK, EH9 3HJ.

Abstract. NGC3603 is one of the youngest massive clusters of the Milky Way which uniquely enables studying the interplay between massive star feedback and the surrounding interstellar medium. Yet, a deep infrared (IR) view of the cluster is missing. We present guaranteed time observations of NGC3603 consisting of near infrared spectroscopy taken with VLT-KMOS. This data set will provide a first, rather complete IR census.

Keywords. stars: early-type, galaxies: star clusters, open clusters and associations: individual (NGC3603)

1. Observations and data reduction

NGC3603 is considered the Galactic counterpart to NGC2070 core (R136), around 2.5 times smaller in massive star content (Drew et al. 2019). Due to its proximity in the Carina nebula (Kalari et al. 2019), this cluster is an ideal test bed for starbursts studies and may be used as a proxy to understand this phenomenon in more distant galaxies where the stellar content is impossible to resolve. The program consisted of observations of the cluster with VLT-KMOS multi IFU infrared spectrograph. Target selection was based on previous spectroscopic and photometric identifications of the massive star population in NGC3603 (Moffat 1983; Melnick et al. 1989; Melena et al. 2008; Roman-Lopes et al. 2016). The magnitudes of the stars included in this study range from bright stars having a 9th magnitude in the J band to the weakest stars having magnitudes up to 12th (Harayama et al. 2008).

To perform the most effective telluric correction, two methodologies have been tested using molecfit (Smette et al. 2015) to obtain a model of the atmospheric conditions. On the one hand, molecfit has been run over A0 telluric standard star spectra observed subsequently to the science objects. On the other hand, molecfit has been executed over the science stars itself. The telluric correction resulting from the two different methodologies is shown in Fig. 1 using two different science stars as examples. We find that the most

†The author is deceased.

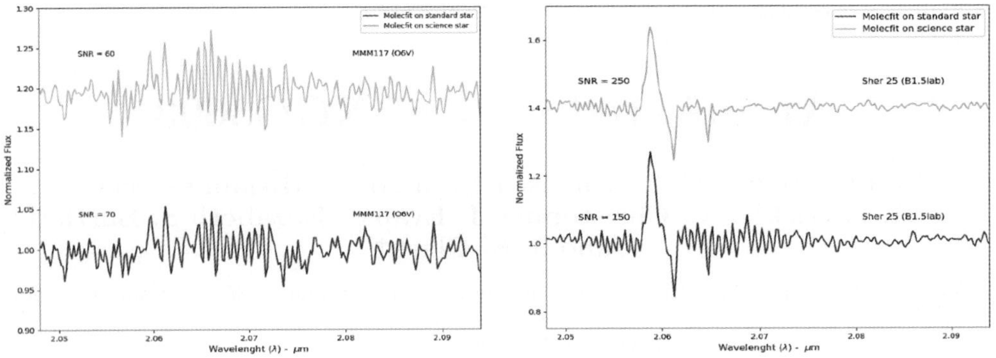

Figure 1. Telluric spectra of stars MMM117 (S/N = 70, **left**) and Sher25 (S/N = 250, **right**) using molecfic on standard (**bottom**) and molecfit on source (**top**) methods.

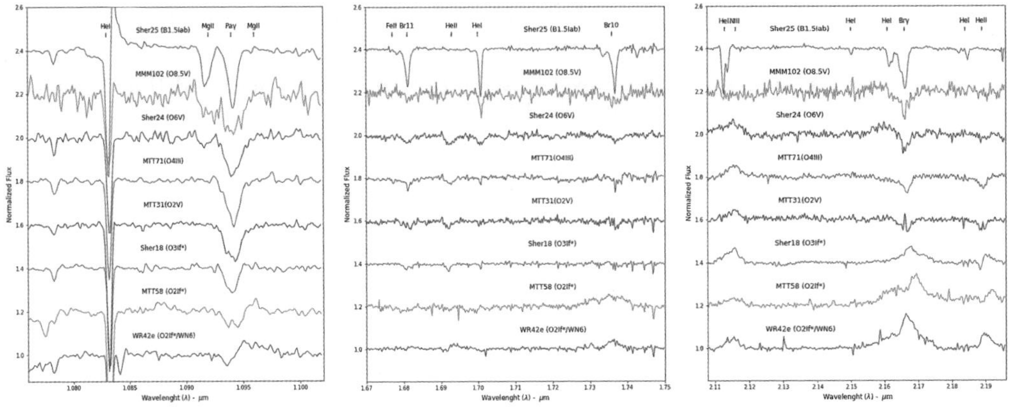

Figure 2. Reduced spectra of a selection of stars in the J, H, and K bands.

accurate results are achieved when molecfit is applied directly to the science objects, unless the observed spectrum of the standard star exhibits a signal to noise ratio (S/N) at least a factor 2.5 higher than the science star spectrum.

2. Preliminary results

As a result of this work, we have obtained an infrared census of 40 stars in the cluster. We have spectra in the J, H, and K bands for 34 of them, while for the remaining 6 we only have spectra in the K band. This census includes O-type stars, and a B supergiant star previously identified(Moffat 1983; Melena et al. 2008; Roman-Lopes et al. 2016). Some of these spectra are shown in Fig. 2. The sample will be analyzed with CMFGEN (Hillier & Miller 1998) model atmospheres to derive stellar and cluster properties.

Acknowledgements

This research has been funded by Spanish grants PID2019-105552RB-C41 and Unidad de Excelencia "Maria de Maeztu" (MDM-2017-0737-19-3).

References

Drew, J. E., Monguió, M., & Wright, N. J. 2019, MNRAS, 486, 1034
Harayama, Y., Eisenhauer, F., & Martins, F. 2008, ApJ, 675, 1319

Hillier, D. J., & Miller, D. L. 1998, ApJ, 496, 407
Kalari, V. M., Vink, J. S., de Wit, W. J., Bastian, N. J., & Méndez, R. A. 2019, A&A, 625, L2
Melena, N. W., Massey, P., Morrell, N. I., & Zangari, A. M. 2008, AJ, 135, 878
Melnick, J., Tapia, M., & Terlevich, R. 1989, A&A, 213, 89
Moffat, A. F. J. 1983, A&A, 124, 273
Roman-Lopes, A., Franco, G. A. P., & Sanmartim, D. 2016, ApJ, 823, 96
Smette, A., et al. 2015, A&A, 576, A77

Massive Stars Near and Far
Proceedings IAU Symposium No. 361, 2024
J. Mackey, J. S. Vink, & N. St-Louis, eds.
doi:10.1017/S1743921322002496

The largest and most ambitious quantitative spectroscopic analysis of B-type supergiants in the Milky Way

Abel de Burgos[1,2], Sergio Simon-Díaz[1,2] and Miguel A. Urbaneja[3]

[1]Universidad de La Laguna, Dpto. Astrofísica, E-38206 La Laguna, Tenerife, Spain
email: astroabelink@gmail.com

[2]Instituto de Astrofísica de Canarias, Avenida Vía Láctea, E-38205 La Laguna, Tenerife, Spain

[3]Institut für Astro- und Teilchenphysik, Universität Innsbruck, Technikerstr. 25/8, A-6020 Innsbruck, Austria

Abstract. Empirical constraints are master keys for testing theoretical evolutionary model predictions. In massive stars, the region in the Hertzsprung-Russell diagram (HRD) in which the Blue Supergiants (BSGs) are located sets several important constraints to the models, and in particular to the theoretical end of the main sequence (MS). So far, we are missing from a full quantitative spectroscopic analysis (QSA) of a sample of BSGs large enough to be statistically significant and without observational biases. We present results from a QSA of a sample of ~700 Galactic BSGs for which we have high-resolution multi-epoch optical spectra.

Keywords. stars: fundamental – stars: evolution – stars: rotation – stars: distances

1. Introduction

BSGs are very important to understand the evolution of massive stars. Evolutionary models predict a gap in the HRD between the MS and the core helium-burning phase. However, this gap is populated by BSGs in the observational diagrams (Fitzpatrick & Garmany 1990; Castro et al. 2014). This area could be filled with stars coming from a RSG (Ekström et al. 2012) and/or that have undergone some type of binary interaction (Langer 2012). Alternatively, the theoretical MS could be wider than traditionally considered (e.g. Vink et al. 2010; Castro et al. 2014). In each of those situations, these different populations should have different physical, chemical and pulsational properties. Aiming at solving this situation, we have gathered a homogeneous high-resolution spectroscopic data-set of ~700 Galactic BSGs, increasing by a factor 25-30 the number of sources considered in previous studies (e.g. Crowther et al. 2006). By using the same analysis strategy, we are able to perform a throughout QSA deriving spectroscopic parameters.

2. Methods

Starting with a sample of ~1500 Galactic O9-B9 stars from the IACOB database (Simón-Díaz et al. 2020), we have used the effect that gravity has on the shape of the Hβ line profile to isolate the descendants of the O-type stars with $M_\odot \gtrsim 20\,M_\odot$. We have used IACOB-broad (Simón-Díaz & Herrero 2007), FASTWIND models (Puls et al. 2005) and MAUI, a Gaussian Process based statistical emulator coupled with exploration of the

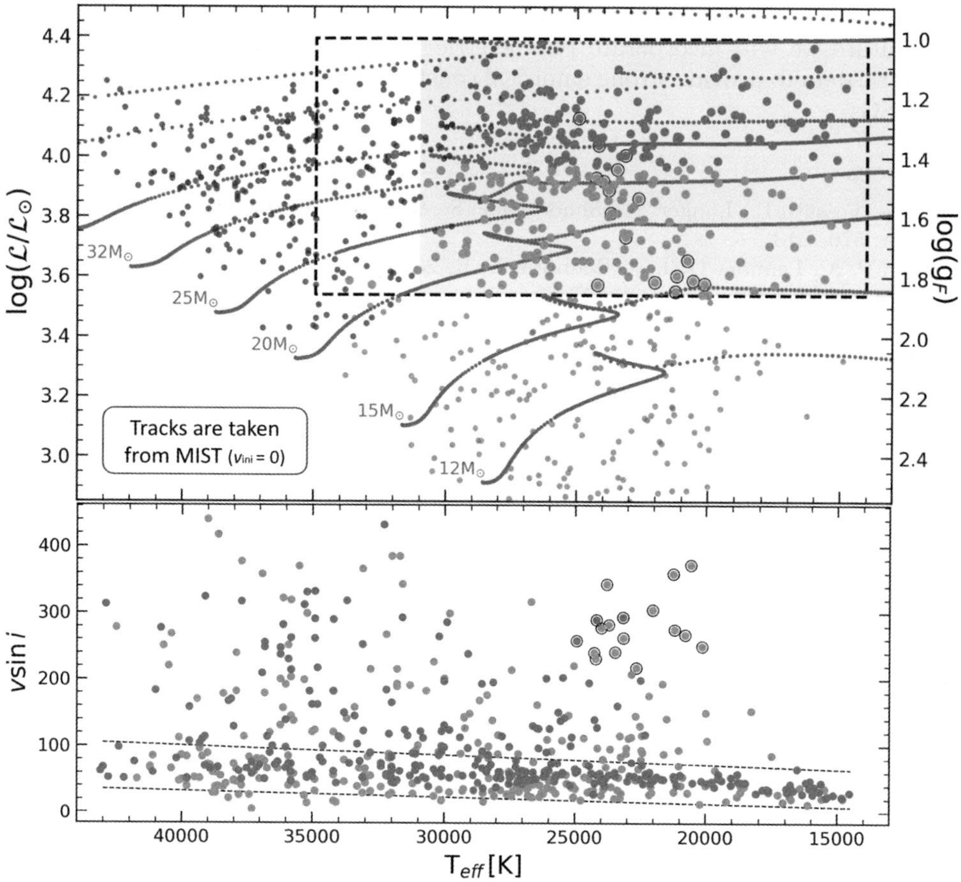

Figure 1. Location of the sample stars in the sHRD (top) and $v \sin i - T_{\rm eff}$ diagram (bottom). The purple dots are results for O-type stars from Holgado et al. (2020). Blue and orange dots in the top panel are BSGs with $3.55 \leq \log(\mathcal{L}/\mathcal{L}_\odot) \leq 4$ and $\log(\mathcal{L}/\mathcal{L}_\odot) \geq 4$, respectively. Gray dots are additional results of stars from IACOB database included for context. The dashed rectangle shows the limits of the MAUI emulator. Bottom panel includes only stars above $\log(\mathcal{L}/\mathcal{L}_\odot) \geq 3.55$. With open circles, location of fast rotating stars with $T_{\rm eff} < 25000$ K.

parameter space by Markov chains (Urbaneja et al., in prep.), to derive the spectroscopic parameters for most of these stars.

3. Results

Fig. 1 shows the sHRD (top) and $v \sin i - T_{\rm eff}$ diagram (bottom) for ~700 B-type stars analyzed using MAUI and ~300 O-type stars from Holgado et al. (2020). We see a group of stars with $v \sin i \gtrsim 100\,{\rm km\,s^{-1}}$and a much denser group with $v \sin i \lesssim 100\,{\rm km\,s^{-1}}$located between the two dotted lines. We also see a slow decrease in $v \sin i$ for all the stars (caused by natural evolution) except for the fast-rotators at ~20000 K where there is a noticeable drop in $v \sin i$.

4. Conclusions

Our results clearly show a high density of stars in the sHRD down to $T_{\rm eff} \sim 21000$ K. We provide empirical evidence that the early B-type SGs could be in the MS. This is supported by the fact that we also observe a drop in the number of fast-rotating stars at the same position where the drop in density occurs and independently from $\log(\mathcal{L}/\mathcal{L}_\odot)$. In the future we will incorporate photometric variability from *TESS* and *Gaia* DR3 astrometric data to provide strong empirical constraints to aid at finally unveil the nature of the BSGs.

References

Castro, N., Fossati, L., Langer, N., Simón-Díaz, S., Schneider, F. R. N., & Izzard, R. G. 2014, A&A, 570, L13

Crowther, P. A., Lennon, D. J., & Walborn, N. R. 2006, A&A, 446, 279

Ekström, S., et al. 2012, A&A, 537, A146

Fitzpatrick, E. L., & Garmany, C. D. 1990, ApJ, 363, 119

Holgado, G., et al. 2020, A&A, 638, A157

Langer, N. 2012, ARA&A, 50, 107

Puls, J., Urbaneja, M. A., Venero, R., Repolust, T., Springmann, U., Jokuthy, A., & Mokiem, M. R. 2005, A&A, 435, 669

Simón-Díaz, S., Britvaskiy, N., Castro, N., & Holgado, G. 2020, in XIV.0 Scientific Meeting (virtual) of the Spanish Astronomical Society, 186

Simón-Díaz, S., & Herrero, A. 2007, A&A, 468, 1063

Vink, J. S., Brott, I., Gräfener, G., Langer, N., de Koter, A., & Lennon, D. J. 2010, A&A, 512, L7

Massive Stars Near and Far
Proceedings IAU Symposium No. 361, 2024
J. Mackey, J. S. Vink, & N. St-Louis, eds.
doi:10.1017/S1743921322002654

Apparently isolated massive stars within Galactic HII regions

Diego de la Fuente[1,†], Amparo Marco[1], Ignacio Negueruela[2], Lee R. Patrick[2], Artemio Herrero[3,4] and Stephen S. Eikenberry[5]

[1]Departamento de Física, Ingeniería de Sistemas y Teoría de la Señal, Universidad de Alicante, San Vicente del Raspeig, Alicante, Spain
email: diego.delafuente@ua.es

[2]Departamento de Física, Aplicada, Universidad de Alicante, San Vicente del Raspeig, Alicante, Spain

[3]Instituto de Astrofísica de Canarias, La Laguna, Tenerife, Spain

[4]Departamento de Astrofísica, Universidad de La Laguna, La Laguna, Tenerife, Spain

[5]CREOL, The College of Optics and Photonics, University of Central Florida, Orlando, FL, USA

Abstract. The Alicante Survey of MAssive Stars in HII Regions (A-SMASHeR) is aimed at finding the ratio of massive stars that are born in isolation. We present LIRIS/WHT images and EMIR/GTC spectra of the massive stellar content in A-SMASHeR regions. Our preliminary analysis yields $\sim 20\%$ of regions hosting relatively (or truly) isolated massive stars.

Keywords. stars: formation, HII regions, infrared: stars, surveys, techniques: spectroscopic, techniques: photometric, open clusters and associations: general

---

1. Introduction

Massive stars mostly belong to clusters or OB associations, and it has been suggested that even field massive stars are born in clustered enviroments from which they have been ejected (Gvaramadze *et al.* 2012). Investigating whether these isolated objects can be formed in situ is crucial for determining which is the dominant mechanism (competitive collapse vs. monolithic accretion) of massive star formation (Zinnecker & Yorke 2007).

Massive stars that are still embedded in the cloud where they were born can be found in studies of the stellar content of HII regions. Following this approach, Oey *et al.* 2013 found such objects, some of them isolated, in the Small Magellanic Cloud (SMC). However, infrared studies of HII regions in the Milky Way are required to rule out compact clusters of massive stars, because such groups may be hard to resolve in the SMC.

2. The A-SMASHeR data

A-SMASHeR (Marco *et al.* 2019) is a homogeneous survey of Galactic HII regions whose objective is estimating the fraction of massive stars that are born in isolation. Further details are provided by Marco *et al.* (these proceedings).

Fig. 1 displays two representative examples of the A-SMASHeR content: a compact cluster of candidate massive stars; and a relatively isolated ionizing source. On the other hand, Fig. 2 shows part of the spectra of three A-SMASHeR point sources. Despite the dominance of nebular lines, we find features that allow us to constrain the nature and

[†]Visiting scientist at the University of Central Florida, USA

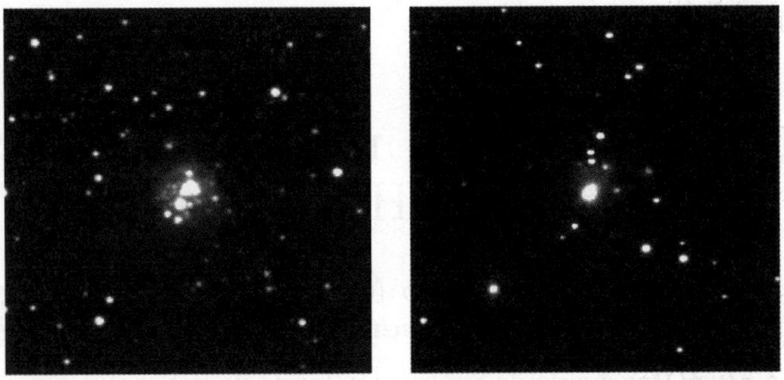

Figure 1. LIRIS/WHT infrared JHK$_S$ images, sized $1' \times 1'$, of two A-SMASHeR targets: a compact cluster *(left)* and a relatively isolated central star surrounded by sparsely distributed faint objects *(right)*.

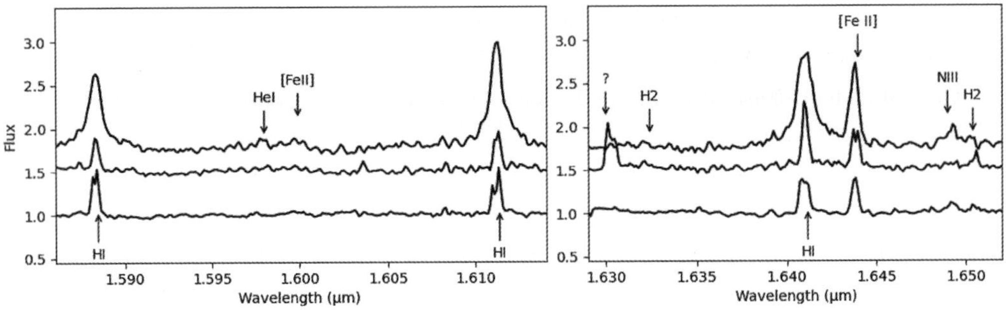

Figure 2. Selected portions of EMIR/GTC H-band spectra of A-SMASHeR targets. Although not being shown here, the K band is also observed by EMIR/GTC.

spectral type of the underlying star. For example, the double-peaked emission lines that point at a circumstellar disk, and the presence or absence of certain metallic lines.

3. Preliminary results and future work

Preliminary analysis yields $\sim 20\%$ of A-SMASHeR regions being ionized by a relatively isolated massive star, although most of them are part of more extended star-forming regions hosting other young objects. Final results will be produced after our planned follow-up multi-object spectroscopy with MIRADAS/GTC (Eikenberry *et al.* 2016), whose multiplexing capability is crucial for the A-SMASHeR variety of regions, ranging from compact clusters to truly isolated stars.

Acknowledgements. Financial support for this work has been provided by Generalitat Valenciana through grants APOSTD/2020/228 and APOSTD/2020/247, and by the Spanish Government projects PGC2018-093741-B-C21/C22.

References

Eikenberry, S. S., Raines, S. N., Stelter, R. D., *et al.* 2016, *Proceedings of the SPIE*, 9908, id. 99081L

Gvaramadze., V., Weidner, C., Kroupa, P., Pflamm-Altenburg, J. 2012, *MNRAS*, 424, 4

Marco, A., Negueruela, I.,Eikenberry, S. S., Herrero, A. 2019, *Highlights on Spanish Astrophysics X*, 353

Oey, M. S., Lamb, J. B., Kushner, C. T., Pellegrini, E. W, Graus, A. S. 2013, *AJ*, 768, 66

Zinnecker, H, Yorke, H. W. 2007, *ARAA*, 45,481

Massive Stars Near and Far
Proceedings IAU Symposium No. 361, 2024
J. Mackey, J. S. Vink, & N. St-Louis, eds.
doi:10.1017/S1743921322002290

MAGIC Intensity Interferometer as a powerful tool to understand massive stars

Irene Jiménez Martínez[1], V. A. Acciari[2], E. Colombo[2], J. Cortina[1], C. Delgado[1], C. Díaz[1], D. Fink[3], M. Fiori[5], D. Guberman[6], T. Hassan[1], E. Lyard[4], S. Mangano[1], M. Mariotti[5], G. Martínez[1], R. Mirzoyan[3], G. Naletto[5], T. Njoh Ekoume[2], N. Produit[4], M. Polo[1], J. J. Rodríguez[1], T. Schweizer[3], R. Walter[4], C. Wunderlich[6] and L. Zampieri[7] for the MAGIC collaboration

[1]CIEMAT, Madrid, Spain
email: `irene.jimenez@ciemat.es`

[2] IAC & University of La Laguna, La Laguna, Spain

[3]Max-Planck-Institut for Physics, Munich, Germany

[4]University of Geneva, Department of Astronomy, Geneva Switzerland

[5]INFN-Padova & Padova University, Department of Physics and Astronomy Padova, Italy

[6]INFN-Pisa & Siena University, Pisa, Italy

[7]INAF-Astronomical Observatory of Padova, Padova, Italy

Abstract. The Intensity Interferometry technique consists of measuring the spatial coherence (visibility) of an object via its intensity fluctuations over a sufficient range of telescope separations (baselines). This allows us to study the size, shape and morphology of stars with an unprecedented resolution. Cherenkov telescopes have a set of characteristics that coincidentally allow for Intensity Interferometry observations: very large reflective surfaces, sensitivity to individual photons, temporal resolution of nanoseconds and the fact that they come in groups of several telescopes. In the recent years, the MAGIC Collaboration has developed a deadtime-free Intensity Interferometry setup for its two 17 m diameter Cherenkov telescopes that includes a 4-channel GPU-based real-time correlator, 410-430 nm filters and new ways of splitting its primary mirrors into submirrors using Active Mirror Control (AMC). With this setup, MAGIC can operate as a long-baseline optical interferometer in the baseline range 40-90 m, which translates into angular resolutions of 0.5-1 mas. Additionally, thanks to its AMC, it can simultaneously measure the zero-baseline correlation or, by splitting into submirrors, access shorter baselines under 17 m in multiple u-v plane orientations. The best candidates to observe with this technique are relatively small and bright stars, in other words, massive stars (O, B and A types). We will present the science cases that are currently being proposed for this setup, as well as the prospects for the future of the system and technique, like the possibility of large-scale implementation with CTA.

Keywords. instrumentation: interferometers, techniques: interferometric

1. Introduction

IACTs (Imaging Atmospheric Cherenkov Telescopes), are telescopes dedicated to observe Extensive Atmospheric Showers (EAS), which emit Cherenkov light. The characterization of this light allows to reconstruct and study the nature, direction and energy

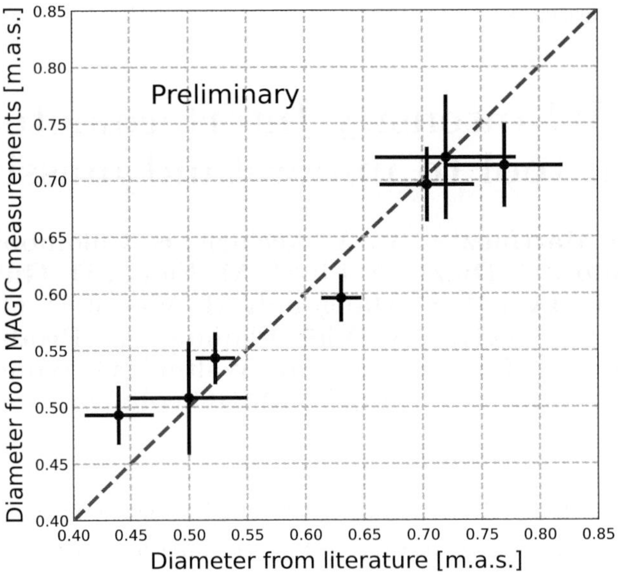

Figure 1. Measured angular diameter of 7 stars by MAGIC compared to the literature values.

of the primary of electromagnetic showers. Cherenkov telescopes are serendipitously very well suited for intensity interferometry observations: they have large reflective surfaces (to have enough statistics), are sensitive to single photo-electrons (to measure the coherence of individual photons), have nanosecond time resolution (to resolve the correlation peak) and they come in groups (several baselines).

The intensity interferometry technique is based on the Van Cittert-Zernike Theorem: a bright object of angular size θ consists of many incoherent regions that produce a pattern of size λ/θ. If the distance between telescopes 1 and 2 is $<< \lambda/\theta$, then they observe the same fluctuations.

As its name indicates, intensity interferometry makes use of the intensity of light from the source detected at different locations to measure the correlation of these fluctuations (Acciari et al. 2020). Since this measurement provides morphological information about the thermal emission of the observed source, it is a valuable tool in a broad range of stellar science cases that requires of such information: angular diameter of massive stars, limb darkening of large bright stars, fast rotators, Be stars, asteroseismology, novae and binary systems.

2. Results

To assess the performance of MAGIC's Intensity Interferometry setup (Delgado et al. 2021) we measured the spatial correlation over a range of baselines for stars with published angular sizes. The preliminary calibration of the setup already provides star diameter measurements close to literature values, as Fig. 1 shows, compatible within errors.

3. Conclusions

In the last few years it has been proved that is possible to apply the intensity interferometry technique with the MAGIC telescopes. Adding LST-1 to the MAGIC setup requires only minor modifications, and can be done as soon as in 2022. MAGIC+LST-1 setup can be further extended to future LSTs, providing large boost in performance, due to linear improvement of the sensitivity with the number of telescopes and mirror area.

Supplementary material

To view supplementary material for this article, please visit http://dx.doi.org/10.1017/S1743921322002290.

References

Acciari, V., et al. 2020, Monthly Notices of the Royal Astronomical Society, 491, 1540

Delgado, C., et al. 2021, in 37th International Cosmic Ray Conference (ICRC 2021), Vol. 395, SISSA, 693

Massive Stars Near and Far
Proceedings IAU Symposium No. 361, 2024
J. Mackey, J. S. Vink, & N. St-Louis, eds.
doi:10.1017/S1743921322003258

Fundamental parameters for 222 B-type stars for the X-shooter Spectral Library

C. J. K. Larkin[1], **Kristiina Verro**[1], **Anaïs Gonneau**[2] **and S. C. Trager**[1]

[1]Kapteyn Astronomical Institute, University of Groningen, Landleven 12, 9747 AD Groningen, The Netherlands
email: `larkin@astro.rug.nl`

[2]Institute of Astronomy, University of Cambridge, Madingley Road, Cambridge CB3 0HA, UK

Abstract. We use the MIDE3700 code to find effective temperatures ($T_{\rm eff}$) and surface gravities ($\log g$) via the Barbier Chalonge Divan (BCD) method for 222 B-type stars in the ESO archive in preparation for their inclusion as an extension to the X-shooter Spectral Library (XSL). We find agreement of $\Delta \log T_{\rm eff} \sim 0.1\sigma$ and $\Delta \log g \sim 0.25\sigma$ of our results with a sample of literature stars. We populate a previously bare region of the XSL Kiel diagram in the ranges $9000 \leq T_{\rm eff} \leq 23000\,{\rm K}$ and $2.8 \leq \log g \leq 4.0\,{\rm dex}$, and thereby extend the lower age limit of XSL stellar population models by up to a factor ~ 10 at $[{\rm Fe/H}] = -1.2\,{\rm dex}$, and by a factor ~ 2 at Solar metallicity.

Keywords. atlases–stars: fundamental parameters–methods: data analysis–techniques: spectroscopic

1. Introduction

Stellar spectral libraries are a crucial component for modelling Simple Stellar Populations (SSPs) (Conroy 2013). One of the most notable empirical spectral libraries in recent years is the X-shooter Spectral Library (XSL, see Verro et al. 2022b and references therein), which in its most recent data release consists of 830 spectra of 683 unique stars with $R \sim 10000$ over the wavelength range 3000–24500 Å. The XSL covers a large parameter space in $T_{\rm eff}$ ($2900 < T_{\rm eff} < 38000\,{\rm K}$), surface gravity $\log g$ ($0 < \log g < 5.7\,{\rm dex}$) and iron abundance [Fe/H] ($-2.5 < [{\rm Fe/H}] < +1.0\,{\rm dex}$) (Gonneau et al. 2016; Arentsen et al. 2019; Verro et al. 2022b). In Figure 1 we see that there is a relative lack of B-type and earlier stars in the Kiel diagram. This effectively sets a lower limit on ages of stellar populations that can be probed with models based on XSL (currently $\sim 500\,{\rm Myr}$ for $[{\rm Fe/H}] = -1.2\,{\rm dex}$ and $\sim$ 50 Myr for Solar metallicity, Verro et al. 2022a). To push this age limit lower, in this work we add B-type spectra for 222 stars originally taken as calibration data for XSL and other ESO programs.

2. Method

We selected observations with identical settings to the XSL from the ESO X-shooter archive (`http://archive.eso.org/wdb/wdb/eso/xshooter/form`). We obtained one or more spectra for 222 unique stars. We use ESO Phase 3 data, meaning the spectra are already reduced, extracted to 1D and telluric corrected for the most important features in the UVB arm.

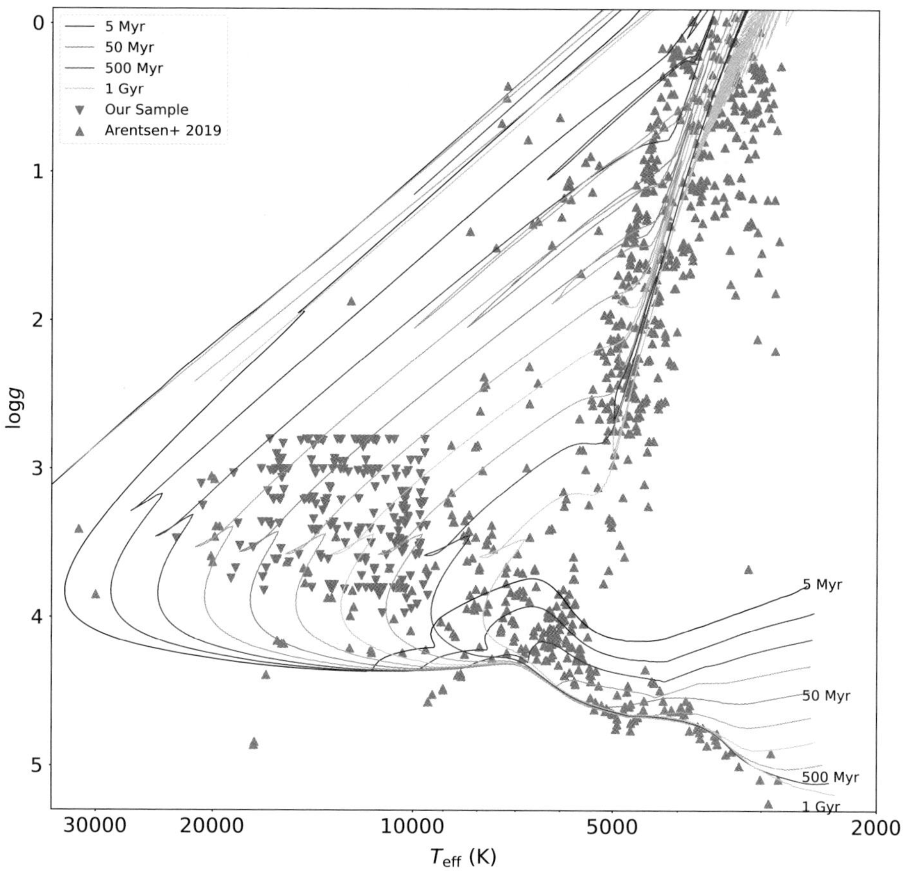

Figure 1. Kiel diagram for the XSL based on Arentsen et al. (2019) (orange), and in this work (light blue), with PARSEC/COLIBRI stellar isochrones at Solar metallicity overlaid.

We used the Barbier–Chalonge–Divan (BCD) method (see Zorec et al. 2009, and references therein) in this work, which uses the parameters (λ_1, D) to determine temperatures and gravities. The BCD parameters are related to the Balmer discontinuity: D is the height of the Balmer discontinuity at $\lambda = 3700$ Å and is sensitive to $T_{\rm eff}$ through the amount of absorption from the Balmer lines, and λ_1 is the mean spectral position of the Balmer discontinuity and is sensitive to $\log g$ through the density of the atmosphere. We use the `MIDE3700` code (Aidelman 2017) to perform the BCD fitting. We note that the BCD method is not sensitive to stellar rotation, which is important as B-type stars are typically fast rotators (e.g. Huang et al. (2010)); it is not sensitive to circumstellar or interstellar extinction; it is largely insensitive to the physics of the star's upper atmosphere as the continuum originates deeper in the photosphere; and it can also be applied to various classes of peculiar B-type stars (Zorec et al. 2009).

3. Results

Our final values of $T_{\rm eff}$ and $\log g$ are shown in Figure 1 along with the parameters of XSL stars from Arentsen et al. (2019). Compared with their results, we now populate a previously sparsely covered parameter space at moderate values of $\log g$ (2.8–4 dex) and high temperatures (9000–20000 K). We compared our results to literature values, in particular to Huang et al. (2010), who inferred values of $T_{\rm eff}$ and $\log g$ for 14 stars in our sample. We find the standard deviation in $\sigma(\Delta \log T_{\mathrm{eff}}) = 0.106$ and $\sigma(\Delta \log g) = 0.254$ dex.

4. Conclusions

We have measured accurate temperatures and gravities for B stars with $9000 \leq T_{\mathrm{eff}} \leq 23000$ K and $2.8 \leq \log g \leq 4.0$ dex for inclusion in XSL. The BCD system does not have calibrations for values of $\log g \leq 2.8$ dex, which causes a visible bottleneck effect in Figure 1. Furthermore, the interpolation method used in `MIDE3700` causes visible streaks in $\log g$ values. However, given the uncertainties we find, the temperature and gravities are acceptable for these spectra to be used in the XSL and other SSP models.

Supplementary material

To view supplementary material for this article, please visit http://dx.doi.org/10.1017/S1743921322003258.

References

Aidelman, Y. 2017, Bol. Asoc. Argentina de Astr. de La Plata, 59, 13
Arentsen, A., et al. 2019, A&A, 627, A138
Conroy, C. 2013, ARA&A, 51, 393
Gonneau, A., et al. 2016, A&A, 589, A36
Huang, W., Gies, D. R., & McSwain, M. V. 2010, ApJ, 722, 605
Verro, K., et al. 2022a, A&A, 661, A50
—. 2022b, A&A, 660, A34
Zorec, J., Cidale, L., Arias, M. L., Frémat, Y., Muratore, M. F., Torres, A. F., & Martayan, C. 2009, A&A, 501, 297

Massive Stars Near and Far
Proceedings IAU Symposium No. 361, 2024
J. Mackey, J. S. Vink, & N. St-Louis, eds.
doi:10.1017/S1743921322002368

Alicante Survey of Massive Stars in H II regions (A-SMASHER)

A. Marco[1], I. Negueruela[2], D. de la Fuente[1,5], L. R. Patrick[2], A. Herrero[3,4] and S. S. Eikenberry[5]

[1]Departamento de Física, Ingeniería de Sistemas y Teoría de la Señal, EPS, Universidad de Alicante, Carretera de San Vicente s/n, E03690, San Vicente del Raspeig, Spain
email: amparo.marco@ua.es

[2]Departamento de Física Aplicada, Facultad de Ciencias, Universidad de Alicante, Carretera de San Vicente s/n, E03690, San Vicente del Raspeig, Spain

[3]Instituto de Astrofísica de Canarias, Vía Láctea s/n, E38200, La Laguna, Tenerife, Spain

[4]Departamento de Astrofísica, Universidad de la Laguna, Avenida Astrofísico Francisco Sánchez, s/n, Facultad de Ciencias, E38200, La Laguna, Tenerife, Spain

[5]CREOL, The College of Optics and Photonics, University of Central Florida, USA

Abstract. The aim of this survey is the homogeneous characterization of a large sample of H II regions with active star formation in order to detect observational trends supporting the two main models of massive star formation.

Keywords. stars: formation, stars: massive, stars: fundamental parameters, – open clusters and associations – Hertzsprung-Russell and colour-magnitude diagrams

1. Introduction

Two main theories exist to explain the formation of massive stars: (a) **Monolithic core accretion**, which is a scaled-up version of classical low-mass formation theories, where very high opacities allow infalling material to overcome the radiation pressure (Krumholz et al. 2009) and (b) **Competitive accretion** where massive stars are formed in cluster cores, benefiting from the gravitational potential of the whole cluster to accrete more material (Bonnell & Bate 2006).

2. Methodology

The sample was carefully selected from existing radio catalogues and consists of targets that have not been previously studied and contain near infrared point sources consistent with a massive star. We obtained deep near infrared photometry of one hundred such H II regions with WHT/LIRIS and we have carefully analyzed the corresponding color-magnitude diagrams.

3. Results

An initial separation has been made between regions containing obvious clusters and those that seem to include single sources (about 25%). For this latter subsample, we have already obtained intermediate-resolution spectroscopy with GTC/EMIR, while the clusters will be observed with multi-object configurations. Examples of the preliminary

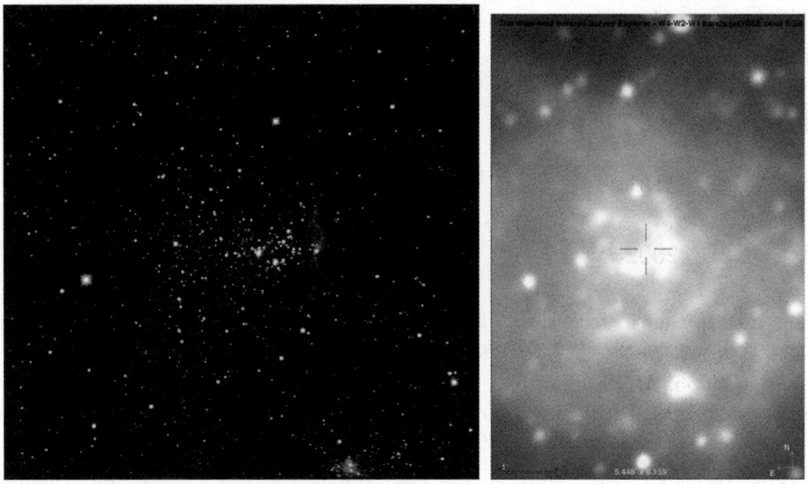

Figure 1. *Left*: J, H and K_S composite image (WHT/LIRIS) for a large cluster. *Right*: WISE image for the large cluster.

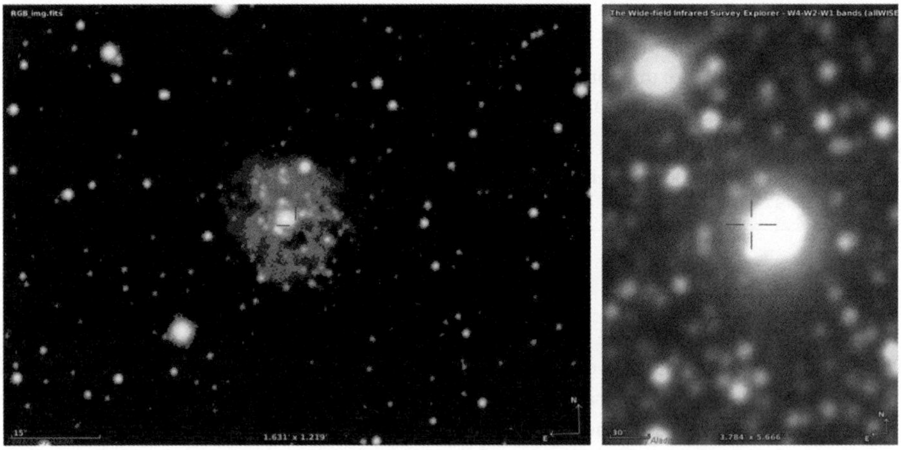

Figure 2. *Left*: J, H and K_S composite image (WHT/LIRIS) for a small cluster. *Right*: WISE image for the small cluster.

study of four H II regions with single sources can be seen in de la Fuente et al. 2022 in these proceedings.

4. Conclusions

The statistical study of the properties of the populations contained within the H II regions in relation to the physical characteristics of the region itself and Galactic environment will provide novel insights into the conditions leading to (relatively) isolated or clustered massive star formation. Below, we show two examples corresponding to two very different H II regions, one containing a large population of massive stars (Figure 1) and the other, only a few members (Figure 2). There are two images for each object: a composite J, H, K_S image (WHT/LIRIS), where we can see the stars inside the H II region (left), and the image from WISE enhancing the content of gas (right).

Acknowledgements

This research is partially supported by the Spanish Government under grants PGC2018-093741-B-C21/C22 (MICIU/AEI/FEDER, UE), and by the Generalitat Valenciana under grant PROMETEO/2019/041.

Supplementary material

To view supplementary material for this article, please visit http://dx.doi.org/10.1017/S1743921322002368.

References

Krumholz et al. 2009, *Science*, 323, 754
Bonnell & Bate, 2006, *MNRAS*, 370, 488
de la Fuente et al. 2022, Proceedings IAU Symposium No. 361, 2022

Massive Stars Near and Far
Proceedings IAU Symposium No. 361, 2024
J. Mackey, J. S. Vink, & N. St-Louis, eds.
doi:10.1017/S1743921322003052

Infrared studies of the massive stellar population at the Galactic Center and the inner disk

**Francisco Najarro[1], J. S. Clark[2]†, Marcus Lohr[2]
Lee R. Patrick[1,2,3], Miriam García[1], Raúl Castellanos[1],
Gonzalo Muñoz-Sánchez[4] and Diego de la Fuente[3]**

[1]Centro de Astrobiología, CSIC-INTA, Carretera de Ajalvir km4, 28850 Torrejón de Ardoz, Madrid, Spain.
email: najarro@cab.inta-csic.es

[2]School of Physical Sciences, The Open University, Walton Hall, Milton Keynes MK76AA, UK.

[3]Departamento de Física Aplicada, Universidad de Alicante, E-03690 San Vicente del Raspeig, Alicante, Spain.

[4]National Observatory of Athens, GR-15236, Penteli, Greece.

Abstract. We present results from our ongoing infrared spectroscopic studies of the massive stellar content at the Center of the Milky Way (GC) and across the obscured Galactic disk. Together with the full characterization of these clusters, we seek to obtain a present day metallicity 2-D map of the inner Galaxy and characterize the influence on the bar in the chemical evolution. We will also constrain the clusters IMFs, infer the presence of possible top-heavy recent star formation histories and test massive star formation channels: clusters vs isolation.

Keywords. stars: early-type, (Galaxy:) open clusters and associations: general, Galaxy: center, Galaxy: abundances

1. Program description

Massive stars in young massive clusters are ideal tracers of the present day metallicity of the Galactic region where they are located. Since α-element and Fe enrichment follows different routes and timescales, measuring the α vs Fe ratio constrains the past star formation history. The infrared allows us to peer through the strong extinction veil towards the inner regions of the galactic disk and hence characterize these clusters. We have shown (Najarro et al. 2004, 2009) that quantitative NIR spectroscopy of high-mass stars using CMFGEN (Hillier & Miller 1998) provides estimates of both absolute abundances and abundance ratios. LBVs and B-hypergiants yield Fe and α-element (Mg, Si) estimates, direct oxygen abundances are obtained from O-supergiants, while indirect oxygen estimates and direct Si values are extracted from WNh stars. Our program has obtained near-infrared moderate-resolution (R 3000-5000) spectroscopic observations of several massive clusters distributed across the inner Galactic disk and the Galactic Center (Arches, Quintuplet and isolated stars) using GNIRS and NIFS at GEMINI, ISAAC, SINFONI and KMOS at VLT and EMIR at GTC.

2. Preliminary results

Arches cluster: a deep spectroscopic survey (Clark et al. 2019) has revealed the presence of O9.5V stars and an apparent turn-off at O4-5V. Using our modeled stellar

†The author is deceased.

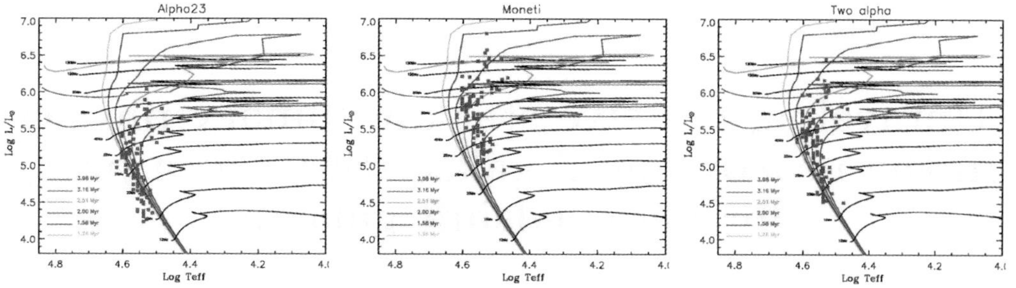

Figure 1. Testing the extinction towards the Arches. Recent (α~2.3,**left**) and previous (Cardelli, **center**)) extinction laws severely underestimate/overestimate the luminosities for the Arches main sequence stars. A two-α law yields better results (**right**).

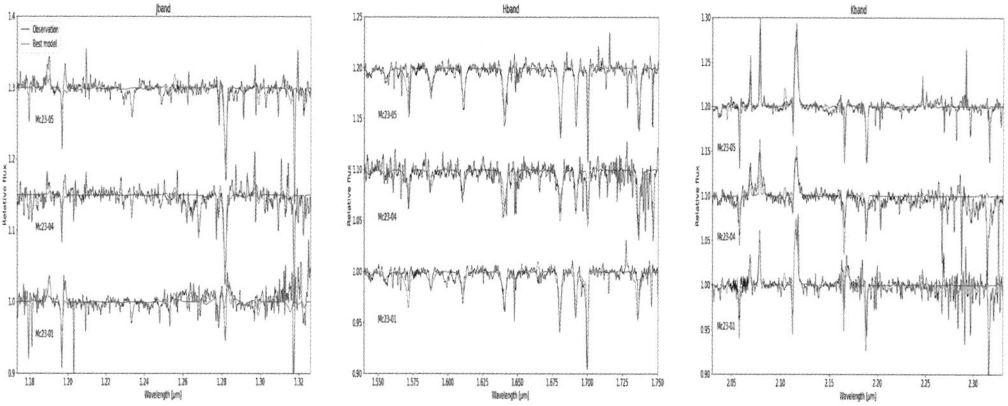

Figure 2. Fits to stars in Mercer23 in the J- (**left**), H- (**center**), and K-bands (**right**).

properties and placing the objects in the HR-diagram we may test the validity of the extinction law. Current best value, α=2.3, from Red Clump stars (Nogueras-Lara et al. 2019) severely underestimates the resulting luminosities for the Arches main sequence stars. We find a degeneracy that is solved by means of a two-α law (see Fig. 1).

Galactic disk: following the strategy presented in previous studies (Najarro et al. 2004, 2009), we are currently analyzing the stellar population of other Galactic disk clusters. As an example, we show in Fig. 4 model fits to a selection of objects in Mercer 23 observed in the J, H and K bands with EMIR/VLT. Preliminary results indicate that, as for the GC population and Mercer 81, this cluster is characterized by an enhanced oxygen abundance. So far, most of the inner disk clusters show α-enhancement and only Mercer 30, placed at a galacto-centric distance of 11 kpc is consistent with solar abundances. Whether the bar is able to create an effective mixing and a flattened gradient in the chemical enrichment within the inner regions of the disk needs to be further investigated.

This work is dedicated to the memory of our friend and colleague Simon Clark. F.N., R.C. and M.G. acknowledge funding by grants PID2019-105552RB-C41 and MDM-2017-0737-19-3 Unidad de Excelencia "María de Maeztu".

References

Clark, J.S., Lohr, M.E, Patrick, L.R. & Najarro, F., 2019, *A&A*,623A, 84C
Hillier, D. J., & Miller, D. L. 1998, *ApJ*, 496, 407
Najarro et al. 2004, *ApJL*, 611, L105
Najarro et al. 2009, *ApJ*, 691, 1816
Nogueras-Lara et al. 2019, *A&A*, 630, L3

Massive Stars Near and Far
Proceedings IAU Symposium No. 361, 2024
J. Mackey, J. S. Vink, & N. St-Louis, eds.
doi:10.1017/S1743921322001867

Unraveling massive star and binary physics in the nearby low-Z galaxy SMC as a proxy for high-redshift galaxies

Daniel Pauli

Institut für Physik und Astronomie, Universität Potsdam, Karl-Liebknecht-Str. 24/25, 14476 Potsdam, Germany
email: dpauli@astro.physik.uni-potsdam.de

Abstract. Low metallicity (Z) massive stars are among the main feedback agents in the early Universe and in present-day blue dwarf galaxies. The nearby star-forming SMC galaxy offers conditions which resemble those at redshift $z \sim 2$, i.e. where modern galaxies formed and star formation peaked. Here we present the recent results about the nature of the eclipsing O-type binary in the SMC, AzV 476, to gain insights on the properties of massive stars and binaries at earlier cosmic epochs. We find that the primary has surprisingly low mass while being much brighter and hotter than the secondary. To place the measured stellar properties in the evolutionary context we modeled the system and confirm that AzV 476 is a post-interaction binary with the primary already being core helium (He) burning, while still having a hydrogen-rich (H-rich) envelope. These results constrain massive binary evolutionary scenarios and guide the searches of stripped stars in low-Z environments.

Keywords. binaries: close, binaries: eclipsing, binaries: spectroscopic, stars: early-type, stars: fundamental parameters, stars: individual: AzV 476, Magellanic Clouds

---

1. Introduction

In the standard picture of binary interactions, it is believed that during stable mass transfer via Roche lobe overflow, more than 70% of the H-rich ($X_H > 0.6$) envelope of the donor star is removed. After the mass-transfer phase, the donor star is expected to evolve quickly into a helium- or WR-star as its remaining H-poor ($X_H \lesssim 0.4$) envelope is quickly removed by powerful stellar winds. In the Hertzsprung-Russell diagram (HRD) these stars can be found on the left side of the zero-age main-sequence (ZAMS). However, recent theoretical predictions of Klencki et al. (2021) suggest that the stripping process via Roche lobe overflow in low metallicity (Z) environments does only partially strip of the hydrogen-rich envelope. Hence, on the HRD, these core He-burning stars are located on the right side of ZAMS, in the same region as core H-burning OB stars.

In our study on the eclipsing binary system AzV 476 Pauli et al. (2022) we might have found one of these stars. Here we present the key results of this recent study, and briefly discuss their impact on our the understanding of massive star evolution at low Z.

2. Overview

AzV 476 harbors an O4 IV-III((f))p and an O9.5: Vn star. If treated as single stars, according to their spectra type, they should have $M_{\rm exp,1} = 49^{+7}_{-6}\,M_\odot$, and $M_{\rm exp,2} = 16^{+7}_{-3}\,M_\odot$ Weidner & Vink (2010). Yet, our orbital analysis revealed that the mass of the

Table 1. Fundamental stellar parameters.

	primary	secondary
$M_{\rm spec}$ $[M_\odot]$	29^{+17}_{-11}	22^{+13}_{-8}
$T_{\rm eff}$ [kK]	42^{+3}_{-3}	32^{+4}_{-4}
$\log\ L$ $[L_\odot]$	$5.65^{+0.2}_{-0.2}$	$4.75^{+0.2}_{-0.2}$
$\log\ \dot{M}$ $[M_\odot\ {\rm yr}^{-1}]$	$-6.1^{+0.2}_{-0.2}$	$-8.8^{+0.5}_{-0.5}$
$v \sin i$ $[{\rm km\,s}^{-1}]$	140^{+10}_{-10}	425^{+25}_{-25}
$X_{\rm H}$ (by mass)	0.73	0.73
$X_{\rm C}/10^{-5}$ (by mass)	2^{+2}_{-1}	21^{a}
$X_{\rm N}/10^{-5}$ (by mass)	45^{+5}_{-10}	3^{a}
$X_{\rm O}/10^{-5}$ (by mass)	80^{+10}_{-20}	110^{a}

Notes: [a]Fixed to standard CNO abundances, as no metal-lines of the secondary can be seen in the spectrum.

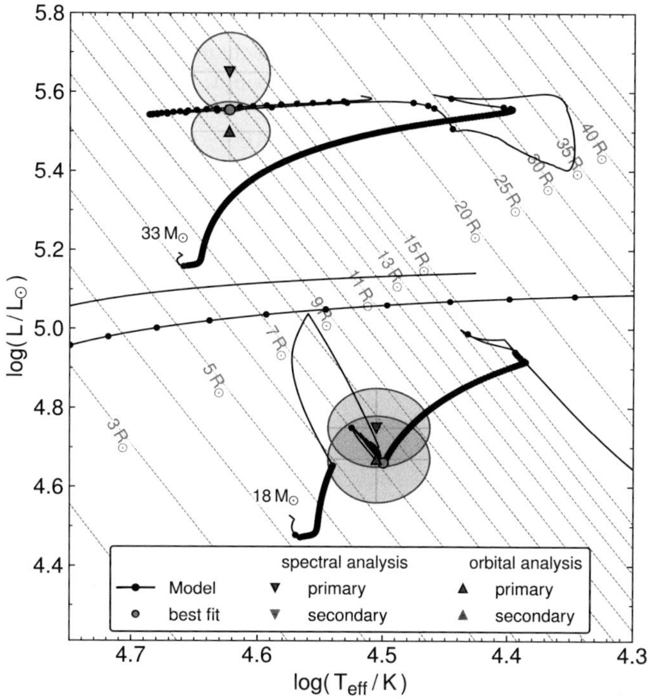

Figure 1. HRD showing the track of the binary evolutionary model (black line). The black dots on the tracks correspond to equidistant time-steps of 0.1 Myr to emphasize the most probable observable phases.

primary is only $M_{\rm orb,1} = 20.2^{+2.0}_{-2.0}\ M_\odot$, and thus a factor of 2.4 less massive than expected! The orbital mass of the secondary $M_{\rm orb,\,2} = 18.0^{+1.8}_{-1.8}\ M_\odot$ agrees with the expectations.

The fundamental stellar parameters of the individual binary components are given in Tab. 2. Note, that the mass-loss rate of the primary, due to the unusual high L/M ratio, is 10 times higher than expected from empirical relations of single stars (e.g. Ramachandran et al. 2019).

The low stellar mass, the unusually high mass-loss rate, and the altered CNO abundances of the primary as well as the fast rotation of the secondary strongly suggest that the binary has undergone a mass-transfer phase in the recent past. We calculated binary evolutionary models with the MESA code. Our best fitting model is shown in Fig. 1. It reproduces the observed stellar and orbital parameters and predicts that the primary of the system is already core He-burning while still being on the right side of the ZAMS in the HRD, hiding among the OB population. These findings, question the standard picture of post-interacting binary.

3. Implications

Uncovering core He-burning massive star in a early O-type binary in the SMC, greatly impact our understanding of stellar feedback in low Z environments, and thus of the early Universe. Binary stripped stars which do not lose their H-rich envelope, have weaker ionizing fluxes than those who do. In addition to that also their mechanical feedback is expected to be lower. The question remains if the system AzV 476 is unique or just the tip of the iceberg of an undiscovered population of partially stripped stars.

References

Klencki, J., Istrate, A. G., Nelemans, G., & Pols, O. 2021, *arXiv*, 2111.10271

Pauli, D., and Oskinova, L. M., Hamann, W.-R., Ramachandran, V., Todt, H., Sander, A. A. C., Shenar, T., Rickard, M., Maíz Apellániz, J. & Prinja, R. 2022, *A&A*, A9, 659

Ramachandran, V., Hamann, W.-R., Oskinova, L. M., Gallagher, J. S., Hainich, R., Shenar, T., Sander, A. A. C., Todt, H. & Fulmer, L. 2019, *A&A*, A104, 625

Weidner, C. & Vink, J. S. 2010, *A&A*, A98, 524

Massive Stars Near and Far
Proceedings IAU Symposium No. 361, 2024
J. Mackey, J. S. Vink, & N. St-Louis, eds.
doi:10.1017/S1743921322002587

Revisiting the Cygnus OB associations

Alexis L. Quintana and Nicholas J. Wright

Astrophysics Group, Keele University, Keele ST5 5BG, UK
email: a.l.p.quintana.isasi@keele.ac.uk

Abstract. We have studied the validity of the historical Cygnus OB associations and have found that many do not show the kinematic coherence expected for true OB associations. We have revisited these groups by photometrically identifying thousands of OB stars across the region with an SED fitting process which combines photometry, astrometry, spectral and evolutionary models. We applied a flexible clustering method and identified seven kinematically-coherent new OB associations. We observe a distinct correlation between position and velocity for two sets of these associations that suggests an expansion pattern. Tracing the motion of the stars back in the past we find that the sets were at their closest around 7.9 and 8.5 Myr ago. We discuss whether this expansion is a natural by-product of the commonly observed size - velocity dispersion relation of molecular clouds, or requires feedback to initiate the dispersal.

Keywords. stars: early-type, (Galaxy:) open clusters and associations: general, Galaxy: kinematics and dynamics

1. Introduction

OB associations are unbound and low-density ($< 0.1 M_\odot$ pc^{-3}) groups of young stars containing many OB stars (Ambartsumian 1947). As an intermediate step between the star-forming regions and the field population of stars (Wright 2020), they are fundamental to understand stellar formation history and how the Galactic field is built up (Armstrong et al. 2020). They have been listed and analysed for decades (see e.g. Humphreys 1978), but most of them remain poorly characterised. This motivates us to revisit them with modern data and survey. To do so we start with the Cygnus OB associations, located within a massive star-forming region (Reipurth & Schneider 2008), with the reddened Cyg OB2 as most prominent (Wright et al. 2015; Berlanas et al. 2019).

2. New OB associations

Most of the historical associations in Cygnus lack kinematic coherence that genuine associations should have (Quintana & Wright 2021, 2022a). We thus built an SED fitting process that we apply on SEDs compiled from photometry and astrometry from several surveys in order to estimate physical stellar parameters. In doing so we identify over 4000 candidate OB stars across Cygnus. Subsequently, we apply a flexible clustering algorithms and find seven new kinematically-coherent OB associations. All associations are expanding in the l direction and three are expanding in the b direction. We observe two large-scale kinematic patterns between position and proper motion in Galactic longitude exhibited by three associations each. This modulation is similar to the pattern observed by Drew et al. (2021) in Carina, with a velocity-scale of $\sim$ 25 km s^{-1} and a length-scale of $\sim$ 150 pc.

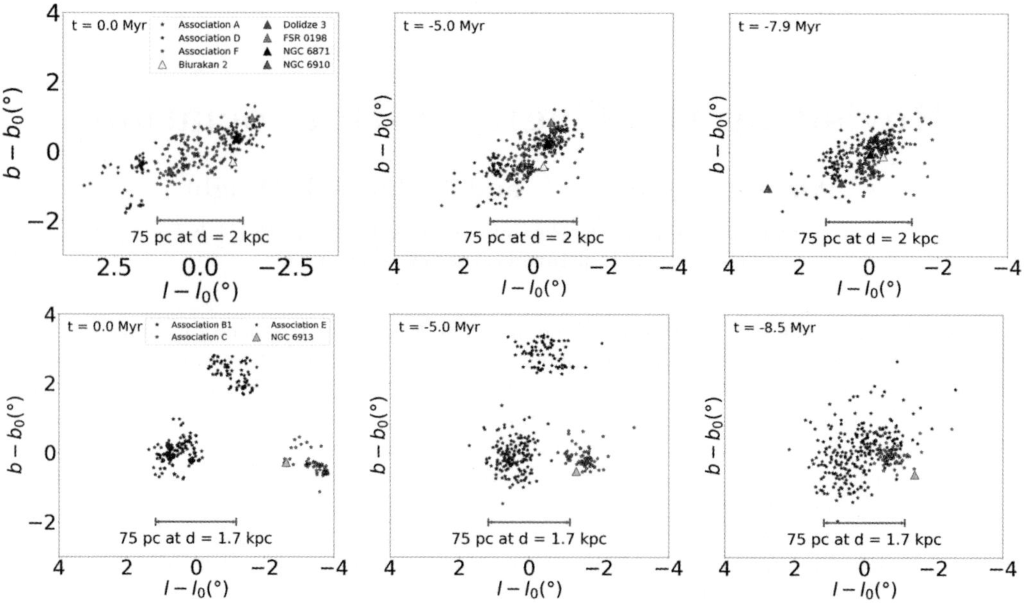

Figure 1. Relative galactic coordinates of all stars and open clusters at various times in the past derived from traceback calculations in their reference frame. Presents results from Quintana & Wright (2022b).

3. Kinematic traceback

We perform a trace back on the associations and possibly related open clusters using 3D kinematics and the epicycle approximation from Fuchs et al. (2006). The two sets of three associations respectively reached their most compactness $7.9^{+3.0}_{-1.8}$ and $8.5^{+0.8}_{-2.8}$ Myr ago, as shown in Fig. 1. A possible explanation for the driver of this expansion is the feedback from a previous generation of stars. This scenario has been notably suggested by Kounkel (2020) and Großschedl et al. (2021) in Orion. However, we find that the observed expansion velocities are completely consistent with arising from the turbulent motions in primordial cloud, as previously observed by Larson (1981); Solomon et al. (1987), and others in the form of the size - linewidth relation. A similar conclusion is reached for the motions of stars in Orion. This implies that, while it is tempting to attribute these kinematic patterns to feedback, turbulence is probably the most likely scenario.

Supplementary material

To view supplementary material for this article, please visit http://dx.doi.org/10.1017/S1743921322002587.

References

Ambartsumian V. A., 1947, The evolution of stars and astrophysics. Armenian SSR Academy of Sciences Press
Armstrong J. J., Wright N. J., Jeffries R. D., Jackson R. J., 2020, MNRAS, 494, 4794
Berlanas S. R., Wright N. J., Herrero A., Drew J. E., Lennon D. J., 2019, MNRAS, 484, 1838
Drew J. E., Monguió M., Wright N. J., 2021, MNRAS, 508, 4952
Fuchs B., Breitschwerdt D., de Avillez M. A., Dettbarn C., Flynn C., 2006, MNRAS, 373, 993
Großschedl J. E., Alves J., Meingast S., Herbst-Kiss G., 2021, A&A, 647, A91
Humphreys R. M., 1978, ApJS, 38, 309
Kounkel M., 2020, ApJ, 902, 122

Larson R. B., 1981, MNRAS, 194, 809
Quintana A. L., Wright N. J., 2021, MNRAS, 508, 2370
Quintana A. L., Wright N. J., 2022a, MNRAS, 511, 1224
Quintana A. L., Wright N. J., 2022b, MNRAS, 515, 687
Reipurth B., Schneider N., 2008, Star Formation and Young Clusters in Cygnus. The Northern Sky ASP Monograph Publications, University of Hawaii, p. 36
Solomon P. M., Rivolo A. R., Barrett J., Yahil A., 1987, ApJ, 319, 730
Wright N. J., 2020, New Astron. Rev., 90, 101549
Wright N. J., Drew J. E., Mohr-Smith M., 2015, MNRAS, 449, 741

Massive Stars Near and Far
Proceedings IAU Symposium No. 361, 2024
J. Mackey, J. S. Vink, & N. St-Louis, eds.
doi:10.1017/S1743921322002484

An exploration of Cygnus OB2 and perspectives for the upcoming WEAVE High-Resolution Cygnus Survey

Sara R. Berlanas[1,2] and **Artemio Herrero**[3,4]

[1]Dept. de Fsica Aplicada, Universidad de Alicante, E-03 690 San Vicente del Raspeig, Spain
email: sara.rb@ua.es

[2]Astrophysics Group, Keele University, Keele ST5 5BG, Staffordshire, UK

[3]Instituto de Astrofsica de Canarias, E-38 200 La Laguna, Tenerife, Spain

[4]Dept. de Astrofsica, Universidad de La Laguna, E-38 205 La Laguna, Tenerife, Spain

Abstract. During the last years we have carried out different studies in the Cygnus OB2 association based on new spectroscopic data and benefiting from the unprecedented Gaia astrometry. They include membership, chemical and structure studies, that allowed us to discern for the first time ever between two stellar groups separated by several hundred parsecs within the association and find at least two star-forming bursts at ∼3 and ∼5 Myr. Using these studies as a template and combining upcoming spectroscopic WEAVE data and the expected accuracy that Gaia will reach in the Cygnus-X area (DR3 and forthcoming releases), we will be able to perform the deepest multi-dimensional study ever done before in a massive star-forming complex. The results of this project will lead to an important improvement of our knowledge of star formation and evolution of star-forming regions and clusters, including our understanding of the dynamics and kinematics of OB associations and stellar groups.

Keywords. stars: massive – stars: early-type – stars: abundances – stars: fundamental parameters – open clusters and associations: individual: Cygnus-X

1. Previously in Cygnus OB2

The Cygnus-X complex represents the most powerful star-forming region at less than 2 kpc from us (Schneider et al. 2006). Its core, Cygnus OB2, contains nearly 100 O-stars (Knödlseder 2000) and can be used as a laboratory for massive star evolution and a proxy for massive clusters. Its distance allows observations at all wavelengths and accurate Gaia measurements. During the last years we have carried out different studies in the Cygnus OB2 association based on new spectroscopic data and benefiting from the unprecedented Gaia astrometry. We have (I) updated the Cygnus OB2 massive OB-star census (Berlanas et al. 2018a); (II) looked for self-enrichment processes and studied the implications of its abundance gradient (Berlanas et al. 2018b); (III) performed the first study of its spatial substructure using Gaia DR2 data (Berlanas et al. 2019); (IV) combined the DR2 data with our spectroscopic analyses to determine the main stellar parameters of the O-type population and explore the recent star formation history (Berlanas et al. 2020).

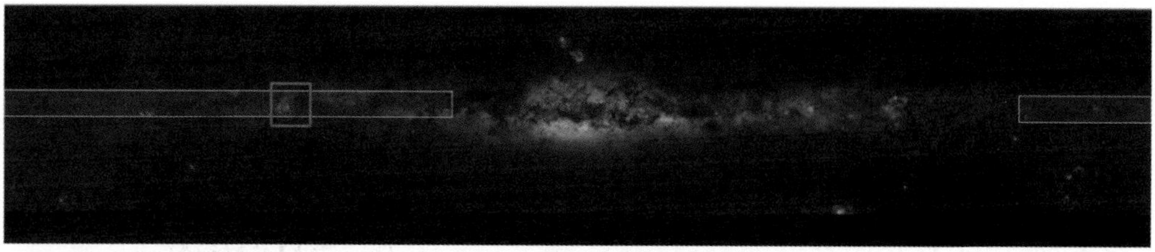

Figure 1. All-sky Milky Way map as seen by GALACTICA (Astroaula Project, IAC). Yellow and red lines indicate the Galactic plane LoRes and the Cygnus HiRes areas covered by WEAVE, respectively.

2. Coming soon

WEAVE is the next multi-object spectrograph at the 4.2m William Herschel Telescope, whose first light is planned for early summer 2022. It will provide high-quality spectra over the coming years for thousands of massive stars in the Galactic plane (LoRes Survey, R=5000) and, specifically, in several rich Cygnus OB associations (HiRes Survey, R=20000, S/N > 120, see Fig. 1). The HiRes Cygnus project is driven by the study of the massive star population of Cygnus-X to explore different scenarios of single and massive star formation and evolution. Our aim is to secure high S/N massive star spectra at the highest WEAVE resolution in the star-forming Cygnus region to:

• obtain rotational velocities and their distributions, focusing on the low vsini region.

• determine binary fractions and stellar multiplicity.

• determine accurate stellar parameters, particularly gravity, improving those obtained from the LoRes survey and allowing more accurate radii and masses with the help of Gaia.

• obtain accurate abundances and spatial abundance patterns for O3-B9 stars in the region for targets with 11<B<16.5.

• determine the kinematical and dynamical status of the stars in the region, checking individual proper motions in order to identify runaway stars or co-moving stellar groups.

Using the studies presented in Sect. 1 as a template and combining upcoming spectroscopic WEAVE data and the expected accuracy that Gaia will reach in the Cygnus-X area (DR3 and forthcoming releases) we will be able to perform the deepest multi-dimensional study ever done before in a massive star-forming complex. Thanks to its close distance, Gaia will provide a complete 3D dynamical plus a 3D spatial picture of the region, being the first time ever we can obtain such a precise spatial resolution along the line of sight for any of the main massive star-forming regions in the Milky Way.

Acknowledgements

S.R.B. thanks financial support by MCIN/AEI/10.13039/501100011033 (contract FJC 2020-045785-I), NextGeneration EU/PRTR and MIU (UNI/551/2021) trough grant Margarita Salas-ULL and MICIU/AEI/FEDER,UE trough grant PGC-2018-093741-B-C21. A.H. thanks support by the Spanish MCI trough grant PGC-2018-0913741-B-C22 and the Severo Ochoa Program trough CEX2019-000920-S.

Supplementary material

To view supplementary material for this article, please visit http://dx.doi.org/10.1017/S1743921322002484.

References

Berlanas, S. R., Herrero, A., Comerón, F., Pasquali, A., Bertelli Motta, C., & Sota, A. 2018a, A&A, 612, A50

Berlanas, S. R., Herrero, A., Comerón, F., Simón-Díaz, S., Cerviño, M., & Pasquali, A. 2018b, A&A, 620, A56
Berlanas, S. R., Wright, N. J., Herrero, A., Drew, J. E., & Lennon, D. J. 2019, MNRAS, 484, 1838
Berlanas, S. R., et al. 2020, A&A, 642, A168
Knödlseder, J. 2000, A&A, 360, 539
Schneider, N., Bontemps, S., Simon, R., Jakob, H., Motte, F., Miller, M., Kramer, C., & Stutzki, J. 2006, A&A, 458, 855

Chapter II:
Stellar atmospheres and winds

Massive Stars Near and Far
Proceedings IAU Symposium No. 361, 2024
J. Mackey, J. S. Vink, & N. St-Louis, eds.
doi:10.1017/S174392132200343X

Stellar Atmospheres and Supernovae: Systematic Errors

D. John Hillier

Department of Physics and Astronomy & Pittsburgh Particle Physics, Astrophysics and Cosmology Center (PITT PACC), University of Pittsburgh, 3941 O'Hara Street, Pittsburgh, PA 15260, USA.
email: hillier@pitt.edu

Abstract. Over the last two decades there have been considerable advances in modelling the spectra of massive stars and supernovae (SNe). Despite this progress, there are still numerous uncertainties that affect the accuracy of models. For massive stars, convection, instabilities, clumping, and our inability to model stellar winds self-consistently likely introduce systematic errors into our analyses. For SNe, and particularly for core-collapse SNe, departures from spherical symmetry strongly affect observed spectra and need to be taken into account. There are also issues with clumping, and mixing processes (both in the progenitor and the SN explosion) that need to be resolved. For both massive stars and SNe, the accuracy and availability of atomic data continues to be an ongoing issue influencing analyses.

Keywords. line: formation, radiative transfer, stars: atmospheres, supernovae: general, stars: winds, outflows etc.

1. Introduction

Systematic errors are a major issue in many areas of astrophysics. They can be difficult to identify, and unfortunately they are generally not reduced by the addition of more data. Systematic errors can arise from many sources: Model assumptions, errors in the analysis code, systematic errors in the observational data, errors in the atomic data, and personal biases.

Some of the most infamous cases of systematic errors relate to distances. For example, early determinations of the size of the Galaxy were biased because it was not appreciated that there were two types of pulsating variables: classical cepheids (associated with Pop I stars) and RR Lyrae stars (associated with Pop II stars). Determinations of the Hubble constant have been plagued by systematic errors, and it is possible that the current difference between the Hubble constant determined via Planck, and that found using Cepheid variables, could be due to systematic errors (e.g., Freedman et al. 2019; Freedman 2021 but see Riess et al. 2022).

With respect to stars, we highlight issues with fitting the He I singlet and triplet lines, particularly in O supergiants. For a long time it was generally argued that the problem was with the triplet lines, and it was thought to arise from a dilution effect (e.g., Voels et al. 1989). However, it is now known that the problem is with the singlet lines. and occurs because weak Fe lines overlap with the He I singlet resonance transition at 584.334 Å (Najarro et al. 2006). The Fe lines remove photons from the singlet transition, lowering the 2p $^1P^o$ population, and hence altering the strength of singlet transitions – particularly those ending on the 2p $^1P^o$ state (Najarro et al. 2006).

A more minor effect is associated with rotational broadening. Generally rotational broadening is taken into account using a convolution technique (e.g., Gray 2008), and for most absorption features in O stars this is an excellent approximation. However this technique does not work for wind lines (since the rotation velocity falls with distance from the star, and since the outflow is not purely radial) and can fail for some photospheric lines. For example, in ζ Puppis several weak emission lines show double-peaked emission profiles (e.g., N IV λ 4058). These cannot be explained using the normal convolution technique, and their existence was used to invoke large-scale co-rotating structures (De Becker et al. 2009). However the double-peaked profiles are automatically produced when rotational broadening is computed from first principles. It occurs in ζ Puppis because non-LTE effects cause these weak photospheric emission lines to be limb brightened (Hillier et al. 2012).

This paper is organized as follows. We briefly discuss some issues with the analysis of massive stars and their winds in §2. An alternative approach for treating clumping in stellar wind models is described in §2.1, while in §2.2 we discuss mass-loss rate determinations. In §3 we turn our attention to core-collapse SNe, and discuss three sources of systematic errors – progenitors (§3.1), chemical mixing (§3.2), and molecular cooling (§3.3). Conclusions can be found in §4.

2. Massive stars & stellar winds: Systematic errors

In classical analyses of massive stars, the star and its wind are generally assumed to be spherical. The wind, originally assumed to be homogeneous, is now understood to be clumped, and this is treated by approximate methods. The presence of significant rotation in many stars immediately indicates that the winds are unlikely to be spherically symmetric. This can be important for interpreting P Cygni profiles since the absorption component samples only along our line of sight to the star (and hence samples only a small volume of wind) while the emission component samples most of the wind. When discrepancies in fitting P Cygni profiles occur, they are normally attributable to wind clumping and porosity.

Both observational (e.g., Eversberg et al. 1998; Prinja & Massa 2010; Hillier & Miller 1999; Crowther et al. 2002) and theoretical studies (e.g., Owocki et al. 1988; Sundqvist & Owocki 2013) indicate that the winds are inhomogeneous with significant spatial and angular variations in wind speed and density. The inhomogeneities are generally thought to arise from the line-driven instability, but other processes, such as convection and non-linear pulsations, may act as seeds.

Since different radiative processes have a different functional dependence on density, the influence of clumping is not uniform. In particular, electron scattering, and the strength of unsaturated P Cygni profiles arising from the dominant ionization state, will scale with density, while recombination processes and free-free emission scale with the square of the density.

In addition, inhomogeneities mean that wind porosity (Oskinova et al. 2007) and vorosity (Owocki 2008) need to be included in models. Porosity arises from the spatial dependence of clumping, and occurs when some sight lines have a significantly lower optical depth because they primarily pass through material of low density. Vorosity (a term coined by Stan Owocki) occurs because some directions may have very little material at certain velocities, and hence this material is transparent to line radiation (Owocki 2008) at these velocities. Porosity affects both lines and the continuum, whereas vorosity only affects line transfer. Due to vorosity, clumps can introduce structure into the absorption component along a single sight line (Fig. 1). The structure can/will disappear when an average is taken over many sight lines but it can affect doublet line ratios

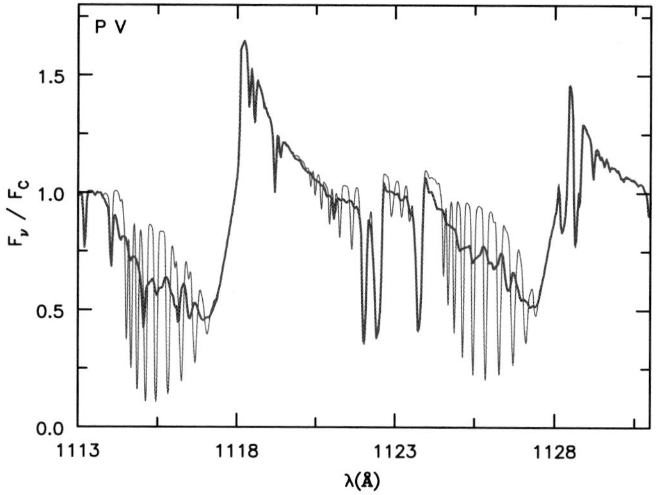

Figure 1. An illustration of the P V λλ1117, 1128 profile in a model for AzV83. The red curve shows the profile computed using the shell model. Because the shells have a distinct velocity, the P Cygni absorption profile presents a comb-like structure which is not observed. However, when we allow the shells to have a sightly different radial velocity along different sight lines we get a smooth profile, much like that observed. Reproduced from: "Using Shell models to investigate clumping in the wind of the O7Iaf+ supergiant AzV83" (figure 4), Flores & Hillier (2021).

(e.g. Massa et al. 2003). Through the study of doublet ratios it is possible to place constraints on the amount of wind porosity/vorosity (Prinja & Massa 2013).

Another important, potential cause of systematic errors is hot gas and its X-ray emission. O stars are significant X-ray emitters with $L_x/L_{\rm BOL} \sim 10^{-7}$ to 10^{-6} (e.g., Cassinelli & Swank 1983; Chlebowski 1989). These X-rays, through Auger ionization, affect the ionization structure of the wind, and are responsible for the so-called superion† P Cygni profiles (e.g., O VI) in O stars (e.g., Cassinelli & Olson 1979). The X-ray emission is thought to arise from shocks produced by the line-driven instability (Lucy & White 1980; Feldmeier et al. 1997). In most modelling the X-ray emission is simply added as an extra wind-emission source, with several ad hoc parameters defining the shock temperature and emissivity as a function of location. The shock structure, and the coupling between the shocks and their associated hot gas, and the cooler wind, is ignored.

2.1. *Clumping*

In most studies clumping is treated using a volume-filling-factor (VFF) approach. A fraction, f, of the spherical wind is assumed to consist of small clumps while the remaining fraction is devoid of material. With this assumption it is easy to modify standard radiative transfer codes to incorporate clumping. However the approach assumes that the interclump medium is void, that the density at a given radius is homogeneous, and that the clumps are small – all assumptions are invalid "at some level", but their influence on results is generally unknown. Further, the radial dependence of f is described by several parameters, which are difficult to constrain.

Extensions to the VFF approach have been developed that allow for an interclump medium, and both porosity and vorosity (e.g., Sundqvist et al. 2014). These techniques are also approximate, require additional free parameters, and can still lead to systematic errors in inferred wind parameters.

† An ion whose presence cannot be explained by the photospheric radiation field.

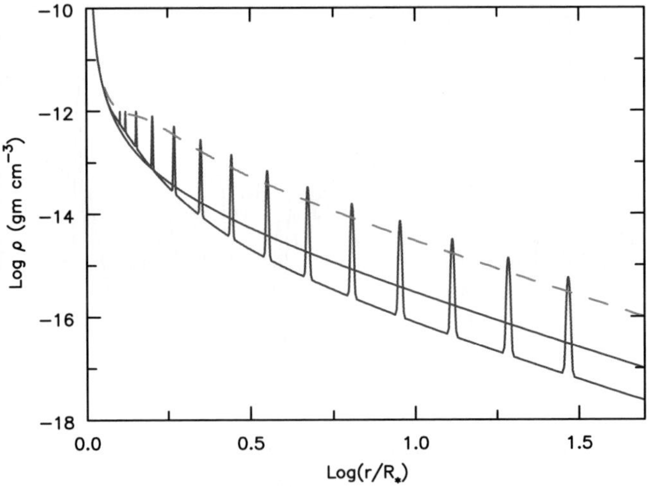

Figure 2. An illustration of the density distribution in three models for AzV83 – all models have the same average mass-loss rate. The red solid curve is for a smooth density distribution and the blue solid curve is for the shell model with an average wind VFF of 0.1. The green dashed curve illustrates the clump density in the VFF model (but recall that the interclump medium is void) with VFF= 0.1. By adjusting model parameters it is possible to vary the number of clumps, clump size and the density of the interclump medium. Reproduced from: "Using Shell models to investigate clumping in the wind of the O7Iaf+ supergiant AzV83" (figure 3), Flores & Hillier (2021).

One way to test the influence of assumptions is to invoke an alternative model which allows for clumping but is based on different assumptions. One such approach is to assume that the star consists of spherical dense shells (shell model) (e.g., Hillier 1991a; Flores & Hillier 2021; Flores & Hillier 2022), as illustrated in Fig. 2. In this approach the finite size of the clumps, the interclump medium, and vorosity can be treated. However it is also unlikely to be valid – it does not allow for non-monotonic velocity fields and the assumption of spherical shells is invalid – they will be broken up on some angular scale.

The shell model has been applied to two stars – AzV83 (an O7Iaf+ supergiant) (Flores & Hillier 2021) and HD 50896 (a classical WN4 star) (Flores & Hillier 2022). In these models the shell structure is set to yield a pre-specified VFF, with the number of shells a free parameter. Typically the shells have a radial thickness of order the Sobolev length ($=rV_{\rm D}/V$ where $V_{\rm D}$ is the line Doppler width).

For AzV83 results obtained with the shell model, and the VFF approach, are similar (Flores & Hillier 2021). A slightly higher mass loss ($\sim 30\%$) is needed in the shell model, an effect also seen in an earlier study that used a more realistic treatment of clumping (Sundqvist et al. 2011). The shells produce highly structured P Cygni absorption components, but the structure averages out if we assume that the shells have small angular variations in velocity (effectively equivalent to breaking the shells up in angle, and performing a slight shift in radial location) when we do a formal solution to compute observed spectra. As the model allows for the interclump medium, the shell model, unlike the VFF model, reproduces the O VI $\lambda\lambda$ 1032, 1038 line in AzV83 for a reasonable X-ray flux. The importance of the interclump medium for explaining superions in O star spectra was previously demonstrated by Zsargó et al. (2008).

By contrast, when applied to HD 50896 the shell models failed to match the He II line strengths – the emission profiles contain a central absorption component and appear double-peaked (Fig. 3). This is attributable to optical depth effects. For lines of sight passing through the shell center, photon escape in a line is substantially influenced by

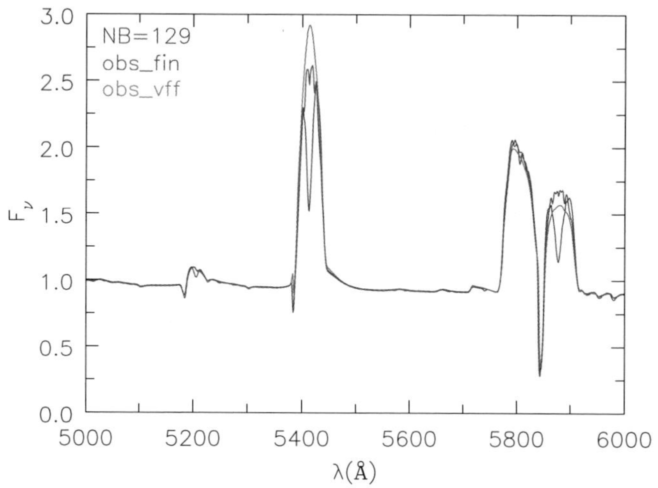

Figure 3. Illustration of how a pure shell model with complete shells (purple curve) fails to produce the He II emission strengths. When the shells are broken up (blue curve), there is much better agreement with the VFF results (pink curve). Full details can be found in Flores & Hillier (2022).

the width of the shell. However for lines of sight through the limb of the shell, the path length is much longer, minimizing the influence of the shell thickness on photon escape, and greatly reducing photon escape in the lines. Continuum formation is much less affected since the scale length for a continuum photon to escape is much larger than for photon escape in a line.

Better consistency can be achieved with the VFF approach if we break the shells up in angle, and randomly distribute each "partial" shell over a radial distance of approximately $\delta r/2$ where δr is the shell separation. Our results argue that the lateral scale of the shells must be small – less than 1°.

2.2. *Mass-loss rates and the velocity law*

Typically the mass-loss rate and velocity law are assumed, with the velocity law (above the sonic point) parameterized by several free parameters. The mass-loss rate and velocity law are then constrained through spectral fitting, although in many cases the constraints on the shape of the velocity law are weak. Unfortunately, it is difficult to determine the mass-loss rate uniquely, since assumptions about both clumping and porosity/vorosity will affect its determination. When porosity/vorosity effects are ignored, f can be constrained by fitting both Hα and P V $\lambda\lambda$1117, 1128 in O stars, while in Wolf-Rayet (WR) stars constraints on f come from a comparison of emission line strengths with their associated electron scattering wings. Obviously our biases about clumping, and (in many cases) the neglect of porosity/vorosity effects will introduce systematic errors into mass-loss rate determinations – the significance of such errors remains to be seen.

A major issue for Wolf-Rayet stars is connecting derived radii and effective temperatures with those of evolutionary models (e.g., Groh et al. 2014). For most WR stars the observed radiation originates in the wind, and is thus insensitive to the radius and effective temperature of the star defined at the sonic point (e.g., Hillier 1991b). Consequently, observationally derived values depend on the assumed velocity law.

Ideally we would like to deduce the velocity law and mass-loss rate from hydrodynamic calculations. Such calculations would be 3D, allow for stellar rotation, and be time dependent (e.g., Moens et al. 2022). Work towards this goal is in progress, but researchers face many hurdles. One-D hydrodynamical models are routinely constructed, but these typically assume a clumping structure. For some stars these 1D solutions appear to give reasonable answers while for others there appears to be significant discrepancies between the mass-loss rates and terminal velocities deduced from 1D hydrodynamic solutions, and those inferred from an analysis of the observations (e.g., Sundqvist et al. 2019; Gormaz-Matamala et al. 2021).

3. Core-collapse supernovae: The final stage of massive star evolution

An understanding of core-collapse SNe is crucial for many areas of astrophysics. SNe influence the interstellar and intergalactic medium through the deposition of momentum, energy, and chemically enriched material and thus influence star formation and galactic evolution, and create/distribute the elements necessary for life. Many questions remain unanswered, particularly related to how core-collapse SNe explode. It is only with the development of complex 3D models that explosions of massive stars can be fully explored (e.g., Wongwathanarat 2017).

Whatever the exact mechanism, core-collapse SNe are intrinsically aspherical. Direct evidence for asphericity comes from polarization observations (e.g., Cropper et al. 1988; Wang et al. 1996; Kawabata et al. 2002), and line profile analyses (e.g., Hanuschik & Dachs 1987; Chugai 2007). Observed SN remnants also show evidence for strong asymmetric explosions. However most modelling of SN light curves and spectra assumes spherical symmetry. The primary reason is computational – it is prohibitively expensive to compute a significant set of time-dependent non-LTE models in 3D, even with approximations. It should also be stressed that the departures from sphericity are unlikely to be described by a simple axi-symmetric model. Theoretical models show complex asymmetries (e.g., Hammer et al. 2010; Janka et al. 2007).

3.1. *Progenitor issues*

Understanding SN explosions requires a detailed progenitor model. However the evolution of massive stars is in itself uncertain, and this will translate into uncertainties in explosion models, and the deduced supernova properties. Uncertainties in massive star properties were recently emphasized in a study by Agrawal et al. (2022) who found that the differences in model results increase with stellar mass, primarily as a consequence of how instabilities associated with the star's proximity to the Eddington limit are treated. An illustration of the different evolutionary tracks they obtained with different evolutionary codes is shown in Fig. 4. Other major uncertainties affecting stellar evolution calculations arise from the adopted mass-loss rates, and the treatment of convection and rotation.

Given the uncertainties in the theoretical treatment of stellar evolution, and uncertainties in mass-loss rates, inferring the progenitor properties, such as its mass on the zero-age main sequence, from the SN ejecta properties is very difficult (e.g., Dessart & Hillier 2019). While mass is obviously the most fundamental parameter that influences a star's evolution, other factors such as metallicity, initial rotation rate, and whether the star's evolution has been influenced by interactions with a companion (which may no longer be present in the system) are also crucial (e.g., Farrell et al. 2020). In principle, mass loss, which can substantially influence a star's evolution, can be deduced from stellar properties at that instant, but in practice evolutionary codes use descriptive formulae with several free parameters. These, of course, introduce systematic errors into an evolutionary calculation. In massive stars one of the more problematic issues is eruptive mass loss. This phenomenum is not understood, and is generally ignored.

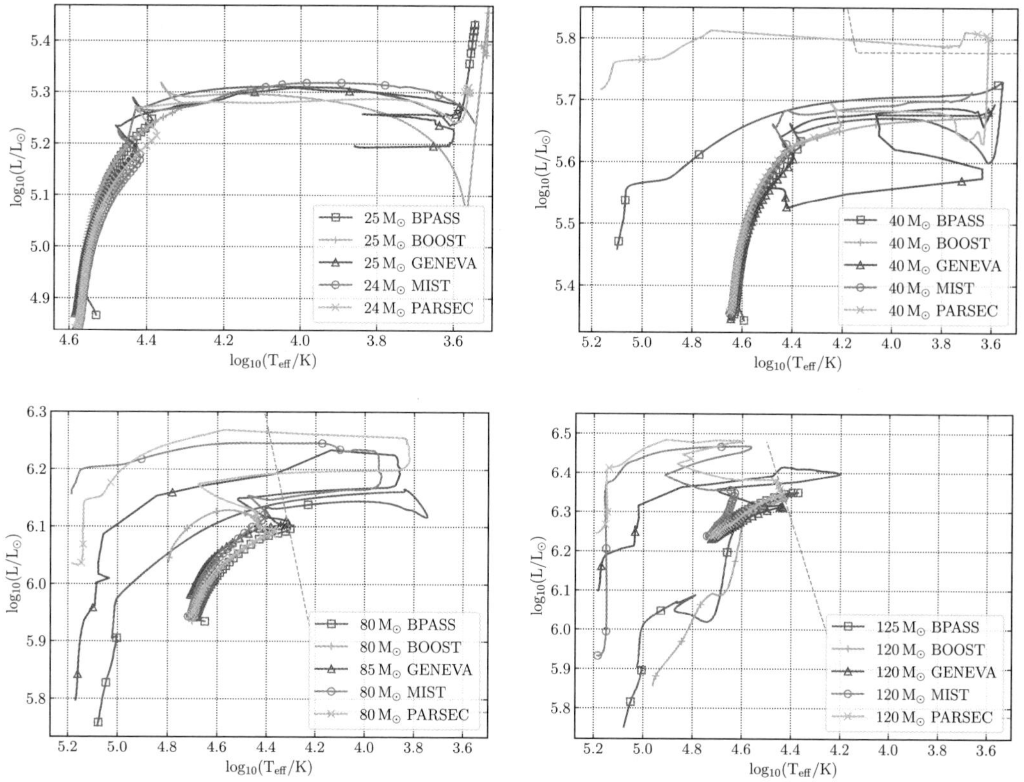

Figure 4. An illustration of the different evolutionary tracks for four ZAMS masses computed using different evolutionary codes. As clearly evident, there is a great deal of dispersion in the evolutionary tracks, particularly at large masses. Reproduced from: "Explaining the differences in massive star models from various simulations" (figure 1), Agrawal et al. (2022).

3.2. *Mixing of Ca into the O zone*

A major problem in modelling core-collapse SNe with 1D models relates to their chemical composition. Stellar evolution calculations for the massive progenitors of core-collapse SNe predict an "onion-like" structure for the chemical composition (see Fig. 6). However in 3D calculations of SN explosions, significant radial mixing of the material occurs (e.g., Janka et al. 2007; Hammer et al. 2010). This mixing is macroscopic – the radially mixed material retains its initial chemical composition. Thus, for example, there is no H in the O-rich material even though their distributions in velocity (or equivalently, radius) overlap.

In 1D modeling, a naive approach for treating mixing would be to run a smoothing function through the material. This causes microscopic mixing and that will/can introduce severe biases into spectroscopic modelling. This occurs because the cooling of the ejecta is determined by its chemical composition, and the cooling efficiency varies widely between different chemical species (and between different ionization stages). A striking illustration of the effect occurs with Ca and Si cooling.

With the smoothing approach, we microscopically mix Ca from the Si-rich shell into the O-rich shell. However, Ca II $\lambda\lambda$ 7291, 7233 is a much more efficient coolant than O I $\lambda\lambda$ 6300, 6364, and consequently a small amount of Ca in the O zone can dramatically weaken the strength of O I $\lambda\lambda$ 6300, 6364 (e.g., Fransson & Chevalier 1989; Dessart & Hillier 2020a). The effect, illustrated in Fig. 5, is of crucial importance since the mass of O in

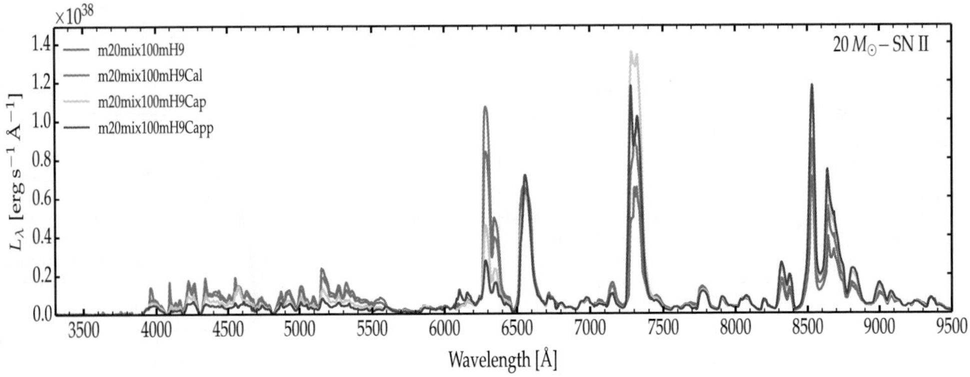

Figure 5. An illustration of the influence of mixing Ca into the O zone of nebular phase spectra. Successive models vary by a factor of 10 in mass fraction and have X(Ca)$= 7 \times 10^{-5}$ (blue), 7×10^{-4} (red), 7×10^{-3} (yellow) and 7×10^{-2} (green). As the Ca to O ratio increases, the O I $\lambda\lambda$ 6300, 6364 emission is depressed, and the Ca II $\lambda\lambda$ 7291, 7233 emission enhanced. Reproduced from: "Radiative-transfer modeling of nebular-phase type II supernovae. Dependencies on progenitor and explosion properties" (figure 7), Dessart & Hillier (2020a).

the ejecta varies strongly with the progenitor mass (Woosley et al. 2002), and hence is used as a diagnostic probe of the progenitor mass (e.g., Jerkstrand et al. 2012).

The influence of Ca on the O line strengths has been known for a long time. In some Kepler models (Ensman & Woosley 1988), pollution of the O-rich shell by the Si-rich shell meant that the models could not explain the observed line strengths in SN 1987A (Fransson & Chevalier 1989). An extensive study of convective oxygen and silicon burning shells in supernova progenitors, and shell mixing, was undertaken by Collins et al. (2018). This problem of shell merging and mixing in progenitor modelling has not been fully solved – it still occurs in some evolutionary calculations (although it can be artificially suppressed), and it is unknown as to what extent it occurs in nature. In some cases, shell merging can occur just prior to collapse, and it is possible that the short time-scale (e.g., 10 minutes before collapse) means that the merging is incomplete when core collapse occurs (Collins et al. 2018).

To treat macroscopic mixing Dessart & Hillier (2020b) developed an approach in which the shells in the original model are broken up into several pieces, and these are then distributed in radius/velocity (Fig. 6). Importantly, this approach preserves the distinct chemical composition of the shells.

3.3. *Cooling by Molecules*

Molecular cooling in core-collapse SN ejecta is known to be important (e.g., Cherchneff & Sarangi 2011; Matsuura 2017; Rho et al. 2021). In SN 1987A, emission in the first overtone of CO was seen in spectra at day 137 and the fundamental band was a contributor to excess emission seen in the M band (Catchpole et al. 1988). Modelling of CO emission in SN ejecta is difficult since it is necessary to solve the chemical reaction equations producing and destroying CO (i.e., chemical equilibrium cannot be assumed). Unfortunately many of the reaction rates are poorly known, or only available at temperatures not relevant to SN ejecta. It is also likely that the vibrational levels are not in local thermodynamic equilibrium, requiring that their populations be deduced from the solution of the kinetic equations. To construct the kinetic equations we need collision rates, which are also poorly known.

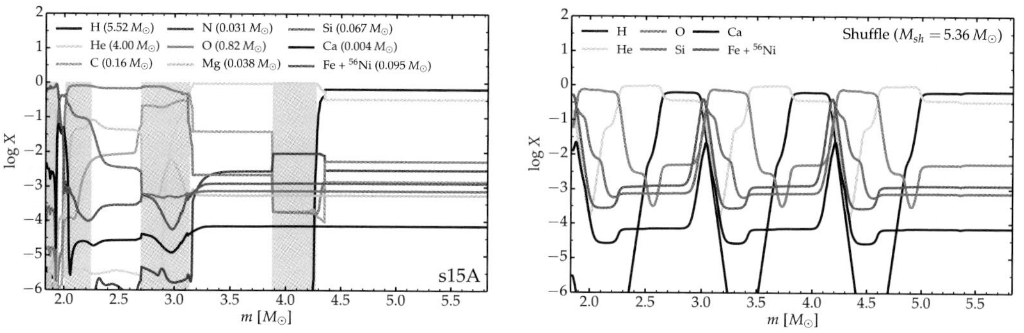

Figure 6. An illustration of the treatment of macroscopic mixing in 1D. On the left we see the chemical composition of the original SN model (based on the S15A ejecta model of Woosley & Heger (2007)) – the onion-like chemical structure of the ejecta is clearly evident. On the right the distinct chemical zones have been broken into three fragments, and these have then been shifted in radius. See Dessart & Hillier (2020b) for further details. Reproduced from: "Radiative-transfer modeling of supernovae in the nebular-phase. A novel treatment of chemical mixing in spherical symmetry" (figure 1), Dessart & Hillier (2020b).

CO formation has been introduced (Liljegren et al. 2020) into SUMO, the SN modelling code of Anders Jersktrand (Jerkstrand et al. 2011). More recently SiO was included, and the code modified to allow the vibrational levels to be treated in non-LTE (Liljegren et al. 2022).

PhD student, Collin McLeod, is currently installing molecules into CMFGEN. For the solution of the chemical equations we treat (in addition to the associated atomic species) CO, CO^+, C_2, C_2^+, O_2, O_2^+, CO_2, CO_2^+, C_2O, and C_2O^+, however only the opacities and emissivities of CO are allowed for when determining the temperature structure of the ejecta. We currently treat CO in LTE, but we plan to relax this assumption in the near future – we also plan to include SiO.

Due to the complexities of cooling in SN ejecta, we have had difficulties in obtaining converged models. Initial results confirm the importance of CO cooling, and its potential impact on SN spectra. In Fig. 7 we show the influence of CO cooling on the optical spectrum for a model roughly applicable to SN 1987A. While this model overemphasizes the importance of CO cooling, it does demonstrate that significant potential errors may occur in fitting optical spectra if cooling by CO and other molecules is neglected. In turn, these erroneous fits may lead to an incorrect estimate of the oxygen mass, and hence the mass of the progenitor that gave rise to the SN.

4. Conclusions

Systematic errors are hard to identify, are hard to quantify, and arise from many sources. Before drawing any conclusions, it is important to investigate possible sources of systematic errors that may influence conclusions.

An obvious potential source of systematic errors arise in codes designed to perform theoretical modeling. All these codes have underlying assumptions that can have a profound influence on their results. In many cases these assumptions are necessary, but they must be considered when drawing conclusions, and tests should be made to examine their applicability. A potential problem with today's codes is that many have an extensive development history. Assumptions valid at one time may no longer be valid when changes are made to the code. We found such issues when installing molecules into CMFGEN.

When using public codes please provide feedback to the developers. Such feedback is important for further code development.

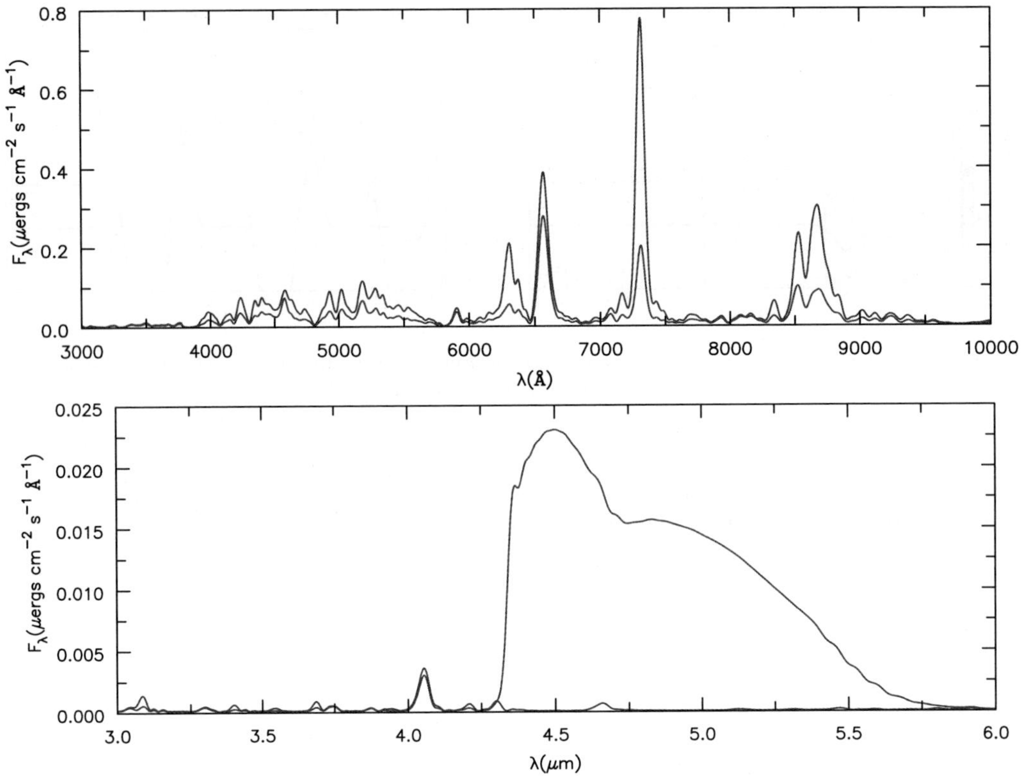

Figure 7. Spectra for a model of a supernova similar to SN 1987A. The model has classical mixing (i.e., not the shell approach). The blue line shows the spectrum at an age of 330 days without cooling, while the (preliminary) model shown in red has CO cooling. CO cooling, primarily by the fundamental band, weakens optical features such as O I λλ 6300, 6364, Ca II λλ 7291, 7233, and Ca II λλ 8498, 8542, 8662. Longward of 4.3 μm we see the strong emission by the fundamental band of CO which completely dominates this spectral region.

References

Agrawal, P., Szécsi, D., Stevenson, S., Eldridge, J. J., & Hurley, J. 2022, MNRAS, 512, 5717
Cassinelli, J. P. & Olson, G. L. 1979, ApJ, 229, 304
Cassinelli, J. P. & Swank, J. H. 1983, ApJ, 271, 681
Catchpole, R. M., Whitelock, P. A., & Feast, M. W., et al. 1988, MNRAS, 231, 75P
Cherchneff, I. & Sarangi, A. 2011, in The Molecular Universe, ed. J. Cernicharo & R. Bachiller, Vol. 280, 228–236
Chlebowski, T. 1989, ApJ, 342, 1091
Chugai, N. N. 2007, in AIP Conf. Ser., Vol. 937, Supernova 1987A: 20 Years After: Supernovae and Gamma-Ray Bursters, ed. S. Immler, K. Weiler, & R. McCray, 357–364
Collins, C., Müller, B., & Heger, A. 2018, MNRAS, 473, 1695
Cropper, M., Bailey, J., McCowage, J., Cannon, R. D., & Couch, W. J. 1988, MNRAS, 231, 695
Crowther, P. A., Hillier, D. J., & Evans, C. J., et al. 2002, ApJ, 579, 774
De Becker, M., Rauw, G., & Linder, N. 2009, ApJ, 704, 964
Dessart, L. & Hillier, D. J. 2019, A&A, 625, A9
—. 2020a, A&A, 642, A33
—. 2020b, A&A, 643, L13
Ensman, L. M. & Woosley, S. E. 1988, ApJ, 333, 754
Eversberg, T., Lépine, S., & Moffat, A. F. J. 1998, ApJ, 494, 799
Farrell, E. J., Groh, J. H., Meynet, G., & Eldridge, J. J. 2020, MNRAS, 494, L53

Feldmeier, A., Kudritzki, R. P., Palsa, R., Pauldrach, A. W. A., & Puls, J. 1997, A&A, 320, 899
Flores, B. L. & Hillier, D. J. 2021, MNRAS, 504, 311
—. 2022, MNRAS, submitted
Fransson, C. & Chevalier, R. A. 1989, ApJ, 343, 323
Freedman, W. L. 2021, ApJ, 919, 16
Freedman, W. L., Madore, B. F., & Hatt, D., et al. 2019, ApJ, 882, 34
Gormaz-Matamala, A. C., Curé, M., & Hillier, D. J., et al. 2021, ApJ, 920, 64
Gray, D. F. 2008, The Observation and Analysis of Stellar Photospheres (Cambridge Uni. Press)
Groh, J. H., Meynet, G., Ekström, S., & Georgy, C. 2014, A&A, 564, A30
Hammer, N. J., Janka, H.-T., & Müller, E. 2010, ApJ, 714, 1371
Hanuschik, R. W. & Dachs, J. 1987, A&A, 182, L29
Hillier, D. J. 1991a, A&A, 247, 455
Hillier, D. J. 1991b, in IAU Symposium, Vol. 143, Wolf-Rayet Stars and Interrelations with Other Massive Stars in Galaxies, ed. K. A. van der Hucht & B. Hidayat, 59
Hillier, D. J., Bouret, J.-C., Lanz, T., & Busche, J. R. 2012, MNRAS, 426, 1043
Hillier, D. J. & Miller, D. L. 1999, ApJ, 519, 354
Janka, H.-T., Langanke, K., Marek, A., Martínez-Pinedo, G., & Müller, B. 2007, Phys. Rep., 442, 38
Jerkstrand, A., Fransson, C., & Kozma, C. 2011, A&A, 530, A45+
Jerkstrand, A., Fransson, C., Maguire, K., Smartt, S., Ergon, M., & Spyromilio, J. 2012, A&A, 546, A28
Kawabata, K. S., Jeffery, D. J., Iye, M., & et al. 2002, ApJL, 580, L39
Liljegren, S., Jerkstrand, A., Barklem, P. S., Nyman, G., Brady, R., & Yurchenko, S. N. 2022, arXiv e-prints, arXiv:2203.07021
Liljegren, S., Jerkstrand, A., & Grumer, J. 2020, A&A, 642, A135
Lucy, L. B. & White, R. L. 1980, ApJ, 241, 300
Massa, D., Fullerton, A. W., Sonneborn, G., & Hutchings, J. B. 2003, ApJ, 586, 996
Matsuura, M. 2017, Dust and Molecular Formation in Supernovae, ed. A. W. Alsabti & P. Murdin (Springer, Cham.), 2125
Moens, N., Poniatowski, L. G., Hennicker, L., Sundqvist, J. O., El Mellah, I., & Kee, N. D. 2022, arXiv e-prints, arXiv:2203.01108
Najarro, F., Hillier, D. J., Puls, J., Lanz, T., & Martins, F. 2006, A&A, 456, 659
Oskinova, L. M., Hamann, W.-R., & Feldmeier, A. 2007, A&A, 476, 1331
Owocki, S. P. 2008, in Clumping in Hot-Star Winds, ed. W.-R. Hamann, A. Feldmeier, & L. M. Oskinova (Potsdam: Univ.-Verl), 121
Owocki, S. P., Castor, J. I., & Rybicki, G. B. 1988, ApJ, 335, 914
Prinja, R. K. & Massa, D. L. 2010, A&A, 521, L55
—. 2013, A&A, 559, A15
Rho, J., Evans, A., Geballe, T. R., & Banerjee, D. P. K., et al. 2021, ApJ, 908, 232
Riess, A. G., Yuan, W., & Macri, Lucas M., et. al. 2022, ApJL, 934, L7
Sundqvist, J. O., Björklund, R., Puls, J., & Najarro, F. 2019, A&A, 632, A126
Sundqvist, J. O. & Owocki, S. P. 2013, MNRAS, 428, 1837
Sundqvist, J. O., Puls, J., Feldmeier, A., & Owocki, S. P. 2011, A&A, 528, A64+
Sundqvist, J. O., Puls, J., & Owocki, S. P. 2014, A&A, 568, A59
Voels, S. A., Bohannan, B., Abbott, D. C., & Hummer, D. G. 1989, ApJ, 340, 1073
Wang, L., Wheeler, J. C., Li, Z., & Clocchiatti, A. 1996, ApJ, 467, 435
Wongwathanarat, A. 2017, in Supernova 1987A:30 years later - Cosmic Rays and Nuclei from Supernovae and their Aftermaths, ed. A. Marcowith, M. Renaud, G. Dubner, A. Ray, & A. Bykov, Vol. 331, 101–106
Woosley, S. E. & Heger, A. 2007, Phys. Rep., 442, 269
Woosley, S. E., Heger, A., & Weaver, T. A. 2002, Reviews of Modern Physics, 74, 1015
Zsargó, J., Hillier, D. J., Bouret, J.-C., Lanz, T., Leutenegger, M. A., & Cohen, D. H. 2008, ApJL, 685, L149

Acknowledgements

Partial support from STScI theory grant HST-AR-16131.001-A (DJH) and Astrophysical theory grant 80NSSC20K0524 is acknowledged. STScI is operated by the Association of Universities for Research in Astronomy, Inc., under NASA contract NAS 5-26555. Special thanks to my students, Brian Flores and Collin McLeod, for allowing me to use their work in this review.

Discussion

Unknown: In the supernova model, how much Ca has been mixed into the O zone?

D. John Hillier: The 1987A model is an old model and uses microscopic mixing. The O/Ca ratio is roughly 300 to 1 in the O-rich material, and it is probable that the level of mixing used causes an overestimate of the importance of CO cooling.

Unknown: You mentioned shocks and X-rays in your talk. A lot of supernovae that appear normal may have shocks arising from the interaction with circumstellar material. Is it possible to incorporate a little bit of X-ray emission into your ejecta?

D. John Hillier: It is easy to include the influence of X-rays if we simply specify a level of X-ray emission through the ejecta – this is already done in models for O and Wolf-Rayet stars. Low levels of X-rays are probably not crucial since we already have γ-rays in the model, and we account for the influence on level populations by the non-thermal electrons that are produced by these γ-rays. We also have run models where we have used CMFGEN to post process (with a fixed temperature) hydrodynamic models where the ejecta has interacted with circumstellar material. More recently we are considering models in which we simply dump energy into the the outer region of the ejecta. None of these approaches are fully satisfactory, but they do give insights into the fluxes and spectra produced by interstellar interactions.

Massive Stars Near and Far
Proceedings IAU Symposium No. 361, 2024
J. Mackey, J. S. Vink, & N. St-Louis, eds.
doi:10.1017/S1743921322002125

Spectroscopic analyses of massive stars at different metallicities

Wolf-Rainer Hamann

Institute for Physics and Astronomy, University Potsdam, D-14476 Potsdam, Germany
email: wrh@astro.physik.uni-potsdam.de

Abstract. Adequate stellar atmosphere models are prerequisite to derive robust stellar parameters from spectroscopic analyses. I will briefly review recent results obtained with the Potdam Wolf-Rayet (PoWR) model atmosphere code, which is applicable to all types of hot stars. Using multi-wavelength observations including the UV, we analyzed large samples of massive stars at various metallicities, gaining important insights on their cosmic role and the feedback to their environment.

A recent extension of PoWR allows to compose the model atmosphere from two zones. A rapidly rotating star, e.g., might possess a cooler equatorial region with a slow wind, and two polar cones with higher photospheric temperature and fast wind. For two examples of rapidly rotating O-type stars, we demonstrate that such model can reproduce wind-line profiles which otherwise would stay inconsistent. Fast rotation, which prevails in particular at low metallicities, thus might bias empirically derived parameters, having implications for feedback as well as for angular-momentum losses of SN and GRB progenitors.

Keywords. stars: atmospheres; stars: winds, outflows

1. Introduction

Detailed information on massive stars and their winds are obtained by comparing spectroscopic observations with synthetic spectra. Hot-star atmospheres and winds can be simulated with, e.g., the *Potsdam Wolf-Rayet* (PoWR) model atmosphere code. It calculates the non-LTE radiative transfer in hot-star atmospheres and winds in spherical symmetry. Complex model atoms and iron-line blanketing are taken into account. Various grids of models for Wolf-Rayet and O-type stars at different metallicities are accessible via the web interface `http://www.astro.physik.uni-potsdam.de/PoWR/`.

Originally focusing on Wolf-Rayet (WR) stars, the hydrogen-deficient nature of WN, WC and WO (nitrogen, carbon, and oxygen) stars was confirmed, establishing the distinction between very luminous "late-subtype" WN stars (WNL) which typically exhibit some hydrogen in their atmosphere, while the hotter "early" subtypes (WNE) are hydrogen-free (Hamann et al. 1995, 2019). A similar pattern was found in the Large Magellanic Cloud (Hainich et al. 2014), while in the Small Magellanic Cloud (SMC) *all* WN stars are hotter than the hydrogen main sequence (i.e. WNE), but at the same time are displaying hydrogen in their atmospheric composition (Hainich et al. 2015). This has been discussed as indication of strong internal mixing and quasi-homogeneous evolution under the condition of the SMC's low metallicity. Generally, the WR analyses have revealed many open issues with stellar evolution modeling, including the role of mass-loss, rotation, internal mixing, overshooting and close-binary evolution.

Applying PoWR models to O- and B-type stars, large samples at different metallicities have been studied. Figure 1 displays a Hertzsprung-Russel diagram (HRD) for massive

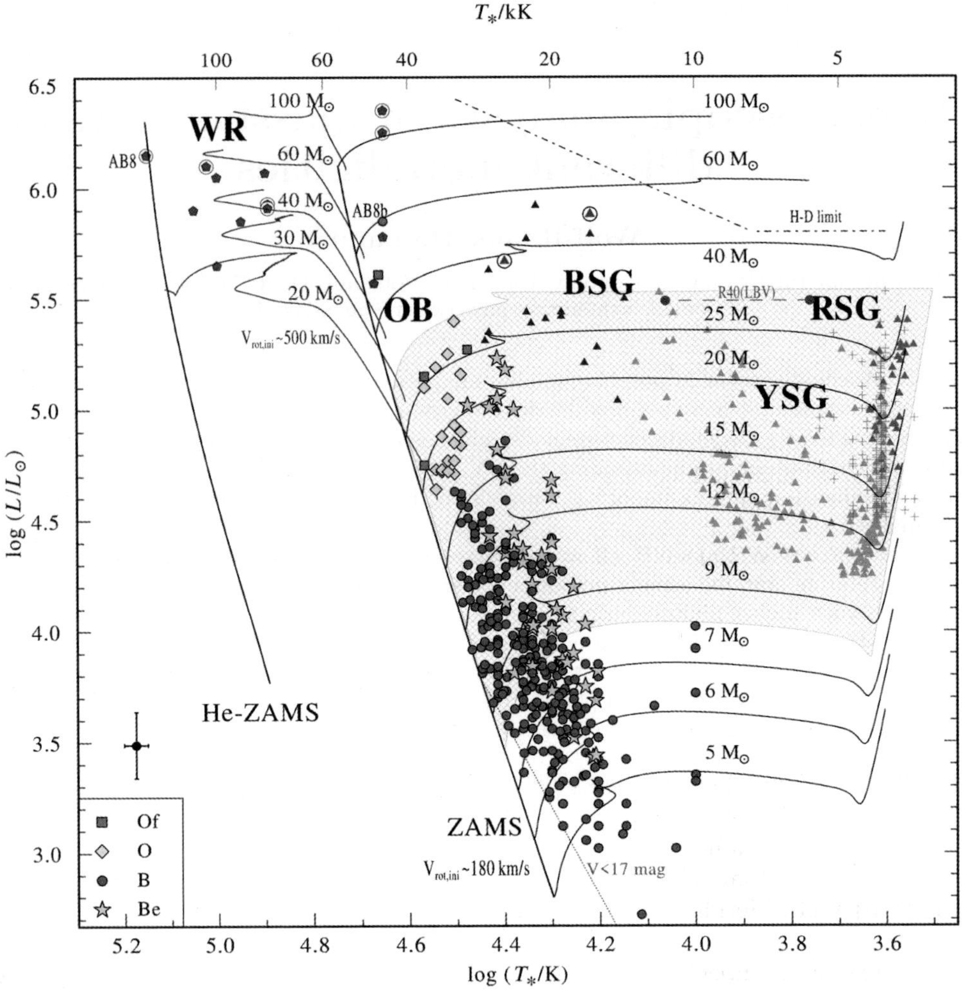

Figure 1. Hertzsprung-Russell-diagram of massive stars in the SMC (from Ramachandran et al. 2019). The OB stars analyzed in that work (symbols as specified in the inset) reside in the SMC's "Wing", while other symbols denote stars from all over the SMC (for more details see the original paper). Known binaries are encircled. Single-star evolutionary tracks shown for comparison are for modest rotation (black lines, labeled with their initial mass), while the blue lines are for very rapid rotation.

stars in the SMC (from Ramachandran et al. 2019). The most-luminous stars are located close to the hydrogen zero-age main sequence (H-ZAMS), or even to its hot side (WRs). These stars would be explainable with quasi-homogeneous evolution. The striking pattern is the sparsity, or even absence, of any other stars that would correspond to an initial mass $> 30\, M_\odot$ and non-homogeneous evolution. This dichotomy could be a consequence of rapid rotation because of reduced angular-momentum losses from weaker stellar winds due to low metallicity. If, alternatively, this pattern could just result from the SMC's particular star formation history plus observational biases remains to be shown.

2. Rotation

Stellar rotation broadens spectral lines by the Doppler effect. If a line forms in layers that are not extended, but can be approximated as the surface of a sphere, rotational

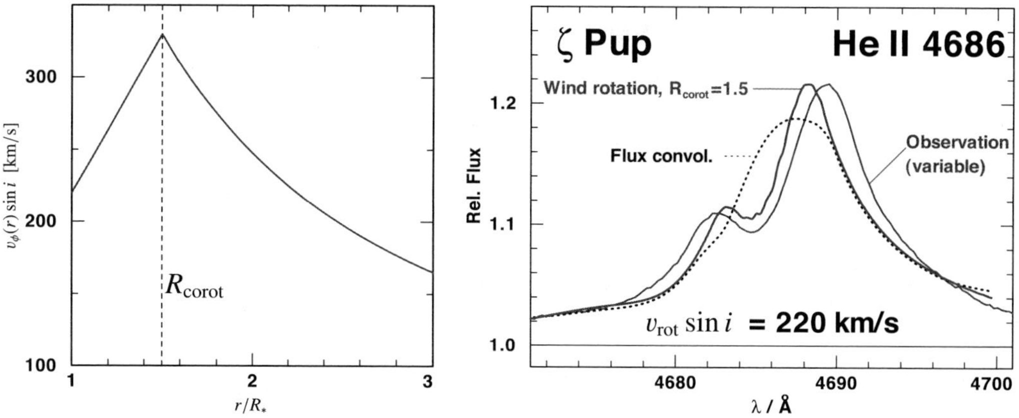

Figure 2. *Left panel:* Projected azimuthal velocity component in the equatorial plane as function of radial distance in units of the stellar radius R_*, assuming rigid-body rotation inside a co-rotation radius of $R_{\rm corot} = 1.5 R_*$ and angular momentum conservation outside. *Right panel:* Profile of the He II line at 4686 Å in the spectrum of ζ Puppis; the observation (blue line) is compared to the PoWR model simulation with wind rotation (red line). The black dotted line is the synthetic profile, but rotationally broadened by convolution of the emergent flux.

broadening can be easily simulated by convolving the emergent spectrum with a semi-ellipse, or a similar function when limb darkening is taken into account. The broadening function depends only $v_{\rm rot} \sin i$, with $v_{\rm rot}$ being the rotation velocity at the equator and i the inclination angle of the rotation axis against the line of sight.

In case of spectral lines formed in a stellar wind, this approximation does not hold. In order to specify how the atmosphere follows the stellar rotation, we assume that up to some co-rotation radius $R_{\rm corot}$ the atmosphere rotates like a solid body, while outside of $R_{\rm corot}$ the azimuthal velocity component v_ϕ drops conserving angular momentum (see Fig. 2, left panel). Meridional velocity components are neglected in PoWR (Shenar et al. 2014).

The remarkable impact of such wind-rotation is demonstrated in the right panel of Fig. 2 for the He II line at 4686 Å of the prototypical supergiant ζ Pup (spectral type O4I(n)fp). The observation, albeit highly variable, mostly exhibits a double-peak profile as in this particular observation shown here.

Convolution of the emergent model spectrum with rotational broadening totally fails to reproduce the line shape. The wind rotation simulation, however, qualitatively reproduces the observed double-peak profile, if some co-rotation of the wind is assumed. One can now vary the co-rotation radius and the velocity gradient (via the β-law exponent) in order to fit in detail the (variable) observed profile. In any case, the co-rotation of the inner wind provides indirect evidence for the existence of magnetic fields that are sufficiently strong to enforce co-rotation, but must not be globally organized for not violating the upper limit from Zeeman polarization studies.

3. Non-spherical winds

In case of very fast rotation that becomes comparable with critical (break-up) rotation, stars and winds deviate from spherical symmetry. For an approximate treatment, PoWR allows to combine two different models, e.g. one for the polar cones and one for the remaining, lower latitudes. Both models are first calculated separately, thus neglecting the mutual radiative interaction between the two domains, but combined in the formal integration of the intensities emerging towards the observer.

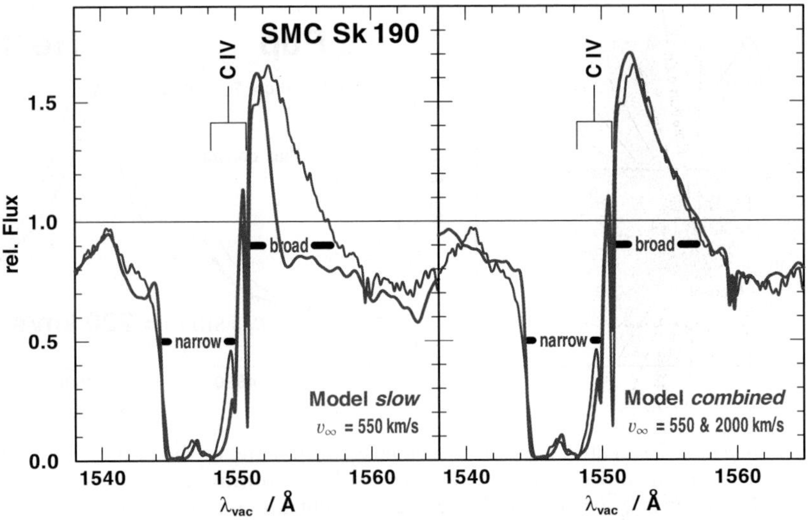

Figure 3. C IV resonance doublet in the spectrum of the SMC star Sk 190 (blue: observation). *Left panel:* The absorption feature of this P Cygni profile is narrow and fits with a PoWR model for a slow wind ($v_\infty = 550$ km/s); however, the width of the emission part of the observed profile is much broader. *Right panel:* The whole profile can be consistently reproduced by a model that combines a fast polar wind ($v_\infty = 2000$ km/s) with a slower outflow at low latitudes. The polar cone has a half opening angle of 60° and an effective temperature of 36 kK, while the model for the equatorial zone is somewhat cooler (30 kK) due to gravity darkening. A high inclination angle has been adopted ($i = 80°$).

As one example, such model is applied for simulating the C IV resonance doublet in the UV spectrum of Sk 190 in the SMC (spectral type O7.5(f)np). Photospheric lines indicate rapid rotation with $v \sin i = 300$ km/s. Figure 3 (left panel) shows the observation with HST-STIS (blue line) together with a PoWR model (red line) adopting a terminal wind velocity of $v_\infty = 400$ km/s, which is just adequate to fit the narrow absorption part of the P-Cygni profile. However, this model predicts a much narrower emission part than observed. The right panel shows the synthetic spectrum as obtained from combining a fast (~2000 km/s) polar wind with a slow equatorial wind, now perfectly matching the whole profile. The adopted geometry is sketched in the left panel of Fig. 4. Rotational broadening is treated with wind rotation as described above.

Another obvious application of such two-zonal model is the Galactic O-type star ζ Oph. This naked-eye object has been classified as spectral type O9.5V(e)nn, where the letter "e" reflects that the star has been observed at some epochs with double-peaked emission lines as evidence for a stellar disk (Ebbets 1981). Its rotation velocity is extremely high, $v \sin i = 400$ km/s.

When fitting the C IV resonance doublet in the UV spectrum, a notorious mismatch is encountered (e.g. Marcolino et al. 2009). The observed profile exhibits a significant extra absorption in the line center at about zero Doppler shift, which cannot be reproduced with any standard wind model (pink-colored area in Fig. 5, left panel). Rotational broadening is treated with wind rotation as described above.

But when adopting an equatorial region with basically zero radial expansion velocity, this disk-like structure partially obscures part of the photosphere as seen from the observer (see the sketch in the right panel of Fig. 4), and the observed profile is roughly reproduced (Fig. 5, right panel). The sharp "narrow absorption components" observed at a Doppler blue-shift of about $-v_\infty$ are not explainable with any smooth-wind models, as has been discussed in the literature.

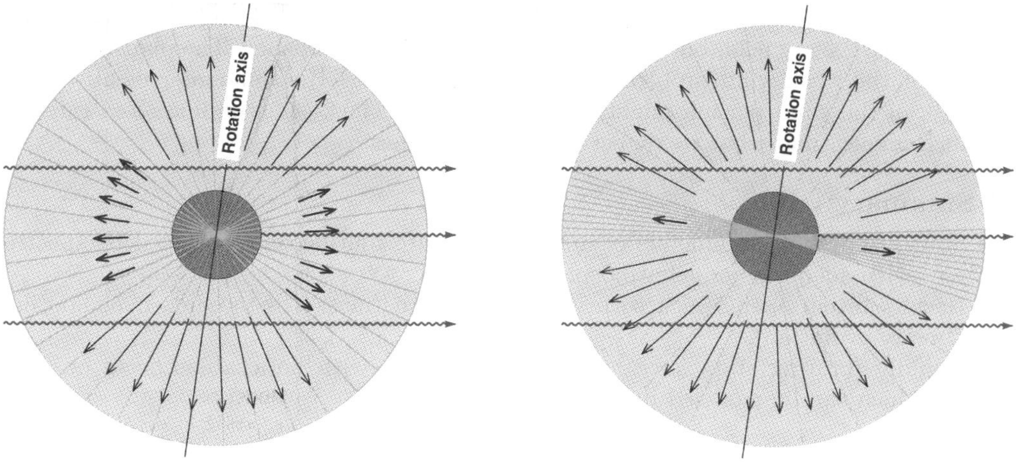

Figure 4. *Left panel:* Sketch of the geometry of the two-zone model for Sk 190. From the observer located to the right, nearly the whole stellar surface is seen through the slow wind of the equatorial zone. Therefore, the absorption part of the P-Cygni profile is narrow, according to a maximum blue-shift of $-v_\infty$. Those rays which are missing the stellar disk, however, collect emission from the fast polar wind, including matter in the back hemisphere that shows large red-shift leading to the broad emission part. *Right panel:* Sketch of the geometry of the two-zone model for ζ Oph. As seen from the observer located to the right, the disk-like region obscures roughly half of the stellar surface, causing an extra absorption at about zero Doppler shift.

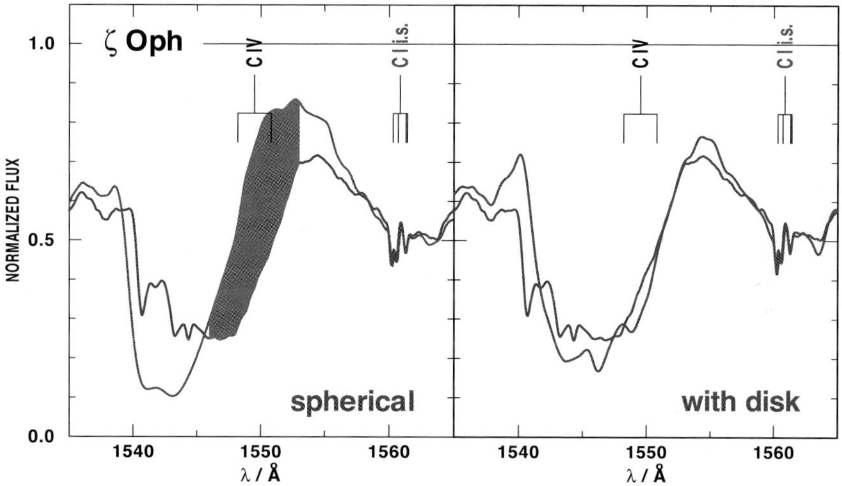

Figure 5. C IV resonance doublet in the UV spectrum of the bright Galactic O-type star ζ Oph. *Left panel:* HST-STIS observation (blue) compared to a standard (i.e. spherically symmetric) wind model ($T_* = 31.8$ kK, $v_\infty = 1600$ km/s). The observed absorption in the line center is not reproduced (pink area). *Right panel:* The central absorption is reproduced when the wind model from the left panel is combined with a disk-like equatorial region that shows no significant radial expansion. In this calculation, this disk is assumed to fill the latitudes within $\pm 5^\circ$ and to be seen edge-on ($i = 90^\circ$). From Henggeng Han (priv. comm.)

Apart from the excess of un-shifted absorption, ζ Oph is an example for the *weak wind problem.* The mass-loss rate, $\log \dot{M} = -9.0$ $[M_\odot/\mathrm{yr}]$ in the shown models, is by orders of magnitude lower than predicted theoretically for a radiation-driven wind from such a luminous star ($\log L/L_\odot = 4.86$) with solar metallicity – for instance, the recipe by Vink et al. (2000) yields a 50 times higher value for $\dot{M}$. In the calculation of the wind profiles we took already into account that the population of C IV is somewhat reduced by the (observed) X-ray field, but this does not resolve the discrepancy. The plausible explanation for the weak-wind problem has been suggested by Huenemoerder et al. (2012): the bulk of the wind plasma might be shock-heated. Thus, the true mass-loss rate can be much higher than what we detect via UV wind lines.

4. Summary

- Stellar and wind parameters of massive stars are routineously obtained by spectral analyses with adequate model atmospheres.
- Results reveal open issues with stellar evolution (mass-loss, mixing, overshooting, close-binary evolution).
- Rapidly rotating stars have non-spherical winds, with noticeable effects on their spectra which can be modeled in first approximation.

References

Ebbets, D. 1981, PASP, 93, 119
Hainich, R., Pasemann, D., Todt, H., et al. 2015, A&A, 581, A21
Hainich, R., Rühling, U., Todt, H., et al. 2014, A&A, 565, A27
Hamann, W. R., Koesterke, L., & Wessolowski, U. 1995, A&A, 299, 151
Hamann, W. R., Gräfener, G., Liermann, A., et al. 2019, A&A, 625, A57
Huenemoerder, D. P., Oskinova, L. M., Ignace, R., et al. 2012, ApJL, 756, L34
Marcolino, W. L. F., Bouret, J. C., Martins, F., et al. 2009, A&A, 498, 837
Ramachandran, V., Hamann, W. R., Oskinova, L. M., et al. 2019, A&A, 625, A104
Shenar, T., Hamann, W. R., & Todt, H. 2014, A&A, 562, A118
Vink, J. S., de Koter, A., & Lamers, H. J. G. L. M. 2000, A&A, 362, 295

Massive Stars Near and Far
Proceedings IAU Symposium No. 361, 2024
J. Mackey, J. S. Vink, & N. St-Louis, eds.
doi:10.1017/S1743921322002800

Next generation spectroscopic analysis for large samples of massive stars

Joachim M. Bestenlehner

Department of Physics & Astronomy, Hounsfield Road,
University of Sheffield, S3 7RH, UK
email: j.m.bestenlehner@sheffield.ac.uk

Abstract. Upcoming large-scale spectroscopic surveys such as WEAVE and 4MOST will provide thousands of spectra of massive stars, which need to be analysed in an efficient and homogeneous way. Studies on massive stars are usually based on samples of a few hundred objects which pushes current spectroscopic analysis tools to their limits because visual inspection is necessary to verify the spectroscopic fit.

The novel spectroscopic analysis pipeline takes advantage of the statistics that large samples provide, and determines the model error to account for imperfections in stellar atmosphere codes due to simplified, wrong or missing physics. Considering observational plus model uncertainties improve spectroscopic fits. The pipeline utilises the entire spectrum rather than selected diagnostic lines allowing a wider range of temperature from B to early O stars to be analysed. A small fraction of stars like peculiar, contaminated or spectroscopic binaries require visual inspection, which are identified through their larger uncertainties.

Keywords. methods: data analysis, stars: early-type, stars: massive, stars: fundamental parameters, stars: abundances

1. Introduction

Historically and still today, the most common way to analyse massive stars is by "eye", which limits the sample size to 10s of stars. This means that stellar parameters as well as uncertainties are estimated rather than determined. Sample of a couple of hundreds of stars are usually analysed with a χ^2-minimisation algorithm. Multi-dimensional probability distribution functions are obtained depending on the number of free parameters and uncertainties are then defined on confidence intervals rather than Gaussian standard deviations (Fig. 1). Those uncertainties can be highly asymmetric and very large in the case of degenerated parameters. In the massive star community there are 2 main flavours of χ^2-minimisation algorithms, grid based (e.g. Simón-Díaz et al. (2011); Castro et al. (2012); Bestenlehner et al. (2014)) and Generic Algorithm on the basis of natural selection (e.g. Mokiem et al. (2007); Brands et al. (2022)). All those pipelines use a pre-defined selection of spectral lines for their analysis.

However, the drawback of the χ^2-method is that parameter estimates can be strongly distorted by minimizing differences between synthetic spectra and real observations. This requires human verification of the spectroscopic fit, which is not feasible for large samples in excess of a couple of hundreds stars. In addition, synthetic spectra based on stellar atmosphere models are imperfect due to missing or simplified physics, insufficient atomic data, and so forth. Therefore, model uncertainties should also be budgeted into the parameter determination.

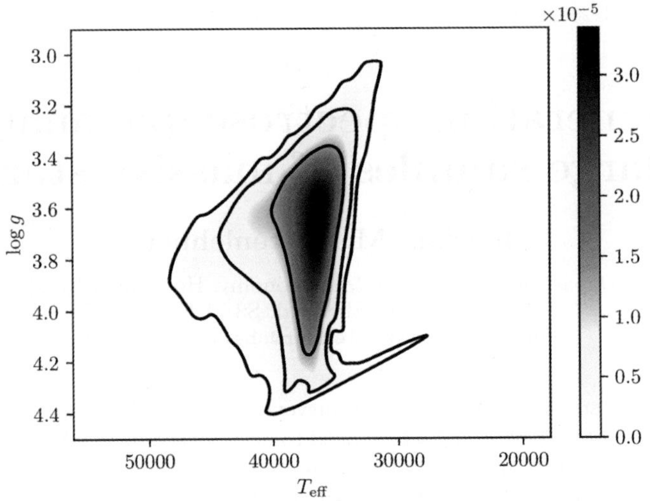

Figure 1. Probability heat map of surface gravity vs. effective temperature. Contours indicate confidence intervals of 68%, 95% and 99.7%.

2. Method

The pipeline presented here is a grid-based χ^2-minimisation algorithm, but uses information of the entire available spectral range instead of specific spectral lines. The observational error spectrum is included into the error matrix N and the χ^2 has the following form

$$\chi^2 = (\vec{d} - \mathrm{R}\vec{s})^{\mathrm{T}} \mathrm{N}^{-1} (\vec{d} - \mathrm{R}\vec{s}) \tag{1}$$

with data $\vec{d}$, observed spectra, response matrix R and idealised synthetic model spectrum $\vec{s}$. Synthetic spectra are de-idealised by asking the reverse question: for a given model spectrum $\vec{s}$, which data $\vec{d}$ can reproduce the model. This is implemented by substituting $\vec{s}$ with $\vec{t}^{[p]} + \vec{u}$. $\vec{t}^{[p]}$ is the idealised model spectrum defined by stellar parameters p and additional error term $\vec{u}$. The model error of a single star does not contain much information, but should be determined utilising the statistic a large sample of homogeneously observed stars provides.

To test and verify the methodology we used the VLT/MUSE observation of ~ 250 OB stars from Castro et al. (2018). The VLT/MUSE data cover the wavelength range between 4600 to 9300 Å at spectral resolution of 2000 to 4000. The spectrum normalisation is fully automated simulating the work-flow of the analysis of a large dataset.

The grid of synthetic spectra was computed with the non-LTE stellar atmosphere and radiative transfer code FASTWIND (Santolaya-Rey et al. (1997); Puls et al. (2005); Rivero González et al. (2011)). The VLT/MUSE data cover a less used wavelength range for hot star and we extended the line list to longer wavelength ($\sim$ 10 000 Å) beyond the tested range of FASTWIND (4000 to 7000 Å). Around 150 000 stellar atmosphere models were computed with varying stellar parameters and chemical composition of He and CNO. Of those models $\sim$20% did not converged properly and were removed. Still the number was too large to individually check all stellar models and we trust that the majority of those remaining are okay.

3. Results

The following parameters were derived: effective temperature T_{eff}, surface gravity $\log g$, helium composition Y and estimates on mass-loss rate $\dot{M}$ and CN chemical abundances.

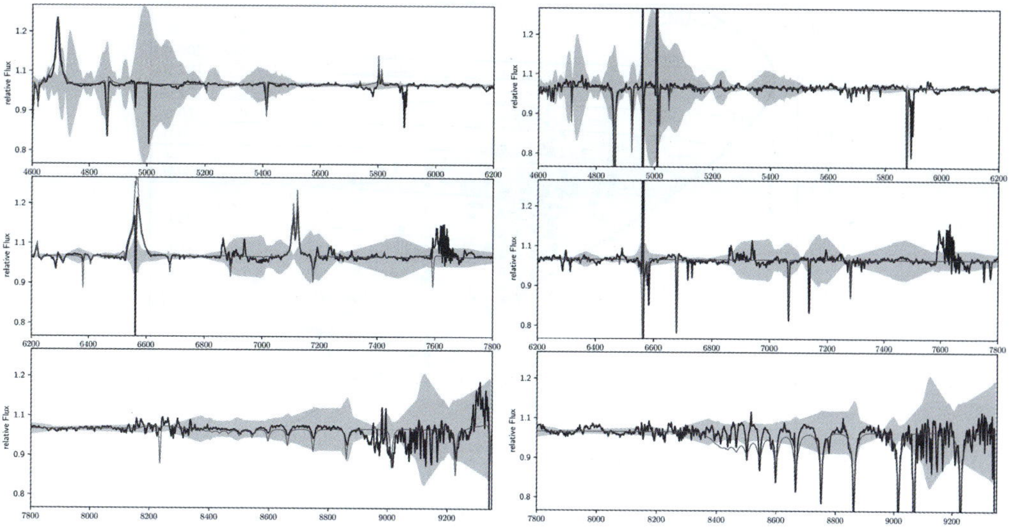

Figure 2. Spectroscopic fit of an early Of supergiant (left) and a early-mid B dwarf (right). Blue solid line is the observation, red solid line the synthetic spectrum and the grey shaded area is the square-root of the diagonal elements of the covariant-matrix calculated by the pipeline.

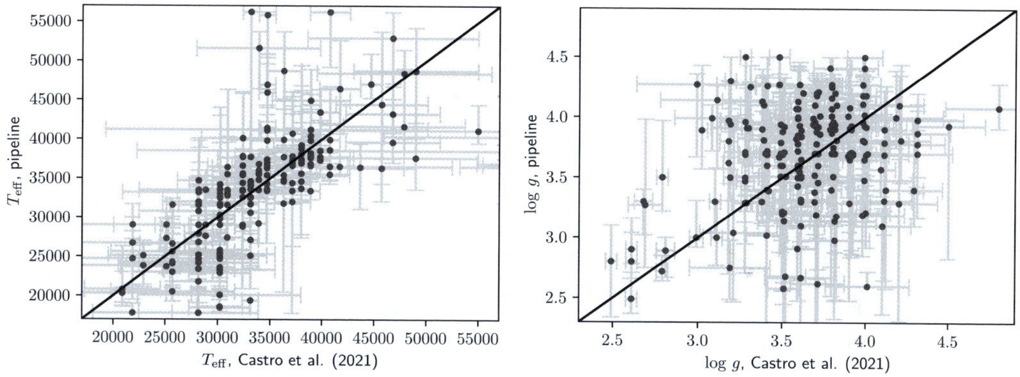

Figure 3. Effective temperatures (left) and surface gravities (right) determined by the pipeline vs. the results from Castro et al. (2021).

Figure 2 shows the spectroscopic fit of an Of supergiant and a B dwarf with $\Delta T_{\rm eff} \approx 25\,000$ K and $\Delta \log \dot{M} \gtrsim 2.5$ dex. This highlights that stars covering of large spectral type range can be successfully and reliably analysed with a single pipeline set up at the same time. However, issues occur for low signal to noise spectra (S/N $\lesssim 10$ to 15) and spectra with strong nebular lines.

In Fig. 3 we compare our results with the ones of Castro et al. (2021) which is based on the ionisation balance of selected HeI and HeII and the wings of H_β. In contrast, we used all H, He plus CNO, Si and Mg metal lines available in the VLT/MUSE wavelength range. In general there is a good agreement for the effective temperature, but the pipeline predicted in correct temperatures for spectra with strong nebular contaminations. The picture is inconclusive for the surface gravity. Even though we also utilised the Paschen lines, $\log g$ is basically unconstrained in both studies.

The Hertzprung-Russell diagram (HRD, Fig. 4) shows that as expected most stars are populated near and to the cool side of the zero-age main-sequence (ZAMS). There are a

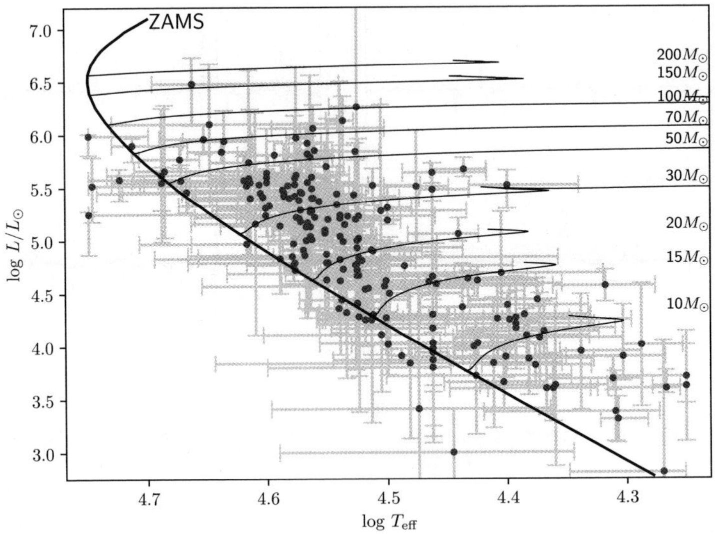

Figure 4. Hertzsprung-Russell diagram of the analysed stars using the VLT/MUSE data from Castro et al. (2018). Thin black lines are stellar evolutionary tracks by Brott et al. (2011) and Köhler et al. (2015).

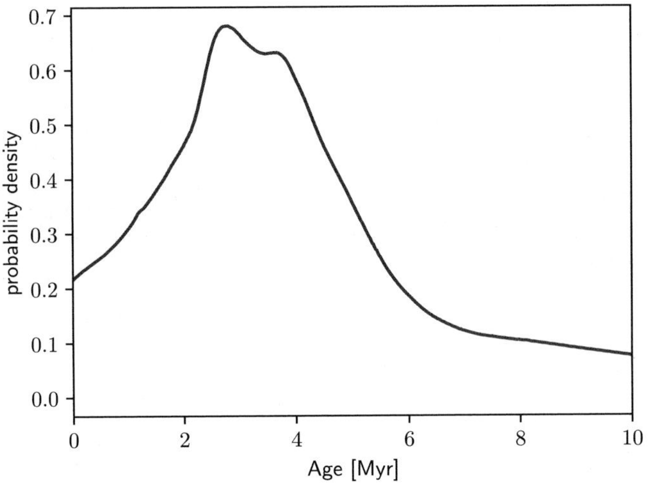

Figure 5. Probability density functions of ages.

couple of exceptions but their uncertainties do not exclude a cooler location in agreement with the majority of the sources. This can be improved by including a meaningful prior into the analysis, e.g. based on evolutionary tracks, and could increase the accuracy of the results, as we used only a flat prior. For example, only hydrogen deficient stars are found to be on the hot side of the ZAMS. A prior would give the star a higher probability to be found either on the hot or cool side of the ZAMS depending on its helium composition.

This can be done as part of the analysis or in the post-processing, e.g. with BONNSAI (Schneider et al. (2014)), when determining stellar masses and ages for our sample. Figure 5 shows the age probability density of our sample based on ages calculated with BONNSAI on the bases of the stellar evolutionary tracks of Brott et al. (2011) and Köhler et al. (2015). The probability density is comparable to the findings of

Schneider et al. (2018), but features are smoother and less prominent as a result of the overall larger uncertainties. However, taking into account that our analysis was completed in less then 2 weeks while the study of Schneider et al. (2018) was based on 4 PhD theses with a combined effort of more than 10 years.

4. Conclusion

With advent of large spectroscopic surveys such as WEAVE† and 4MOST‡ 10 000s of massive stars are going to be observed and need to be analysed. We will require fully automated spectroscopic analysis tools which reduce the human interaction to a minimum to cope with the amount of data. The pipeline presented here analyses $\sim$ 250 stars in less than half a day. Overall the quality of the spectroscopic fits is good, but around 15% of the stars need additional attention as a result of strong nebular contamination, low S/N, multiplicity et cetera. Still this will provide more time for doing science rather than data analysis.

Future developments will replace the χ^2-minimiser with machine learning approach including observational and model errors. After training the neural network the analysis process should be faster, less resource hungry and increasing the accuracy.

Acknowledgements

JMB is supported by the Science and Technology Facilities Council research grant ST/V000853/1 (PI. V. Dhillon).

References

Bestenlehner, J. M., Gräfener, G., Vink, J. S., Najarro, F., de Koter, A., Sana, H., Evans, C. J., Crowther, P. A., Hénault-Brunet, V., Herrero, A., Langer, N., Schneider, F. R. N., Simón-Díaz, S., Taylor, W. D., Walborn, N. R. 2014, *A&A*, 570, A38

Brands, S. A., de Koter, A., Bestenlehner, J. M., Crowther, P. A., Sundqvist, J. O., Puls, J., Caballero-Nieves, S. M., Abdul-Masih, M., Driessen, F. A., García, M., Geen, S., Gräfener, G., Hawcroft, C., Kaper, L., Keszthelyi, Z., Langer, N., Sana, H., Schneider, F. R. N., Shenar, T., Vink, J. S. 2022, *A&A accepted*, arXiv:2202.11080

Brott, I., de Mink, S. E., Cantiello, M., Langer, N., de Koter, A., Evans, C. J., Hunter, I., Trundle, C., Vink, J. S. 2011, *A&A*, 530, A115

Castro, N., Urbaneja, M. A., Herrero, A., Garcia, M., Simón-Díaz, S., Bresolin, F., Pietrzyński, G., Kudritzki, R. -P., Gieren, W. 2012, *A&A*, 542, A79

Castro, N., Crowther, P. A., Evans, C. J., Mackey, J., Castro-Rodriguez, N., Vink, J. S., Melnick, J., Selman, F. 2018, *A&A*, 614, A147

Castro, N., Crowther, P. A., Evans, C. J., Vink, J. S., Puls, J., Herrero, A., Garcia, M., Selman, F. J., Roth, M. M., Simón-Díaz, S. 2021, *A&A*, 648, A65

Köhler, K., Langer, N., de Koter, A., de Mink, S. E., Crowther, P. A., Evans, C. J., Gräfener, G., Sana, H., Sanyal, D., Schneider, F. R. N., Vink, J. S. 2015, *A&A*, 573, A71

Mokiem, M. R., de Koter, A., Puls, J., Herrero, A., Najarro, F., Villamariz, M. R. 2005, *A&A*, 441, 711

Puls, J., Urbaneja, M. A., Venero, R., Repolust, T., Springmann, U., Jokuthy, A., Mokiem, M. R. 2005, *A&A*, 435, 669

Rivero González, J. G., Puls, J., Najarro, F. 2011, *A&A*, 536, A58

Santolaya-Rey, A. E., Puls, J., Herrero, A. 1997, *A&A*, 323, 488

Schneider, F. R. N., Izzard, R. G., de Mink, S. E., Langer, N., Stolte, A., de Koter, A., Gvaramadze, V. V., Hußmann, B., Liermann, A., Sana, H. 2014, *ApJ*, 780, 117

† `https://www.ing.iac.es/weave`
‡ `https://www.4most.eu`

Schneider, F. R. N., Ramírez-Agudelo, O. H., Tramper, F., Bestenlehner, J. M., Castro, N., Sana, H., Evans, C. J., Sabín-Sanjulián, C., Simón-Díaz, S., Langer, N., Fossati, L., Gräfener, G., Crowther, P. A., de Mink, S. E., de Koter, A., Gieles, M., Herrero, A., Izzard, R. G., Kalari, V., Klessen, R. S., Lennon, D. J., Mahy, L., Maíz Apellániz, J., Markova, N., van Loon, J. Th., Vink, J. S., Walborn, N. R. 2018, *A&A*, 618, A73

Simón-Díaz, S., Castro, N., Herrero, A., Puls, J., Garcia, M., Sabín-Sanjulián, C. 2011, *JPhCS*, 328, 012021

Discussion

FABRY: As you mentioned the $\log g$ uncertainty is driven by normalisation issues (Paschen lines) or uncertainties on the normalisation. Do you have suggestions, how to improve on this or will machine learning solve this?

BESTENLEHNER: You can use machine learning to normalise your spectra, but if your analysis pipeline uses machine learning the normalisation is less of an issue as preliminary tests have shown. Here we use a simple normalisation routine to simulate the analysis work-flow of 4MOST. However, if spectra are flux calibrated and the temperature is reasonably well known, they can be normalised by dividing the observation by the theoretical model energy distribution (SED). In this way the Paschen jump is better normalised out and the accuracy of $\log g$ improves.

NAJARRO: Large errors on logg comes from high Paschen series which are not well fitted. In comparison with VFTS, where the $\log g$ are based on low order Balmer series, $\log g$ on the Paschen series is unconstrained.

BESTENLEHNER: The Paschen lines contain a lot of information on $\log g$, but the line profiles overlap and become not only very sensitive to the normalisation but also to the line broadening parameters (macro-turbulent and projected rotational velocities). In addition, the Paschen line profiles are less verified in stellar atmosphere codes than the Balmer series. The machine learning approach might improve on this in the future, but I would like to understand first the issues and how well my current approach works before using a new approach where I want to be able to follow step by step the analysis process like using a black-box.

Massive Stars Near and Far
Proceedings IAU Symposium No. 361, 2024
J. Mackey, J. S. Vink, & N. St-Louis, eds.
doi:10.1017/S1743921322002186

MUSE 3D spectroscopy of BA-type supergiants in NGC 300

Gemma González-Torà[1,2,3], Miguel A. Urbaneja[3], Norbert Przybilla[3], Stefan Dreizler[4], Martin M. Roth[5], Sebastian Kamann[2] and Norberto Castro[5]

[1]European Southern Observatory (ESO), Karl-Schwarzschild-Str. 2, 85748 Garching bei München, Germany
email: Gemma.GonzaleziTora@eso.org

[2]Astrophysics Research Institute, LJMU, 146 Brownlow Hill, Liverpool L3 5RF, UK

[3]Institut für Astro- und Teilchenphysik, Universität Innsbruck, Technikerstr. 25/8, 6020 Innsbruck, Austria

[4]Institute for Astrophysics, University of Göttingen, Friedrich-Hund-Platz 1, 37077 Göttingen, Germany

[5]Leibniz-Institut für Astrophysik (AIP), An der Sternwarte 16, 14482 Postdam, Germany

Abstract. We present the results obtained using spectroscopic data taken with the intermediate-resolution Multi Unit Spectroscopic Explorer (MUSE) of B and A-type supergiants and bright giants in the Sculptor Group galaxy NGC 300. For our analysis, a hybrid local thermodynamic equilibrium (LTE) line-blanketing+non-LTE method was used to improve the previously published results for the same data. In addition, we present some further applications of this work, which includes extending the flux-weighted gravity luminosity relationship (FGLR), a distance determination method for supergiants. This pioneering work opens up a new window to explore this relation, and also demonstrates the enormous potential of integral field spectroscopy (IFS) for extragalactic quantitative stellar studies.

Keywords. Galaxies: individual (NGC 300) – Galaxies: distances and redshifts – Stars: early-type – Stars: fundamental parameters – supergiants

1. Introduction

Resolving individual stars beyond the Local Group has been a turning point for astronomy (e.g., Hubble 1929; Baade 1944). The best candidates for such observations are massive BA-type supergiants, as they are the brightest objects in optical light, reaching absolute visual magnitudes of $M_V \simeq -9.5$ (e.g. Humphreys & Davidson 1979; Humphreys & Aaronson 1987). The spiral galaxy NGC 300 is located at the Sculptor Group, close to the Galactic southern pole. These galaxies are the least affected by Galactic extinction and therefore convenient targets to study. Moreover, NGC 300 is oriented face-on, being a great candidate for quantitative multi-object slit spectroscopy (e.g., Bresolin et al. 2002, 2004; Urbaneja et al. 2005). An extremely powerful tool for this purpose is the so-called integral field spectroscopy (IFS). This technique allows for a spectrum for each pixel across an image to be obtained simultaneously. The Multi Unit Spectroscopic Explorer (MUSE, Bacon et al. 2014) on the European Southern Observatory Very Large Telescope (ESO VLT) is groundbreaking in this context, as it combines a wide field of view with high spatial sampling. Spectroscopic analysis of the

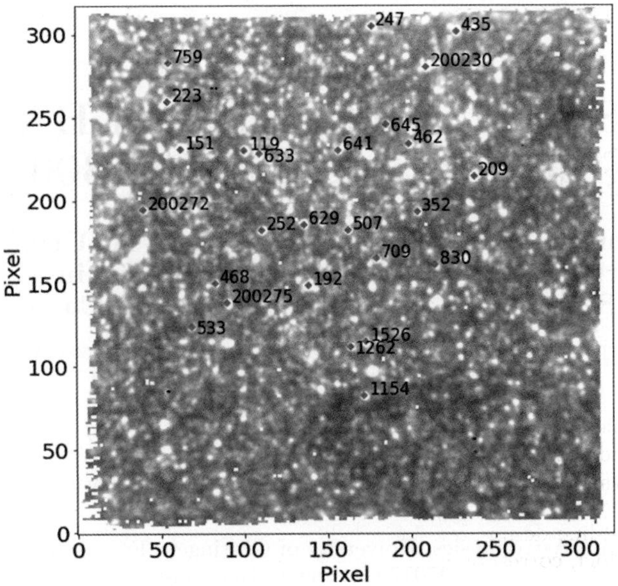

Figure 1. Chart of NGC 300 field (i) in spaxel coordinates with the programme stars marked in red and their ID numbers marked in black (according to Paper I).

brightest objects in nearby galaxies, i.e. supergiant stars, can be performed to determine their stellar parameters.

The present study based on González-Torà et al. (2022) provides a detailed analysis of MUSE spectra of BA-type supergiants and bright giants in NGC 300 in the field seen in Figure 1 from (Roth et al. 2018, henceforth Paper I). The analysis is based on synthetic spectra accounting for deviations from local thermodynamic equilibrium (LTE).

2. Methods

The spectroscopic data were obtained using MUSE (Bacon et al. 2014), in the wide field mode (WFM) with $1'\times1'$ spatial coverage and $0.2''$ sampling. The pointing observed under the best seeing conditions (FWHM $= 0.47''$-$0.59''$, measured from the data) was investigated.

The initial reduction was achieved with the MUSE pipeline V1.0 (see Paper I for details Weilbacher et al. 2020). The final data were produced in the form of a datacube, and the spectra of 606 individual sources were extracted using the PampelMUSE software (Kamann et al. 2013).

Out of these 606 extracted sources, 26 were classified as late-B to early-A supergiants or bright giants in Paper I with a $7 < \mathrm{S/N} \lesssim 20$, the minimum was demanded for our analysis to provide valid results. They are identified in Fig. 1. Due to model restrictions, we ended up with 16 objects that fulfilled the criteria for a quantitative analysis (see González-Torà et al. 2022, for more details).

We consider a grid of model atmospheres covering the parameter space defined by the effective temperature $T_{\rm eff}$ and surface gravity $\log g$. We adopt the modelling methodology by Przybilla et al. (2006) for the analysis of our final sample of 16 stars. Very briefly, the method employs a combination of model atmosphere structures calculated under the assumption of LTE+line-blanketing and a detailed non-LTE level population as well as line-formation calculations. The reader is referred to Przybilla et al. (2006) for the advantages and drawbacks of this hybrid approach, as well as its limitations.

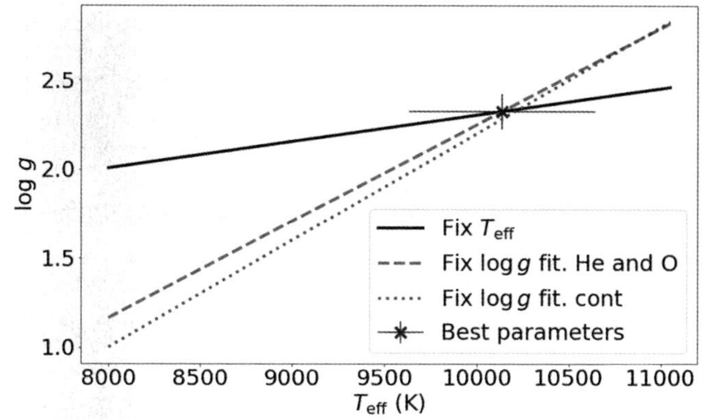

Figure 2. Example of the method used to determine the temperature and surface gravity for star #151. The solid black line shows the locus obtained by varying $\log g$ (in cgs units) for each fixed model-grid $T_{\rm eff}$ fitting the hydrogen lines. As red-dashed and green-dotted lines, we show the loci obtained by varying $T_{\rm eff}$ at each fixed model-grid $\log g$ best fitting the He I and O I lines, and the metal-line dominated 4950-5600 Å region, respectively. The blue cross marks the intersection of the loci, corresponding with the adopted atmospheric parameter values.

We followed a well-established methodology to find the best solution for each object. First, by adopting $T_{\rm eff}$, we found the model that best reproduces the features sensitive to gravity changes (the hydrogen Balmer lines); this step was repeated for different adopted $T_{\rm eff}$ values, hence allowing us to define the locus of models for which the Balmer lines are equally well represented. This is referred as the $\log g$ locus. Similarly, but adopting the surface gravity, we identified the model that best reproduces the $T_{\rm eff}$ sensitive features (in our case either the metal lines He I and O I lines, or the metal-line dominated 4950-5600 Å region). We repeated this step for different values of $\log g$, defining the $T_{\rm eff}$ locus. Finally, the intersection of both lines represents the best possible solution for a given object in the $T_{\rm eff}$–$\log g$ plane (see Fig. 2).

A catalogue of Johnson B- and V-band magnitudes for sources in NGC 300, based on the work by Pietrzyński et al. (2001), was kindly provided by F. Bresolin (private communication). It allowed to calculate bolometric magnitudes $M_{\rm bol}$ from the extinction-corrected apparent V-band magnitudes, the distance to NGC 300 of $d = 1.86 \pm 0.07$ Mpc as determined by Rizzi et al. (2006)., and the bolometric corrections ($B.C.$) for each object individually, calculated from tailored models.

3. The flux-weighted gravity luminosity relationship (FGLR)

The FGLR was first derived by Kudritzki et al. (2003) as a new method for distance determination of supergiants:

$$-M_{\rm bol} = a(\log g_{\rm F} - 1.5) + b \tag{1}$$

with a determined by the mass-luminosity relation ($L \propto M^x$) exponent x to $a = 2.5x/(1-x)$, and $g_{\rm F}$ the so-called flux-weighted gravity, defined as $g_{\rm F} = g \times \left(T_{\rm eff}/10^4\right)^{-4}$.

This relation holds for all supergiants and bright giants that have a constant luminosity track when they move to the right of the Hertzsprung-Russell diagram (HRD). Therefore, this relation can be used to estimate the bolometric magnitudes, luminosities, and distances of supergiants and bright giants for which only spectral information is available. If we are able to resolve massive stars in distant galaxies, this can be a very powerful tool to determine extragalactic distances.

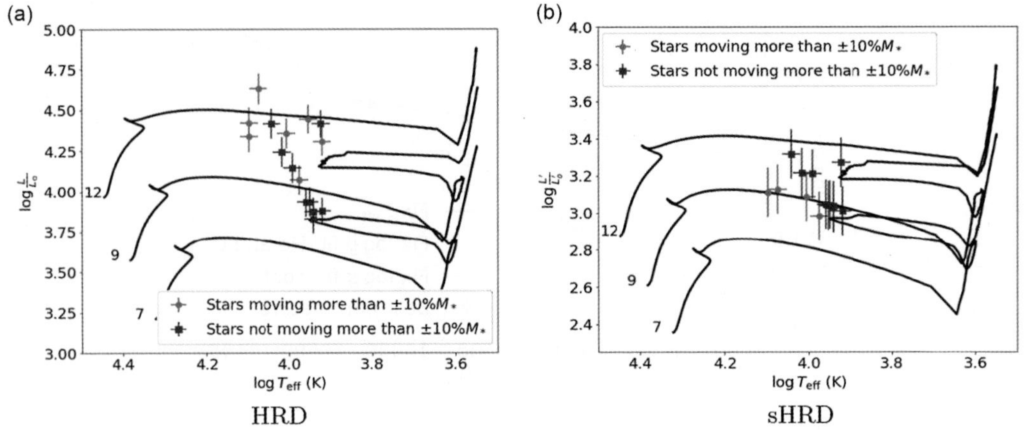

Figure 3. a) HRD with evolutionary tracks for different masses (black solid lines, in $M_\odot$) accounting for rotation (Ekström et al. 2012). The red dots represent sample stars that move by more than 10% their evolutionary mass track position with respect to the sHRD in Fig. 3b, and the blue squares are those that move less than 10%. b) Same as Fig. 3a, but for the sHRD. The L' is defined as the inverse of the flux-weighted gravity.

To study the FGLR, the stars need to be at the correct evolutionary stage. To verify that, we plotted the stars in relation to evolutionary tracks from Ekström et al. (2012) in the regular HRD (Fig. 3a) and compared their position with respect to the same tracks in the sHRD (Fig. 3b). The sHRD (Langer & Kudritzki 2014) shows L' which is the inverse of $g_{\rm F}$, with respect to $T_{\rm eff}$. Using the sHRD, we can place the stars with only their spectroscopic information and without any knowledge of their distance or brightness. The advantage of the sHRD is that the stars fall into different iso-gravity lines, enabling to discriminate stars with different radii as well as multiple sytems.

The red dots in Figs. 3a and 3b represent stars that move more than the 10% threshold (corresponding to their mass error), which indicates that they are not well-behaved objects. These stars show significantly larger spectroscopic than evolutionary masses that are derived from comparison with evolutionary tracks. The blue squares in Figs. 3a and 3b do not move by more than the threshold, and we can assume a good correspondence between their spectral information and their true evolutionary stage and their single star status. The latter stars are certain to be in the supergiant stage and therefore the FGLR would hold.

To prove our last point, we considered our 16 stars along with the objects previously studied by Kudritzki et al. (2008) to derive the FLGR. As we can see in Fig. 4, the blue squares follow the old FGLR within their error limits. The initial FGLR gives the old parameters determined by Kudritzki et al. (2008): $a_{\rm old} = -3.52$ and $b_{\rm old} = 8.11$. Adding the contribution of our newly found supergiants (blue squares in Fig. 4), we obtained $a = -3.40 \pm 0.04$ and $b = 8.02 \pm 0.14$. The results from this work are also in accordance with the new FGLR distance to NGC 300 $(m - M)_{\rm FGLR} = 26.34 \pm 0.06$ by Sextl et al. (2021).

3.1. *Discrepant cases*

As seen in Fig. 4, stars depicted by blue squares follow – within the uncertainties – the trend defined by the FGLR, while the ones represented by red dots deviate to some extent. For the stars depicted by red circles, either the derived luminosity is too high, or the flux-weighted gravity is too large (or both). The main reason for the luminosity (bolometric magnitude) to be too high would be for the apparent magnitude to be too

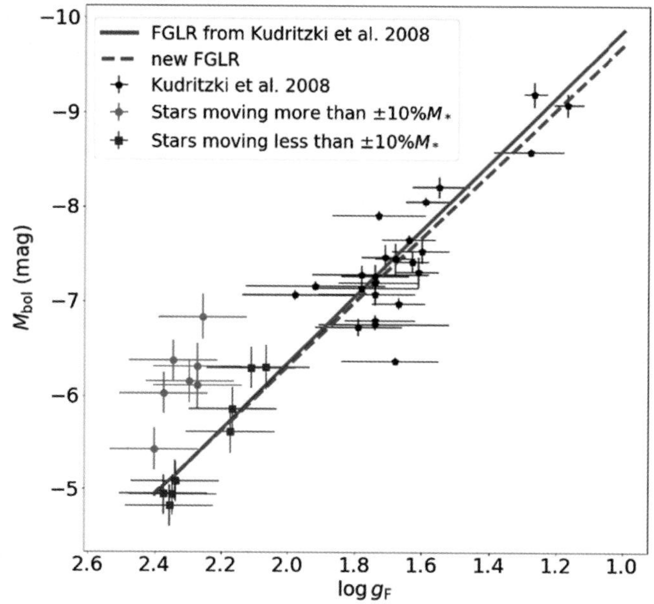

Figure 4. FGLR for the stars in NGC 300. The stars studied by Kudritzki et al. (2008) are shown in black pentagons, with the corresponding FGLR regression line marked in solid green. Blue squares and red circles mark the stars analysed in this work, with the magenta-dashed line representing the regression line derived from the black and blue symbols.

high. This could mean that what is seen as a single star is a combination of several unresolved sources instead. Inspecting the MUSE datacube (see Fig. 1), none of these objects show signs of being an extended source, which makes it unlikely that they are large star clusters. On the other hand, it cannot be ruled out that they are small stellar aggregates that are not resolved at the distance of NGC 300.

An alternative explanation for this deviation could be the following: as we increase the $\log g_{\rm F}$ and move to the bottom left of the FGLR, the stars decrease in mass. Population simulations predict that the FGLR will get wider for lower masses, as discussed by Meynet et al. (2015). Because of the so-called initial mass function (IMF) effect, we always expect to find a higher number of low mass stars than of massive stars (e.g. Salpeter 1955; Kroupa 2001), widening the FGLR because of the increased scatter.

4. Conclusions

We performed a quantitative spectroscopic analysis of 16 BA-type supergiants and bright giants in NGC 300, based on VLT/MUSE IFS data. Our focus lied on determining basic atmospheric and fundamental stellar parameters. This allowed us to extend the FGLR towards less luminous stars than studied before. However, the study has faced limitations by the relatively low $S/N \lesssim 20$ of the spectra. For future work, the S/N and spatial resolution should be improved by taking advantage of the adaptive optics mode of MUSE and longer exposure times. This would not only reduce the uncertainties for similar studies as this work, but would also help to determine metallicities and likely elemental abundances for selected individual chemical elements. Therefore, demonstrating the full potential of MUSE for extragalactic stellar astrophysics. In addition, BlueMUSE is the new proposed medium-resolution IFS instrument at the VLT. Optimized for the optical blue, BlueMUSE will be the perfect instrument to study hot massive stars as most of their spectral features are located in its spectral range - a Highlight Science Case outlined in the BlueMUSE White Paper (Richard et al. 2019).

References

Baade, W. 1944, ApJ, 100, 137
Bacon, R., et al. 2014, The Messenger, 157, 13
Bresolin, F., Gieren, W., Kudritzki, R.-P., Pietrzyński, G., & Przybilla, N. 2002, ApJ, 567, 277
Bresolin, F., Pietrzyński, G., Gieren, W., Kudritzki, R.-P., Przybilla, N., & Fouqué, P. 2004, ApJ, 600, 182
Ekström, S., et al. 2012, A&A, 537, A146
González-Torà, G., Urbaneja, M. A., Przybilla, N., Dreizler, S., Roth, M. M., Kamann, S., & Castro, N. 2022, A&A, 658, A117
Hubble, E. P. 1929, ApJ, 69, 103
Humphreys, R. M., & Aaronson, M. 1987, AJ, 94, 1156
Humphreys, R. M., & Davidson, K. 1979, ApJ, 232, 409
Kamann, S., Wisotzki, L., & Roth, M. M. 2013, A&A, 549, A71
Kroupa, P. 2001, MNRAS, 322, 231
Kudritzki, R. P., Bresolin, F., & Przybilla, N. 2003, ApJL, 582, L83
Kudritzki, R. P., Urbaneja, M. A., Bresolin, F., Przybilla, N., Gieren, W., & Pietrzyński, G. 2008, ApJ, 681, 269
Langer, N., & Kudritzki, R. P. 2014, A&A, 564, A52
Meynet, G., Kudritzki, R. P., & Georgy, C. 2015, A&A, 581, A36
Pietrzyński, G., Gieren, W., Fouqué, P., & Pont, F. 2001, A&A, 371, 497
Przybilla, N., Butler, K., Becker, S. R., & Kudritzki, R. P. 2006, A&A, 445, 1099
Richard, J., et al. 2019, arXiv e-prints, arXiv:1906.01657
Rizzi, L., Bresolin, F., Kudritzki, R. P., Gieren, W., & Pietrzyński, G. 2006, ApJ, 638, 766
Roth, M. M., et al. 2018, A&A, 618, A3 (Paper I)
Salpeter, E. E. 1955, ApJ, 121, 161
Sextl, E., Kudritzki, R.-P., Weller, J., Urbaneja, M. A., & Weiss, A. 2021, ApJ, 914, 94
Urbaneja, M. A., et al. 2005, ApJ, 622, 862
Weilbacher, P. M., et al. 2020, A&A, 641, A28

Discussion

QUESTION: How do you know that the stars are in the right evolutionary stage?

GONZÁLEZ-TORÀ: We plot all the targets both in the HRD and the sHRD. The blue dots in Figures 3a, 3b are the stars in the supergiant phase, since their photometric and spectroscopic information coincide *(see more in Section 3, 3rd and 4th paragraphs)*.

Massive Stars Near and Far
Proceedings IAU Symposium No. 361, 2024
J. Mackey, J. S. Vink, & N. St-Louis, eds.
doi:10.1017/S1743921322002423

SPAMMS: applications and use cases for the 3D spectroscopic analysis technique to study deformed massive stars

Michael Abdul-Masih

European Southern Observatory, Alonso de Cordova 3107, Vitacura,
Casilla 19001, Santiago de Chile, Chile
email: Michael.Abdul-Masih@eso.org

Abstract. Whether it be due to rapid rotation or binary interactions, deviations from spherical symmetry are common in massive stars. These deviations from spherical symmetry are known to cause non-uniform distributions of various parameters across the surface including temperature, which can drive internal mixing processes within the envelopes of these massive stars. Despite how common these 3D distortions are, they are often neglected in spectroscopic analyses. We present a new spectral analysis code called SPAMMS (Spectroscopic PAtch Model for Massive Stars) specifically designed to analyze non-spherical systems. We discuss how the code works and discuss its assumptions. Furthermore, we demonstrate how SPAMMS can be applied to a variety of different types of systems and we show how it can model 3D effects in a way that current analysis techniques are not able to.

Keywords. techniques: spectroscopic, stars: rotation, stars: fundamental parameters, stars: abundances,, binaries: close

1. Introduction

Deviations from spherical symmetry are common throughout the lifetime of massive stars (Langer 2012). Nearly a quarter of all massive stars will merge with a companion during their lifetimes, and this coalescence is preceded by a highly non-spherical contact phase (Pols 1994; Sana et al. 2012). The internal processes occurring during the contact phase can have large effects on the eventual end-products of these systems (e.g., Shao & Li 2014; de Mink & Mandel 2016; Mandel & de Mink 2016; Marchant et al. 2016; Schneider et al. 2019). Even in massive single stars, deviations from spherical symmetry are common due to rotational distortion (Ramírez-Agudelo *et al.* 2015; Bodensteiner et al. 2020). The three-dimensional geometry of these non-spherical systems is, however, typically not taken into account in conventional tools available for the spectroscopic analysis of massive stars.

Due to the temperature regime where massive stars are found, NLTE effects become very important to accurately model the spectral lines. Thus, radiative transfer codes suitable for massive stars are typically computed in 1D because the NLTE calculations are computationally expensive (e.g., Hillier & Miller 1998; Gräfener et al. 2002; Puls et al. 2005). This imposes the assumption of spherical symmetry on the system, and ignores any effects arising from the surface variations resulting from the 3D geometry. Here, we present a new spectral analysis code called /textscspamms, which is specifically designed to analyze these types of distorted systems, and we demonstrate its ability to handle various geometries. We then apply it to a sample of massive overcontact binaries and compare its performance to traditional spectroscopic binary analysis techniques. Additionally, we

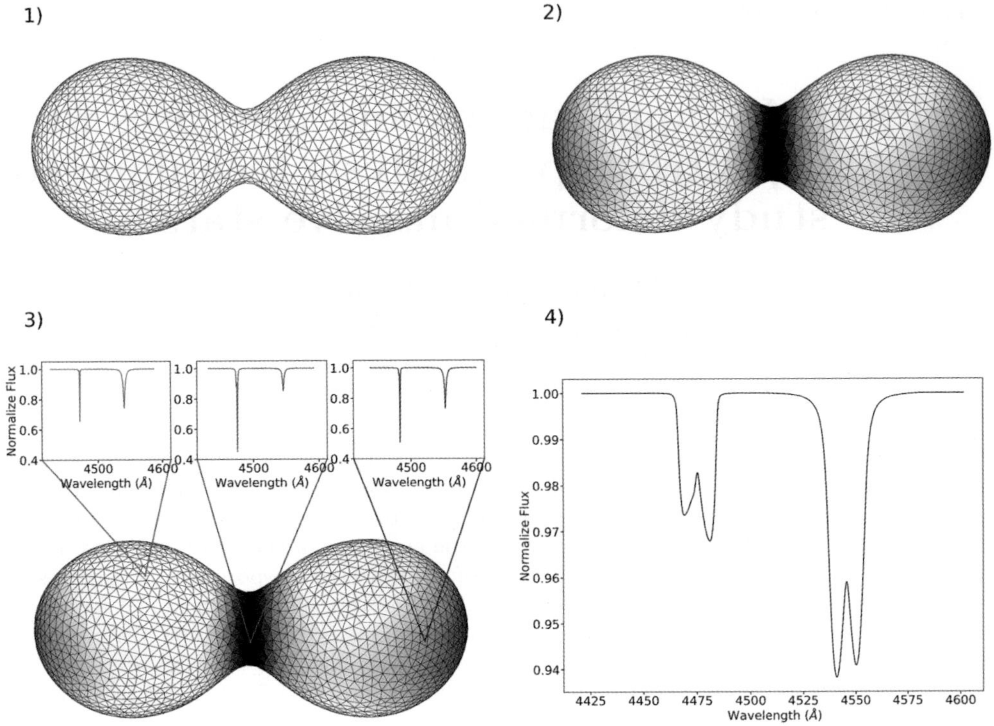

Figure 1. Graphical representation of the four step SPAMMS model computation as described in Sec. 2. 1) represents the generation of the mesh. 2) represents the population of the mesh with local parameters. Here the surface temperature distribution is shown as the face colors of each patch triangle. 3) represents the assigning of FASTWIND models to each patch. 4) shows the final integrated line profile for the given system at the given phase. This figure originally appeared in Abdul-Masih et al. (2020)

demonstrate how the code can be used to investigate the degree to which the 3D geometry affects the observed temperature in rapidly rotating systems. Finally, we show how the code can model various 3D effects that occur in spherical binary systems as well.

2. SPAMMS

In order to properly model distorted massive stars, we require a code that can simultaneously model the 3D geometry (and resulting surface parameter distributions) and model the NLTE effects in the atmosphere. For this purpose, we present the Spectroscopic PAtch Model for Massive Stars (SPAMMS; Abdul-Masih et al. 2020). This code combines the 3D modelling capabilities of PHOEBE (Prša et al. 2016) with the NLTE radiative transfer code FASTWIND (Puls et al. 2005) to create a spectroscopic patch model across the surface of a three dimensional system.

SPAMMS works in 4 steps. First, given the physical stellar or binary parameters of the system (radius, mass, inclination, rotation rate, period, phase, etc.), a 3D triangulated mesh is generated accounting for Roche or rotational distortions. Next, the mesh is populated with local parameters such as temperature, surface gravity, radius and radial velocity. Third, based on the local parameters, FASTWIND synthetic line profiles are assigned to each patch across the surface. Finally, the visible surface is integrated over, accounting for surface area projection effects, emergent angles and the intensity contributions from each patch. This results in a final synthetic line profile that accounts for the geometry and the surface effects of the system. A graphical representation of these steps can be found in Fig. 1.

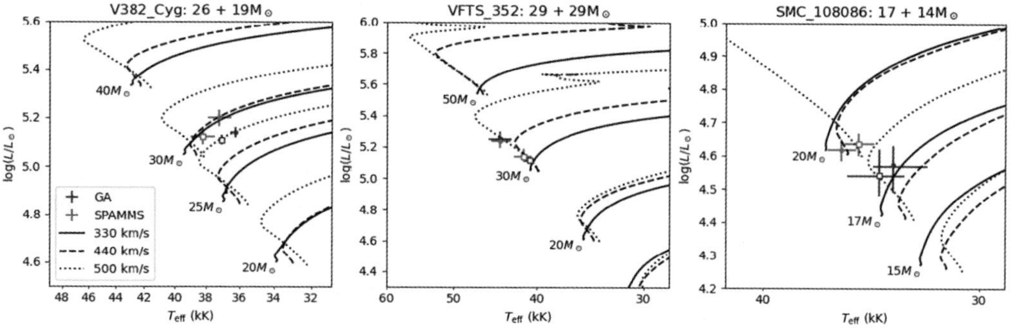

Figure 2. HRDs for the three overcontact systems that we analyzed. The positions of the primary and secondary are indicated with closed and open points respectively. The SPAMMS fit is indicated in red while the fit from the 1D techniques are indicated in blue. For reference, rotating evolutionary tracks for various masses from (Brott et al. 2011) are plotted. This figure is adapted from Abdul-Masih et al. (2021)

3. Applications

3.1. *Overcontact binaries*

The overcontact phase is expected to occur in about a quarter of all massive stars (Sana et al. 2012) and is characterized by a highly distorted geometry (see e.g., Fabry et al. 2022). This evolutionary phase is of particular importance in the context of mergers and gravitational wave progenitors, so an accurate determination of surface parameters can be vitally important in understanding the future evolution of these systems. Traditional fitting techniques involve first disentangling the observed spectra and then fitting each component with 1D models (e.g., Abdul-Masih et al. 2019). This imposes the spherical assumption several times throughout the process and can bias the derived stellar parameters.

To investigate how big of a difference accounting for the 3D geometry has on the derived solution, we modeled a sample of three massive overcontact systems using both SPAMMS and traditional fitting techniques. Since the three overcontact systems had been characterized photometrically in the past (Hilditch et al. 2005; Almeida et al. 2015; Martins et al. 2017), we use the binary solution from these works as the input for the SPAMMS model to constrain the geometry. As shown in Abdul-Masih et al. (2021), we are able to reproduce the observed spectra to a higher accuracy with significantly fewer free parameters using SPAMMS than with the traditional fitting techniques (8 and 22 free parameters respectively).

In addition to better morphological agreements between the observed spectra and the models, we also found a difference in the derived parameters using both techniques. The most significant and important of these differences involves the measured temperature and luminosities. We find that for the systems with inclinations close to 90, SPAMMS measures a higher temperature and luminosity than the 1D techniques. On the other hand, for the system with an inclination close to 60, the measured temperatures and luminosities agree quite well. This can be seen in Fig. 2. This is significant, because the difference in the measured temperatures and luminosities has a large impact on our understanding of the internal mixing processes for these systems, and thus their future evolution.

3.2. *Rapid rotators*

Motivated by our initial benchmarking of SPAMMS (see Sec. 3 of Abdul-Masih et al. 2020) as well as the differences in measured temperature found for the overcontact systems when using SPAMMS and 1D techniques, we investigated how the geometry of rapidly rotating systems affects the stellar parameters. As with overcontact systems, the geometry of rapidly rotating systems can deviate quite strongly from spherical symmetry, and the temperature range across the surface is expected to be more extreme for systems approaching critical rotation than in the case of overcontact systems.

Using SPAMMS, we generate synthetic spectra for a system rotating at 90% of critical velocity at different inclinations. We then perform a Monte Carlo simulation where we simulate a signal to noise of 300 and fit the synthetic SPAMMS spectra with 1D models to characterize how the measured temperature varies with inclination. We find that at most inclinations, the measured 1D temperature is more than 3σ away from the input temperature. At lower inclinations, a hotter temperature is measured, while at higher inclinations a coolet temperature is measured, with equivalent temperatures being found at an inclination of $\sim 60°$. Our findings indicate that not only does the temperature change by up to 5000 K based on the inclination, but also that the measured helium abundance also changes by a significant amount.

3.3. *Binary effects*

In addition to the two classes of objects discussed above, SPAMMS has also been applied to semi-detached and fully detached binary systems. In both of these cases, we find that SPAMMS is able to reproduce interesting 3D effects.

One such effect is known as the Struve-Sahade effect (Struve 1950; Sahade 1959; Linder et al. 2007). This effect is characterized by a strengthening of spectral lines when they are blue-shifted and a weakening when they are red-shifted. This effect is typically associated with semi-detached binaries and work by Palate & Rauw (2012) demonstrated that this effect could be reproduced using a similar approach to SPAMMS. In Abdul-Masih et al. (2020), we demonstrated that SPAMMS can also reproduce this effect, however in this case, we saw it when modeling an overcontact system. It is still unclear what causes this effect, but the fact that it can be produced by codes such as SPAMMS indicate that at least some of the effect is geometrical in nature.

Another effect that SPAMMS can reproduce is known as the Rossiter Mclaughlin effect (Rossiter 1924; McLaughlin 1924). This effect occurs when a rotating star is eclipsed by a companion. Since the rotating star has regions of the surface that are red-shifted and others that are blue-shifted, when the star is eclipsed by its companion, the spectral lines of the rotating system get skewed in different ways depending on which region of the star is being eclipsed. This is demonstrated in Fig. 3 This effect is often observed when a planet eclipses its host star and is very important for determining if there is a spin-orbit misalignment.

4. Conclusions and future prospects

Here, we have presented SPAMMS, a spectroscopic analysis code specifically designed for distorted stars. We have explained how the code works, and how it is able to reproduce the 3D effects associated with non-spherical geometries. We then described various ways in which SPAMMS has been applied, and we have shown some of the results to have come from this code so far.

While we have gone through some examples, this is by no means the limit of what SPAMMS can model. In addition to the cases shown, SPAMMS can spectroscopically model anything that PHOEBE can create a mesh for. This means that SPAMMS benefits from all of the advanced physical prescriptions and phenomena that are already included in PHOEBE including things such as reflection effects, spots, Doppler boosting, and many more. Additionally, as PHOEBE evolves and includes more features, so will SPAMMS. Some

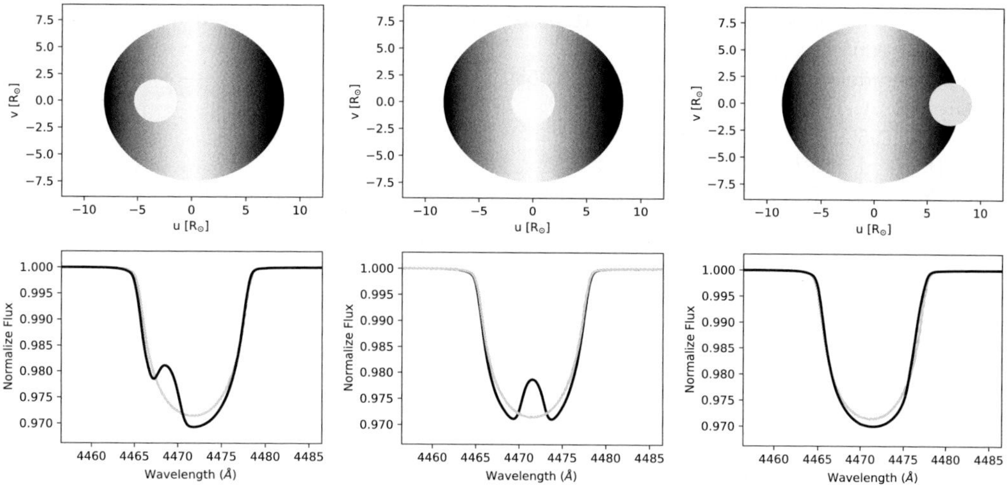

Figure 3. Example of the Rossiter Mclaughlin effect. The top panels show a rotating star at various phases during the eclipse with the face color representing the radial velocity profile across the surface. The bottom panels show how the line profile changes during the eclipse with the out of eclipse line profile plotted in grey for reference. This figure originally appeared in Abdul-Masih et al. (2020)

notable features that are currently in development that will be particularly interesting for SPAMMS include pulsations, alternative gravity darkening prescriptions and the ability to model higher order multiple system.

SPAMMS represents an exciting step forward in our understanding and ability to model massive star systems. With each new SPAMMS release, we will be able to model new and different types of systems that up until now we have not had the ability to model in 3D.

References

Abdul-Masih, M., Sana, H., Conroy, K. E., Sundqvist, J., Prša, A., Kochoska, A., & Puls, J. 2020, A&A, 636, A59

Abdul-Masih, M., et al. 2019, ApJ, 880, 115

—. 2021, A&A, 651, A96

Almeida, L. A., et al. 2015, ApJ, 812, 102

Bodensteiner, J., et al. 2020, A&A, 634, A51

Brott, I., et al. 2011, A&A, 530, A116

de Mink, S. E., & Mandel, I. 2016, MNRAS, 460, 3545

Fabry, M., Marchant, P., & Sana, H. 2022, A&A, 661, A123

Gräfener, G., Koesterke, L., & Hamann, W. R. 2002, A&A, 387, 244

Hilditch, R. W., Howarth, I. D., & Harries, T. J. 2005, MNRAS, 357, 304

Hillier, D. J., & Miller, D. L. 1998, ApJ, 496, 407

Langer, N. 2012, ARA&A, 50, 107

Linder, N., Rauw, G., Sana, H., De Becker, M., & Gosset, E. 2007, A&A, 474, 193

Mandel, I., & de Mink, S. E. 2016, MNRAS, 458, 2634

Marchant, P., Langer, N., Podsiadlowski, P., Tauris, T. M., & Moriya, T. J. 2016, A&A, 588, A50

Martins, F., Mahy, L., & Hervé, A. 2017, A&A, 607, A82

McLaughlin, D. B. 1924, ApJ, 60, 22

Palate, M., & Rauw, G. 2012, A&A, 537, A119

Pols, O. R. 1994, A&A, 290, 119

Prša, A., et al. 2016, ApJS, 227, 29

Puls, J., Urbaneja, M. A., Venero, R., Repolust, T., Springmann, U., Jokuthy, A., & Mokiem, M. R. 2005, A&A, 435, 669
Ramírez-Agudelo, O. H., et al. 2015, A&A, 580, A92
Rossiter, R. A. 1924, ApJ, 60, 15
Sahade, J. 1959, PASP, 71, 151
Sana, H., et al. 2012, Science, 337, 444
Schneider, F. R. N., Ohlmann, S. T., Podsiadlowski, P., Röpke, F. K., Balbus, S. A., Pakmor, R., & Springel, V. 2019, Nature, 574, 211
Shao, Y., & Li, X.-D. 2014, ApJ, 796, 37
Struve, O. 1950, Stellar evolution, an exploration from the observatory.

Discussion

F. Martins: You showed for rapidly rotating stars that these effects are important. I was wondering if you can you say how close binaries need to be for these effects to be important

M. Abdul-Masih: These effects arise from the temperature distribution across the surface, so in general, the more deformed your system is, the larger these effects are. I would say that the deformations due to rotation and the Roche geometry are more important than how physically close the two stars are, so I can't really cite a specific value for this, but we do plan to investigate what level of distortion is needed for these effects to become important in binary systems.

A. Herrero: I have two short questions, the first is concerning the surface gravity. Do you see the same effects on the surface gravity that you do on the temperatures and if so what is the range? And what is the effect when you change the temperature, do you see the same effects at lower and higher temperatures?

M. Abdul-Masih: We do see the same effect for surface gravities, you would see lower surface gravities when looking at an inclination of 90. The change in surface gravity with inclination depends quite heavily on how rapidly rotating your star is, but at close to critical, you can have differences in $\log g$ of ~ 0.5 or more.

When we change the temperature we still see the same inclination based effect. In this study, we focused on helium lines as temperature diagnostics, but as you shift to higher or lower temperatures you can use the ionization balance of other species to do the same thing.

Massive Stars Near and Far
Proceedings IAU Symposium No. 361, 2024
J. Mackey, J. S. Vink, & N. St-Louis, eds.
doi:10.1017/S1743921322002812

Mass loss and the Eddington parameter

Joachim M. Bestenlehner

Department of Physics & Astronomy, Hounsfield Road, University of Sheffield, S3 7RH, UK
email: j.m.bestenlehner@sheffield.ac.uk

Abstract. Mass loss through stellar winds plays a dominant role in the evolution of massive stars. Very massive stars (VMSs, $> 100 M_\odot$) display Wolf-Rayet spectral morphologies (WNh) whilst on the main-sequence. Bestenlehner (2020) extended the elegant and widely used stellar wind theory by Castor, Abbott & Klein (1975) from the optically thin (O star) to the optically thick main-sequence (WNh) wind regime. The new mass-loss description is able to explain the empirical mass-loss dependence on the Eddington parameter and is suitable for incorporation into stellar evolution models for massive and very massive stars. The prescription can be calibrated with the transition mass-loss rate defined in Vink & Gräfener (2012). Based on the stellar sample presented in Bestenlehner et al. (2014) we derive a mass-loss recipe for the Large Magellanic Cloud using the new theoretical mass-loss prescription of Bestenlehner (2020).

Keywords. stars: early-type, stars: massive, stars: mass loss, stars: winds, outflows, stars: atmospheres, stars: evolution

1. Introduction

The evolution of massive stars is dominated by mass loss and significantly influences their potential evolutionary phases (e.g. Langer 2012). In particular, stars with masses greater than $60\,M_\odot$ the evolution is dominated by mass loss (Vink 2015). In stellar structure calculations the theoretical mass-loss descriptions by Vink et al. 2000, 2001 are usually used for massive main-sequence stars. This mass-loss recipe under-predicts mass-loss rates for the most massive stars (e.g. Bestenlehner et al. 2014), but over-predicts at lower masses, which is also known as the weak-wind problem (e.g. Puls et al. 2008).

Mass-loss rates depend on metallicity and the Eddington parameter Γ_e, considering only the electron opacity, (e.g. Gräfener & Hamann 2008). Vink et al. (2011) predicted a steep increase of the mass-loss rate at the transition region from optically thin to optically thick which was empirically confirmed by Bestenlehner et al. (2014, Fig. 1). Based on the line driven wind theory Castor, Abbott & Klein (1975, hereafter CAK) derive a mass-loss prediction for massive stars as function of stellar mass M, Γ_e and CAK parameters k and α (see their equ. 46). However, the dependence of $\dot{M}$ on Γ_e was too weak to reproduce the steep increase of $\dot{M}$ when approaching the Eddington limit (Fig. 9 of Bestenlehner et al. 2014).

2. New mass-loss description for massive and very massive stars

The mass-luminosity relation of massive stars changes from $L \propto \mu^4 M^3$ to $L \propto M$ for $\Gamma_e \rightarrow 1$ with luminosity L and mean molecular weight μ (e.g. Yusof et al. 2013). This behaviour is not represented in M of the CAK mass-loss prediction. As shown in Bestenlehner (2020) the mass-term in the CAK mass-loss prediction can be replaced with

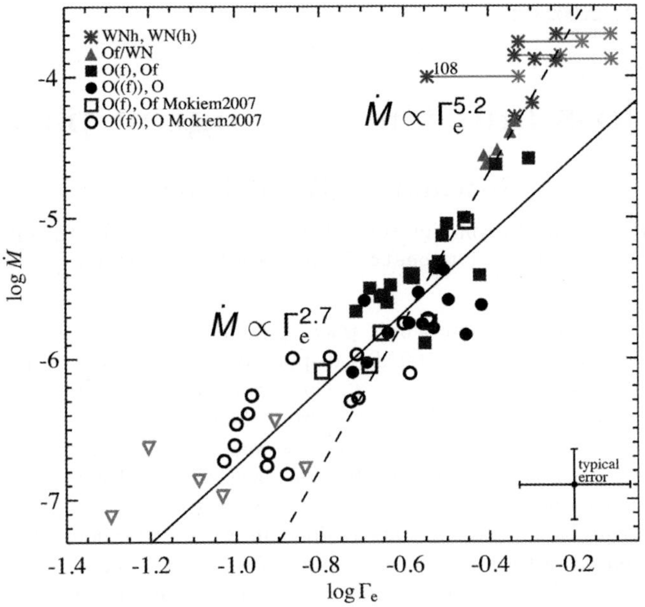

Figure 1. $\dot{M}$ vs. Γ_e for several spectral types. Figure adapted from Bestenlehner et al. (2014).

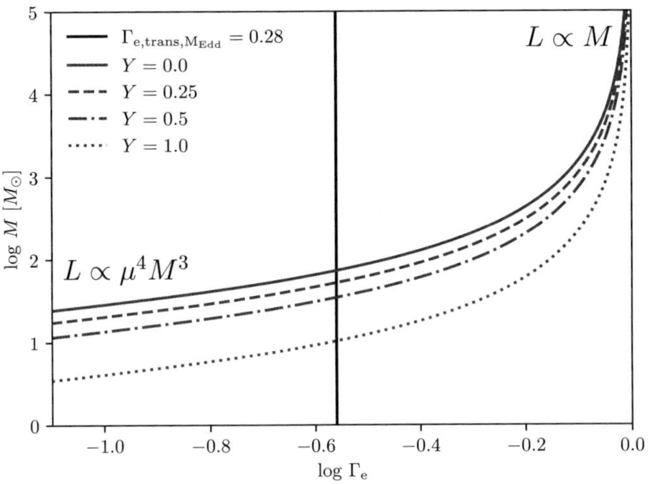

Figure 2. Stellar mass vs. Eddington parameter for several mean molecular weights. The transition to a linear mass-luminosity relation is independent of the mean molecular weight.

the $M - \Gamma_e$ relation obtained under the assumption of a fully radiative star (Eddington stellar model), which fullfills the condition $\Gamma_e \ll 1 : L \propto \mu^4 M^3, \Gamma_e \to 1 : L \propto M$:

$$M \propto \frac{1}{\mu^2} \frac{\Gamma_e^{1/2}}{(1-\Gamma_e)^2} \tag{1}$$

with $\Gamma_e \propto L/M$. The transition occurs at the location where $\Gamma_e^{1/2} = (1-\Gamma_e)^2$. The transition Eddington parameter $\Gamma_{e,trans,M_{Edd}} \approx 0.28$ is independent of the mean molecular weight. As shown in Fig. 2 the transition mass is strongly dependent on the mean molecular weight. For a pure hydrogen star the transition point occurs at $\sim 75\,M_\odot$ while for pure helium star the transition mass is only $\sim 10\,M_\odot$. However, stars close to the Eddington limit ($\Gamma_\mathrm{e} \rightarrow 1$) have hardly any dependence on μ.

Substituting M of the CAK mass-loss prediction with the $M - \Gamma_\mathrm{e}$ relation (Equation 1) leads to a new CAK-type mass-loss description of the form

$$\dot{M} \propto \frac{1}{\mu^2} \frac{\Gamma_\mathrm{e}^{1/\alpha+1/2}}{(1-\Gamma_\mathrm{e})^{(1-\alpha)/\alpha+2}} \tag{2}$$

as derived in Bestenlehner (2020). By using the logarithmic form of Equation 2 and absorbing all constants into $\dot{M}_0$ we obtain

$$\log \dot{M} = \log \dot{M}_0 + \left(\frac{1}{\alpha} + 0.5\right) \log(\Gamma_\mathrm{e}) - \left(\frac{1-\alpha}{\alpha} + 2\right) \log(1-\Gamma_\mathrm{e}). \tag{3}$$

For $\Gamma_\mathrm{e} \ll 1$, $\dot{M} \propto \Gamma_\mathrm{e}^{1/\alpha+1/2}$ and $\Gamma_\mathrm{e} \rightarrow 1$, $\dot{M} \propto 1/(1-\Gamma_\mathrm{e})^{(1-\alpha)/\alpha+2}$. At the transition, where $(\frac{1}{\alpha} + 0.5)\log(\Gamma_\mathrm{e}) = (\frac{1-\alpha}{\alpha} + 2)\log(1-\Gamma_\mathrm{e})$, we defined the transition Eddington parameter $\Gamma_\mathrm{e,trans}$. The CAK parameter α is metallicity dependent (e.g. Puls et al. 2000). The new mass-loss rate description predicts a stronger dependence on Γ_e, larger exponent, at lower metallicity (smaller α) and vice versa which can be observationally verified.

3. Discussion

To determine the CAK parameter α and $\dot{M}_0$ we fitted Equation 3 to the data of Bestenlehner et al. (2014) and Mokiem et al. (2007). The result is shown in Fig. 3 where we plot unclumped $\dot{M}$ against Γ_e. We are able to reproduce not only the mass-loss rates of O stars but also the enhanced $\dot{M}$ of the Of/WN and WNh stars. In addition, the fit follows closely the relations derived by Bestenlehner et al. (2014) and we obtain $\alpha = 0.50 \pm 0.04$ and $\log \dot{M}_0 = -4.44 \pm 0.10$. From this it follows that $\dot{M} \propto \Gamma_\mathrm{e}^{2.50^{+0.17}_{-0.15}}$ for $\Gamma_\mathrm{e} \ll 1$, which is in good agreement with the findings by Bestenlehner et al. (2014), $\dot{M} \propto \Gamma_\mathrm{e}^{2.73 \pm 0.43}$. The transition, where the $1/(1-\Gamma_\mathrm{e})^{3.00^{+0.17}_{-0.15}}$ term starts to dominate, occurs at $\Gamma_\mathrm{e,trans} = 0.47$ (Fig. 3).

Mass-loss rates in Fig. 3 are unclumped under the assumption of homogeneous stellar winds. The transition mass-loss rate $\dot{M}_\mathrm{trans}$, introduced by Vink & Gräfener (2012), is defined at the transition from optically thin O-type star to optically thick WR star winds, where the wind efficiency η is equal to the optical depth τ and 1, $\eta = \tau = 1$. The mass-loss rates of O and WR stars can be calibrated by utilising Equation 12 of Vink & Gräfener (2012): $\dot{M} = f\, L_\mathrm{trans}/(v_\infty c) \simeq 0.6 \dot{M}_\mathrm{trans}$ with correction factor $f \sim 0.6$, transition luminosity L_trans, terminal velocity v_∞ and light speed c.

The transition of optically thin to optically thick winds occurs between stars of spectral type Of and Of/WN (Fig. 1) with the Of/WN stars the desired transition objects (Vink & Gräfener (2012)). The sample of Bestenlehner et al. (2014) contains 6 Of/WN stars. Due to peculiarities we exclude VFTS 457 (Bestenlehner et al. 2014, for more details). Based on those 5 stars a transition $\Gamma_\mathrm{e,trans,vink} \approx 0.42$ was determined with corresponding $\log L_\mathrm{trans} = 6.31 \pm 0.12$ and $\log \dot{M}_\mathrm{trans}\,[M_\odot/\mathrm{yr}] = -4.48 \pm 0.12$ for unclumped winds. With L_trans and the averaged terminal velocity $v_\infty \approx 2700$km/s we derived an $\log \dot{M}_\mathrm{trans}\,[M_\odot/\mathrm{yr}] \approx -5.04 \pm 0.14$. This corresponds to a wind clumping factor $D = 13.2^{+17.0}_{-7.4}$ or volume filling factor $f_\mathrm{v} = 0.08^{+0.10}_{-0.04}$, which is consistent to the empirical clumping factor $D = 10$ found for the Of/WN stars in Bestenlehner et al. (2014)

On the basis of this result we calibrate $\dot{M}_0$ of Equ. 3 and obtain the mass-loss description for hot, massive stars in 30 Doradus in the Large Magellanic Cloud:

$$\log \dot{M} = -5.00 \pm 0.21 + 2.50^{+0.17}_{-0.15} \log(\Gamma_\mathrm{e}) \quad 3.00^{+0.17}_{-0.15} \log(1-\Gamma_\mathrm{e}). \tag{4}$$

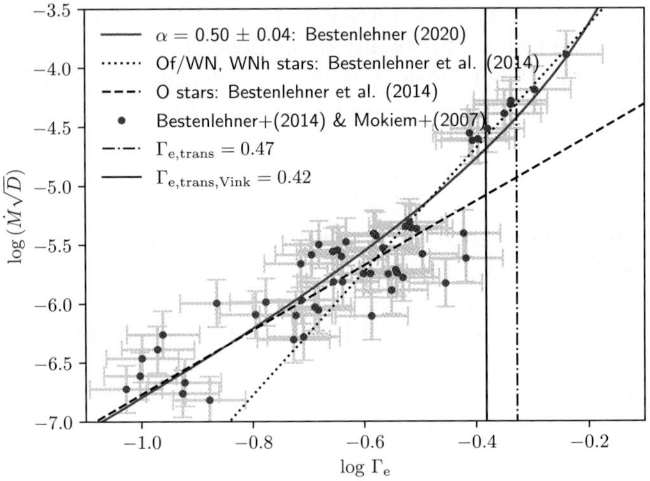

Figure 3. Wind clumping corrected mass-loss rate vs. Eddington parameter. The derived mass-loss description by Bestenlehner (2020) well reproduce the slopes obtained by Bestenlehner et al. (2014). The two vertical lines indicate the transitions discussed in the text.

This well agrees with the study of the cluster R136 in the centre of 30 Doradus by Brands et al. (2022), who found $\log \dot{M} = -5.19 + 2.69\ \log(\Gamma_{\rm e}) - 3.19\ \log(1 - \Gamma_{\rm e})$ on the basis of an ultraviolet+optical spectroscopic analysis.

Acknowledgements

JMB is supported by the Science and Technology Facilities Council research grant ST/V000853/1 (PI. V. Dhillon).

References

Bestenlehner, J. M., Gräfener, G., Vink, J. S., Najarro, F., de Koter, A., Sana, H., Evans, C. J., Crowther, P. A., Hénault-Brunet, V., Herrero, A., Langer, N., Schneider, F. R. N., Simón-Díaz, S., Taylor, W. D., Walborn, N. R. 2014, *A&A*, 570, A38

Bestenlehner, J. M. 2020, *MNRAS*, 493, 3938

Brands, S. A., de Koter, A., Bestenlehner, J. M., Crowther, P. A., Sundqvist, J. O., Puls, J., Caballero-Nieves, S. M., Abdul-Masih, M., Driessen, F. A., García, M., Geen, S., Gräfener, G., Hawcroft, C., Kaper, L., Keszthelyi, Z., Langer, N., Sana, H., Schneider, F. R. N., Shenar, T., Vink, J. S. 2022, *A&A accepted*, arXiv:2202.11080

Castor, J. I., Abbott, D. C., Klein, R. I. 1975, *ApJ*, 195, 157

Gräfener, G., Hamann, W.-R. 2008, *A&A*, 482, 945

Langer, N. 2012, *ARA&A*, 50, 107

Mokiem, M. R., de Koter, A., Evans, C. J., Puls, J., Smartt, S. J., Crowther, P. A., Herrero, A., Langer, N., Lennon, D. J., Najarro, F., Villamariz, M. R., Vink, J. S. 2007, *A&A*, 465, 1003

Pauldrach, A., Puls, J., Kudritzki, R. P. 1986, *A&A*, 164, 86

Puls, J., Springmann, U., Lennon, M. 2000, *A&AS*, 141, 23

Puls, J., Vink, J. S., Najarro, F. 2008, *A&ARv*, 16, 209

Sander, A. A. C., Vink J. S. 2020, *MNRAS*, 499, 873

Vink J. S., de Koter A., Lamers H. J. G. L. M. 2000, *A&A*, 362, 295

Vink J. S., de Koter A., Lamers H. J. G. L. M. 2001, *A&A*, 369, 574

Vink, J. S., Muijres, L. E., Anthonisse, B., de Koter, A., Grfener, G., Langer, N. 2011, *A&A*, 531, A132

Vink, J. S., Gräfener, G. 2012, *ApJL*, 751, L34

Vink, J. S. 2015, *ASSL*, 412, 77
Yusof, N., Hirschi, R., Meynet, G., Crowther, P. A., Ekström, S., Frischknecht, U., Georgy, C., Abu Kassim, H., Schnurr, O. 2013, *MNRAS*, 433, 1114

Discussion

PULS: This approach has one big advantage, namely that they no longer differentiate the wind features as a function of the optical depth, which more or less always refers to the continuum and not to the lines. In the original CAK version the question was what has the continuum to do with the different kinds of line driving for optically thick and thin winds. As far as I have seen it here this is not the question anymore. We have Γ_e from the original CAK as well as the different relation between luminosity and mass. So this is the big advantage, but just one warning: this all assumes that in these extreme massive O stars and main-sequence WR stars have the same mass driving as for normal stars, so that the mass loss is not related to an iron opacity bump or different mechanism to the CAK line driving as in the classical WR stars. If all this would be true then all these stars (massive O and main-sequence WR stars) are just normal and have usual O star winds. Do you agree?

BESTENLEHNER: Yes, I agree, but there might be slightly different values of α at the optically thin and thick wind regime, but if the iron-peak opacity would be the major driver of the mass loss this parametrisation should break down. Looking at the quality of the fit (Fig. 3) there is no clear indication and an effective α seemed to be representative for all those stars. Regarding classical WR stars, Sander & Vink (2020) show that this simple parametrisation works for some WR stars but indeed breaks down for the less luminous ones.

MURPHY: Do you have a relation how α changes with metallicity? You say, you can determine α from observations, but for very low metallicities or even zero, where you do do not have those observations, is there a relation that you can use there?

BESTENLEHNER: There is not an empirical relation, but Puls et al. (2000) lists theoretical αs for several metallicities. α decreases with decreasing metallicity. With ULLYSES we will be able to derive empirical αs for LMC, SMC and maybe Sextans A metallicities as only a few OB stars are required, but this is far from zero metallicity. Once you know α you are able to calibrate the mass-loss descriptions with the transition mass-loss rate.

PONIATOWSKI: You mentioned that α depends on metallicity, but α is also something that depends on the general distribution of lines, what species you have, how they are excited and so on, which can be computed, but now you can also fit that. How does this compared to general values that are normally used in literature? For instance, one typical number is $\alpha = 2/3$.

BESTENLEHNER: The α, you are referring to, is α' which is corrected for the finite cone angle, $\alpha = \alpha' - \delta$ with $\delta \approx 0.1$, e.g. Pauldrach et al. (1986). In addition, the α here is an effective alpha based on all types of stars and is therefore an average value. The value presented here ($\alpha = 0.5$) is still slightly lower, but derived for LMC metallicity. Using the modified CAK-theory by Pauldrach et al. (1986) would have made equation 2 much more complex. The beauty and simplicity of this mass-loss description would have not been given anymore, even though it is sufficient to reproduce the observations.

Massive Stars Near and Far
Proceedings IAU Symposium No. 361, 2024
J. Mackey, J. S. Vink, & N. St-Louis, eds.
doi:10.1017/S1743921322003234

New empirical mass-loss rates and clumping properties of massive stars

Calum Hawcroft[1], Hugues Sana[1], Laurent Mahy[2,1], J. O. Sundqvist[1], M. Abdul-Masih[1], J. C. Bouret[4], S. A. Brands[3], A. de Koter[3], F. A. Driessen[1] and J. Puls[5]

[1]Institute of Astronomy, KU Leuven, Celestijnenlaan 200D, 3001, Leuven, Belgium
email: calum.hawcroft@kuleuven.be

[2]Royal Observatory of Belgium, Avenue Circulaire 3, B-1180 Brussels, Belgium

[3] Astronomical Institute Anton Pannekoek, Amsterdam University, Science Park 904, 1098 XH Amsterdam, The Netherlands

[4]Aix Marseille Univ, CNRS, CNES, LAM, Marseille, France

[5]LMU München, Universitätssternwarte, Scheinerstr. 1, 81679 München, Germany

Abstract. Hot, massive stars are known to host unstable, radiation-driven outflowing winds, giving rise to dense clumps of material which severely affect the diagnostic techniques used to derive wind properties of massive stars. Most of the current diagnostic models account for wind inhomogeneities by assuming a one-component medium consisting of optically thin clumps, and maintaining a smooth velocity-field. However, this neglects important light-leakage effects through porous channels in-between the clumps. These light-leakage effects have recently been incorporated in the stellar atmosphere modelling code FASTWIND, and here we will present quantitative mass-loss results from a combined Ultraviolet-Optical wind analysis of O-supergiants in the Galaxy. Using a genetic-algorithm fitting-approach, we systematically investigate the impact the wind physics has on derived stellar and wind parameters, and how this depends on metallicity and spectral type. We compare our findings with earlier results (which do not take into account such light-leakage effects), to standard mass-loss rates usually included in evolution model studies of massive stars, and with theoretical predictions of clumping properties. We will also present the first systematic empirical constraints on the new wind parameters, associated with light-leakage, and compare these with theoretical predictions.

Keywords. stars: mass loss, stars: early-type, stars: atmospheres

---

1. Introduction

Massive stars ($M_{\rm ini} > 8 M_{\odot}$) are important objects in the context of the evolution of their host environments and galaxies at large Bresolin et al. (2008). Through strong, radiation-driven outflows these stars enrich the interstellar medium with heavy elements, as well as imparting significant amounts of ionising radiation and kinetic energy (Chiosi & Maeder (1986), Matteucci (2008), Geen et al. (2020)). These stellar winds are also one of the most critical factors in understanding the evolution of massive stars as the magnitude of the mass lost through the stellar lifetime affects the evolutionary pathway Langer (2012). Typical uncertainties on observed mass-loss rates are a around a factor of three in the evolutionary stages near the main sequence Puls, Vink, & Najarro (2008). This uncertainty is a result of a wind structure composed of slow, dense clumps of material and fast, tenuous flows developing in the winds (see e.g. Owocki, Castor, & Rybicki

(1988); Feldmeier (1995); Dessart & Owocki (2003); Sundqvist & Owocki (2013); Sundqvist, Owocki, & Puls (2018); Driessen, Sundqvist, & Kee (2019)) due to the inherent instability of radiation line driving (MacGregor, Hartmann, & Raymond (1979), Carlberg (1980), Owocki & Rybicki (1984)), which affects observational diagnostics of mass-loss rates (e.g. Oskinova, Hamann, & Feldmeier (2007), Sundqvist, Puls, & Feldmeier (2010), Šurlan et al. (2013)). This level of uncertainty in mass loss is also important for understanding the supernova explosion and resulting chemical yields Renzo et al. (2017), as well as the angular momentum transport in the stellar interior Keszthelyi, Puls, & Wade (2017).

In order to reduce the uncertainties on empirical mass-loss rates we must consider the wind structure in simultaneous fitting of UV and optical wind diagnostics. In this work we achieve this by performing a combined UV and optical fit of a sample of well-studied O-supergiant stars in the Milky Way, using the NLTE stellar atmosphere and wind modelling software FASTWIND which has recently incorporated the effects of optically thick clumping through a statistical approach.

2. Observations

The sample of eight Galactic O-supergiants analysed in this study is the same as the full sample of Bouret et al. (2012). We utilise the spectra as presented in Bouret et al. (2012), without adjusting the normalisation. For each star there is FUV, UV and optical coverage, the details of which are presented in Bouret et al. (2012) and Hawcroft et al. (2021). The spectra include a multitude of diagnostic lines, of which we focus on a subset of hydrogen, helium, carbon, nitrogen, oxygen, silicon and phosphorus features.

3. Methods

We determine stellar and wind parameters we compare the observed spectra with synthetic spectra, produced using the code FASTWIND (v10.3, Santolaya-Rey, Puls, & Herrero (1997), Puls et al. (2005), Rivero González, Puls, & Najarro (2011), Carneiro et al. (2016), Sundqvist & Puls (2018)). We employ a genetic-algorithm (GA) fitting technique to simultaneously optimise all the input parameters of the FASTWIND models, including the effective temperature $T_{\rm eff}$, surface gravity $\log g$, rotational and macroturbulent velocities ($v \sin i$, $v_{\rm macro}$), the wind β acceleration parameter, terminal wind speed v_∞, mass-loss rate $\dot{M}$, clumping factor $f_{\rm cl}$, surface abudances of helium, carbon, nitrogen and oxygen. We also include parameters associated with optically thick clumping, the velocity filling factor $f_{\rm vel}$, the interclump density factor $f_{\rm ic}$ and the clumping onset velocity $v_{\rm cl}$. The GA is based on the principles of genetic evolution and built upon the framework of Charbonneau (1995) and Mokiem et al. (2005). GA fitting has been shown to be effective in producing statisically unique solutions in fits of massive early-type stars in a number of studies (Mokiem et al. (2006); Mokiem et al. (2007); Tramper et al. (2011); Tramper et al. (2014); Ramírez-Agudelo et al. (2017); Abdul-Masih et al. (2019); Abdul-Masih et al. (2021), Hawcroft et al. (2021), Brands et al. (2022)). Figure 1 shows a spectral fit for the star HD14947, and Fig. 2 the reduced χ^2 distributions of models with respect to various parameters.

4. Results

We determine new mass-loss rates while including the effects of optically thick clumping. Figure 3 shows that the best-fit mass-loss rates are, on average, a factor of 3.6 below the predictions of Vink, de Koter, & Lamers (2000), and within a factor of 1.4 of the

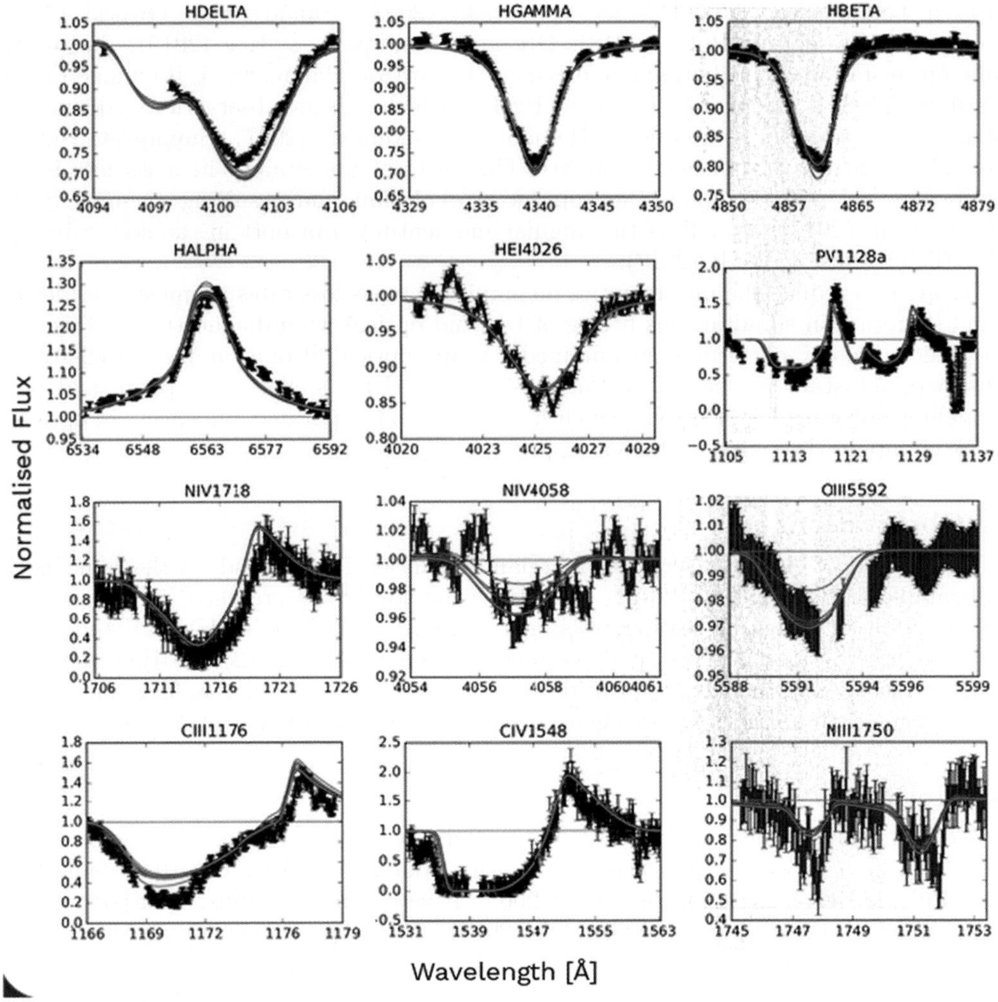

Figure 1. Subset of diagnostic lines with best-fit for HD14947 from the GA fit with optically thick clumping. Observed spectrum is shown in the dashed line, solid lines are statistically equivalent best-fit models within the confidence interval. Figure adapted from Hawcroft et al. (2021).

predictions of Björklund et al. (2021). To compute these predictions, best-fit parameter values from Hawcroft et al. (2021) are used as inputs for the recipes provided by the respective theoretical works. The empirical mass-loss rates we find are, on average, within 0.3 dex of the mass-loss rates found by Bouret et al. (2012), without the need to change the phosphorus abundance from solar, as the additional light leakage offered by the reduced effective opacity due to the inclusion of optically thick clumping allows us to reproduce the UV P-Cygni PV profiles. Bouret et al. (2012) do not include optically thick clumping and so must reduce the phosphorus abudance to reproduce the PV profiles. We find empirical constraints on parameters associated with optically thick clumping of: $f_{\rm cl} \sim 25$, $f_{\rm vel} \sim 0.5$ and $f_{\rm ic} \sim 0.1 - 0.3$.

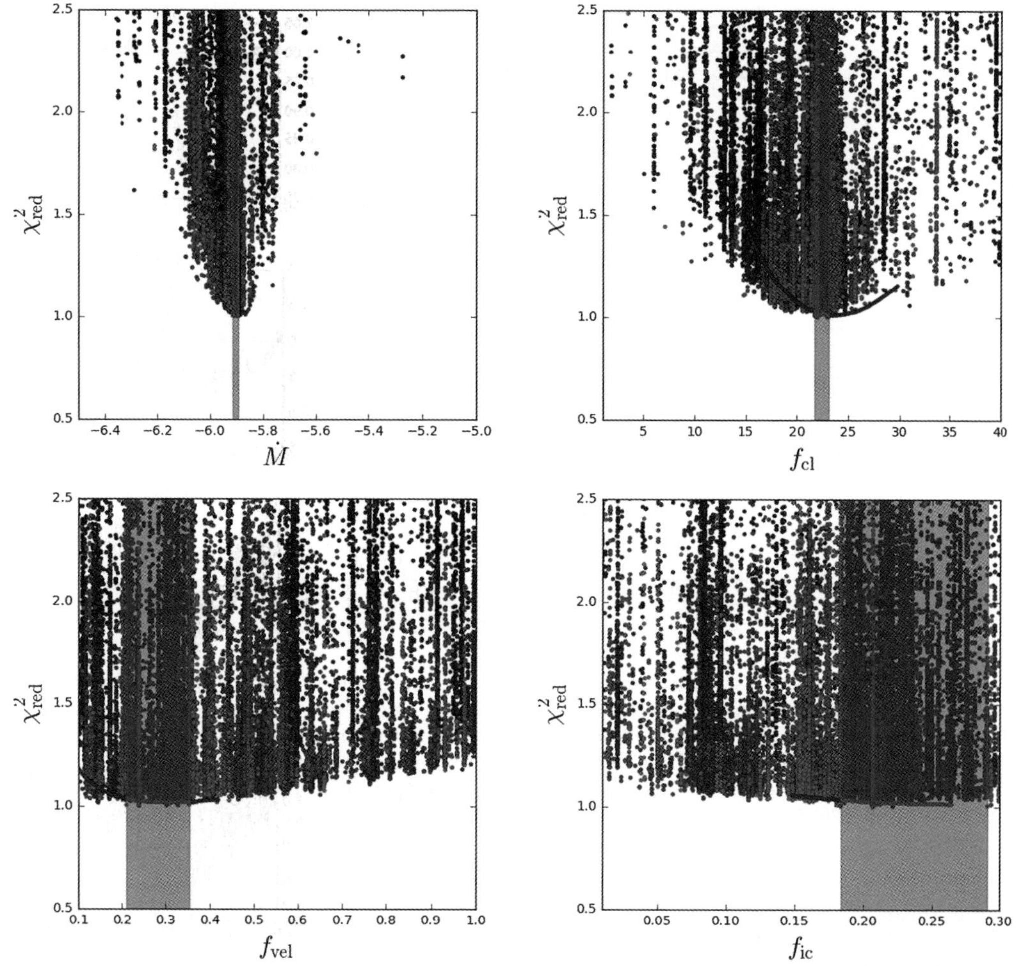

Figure 2. Subset of χ^2 distribution per parameter for HD14947 including mass-loss rate $\dot{M}$ (upper left), clumping factor $f_{\rm cl}$ (upper right), velocity filling factor $f_{\rm vel}$ (lower left) and inter-clump density factor $f_{\rm ic}$ (lower right). Models in blue are those produced by the GA, red points are models created to define the confidence interval around the GA best-fit. Region marked in red is the confidence interval.

5. Conclusions

We carried out a simultaneous UV and optical spectroscopic analysis of 8 early O-supergiant stars. By including a description of the span of the outflowing clumps in velocity space and allowing the clumps to become optically thick we have constrained two new parameters associated with wind structure for the first time. These are the fractional filling of velocity space by clumps and the density of interclump material. We are also able to constrain the degree of the overdensity in the clumps and the conditions for the onset of clumping. We estimate that, on average, 50% of the velocity field must be covered by clumps and the interclump material is 10-30% less dense than the mean outflow. We find that the clumps are, on average, 20 times more dense than the average wind. With these considerations we are able to fit UV and optical wind features with consistent mass-loss rates and abundances. The new empirical mass-loss rates are roughly three times lower than those commonly implemented in stellar evolution calculations.

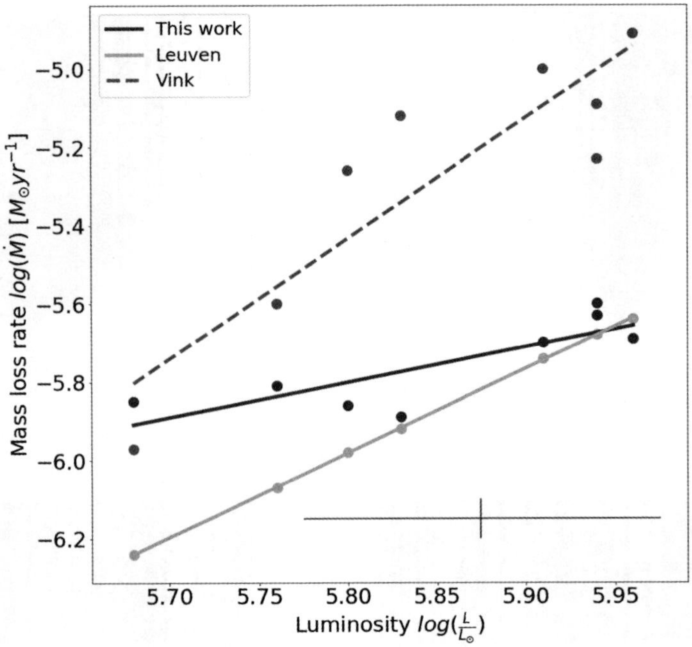

Figure 3. Mass-loss rates from the GA best-fits of Hawcroft et al. (2021), compared to the pre- dictions made by Leuven (Björklund et al. 2021) and Vink (Vink, de Koter, & Lamers 2000) for each star in our sample. The GA mass-loss rates are those found from the best-fit with optically thick clumping. Black cross in bottom right shows minimum error on mass-loss from this analysis and error on luminosity ±0.1 from Bouret et al. (2012). Figure adapted from Hawcroft et al. (2021).

References

Abdul-Masih M., Sana H., Sundqvist J., Mahy L., Menon A., Almeida L. A., De Koter A., et al. 2019, *ApJ*, 880, 115

Abdul-Masih M., Sana H., Hawcroft C., Almeida L. A., Brands S. A., de Mink S. E., Justham S., et al. 2021, *A&A*, 651, A96

Bouret J.-C., Hillier D. J., Lanz T., Fullerton A. W. 2012, *A&A*, 544, A67

Björklund R., Sundqvist J. O., Puls J., Najarro F. 2021, *A&A*, 648, A36

Brands S. A., de Koter A., Bestenlehner J. M., Crowther P. A., Sundqvist J. O., Puls J., Caballero-Nieves S. M., et al. 2022, arXiv, arXiv:2202.11080

Bresolin, F., Crowther, P. A., Puls, J. 2008, *Proceedings of the International Astronomical Union*, 250, 273

Carlberg R. G. 1980, *ApJ*, 241, 1131

Carneiro L. P., Puls J., Sundqvist J. O., Hoffmann T. L. 2016, *A&A*, 590, A88

Charbonneau P. 1995, *ApJS*, 101, 309

Chiosi, Cesare & Maeder, André 1986, *Annual Review of Astronomy and Astrophysics*, 24, 329–375

Dessart L., Owocki S. P. 2003, *A&A*, 406, L1

Driessen F. A., Sundqvist J. O., Kee N. D. 2019, *A&A*, 631, A172

Feldmeier A. 1995, *A&A*, 299, 523

Geen S., Pellegrini E., Bieri R., Klessen R. 2020, *MNRAS*, 492, 915

Hawcroft C., Sana H., Mahy L., Sundqvist J. O., Abdul-Masih M., Bouret J. C., Brands S. A., de Koter, A., Driessen, F. A., Puls, J. 2021, *A&A*, 655, A67

Keszthelyi Z., Puls J., Wade G. A. 2017, *A&A*, 598, A4

Langer N. 2012, *ARA&A*, 50, 107

MacGregor K. B., Hartmann L., Raymond J. C. 1979, *ApJ*, 231, 514

Matteucci F., 2008 *Proceedings of the International Astronomical Union*, 250, 391

Mokiem M. R., de Koter A., Puls J., Herrero A., Najarro F., Villamariz M. R. 2005, *A&A*, 441, 711
Mokiem M. R., de Koter A., Evans C. J., Puls J., Smartt S. J., Crowther P. A., Herrero A., et al. 2006, *A&A*, 456, 1131
Mokiem M. R., de Koter A., Evans C. J., Puls J., Smartt S. J., Crowther P. A., Herrero A., et al. 2007, *A&A*, 465, 1003
Oskinova L. M., Hamann W.-R., Feldmeier A. 2007, *A&A*, 476, 1331
Owocki S. P., Rybicki G. B. 1984, *ApJ*, 284, 337
Owocki S. P., Castor J. I., Rybicki G. B. 1988, *ApJ*, 335, 914
Puls J., Urbaneja M. A., Venero R., Repolust T., Springmann U., Jokuthy A., Mokiem M. R. 2005, *A&A*, 435, 669
Puls J., Vink J. S., Najarro F. 2008, *A&ARv*, 16, 209
Ramírez-Agudelo O. H., Sana H., de Koter A., Tramper F., Grin N. J., Schneider F. R. N., Langer N., et al. 2017, *A&A*, 600, A81
Renzo M., Ott C. D., Shore S. N., de Mink S. E. 2017, *A&A*, 603, A118
Rivero González J. G., Puls J., Najarro F. 2011, *A&A*, 536, A58
Santolaya-Rey A. E., Puls J., Herrero A. 1997, *A&A*, 323, 488
Sundqvist J. O., Puls J., Feldmeier A. 2010, *A&A*, 510, A11
Sundqvist J. O., Owocki S. P. 2013, *MNRAS*, 428, 1837
Sundqvist J. O., Puls J. 2018, *A&A*, 619, A59
Sundqvist J. O., Owocki S. P., Puls J. 2018, *A&A*, 611, A17
Šurlan B., Hamann W.-R., Aret A., Kubát J., Oskinova L. M., Torres A. F. 2013, *A&A*, 559, A130
Tramper F., Sana H., de Koter A., Kaper L. 2011, *ApJL*, 741, L8
Tramper F., Sana H., de Koter A., Kaper L., Ramírez-Agudelo O. H. 2014 *A&A*, 572, A36
Vink J. S., de Koter A., Lamers H. J. G. L. M. 2000, *A&A*, 362, 295

Discussion

J. VINK: The luminosity of the stars in your sample approach $10^6 L_\odot$ which is close to the transition mass-loss where the Björklund et al. (2021) mass-loss rates are a factor of three too low and the Vink, de Koter, & Lamers (2000) rates agree better with the observations, and this doesn't depend on clumping or porosity. Could it be that the clumping prescription in the models is incorrect which is affecting the mass-loss rates, for example the onset of clumping?

C. HAWCROFT: Indeed, there could be issues with the clumping prescriptions as they are currently implemented. For example, we find that the clumping onset, which currently has a variable onset velocity and increases linearly to the maximum, is degenerate with the wind acceleration parameter β, so the shape of the wind and clumping acceleration profile is uncertain. To try to get a better understanding of this we have tried to compared to theoretical predictions, specifically the LDI simulations of Driessen, Sundqvist, & Kee (2019), these authors find onset velocities of clumping which are consistent with the velocities determined through the GA fits. However, these are 1D prescriptions and we may see changes in the clumping parameters and prescriptions as the simulations are computed in higher dimensions.

M. WILLIAMSON OR M. GAGNE: In which wavelength ranges are the twenty times denser clumps optically thick?

C. HAWCROFT: It is difficult to give an exact answer to this across the full spectrum from this work as we mainly focus on UV and optical diagnostics, but in this parameter regime the clumps are mainly optically thick at UV wavelengths and less so at optical wavelengths.

Massive Stars Near and Far
Proceedings IAU Symposium No. 361, 2024
J. Mackey, J. S. Vink, & N. St-Louis, eds.
doi:10.1017/S1743921322002101

ISOSCELES: Grid of stellar atmosphere and hydrodynamic models of massive stars

Ignacio Araya[1], Natalia Machuca[2], Michel Curé[2] and Catalina Arcos[2]

[1]Vicerrectoría de Investigación, Universidad Mayor, Chile
email: ignacio.araya@umayor.cl

[2]Instituto de Física y Astronomía, Universidad de Valparaíso, Chile

Abstract. Spectroscopy can decode the radiation from stars in an appropriate way and derive many properties of different stellar objects. In this work we seek to derive simultaneously stellar and wind parameters of massive stars. To model the data we use the radiative transport code FASTWIND with the hydrodynamic solutions derived using our stationary code HYDWIND as input, instead of the β-law. Then, ISOSCELES, our grid of stellar atmosphere and hydrodynamic models of massive stars, is used to derive the physical properties of the observed spectra through spectral line fittings. This quantitative spectroscopic analysis provide an estimation about the line–force parameters, whose theoretical calculations are complex. In addition, we expect to confirm that the hydrodynamic δ-slow solutions, describe quite reliable the radiation line-driven winds of A and late B supergiant stars and, at the same time, explain disagreements between observational data and theoretical models for the Wind–Momentum Luminosity Relationship (WLR).

Keywords. stars: early-type, stars: atmospheres, stars: mass loss

1. Introduction

ISOSCELES is the first grid of synthetic data for massive stars that involves both, the m-CAK hydrodynamics and the NLTE radiative transport. To produce the grid of synthetic line profiles, we first computed a grid of hydrodynamic wind solutions with our stationary code HYDWIND (Curé 2004). These hydrodynamic wind solutions, based on the CAK theory and its improvements (Castor et al. 1975; Friend & Abbott 1986; Curé 2004; Curé et al. 2011) are used as input in the NLTE radiative transport code FASTWIND (Puls et al. 2005). Each HYDWIND model is described by six parameters: T_{eff}, logg, R_*, α, k, and δ. All these models consider, for the optical depth, the boundary condition $\tau = 2/3$, at the stellar surface. In the case of the FASTWIND grid, we calculated a total of 573 433 models. From these models line profiles of H, He, and Si elements were calculated in the optical and infrared range. Also, it is important to notice, that all our models are calculated without stellar rotation.

2. Method and Results

The observational data was pre-processed using the *iacob broad* tool Simón-Díaz & Herrero to derive the projected rotational and macroturbulent speeds. Then, these values were used in our search code to perform the spectral fitting with the purpose to obtain stellar and wind parameters. First, the code reads an input file that contains the information of the observational data, the convolution parameters and (optional) the

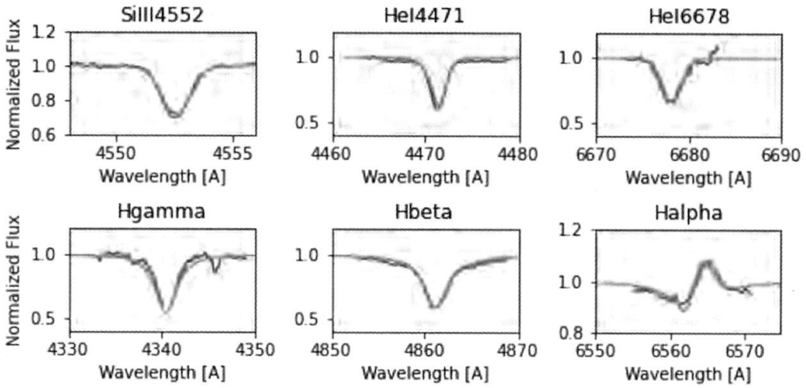

Figure 1. Spectral line fitting for HD 99953. Blue solid line shows the spectra retrieved from Haucke et al. (2018). Orange solid line corresponds to a model with the following parameters: $T_{\rm eff} = 18500$ K, $\log g = 2.4$, $\alpha = 0.45$, $k = 0.25$, $\delta = 0.32$, $v_{\rm micro} = 15$ km/s and $\log \epsilon_{\rm Si} = 7.81$.

type of solution we are searching for, fast or δ-slow. In a second step, it uses multiprocessing tools to search through the grid. The line profiles are rotationally convolved and interpolated with the observed line. Then a χ^2 test is performed. Finally, the code collect all these results and sort them from lower to higher χ^2 values, selecting and returning the one that most resembles to the observational data. By way of example, we present the results for the star HD 99953 considering six line profiles (see Figure 1). We derived stellar parameters similar to the ones obtained by Haucke et al. (2018). Regarding to the wind parameters, they found $\dot{M} = 0.08 \times 10^{-6}\ M_{\odot}$/year and $v_\infty = 250$ km/s, with $\beta = 2$, for the velocity profile. We obtained line force parameters that corresponds to a δ-slow solution, with the following wind parameters: $\dot{M} = 0.243 \times 10^{-6}\ M_{\odot}$/year and $v_\infty = 254$ km/s.

3. Conclusions

According to our experience in spectral line fittings, we predict that the models with values of $\beta \gtrsim 1.5$ can be properly reproduced with the hydrodynamic solution δ-slow. Further studies with more stars and analysis of more spectral lines will give us more precise estimates of the stellar and wind parameters, leading us to calibrate the WLR accurately. As future work, we expect to analyze the χ^2 distributions and then use it to compute the uncertainties for each derived parameter, also, we are working in a statistical study of the theoretical line profiles shapes in contrast with stellar and wind parameters from ISOSCELES in order to study the influence of the wind in the observed line profiles.

Acknowledgements

M.C., C.A., & I.A. are grateful with the support from FONDECYT project 1190485. I.A. & C.A. also thanks the support from FONDECYT projects 11190147 and 11190945, respectively. This project receive funding from the European Union's Framework Programme for Research and Innovation Horizon 2020 (2014–2020) under the Marie Skłodowska–Curie grant Agreement No. 823734.

References

Castor, J. I., Abbott, D. C., & Klein, R. I. 1975, ApJ, 195, 157
Curé, M. 2004, ApJ, 614, 929
Curé, M., Cidale, L., & Granada, A. 2011, ApJ, 737, 18
Friend, D. B., & Abbott, D. C. 1986, ApJ, 311, 701

Haucke, M., Cidale, L. S., Venero, R. O. J., Curé, M., Kraus, M., Kanaan, S., & Arcos, C. 2018, A&A, 614, A91

Puls, J., Urbaneja, M. A., Venero, R., Repolust, T., Springmann, U., Jokuthy, A., & Mokiem, M. R. 2005, A&A, 435, 669

Simón-Díaz, S., & Herrero, A. 2014, A&A, 562, A135

Massive Stars Near and Far
Proceedings IAU Symposium No. 361, 2024
J. Mackey, J. S. Vink, & N. St-Louis, eds.
doi:10.1017/S1743921322003301

Access platform to our database of CMFGEN spectra for atmospheric studies of massive stars

A. Payen-Sandoval[1], L. Arias[1], J. Zargó[2], A. Arrieta[1], C. R. Fierro-Santillán[3] and J. Klapp[3,4]

[1]Universidad Iberoamericana
email: lorena.arias@ibero.mx

[2]Escuela Superior de Física y Matemáticas, IPN

[3]ABACUS-CINESTAV

[4]Instituto Nacional de Investigaciones Nucleares

Abstract. We have collected a database of more than 43,000 spectra of atmospheres of massive stars. These spectra have been generated with the CMFGEN code of Hillier & Miller (1998) by systematically varying stellar parameters: effective temperature, luminosity, metallicity and mass loss rate for stars from 9 to 120 solar masses (Zsargó et al. 2020) In this work we present a web-based platform for accessing the database. The platform allows an online comparison between an observed and a synthetic spectrum to quickly assess the stellar and wind parameters. The platform will be available without cost to the astronomical community and will be hosted on servers shared between Mexican Universities.

Keywords. stars: atmospheres, stars:mass loss, catalogs.

1. Our atmosphere grid: 43,340 synthetic spectra

Synthetic spectra were generated with CMFGEN (Hillier & Miller 1998, 1999), a non-LTE radiative transfer, fully line-blanketed atmosphere code, designed to solve the statistical equilibrium and radiative transfer equations in spherical geometry. In order to properly constrain the input parameters, we used stellar evolutionary tracks and isochrones for both rotating and stationary stellar models reported by Ekström et al. (2012). Considered parameters are luminosity L, surface temperature T_{eff}, mass M, mass-loss rate $\dot{M}$ and surface chemical composition. Nine evolutionary tracks were included (initial mass from 12 to 120 $M_{\odot}$) as well as eight isochrones (log age in years from 6.5 to 7.2). To describe the wind structure, we estimated V_{∞} using V_{esc}. For each stellar model we generated 28 atmosphere models using combinations of the wind clumping factor $F_{cl} = 0.05$, 0.3, 0.6 and 1.0, and the type velocity law of the wind $\beta = 0.5$, 0.8, 1.1, 1.4, 1.7, 2.0 and 2.3. Chemical elements included in our models are H, He, C, N, and O in initial abundances as reported by Ekström et al. (2012). The solar metallicity assumed was the one reported by Asplund et al. (2009) for Si, P, S, and Fe, in all models. A database has been constructed with all synthetic spectra.

2. User interface

Written using the Python web development framework *Dash*, the application will allow users to search the database for stars with parameters within the inputs, search for stars

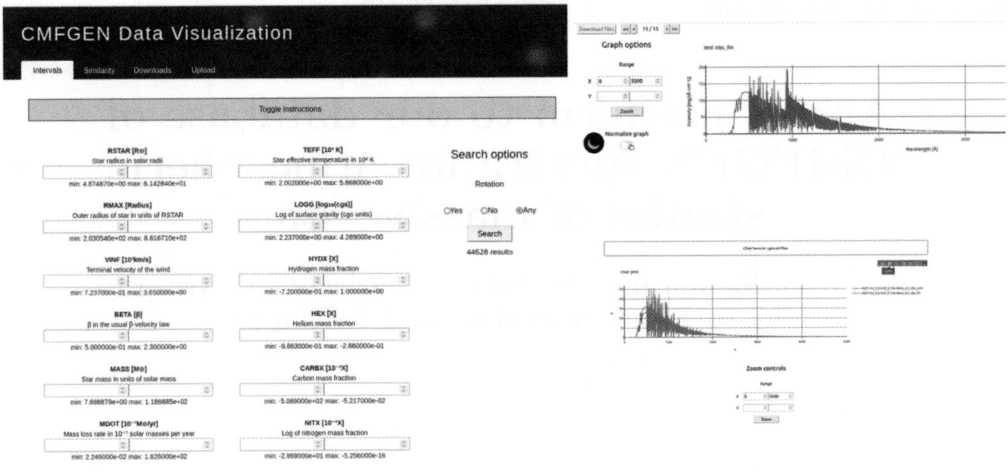

Figure 1. Online appearance of the access platform to the database.

by how similar they are to the parameters entered, upload multiple files with tabular data to compare their graphs in a single window, visualize the spectra of every star in our database, and even download the data. Figure 1 shows the online appearance of the access platform to the database:

- Entering values in boxes, user can search for stars with physical parameters between those values (Figure 1, left).
- The number of results will be displayed. When looking at the results of a search, the spectrum can be visualized as is or in normalized form in a fully interactive graph (Figure 1, top right). The figure can be saved as a .png image and zoomed in and out. The graph can be switched by selecting another star; by selecting stars in the table and pressing the download button, their data can be downloaded. When searching stars by similarity, the parameters by the user are transformed into a vector, normalized alongside the database data, and compared with every other star by taking their differences. The star with the highest similarity score is returned first, with all other stars following by order of similarity score.
- By uploading up to three files containing tabular data, their graphs can be visualized in a single window (Figure 1, bottom right). As in the case of the search results page, this graph is also fully interactive, and its figure can be saved in .png format.

The database will be hosted on servers at our institutions in México (Instituto Politécnico Nacional, Universidad Iberoamericana and CINVESTAV) and will be available soon.

References

Asplund, M., Grevesse, N., Sauval, A., Jacques, A. & Scott, P. 2009, *ARAA*, 47, 481
Ekström, S., et al. 2012, *A&A*, 537, A146
Hillier, J., & Miller D.L. 1998, *ApJ*, 496, 407
Hillier, J., & Miller D.L. 1999, *ApJ*, 519, 354
Zsargó, J., et al. 2020, *A&A*, 643, A88

Massive Stars Near and Far
Proceedings IAU Symposium No. 361, 2024
J. Mackey, J. S. Vink, & N. St-Louis, eds.
doi:10.1017/S1743921323004404

CMD for atmospheres of massive stars created by using CMFGEN synthetic spectra

Anabel Arrieta[1], **Agustín Payen-Sandoval**[1], **Lorena Arias**[1], **Janos Zsargo**[2], **Celia Fierro-Santillán**[3] **and Jaime Klapp**[3,4]

[1]Universidad Iberoamericana, Prolongación Paseo de la Reforma 880, Lomas de Santa Fe, 01219 Ciudad de México, México
email: anabel.arrieta@ibero.mx

[2]Escuela Superior de Física y Matemáticas, IPN

[3]ABACUS-CINESTAV

[4]Instituto Nacional de Investigaciones Nucleares

Abstract. We present Color Magnitude Diagrams (CMDs) created by using our database of 43,340 synthetic CMFGEN spectra.For each calculated spectra we measure the absolute flux by Johnson and Gaia photometry filters and build synthetic CMDs.

Keywords. Color Magnitude Diagram, massive star, Hertzsprung-Russell Diagram, stellar wind

---

1. Introduction

GaiaData Release provides very high-precision astrometry and photometry for a large number of stars. Now very precise Hertzsprung-Russell Diagrams (HRDs) have been constructed and extremely fine structures for open and globular clusters have been reported (Gaia Collaboration et al. 2018). A detailed study of these fine structures will provide important constraints for stellar structure and evolutionary studies.

2. Our atmosphere grids

To calculate the synthetic spectra we use CMFGEN (Hillier & Miller 1998, 1999) non local thermal equilibrium stellar atmosphere code.

In order to properly constrain the input parameters, we used stellar evolutionary tracks and isochrones for both rotating and stationary stellar models reported by Ekström et al. (2012). The luminosity L, surface temperature T_{eff}, mass $M_{\odot}$ and mass-loss rates $\dot{M}$ and surface chemical composition for 9 evolutionary tracks (initial masses between 12 to 120 $M_{\odot}$) as well 8 isochrones (log age 6.5 to 7.2 yrs) were taken from Ekström et al. (2012).

To describe the wind structure we estimate $V_{\rm inf}$ using V_{esc}. For each stellar model we calculated 28 atmosphere models using combinations of $F_{cl} = 0.05$, 0.3, 0.6 and 1.0, and the type of velocity law of the wind $\beta = 0.5$, 0.8, 1.1, 1.4, 1.7, 2.0 and 2.3.

The chemical elements included in our models are H, He, C, N, O reported by Ekström et al. (2012) and we assumed solar metallicity as reported by Asplund et al. (2009) for Si, P, S, and Fe in all models.

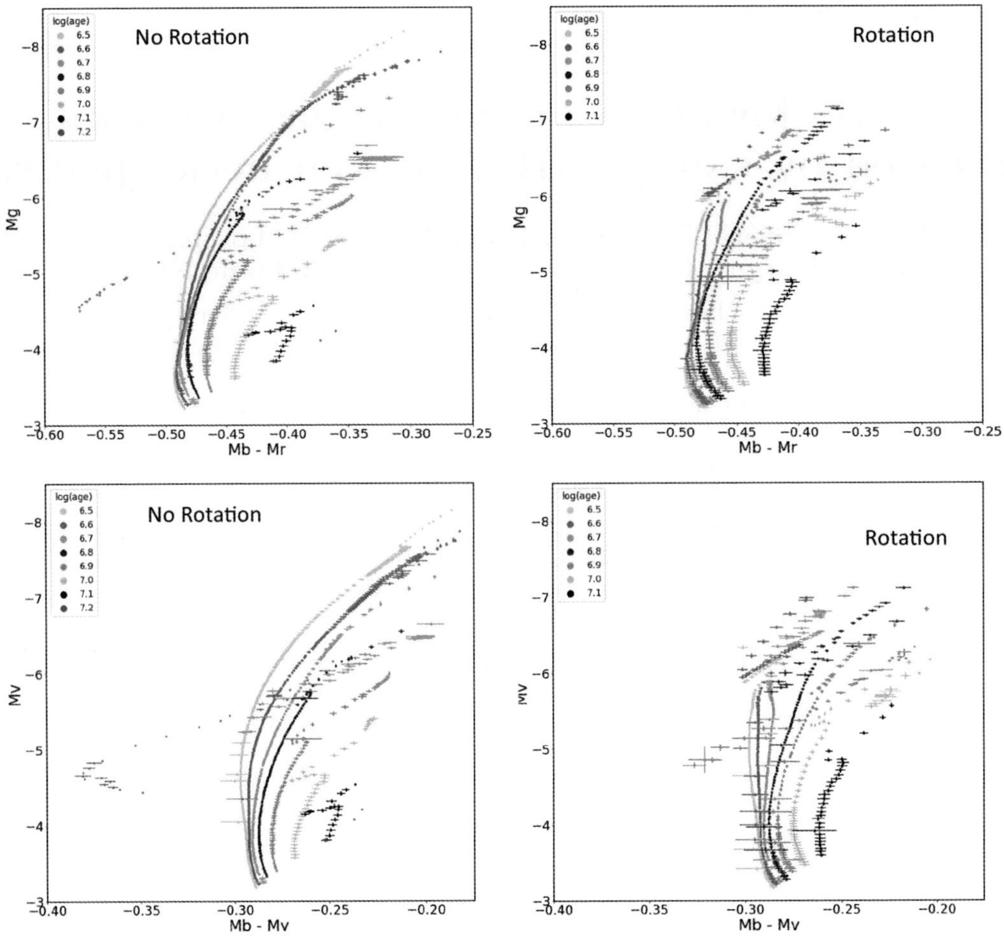

Figure 1. Syntetic color-magitude diagrams for Gaia (top) and Johnson (bottom) filters for evolutionary tracks.

3. Color-Magnitude Diagrams

Synthetic photometry for the grid of models has been calculated in order to obtain absolute magnitudes and colors in the BV Johnson and Gaia systems. For each broadband filter the computed magnitude and colors are:

$$M_i = -2.5\frac{\int F(\lambda)S_i(\lambda)d\lambda}{\int F_0(\lambda)S_i(\lambda)d\lambda} + C_1 \tag{1}$$

and

$$(M_i - M_j)_{obs} = (M_i - M_j)_{calc} + C_2. \tag{2}$$

Where $F(\lambda)$ is the CMFGEN flux computed for a distance of 1 kpc from the star, $S_i(\lambda)$ is the filter response in the i band, $F_0(\lambda)$ is the Vega observed flux and C_1 and C_2 are constants needed to fix the zero point. Due to the characteristics of CMFGEN it is not possible to generate a model for a Vega type star, for this reason stars for which there is a fitted CMFGEN model and whose colors are known have been used (Fierro-Santillán et al. (2018)).

A CMD was constructucted for Johnson and Gaia photometry filters. For different wind parameters in a same stellar model, the color index and absolute magnitude show a dispersion. We found that the position in color diagram have an important dependence not only on temperature and abundance, but with wind parameters as well. In order to visualize the dispersion, Figure 1 shows the average and standard deviation in the position for each stellar parameter model calculated with different wind parameters.

References

Asplund, M., Grevesse, N., Sauval, A., Jacques, A. & Scott, P. 2009, *ARAA*, 47, 481
Fierro-Santillán, et al. 2018, *ApJS*, 236, 38
Ekström, S., et al. 2018, *A&A*, 537, A146
GAIA Collaboration 2018, *A&A*, DR2HRD
Hillier, J., & Miller D.L. 1998, *ApJ*, 496, 407
Hillier, J., & Miller D.L. 1999, *ApJ*, 519, 354
Zsargó, J., et al. 2020, *A&A*, 643, A88

Massive Stars Near and Far
Proceedings IAU Symposium No. 361, 2024
J. Mackey, J. S. Vink, & N. St-Louis, eds.
doi:10.1017/S174392132300114X

Clumping and X-rays in cool B-Supergiants

Matheus Bernini-Peron[1], **W. L. F. Marcolino**[2] **and A. A. C. Sander**[1]

[1]Zentrum für Astronomie der Universität Heidelberg, Astronomisches Rechen-Institut
Mönchhofstr. 12-14, 69120, Heidelberg, Germany
email: `matheus.bernini@uni-heidelberg.de`

[2]Observatório do Valongo, Federal University of Rio de Janeiro,
Ladeira do Pedro Antônio, 20080-090, Rio de Janeiro, Brazil

Abstract. B supergiants (BSGs) lie on the cool end of line-driven wind regime, such that the study of their atmospheres can help us to understand the physics of line-driven winds. So far key features of their spectra, especially in the UV region, could not be reproduced consistently with atmosphere models. This represents a significant gap in our knowledge of their physical properties and behavior, which is particularly striking for BSGs on the cool side of the Bi-Stability Jump (cooler than B1). To address this problem, we analysed a sample of Galactic cool BSGs, with sufficient UV and optical coverage. None of our targets are detected in X-rays with only upper limits existing for some of them.

Keywords. massive stars, stellar winds, B-supergiant, clumping, X-ray, Bi-stability

---

Introduction and Methods: B supergiants (BSGs) are evolved stars with effective temperatures between ~12 and ~29 kK, lying on the cool end of the line-driven wind regime. Studying them can help us understanding high-mass stellar evolution, the physics of BSGs winds, and phenomena like the Bi-stability Jump (Lamers et al. 1995). Of major concern are in particular the UV P-Cygni profiles, which could not be reproduced consistently in previous studies (Crowther et al. 2006, Searle et al. (2008)). This leads to considerable uncertainties in our knowledge of BSG wind properties, in particular for stars with a spectral type cooler than B1.

To address this issue, we analysed a sample of four Galactic cool BSGs with sufficient UV and optical coverage. None of our targets are detected in X-rays. In our analysis, we used CMFGEN (and PoWR) to verify whether the inclusion of (micro- and macro-) clumping and X-rays in the wind could improve the agreement between models and observations. As a first step we retrieved spectra (UV and optical), magnitudes (from UV to IR) and distances (GAIA eDR3 and Hipparcos). From fitting the optical spectrum we obtained the stellar physical properties, whereas SED fitting and distances yielded the luminosity and reddening. We then obtained the wind properties by fitting the UV and investigated the impacts of clumping and X-rays in the wind using CMFGEN. PoWR was then used to test the impact of optically thick clumping on Hα and UV lines which were not well modelled with microclumping only.

Results and Discussion: We found a reasonable agreement between synthetic and observed spectra (Fig. 1) for our sample stars – both in the optical and the UV. We obtained stellar properties compatible with evolved objects: (i) being far from the ZAMS and (ii) displaying altered chemical compositions (↓C, ↑N, ↓O) relative to the Sun (Asplund et al. 2009). We found spectroscopic masses discrepant with rotating (single star) evolution models – which deserves deeper investigation in the context of the mass

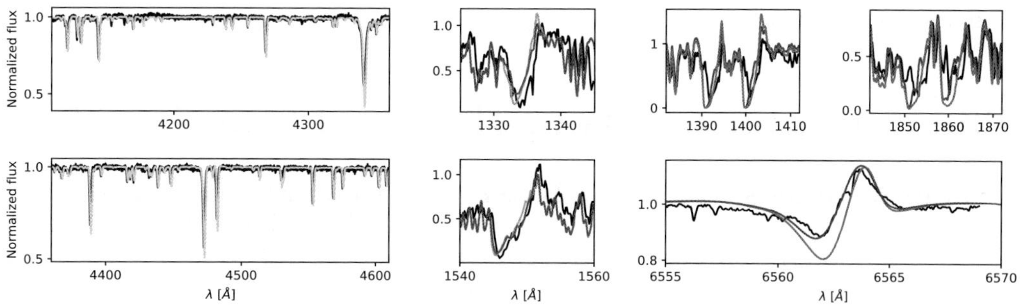

Figure 1. Spectral analysis of HD53138 (B3 Ia): In all panels the observed spectra are depicted as a black line. *Left panels:* Optical fitting. *Middle Panels:* Fitting of C II and C IV profiles including clumping and X-rays – green line: model with $f_\infty = 0.5$, $L_x/L = 10^{-7.5}$ and $v_x = 0.8\, v_\infty$ (where v_x is the onset of X-rays); yellow: model with $f_\infty = 0.9$, $L_x/L = 10^{-8.0}$ and $v_x = 0.05\, v_\infty$. *Right panels:* Models with (red) and without (blue) macroclumping.

discrepancy problem (see, e.g., Cantiello et al. 2009). In the UV, we reproduced key wind lines, where previous studies were unsuccessful. Our results show that both X-rays and clumping need to be included in order to fit C II and C IV lines simultaneously – and to not overestimate Hα. However, we required different values than those typically obtained for hotter stars ($L_x/L = 10^{-7}$; $f_\infty = 0.1$), namely $f_\infty > 0.5$. We also inferred $\log(L_x/L)$ of about -7.5 to -8. Additionally, we had problems in fitting Hα, Si IV and Al III lines. Thus, motivated by previous work (Prinja & Massa 2010, Petrov et al. 2014), we investigated the impact of optically thick clumping on HD53138 using PoWR (Oskinova et al. 2007). We obtained an improvement for Hα, Si IV and Al III, although yet insufficient to totally remove the discrepancy. We thus draw the following conclusions from our study, the first to investigate clumping and X-rays in cool BSGs modelling consistently UV and optical:

- Our results point towards weaker clumping and X-ray emission in cool BSGs compared to their hotter analogs. This is in line with recent theoretical predictions (Driessen et al. 2019) and observational results in the infrared (Rubio-Díez et al. 2022).
- The behavior of X-ray emission in such stars is still unclear (Berghoefer et al. 1997, Nazé 2009). In this study we show that a certain amount of X-Rays are necessary to explain the observed superionization in the UV.
- The lower amount of clumping obtained would imply a smaller reduction of the mass-loss rates compared to the Vink et al. (2000) prediction than on the hot side of the Bi-Stability Jump.

References

Asplund, M., Grevesse, N., Sauval, A. J., & Scott, P. 2009, *ARA&A*, 47, 481
Berghoefer, T. W., Schmitt, J. H. M. M., Danner, R., & Cassinelli, J. P. 1997, *A&A*, 322, 167
Cantiello, M., Langer, N., Brott, I., et al. 2009, *A&A*, 499, 279
Crowther, P. A., Lennon, D. J., & Walborn, N. R. 2006, *A&A*, 446, 279
Driessen, F. A., Sundqvist, J. O., & Kee, N. D. 2019, *A&A*, 631, A172
Lamers, H. J. G. L. M., Snow, T. P., & Lindholm, D. M. 1995, *ApJ*, 455, 269
Nazé, Y. 2009, *A&A*, 506, 1055
Oskinova, L. M., Hamann, W. R., & Feldmeier, A. 2007, *A&A*, 476, 1331
Petrov, B., Vink, J. S., & Gräfener, G. 2014, *A&A*, 565, A62
Prinja, R. K. & Massa, D. L. 2010, *A&A*, 521, L55
Rubio-Díez, M. M., Sundqvist, J. O., Najarro, F., et al. 2022, *A&A*, 658, A61
Searle, S. C., Prinja, R. K., Massa, D., & Ryans, R. 2008, *A&A*, 481, 777
Vink, J. S., de Koter, A., & Lamers, H. J. G. L. M. 2000, *A&A*, 362, 295

Massive Stars Near and Far
Proceedings IAU Symposium No. 361, 2024
J. Mackey, J. S. Vink, & N. St-Louis, eds.
doi:10.1017/S1743921322002277

The clumped winds of the most massive stars

Sarah A. Brands[1], Alex de Koter[1,2], Joachim M. Bestenlehner[3], Paul A. Crowther[3], Jon O. Sundqvist[2], Joachim Puls[4], Saida M. Caballero-Nieves[5], Michael Abdul-Masih[2,6], Florian A. Driessen[2], Miriam García[7], Sam Geen[1], Götz Gräfener[8], Calum Hawcroft[2], Lex Kaper[1], Zsolt Keszthelyi[1], Norbert Langer[8], Hugues Sana[2], Fabian R. N. Schneider[9,10], Tomer Shenar[1] and Jorick S. Vink[11]

[1]Anton Pannekoek Institute for Astronomy, University of Amsterdam, 1090 GE Amsterdam, The Netherlands
email: s.a.brands@uva.nl

[2]Institute of Astronomy, KU Leuven, Celestijnenlaan 200D, 3001 Leuven, Belgium

[3]Department of Physics and Astronomy, University of Sheffield, Sheffield, S3 7RH, UK

[4]LMU München, Universitätssternwarte, Scheinerstr. 1, 81679 München, Germany

[5]Department of Aerospace, Physics and Space Sciences, Florida Institute of Technology, 150 W. University Boulevard, Melbourne, FL 32901, USA

[6]European Southern Observatory, Alonso de Córdova 3107, Vitacura, Santiago, Chile

[7]Centro de Astrobiología, CSIC-INTA. Crtra. de Torrejón a Ajalvir km 4. 28850 Torrejón de Ardoz (Madrid), Spain

[8]Argelander-Institut für Astronomie, Universität Bonn, Auf dem Hügel 71, 53121 Bonn, Germany

[9]Heidelberger Institut für Theoretische Studien, Schloss-Wolfsbrunnenweg 35, 69118 Heidelberg, Germany

[10]Astronomisches Rechen-Institut, Zentrum für Astronomie der Universität Heidelberg, Mönchhofstr. 12-14, 69120 Heidelberg, Germany

[11]Armagh Observatory, College Hill, Armagh BT61 9DG, UK

Abstract. The core of the cluster R136 in the Large Magellanic Cloud hosts the most massive stars known. The high mass-loss rates of these stars strongly impact their surroundings, as well as the evolution of the stars themselves. To quantify this impact accurate mass-loss rates are needed, however, uncertainty about the degree of inhomogeneity of the winds ('wind clumping'), makes mass-loss measurements uncertain. We combine optical and ultraviolet HST/STIS spectroscopy of 56 stars in the core of R136 in order to put constraints on the wind structure, improving the accuracy of the mass-loss rate measurements. We find that the winds are highly clumped, and use our measured mass-loss rates to test theoretical predictions. Furthermore we find, for the first time, tentative trends in the wind-structure parameters as a function of mass-loss rate, suggesting that the winds of stars with higher mass-loss rates are less clumped than those with lower mass-loss rates.

Keywords. stars: early-type, stars: atmospheres, stars: mass loss, galaxies: Magellanic Clouds

1. Introduction

The young, massive star cluster R136 inside the Large Magellanic Cloud (LMC) hosts a rich population of massive stars, including the most massive stars known

(Bestenlehner et al. 2020; Crowther et al. 2010). Nine stars within this cluster have a mass around or exceeding 100 solar masses, reaching 300 solar masses for the most massive one. These extremely massive stars strongly impact their surroundings through their high ionising fluxes and powerful stellar winds (e.g. Weaver et al. 1977). Furthermore, the mass-losses resulting from these winds impact the stars themselves by altering their evolution and subsequent endpoints (e.g. Langer 2012).

Accurate measurements of the mass-loss rates are crucial for understanding the impact of the stellar winds. However, measurements of these rates are uncertain, in large part due to uncertainty about the degree of inhomogeneity of the winds, often referred to as 'wind clumping' (see e.g. Puls et al. 2008, for a review). In this work, we analyse optical and ultraviolet (UV) spectroscopy of R136, allowing us to study in detail the mass-loss rates and wind structure. For the first time, we study wind structure properties over a wide range of stellar properties, covering late to early O-stars, and three WNh stars.

2. Observations

R136 lies inside the Tarantula Nebula, a region extensively studied in the VLT-FLAMES Tarantula Survey (Evans et al. 2011), in which optical spectra of about 800 massive stars were taken and analysed. However, due to crowding, the cluster R136 has been excluded from this survey. To complement the VFTS survey, R136 has been observed in optical and UV with HST/STIS (PI: Crowther, Proposal ID: 12465). These observations were previously analysed by Crowther et al. (2016), who focus their analysis on the UV part of the data set and in particular on the He II λ1640 emission, and by Bestenlehner et al. (2020), who use the optical part of the data only, deriving detailed stellar parameters, ages, abundances and masses. The present study combines the optical and UV spectra and focuses on the wind properties.

3. Methods

We analyse the spectra of 56 stars using the model atmosphere code FASTWIND, version 10.3.1 (Santolaya-Rey et al. 1997; Puls et al. 2005; Rivero González et al. 2012; Carneiro et al. 2016; Sundqvist & Puls 2018). FASTWIND is a code that is tailored to hot stars with winds. It stands out in terms of speed, allowing for the computation of many models; one model can be computed in about 15-45 minutes on a single CPU.

In our analysis, we take into account a non-smooth wind, in which the clumps (regions of over-density) can become optically thick. This can lead to porosity effects, which may affect the line profiles and hence affect the mass-loss rate measurements. In FASTWIND the clumped wind is implemented by means of a two-component formalism. Fig. 1 contains a schematic explanation of the different wind structure properties.

For each star, we fit the temperature, surface gravity, helium abundance, projected rotational velocity, wind acceleration parameter β, terminal velocity, mass-loss rate and clumping factor $f_{\rm cl}$, a measure for the degree of inhomogeneity of the wind. If the data quality allowed (which was the case for 39 out of 56 stars), we also fitted the nitrogen en carbon abundance, as well as detailed wind structure parameters: the clumping onset velocity $v_{\rm cl,onset}$, the interclump density contrast $f_{\rm ic}$, the velocity-porosity $f_{\rm fvel}$ and the wind turbulence $v_{\rm windturb}$.

We do not obtain these parameters one by one, but fit them simultaneously. With up to 14 free parameters for one fit, this requires an efficient exploration of the parameter space. To this end, we developed a genetic algorithm called KIWI-GA†, using elements of the algorithms of Charbonneau (1995), Mokiem et al. (2005), and Abdul-Masih et al. (2021). Genetic Algorithms operate based on the principles of natural selection and survival of the fittest, and they are particularly suitable for efficient search in large parameter spaces.

† `https://github.com/sarahbrands/Kiwi-GA`

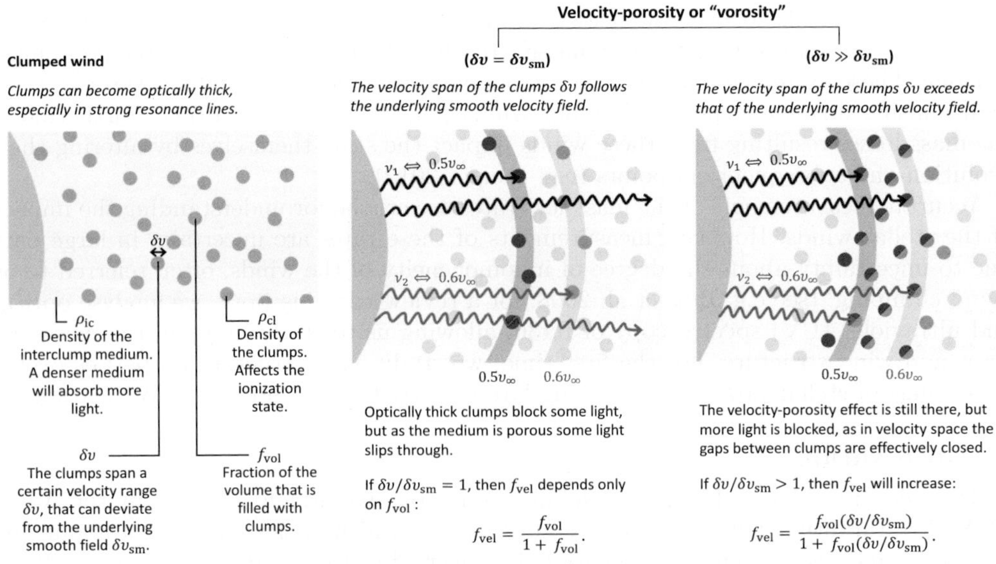

Figure 1. Illustration of the effects of clumping, porosity and velocity-porosity. Credit: Brands et al. (A&A, 663, A36, 2022), reproduced with permission © ESO.

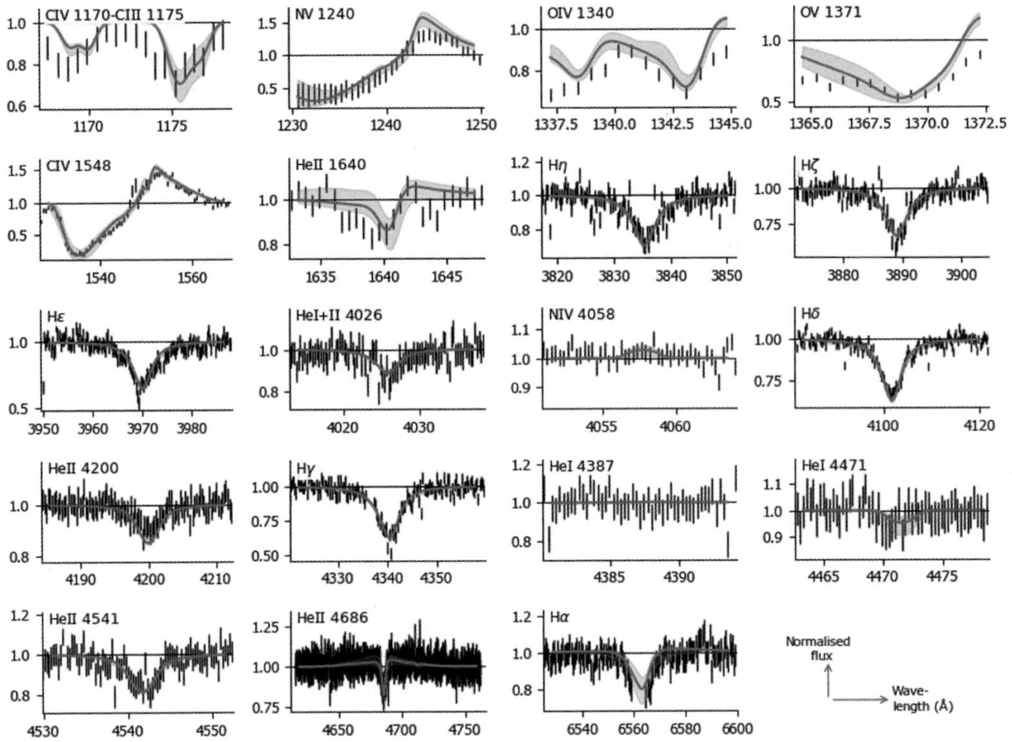

Figure 2. Diagnostic lines in the UV and optical of the star H35. Overlaid are the best fitting model (solid line) and the family of best fitting models (corresponding to 95% confidence intervals on the fitted parameters, shaded region). Credit: Brands et al. (A&A, 663, A36, 2022), reproduced with permission © ESO; figure adapted.

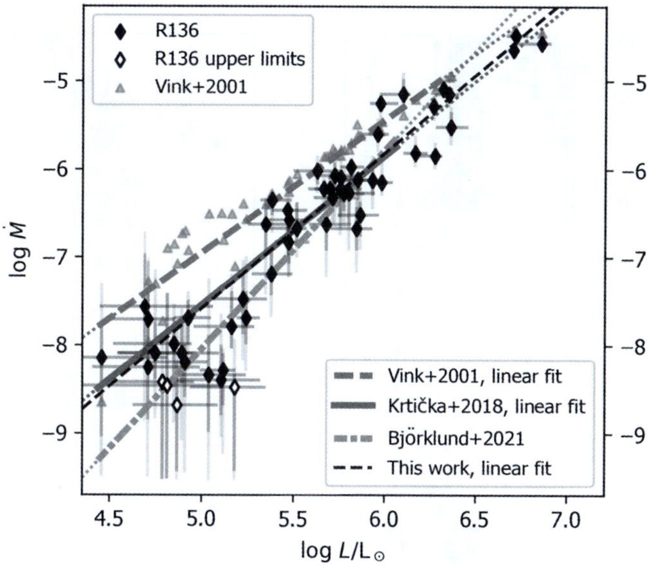

Figure 3. Observed and predicted mass-loss rates as a function of luminosity. Diamonds show our observed values, with light and dark error bars corresponding to 67% and 95% confidence intervals, respectively, thin dashed line the best fit to these values. Thick dashed, solid, and dot-dashed lines show theoretical predictions of Vink et al. (2001), Krtička & Kubát (2018), and Björklund et al. (2021), respectively. Credit: Brands et al. (A&A, 663, A36, 2022), reproduced with permission © ESO; figure adapted.

With the combination of FASTWIND and KIWI-GA we analyse all stars in the sample and constrain the free parameters listed above, including mass-loss rates and wind structure parameters. Fig. 2 shows the spectrum and best fitting models of H35 (spectral type: O3 V), for which we fitted 12 free parameters.

4. Mass-loss rates & wind structure parameters

Figure 3 shows a comparison of the observed mass-loss rates to the theoretical predictions of Vink et al. (2000, 2001), Krtička & Kubát (2018), and Björklund et al. (2021). Overall, the prediction of Krtička & Kubát (2018) matches our observations best, outperforming the other two in the regime where mass-loss impacts evolution the most ($\log L/L_\odot > 5.3$). The observed values in the intermediate luminosity regime ($5.3 < \log L/L_\odot \leq 6.25$) are about a factor two lower than the often used Vink et al. (2001) prescription. This is in line with other studies (Bouret et al. 2005, 2012; Fullerton et al. 2006; Puls et al. 2006; Cohen et al. 2010; Sundqvist et al. 2011; Šurlan et al. 2012; Rauw et al. 2015).

Simultaneously with mass-loss rate, we measure the wind structure parameters $f_{\rm cl}$, $f_{\rm fic}$, $f_{\rm vel}$ and $v_{\rm cl,onset}$. For the first time, these quantities are measured for a wide range of spectral types. Figure 4 shows the obtained values for these quantities against mass-loss rate. Tentative trends in the wind structure parameters as a function of mass-loss rate can be seen: the stars with the highest mass-loss rates seem to have smoother, albeit still clumpy, winds. Comparing these with the measurements of Hawcroft et al. (2021), who study a Galactic sample, we find that they generally agree well. However, the mass-loss rates of the sources studied by Hawcroft et al. (2021) span only a small range, and therefore their study confirms nor contradicts the trends we find. The trends we find thus need to be further investigated in the future.

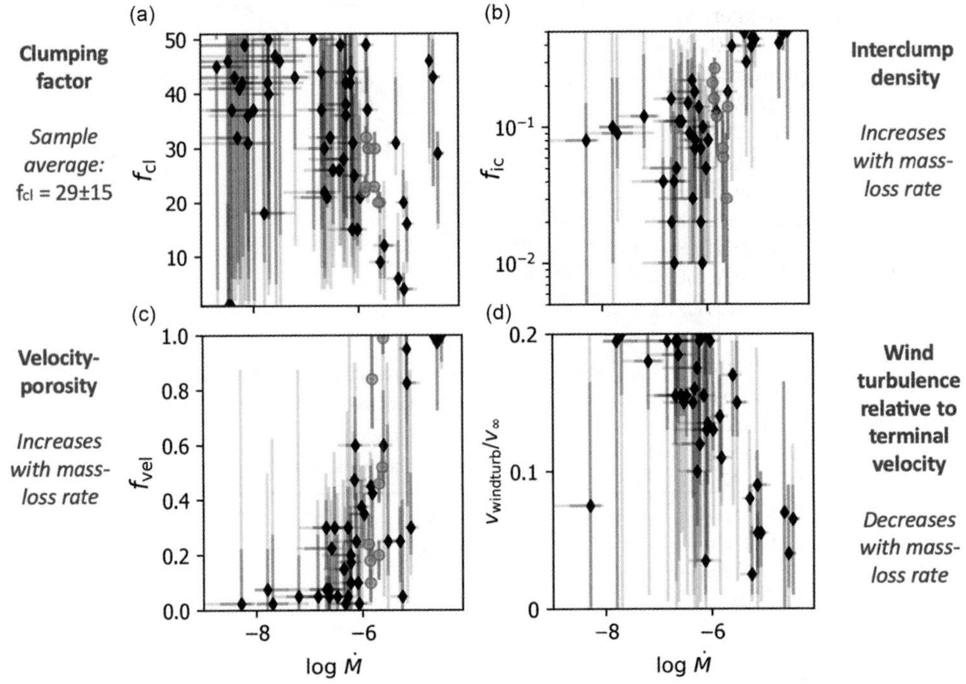

Figure 4. Measured wind structure parameters as a function of mass-loss rate. Diamonds are the values of R136 obtained in this study, with light and dark error bars corresponding to 67% and 95% confidence intervals, respectively. Circles Galactic observations of Hawcroft et al. (2021), error bars corresponding to 95% confidence intervals. Credit: Brands et al. (A&A, 663, A36, 2022), reproduced with permission © ESO; figure adapted.

5. Summary & Outlook

We simultaneously analysed the optical and UV spectra of 56 massive stars in the core of R136, and investigated the mass-loss rates and wind structure for a wide range of spectral types, covering the full O-star range and three WNh stars. We find that the mass-loss rates are in good agreement with the prescription of Krtička & Kubát (2018). We find high clumping factors for most of our stars. Furthermore we observe, for the first time, tentative trends in the wind structure parameters as a function of mass-loss rate. These trends require further investigation; the UV and optical spectra of the ULLYSES and XshootU projects are very well suited for this purpose, and analyses are currently being carried out.

Supplementary material

To view supplementary material for this article, please visit http://dx.doi.org/10.1017/S1743921322002277.

References

Abdul-Masih, M., et al. 2021, A&A, 651, A96
Bestenlehner, J. M., et al. 2020, MNRAS, 499, 1918
Björklund, R., Sundqvist, J. O., Puls, J., & Najarro, F. 2021, A&A, 648, A36
Bouret, J. C., Hillier, D. J., Lanz, T., & Fullerton, A. W. 2012, A&A, 544, A67
Bouret, J. C., Lanz, T., & Hillier, D. J. 2005, A&A, 438, 301
Carneiro, L. P., Puls, J., Sundqvist, J. O., & Hoffmann, T. L. 2016, A&A, 590, A88
Charbonneau, P. 1995, ApJS, 101, 309

Cohen, D. H., Leutenegger, M. A., Wollman, E. E., Zsargó, J., Hillier, D. J., Townsend, R. H. D., & Owocki, S. P. 2010, MNRAS, 405, 2391
Crowther, P. A., Schnurr, O., Hirschi, R., Yusof, N., Parker, R. J., Goodwin, S. P., & Kassim, H. A. 2010, MNRAS, 408, 731
Crowther, P. A., et al. 2016, MNRAS, 458, 624
Evans, C. J., et al. 2011, A&A, 530, A108
Fullerton, A. W., Massa, D. L., & Prinja, R. K. 2006, ApJ, 637, 1025
Hawcroft, C., et al. 2021, A&A, 655, A67
Krtička, J., & Kubát, J. 2018, A&A, 612, A20
Langer, N. 2012, ARA&A, 50, 107
Mokiem, M. R., de Koter, A., Puls, J., Herrero, A., Najarro, F., & Villamariz, M. R. 2005, A&A, 441, 711
Puls, J., Markova, N., Scuderi, S., Stanghellini, C., Taranova, O. G., Burnley, A. W., & Howarth, I. D. 2006, A&A, 454, 625
Puls, J., Urbaneja, M. A., Venero, R., Repolust, T., Springmann, U., Jokuthy, A., & Mokiem, M. R. 2005, A&A, 435, 669
Puls, J., Vink, J. S., & Najarro, F. 2008, A&A Rev., 16, 209
Rauw, G., et al. 2015, ApJS, 221, 1
Rivero González, J. G., Puls, J., Massey, P., & Najarro, F. 2012, A&A, 543, A95
Santolaya-Rey, A. E., Puls, J., & Herrero, A. 1997, A&A, 323, 488
Sundqvist, J. O., & Puls, J. 2018, A&A, 619, A59
Sundqvist, J. O., Puls, J., Feldmeier, A., & Owocki, S. P. 2011, A&A, 528, A64
Vink, J. S., de Koter, A., & Lamers, H. J. G. L. M. 2000, A&A, 362, 295
—. 2001, A&A, 369, 574
Šurlan, B., Hamann, W. R., Kubát, J., Oskinova, L. M., & Feldmeier, A. 2012, A&A, 541, A37
Weaver, R., McCray, R., Castor, J., Shapiro, P., & Moore, R. 1977, The Astrophysical Journal, 218, 377

Discussion

BRANKICA KUBÁTOVÁ: Did you check uniqueness of the obtained results, concerning clumping parameters? Could it happen in your calculation that combinations of different clumping parameters gives similar good fit with observations?

SARAH BRANDS: I checked this to some extent by running single models and comparing them, and each of the clumping parameters seems to have a different behaviour for different lines, so that degeneracy can be broken. But the signal to noise of this data is moderate, so I think it would be good to study this further with (higher signal to noise) ULLYSSES data, to be sure of this.

BRANKICA KUBÁTOVÁ: I think that is really important. One more question, which elements do you include in the FASTWIND calculation? When you talk about abundances you mention only carbon and nitrogen, but did you use some more elements?

SARAH BRANDS: Carbon and nitrogen are the ones that I have as free parameter. Helium is fixed but I fitted that from optical only spectra. The other elements are in there but are fixed. This includes also iron.

PACO NAJARRO: I have a question regarding the clumping. Do you use a structure for the clumping, in the sense that it is variable throughout the wind, or is it constant?

SARAH BRANDS: Clumping starts at a certain point in the wind, which is also a free parameter, and then it increases linearly until 0.3 of the terminal velocity. That is how it is implemented.

PACO NAJARRO: So it goes linearly, and doesn't grow exponentially with the velocity?

SARAH BRANDS: Indeed, it is a linear behaviour.

Massive Stars Near and Far
Proceedings IAU Symposium No. 361, 2024
J. Mackey, J. S. Vink, & N. St-Louis, eds.
doi:10.1017/S1743921322002605

New self-consistent wind parameters to fit optical spectra of O-type stars

Alex Camilo Gormaz-Matamala[1,2]

[1]Departamento de Ciencias, Facultad de Artes Liberales, Universidad Adolfo Ibáñez, Av. Padre Hurtado 750, Viña del Mar, Chile
email: alex.gormaz@uv.cl

[2]Instituto de Astrofísica, Facultad de Física, Pontificia Universidad Catlica de Chile
782-0436 Santiago, Chile

Abstract. We perform spectral fittings for O-type stars based on self-consistent wind solutions, providing $\dot{M}$ and $v(r)$ directly derived from the initial stellar parameters. We introduce our two methods: m-CAK prescription and Lambert-procedure.

The Lambert-procedure allows the calculation of consistent $v(r)$ that reduce the number of free parameters when a spectral fitting using CMFGEN is performed, even without recalculation of the $\dot{M}$. Spectra calculated from our Lambert-solutions show significant differences compared to the initial β-law CMFGEN models. For m-CAK prescription, self-consistent solutions provide values for theoretical $\dot{M}$ on the order of the most recent predictions from other studies. Later, we find a global fit with the RT code FASTWIND. This is an important step towards the determination of stellar and wind parameters without using β-law. Our m-CAK prescription is valid for the O-type stars with $T_{\rm eff} \geq 30$ kK and $\log g \geq 3.2$.

We expect that solutions introduced here to be extended to numerous studies about massive stars in future.

Keywords. Early-type stars, Hydrodynamics, Stellar winds

1. The Lambert-procedure

Given the radiative acceleration $g_{\rm rad}(r)$ calculated by CMFGEN, we have the wind equation of motion

$$v\frac{dv}{dr} = -\frac{1}{\rho}\frac{dp}{dr} - \frac{GM_*}{r^2} + g_{\rm rad}(r)\,, \tag{1}$$

new velocity profile $v(r)$ is analytically calculated

$$\Rightarrow -\hat{v}^2 e^{-\hat{v}^2} = -\left(\frac{\hat{r}_c}{\hat{r}}\right)^4 \exp\left[-1 - 2\,\hat{v}_{\rm crit}^2\left(\frac{1}{\hat{r}} - \frac{1}{\hat{r}_c}\right) - 2\int_{\hat{r}_c}^{\hat{r}} \hat{g}_{\rm line}\, d\hat{r}\right] \tag{2}$$

by implementing the Lambert W-function as

$$\hat{v}(\hat{r}) = \sqrt{-W(x(\hat{r}))}\;. \tag{3}$$

This is iteratively implemented, until convergence is achieved. Mass-loss rate $\dot{M}$ is a free parameter for the Lambert-procedure, but it needs to be constrained to accurately satisfy the full equation of motion in CMFGEN, including clumping.

$$\left(v - \frac{a^2}{v}\right)\frac{dv}{dr} = g_{\rm rad} - \frac{GM_*}{r^2} + 2\frac{a^2}{r} + \frac{a^2}{f}\frac{df}{dr}\,. \tag{4}$$

Results of this procedure are shown in Table 1.

Table 1. Results for our modelled stars using the Lambert-procedure (Gormaz-Matamala et al. 2021)

	ζ-Puppis	**HD 163758**	**α-Cam**
$T_{\rm eff}$ [kK]	41	34.5	28.2
$\log g$	3.6	3.4	2.975
$R_*/R_\odot$	17.9	19.1	30.3
v_∞ [km s^{-1}]	2 740	2 400	2 650
$\dot{M}$ [$10^{-6} M_\odot$ yr^{-1}]	2.7	1.2	0.85
f_∞ [$1/D_\infty$]	0.10	0.05	0.05

Table 2. Stellar and wind parameters for HD 192639, obtained using the m-CAK prescription (Gormaz-Matamala et al. 2022a)

		HD 192639	
$T_{\rm eff}$ (kK)	34.0	$(k, \alpha, \delta)_{\rm sc}$	(0.047,0.694,0.089)
$\log g$	3.25	$\log \dot{M}_{\rm sc}$ ($M_\odot$ yr^{-1})	$-5.783\pm.090$
$R_*/R_\odot$	19.8	v_∞ (km s^{-1})	$1\,460\pm160$
$M_*/M_\odot$	25.4	$f_{\rm cl}$	6.25
$L_*/L_\odot$	4.73×10^5	$v_{\rm rot}$ (km s^{-1})	100
[He/H]	0.10	$\log D_{\rm mom}$	28.83

2. The m-CAK prescription

Equation of motion is solved using the line-force parameters k, α and δ (Gormaz-Matamala et al. 2019)

$$v\frac{dv}{dr} = -\frac{1}{\rho}\frac{dp}{dr} - \frac{GM_{\rm eff}}{r^2} + g_{\rm es}(r)\, k\, t^{-\alpha}\left(\frac{N_{\rm e}}{W}\right)^{\delta} . \qquad (5)$$

Line-force parameters (k, α, δ) are self-consistently calculated with the wind hydrodynamics, by means of the force-multiplier ($\mathcal{M}(t)$) which is defined as a function of the independent optical depth (t)

$$\mathcal{M}(t) = k\, t^{-\alpha}\left(\frac{N_{\rm e}}{W}\right)^{\delta} = \frac{g_{\rm line}}{g_{\rm es}}\ , \text{ with } t = \sigma_{\rm es} v_{\rm th}\, \rho(r)\left(\frac{dv}{dr}\right)^{-1} , \qquad (6)$$

We generate a large grid with of solutions for the line-force parameters, for different values of effective temperatures and surface gravities. Self-consistent solution is delimited to a specific range of t, which depends on the hydrodynamic conditions such as incorporation of temperature structure $T(r)$. New self-consistent values for mass-loss rate ($\dot{M}_{\rm sc}$) and terminal velocity ($v_{\infty,\rm sc}$) are obtained, which are used to perform synthetic spectra and determine new stellar and wind parameters. Results for the star HD 192639 are shown in Table 2.

Because wind parameters $\dot{M}_{\rm sc}$ and $v_{\infty,sc}$ now depend on stellar parameters, we do our spectral fittings reducing the number of free parameters. Mass-loss rates derived from the m-CAK prescription can be applied for the generation of evolutionary tracks for massive stars (Gormaz-Matamala et al. 2022b, accepted in A&A).

References

Gormaz-Matamala A. C., Curé M., Lobel A., Panei J. A., Cuadra J., Araya I., Arcos C., et al., 2022a, *A&A*, 661A, 51G

Gormaz-Matamala A. C., Curé M., Hillier D. J., Najarro F., Kubátová B., Kubát J., 2021, *ApJ*, 920, 64G

Gormaz-Matamala A. C., Curé M., Cidale L. S., Venero R. O. J., 2019, *ApJ*, 873, 131G

Massive Stars Near and Far
Proceedings IAU Symposium No. 361, 2024
J. Mackey, J. S. Vink, & N. St-Louis, eds.
doi:10.1017/S1743921322002289

Structure in stellar winds of early B type supergiants

Timothy N. Parsons[1], Raman K. Prinja[1], Lidia M. Oskinova[2] and Derck L. Massa[3]

[1]Department of Physics and Astronomy, University College London, Gower Street, London WC1E 6BT, UK
email: timothy.parsons.15@ucl.ac.uk, rkp@star.ucl.ac.uk

[2]Institut für Physik und Astronomie, Universität Potsdam, Karl-Liebknecht-Str. 24/25, 14476 Potsdam, Germany
email: lida@astro.physik.uni-potsdam.de

[3]Space Science Institute, 4750 Walnut Street, Suite 205, Boulder, CO 80301, USA
email: dmassa@SpaceScience.org

Abstract. One significant difficulty in reliable quantification of the rates of mass-loss from hot, massive stars lies in uncertainties associated with quantifying temporal and spatial variability within stellar winds. The consequences of low-metallicity conditions for wind structure also merit continued investigation. We present initial results from ULLYSES data with the aim of identifying structure within the stellar winds of early B type supergiants with sub-solar metallicities in the Large and Small Magellanic Clouds. We demonstrate how single-epoch ULLYSES data can be used to investigate significant wind structure for these stars.

Keywords. stars: winds, outflows, supergiants, ultraviolet: stars, techniques: spectroscopic

1. Background and Technique

Mass-loss rates are critical for understanding the evolution of hot, massive stars and their feedback into the interstellar medium. These stars exhibit structured and variable stellar winds and diagnostics of these features are reflected in the UV wind line profiles. Through use of the HST ULLYSES UV data sets, we have the opportunity to probe wind structure of these stars at sub-solar metallicity.

Earlier studies, such as Prinja & Massa (2010) have sought to identify observational signatures of structures within the stellar winds of OB type stars, based on individual UV spectra, including the effect of wind clumping on doublet line ratios. We focus here on evidence for structure in winds of B type supergiant stars within the Large and Small Magellanic Clouds (LMC, SMC).

In smooth winds, the ratios of the optical depths of the doublet components will be the same as the ratios of the doublet oscillator strengths (f). In structured winds however, the doublet component optical depths are determined by the coverage fraction, which itself describes the fraction of the wind structure that is optically thick. In such a case therefore, the optical depth ratio of the doublet components will depart from the f-value ratio and approach unity.

We consider evidence for such departure by examining each element of the Si IV 1394, 1403 Å feature (f-value ratio $= 2$) in the UV spectra of target stars. These doublet components are well-separated for moderate terminal wind velocities, so can be treated as radiatively decoupled, permitting them to be separately fitted using a Sobolev with exact integration method (SEI) model (Lamers et al. (1987), Massa et al. (2003)).

Table 1. Tabulation of spectral types of initial target stars and mean radial optical depth ratios, across a normalised velocity (v/v_∞) range of 0.7-0.35, for blue/red components of the Si IV 1394, 1403 Å doublet feature (f-value ratio = 2). The final two columns show the standard deviation and standard error for each mean optical depth ratio calculation.

Star	Spectral type	$<\tau_{\rm blue}/\tau_{\rm red}>$	Std dev.	Std error
AzV 210	B1.5 Ia	1.55	0.27	0.10
AzV 242	B0.7 Iaw	1.59	0.25	0.09
AzV 266	B1 I	1.68	0.57	0.20
Sk-68 140	B0.7 Ib-Iab Nwk	1.59	0.46	0.16
NGC2004 ELS 3	B5 Ia	1.92	0.20	0.07

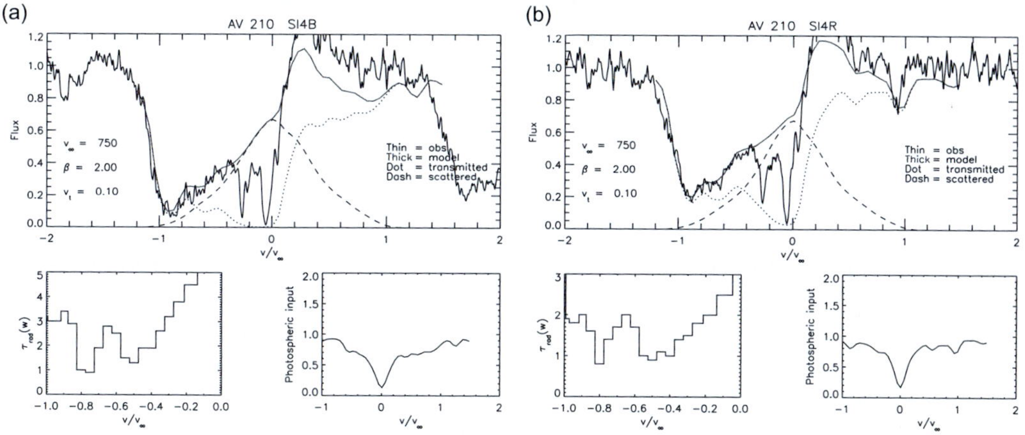

Figure 1. SEI-derived model fits for (a) blue and (b) red components of the Si IV doublet feature in the UV spectrum of SMC star AzV 210 (spectral type B1.5 Ia).

2. Observations, Results and Future Work

The target stars initially examined are all supergiant stars, principally early B types (see Table 1). Fig. 1 shows an example, for the SMC blue supergiant AzV 210, of the separate SEI model fit for the blue and red components of the Si IV doublet (thick red line), together with radial optical depths across normalised wind velocities and the photospheric contribution to the spectrum, from a normalised TLUSTY model fit (Hubeny (1988) and later guides).

We present initial results, showing evidence for wind structure around the SMC stars, AzV 210 and AzV 242 and a lack of significant wind structure around the later-type LMC star, NGC2004 ELS 3. Other results are inconclusive. Further early B type supergiant stars from the ULLYSES database are being similarly examined.

Supplementary material

To view supplementary material for this article, please visit http://dx.doi.org/10.1017/S1743921323001771.

References

Hubeny, I. 1988, *Comp Phys Comm*, 52, 103
Lamers, H.J.G.L.M., Cerruti-Sola, M. & Perinotto, M. 1987, *ApJ*, 314, 726
Massa, D.L., Fullerton, A.W., Sonneborn, G. & Hutchings, J.B. 2003, *ApJ*, 586, 996
Prinja, R.K. & Massa, D.L. 2010, *A&A*, 521, L55

Massive Stars Near and Far
Proceedings IAU Symposium No. 361, 2024
J. Mackey, J. S. Vink, & N. St-Louis, eds.
doi:10.1017/S1743921322003271

Upper Mass-Loss Limits and Clumping in the Intermediate and Outer Wind Regions of OB stars

M. M. Rubio-Díez[1,2], J. O. Sundqvist[3], F. Najarro[2], A. Traficante[4], J. Puls[5], L. Calzoletti[4] and D. Figer[6]

[1]Facultad de Ciencias, Universidad de Alcalá, Alcalá de Henares, Madrid, Spain
email: mmar.rubio@uah.es

[2]Centro de Astrobiología, CSIC-INTA, Torrejón de Ardoz, Madrid, Spain

[3]Instituut voor Sterrenkunde, KU Leuven, Leuven, Belgium

[4]IAPS-INAF, Roma, Italy

[5]Universitäts-Sternwarte München, München, Germany

[6]Center for Detectors, Rochester Institute of Technology, Rochester, New York, USA

Abstract. Mass-loss is a key parameter throughout the evolution of massive stars. In this work we probe the radial clumping stratification of OB stars in the intermediate and outer wind regions ($r \gtrsim 2R_*$; r, radial distance to photosphere), derive upper limits for mass-loss rates, $\dot{M}_{\rm max}$, and compare them to current theoretical mass-loss recipes implemented in evolutionary models. A key conclusion of our analysis regards the derived upper-limit mass-loss rates of B supergiants, independently of clumping, which calls for an urgent revision of the role recombination of iron-like elements plays in determining the mass-loss rates of objects that cross the bi-stability region, and a careful analysis of corresponding effects for stellar evolution models.

Keywords. stars: early-type, stars: atmospheres, stars: mass loss

---

It is well known that the presence of inhomogeneities in stellar winds (*clumping*; see Puls et al. (2008) and references within) leads to severe discrepancies among different mass-loss rate diagnostics, and between empirical estimates and theoretical predictions (e.g. Fullerton et al. (2006)).

In this work, we derive the minimum radial stratification of the clumping, $f_{\rm cl}^{\rm min}(r)$, through the stellar wind of 25 OB starts, by using density-squared diagnostics, and the corresponding maximum mass-loss rate, $\dot{M}_{\rm max}$, normalising clumping factors to the outermost wind region ($f_{\rm cl}^{\rm far} = 1$) (Puls et al. (2006)). Finally, we compare the obtained $\dot{M}_{\rm max}$ to theoretical mass-loss rate recipes (Vink et al. (2000), 2001; hereafter V00 and V01) commonly implemented in current stellar evolution codes (MESA-Bonn, Geneva).

We find that the clumping degree for r $\gtrsim 2R_*$ decreases or stays constant with increasing radius, regardless of luminosity class or spectral type (22/25 in our sample). However, a dependence of the clumping degree on luminosity class and spectral type at the intermediate region relative to the outer ones has been observed: O supergiants present, on average, a factor 2 larger clumping factors than B supergiants (Fig. 1). In addition, we

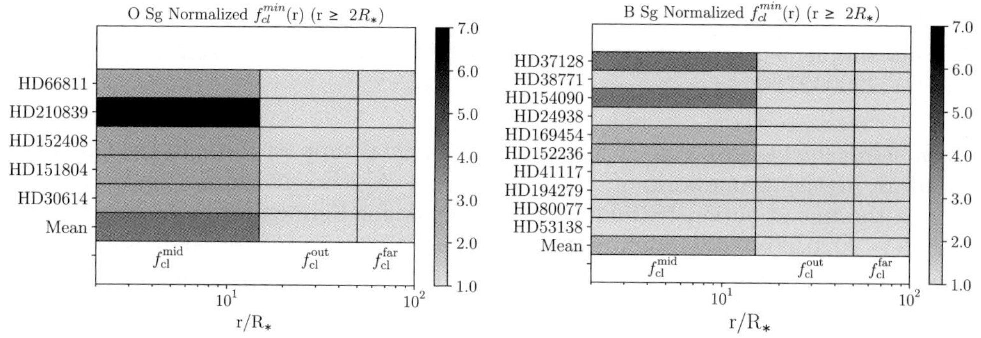

Figure 1. Derived individual minimum and average values of the clumping factors through the stellar wind for $r \gtrsim 2R_*$ for the O (left) and B (right) supergiants in our sample.

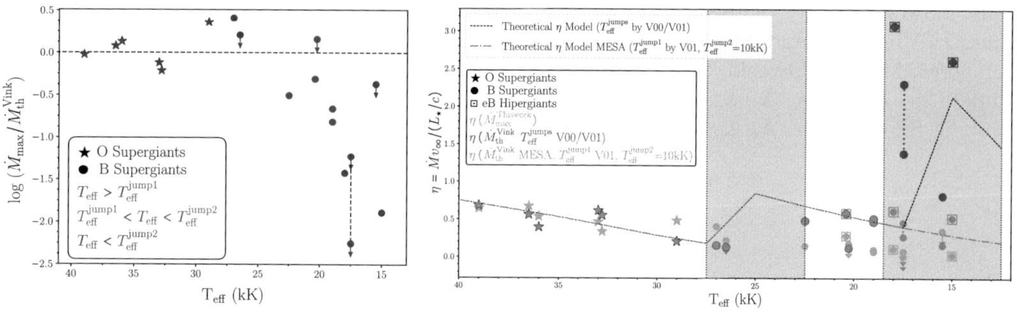

Figure 2. *Left:* this work empirical-maximum ($\dot{M}_{\rm max}$) to theoretical ($\dot{M}_{\rm th}^{\rm Vink}$) mass-loss rates ratio as a function of $T_{\rm eff}$ for the OB supergiants in our sample. $\dot{M}_{\rm th}^{\rm Vink}$ was computed via the recipes by V00 and V01, depending on the temperatures of the jumps (see legend). Arrows indicate upper limits and a dotted line joints two possible values for a given source. *Right:* empirical-maximum (clear blue) vs. theoretical (magenta and orange) wind performance numbers, η, as a function of $T_{\rm eff}$, for OB supergiants. Theoretical η where obtained using $\dot{M}_{\rm th}^{\rm Vink}$ recipes as they are implemented in Geneva and MESA evolution model and code, respectively. Dashed and dotted-dashed lines correspond to theoretical predictions for a source with log L/L$_\odot = 5.75$ and $M_* = 45 M_\odot$ for Z$_\odot$ from Geneva and MESA implementations, respectively. The shadowed regions are the first and second bi-stability jump zones as defined by V00.

find that the estimated $\dot{M}_{\rm max}$ for B supergiants is at least one order of magnitude (before finally decreasing) lower than the values usually adopted by stellar evolution models, whereas for O supergiants upper observational limits and predictions agree within errors (Fig. 2). This implies large reductions of mass-loss rates applied in evolution-models for B supergiants, independently of the actual clumping properties of these winds. However, hydrodynamical models of clumping suggest absolute clumping factors in the outermost radio-emitting wind on the order $f_{\rm cl}^{\rm far} \approx 4-9$ (Runacres & Owocki (2002)); assuming these values would imply a reduction of mass-loss rates included in stellar evolution models by a factor $2-3$ for O supergiants (above $T_{\rm eff} \sim 26.5$ kK) and by factors $6-200$ for B supergiants below the so-called first bi-stability jump (below $T_{\rm eff} \sim 22$ kK). While such reductions agree well with new theoretical mass-loss calculations for O supergiants (Björklund et al. (2021)), our empirical findings call for a thorough re-investigation of B supergiants mass-loss rates and their associated effects on stellar evolution.

Supplementary material

To view supplementary material for this article, please visit http://dx.doi.org/10.1017/S1743921322003271.

Acknowledgements.- M.M.R-D acknowledge financial support through the Community of Madrid in the framework of the Pluriannual Agreement with the University of Alcalá in the line of action "Stimulus to Excellence for Permanent University Lecturers (EPU-INV/2020/008)", through Spanish grants ESP2017-86582-C4-1-R and PID2019-105552RB-C41 (MINECO/MCIU/AEI /FEDER), and from the Spanish State Research Agency (AEI) through the Unidad de Excelencia "María de Maeztu"-Centro de Astrobiología (CSIC-INTA) project No. MDM-2017-0737.

References

Björklund, R., Sundqvist, J. O., Puls, J., & Najarro, F., 2021, *A&A*, 648, A36
Fullerton, A.W., Massa, D.L. & Prinja, R.K., 2006, *ApJ*, 637, 1025
Puls, J., Markova, N., Scuderi, S., et al., 2006, *A&A*, 454, 625
Puls, J., Vink, J.S. and Najarro, F., 2008, *A&A Rev*, 16, 209
Runacres, M. C. & Owocki, S. P., 2002, *A&A*, 381, 1015
Vink, J. S., de Koter, A., & Lamers, H. J. G. L. M., 2000, *A&A*, 362, 295
Vink, J. S., de Koter, A., & Lamers, H. J. G. L. M., 2001, *A&A*, 369, 574

Massive Stars Near and Far
Proceedings IAU Symposium No. 361, 2024
J. Mackey, J. S. Vink, & N. St-Louis, eds.
doi:10.1017/S1743921322003295

Astro+: Design, construction, and scientific exploitation of a large-scale massive star spectroscopic database

Klaus Rübke[1], Amparo Marco[1], Ignacio Negueruela[2], Artemio Herrero[3,4], Sergio Simon-Diaz[3,4], Hugo Tabernero[5] and Lee Patrick[2,6]

[1]Departamento de Física, Ingeniería de Sistemas y Teoría de la Señal, Universidad de Alicante, Carretera de San Vicente s/n, E03690, San Vicente del Raspeig, Spain
email: klaus.rubke@ua.es

[2]Departamento de Física Aplicada, Facultad de Ciencias, Universidad de Alicante, Carretera de San Vicente s/n, E03690, San Vicente del Raspeig, Spain

[3]Instituto de Astrofísica de Canarias, Vía Láctea s/n, E38200, La Laguna, Tenerife, Spain

[4]Universidad de La Laguna, Departamento de Astrofísica, E38206 La Laguna, Tenerife, Spain

[5] Centro de Astrobiología, CSIC-INTA. Campus ESAC. C. bajo del castillo s/n. E-28 692 Villanueva de la Cañada, Madrid, Spain

[6]School of Physical Sciences, The Open University, Walton Hall, Milton Keynes MK7 6AA, UK

Abstract. Massive stars are the objects that condition the evolution of the interstellar medium by the amount of energy released during their lives and especially by their death as a supernova explosion. All the data provided by the previous and ongoing missions of ground and space telescopes have saturated us with the amount of information, which is no longer easy to process daily by human routines. To this end, we present the development of a massive star spectroscopic interactive database designed for scientific research.

Keywords. Database, Massive-stars, Automatic-programs, parameter-determination.

1. Introduction

We have created a web application Astro+, a dedicated massive star spectra storage site, to have a centralised place for all the spectra available from, 1.the users who want to contribute, 2.the data already available and 3.the data coming from the different new missions. Our application is divided into three main parts, the upload, the search and the science modules. Within the first, we find the upload process, where the user will be able to upload spectra in both FITS and ASCII format.

2. Upload method

FITS has different configurations (array, binary data, etc.),and so we allow the user options to set the FITS configuration. On the other hand, the ASCII upload is quite simple, a separate header and spectrum need to be provided, following the standard template required. Once the upload is done, the user can track the upload and correct any errors by using the log we provide. The entered parameters can also be checked, and the user can reprocess the already uploaded file without the need to re-upload the files and then submit the upload for review.

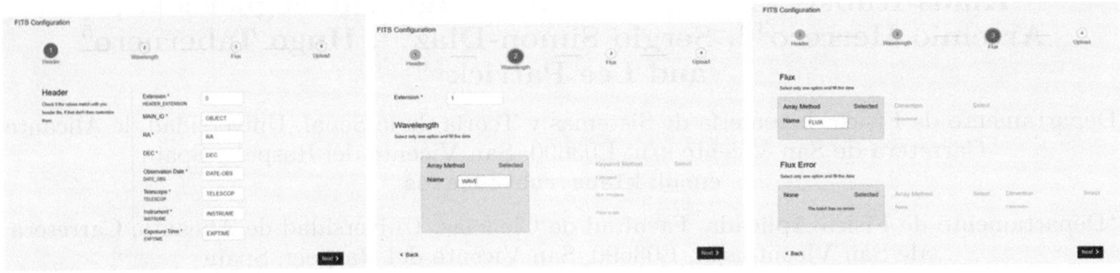

3. Visualization/Search

To inspect the uploaded spectrum, we provide a friendly and easy way to visualize the spectrum, complemented with visual identification using the Aladin API. We add the data from Simbad and Gaia databases through a correlation based on ID or coordinates to check if the star was correctly identified. From Simbad we also adopt the name of the star as the main identification ID to our database.

The search tool,can be used to search through the available spectra in the database by ID, coordinates or by using SQL syntax for more specific star types.

4. Science/AI

By a battery of programs created in Python, based on the best model determination methods (Simón-Díaz & Herrero 2007), (Simón-Díaz et al. 2011), (Simón-Díaz & Herrero 2014) and (Simón-Díaz et al. 2015) on a grid FASTWIND (Santolaya-Rey et al. 1997), (Puls et al. 2005), we can obtain the most critical parameters, such as $V_{\rm rad}$, $V \sin i$, $T_{\rm eff}$ and $\log g$, in a completely automatic way, for early or late massive stars. We are now contrasting preliminary results for Blue/Hot stars with those obtained by (Holgado et al. 2018) from which we are finding errors within the expected, $T_{\rm eff} \sim 1000$[K] and $\log\, g \sim 0.3$[dex].

Acknowledgements

This research is partially supported by the Spanish Government Ministerio de Ciencia, Innovación y Universidades under grant PGC2018-093741-B-C21/22 (MICIU/AEI/FEDER, UE) and the Generalitat Valenciana under grant PROMETEO/2019/041.

Supplementary material

To view supplementary material for this article, please visit http://dx.doi.org/10.1017/S1743921322003295.

References

Holgado, G., et al. 2018, A&A, 613, A65

Puls, J., Urbaneja, M. A., Venero, R., Repolust, T., Springmann, U., Jokuthy, A., & Mokiem, M. R. 2005, A&A, 435, 669

Santolaya-Rey, A. E., Puls, J., & Herrero, A. 1997, A&A, 323, 488
Simón-Díaz, S., Castro, N., Garcia, M., Herrero, A., & Markova, N. 2011, Bulletin de la Societe Royale des Sciences de Liege, 80, 514
Simón-Díaz, S., & Herrero, A. 2007, A&A, 468, 1063
—. 2014, A&A, 562, A135
Simón-Díaz, S., et al. 2015, in Highlights of Spanish Astrophysics VIII, 576–581

Massive Stars Near and Far
Proceedings IAU Symposium No. 361, 2024
J. Mackey, J. S. Vink, & N. St-Louis, eds.
doi:10.1017/S1743921322002046

Quantitative spectroscopy of mid B-type supergiants

David Weßmayer[1], Norbert Przybilla[1] and Keith Butler[2]

[1]Institut für Astro- und Teilchenphysik, Universität Innsbruck, Technikerstr. 25/8, 6020 Innsbruck, Austria
email: david.wessmayer@uibk.ac.at

[2]LMU München, Universitätssternwarte, Scheinerstr. 1, 81679 München, Germany

Abstract. B-type supergiants show enormous potential as resourceful tools to address a wide range of astrophysical questions concerning stellar atmospheres, stellar and galactic evolution and the cosmic distance scale. For the purposes of a comprehensive analysis of these objects we test a hybrid non-LTE approach – line-blanketed model atmospheres computed under the assumptions of local thermodynamic equilibrium (LTE) in combination with non-LTE line-formation calculations. An observational sample of 14 Galactic B-type supergiants with masses below about 30 $M_\odot$ is investigated on the basis of high-resolution Echelle spectra. The results of this analysis – atmospheric and fundamental stellar parameters, the characterisation of the interstellar sightlines to the objects, as well as derived spectroscopic distances and multi-species abundances – are subjected to multiple tests of consistency.

Keywords. Stars: abundances, Stars: atmospheres, Stars: early-type, Stars: evolution, supergiants

---

With their strong stellar winds, ionizing radiation and inevitable demise in supernova explosions, massive stars contribute crucially to the evolution of galaxies. Overall, the evolutionary path of massive stars in the Herzsprung-Russell diagram (HRD) is split into two domains: stars with masses larger than about 40 $M_\odot$ start their lives as early- and mid-O type stars on the main sequence before becoming early B-type supergiants. They lose their envelopes via strong stellar winds, causing them to remain blue objects throughout their lives. Their less massive counterparts begin as late O- and early B-type main-sequence stars, and evolve past the B-type supergiant stage to become red supergiants at least for some time in their later evolution. Though multiple aspects of these evolutionary pathways – both single and binary – are understood, some details are yet to be resolved (e.g. Maeder & Meynet 2012). A considerable reduction in the observational uncertainties is crucial to further our understanding of these objects. For this purpose, 14 high S/N, high-resolution Echelle spectra of B-type supergiants with masses below about 30 $M_\odot$ are analysed employing a hybrid non-LTE approach as outlined by Przybilla et al. (2006). In this methodology, static, plane-parallel line-blanketed LTE model atmospheres are computed with ATLAS12 (Kurucz 2005), which serve as the basis for non-LTE line-formation calculations with DETAIL (Giddings 1981) and SURFACE (Butler & Giddings 1985). For a thourough chemical analysis, state-of-the-art model atoms for the 12 most abundant chemical species (H, He, C, N, O, Ne, Mg, Al, Si, S, Ar and Fe) are utilised. The interstellar sightlines to the objects were characterised by fitting the model spectral energy distributions to archival photometric data and UV-spectrophotometry for all objects.

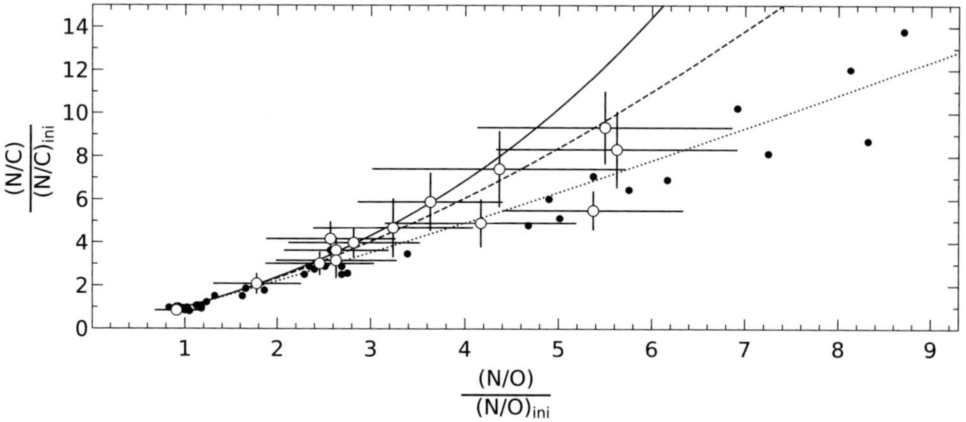

Figure 1. Nitrogen-to-carbon ratio vs. nitrogen-to-oxygen ratio, normalised to initial values corresponding to cosmic abundance standard values (Nieva & Przybilla 2012). Models of stellar evolution – $15\,M_{\odot}$, with metallicity $Z = 0.014$ and angular velocity $\Omega_{\rm rot} = 0.95\Omega_{\rm crit}$ (Georgy et al. 2013, solid line), $15\,M_{\odot}$, $Z = 0.014$, $\Omega_{\rm rot} = 0.568\Omega_{\rm crit}$ (Ekström et al. 2012, dashed line) and $25\,M_{\odot}$, $Z = 0.014$, $\Omega_{\rm rot} = 0.568\Omega_{\rm crit}$ (Ekström et al. 2012, dotted line) – are compared to objects from this work (open circles), B-type main-sequence stars (Nieva & Przybilla 2012) and BA-type supergiants (Przybilla et al. 2010, black dots).

Our methodology can be shown to produce results equivalent to hydrostatic full non-LTE models for the deeper photospheric layers of the B-type supergiants investigated. In particular, we show that a precise and accurate determination of stellar parameters is possible, i.e. effective temperatures can be constrained to 2-3% uncertainty, surface gravities to better than 0.07 dex, masses to about 5%, radii to about 10% and luminosities to better than 25%. Qualitatively, the synthetic spectra corresponding to the derived parameters tightly and reliably reproduce essentially the entire observed optical to near-IR spectra. The determined parameters are further substantiated in a comparison of derived spectroscopic distances with those derived from Gaia EDR3 parallaxes, which shows good agreement for the entire sample, spanning distances from 400 pc to about 6 kpc. The abundances of all analysed species have statistical uncertainties of 0.05 to 0.10 dex, generally. A diagnostic diagram displaying the CNO-abundances of the sample objects and evolution models for rotating stars is shown in Fig. 1. It is apparent that most sample stars are considerably enriched in CN-processed matter, though not as much as some BA-type supergiants in a previous analysis (Przybilla et al. 2006). The close reproduction of the path of nuclear enrichment set out by the models of stellar evolution serves as yet another quality assessment of the derived atmospheric parameters and indicates a low level of systematic uncertainty (Maeder et al. 2014).

Finally, tests are performed for application to intermediate-resolution spectra of B-type supergiants, which can be obtained over extragalactic distances, even in galaxies beyond the Local Group (e.g. Kudritzki et al. 2016). This is in order to facilitate homogeneous analyses of blue supergiants over the full range of the B and A spectral types below the $\sim$30 $M_{\odot}$ limit.

Acknowledgements

D.W. and N.P. gratefully acknowledge support from the Austrian Science Fund FWF project DK-ALM, grant W1259-N27.

Supplementary material

To view supplementary material for this article, please visit http://dx.doi.org/10.1017/S1743921322002046.

References

Butler, K., & Giddings, J. R. 1985, Newsletter of Analysis of Astronomical Spectra, 9 (Univ. London)
Ekström, S., Georgy, C., Eggenberger, P., et al. 20212 *A&A*], 537, A146
Georgy, C., Ekström, S., Granada, A., et al. 2013, *A&A*, 553, A24
Giddings, J. R. 1981, PhD thesis (Univ. London)
Kudritzki, R. P., Castro, N., Urbaneja, M. A., et al. 2016, *ApJ*, 829, 70
Kurucz, R. 2005, *Mem. Societa Astronomica Italiana Suppl.*, 8, 14
Maeder, A. & Meynet, G. 2012, *Rev. Mod. Phys.*, 84, 25
Maeder, A., Przybilla, N., Nieva, M. F., et al. 2014, *A&A*, 565, A39
Nieva, M. F., & Przybilla, N. 2012, *A&A*, 539, A143
Przybilla, N., Butler, K., Becker, S. R., & Kudritzki, R. P. 2006, *A&A*, 445, 1099
Przybilla, N., Firnstein, M., Nieva, M. F., Meynet, G., & Maeder, A. 2010, *A&A*, 517, A38

Chapter III:
Physical processes in massive stars

Massive Stars Near and Far
Proceedings IAU Symposium No. 361, 2024
J. Mackey, J. S. Vink, & N. St-Louis, eds.
doi:10.1017/S174392132200285X

Wind line variability and intrinsic errors in observational mass loss rates

Derck Massa[1] **and Raman Prinja**[2]

[1]Space Science Institute, 4750 Walnut Street Suite 205 Boulder, Colorado 80301, USA
email: dmassa@spacescience.org

[2]Department of Physics & Astronomy, University College London,
Gower Street, London WC1E 6BT, UK
email: rkp@star.ucl.ac.uk

Abstract. UV wind line variability in OB stars appears to be universal. We review the evidence that the variability is due to large, dense, optically thick structures rooted in or near the photosphere. Using repeated observations and a simple model we translate observed profile variations into optical depth variations and, consequently, variations in measured mass loss rates. Although global rates may be stable, measured rates vary. Consequently, profile variations infer how mass loss rates determined from UV wind lines vary. These variations quantify the intrinsic error inherent in any mass loss rate derived from a single observation. These derived rates can differ by factors of 3 or more. Our results also imply that rates from non-simultaneous observations (such as UV and ground based data) need not agree. Finally, we use our results to examine the nature of the structures responsible for the variability.

Keywords. stars: winds, outflows, stars: mass loss

1. Introduction

Wind line variability, as seen in Hα, has been known for many years (e.g., Ebbets 1982). With the advent of *IUE*, it became clear that well developed but unsaturated UV wind lines in OB stars were also highly variable (e.g., Prinja & Howarth 1986). It is important to study this wind line variability for the following reasons: 1) It appears to be a universal property of radiatively driven winds. 2) Currently there are four ways to measure mass loss rates ($\dot{M}$): wind lines, IR/Radio excesses, X-rays and bow shocks. They do not all agree, due to structures (clumping) in the winds (e.g., Fullerton *et al.* 2006; Massa et al. 2017; Kobulnicky et al. 2019 AJ, 158, 73) and wind line variability *presents an opportunity to characterize these wind structures and provide feedback for modelers.* 3) Wind line variability *determines the intrinsic accuracy of a single measurement of* $\dot{M}$.

In the following, we discuss what is known about the nature and origin of the variability. We then introduce a simple model, use it to analyze the variability in a few times series and then summarize our results.

2. Observations of wind line variability

In this section we describe what has been learned from the morphology of wind line variability. Figure 1 shows examples of UV wind line variability in 4 stars with a range in spectral types, but all of which have well developed, but unsaturated Si IV λ1400 resonance doublets. Figure 2 is the dynamic spectrum (a gray scale of all the spectra in a series divided by the mean) of a time series of the B0.5 Ib star, HD 64760, described by Massa *et al.* (1995). Such data sets have lead to the following conclusions.

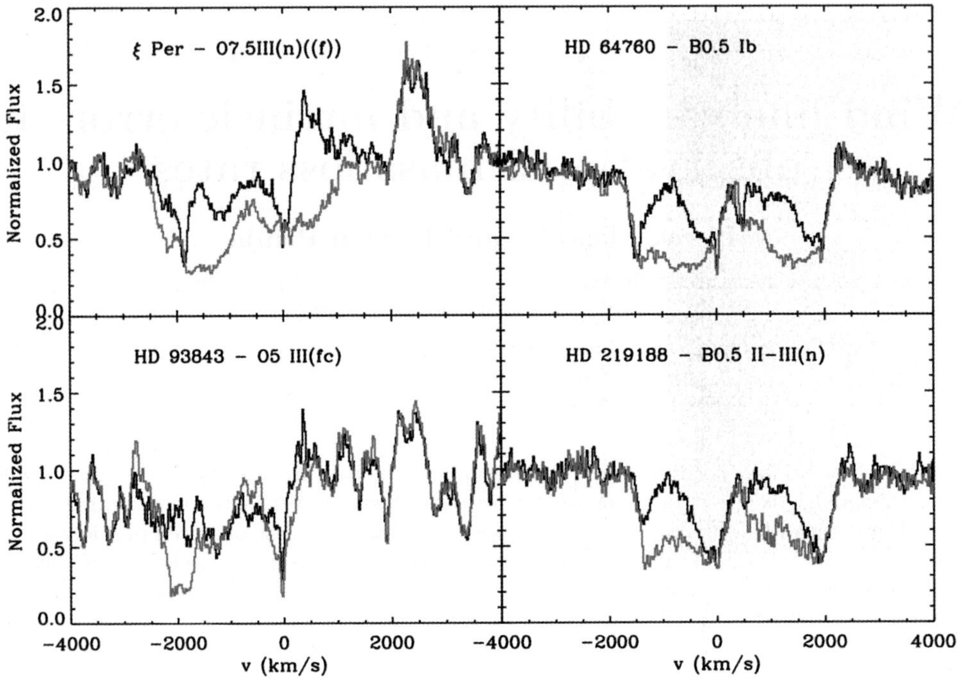

Figure 1. Examples of Si IV$\lambda\lambda$1400 variability in four stars with a range in spectral types.

Wind variability is universal: Wind line variability is seen in repeated observations of just about every OB star with a well developed but unsaturated wind line (e.q. Prinja & Howarth 1986). Further, it has also been observed in LMC and SMC OB stars (Massa *et al.* 2000) and the central stars of planetary nebulae (Prinja *et al.* 2012).
Spiral structures cause the variability: Time series of UV wind lines revealed bow shaped patterns in the dynamic spectra OB Stars (see Figure 2). Analysis of these patterns lead Cranmer & Owocki (1996) to interpret them in terms of co-rotating interaction regions (CIRs). However, any phenomenon that produces a spiral structure in the wind will do. To see the classic bow shape, one must view the star with $\sin i \simeq 1$ and the origin of the spot must be near the equator. Consequently, it is not surprising that the pattern is not seen in some stars with lower $v_{eq} \sin i$ s, although regularly repeating patterns are (Prinja *et al.* 2002).
The spiral structures are optically thick: Prinja & Massa (2010) applied a simple model to show that the Si IV λ1400 doublet wind line profiles of a large sample of B supergiants are best fit using doublet ratios that are between unity and the actual ratio of the oscillator strengths of the doublet, $f_B/f_R \simeq 2$. This indicates that the line of sight to the stellar disk is partially covered by very optically thick structures, and by optically thin material. The result is a doublet absorption that appears unsaturated (because it does not go to zero) but whose relative strength is close to unity (since most of the absorption is from the strongly saturated optically thick component).
The spiral structures reach the stellar surface and are large: The fact that the variability appears to extend to $v \simeq 0$, does not necessarily mean that the physical structures extend to $r \simeq R_\star$, since this could simply reflect a non-monotonic velocity law. However, Massa & Prinja (2015) showed that the absorption of the excited state line N IV 1718 blends smoothly with the high velocity absorption in resonance lines (see, Fig 3). Since N IV 1718 cannot exist without the strong photospheric radiation field,

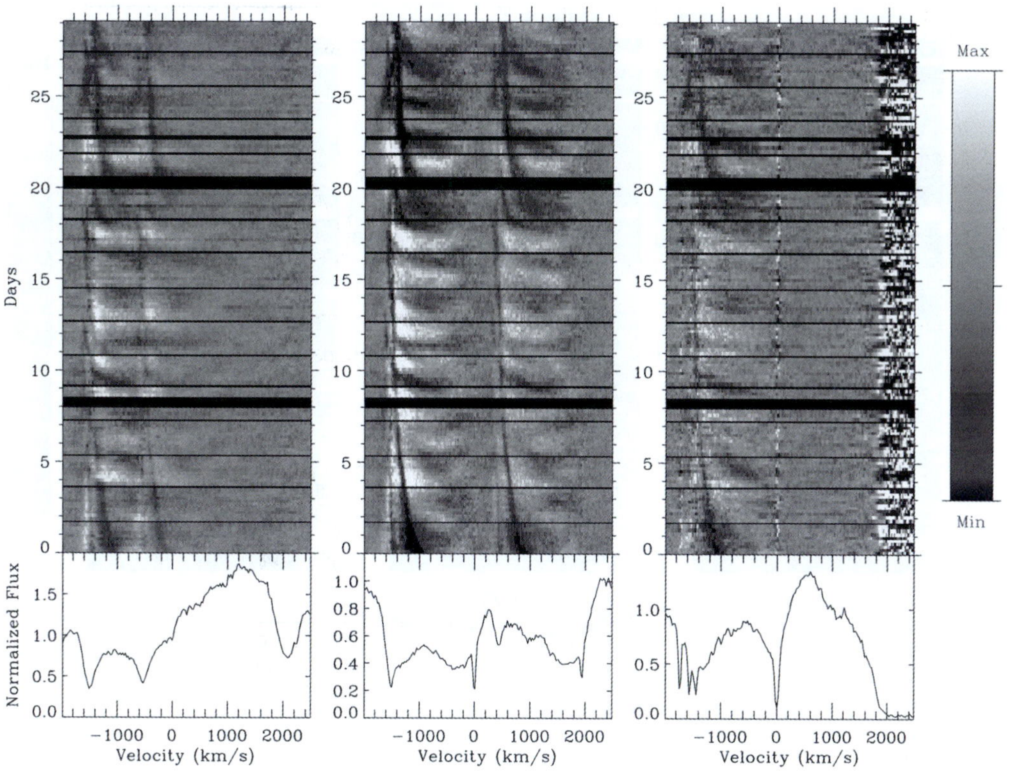

Figure 2. Dynamic spectra of (left to right) N V, Si IV and Si III for HD 64760 (B0.5 Ib, $v \sin i = 216$ km s^{-1}. Massa *et al.* (1995).

this analysis shows that the features must originate near $r = R_\star$. Furthermore, since the features cause detectable absorption near $R_\star$, they must occult a significant portion of the stellar disk.

The spiral structures are denser than ambient wind: A large change in the optical depth can indicate either a change in density, a flattening of the velocity law, or a change in the ionization (see, §3). It is, therefore, important to examine observational diagnostics that respond only to density. X-ray fluxes are one such diagnostic. It has been known for some time that the X-ray fluxes in OB stars are variable (see Oskinova *et al.* 2001, Nazé *et al.* 2013, Nazeé *et al.* 2018). Most recently, Massa *et al.* (2019) obtained contemporaneous *XMM* X-ray and *HST* STIS UV spectra of the O7.5 III(n)((f)) star ξ Per. They were able to show that the X-rays are modulated by the spiral patterns, implying that they are significantly denser than the rest of the wind (the exact amount is model dependent).

3. Modeling wind line variability

In order to translate flux variations into the physical parameters needed to estimate the intrinsic error in a single $\dot{M}$ measurement, a model is required. We use the SEI Sobolev model as formulated by Lamers *et al.* (1987) and modified by Massa *et al.* (2003). The calculation of an SEI profile requires the following parameters:

1) A value for v_∞ (determined by a grid search),
2) A velocity law: typically a β-law, of the form $v = v_\infty(1 - a/x)^\beta$, where $x = r/R_\star$.

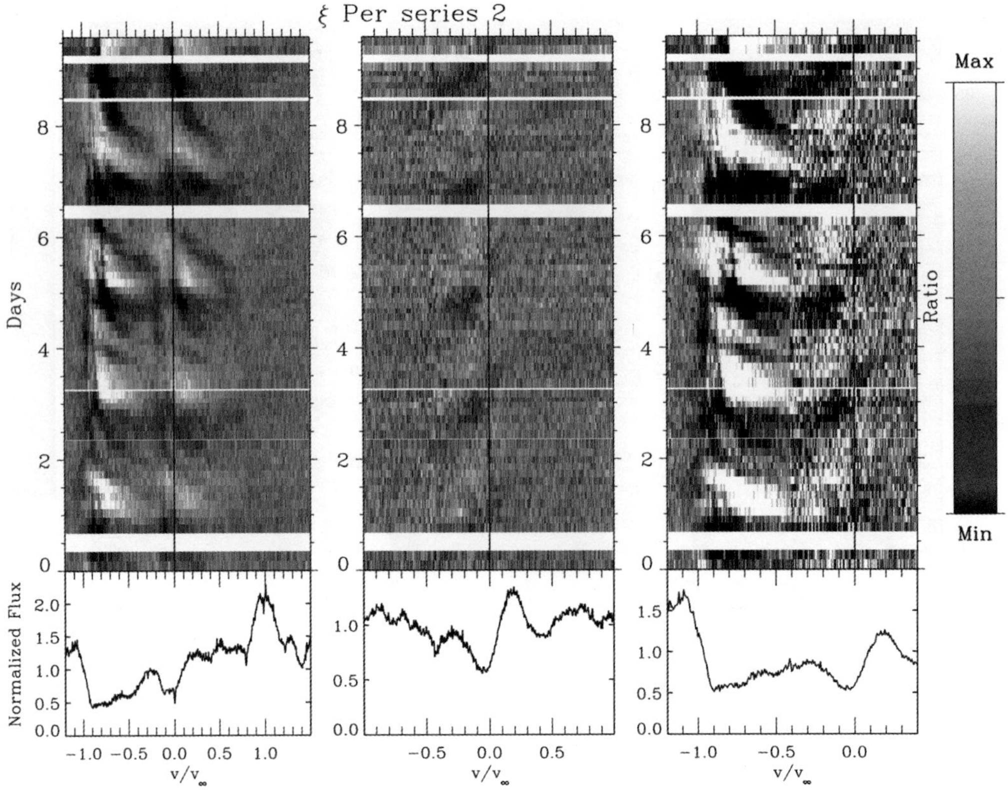

Figure 3. Dynamic spectra of Si IV, N IV, and the two spliced together for ξ Per (v_∞= 2450 km s^{-1}, Massa & Prinja (2015). Shows that structures are tied to the base of the wind where the radiation field is intense.

3) The radial (Sobolev) optical depth of the wind ($w = v/v_\infty$):

$$\tau_{rad}(w) = Const \frac{\dot{M}}{R_\star v_\infty^2} q_i(w) \left(x^2 w \frac{dw}{dx} \right)^{-1}$$

where $Const$ contains atomic parameters and q_i is the ionization fraction. $\tau_{rad}(w)$ is modeled by 20 velocity bins adjusted to obtain the best fitting profile. This approach can be viewed as an inversion of the profile to obtain $\tau_{rad}(w)$. *Note that* τ_{rad} *variations are proportional to derived* $\dot{M}q$ *variations.*

In addition to the usual parameters, we also allow the ratio of the optical depths of the doublets, f-ratio $\equiv f_B/f_R$, to vary. This is a well known means to mimic the effect of optically thick structures partially covering the stellar disk (Prinja & Massa 2015). It also provides an additional diagnostic of how the portion of the wind structures in front of the star varies with time. We also note that it has little affect on τ_{rad}. Its main effect is to improve the SEI fit to the red component of the wind line. All of these parameters are determined by a non-linear least squares fit to the observed profile.

4. Results

Figure 4 shows the quality of the fits that can be achieved. It shows non-linear least squares fits to the Si IV and N V profiles in ρ Leo. The model used v_∞= 1150 km s^{-1}, and a TLUSTY Lanz & Hubeny (2003) model with solar abundance and $(T_{eff}, \log g) = (25kK, 2.5)$ for the photospheric spectrum. The derived τ_{rad}s are shown

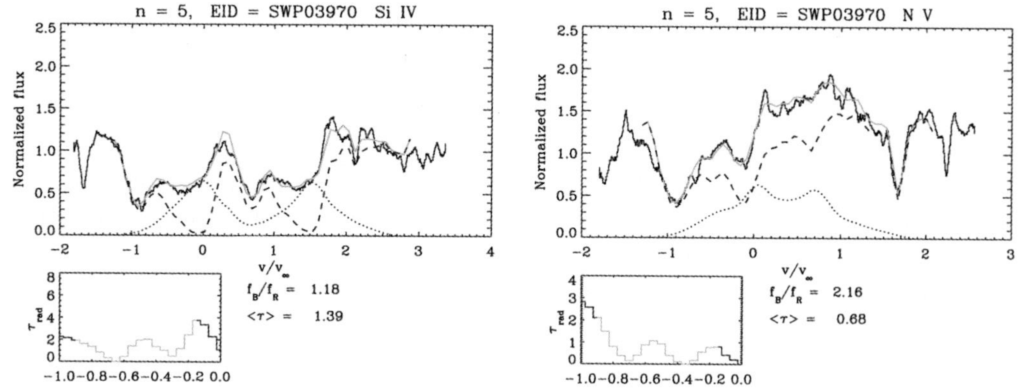

Figure 4. Fits to a Si IV and N V profiles in ρ Leo with derived τ_{rad}s below. Observed = solid black, fit = gray, absorbed photospheric profile = dashed and emission = dotted.

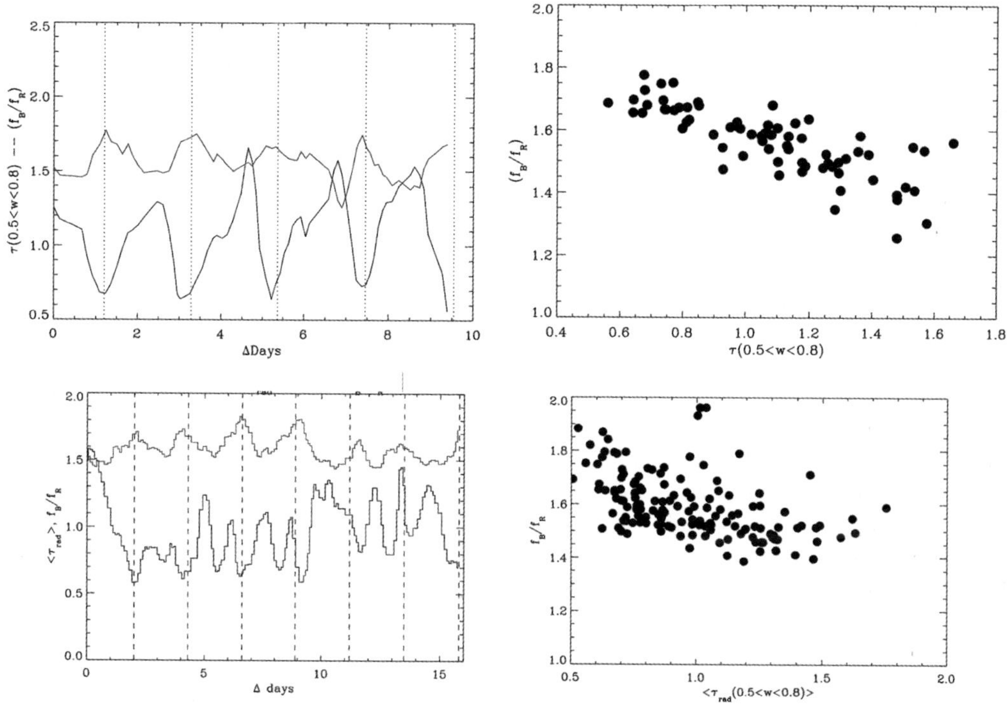

Figure 5. **Top row:** 70 Si IV spectra of ξ Per. Left: Variations of τ_{rad} averaged over $0.5 \leq v/v_\infty \leq 0.8$ (lower curve) and f-ratios (upper curve) versus days from first spectrum. Dotted vertical lines are the 2.086 day period determined by de Jong *et al.* (2001). Right: The same two quantities plotted against each other. **Bottom row:** 146 Si IV spectra of HD 64760. Left: Mean τ_{rad} s (lower curve) and f-ratios (upper curve) versus days from the first spectrum. Dotted vertical lines are a 2.30 Day period. Right: The same two quantities plotted against each other.

below each spectrum, and the portion used to determine $< \tau >$, $0.2 \leq v/v_\infty < 0.9$, is shaded gray. The best fit f-ratio and $< \tau >$ values are also listed. Notice that the fit with the smaller $< \tau >$ has the larger f-ratio.

As part of an ongoing program to analyze all repeated *IUE* observations of normal stars with well developed but unsaturated lines, we have begun with an analysis of a few time series. Figure 5 summarizes the results for two time series, one consisting of

70 spectra of ξ Per and one for 146 spectra of HD 64760. The left hand panels show the mean τ_{rad} s and f-ratios versus time, while the right hand panels show the two quantities plotted against each other.

5. Analysis

The previous results can be used to address two distinct issues: the intrinsic error present in any $\dot{M}$ derived from a single observation, and the nature of the structures responsible for the variations.

The intrinsic error: Recall that τ_{rad} is directly proportional to $\dot{M}$, so variations in τ_{rad} are a surrogate for variations in $\dot{M}$. Examination of the right panels of Figure 5 shows that the ranges of $< \tau_{rad} >$ can be a factor of 3 or more. The means and variances of $< \tau_{rad} >$ for the ξ Per series are $= 1.09$ and 0.27 and for the HD 64760 series they are 1.05 and 0.26. Both cases suggest the best accuracy one can expect is about $\pm 25\%$. However, this may be a lower limit since the distributions are highly non-Gaussian and the fact that the observations were clustered in time may introduce an additional bias. Nevertheless, a reasonable first estimate is $\sigma(\dot{M})/\dot{M} \simeq 1.25$, although this is probably a lower limit.

Constraints on the structures: It is interesting that, for the two series analyzed, the f-ratio varies periodically, as distinctly as the $< \tau_{rad} >$ or even more so. Further, the f-ratio decreases as $< \tau_{rad} >$ increases, suggesting that the change in the apparent optical depth is actually due to a larger fraction of the stellar surface being covered by optically thick material. These results are intriguing, but we emphasize that they are very preliminary.

6. Summary

We reviewed the evidence that wind line variability is universal and probably due to large, spatially coherent spiral structures. These structures extend to the base of the wind and cover much of the stellar disk. Further, they are denser than their surroundings, and because they are optically thick, *their geometry matters.* We then showed that the variability limits the intrinsic accuracy of a single $\dot{M}$ measurement derived from a resonance line to roughly 25%, and that individual measurements can vary by as much as a factor of 3. This variability also compromises the consistency of non-simultaneous measurements. Finally, we discussed how our results can be used to constrain the nature of the structures responsible for the variability.

References

Cranmer, S.R. & Owocki, S.P. 1996, *Ap.J.*, 462, 469
Ebbets, D. 1982, *Ap.J. Supp.*, 48, 399
de Jong, J.A., Henrichs, H.F., Kaper, L., et al 2001, *A&A*, 368, 601
Fullerton, A.W., Massa, D.L., & Prinja, R.K. 2006, *Ap.J.*, 637, 1025
Lamers, H.J.G.L.M., Cerruti-Sola, M., & Perinotto, M. 1987, *Ap.J.*, 314, 726
Lanz, T. & Hubeny, I. 2003, *Ap.J. Supp.*, 146, 417
Massa, D., Fullerton, A.W., Nichols, J.S., et al. 1995, *Ap.J. Letters*, 452, L53
Massa, D., Fullerton, A.W., Hutchings, J.B., et al. 2000, *ApJL*, 538, L47
Massa, D. & Prinja, R.K. 2015, *ApJ*, 809, 12
Massa, D., Oskinova, L., Prinja, R. & Ignace, R. 2019, *Ap.J.*, 873, 81
Massa, D., Fullerton, A.W., Sonneborn, G., et al. 2003, *Ap.J.*, 586, 996
Nazé, Y., Oskinova, L.M., & Gosset, E. 2013, *Ap.J.*, 763, 143
Nazé, Y., Ramiaramanantsoa, T., Stevens, I.R., et al 2018, *A&A*, 609, A81
Oskinova, L.M., Clarke, D., & Pollock, A.M.T. 2001, *A&A*, 378, L21
Fullerton, A.W., Massa, D.L., Prinja, R.K., et al 1997, *A&A*, 327, 699

Prinja, R.K. & Howarth, I.D. 1986, *Ap.J. Supp.*, 61, 357
Prinja, R.K., Massa, D.L., Urbaneja, M.A., et al 2012, *M.N.R.A.S.*, 422, 3142
Prinja, R.K., Massa, D., & Fullerton, A.W. 2002, *A&A*, 388, 587
Prinja, R.K. & Massa, D.L. 2010, *A&A*, 521, L55

Discussion

QUESTION: Can the repetition time of the absorption variability be used to infer the rotation period of the star?

ANSWER: That is possible, however it is complicated. The variability occurs on the rotation time scale and it often appears as if it repeats twice per rotation. However, the variability is not "locked" to the rotation period. If one observes the same star at a different time, the variability time scale may be the same, but the phase will have wandered. So I guess the bottom line is I don't know.

QUESTION: Could a magnetic spot be responsible for the variability of the absorption?

ANSWER: That is certainly possible. However, we have to keep in mind that many of the stars which show the variability are rapid rotators. So if a magnetic field is responsible, it cannot be too strong – otherwise the star would have spun down due to magnetic breaking.

Massive Stars Near and Far
Proceedings IAU Symposium No. 361, 2024
J. Mackey, J. S. Vink, & N. St-Louis, eds.
doi:10.1017/S1743921322002794

Magnetic massive stars from stellar mergers

Fabian R. N. Schneider[1], Sebastian T. Ohlmann[2], Philipp Podsiadlowski[3], Friedrich K. Röpke[4], Steven A. Balbus[5], Rüdiger Pakmor[6] and Volker Springel[6]

[1]Heidelberger Institut für Theoretische Studien,
Schloss-Wolfsbrunnenweg 35, 69118 Heidelberg, Germany
Zentrum für Astronomie der Universität Heidelberg, Astronomisches Rechen-Institut,
Mönchhofstr. 12-14, 69120 Heidelberg, Germany
email: fabian.schneider@h-its.org

[2]Max Planck Computing and Data Facility, Gießenbachstr. 2, 85748 Garching, Germany
email: sebastian.ohlmann@mpcdf.mpg.de

[3]University of Oxford, St Edmund Hall, Oxford, OX1 4AR, United Kingdom
email: podsi@hotmail.com

[4]Heidelberger Institut für Theoretische Studien,
Schloss-Wolfsbrunnenweg 35, 69118 Heidelberg, Germany
Zentrum für Astronomie der Universität Heidelberg, Institut für Theoretische Astrophysik,
Philosophenweg 12, 69120 Heidelberg, Germany
email: friedrich.roepke@h-its.org

[5]Department of Physics, University of Oxford, Keble Rd, Oxford OX1 3RH, United Kingdom
email: steven.balbus@physics.ox.ac.uk

[6]Max-Planck-Institut für Astrophysik, Karl-Schwarzschild-Str. 1, 85748 Garching, Germany
email: rpakmor@MPA-Garching.MPG.DE, email: vspringel@MPA-Garching.MPG.DE

Abstract. The first magnetic field in a star other than the Sun was detected in 1947 in the star 78 Vir. Today, we know that about 10% of these intermediate-mass and high-mass stars have strong, large-scale surface magnetic fields whose origin has remained a mystery till today. It has been suggested that merging of main-sequence and pre-main-sequence stars could produce such strong fields. The massive star τ Sco is a well-known member of the group of magnetic stars and is a blue straggler given its apparently young age compared to that of other members of the Upper Scorpius association. Here, we present 3D magnetohydrodynamic simulations of the coalescence of two massive main-sequence stars and 1D stellar evolution computations of the subsequent evolution of the merger product that can explain τ Sco's magnetic field, apparent youth and other observed characteristics. We argue that field amplification in stellar mergers is a general mechanism to form strongly-magnetised massive stars. Such stars are promising progenitors of magnetars, which may give rise to some of the enigmatic fast radio bursts, and their supernova explosions may be affected by the strong magnetic fields.

Keywords. MHD, binaries: general, blue stragglers, stars: evolution, stars: magnetic fields

1. Introduction

Magnetic fields are ubiquitous in the Universe. They were discovered in our Sun in 1908 by Hale, and some 40 years later also in the intermediate-mass, A-type star 78 Vir (Babcock 1947). Surveys have shown that about 7% of all intermediate- and high-mass stars have strong, large-scale surface magnetic fields (Donati & Landstreet 2009; Grunhut et al. 2017; Schöller et al. 2017). Contrarily to solar-like and other low-mass stars, the magnetic field strengths are not found to correlate with mass of the star or rotational velocity (Kochukhov & Bagnulo 2006). This led to the interpretation that the fields are "fossil", i.e. that they are frozen into the stellar structure after they were created by some, as yet unknown process. This unknown process must be able to explain the existence of magnetic fields in a small subset of main-sequence and pre-main-sequence stars, and it must also explain the observed lack of magnetic stars in close binaries (Carrier et al. 2002; Neiner et al. 2015). Also, there are hardly any binaries known with two magnetic components (see Shultz et al. 2015, for one such example). Magnetic dynamo models or processes acting during the star-formation process have difficulties explaining why only a fraction of stars obtain strong surface magnetic fields and why this is even lower in close binaries. It has thus been suggested that the merging of stars may explain the strong magnetic fields in some stars (Ferrario & Wickramasinghe 2005; Ferrario et al. 2009; Langer 2012; Wickramasinghe et al. 2014) and this idea is further supported by identifying some magnetic stars as blue stragglers (Schneider et al. 2016). In this proceeding, we present our findings that merging of massive main-sequence stars indeed lead to strong magnetic-field amplification and that observed key properties of the blue straggler star τ Sco appear consistent with our merger models (Schneider et al. 2019, 2020).

2. Magnetic-field amplification in a stellar merger

From the apparent age of τ Sco in comparison to the age of its host environment, the Upper Scorpius Association, Schneider et al. (2016) were able to identify possible binary-progenitor configurations that may lead to a star resembling τ Sco via a stellar merger. In Schneider et al. (2019), we picked one of these configurations, namely a $9+8\,\mathrm{M}_\odot$ binary merging at an age of 9 Myr (fractional main-sequence age of $\approx$35%). The merger is carried out with the 3D moving-mesh, magnetohydrodynamic code AREPO (Springel 2010; Pakmor et al. 2011; Pakmor & Springel 2013). The divergence of the magnetic field is controlled following Powell et al. (1999). Snapshots of the density evolution and the magnetic-field amplification process are shown in Fig. 1.

The magnetic-field amplification process is complex, highly non-linear and proceeds exponentially, eventually reaching magnetic field strengths of up to 10^7 G. It is difficult to attribute the amplification to individual mechanisms. The magnetic field is first strongly amplified in a shear layer that forms around both stars by mass flows in the initial overcontact binary configuration. We have identified the magneto-rotational instability and Kelvin-Helmholtz-like instabilities to contribute to the exponential build-up of the magnetic field. Saturation is reached at a level larger than expected from equipartition with turbulent energy alone likely because of strong, ordered differential rotation that also feeds into the magnetic energy. While the amplification process first produces fields on local, small scales ($\approx$0.1 $\mathrm{R}_\odot$), they are later redistributed such that most of the magnetic energy is at larger scales of 1–5 $\mathrm{R}_\odot$ (Ohlmann et al. 2022). Magnetic flux freezing from the end of the 3D simulation to a star of τ Sco's radius results in surface magnetic field strengths of a few kG, i.e. in the ball park of values observed in magnetic massive stars.

In the merging process, a massive, rotationally-supported torus forms ($\approx$3 $\mathrm{M}_\odot$) that carries a large fraction of the initial angular momentum ($\approx$60%). It is expected that this torus is quickly accreted onto the central merger remnant and it is during this accretion and co-evolution process that the final, total angular momentum of the merged star is determined (Schneider et al. 2019, 2020). This process is on timescales beyond what can be followed by our simulations such that the final angular momentum of the merged star remains uncertain. At the end of the merger simulation, the interior rotation profile approaches that of a solid body, which smoothly transitions into a Keplerian-rotation profile of the torus (the torus rotates at about 80% of the Keplerian value and is pressure supported).

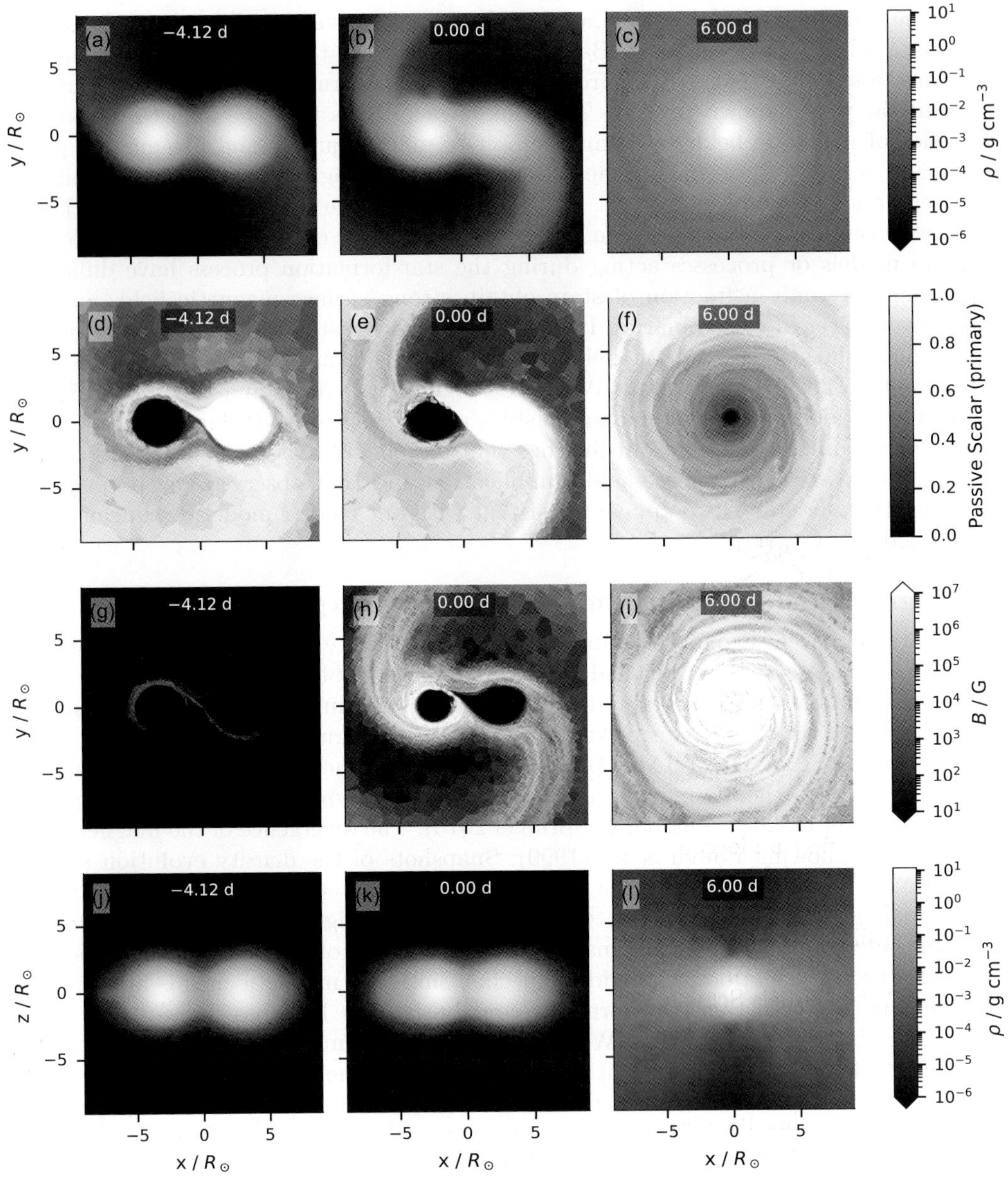

Figure 1. Evolution of density and absolute magnetic field strength in the orbital xy-plane and in its perpendicular, edge-on xz-plane. Figure adopted from Schneider et al. (2019).

3. Evolution of the post-merger star and comparison to τ Sco

Assuming that most of the torus is accreted onto the merger remnant and that the merged star is able to maintain its solid-body rotation profile matched to 80% of Keplerian rotation of the remaining torus material at the new stellar surface, we map the entropy profile of 3D merger simulation into the 1D stellar evolution code MESA (Paxton et al. 2011, 2013, 2015; Paxton et al. 2018) as described in Schneider et al. (2019) and Schneider et al. (2020). The magnetic field is incorporated into the 1D model with a dipole radial scaling, and it contributes to angular-momentum transport and leads to magnetic braking. The merged star starts its evolution at an effective temperature of

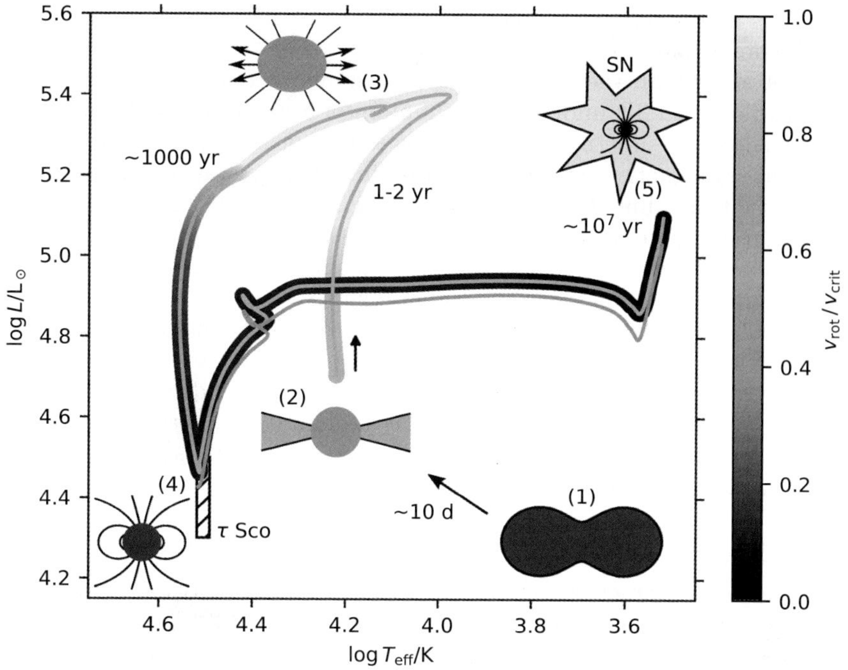

Figure 2. Hertzsprung–Russell diagram of the evolution of the merged star in comparison to a genuine single star of the same initial mass as the merger remnant ($16.9\,M_\odot$). Colour indicates the surface rotational velocity with respect to the critical Keplerian value. The cartoons illustrate key evolutionary steps. Figure adopted from Schneider et al. (2019).

$\approx$14500 K and first undergoes a thermal relaxation phase (Fig. 2): its luminosity reaches a maximum value of $\log L/L_\odot \approx 5.4$ within a few years before it contracts back to the main-sequence within 1000–10,000 yrs. During this relaxation phase, the star actually spins up while expanding and then spins down when contracting. This somewhat anti-intuitive behaviour is explained by a restructuring of the star's interior that modifies the moment of inertia (see Schneider et al. 2020, for more details). Only a negligible amount of angular momentum is lost in this short evolutionary phase despite mass shedding when rotating critically and magnetic braking. At the end of the relaxation phase when the star is back on the main sequence, the merger product resembles the magnetic star τ Sco in terms of its luminosity, effective temperature, surface gravity, apparent youth and also slow spin. Furthermore, the surface magnetic field, assuming magnetic flux freezing, is well in the range of what is observed in τ Sco and other magnetic stars.

4. Final fate of merged stars

The future evolution of the merged star closely follows that of a genuine single star of the same mass with one difference: the merged star has a higher average helium content in its envelope such that it evolves at a higher luminosity than the comparison single star (Fig. 2). The merger product is predicted to explode in a SN IIP upon reaching core collapse as a red supergiant. Ferrario & Wickramasinghe (2008) suggested that magnetic massive stars may be the direct progenitors of magnetars, i.e. neutron stars with the strongest known magnetic fields in the entire Cosmos. Magnetic flux freezing in the inner $1.5\,M_\odot$ of the merger simulation to neutron-star dimensions results in surface magnetic fields of the neutron star of up to 10^{16} G. So even a much reduced magnetic flux could in principle result in magnetar-like field strengths. It should, however, be noted that it is as yet unknown how much (if any) of the magnetic flux might indeed survive until core collapse.

In Varma et al. (2022), the pre-supernova structure of our merger product has been successfully exploded in 3D under the assumption of a highly-magnetised pre-supernova core. A more energetic supernova explosion is found for a highly-magnetised core compared to a weaker magnetisation. Also, the forming neutron star may have magnetar-like surface magnetic fields, but it is too early to draw firm conclusions.

Magnetars are predicted to be born at a rate of $\approx$0.3 per century in the Milky Way (Keane & Kramer 2008) which converts into a magnetar birth fraction of $\approx$15% of all supernovae assuming the inferred core-collapse supernova rate of $\approx$2.0 per century by Diehl et al. (2006). This birth fraction of magnetars appears consistent with the observed incidence of magnetic massive stars of about 10%. While the exploding, merged star may be highly magnetised, we did not expect to find engine-driven explosions such as superluminous supernovae. Our merged star is probably slowly rotating and, even if it was rotating faster, it will likely lose much of its spin angular momentum in stellar winds before reaching the supernova stage.

5. Concluding remarks

Strong dynamical interactions in various binary systems have been found to lead to the formation of strong magnetic fields and thus seem to be an ubiquitous outcome of such phases. This includes stellar mergers of main-sequence stars as discussed here, but also mergers of white dwarfs (e.g. Zhu et al. 2015) and neutron stars (e.g. Price & Rosswog 2006); it may also extend to mergers of pre-main-sequence stars. Also strong magnetic fields emerge in the common-envelope phase where a companion is dragged into the envelope of a (super)giant star (e.g. Ohlmann et al. 2016) and even lead to the formation of fast bipolar outflows that help explain the shapes of (proto-)planetary nebulae (Ondratschek et al. 2022).

Acknowledgements

This work has received funding from the European Research Council (ERC) under the European Union's Horizon 2020 research and innovation programme (Grant agreement No. 945806). This work is supported by the Deutsche Forschungsgemeinschaft (DFG, German Research Foundation) under Germany's Excellence Strategy EXC 2181/1-390900948 (the Heidelberg STRUCTURES Excellence Cluster).

References

Babcock H. W., 1947, Publications of the Astronomical Society of the Pacific, 59, 112
Carrier F., North P., Udry S., Babel J., 2002, A&A, 394, 151
Diehl R., et al., 2006, Nature, 439, 45
Donati J.-F., Landstreet J. D., 2009, ARA&A, 47, 333
Ferrario L., Wickramasinghe D. T., 2005, MNRAS, 356, 615
Ferrario L., Wickramasinghe D., 2008, MNRAS, 389, L66
Ferrario L., Pringle J. E., Tout C. A., Wickramasinghe D. T., 2009, MNRAS, 400, L71
Grunhut J. H., et al., 2017, MNRAS, 465, 2432
Hale G. E., 1908, ApJ, 28, 315
Keane E. F., Kramer M., 2008, MNRAS, 391, 2009
Kochukhov O., Bagnulo S., 2006, A&A, 450, 763
Langer N., 2012, ARA&A, 50, 107
Neiner C., Mathis S., Alecian E., Emeriau C., Grunhut J., BinaMIcS t., MiMeS collaborations 2015, preprint, (`arXiv:1502.00226`)
Ohlmann S. T., Röpke F. K., Pakmor R., Springel V., Müller E., 2016, MNRAS, 462, L121

Ohlmann S. T., Schneider F. R. N., Röpke F. K., Podsiadlowski P., Pakmor R., Springel V., Balbus S., 2022, in prep.

Ondratschek P. A., Röpke F. K., Schneider F. R. N., Fendt C., Sand C., Ohlmann S. T., Pakmor R., Springel V., 2022, A&A, 660, L8

Pakmor R., Springel V., 2013, MNRAS, 432, 176

Pakmor R., Bauer A., Springel V., 2011, MNRAS, 418, 1392

Paxton B., Bildsten L., Dotter A., Herwig F., Lesaffre P., Timmes F., 2011, ApJS, 192, 3

Paxton B., et al., 2013, ApJS, 208, 4

Paxton B., et al., 2015, ApJS, 220, 15

Paxton B., et al., 2018, ApJS, 234, 34

Powell K. G., Roe P. L., Linde T. J., Gombosi T. I., De Zeeuw D. L., 1999, Journal of Computational Physics, 154, 284

Price D. J., Rosswog S., 2006, Science, 312, 719

Schneider F. R. N., Podsiadlowski P., Langer N., Castro N., Fossati L., 2016, MNRAS, 457, 2355

Schneider F. R. N., Ohlmann S. T., Podsiadlowski P., Röpke F. K., Balbus S. A., Pakmor R., Springel V., 2019, Nature, 574, 211

Schneider F. R. N., Ohlmann S. T., Podsiadlowski P., Röpke F. K., Balbus S. A., Pakmor R., 2020, MNRAS, 495, 2796

Schöller M., et al., 2017, A&A, 599, A66

Shultz M., Wade G. A., Alecian E., BinaMIcS Collaboration 2015, MNRAS, 454, L1

Springel V., 2010, MNRAS, 401, 791

Varma V., Mueller B., Schneider F. R. N., 2022, arXiv e-prints, p. arXiv:2204.11009

Wickramasinghe D. T., Tout C. A., Ferrario L., 2014, MNRAS, 437, 675

Zhu C., Pakmor R., van Kerkwijk M. H., Chang P., 2015, ApJL, 806, L1

Massive Stars Near and Far
Proceedings IAU Symposium No. 361, 2024
J. Mackey, J. S. Vink, & N. St-Louis, eds.
doi:10.1017/S174392132200268X

Asteroseismology reveals the near-core magnetic field strength in the early-B star HD 43317

D. M. Bowman[1,2], D. Lecoanet[3,4,2] and T. Van Reeth[1,2]

[1]Institute of Astronomy, KU Leuven, Celestijnenlaan 200D, 3001 Leuven, Belgium
email: dominic.bowman@kuleuven.be

[2]Kavli Institute for Theoretical Physics, University of California, Santa Barbara, CA 93106, USA

[3]Department of Engineering Sciences and Applied Mathematics, Northwestern University, Evanston, IL 60208, USA

[4]CIERA, Northwestern University, Evanston, IL 60201, USA

Abstract. Spectropolarimetic campaigns have established that large-scale magnetic fields are present at the surfaces of approximately 10% of massive dwarf stars. However, there is a dearth of magnetic field measurements for their deep interiors. Asteroseismology of gravity-mode pulsations combined with rotating magneto-hydrodynamical calculations of the early-B main-sequence star HD 43317 constrain its magnetic field strength to be approximately 5×10^5 G just outside its convective core. This proof-of-concept study for magneto-asteroseismology opens a new window into the observational characterisation of magnetic fields inside massive stars.

Keywords. magnetic fields, stars: early-type, stars: oscillations, stars: rotation

1. Introduction

Only about 10% of early-type (i.e. spectral types O, B and A) dwarf stars are known to host a strong, large-scale magnetic field at their surface, which is thanks to dedicated spectropolarimetric campaigns (Wade et al. 2014; Grunhut et al. 2017; Shultz et al. 2019; Sikora et al. 2019). The magnetic fields are predominantly dipolar, typically inclined with respect to the rotation axis, and have strengths between approximately 100 G and a few tens of kG. Large-scale magnetic fields in main-sequence stars are thought to be formed through the star formation process (e.g. Neiner et al. 2015) or through binary star mergers that leave behind a strongly magnetic and apparently single star (e.g. Schneider et al. 2019). Regardless of how they arise, magnetic fields are one of the largest uncertainties in stellar structure and evolution theory, as they strongly impact interior rotation, mixing and angular momentum transport (Maeder & Meynet 2005; Keszthelyi et al. 2019).

However, there are few inferences of the strength and geometry of a magnetic field below the surface of an early-type star, specifically in the deep interior near the convective cores of main-sequence massive stars. Thus any direct inference of interior stellar properties from asteroseismology – the study of stellar structure from pulsations – are extremely valuable in improving stellar structure and evolution theory (Aerts 2021). To date, only a handful of massive stars have undergone forward asteroseismic modelling to constrain their interior properties (see Bowman 2020 for a review).

Amongst massive stars, there are two main types of pulsations that can be excited: pressure and gravity modes, which are standing waves that are restored by the pressure force and buoyancy, respectively. Coherent pulsation modes (i.e. standing waves) are excited by a heat-engine mechanism operating in the iron-nickel opacity bump at 200 000 K, and have pulsation periods of order hours to days (Dziembowski & Pamyatnykh 1993; Dziembowski et al. 1993). Gravity modes in particular are sensitive to the properties (e.g. mass and size) of the convective core in main-sequence massive stars, because they probe the Brunt-Väisälä frequency profile. Furthermore, rotation shifts pulsation mode frequencies such that the interior rotation profiles of massive stars can be deduced from gravity-mode period spacing patterns (Bouabid et al. 2013) and rotationally split multiplets (e.g. Aerts et al. 2003).

In these proceedings, we summarise the methods and results of applying magneto-asteroseismology to the magnetic, pulsating, early-B star HD 43317 published by Lecoanet, Bowman, & Van Reeth (2022), to which the reader is referred for full details.

2. The Magnetic Pulsator HD 43317

HD 43317 is an early-type main-sequence star with spectral type B3.5 V. From dedicated high-resolution spectropolarimetry, a precise rotation period of 0.897673(4) d and a large-scale dipolar surface magnetic field with strength $B_{\rm p} = 1312 \pm 332$ G have been measured (Briquet et al. 2013; Buysschaert et al. 2017). Furthermore, atmospheric modelling demonstrated that HD 43317 is likely to be a single star with solar metallicity, and has an effective temperature of $T_{\rm eff} = 17350 \pm 750$ K, a surface gravity of $\log\, g = 4.0 \pm 0.1$ and a projected surface rotational velocity of $v\, \sin\, i = 115 \pm 9\, {\rm km\, s^{-1}}$ (Pápics et al. 2012).

HD 43317 was observed by the CoRoT space mission (Auvergne et al. 2009), which assembled a light curve spanning 150.5 d with an average cadence of 32 s. Frequency analysis of the CoRoT light curve revealed dozens of significant gravity-mode frequencies (Pápics et al. 2012; Buysschaert et al. 2018). The detection of gravity modes in a strongly magnetic star makes HD 43317 unique among early-type stars, since strong magnetic fields are expected to suppress their excitation (e.g. Saio 2005; Lecoanet et al. 2017).

3. Forward Asteroseismic Modelling

The gravity-mode pulsation mode frequencies of HD 43317 extracted from the CoRoT light curve allowed Buysschaert et al. (2018) to perform forward asteroseismic modelling to ascertain the best-fitting interior structure. As a first step, the effective temperature and surface gravity from spectroscopy were used to delimit the parameter space in the Kiel diagram (see Fig. 1). Within this parameter space, a grid of non-rotating stellar structure models was calculated using the open-source MESA evolution code (Paxton et al. 2011, 2013; Paxton et al. 2015). The free parameters were the mass, M, the central core hydrogen content, $X_{\rm c}$, as a proxy for age on the main sequence, and the amount of convective boundary mixing (CBM) of the convective core, $f_{\rm CBM}$, assuming a diffusive exponential prescription. The metallicity was fixed at solar as indicated by the spectroscopic abundance analysis (i.e. $Z = 0.014$), and a fixed amount of mixing was set in the radiative envelope, specifically $D_{\rm env} = 10\ {\rm cm^2\, s^{-1}}$. Models were calculated from the zero-age main sequence until core hydrogen depletion (i.e. $X_{\rm c} = 0$).

For every structure model, the adiabatic pulsation mode frequencies for dipole (i.e. $\ell = 1$) and quadrupole (i.e. $\ell = 2$) gravity modes and all possible azimuthal orders (i.e. $|m| \leq \ell$) were calculated using the GYRE stellar pulsation code (Townsend & Teitler 2013; Townsend et al. 2018). A uniform (i.e. rigid) interior rotation profile was assumed applying the measured surface rotation period of 0.897673(4) d and the traditional approximation for rotation (TAR) was used to include the effects of the Coriolis force on the pulsation mode frequencies. However, such an approach neglects the horizontal component and the structural deformation due to the centrifugal force, because Buysschaert et al. (2018) used non-rotating structure models.

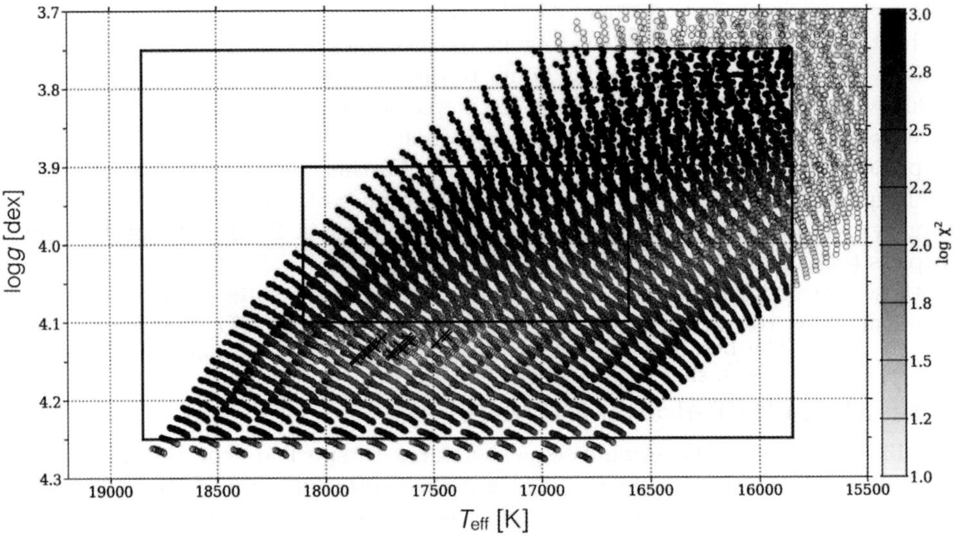

Figure 1. Kiel diagram showing the 20 best-fitting asteroseismic structure models as black crosses for HD 43317. The 1- and 2-σ spectroscopic confidence intervals are denoted by the two black rectangles. Each calculated stellar structure model is shown as a circle, which have been colour-coded by their resulting asteroseismic χ^2 value in the forward asteroseismic modelling methodology. Figure adapted from Buysschaert et al. (2018), their figure 7.

Forward asteroseismic modelling is the quantitative comparison of observed pulsation mode frequencies to those theoretically predicted from a grid of stellar structure models by use of a merit function (Aerts 2021). After eliminating the unlikely mode geometries, Buysschaert et al. (2018) identified the best-fitting structure model and then recalculated a finer grid of MESA models to oversample the parameter space identified by the global χ^2 minimum and repeated the forward asteroseismic modelling exercise. The 20 best-fitting structure models that satisfy the 2σ confidence interval from forward asteroseismic modelling are shown in Fig. 1. The gravity mode frequencies of HD 43317 were identified to have radial orders between $-15 \le n_{\rm pg} \le -1$, and the best overall structure model had a mass of $M = 5.8^{+0.1}_{-0.2}$ M$_\odot$, a core hydrogen content of $X_{\rm c} = 0.54^{+0.01}_{-0.02}$ and a parameterisation of CBM to be $f_{\rm CBM} = 0.004^{+0.014}_{-0.002}$ (Buysschaert et al. 2018).

The results of the forward asteroseismic modelling make HD 43317 particularly interesting for two reasons, both of which support the hypothesis that a strong magnetic field must exist in the deep interior. Firstly, the low amount of measured CBM is plausibly a consequence of a magnetic field suppressing convective motions from 'overshooting' the Schwarzschild boundary of the convective core into the overlying radiative envelope. This is observed in magneto-hydrodynamical (MHD) simulations of magnetised core convection (e.g. Browning et al. 2004; Featherstone et al. 2009; Augustson et al. 2016). Secondly, the gravity-mode frequencies identified in HD 43317 are particularly high frequency and have atypical radial orders for gravity modes in stars with masses between $3 \lesssim M \lesssim 9$ M$_\odot$ (see Pedersen et al. 2021). This lends further support to the presence of a strong interior magnetic field, because, if present, it would interact with and suppress low-frequency gravity modes leaving only high-frequency gravity modes unaffected.

4. Interior Magnetic Field Inference

To test the hypothesis of a strong magnetic field in the deep interior of HD 43317, Lecoanet, Bowman, & Van Reeth (2022) performed a magneto-asteroseismic analysis to

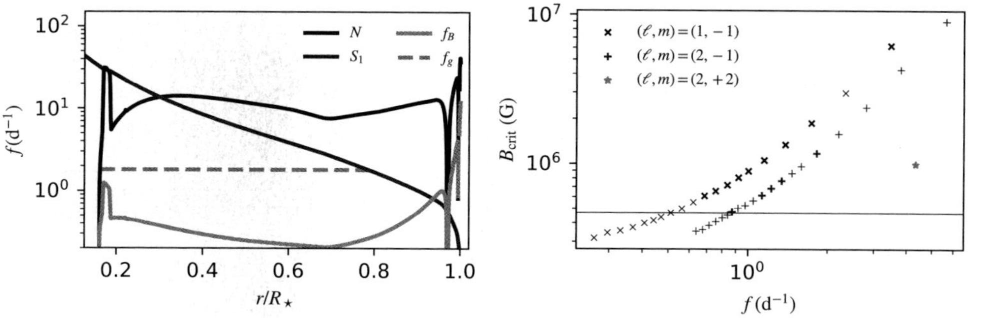

Figure 2. Left: The best fitting structure model from forward asteroseismic modelling by Buysschaert et al. (2018), showing the Lamb (S_l) and Brunt-Väisälä (N) frequency profiles, the lowest observed frequency gravity mode (f_g) and the magnetic interaction frequency (f_B; cf. Eqn. 1). Right: Gravity mode frequencies as a function of critical magnetic field strength needed to suppress their propagation, in which observed gravity modes are shown as thick symbols. The horizontal line denotes the maximum permitted magnetic field strength without suppressing the lowest frequency observed gravity mode. Figures adapted from Lecoanet, Bowman, & Van Reeth (2022), their figures 1 and 3.

infer the critical magnetic field strength needed to suppress non-observed gravity modes. Previous work has found that gravity waves can strongly interact with magnetic fields in stellar interiors (e.g. Rogers & MacGregor 2010; Fuller et al. 2015; Lecoanet et al. 2017). If a gravity wave of frequency f enters a region where

$$f \lesssim f_{\mathrm{B}} = \frac{1}{2\pi}\sqrt{\frac{B_{\mathrm{r}}}{\pi\rho}\frac{N\Lambda}{r}} \tag{1}$$

where B_{r} is the radial magnetic field, ρ is the density, $\Lambda = \sqrt{\ell(\ell+1)}$ with ℓ the spherical harmonic degree of the gravity wave, and r is the local radius, it converts into a magnetic (i.e. Alfvén) wave, which in turn prevents the formation of a gravity mode (i.e. standing gravity wave). The best fitting structure model as found by Buysschaert et al. (2018), as well as the resultant magnetic interaction frequency (f_{B}; cf. Eqn. 1) for HD 43317 are shown in the left panel of Fig. 2. The lowest observed gravity mode frequency of HD 43317 is close to the value of f_{B} in the near-core region at $r = 0.18R_\star$. The critical magnetic field strength required to suppress all of the non-observed low-frequency gravity modes given the structure model of HD 43317 is inferred to be $B_{\mathrm{r}} \simeq 4.7 \times 10^5$ G at $r = 0.18R_\star$, as shown in the right panel of Fig. 2 (Lecoanet, Bowman, & Van Reeth 2022).

To accurately quantify the magnetic field strength in the near-core of HD 43317, Lecoanet, Bowman, & Van Reeth (2022) calculated the linear waves for a rotating and magnetised stellar model taking the best model in Fig. 2 as input and using the Wentzel-Kramers-Brillouin-Jeffreys (WKBJ) approximation with the DEDALUS code (Lecoanet et al. 2019; Vasil et al. 2019; Burns et al. 2020). Spherical geometry, a predominately dipolar magnetic field geometry and uniform interior rotation at the rate used in the prior asteroseismic analysis were assumed. A graphical representation of the results of these DEDALUS calculations showing the azimuthal velocity structure of gravity modes above and below the critical magnetic field threshold is shown in the left panel of Fig. 3. The important conclusion is that frequencies below the lowest observed $n_{\mathrm{pg}} = -15$ gravity mode strongly interact with the interior magnetic field. This places a strong constraint on the maximum magnetic field strength inside HD 43317 corroborating the observed distribution of gravity modes. To check the robustness of their results, Lecoanet, Bowman, & Van Reeth (2022) propagated the uncertainties on the best-fitting structure model into the magnetic field inference and find they have only a small impact,

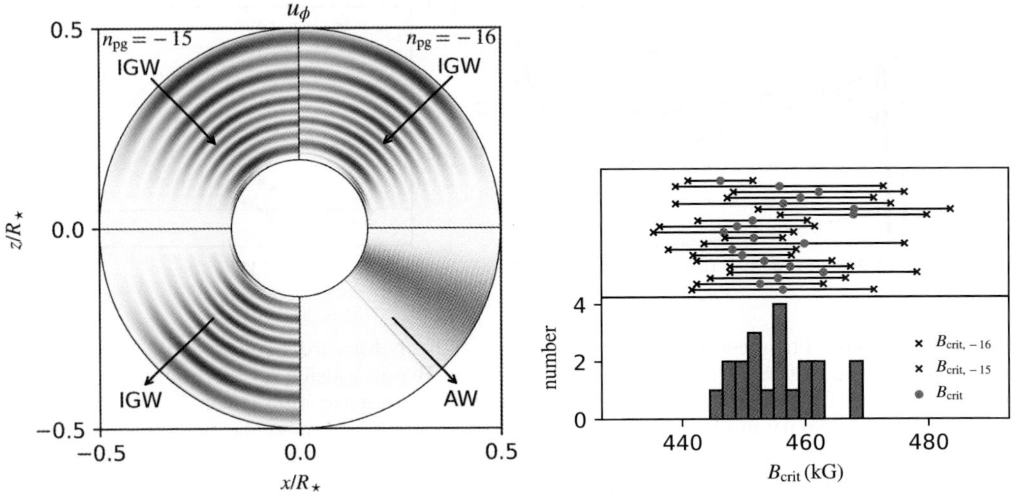

Figure 3. Left: DEDALUS calculation showing the azimuthal velocity structure for the $(\ell, m) = (2, -1)$ gravity wave with $f = 0.872$ d^{-1} and $n_{pg} = -15$ in the left quadrants and $f = 0.840$ d^{-1} and $n_{pg} = -16$ in the right quadrants. The top and bottom quadrants show the incoming and outgoing wave structure, respectively, which in the case of the right quadrants show that the gravity wave is converted into an Alfvén wave, hence not observed. Right: Inferred magnetic field strengths in the near-core region taking the 20 best-fitting structure modelling from Buysschaert et al. (2018) as input. The upper and lower limits are assuming radial orders of $n_{pg} = -15$ and -16 for the lowest frequency observed gravity mode, respectively. Figures adapted from Lecoanet, Bowman, & Van Reeth (2022), their figures 4 and 5.

as shown in the right panel of Fig. 3. Thus the magnetic inference is robust to the asteroseismic uncertainties.

5. Conclusions

The recent results of Lecoanet, Bowman, & Van Reeth (2022) place a strong constraint on the maximum magnetic field strength of $B_r \simeq 4.7 \times 10^5$ G at $r = 0.18 R_\star$ for the magnetic pulsating early-B star HD 43317. The dominant probing region of the gravity modes is just outside the convective core (see Fig. 2). Since HD 43317 has evolved beyond the zero-age main-sequence and its convective core has receded, the current gravity-mode cavity was previously within the convective core. The inferred near-core magnetic field strength is comparable to predictions from MHD simulations containing a fossil field and convective core dynamo in a main sequence massive star (e.g. Browning et al. 2004; Featherstone et al. 2009; Augustson et al. 2016).

HD 43317 is currently the only magnetic star pulsating in gravity modes to have undergone forward asteroseismic modelling (Buysschaert et al. 2018), but serves as a valuable proof-of-concept for magneto-asteroseismology. The ongoing TESS space mission is providing excellent time series data (Ricker et al. 2015), which have revealed a high pulsator fraction among massive stars (e.g. Bowman et al. 2019; Burssens et al. 2020). With future spectropolarimetric and spectroscopic campaigns, such as the CubeSpec mission (Bowman et al. 2022), the future of massive star asteroseismology is bright.

Acknowledgements

DMB and TVR gratefully acknowledge funding from the Research Foundation Flanders (FWO) by means of senior and junior postdoctoral fellowships with grant agreements 1286521N and 12ZB620N, respectively, and FWO long stay travel grants V411621N and

V414021N, respectively. DL is supported in part by the US National Aeronautics and Space Administration (NASA) grant 80NSSC20K1280. Computations were conducted with support by the NASA High End Computing Program through the NASA Advanced Supercomputing (NAS) Division at Ames Research Center on Pleiades with allocation GIDs s2276. This research was supported in part by the National Science Foundation under grant no. NSF PHY-1748958.

References

Aerts et al. 2003, *Science*, 300, 1926-1928
Aerts (2021), *Rev. Mod. Phys.*, 93, 015001
Augustson et al. 2016, *ApJ*, 829, 92
Auvergne et al. 2009, *A&A*, 506, 411
Briquet et al. 2013, *A&A*, 557, L16
Browning et al. 2004, *ApJ*, 601, 512
Bouabid et al. 2013, *MNRAS*, 429, 2500
Bowman et al. 2019, *Nat. Ast.*, 3, 760-765
Bowman et al. 2022, *A&A*, 658, A96
Bowman 2020, *Frontiers Astron. Space Sci.* 7, 70
Burns et al. 2020, *Phys. Rev. Research*, 2, 023068
Burssens et al. 2020, *A&A*, 639, A81
Buysschaert et al. 2017, *A&A*, 605, A104
Buysschaert et al. 2018, *A&A*, 616, A148
Dziembowski & Pamyatnykh 1993, *MNRAS*, 262, 204-212
Dziembowski et al. 1993, *MNRAS*, 265, 588
Featherstone et al. 2009, *ApJ*, 705, 1000
Fuller et al. 2015, *Science*, 350, 423-426
Grunhut et al. 2017, *MNRAS*, 465, 2432
Keszthelyi et al. 2019, *MNRAS*, 485, 5843
Lecoanet et al. 2017, *MNRAS*, 466, 2181-2193
Lecoanet et al. 2019, *Journal of Computational Physics: X*, 3, 100012
Lecoanet, Bowman & Van Reeth 2022, *MNRAS*, 512, L16–L20
Maeder & Meynet 2005, *A&A*, 440, 1041
Neiner et al. 2015, *Proc. IAU Symp* Vol. 305, Kluwer, Dordrecht, p. 61
Pápics et al. 2012, *A&A*, 542, A55
Paxton et al. 2011, *ApJS*, 192, 3
Paxton et al. 2013, *ApJS*, 208, 4
Paxton et al. 2015, *ApJS*, 220, 15
Pedersen et al. 2021, *Nature Astronomy*, 5, 715-722
Ricker et al. 2015, *Journal of Astronomical Telescopes, Instruments, and Systems*, 1, 014003
Rogers & MacGregor 2010, *MNRAS*, 401, 191-196
Saio 2005, *MNRAS*, 360, 1022-1032
Schneider et al. 2019, *Nature*, 574, 211
Sikora et al. 2019, *MNRAS*, 483, 2300
Shultz et al. 2019, *MNRAS*, 490, 274
Townsend & Teitler 2013, *MNRAS*, 435, 3406
Townsend et al. 2018, *MNRAS*, 475, 879
Vasil et al. 2019, *Journal of Computational Physics: X*, 3, 100013
Wade et al. 2014, *Proc. IAU Symp.* Vol. 302, Cambridge University Press, Dordrecht, p. 265

Massive Stars Near and Far
Proceedings IAU Symposium No. 361, 2024
J. Mackey, J. S. Vink, & N. St-Louis, eds.
doi:10.1017/S1743921322003039

Constraining physical processes in pre-supernovae massive star evolution

Erin R. Higgins[1], Jorick S. Vink[1], Andreas Sander[2] and Raphael Hirschi[3,4]

[1]Armagh Observatory and Planetarium, College Hill, Armagh BT61 9DG, N. Ireland
email: erin.higgins@armagh.ac.uk

[2]Zentrum für Astronomie der Universität Heidelberg, Astronomisches Rechen-Institut, Mönchhofstr. 12-14, 69120 Heidelberg, Germany

[3]Astrophysics Group, Keele University, Keele, Staffordshire ST5 5BG, UK

[4]Kavli Institute for the Physics and Mathematics of the Universe (WPI), University of Tokyo, 5-1-5 Kashiwanoha, Kashiwa 277-8583, Japan

Abstract. While we have growing numbers of massive star observations, it remains unclear how efficient the key physical processes such as mass loss, convection and rotation are, or indeed how they impact each other. We reconcile this with detailed stellar evolution models, yet these models have their own drawbacks with necessary assumptions for 3-dimensional processes like rotation which need to be adapted into 1-dimensional models. The implementation of empirical mass-loss prescriptions in stellar evolution codes can lead to the extrapolation of base rates to unconstrained evolutionary stages leading to a range of uncertain fates. In short, there remain many free parameters and physical processes which need to be calibrated in order to align our theory better with upcoming observations. We have tested various processes such as mass loss and internal mixing, including rotational mixing and convective overshooting, against multiple observational constraints such as using eclipsing binaries, the Humphreys-Davidson limit, and the final masses of Wolf-Rayet stars, across a range of metallicities. In fact, we developed a method of disentangling the effects of mixing and mass loss in the 'Mass-Luminosity Plane' allowing direct calibration of these processes. In all cases, it is important to note that a combined appreciation for both stellar winds and internal mixing are important to reproduce observations.

Keywords. stars: evolution, early-type, interiors, mass loss, winds, outflows, Wolf-Rayet

---

1. Introduction

In the first instance, we expect that the effects of internal mixing dominate the evolution of the lower mass range of massive stars (~10-30 $M_\odot$) and mass loss dominates the upper mass range (~30-60 $M_\odot$), however, the picture is much more complex than this. For instance, the effects of stellar winds on the surface properties of massive stars directly impacts the subsurface convective layers, including semiconvective regions which induce mixing in the envelope. Moreover, for the most massive stars it may be the case that internal mixing – in particular very low internal mixing – allows for smaller convective cores which collapse to form heavy black holes on the order of 80-90 $M_\odot$. It is therefore important to note that while in certain regimes we can focus on the effects of mixing *or* mass loss, in reality it will be important to have a real appreciation for how both interact with and counteract each other. This may be key in understanding the nature of how

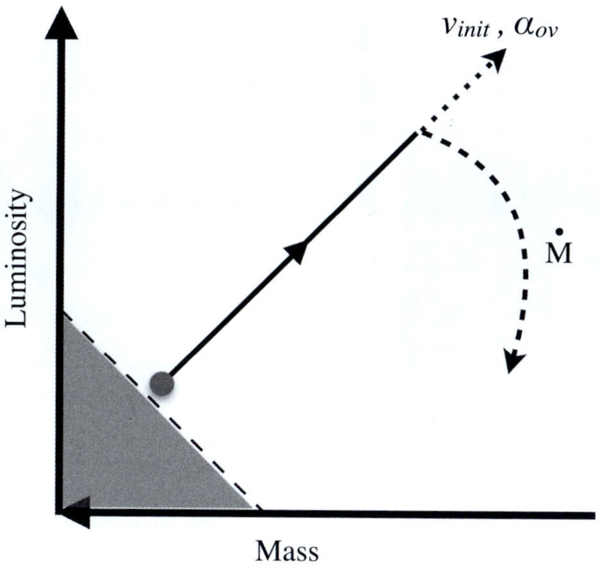

Figure 1. Illustration of the 'Mass-Luminosity' plane with an evolutionary track starting at the ZAMS with the red dot, evolving along the black vector. The dotted line extending the vector suggests how much additional mixing via increased rotation and/or convective overshooting should be applied to extend the $M - L$ vector. The curved dashed line represents the gradient of the vector, dominated by the mass-loss rates. The red solid region represents the boundary set by the mass-luminosity relationship, and as such is forbidden.

and where certain massive star evolutionary channels split towards becoming black holes (BHs), neutron stars (NS), Wolf-Rayets (WRs) and Red supergiants (RSGs).

2. Main Sequence Evolution

We have developed a new method of calibrating various physical processes in Higgins & Vink (2019), called the Mass-Luminosity Plane. From our analysis of a detached eclipsing binary in the Galaxy (HD166734, see Mahy *et al.* 2017) we determined that individual processes such as rotational mixing or core overshooting could not be calibrated in a standard Hertzsprung-Russell diagram (HRD). In this study we used the M-L plane to reproduce the observed stellar parameters, however with continued use and development we have found that it may be an extremely powerful tool in disentangling physical processes.

Just like in a HRD, models evolve to cooler effective temperatures along the vector towards the right-hand side, but we find that the length of the vector can be extended by extra mixing (see Fig.1). In the first instance rotation is the dominant factor, with core overshooting as a secondary extension after rotation has been constrained by, for example, the surface Nitrogen abundances ([N/H]). The key feature is that we can separate these internal mixing effects from the effects of mass loss, whereby the gradient of the vector track shows the effects of winds, with stronger mass-loss rates leading to a shallow gradient and weak winds leading to a steep gradient. This effect will also be seen in different metallicities (Z) since the gradients will be steeper in low Z environments due to Z-dependent winds.

With the M-L plane we found that the eclipsing binary required additional internal mixing by core overshooting of up to $\alpha_{\rm ov}$= 0.5 for the 30-40 $M_{\odot}$ components. While simultaneously we were able to hone in on the extent of mass loss to a factor of unity

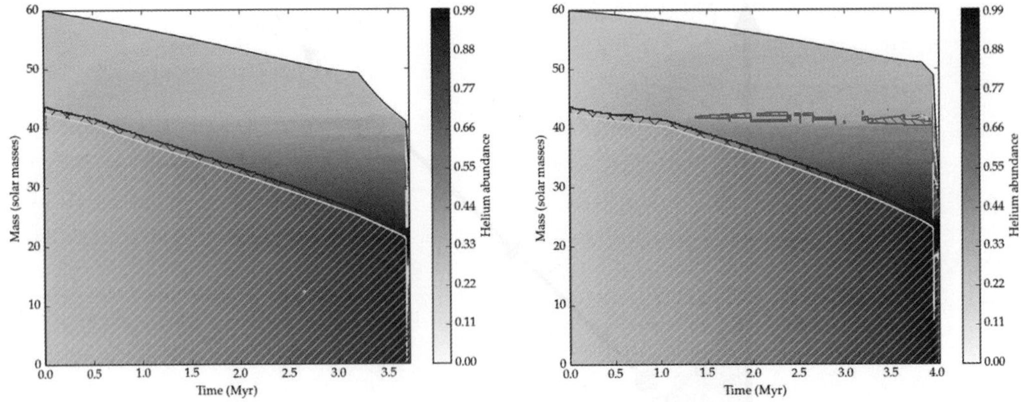

Figure 2. Kippenhahn diagrams illustrating the structure of $60\,M_\odot$ models with $\alpha_{\rm ov}= 0.1$, $\alpha_{\rm semi}= 100$ and $\Omega/\Omega_{\rm crit} = 0.4$. The colourbar represents Helium abundance, while the green hatched region denotes convection. The model which evolves at Galactic metallicity (which experiences increased mass loss due to Z-dependent winds) is shown on the left. The model which has evolved at $Z_{\rm LMC}$ is shown on the right, where red semiconvective regions are seen as a consequence of lower mass-loss rates and subsequently a larger envelope mass.

of the standard Vink *et al.* (2001) rates, excluding for this range of models any large amplifications or reductions (within a factor of two) in the mass-loss rates. We also find in Higgins & Vink (2019) that by calculating a large grid of models with varied core overshooting that due to the relative compactness of our pre-supernovae models, that BHs were more likely to form with lower overshooting, which has remained important for our follow up studies, including Vink *et al.* (2021).

3. Post-Main Sequence Evolution

Considering the idea of 'turning points' in evolutionary channels, we probed the post-Main Sequence (post-MS) evolution towards red and blue supergiants (RSGs and BSGs), considering what we had learnt from the Galactic eclipsing binary and the M-L plane. The upper RSG luminosity or Humphreys-Davidson (HD) limit points towards a tipping point at which stars stop evolving to RSGs and start evolving to hotter temperatures as WRs. This change in evolutionary channels is important for comparing populations of BSGs and RSGs or various SNe progenitors with population synthesis models. But this turning point is also considered to be highly sensitive to mass loss. In Higgins & Vink (2020), we found that while mass loss did play a key role, it was actually in its indirect effect on the envelope structure which prohibited semiconvective regions forming, which would promote blueward evolution above the RSG limit. In this study, it was important that we maintained a low convective core mass to allow for additional, or in this case, semiconvective mixing to take place, and the standard Z-dependent mass loss naturally reproduced the observed HD limit at each Z. However, recent work by Sabhahit *et al.* (2021) suggests that if the HD-limit is in fact Z-independent, then the effects of envelope mixing and Z-dependent winds may balance to produce an upper RSG luminosity of log $L/L_\odot = 5.5$ regardless of host galaxy Z.

Figure 2 showcases the effect of stellar winds on internal mixing processes. Due to increased winds at $Z_\odot$ compared to lower Z environments, such as the LMC (50% $Z_\odot$), the envelope mass is reduced such that semiconvective regions do not develop. We find that increasing the convective core by overshooting has a similar effect on envelope mixing. This shows that while the M-L plane can disentangle each process to better calibrate our prescriptions to reproduce observations, the indirect effects of stellar winds on the interior structure and efficiency of mixing processes also plays a key role in our understanding of massive stars.

Since it was the higher mass models which set the HD limit, and with prior knowledge of having a low core overshooting parameter to produce more BHs from our study of HD166734, we could actually use these results in reproducing heavy BHs. When the announcement of an 85 and 65 $M_\odot$ BH merger was detected by LIGO, we aimed to reproduce the heaviest BH from first generation black holes which have not had previous merger events. In fact, with this prior work on massive star evolution, and in particular a combined appreciation for how mixing can impact mass loss and vice versa, we were able to form an 85 $M_\odot$ BH with a moderate Z (10% solar) and an initial mass of $\sim$100 $M_\odot$ (see Vink *et al.* 2021).

4. Wolf-Rayet stars as black hole progenitors

Now we look at the point at which stars stop evolving to cooler effective temperatures as RSGs, above which the upper mass range is dominated by WR stripped Helium stars. Since these stars dominate the upper mass range, they also will dominate the BH mass spectrum across different metallicities. As a result, it is important to know how much mass they lose over their He-burning lifetime, since the remaining stages are exceedingly short. WR stars have very strong, Z-dependent winds which dominate the final masses of the most massive stars. It is important therefore to correctly measure the extent of these winds if we are to probe the full mass spectrum of stellar BHs.

More recently, in Higgins *et al.* (2021) we have developed a study of Helium star evolution for three commonly-used wind prescriptions, including the recent hydrodynamically-consistent rates by Sander & Vink (2020), at 5 metallicities from Solar down to 2% solar, and Helium ZAMS masses of 20 to 200 $M_\odot$. Our models show a qualitative difference in mass-loss behaviour of the various prescriptions in terms of "convergence" on the upper BH mass. Using the prescription from Nugis & Lamers (2000) the maximum stellar black hole below the pair instability gap is found to converge to a value of 20-30 $M_\odot$, independent of metallicity. We also provide a comparison of the Helium star mass-loss rates utilised in population synthesis codes such as COMPAS, see Belczynski *et al.* (2010). This mass-loss prescription is based on the Hamann *et al.* (1995) rates (denoted here as H95+), which are then reduced by a factor of ten to account for wind clumping and implemented with a Z-scaling from Vink & de Koter (2005), see also Yoon *et al.* (2006).

We find that when implementing the new prescription from Sander & Vink (2020) there is no convergence to a maximum black hole mass value across all Z, but the final mass is larger for larger initial He star mass, in which case the upper black hole limit is set by prior main sequence evolution and mass loss or by the pair instability. Interestingly, the upper black hole mass limit for each Z also corresponds here to the Z-limit of pair instability supernovae, with our results reaching the pair instability gap at a higher metallicity limit (50% $Z_\odot$) than previously predicted. Finally, we show in Fig. 3 that above the second black hole mass gap, the Nugis & Lamers (2000) prescription would not predict any black holes, whereas the most massive He stars with Sander & Vink (2020) winds included, demonstrate a channel for very massive stars to form black holes above the $\sim$130 $M_\odot$ Pair instability supernova limit at $\sim$2% $Z_\odot$ or below.

These studies have shown that in order to predict evolutionary channels for the relevant mass ranges with realistic final masses and fates, we must have a combined understanding of metallicity-dependent mass loss – both on the main sequence and for WRs – and internal mixing efficiencies. In summary, our implementation and understanding of these key processes will be aided with further studies of eclipsing binaries, alongside the advancement of asteroseismology and updated wind prescriptions.

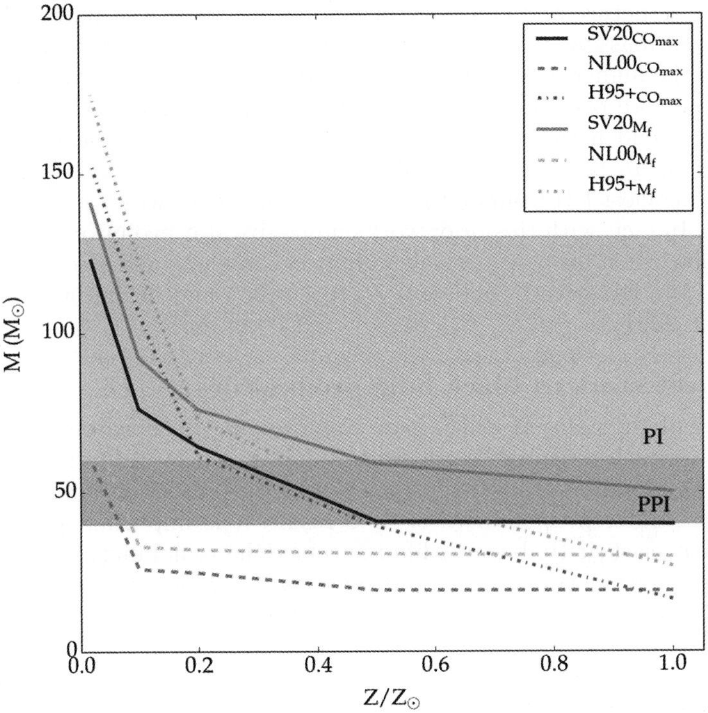

Figure 3. Final CO masses at core He-exhaustion as a function of initial metallicity for each wind recipe. Models which implement Sander & Vink (2020) are represented by solid red lines, H95+ in dashed-dotted green lines and Nugis & Lamers (2000) in dashed blue lines. The maximum final mass (M_f) models, for each metallicity, which include mass-loss rates from SV20 are represented by solid light red lines, H95+ in dashed-dotted light green lines and NL00 in dashed light blue lines. The grey shaded region highlights the region where pulsational pair-instability supernovae occur, with CO core masses above 40 $M_\odot$.

5. Conclusions

These studies have shown that in order to predict evolutionary channels for the relevant mass ranges with realistic final masses and fates, we must have a combined understanding of metallicity-dependent mass loss – both on the MS and for WRs – as well as internal mixing efficiencies.

We find that while high values (α_{ov}=0.5) of core overshooting were needed to reproduce a galactic eclipsing binary with masses of 30-40 $M_\odot$. However, we find that lower values of α_{ov}=0.1 were required for reproducing the 85 $M_\odot$ BH. This means that there may be a variation of internal mixing efficiencies at different masses or metallicities.

The M-L plane is an important tool for calibrating stellar evolution models since it can disentangle the effects of internal mixing and mass loss allowing for constraints on each process. We find that MS mass loss cannot be increased or decreased by extreme factors of the standard Vink *et al.* (2001) rates for the 20-60 $M_\odot$ range due to the resulting gradient in M-L plane which would not be able to reproduce observed luminosities simultaneously with stellar masses. Furthermore, WR mass loss may be the decisive feature in setting the BH mass spectrum across multiple metallicities. Though our studies show that the pre-Helium evolution and Pair instability also play an important role.

We find that stellar winds have a direct impact on the efficiency of mixing in stellar interiors as the envelope is depleted, thereby altering the subsurface regions where mixing such as semiconvection or subsurface convection occurs. It will be important to have a combined view of both mixing and mass loss in reproducing future observations.

In summary, our implementation and understanding of these key processes will be aided by further studies of eclipsing binaries, alongside the advancement of asteroseismology for the most massive stars and updated wind prescriptions, as well as follow up studies of BH progenitors and merger events.

References

Belczynski, K., Dominik, M., Bulik, T., O'Shaughnessy, R., Fryer, C. and Holz, D.E. 2010, *ApJ*, L138-L141, 715

Hamann, W.-R., Koesterke, L. and Wessolowski, U. 1995, *A&A*, 151, 299

Higgins, E. R., Sander, A.A.C., Vink, J. S., and Hirschi, R. 2021, *MNRAS*, 4874-4889, 505

Higgins, E.R. and Vink, J.S. 2020, *A&A*, A175, 635

Higgins, E.R. and Vink, J.S. 2019, *A&A*, A50, 622

Mahy, L., Damerdji, Y., Gosset, E., Nitschelm, C., Eenens, P., Sana, H. and Klotz, A. 2017, *A&A*, 96, 607

Nugis, T. and Lamers,H. J. G. L. M. 2000, *A&A*, 227-244, 360

Sabhahit, G.N., Vink, J.S., Higgins, E.R. and Sander, A.A. C. 2021, *MNRAS*, 4473-4487, 506

Sander, Andreas A.C. and Vink, Jorick S. 2020, *MNRAS*, 873-892, 499

Vink, J.S., Higgins, E.R., Sander, A.A.C., and Sabhahit, G.N. 2021, *MNRAS*, 146-154, 504

Vink, J.S. and de Koter, A. 2005, *A&A*, 587-596, 442

Vink, J.S., de Koter, A. and Lamers, H.J.G.L.M. 2001, *A&A*, 574-588, 369

Yoon, S.-C., Langer, N. and Norman, C. 2006, *A&A*, 199-208, 460

Discussion

TOMER SHENAR: I have a question regarding the HD limit, how do you define the limit in terms of luminosities?

ERIN HIGGINS: Prior to updated observations in 2020, the upper RSG luminosity of the Milky Way was log $L/L_{\odot}=5.8$, but has now been revised to 5.5. While the LMC and SMC maintained an upper RSG luminosity of log $L/L_{\odot}=5.5$.

JON SUNDQUIST: How do you evolve the stars towards WRs at low metallicity rather than evolving stars which inflate towards the red?

ERIN HIGGINS: We evolve our Helium star model grid in two ways, with increased rotation or with increased core overshooting to fully mix the models towards the Helium ZAMS. We also test models with main sequence mass loss, and without. This allows for the entire mass range to be tested at each metallicity.

JON SUNDQUIST: Do you assume the enhanced convective efficiency by MLT++?

ERIN HIGGINS: No, we don't include MLT++ during core He-burning, but to evolve the high mass Hydrogen-burning stars towards the Helium ZAMS we implement MLT++ and then disable this at the He ZAMS.

Massive Stars Near and Far
Proceedings IAU Symposium No. 361, 2024
J. Mackey, J. S. Vink, & N. St-Louis, eds.
doi:10.1017/S1743921322002411

3D effects of rotation on spectroscopic observables

Michael Abdul-Masih

European Southern Observatory, Alonso de Cordova 3107, Vitacura,
Casilla 19001, Santiago de Chile, Chile
email: Michael.Abdul-Masih@eso.org

Abstract. When a star is rapidly rotating, it deviates from spherical symmetry causing non-uniform distributions of the surface gravity and temperature across the surface. These three-dimensional effects lead to an inclination dependence of many observable spectroscopic parameters, however this is often neglected when analyzing rapidly rotating systems. Using SPAMMS, we generate synthetic spectra that account for the 3D geometry of the system and fit them with 1D models to investigate how much the 3D effects can change the derived stellar parameters. We show that these 3D effects can lead to observed temperature differences of thousands of kelvin for the same star viewed at different inclinations, and a systematic underestimation of the helium abundance.

Keywords. stars: rotation, stars: fundamental parameters, stars: abundances, techniques: spectroscopic

1. The problem

Rapid rotation is common among massive stars, and can play a crucial roll in massive star evolution (e.g., Maeder 1987; Brott et al. 2011; Langer 2012). A key aspect of rapid rotation is a deviation from spherical symmetry, which in turn leads to non-uniform surface gravity and surface temperature distributions across the surface of rapidly rotating stars (von Zeipel 1924). The temperature distribution can then lead to a change in the ionization balance across the surface, meaning that different regions of the star will show certain species while others may not. Unfortunately however, these 3D effects are often neglected in spectroscopic analyses of massive rotating systems.

The radiative transfer codes available for the analysis of massive stars are mostly 1D due to the computationally intensive NLTE calculations required in this temperature regime (e.g., Hillier & Miller 1998; Gräfener et al. 2002; Puls et al. 2005). The synthetic spectra generated are then convolved with a rotation kernel to mimic rotational broadening, but this treats all of the lines in the same way and does not account for the temperature and ionization balance distributions across the surface. Neglecting these effects may bias our measurements for these systems, and could have a large impact on our understanding of physical processes like internal mixing efficiencies.

2. How can we address this?

In order to quantify the impact that these 3D effects have on observed stellar parameters, we use the SPAMMS code (Abdul-Masih et al. 2020), which allows us to model the distorted surface geometry and the subsequent surface parameter distributions for these stars. SPAMMS combines the advanced 3D mesh triangulation capabilities of the PHOEBE

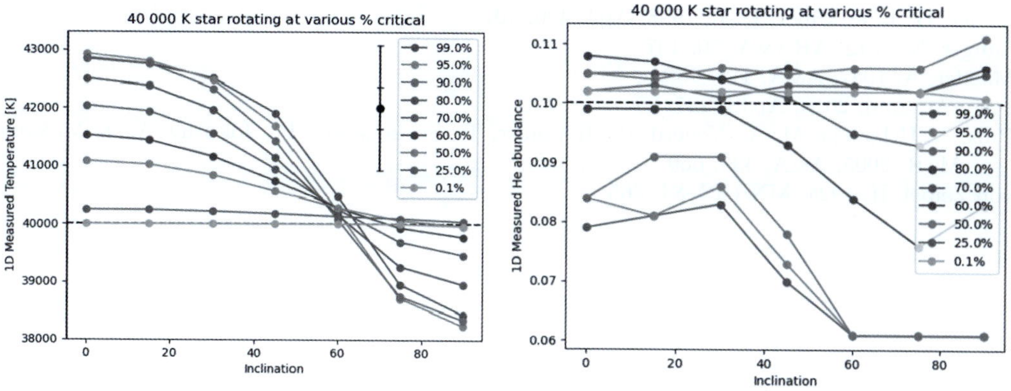

Figure 1. Left: 1D measured temperature as a function of inclination for a 40kK star rotating at different percentages of critical (indicated by the colors). The dashed horizontal line represents the real input temperature of the system. A representative 1σ and 3σ error bar are plotted in black assuming a SNR of 300. Right: 1D measured helium abundance as a function of inclination for the same star.

II code (Prša et al. 2016) with the NLTE radiative transfer code FASTWIND (Puls et al. 2005) to create a surface patch model. Given a set of stellar parameters, SPAMMS generates a mesh that represents the 3D geometry of the photosphere, and based on the local conditions of each mesh triangle, assigns a FASTWIND model. Integrating over the visible surface returns the final synthetic spectrum.

Using SPAMMS, we generate a small grid of synthetic models, where we only vary the rotation rate and the inclination. The rest of the parameters ($T_{\rm eff} = 40$kK, $r_{\rm pole} = 6.0\mathrm{R}_\odot$, $M = 25\mathrm{M}_\odot$) are left constant. We then fit each of these synthetic 3D models with 1D models to see how the observed parameters change when accounting for the 3D geometry.

3. Are these effects important?

We find a strong correlation between the 1D measured temperature and the inclination, with lower inclinations corresponding to higher temperatures and vice versa (Abdul-Masih 2022 in prep; see Fig. 1). The effect can be seen at rotation rates as low as 50%, but gets stronger as the rotation rate increases. We also find a strong deviation between the input and the 1D measured helium abundance. This effect only begins to appear at rotation rates of greater than 70% critical, but once above this, the helium abundance seems to be systematically underestimated via the 1D fits at this temperature (Fig. 1).

Our findings demonstrate that failing to account for the 3D geometry can lead to incorrect measurements of $\mathrm{T}_{\rm eff}$, abundances and presumably many other parameters. This can have a large impact on our understanding of rapidly rotating massive stars and can affect our calibrations of their internal processes such as rotational mixing efficiency.

Supplementary material

To view supplementary material for this article, please visit http://dx.doi.org/10.1017/S1743921322002411.

References

Abdul-Masih, M., Sana, H., Conroy, K. E., Sundqvist, J., Prša, A., Kochoska, A., & Puls, J. 2020, A&A, 636, A59
Brott, I., et al. 2011, A&A, 530, A116
Gräfener, G., Koesterke, L., & Hamann, W. R. 2002, A&A, 387, 244

Hillier, D. J., & Miller, D. L. 1998, ApJ, 496, 407
Langer, N. 2012, ARA&A, 50, 107
Maeder, A. 1987, A&A, 178, 159
Prša, A., et al. 2016, ApJS, 227, 29
Puls, J., Urbaneja, M. A., Venero, R., Repolust, T., Springmann, U., Jokuthy, A., & Mokiem, M. R. 2005, A&A, 435, 669
von Zeipel, H. 1924, MNRAS, 84, 665

Massive Stars Near and Far
Proceedings IAU Symposium No. 361, 2024
J. Mackey, J. S. Vink, & N. St-Louis, eds.
doi:10.1017/S1743921322002174

First results of a magnetic survey of classical Cepheids

James A. Barron[1,2], Gregg A. Wade[2], Colin P. Folsom[3] and Oleg Kochukhov[4]

[1]Department of Physics, Engineering Physics & Astronomy, Queen's University, Canada
email: j.barron@queensu.ca

[2]Department of Physics and Space Science, Royal Military College of Canada

[3]Tartu Observatory, University of Tartu, Estonia

[4]Department of Physics and Astronomy, Uppsala University, Sweden

Abstract. We report recent ESPaDOnS and HARPSpol spectropolarimetric observations from our ongoing magnetic survey of the brightest twenty-five classical Cepheids. Stokes V magnetic signatures are detected in eight of fifteen targets observed to date. The Stokes V profiles show a diversity of morphologies with weak associated longitudinal field measurements of order 1 G. Many of the Stokes V profiles are difficult to interpret in the context of the normal Zeeman effect. They consist of approximately unipolar single or double lobe(s) of positive or negative circular polarization. We hypothesize that these unusual signatures are due to the Zeeman effect modified by atmospheric velocity or magnetic field gradients. In contrast, the Stokes V profiles of Polaris and MY Pup appear qualitatively similar to the complex magnetic signatures of non-pulsating cool supergiants, possibly due to the low pulsation amplitudes of these two stars.

Keywords. Cepheids, stars: magnetic fields, stars: individual (β Dor, δ Cep, MY Pup, Polaris)

1. Introduction

Classical Cepheid variables are essential tools in studying stellar evolution and cosmology due to their radial pulsations and period-luminosity relation (e.g. Riess et al. 2021). As luminous yellow supergiants, Cepheids are the evolutionary descendants of main-sequence intermediate to massive B-type stars. While Cepheids have been studied for over a century, we know surprisingly little about how magnetic fields may influence their evolution and stellar properties.

We have identified a magnitude-limited sample ($V < 6$ mag) of the brightest twenty-five Cepheids to perform a first systematic magnetic study (Fig. 1). Before this survey, the only Cepheid with a confirmed magnetic detection was η Aql (Grunhut et al. 2010). Initial results of this survey were reported by Barron et al. (2022), including new magnetic detections of Polaris, δ Cep, and ζ Gem. Subsequently, we have obtained magnetic detections in four additional Cepheids: β Dor, MY Pup, T Vul, and W Sgr.

2. Results

Fifteen survey targets have been observed to date with ESPaDOnS at the Canada-France-Hawaii Telescope and HARPSpol at the ESO 3.6m Telescope. We apply the multi-line procedure Least Squares Deconvolution (LSD) to the Stokes V spectra to diagnose

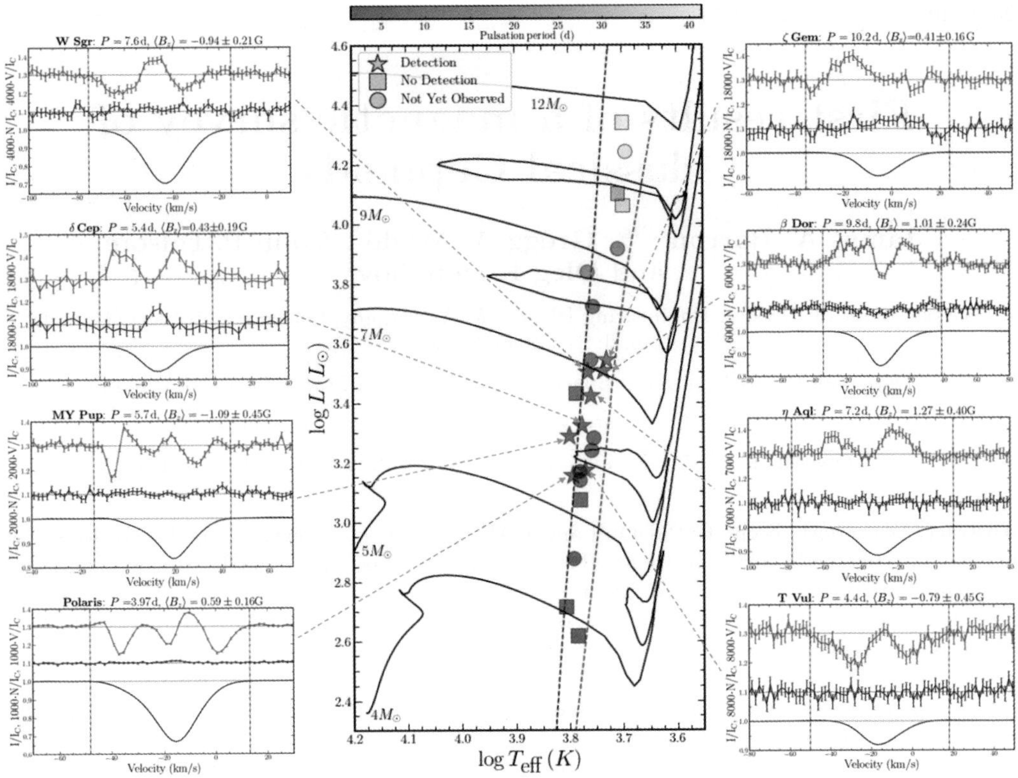

Figure 1. LSD Stokes I (black), V (red) and diagnostic null (blue) profiles for the Cepheid magnetic detections we have obtained to date. Stellar parameters, evolutionary tracks and IS boundaries from Turner (2010), Luck (2018), Georgy et al. (2013) and Anderson et al. (2016).

the presence of a photospheric magnetic field. To detect weak magnetic signatures the Stokes V spectra are obtained at high signal-to-noise ratio (~ 4000 at 500 nm).

We detect Stokes V signatures in eight targets, demonstrating that magnetic fields are frequently detectable in Cepheids when observed with sufficient precision. Many targets show peculiar unipolar positive or negative Stokes V lobes, which are not predicted by standard Zeeman theory (e.g. δ Cep). These Stokes V profiles show similarities to those detected in some Am stars (Petit et al. 2011). We hypothesize that these features are due to the Zeeman effect modified by atmospheric velocity and magnetic field gradients. Polarized radiative transfer modelling of the Stokes V signatures could provide a new and unique probe of Cepheid atmospheric dynamics (see Barron et al. 2022 and references therein).

In contrast, the LSD Stokes V profiles of Polaris and MY Pup appear more similar to those observed in cool non-pulsating supergiants (Grunhut et al. 2010), with some net negative asymmetry. The lesser distortion in the Stokes V profiles may be related to the stars' low amplitude pulsations (Kienzle et al. 1999; Anderson 2019). We have initiated a monitoring campaign of Polaris using ESPaDOnS to map its magnetic field using Zeeman-Doppler Imaging.

Supplementary material

To view supplementary material for this article, please visit http://dx.doi.org/10.1017/S1743921322002174.

References

Anderson, R. I. 2019, A&A, 623, A146
Anderson, R. I., Saio, H., Ekström, S., Georgy, C., & Meynet, G. 2016, A&A, 591, A8
Barron, J. A., Wade, G. A., Evans, N. R., Folsom, C., & Neilson, H. R. 2022, MNRAS, 512, 4021
Georgy, C., et al. 2013, A&A, 553, A24
Grunhut, J. H., Wade, G. A., Hanes, D. A., & Alecian, E. 2010, MNRAS, 408, 2290
Kienzle, F., Moskalik, P., Bersier, D., & Pont, F. 1999, A&A, 341, 818
Luck, R. E. 2018, AJ, 156, 171
Petit, P., et al. 2011, A&A, 532, L13
Riess, A. G., et al. 2021, arXiv e-prints, arXiv:2112.04510
Turner, D. G. 2010, AP&SS, 326, 219

Massive Stars Near and Far
Proceedings IAU Symposium No. 361, 2024
J. Mackey, J. S. Vink, & N. St-Louis, eds.
doi:10.1017/S1743921322003283

A comprehensive empirical investigation of Galactic fast-rotating O-type stars

N. Britavskiy[1], G. Holgado[3], S. Simón-Díaz[2], J. Maíz Apellániz[3], M. Pantaleoni González[3], Y. Nazé[1], S. Burssens[4] and J. J. Eldridge[5]

[1]Université de Liège, Allée du 6 Aout 19c, B-4000 Sart Tilman, Liège, Belgium
email: mbritavskiy@uliege.be

[2]Instituto de Astrofísica de Canarias. E-38200 La Laguna, Tenerife, Spain

[3]Centro de Astrobiología, campus ESAC. Villanueva de la Cañada, E-28692, Spain

[4]Institute of Astronomy, KU Leuven, Celestijnenlaan 200D, 3001 Leuven, Belgium

[5]Department of Physics, University of Auckland, Private Bag 92019, Auckland, New Zealand

Abstract. Rotation is one of the important parameters affecting the evolution and final fate of massive stars but the origin of fast rotators remains unclear (imprint of the star formation process, result of binary interactions). In this work, we aim at investigating the binary status, photometric variability, and runaway status of a statistically meaningful sample of Galactic fast-rotating O stars. We perform a comprehensive multi-epoch analysis of new high-quality spectroscopic observations gathered by the IACOB and OWN surveys. Notably, we find that the total percentage of spectroscopic binaries in the investigated sample range between 25 and 40%, in agreement with previous finding for the case of O-type stars with lower projected rotational velocities. On the contrary, the fraction of runaway stars among fast rotators (∼35-50%) is significantly higher than in the case of slow rotators (∼20-30%). By combining all these observational results we will evaluate each scenario about the origin of fast rotators.

Keywords. Stars: early type, Stars: rotation, Stars: binaries.

1. Introduction

Several investigations of the distribution of projected rotational velocities ($v \sin i$) in the O-type domain, both in the Galaxy and the Magellanic Clouds (see Holgado et al. 2022, and references therein) have shown a clear and ubiquitous bimodal character. According to some theoretical simulations (de Mink et al. 2013, de Mink et al. 2014), different scenarios of binary interaction are needed to describe such $v \sin i$ distributions, and the existence of fast rotators with $v \sin i > 200$ km s^{-1}. The authors emphasize several channels: (i) mergers, (ii) mass transfer with a speed-up gainer and (iii) tidal coupling in a close binary systems. All these scenarios require high-quality observational evidence to distinguish between them. In our work (Britavskiy et al. in prep), we aim to make a comprehensive characterization of fast-rotating stars in the O-type domain. Such an analysis became possible with the help of new multi-epoch spectroscopic observations compiled in the framework of the IACOB project (see Simón-Díaz, these proceedings), proper motion analyses based on the Gaia data, and photometric variability analyses from TESS.

We started our work by using a cross-correlation technique for a sample of 51 fast rotators to find the radial velocity variability defined as peak-to-peak amplitude ($RV_{\rm pp}$). However, significant radial velocity variations cannot serve as a definitive diagnostic

Table 1. Number (and percentage) of identified O-type spectroscopic binaries and runaway stars among 51 fast-rotating stars with $v \sin i > 200$ km s^{-1} and distances up to 2.5 kpc.

	Fast rotators	
SB2:	4	8 %
LPV/SB2?:	6	12 %
SB1:	8	16 %
LPV/SB1?:	2	4 %
LS	31	60 %
LS runaways	11	22 %
LS not runaways	20	39 %

parameter for the presence of a secondary component. Indeed, orbital motion is not the only effect which can led to the detection of large radial velocity variations in O-type stars, especially in the case of fast rotators. The line-profiles of most O- and B-type stars are often subject to various types of variability due to, e.g., stellar pulsations or spots.

Taking into account these ideas, we decided to complement the information from the measured $RV_{\rm pp}$ with the outcome from a visual inspection of the detected variability of the He I λ5875 line-profile in all stars in our program sample, classifying each star as being part of one of the following subgroups: likely single (LS), one- and two-component spectroscopic binary (SB1 and SB2), and unclear spectroscopic binaries "LPV/SB1?", and "LPV/SB2?" which require further investigation.

The Gaia and TESS missions have provided a unique opportunity to have access to very valuable informatison in a homogeneous way. The proper motions delivered by Gaia-EDR3 can be used to identify runaway stars. Some of these runaways are expected to be produced by the dynamical ejection of the surviving companion in a high-mass binary system after a supernova explosion event. Following de Mink et al. (2013), an important fraction of the O-type stars with $v \sin i > 200$ km s^{-1} are the mass gainers in binary systems after Roche-lobe overflow of the initially more massive star. In this sense, it is interesting to know, not only which of the stars in our program sample are identified as a runaway star, but also to compare the percentage of runaways detected in the fast- and slow-rotator samples of O-type stars investigated in Holgado et al. (2022).

In Table 1 we list the number of identified spectroscopic binaries together with a number of runaway stars among our working sample of 51 fast rotators. In addition, we searched for signatures of eclipses in the TESS light curves, identifying four eclipsing binaries in our sample (three of them marked as clear SB1 systems, and the fourth one labeled as LS). Finally, we modeled the observational appearance of fast rotators with a help of Binary Population and Spectral Synthesis v2.2.1 code in order to estimate the binary fraction of different O-star samples that we can expect to be rapidly rotating. These theoretical results appeared to be in agreement with our observational detections of spectroscopic binaries among fast rotators. A detailed discussion of these observational results and their interpretation will be presented in Britavskiy et al. (in prep).

References

Holgado, G, Simón-Díaz, S., Herrero, A., & Barbá, R. 2022, A&A, accepted.
de Mink, S. E., Langer, N., Izzard, R. G., Sana, H., & de Koter, A. 2013, ApJ, 764, 166
de Mink, S. E., Sana, H., Langer, N., Izzard, R. G., & Schneider, F. R. N. 2014, ApJ, 782, 7

Massive Stars Near and Far
Proceedings IAU Symposium No. 361, 2024
J. Mackey, J. S. Vink, & N. St-Louis, eds.
doi:10.1017/S1743921322002617

Can we detect deep axisymmetric toroidal magnetic fields in stars?

Hachem Dhouib[1], Stéphane Mathis[1], Lisa Bugnet[2], Timothy Van Reeth[3] and Conny Aerts[3,4,5]

[1]Université Paris Cité, Université Paris-Saclay, CEA, CNRS, Astrophysique, Instrumentation et Modélisation Paris-Saclay, F-91191, Gif-sur-Yvette, France
email: hachem.dhouib@cea.fr

[2]Flatiron Institute, Simons Foundation, 162 Fifth Ave, New York, NY 10010, USA

[3]Institute of Astronomy, KU Leuven, Celestijnenlaan 200D, 3001 Leuven, Belgium

[4]Department of Astrophysics, IMAPP, Radboud University Nijmegen, PO Box 9010, 6500 GL Nijmegen, the Netherlands

[5]Max Planck Institute for Astronomy, Koenigstuhl 17, 69117 Heidelberg, Germany

Abstract. The current angular momentum (AM) transport models fail to reproduce asterosieimic observations. One of the best candidates to explain this discrepancy is the magnetic field in radiative zones with its various possible topologies, for instance axisymmetric toroidal magnetic field. If such azimuthal field is strong enough, the Tayler's instability could occur which induces a magnetic torque that allows a very efficient transport of AM and could trigger dynamo action in radiative layers. If such important field does not emerge at the surface, spectropolarimetry is blind. In this case, the only way to detect and characterise the field is by using magneto-asteroseismology. It consists in searching for the characteristic signatures of magnetic field in the observed frequency spectra of stellar oscillations.

Keywords. stars: early-type, stars: interiors, stars: magnetic fields

1. Introduction

Asteroseismology has revealed the existence of a strong extraction of angular momentum (AM) operating in stars. One of the plausible candidates to efficiently carry AM is the magnetic field in stellar radiative zones. Since the seminal work by Spruit (2002), a strong interest has been devoted to axisymmetric toroidal magnetic fields. He suggested that such field could trigger a dynamo action in stably stratified layers (the way the dynamo loop is closed has been debated in Zahn et al. (2007) and Fuller et al. (2019)) and that the resulting magnetic torque allows a very efficient transport of AM. However if such field is present in the radiative envelope but does not emerge at the stellar surface, spectropolarimetry will not be able to detect it. Thus, the key question to answer is how they can be detected and characterised. The only diagnostic tool that we have to address this problem is magneto-asteroseismology (Neiner et al. 2015; Mathis et al. 2021). It consists of searching for magnetic signatures in the stellar oscillations frequency spectrum. We here study the impact of general axisymmetric toroidal magnetic fields on gravito-inertial modes in a non-perturbative way. These modes are the oscillations that propagate in rotating stellar radiative zones under the combined action of the buoyancy force and the Coriolis acceleration. In the magnetised case, these modes become magneto-gravito-inertial modes (MGIM) because of the Lorentz force.

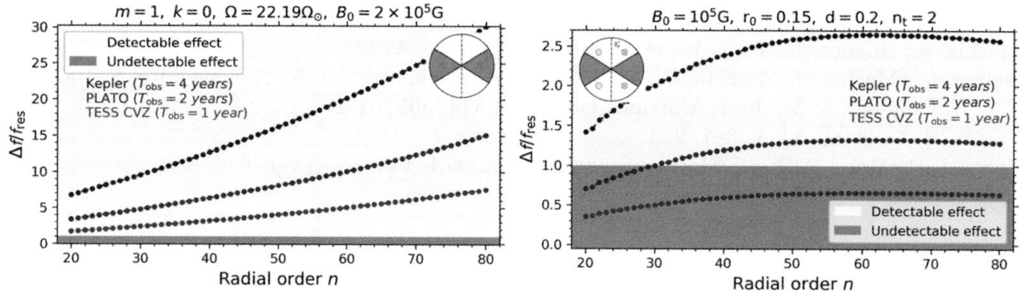

Figure 1. Detectability of the signature of an equatorial (left) and hemispheric (right) toroidal magnetic field on the $\{k=0, m=1\}$ mode as a function of the radial order n using a $1.6M_\odot$ γ Dor model near ZAMS.

2. Magnetic TAR

We generalise the traditional approximation of rotation (TAR) which is intensively used in asteroseismology for the study of low-frequency waves propagating in strongly stratified zones (e.g. Lee & Saio 1997). This approximation allows us to neglect the vertical component of the Coriolis acceleration along the stratification direction because it is dominated by the buoyancy force. The TAR, in its standard version, relies mainly on three assumptions. First, the rotation is assumed to be uniform. Second, the star is assumed to be spherical. Finally, the magnetic field is neglected. To introduce magnetic effects, we derive the magnetic TAR, which includes a general axisymmetric toroidal magnetic field (with a general 2D profile of the Alfvén frequency ω_A) in a non-perturbative way and a general 2D differential rotation Ω. Using this formalism we derive a new generalisation of the Laplace tidal equation (Laplace 1799). The derivation of the generalised magnetic Laplace tidal equation can be found in Dhouib et al. (2022).

As a proof of concept, we applied this new formalism to representative stellar models of typical γ Dor stars during their main-sequence evolution. We choose to investigate the impact of two different configurations of toroidal magnetic field. The first one is an equatorial field localised near the equator. The second is a hemispheric field composed of two magnetic distributions localised in the Northern and Southern hemispheres and vanishing at the equator. As we can see in Fig. 1, the effect of an equatorial magnetic field is in principle largely detectable for all radial orders using nominal TESS CVZ, *Kepler* and PLATO light curves. In opposite, the detection becomes much harder for hemispheric fields because gravito-inertial modes can be equatorially trapped.

3. Conclusion

The inclusion of magnetic effects in stellar oscillation theory and seismic forward modelling will provide important observational constraints on the theory of transport processes in stellar interiors. This would lead to breakthrough in our understanding of the evolution and the magnetism of stars.

Supplementary material

To view supplementary material for this article, please visit http://dx.doi.org/10.1017/S1743921322002617.

References

Dhouib, H., Mathis, S., Bugnet, L., Van Reeth, T., & Aerts, C. 2022, A&A, 661, A133

Fuller, J., Piro, A. L., & Jermyn, A. S. 2019, MNRAS, 485, 3661

Laplace, P. S. 1799, Traité de Mécanique Cëleste (Imprimerie de Crapelet (Paris))

Lee, U. & Saio, H. 1997, ApJ, 491, 839
Mathis, S., Bugnet, L., Prat, V., et al. 2021, A&A, 647, A122
Neiner, C., Mathis, S., Alecian, E., et al. 2015, in Polarimetry, ed. K. N. Nagendra, S. Bagnulo, R. Centeno, & M. Jesús Martínez González, Vol. 305, 61–66
Spruit, H. C. 2002, A&A, 381, 923
Zahn, J. P., Brun, A. S., & Mathis, S. 2007, A&A, 474, 145

Massive Stars Near and Far
Proceedings IAU Symposium No. 361, 2024
J. Mackey, J. S. Vink, & N. St-Louis, eds.
doi:10.1017/S1743921322002356

Apsidal motion in NGC 6231: Sounding the internal structure of massive stars

S. Rosu[1], G. Rauw[1], M.-A. Dupret[1], A. Noels[1] and M. Farnir[2]

[1]STAR Institute, Université de Liège, Allée du 6 août 19c, Bât B5c, 4000 Liège, Belgium
email: sophie.rosu@uliege.be

[2]University of Warwick, Coventry CV4 7AL, UK

Abstract. The young open cluster NGC 6231 hosts a rich population of O-type binary stars. We study several of these eccentric short-period massive eclipsing binaries and assess their fundamental parameters. The properties of these systems make them interesting targets to study tidally induced apsidal motion. The analysis of apsidal motion offers a powerful means to obtain information about the internal structure of the stars. Indeed, since the rate of apsidal motion in a binary system is proportional to the internal structure constants of the stars composing it, its value gives direct insight into the internal structure and evolutionary state of these stars. Stellar evolution models are constructed based on the observationally-determined fundamental parameters and a theoretical rate of apsidal motion is inferred. The results are striking: Adopting standard stellar evolution models yields a theoretical rate of apsidal motion much larger than the observational value. This discrepancy results from the standard models predicting too low an efficiency of internal mixing and thus too homogenous stars in terms of density. By enforcing the theoretical rates of apsidal motion to match the observational values, enhanced mixing is required, through a large overshooting parameter and/or additional turbulent/rotational mixing. Our analysis leads to the conclusion that the chemically mixed cores in those massive stars must be more extended than anticipated from standard models.

Keywords. Stars: early-type, evolution, massive, Binaries: spectroscopic, eclipsing

1. Introduction and motivations

In close binaries, tidal interactions – which are responsible for the non-spherical gravitational fields of the stars – take place between the two components. If the orbit is eccentric, the main effect of these tidal interactions is the precession of the line of apsides with time, which is called apsidal motion. Whilst massive binaries, especially double-line spectroscopic eclipsing ones, are interesting systems as they offer possibilities to constrain the fundamental properties – such as the masses and radii – in a model independent way, massive binaries showing evidence of apsidal motion are even more interesting systems as the apsidal motion allows us to sound the internal structure of the stars (see e.g. Rosu *et al.* 2020a, and references therein). Indeed, the apsidal motion rate depends upon the internal structure constants of the stars, a measure of the density stratification between the core and the external layers of the stars. In this respect, the most promising systems are close short-period eccentric systems for which photometric, spectroscopic, and radial velocity data can be analysed to constrain the fundamental stellar parameters and the apsidal motion rate of the system. Confronting stellar evolution models computed with different prescriptions for the internal mixing and/or the mass-loss rate to the parameters derived from the observations then allows us to constrain the internal mixing processes occurring inside the stars. We focus here on the young and rich open cluster NGC 6231 that hosts

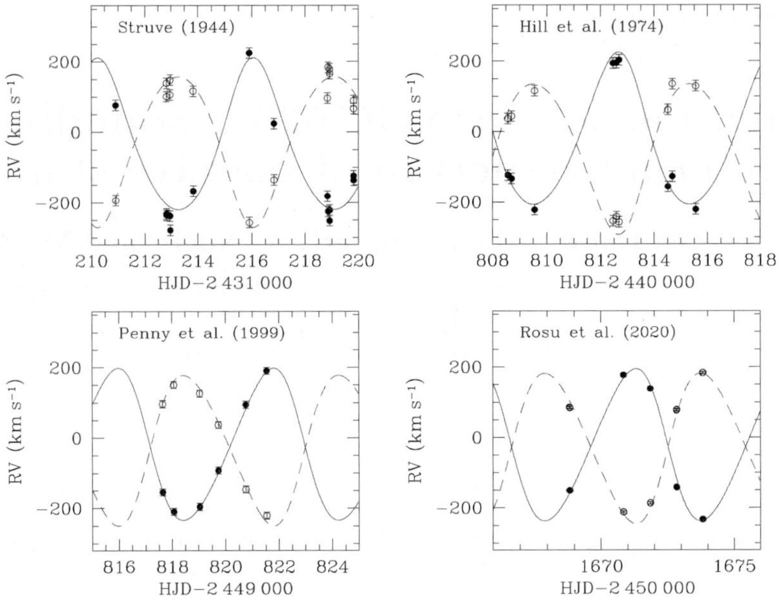

Figure 1. Measured RVs of the primary (filled dots) and secondary (open dots) stars of HD 152248, and best-fit RV curves (blue and red). Figure from Rosu (2021).

a substantial population of O-type binaries which are interesting candidates for such studies, notably HD 152248 (Rosu *et al.* 2020a,b) and HD 152219 Rosu *et al.* (2022).

2. How to measure the apsidal motion rate

The apsidal motion rate in a binary system can be determined using two different methods, depending on whether the binary system is a spectroscopic binary for which radial velocity measurements are available or it is an eclipsing binary for which photometric observations are available.

In the former case, the spectroscopic observations of the binary are analysed to disentangle the contributions of the two stars and to derive the radial velocities at each time of observation. All radial velocity data (either primary, secondary, or equivalent primary) are fit according to the following equation

$$RV(t) = K\,(\cos(\phi(t) + \omega(t)) + e\cos(\omega(t))) + \gamma, \tag{2.1}$$

where K is the semi-amplitude of the radial velocity curve, γ is the apparent systemic velocity, e is the eccentricity of the orbit, ω is the argument of periastron, and ϕ is the true anomaly. This expression explicitly accounts for apsidal motion through the linear variation of ω with time:

$$\omega(t) = \omega_0 + \dot{\omega}(t - T_0), \tag{2.2}$$

where $\dot{\omega}$ is the apsidal motion rate and ω_0 is the argument of periastron at the time of reference T_0. An example of such a radial velocity adjustment is shown in Fig. 1 for the binary system HD 152248 (see also Rauw *et al.* 2016; Rosu *et al.* 2020b, 2022a).

In the latter case, the inclination i of the system has to be such that the eclipses of the two stars are observed during one orbital cycle. In an eccentric binary, both the depths and the phases of the eclipses change with ω. This is illustrated in Fig. 2 where nine different light curves are plotted for nine different values of ω ranging from 0° to 160°, all other parameters identical ($e = 0.134$, $i = 68.6°$, and stars having the same mass m,

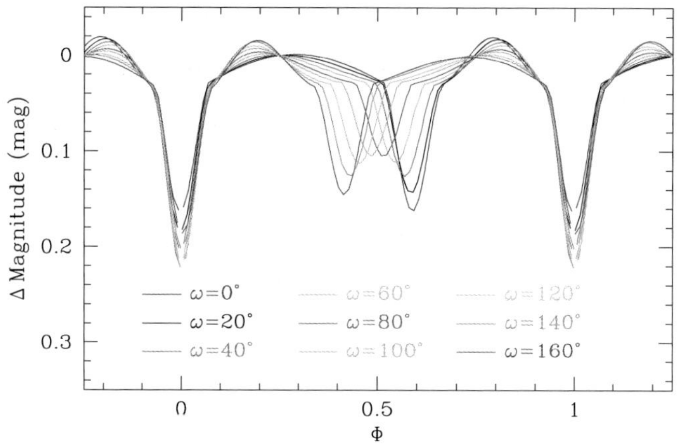

Figure 2. Set of theoretical light curves for different values of ω, all other parameters being identical ($e = 0.134$, $i = 68.6°$, and stars having the same m, R, and $T_{\rm eff}$). Figure from Rosu (2021).

radius R, and effective temperature $T_{\rm eff}$). The method then consists either in directly fitting the light curves taken at different epochs with ω as a free parameter, or in fitting the times of minimum of the eclipses of the corresponding light curves to get values for the phase differences at different epochs that can in turn be fitted to get a value for the apsidal motion rate through the equations of Giménez & Bastero (1995), see also Zasche & Wolf (2019), Baroch *et al.* (2021), and Rosu *et al.* (2022a,b).

3. Apsidal motion equations and internal structure constant

The apsidal motion rate of a system is made, in the two-body problem, of a Newtonian contribution $\dot{\omega}_{\rm N}$, which originates from the non-Keplerian terms of the stellar gravitational potentials, and a general relativistic correction $\dot{\omega}_{\rm GR}$. Their expressions are given by, according to Sterne (1939), where only the contributions arising from the second-order harmonic distorsions of the gravitational potential are considered,

$$\dot{\omega}_{\rm N} = \frac{2\pi}{P_{\rm orb}} \left[15 f(e) \left\{ k_{2,1} q \left(\frac{R_1}{a}\right)^5 + \frac{k_{2,2}}{q}\left(\frac{R_2}{a}\right)^5 \right\} + g(e) \left\{ k_{2,1}(1+q) \left(\frac{R_1}{a}\right)^5 \left(\frac{P_{\rm orb}}{P_{\rm rot,1}}\right)^2 + k_{2,2}\frac{1+q}{q}\left(\frac{R_2}{a}\right)^5 \left(\frac{P_{\rm orb}}{P_{\rm rot,2}}\right)^2 \right\} \right], \tag{3.3}$$

and by, according to Shakura (1985),

$$\dot{\omega}_{\rm GR} = \left(\frac{2\pi}{P_{\rm orb}}\right)^{5/3} \frac{3(G(m_1+m_2))^{2/3}}{c^2(1-e^2)}, \tag{3.4}$$

where a is the semi-major axis, $q = m_2/m_1$ is the mass ratio, $k_{2,j}$ the internal structure constant and $P_{\rm rot,j}$ the rotational period of star j ($j = 1, 2$ for the primary and secondary stars, respectively), $f(e)$ and $g(e)$ are functions of the eccentricity of the orbit which expressions are given in e.g. Rosu (2021), G is the gravitational constant and c is the speed of light. The internal structure constant k_2 is a measure of the density stratification between the core and the external layers of the star. It is an algebraic expression of the η_2-parameter which is the solution, evaluated at the stellar surface, of the Clairaut-Radau differential equation (Hejlesen 1987) that depends upon the density profile inside the

Table 1. Set of observationally-determined properties of the massive binaries HD 152248 (Rosu *et al.* 2020b) and HD 152219 (Rosu *et al.* 2022a).

Parameter	Value		
	HD 152248	**HD 152219**	
		Primary	**Secondary**
Spectral type	O7.5 III-II(f)	O9.5 III	B1-2 V-III
M $(M_\odot)$	$29.5^{+0.5}_{-0.4}$	18.64 ± 0.47	7.70 ± 0.12
R $(R_\odot)$	$15.07^{+0.08}_{-0.12}$	9.40 ± 0.15	3.69 ± 0.06
$T_{\rm eff}$ (K)	$34\,000 \pm 1000$	$30\,900 \pm 1000$	$21\,700 \pm 1000$
$L_{\rm bol}$ $(L_\odot)$	$(2.73 \pm 0.32) \times 10^5$	$(7.26 \pm 0.97) \times 10^4$	$(2.73 \pm 0.51) \times 10^3$
$\bar{k}_2$	0.0010 ± 0.0001	0.00173 ± 0.00052	
$\dot{\omega}$ $(^\circ\,{\rm yr}^{-1})$	$1.843^{+0.064}_{-0.083}$	1.198 ± 0.300	
$P_{\rm orb}$ (d)	$5.816475^{+000085}_{-0.000075}$	$4.24046^{+0.00005}_{-0.00004}$	
e	$0.134^{+0.0007}_{-0.0004}$	$0.072^{+0.004}_{-0.005}$	

star. k_2 takes its maximum value of 0.75 for an homogenous sphere of constant density, and can take values as low as 10^{-4} for massive stars having a dense core and a diluted atmosphere (Rosu *et al.* 2020a). As the star evolves, its core contracts while its enveloppe expands, hence its k_2 decreases with time, rendering this quantity a good indicator of stellar evolution.

All terms appearing in the right-hand side of Eq. (3.4) are known from observations, hence $\dot{\omega}_{\rm GR}$ is also. As $\dot{\omega}$ is known from observations, all terms – except for $k_{2,1}$ and $k_{2,2}$ – appearing in Eq. (3.3) are also. In the special case of a twin system, that is to say, if the two stars share exactly the same properties in terms of m, R, and $T_{\rm eff}$, we can assume their k_2 are identical. An observational value can then be inferred for k_2 through Eq. (3.3). In the general case of two different stars, we define a weighted-average mean of the internal structure constants

$$\bar{k}_{2,\rm obs} = \frac{c_1 k_{2,1} + c_2 k_{2,2}}{c_1 + c_2} = \frac{\dot{\omega}_{\rm N}}{c_1 + c_2}, \tag{3.5}$$

where c_1 and c_2 are known functions of stellar and orbital parameters, whose expressions can be obtained from Eq. (3.3) and which can be inferred from observations.

4. Two concrete cases: `Clés` models

We built stellar structure and evolution models for the massive binaries HD 152248 and HD 152219 with the `Clés` code (Scuflaire *et al.* 2008). The observationally determined properties of the two binaries are summarised in Table 1. Whilst HD 152248 is a twin system, hence the individual k_2 are identical and equal to $\bar{k}_2$, HD 152219 is made of a primary star bigger and more massive than the secondary companion, hence we have $k_{2,1} \leq \bar{k}_2 \leq k_{2,2}$. In the `Clés` models, we included overshooting, implemented as a step-function through the overshooting parameter $\alpha_{\rm ov}$, as well as turbulent diffusion, implemented as a partial mixing process with velocities of chemicals $V_i = -D_T \frac{d \ln X_i}{dr}$, where D_T is the turbulent diffusion coefficient (measured in ${\rm cm^2\,s^{-1}}$). Turbulent diffusion acts as reducing the abundance gradient of the chemicals. Both mechanisms bring additional hydrogen from the external layers to the chemically mixed core, that is to say, fuel for nuclear reactions, and, therefore, increase the lifetime of the star on its main-sequence phase. The additional mixing leads to a reduced mass concentration at a given age. Mass-loss is implemented as in Vink *et al.* (2001). We used the `min-Clés` routine, in which a Levenberg-Marquardt minimisation technique is implemented, to search for best-fit models of the stars in terms of their observational properties (Rosu *et al.* 2020a).

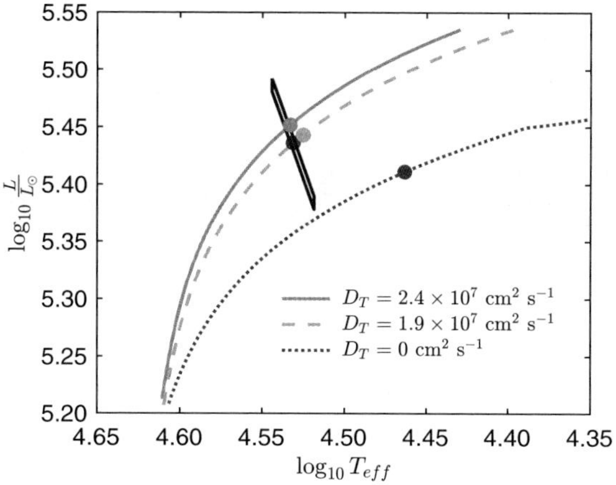

Figure 3. HR diagram for HD 152248: evolutionary tracks of `Clés` models. The dots over-plotted on the tracks correspond to the models that fit the observational k_2. The observational value and its error bars are represented in red.

HD 152248 (see Rosu et al. 2020a). We first requested the `Clés` models to reproduce the stellar mass, radius, and position in the Hertzsprung-Russell (HR) diagram only. To this aim, we built a first best-fit model with $\alpha_{\rm ov} = 0.20$ and $D_T = 0\,{\rm cm^2\,s^{-1}}$. The corresponding evolutionary track is shown in Fig. 3 (purple dotted line) together with the observational value and its error bars (in red). This model does not reproduce the physical properties of the stars, even within the error bars. We then built a second model leaving the turbulent diffusion coefficient as a free parameter of the adjustment and obtained $D_T = 1.9 \times 10^7\,{\rm cm^2\,s^{-1}}$. The mass, radius, and position in the HR diagram are well-reproduced, as the corresponding track indicates (green dashed line in Fig. 3). However, we observe that the model that reproduces the observational k_2-value, symbolised by the dot over-plotted on the track, is located further away on the evolutionary track compared to the location of the best-fit model. It means that our best-fit model has too high a k_2-value, that is to say, too low a density contrast between its core and external layers. We solved this issue by enforcing enhanced turbulent diffusion inside the models. We obtained a best-fit model in terms of the mass, radius, position in the HR diagram, k_2, and the apsidal motion rate that has $D_T = 2.4 \times 10^7\,{\rm cm^2\,s^{-1}}$. The corresponding track is shown in Fig. 3 (orange line). This is a valuable result in the sense that without taking into account the apsidal motion occurring in this binary system, we would not highlight that the standard models predict too low a density contrast between the core and the external layers compared to what is expected from observations.

HD 152219 (see Rosu et al. 2022a). Given that the secondary star has a much smaller radius than the primary star, we can assume that $c_2 \ll c_1$ in Eq. 3.5, and hence, $k_{2,1} \sim \bar{k}_2$. We hence focus here on the primary star only. As for HD 152248, the best-fit model in terms of the mass, radius, and position in the HR diagram has too high a $k_{2,1}$-value, as indicated in Fig. 4 (pink dotted and green dashed lines for $D_T = 0$ and $2.12 \times 10^6\,{\rm cm^2\,s^{-1}}$, respectively). We proceeded as for HD 152248 and obtained a best-fit model in terms of the afore-mentioned parameters and of the internal structure constant, adopting the least stringent constraint on $k_{2,1}$, that is to say, enforcing $k_{2,1} = \bar{k}_2$, that has $D_T = 9.92 \times 10^6\,{\rm cm^2\,s^{-1}}$. The corresponding evolutionary track is shown in Fig. 4 (orange line). The discrepancy in k_2 is lowered at the cost that now the best-fit model does not reproduce the position in the HR diagram anymore.

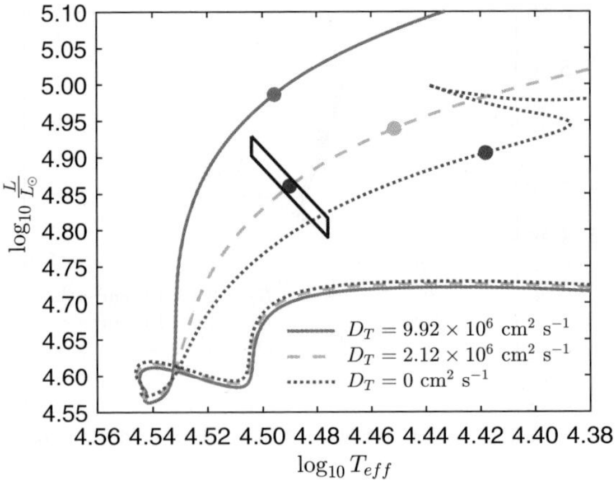

Figure 4. HR diagram for HD 152219: evolutionary tracks of `Clés` models. The dots overplotted on the tracks correspond to the models that fit the observational $\bar{k}_2$. The observational value and its error bars are represented in red.

5. Conclusion

We studied the apsidal motion in massive binary systems from a theoretical point of view. We highlighted that the best candidates for this purpose are the twin systems, for which the apsidal motion equations can be solved for the internal structure constants. We built stellar structure and evolution models for two massive binaries, HD 1522448 and HD 152219, based on the observationally-determined stellar parameters (masses, radii, and effective temperatures). The standard models obtained in this way predicted too high a value of the apsidal motion rate compared to the observational value. We then enforced the models to reproduce the apsidal motion rate of the system through the internal structure constants of the stars. The best-fit models obtained in this way require enhanced mixing through the turbulent diffusion compared to standard ones. We conclude that standard models predict too low an efficiency of internal mixing and thus stars that have too low a density contrast between their core and external layers, and that the chemically mixed cores in these massive stars must be more extended than usually considered in stellar evolution.

Acknowledgements. S. Rosu acknowledges support from the Fonds de la Recherche Scientifique (F.R.S.- FNRS, Belgium). The authors thank the referee for his/her suggestions and comments towards the improvement of the manuscript.

References

Baroch, D., Giménez, A., Ribas, I., *et al.* 2021, *A&A*, 649, A64
Giménez, A., & Bastero, M. 1995, *Ap&SS*, 226, 99
Hejlesen, P. M. 1987 *A&AS*, 69, 251
Rauw, G., Rosu, S., Noels, A., *et al.* 2016, *A&A*, 594, A33
Rosu, S. 2021, *Bulletin de la Société Royale des Sciences de Liège*, 90, 1
Rosu, S., Noels, A., Dupret, M.-A., *et al.* 2020a, *A&A*, 642, A221
Rosu, S., Rauw, G., Conroy, K. E., *et al.* 2020b, *A&A*, 635, A145
Rosu, S., Rauw, G., Farnir, M., Dupret, M.-A., & Noels, A. 2022a, *A&A*, 660, A120
Rosu, S., Rauw, G., Nazé, Y., Gosset, E., & Sterken, C. 2022b, *A&A*, accepted [arXiv:2205.11207]
Scuflaire, R., Théado, S., Montalbán, J., *et al.* 2008, *Ap&SS*, 316, 83
Shakura, N. I. 1985, *Sov. Astron. Lett.*, 11, 224

Sterne, T. E. 1939, *MNRAS*, 99, 451
Vink, J. S., de Koter, A., & Lamers, H. J. G. L. M. 2001 *A&A*, 369, 574
Zasche, P., & Wolf, M. 2019, *AJ*, 157, 87

Discussion

HIGGINS: I was just wondering how do you translate these values of $k_{2,1}$ and $k_{2,2}$ into your mixing. So is it like a model grid and then how fine would the spacing be because I saw that for one component it was like an $\alpha_{\rm ov}$ of 0.3 and other was an $\alpha_{\rm ov}$ of 0.2 so how do you translate your extra mixing into like values so that we can use them for stellar evolution?

ROSU: Basically, I tried several values of $\alpha_{\rm ov}$. I made a grid from 0.1 to 0.4 and got different values of the turbulent mixing so, as we don't really know the value of $\alpha_{\rm ov}$, it translates into the change in turbulent diffusion. But whatever the value of $\alpha_{\rm ov}$ in this range, we get models that are able to reproduce the different parameters in the same way. For instance for the first system, I only presented one value but we can get models that have different values of $\alpha_{\rm ov}$ and so different values of the turbulent diffusion, and the corresponding tracks mostly overlap.

HIGGINS: Yes, that was I was kind of wondering and how you would get like a unique solution because if you have other parameters like the rotation or other efficiencies you know you could have multiple solutions for your k_2-values.

ROSU: Yes, we get several solutions and we also tested the rotational influence. We cannot put rotation inside the `Clés` models but we used `GENEC` models with rotation included and we arrived to the conclusion, for this system at least, that the rotation can be simulated through this turbulent diffusion for instance.

BOWMAN: Maybe I missed the values but can you comment perhaps on this turbulent mixing being mass dependant because I think I saw that your primary needed larger values than the secondary? Is that correct?

ROSU: Exactly. Here, we expect that the turbulent diffusion depends upon the masses of the stars but it might also depend on the rotation of the stars as it might also simulate a bit the rotation of the stars. We expect that if the star is rotating faster than it should have a turbulent diffusion much higher than a star rotating slower.

BOWMAN: I am just thinking about synchronicity though in such short-period systems as well so if they are approximately synchronous and then you still find larger values in the primary that is very interesting.

Massive Stars Near and Far
Proceedings IAU Symposium No. 361, 2024
J. Mackey, J. S. Vink, & N. St-Louis, eds.
doi:10.1017/S1743921322002344

Apsidal Motion in Massive Binaries: CPD-41° 7742, an Extreme Case?

S. Rosu[1], G. Rauw[1], Y. Nazé[1], E. Gosset[1] and C. Sterken[2]

[1]STAR Institute, Université de Liège, Allée du 6 août 19c, Bât B5c, 4000 Liège, Belgium
email: sophie.rosu@uliege.be

[2]Physics Department, Vrije Universiteit Brussel, Pleinlaan 2, 1050 Brussels, Belgium

Abstract. We study the apsidal motion in close eccentric massive binaries. Measuring the rate of apsidal motion in such a system gives insight into the internal structure and evolutionary state of the stars. We focus on CPD-41° 7742, for which independent studies in the past showed large discrepancies in the longitude of periastron of the orbit, hinting at the presence of apsidal motion. We perform a consistent analysis of all observational data to solve this apparent discrepancy and report the first determination of apsidal motion in this system. This study confirms the need for enhanced mixing in the stellar evolution models of the primary star to reproduce the observational properties. This points towards larger convective cores than usually considered.

Keywords. Stars: early-type, evolution, massive, Binaries: spectroscopic, eclipsing

1. Introduction and motivations

The majority of massive stars belong to binary systems: This considerably affects the evolution of the stars and offers tremendous possibilities to constrain their properties in a model independent way. The most interesting systems in this respect are double-line spectroscopic (SB2) eclipsing binaries showing a significant apsidal motion. The apsidal motion is the slow precession of the line of apsides with time. It arises because of the tidal interactions occurring between the stars of a close eccentric binary, interactions which are responsible for the non-spherical gravitational fields of the stars. The rate of apsidal motion is directly related to the internal structure of the stars. Measuring the rate of apsidal motion hence provides a diagnostic of the internal mass-distribution of the stars, which is otherwise difficult to constrain, and also offers a test of our understanding of stellar structure and evolution. We here focus on the massive SB2 eclipsing binary CPD-41° 7742, located in the young open cluster NGC 6231. Large discrepancies in the longitude of periastron (ω) derived for this binary were reported in the past by independent studies (Sana et al. 2003, Sana et al. 2005, Bouzid et al. 2005). These discrepancies hint at the presence of apsidal motion in the system. We carry out the first consistent analysis of all the observational data of CPD-41° 7742 available at the time of this study to enforce a rate of apsidal motion ($\dot{\omega}$) of the system; further details are given by Rosu *et al.* (2022).

2. Methods and results

We used our disentangling code based on the method described by González & Levato (2006) to reconstruct, from a set of optical spectra of CPD-41° 7742, the individual spectra of the stars as well as their radial velocities (RVs) at each time of observation. The reconstructed spectra were analysed with `CMFGEN` (Hillier & Miller 1998) to determine the

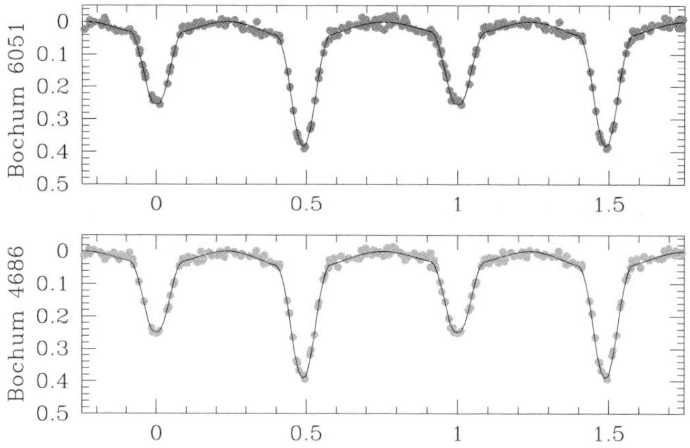

Figure 1. Best-fit to the Bochum light curves.

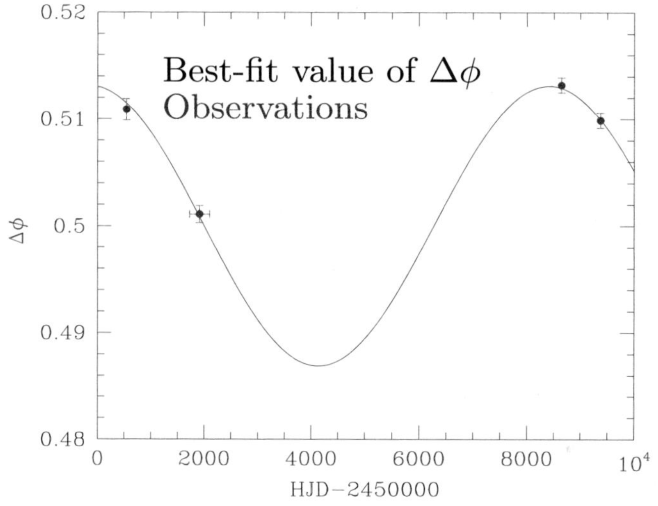

Figure 2. $\Delta\phi$ between primary and secondary minima as a function of time.

primary star's effective temperature. We combined our re-derived RVs to those coming from the literature to infer values for the orbital period. We analysed our set of light curves taken with the 0.6 m Bochum telescope at La Silla observatory (Royer et al. 1998) with `Nightfall` (Wichmann 2011) to determine the orbital inclination (i), stellar radii and masses, and effective temperature of the secondary star (see Fig. 1). We used the times of minima of the photometric observations (Bochum, *uvby* of Bouzid et al. (2005), and TESS taken during sectors 12 and 39, see Ricker et al. 2015) for the determination of $\dot{\omega}$. We used the relations from Giménez & Bastero (1995) that provide the phase difference as a function of ω, i, and the eccentricity e to adjust the curve to the observations, explicitly accounting for the apsidal motion (see Fig. 2). In this way, we derived $\dot{\omega} = 15.^{\circ}38^{+0.42}_{-0.51}\,\mathrm{yr}^{-1}$ and $e = 0.0204 \pm 0.0016$. We computed stellar evolution models for the binary system with the `Clés` code (Scuflaire et al. 2008) adopting different prescriptions for the internal mixing (overshooting and turbulent diffusion). We used the `min-Clés` routine to search for best-fit models of the stars. To reproduce $\dot{\omega}$, a large amount of turbulent diffusion needs to be included in the models of the primary star. The stellar models have thus larger convective cores than standard models.

3. Conclusion

We solved the apparent discrepancies in the longitude of periastron through the coherent analysis of all observational data of the system, explicitly accounting for the apsidal motion, which is determined for the first time. This study suggests that massive stars have larger convective cores than usually considered in stellar evolution codes.

Acknowledgements. S.R., Y.N., & E.G. acknowledge support from the FNRS, Belgium. The authors thank the referee for his/her suggestions and comments towards the improvement of the manuscript.

References

Bouzid, M. Y., Sterken, C., & Pribulla, T. 2005, *A&A*, 437, 769
Giménez, A., & Bastero, M. 1995, *Ap&SS*, 226, 99
González, J. F., & Levato, H. 2006, *A&A*, 448, 283
Hillier, D. J., & Miller, D. L. 1998, *ApJ*, 496, 407
Ricker, G. R., Winn, J. N., Vanderspek, R., et al. 2015, *JATIS*, 1, 014003
Rosu, S., Rauw, G., Nazé, Y., Gosset, E., & Sterken, C. 2022, *A&A*, in press [arXiv:2205.11207]
Royer, P., Vreux, J.-M., & Manfroid, J. 1998, *A&AS*, 130, 407
Sana, H., Antokhina, E., Royer, P., et al. 2005, *A&A*, 441, 213
Sana, H., Hensberge, H., Rauw, G., & Gosset, E. 2003, *A&A*, 405, 1063
Scuflaire, R., Théado, S., Montalbán, J., et al. 2008, *Ap&SS*, 316, 83
Wichmann, R. 2011, Nightfall, Astrophysics Source Code Library, record ascl:1106.016

Massive Stars Near and Far
Proceedings IAU Symposium No. 361, 2024
J. Mackey, J. S. Vink, & N. St-Louis, eds.
doi:10.1017/S1743921322003143

Stability analysis of radial oscillation modes in massive stars

J. P. Sánchez Arias[1], **S. Guha**[1,2] **and M. Kraus**[1]

[1]Astronomical Institute, Czech Academy of Sciences, Fričova 298, 25165 Ondřejov, CZ
email: julieta.sanchez@asu.cas.cz

[2]Matematicko fyzikální fakulta, Univerzita Karlova, 12116 Praha, Czech Republic

Abstract. During the evolution of massive stars, their properties change significantly. But stellar parameters of massive stars have the biggest uncertainties in stellar astrophysics, specifically in the post-main sequence stages where blue supergiant stars are located. These stars experience mass loss events during their evolution which are supposed to be related to strange-modes instabilities. In this work, we explore the stability of oscillation modes in massive stars for different masses, and a range of mass-loss rates, with the aim to provide clues about the connection between strange modes instabilities and mass-loss events.

Keywords. stars: early-type, stars: interiors, stars: mass loss

---

1. Introduction

Numerous theoretical studies suggest the existence of strange modes in massive stars (Gautschy et al. 1990). These are non-linear non-adiabatic modes whose instabilities are not excited by the classical κ-mechanism. To confirm the excitation of strange modes and to determine their final fate, unstable modes in the linear stability analysis need to be followed into the non-linear regime. If the pulsation velocity exceeds the escape velocity, it would be a direct evidence for mass loss. The presence of these modes has been suggested for 55 Cyg as well as their connection with its mass loss (Kraus et al. 2015). With the aim to further study strange mode oscillations in different evolutionary stages and to understand when these oscillations are triggered as well as their connection with mass loss, we initiated a systematic analysis of linear instabilities in massive stars.

2. Evolutionary and pulsational models

Using MESA (Paxton et al. 2019) we developed evolutionary models of stars with a metalicity of 0.02 and with initial masses between 50 and 65 solar masses up to the A-type supergiant regime (see Fig. 1). We adopted Vink recipe (Vink et al. 2001) for the mass loss. We considered different wind efficiencies (η) to account for reductions in mass loss rates due to wind inhomogeneities. Left and right panels in Fig. 1 correspond to η equal to 0.5 and 1, respectively. The numbers indicate the actual mass at two effective temperatures considered. The classical mixing length theory was adopted with $\alpha = 1.5$.

We employed the GYRE code (Townsend & Teitler 2013) to calculate non-adiabatic radial oscillations for models that have reached temperatures of 20 000 K and 10 000 K, corresponding to early- and late-type B supergiants. We restrict to modes between 0.032 c/d and 1 c/d, the typical range for strange-mode pulsations. The unstable modes obtained for $\eta = 0.5$ and 1 are shown in the left and right panels of Fig. 2.

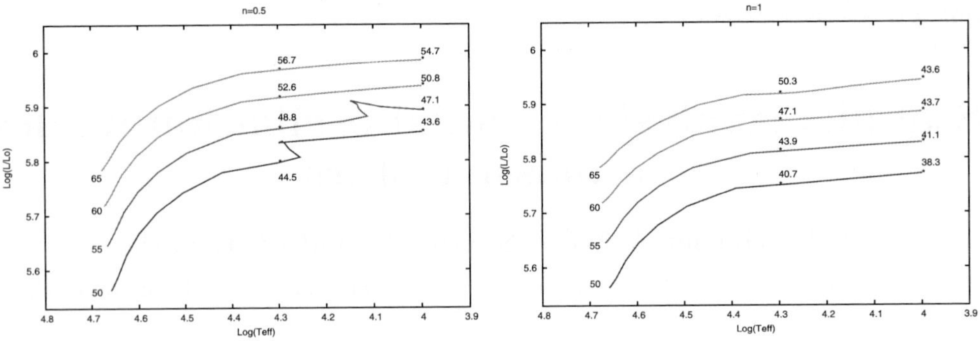

Figure 1. HR diagram for stars with different initial masses and wind efficiency.

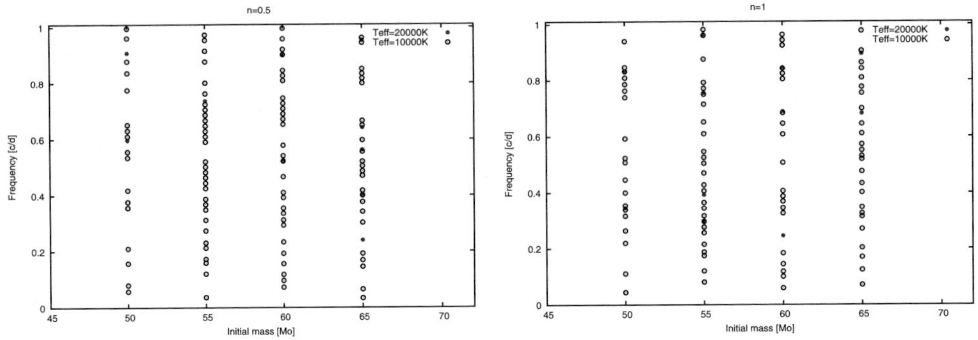

Figure 2. Non-adiabatic radial modes in B supergiants with different initial mass and η.

3. Conclusions

This project is the first step to study radial strange modes in massive stars and their interplay with mass loss through stellar winds. Excited modes are found in all stellar models, and their number significantly increases for late-type B supergiants. For these, the models with lower η also seem to show more excited modes, possibly due to the higher L/M which helps the radiative pressure to act in favor of their excitation. Non-linear stability analysis is required to properly study the link between these unstable modes (possible strange modes) and their contribution to the total mass loss from evolved massive stars.

Acknowledgements

We acknowledge financial support from the Czech Science Foundation (GAČR 20-00150S). The Astronomical Institute Ondřejov is supported by RVO:67985815. This project has received funding from the European Union's Framework Programme for Research and Innovation Horizon 2020 (2014-2020) under the Marie Skłodowska-Curie Grant Agreement No. 823734.

Supplementary material

To view supplementary material for this article, please visit http://dx.doi.org/10.1017/S1743921322003143.

References

Gautschy, A., Glatzel, W., Gautschy, A., & Glatzel, W. 1990, MNRAS, 245, 597
Kraus, M., et al. 2015, A&A, 581, A75
Paxton, B., et al. 2019, ApJS, 243, 10
Townsend, R. H. D., & Teitler, S. A. 2013, MNRAS, 435, 3406
Vink, J. S., de Koter, A., & Lamers, H. J. G. L. M. 2001, A&A, 369, 574

Massive Stars Near and Far
Proceedings IAU Symposium No. 361, 2024
J. Mackey, J. S. Vink, & N. St-Louis, eds.
doi:10.1017/S1743921322002113

Spins of stripped B stars support magnetic internal angular momentum transport

Christoph Schürmann[1,2], Norbert Langer[2,1], Xiao-Tian Xu[1,2] and Chen Wang[3]

[1]Max-Planck-Institut für Radioastronomie, Auf dem Hügel 69, 53121 Bonn, Germany
email: chr-schuermann@uni-bonn.de

[2]Argelander-Institut für Astronomie, Univ. Bonn, Auf dem Hügel 71, 53121 Bonn, Germany

[3]Max-Planck-Institut für Astrophysik, Karl-Schwarzschild-Str. 1, 85748 Garching, Germany

Abstract. The galactic binary star LB-1 contains a recently stripped B-type star. Comparing its properties to detailed binary star models shows that tidal braking and magnetic torques lead to low surface rotational velocities in the stripped donors after Roche-lobe overow. Models without magnetic torques cannot reproduce the observed low surface rotation.

Keywords. stars: rotation, stars: interiors, stars: binaries, stars: magnetic fields

1. Motivation

In order to predict the spins of stellar remnants, we need to understand the evolution of the internal rotation of stars, and to identify at which stage the rotation of the contracting cores of evolved stars decouples from their expanding envelopes. The donor stars of mass transferring binaries lose almost their entire envelope and may thus offer a straight view on their core rotation. While after the mass transfer event they contract and fade rapidly, they are well observable when caught in the short-lived B-star phase. The B-type primary of the galactic binary star LB-1, which was originally suggested to contain a massive black hole (Liu et al. 2019), is well explained as a stripped star (He- and N-rich) accompanied by a fainter Be star (Shenar et al. 2020). The narrow absorption lines in the primary's spectrum signify extremely slow rotation ($v_{\rm rot}\sin i = 7\,{\rm km/s}$), atypical for B-type main-sequence stars (Lennon et al. 2021; Shenar et al. 2020).

2. Methods

We identify a LB-1 progenitor model using a grid of generic MESA (Paxton et al. 2013) binary models (Wang et al. 2022) with initial rotation rates of 60% and 20% critical. We search for post-Roche-lobe-overflow model with $T_{\rm eff}$, $L/M \sim T_{\rm eff}^4/g$ and $Y_{\rm surf}$ similar to the observations (Shenar et al. 2020; Lennon et al. 2021). We find an initial donor mass of about $4\,{\rm M}_\odot$ and an initial period of 16 d.

Three processes can alter the rotation rate of a mass shell in a stellar model. Contraction (expansion) leads to acceleration (deceleration). If a model fills its Roche-lobe, tidal torques force the outer layers in corotation with the orbit (Zahn 1977). At last, angular momentum is transported within the stellar model by torques, which can be induced by convection, hydrodynamic instabilities (Heger et al. 2000), and magnetic fields e.g. due to the Spruit-Tayler dynamo (STD), which grows with the Brunt-Väisälä frequency to the $-4^{\rm th}$ power, wherefore gradients of entropy and mean molecular weight weaken the magnetic torque (Spruit 2002; Heger et al. 2005)

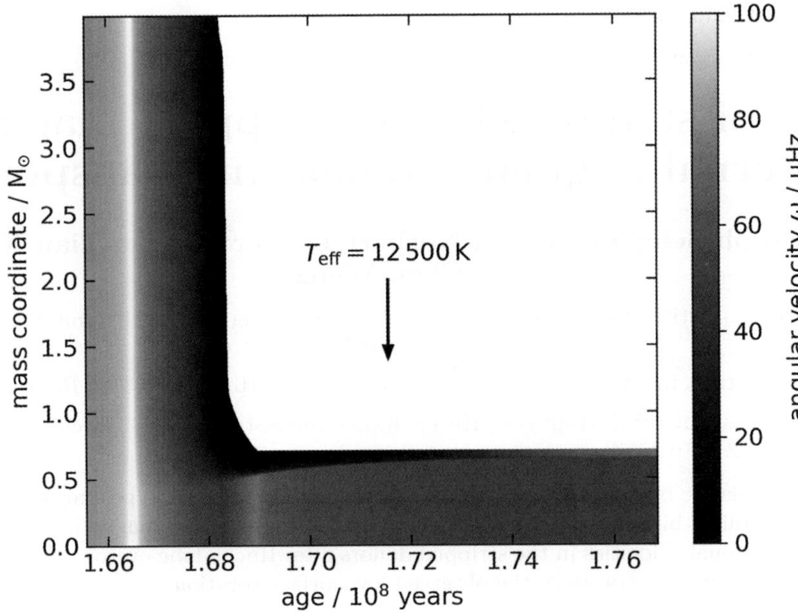

Figure 1. Kippenhahn-type diagram of the mass donor of our fiducial binary model, showing the internal evolution of its rotational frequency (colorbar on the right) after TAMS. The arrow marks the time corresponding to LB-1.

3. Rotational evolution with and without magnetic fields

During hydrogen core burning (until $1.665 \cdot 10^8$ years, Fig. 1) the magnetic torques of the STD keep the model in nearly rigid rotation. After central hydrogen depletion the envelope expands and tides become important, so the surface rotation rate drops. The magnetic torques drain so much angular momentum out of the core that it decelerates in spite of its contraction.

When the star fills its Roche-lobe ($1.68 \cdot 10^8$ years) the surface rotates at the same rate as the orbit. Hydrogen shell burning has increased the gradients of entropy and mean molecular weight so much that the STD becomes ineffective. No angular momentum is drained out of the core anymore and it accelerates due to contraction. Soon after that, helium core burning ignites. The envelope contracts increasing its rotation rate. At an age of $1.75 \cdot 10^8$ years it becomes faster than the core, which could not be archived by the STD. The model will evolve to a subdwarf with a rotational velocity in broad agreement with observations (Wang et al. 2021).

In an alternative model without STD the rotation rate of core and envelope decouple already during hydrogen burning. The core is barely braked during Roche-lobe filling leading to a fast rotating stripped star in contradiction with observation. The results are independent of initial rotation.

4. Conclusions

Tidal braking and magnetic torques lead to low surface rotational velocities in the stripped donor after Roche-lobe overflow. Models without magnetic torques cannot reproduce the low surface rotation of LB-1. The removal of angular momentum from core is supported by earlier works.

Supplementary material

To view supplementary material for this article, please visit http://dx.doi.org/10.1017/S1743921322002113.

References

Heger, A., Langer, N., & Woosley, S. E. 2000, ApJ, 528, 368
Heger, A., Woosley, S. E., & Spruit, H. C. 2005, ApJ, 626, 350
Lennon, D. J., et al. 2021, A&A, 649, A167
Liu, J., et al. 2019, Nature, 575, 618
Paxton, B., et al. 2013, ApJS, 208, 4
Shenar, T., et al. 2020, A&A, 639, L6
Spruit, H. C. 2002, A&A, 381, 923
Wang, C., et al. 2022, Nature Astronomy, 6, 480
Wang, L., Gies, D. R., Peters, G. J., Götberg, Y., Chojnowski, S. D., Lester, K. V., & Howell, S. B. 2021, AJ, 161, 248
Zahn, J. P. 1977, A&A, 500, 121

Massive Stars Near and Far
Proceedings IAU Symposium No. 361, 2024
J. Mackey, J. S. Vink, & N. St-Louis, eds.
doi:10.1017/S1743921322002307

Slowly-rotating nitrogen-rich O stars in 30 Doradus

F. Tramper

Institute of Astronomy, KU Leuven, Celestijnenlaan 200D, 3001 Leuven, Belgium
email: frank.tramper@kuleuven.be

Abstract. The VLT/FLAMES Tarantula Survey (Evans et al. 2011) identified a group of slowly-rotating nitrogen-rich O-type stars that cannot be explained by current evolutionary models. Here we present high-quality VLT/UVES observations of four of these stars that allow a detailed quantitative spectroscopic analysis. We present the analysis of the spectra with a genetic algorithm, and discuss the future steps to be taken to further investigate the cause of the nitrogen enrichment.

Keywords. stars: abundances, stars: atmospheres, stars: chemically peculiar, stars: early-type, stars: evolution, stars: fundamental parameters

1. Introduction

Rotation is a key element affecting the evolution of massive stars. High rotation rates allow the mixing of material between the core and the envelope. As a consequence, more hydrogen becomes available in the core resulting in significantly longer main-sequence lifetimes (up to 30%, e.g., Brott et al. 2011). Simultaneously, CNO processed material is mixed into the envelope, increasing the nitrogen and decreasing the carbon and oxygen surface abundances. While the current generation of evolutionary models differ in the amount of mixing predicted and the rotation rate needed to have a significant impact on massive star evolution, they do agree on two important observational effects:

(1) a strong correlation between surface nitrogen abundance and rotation rate, and
(2) the almost complete absence of nitrogen enrichment for slowly-rotating massive stars.

Contrary to these predictions, Hunter et al. (2008) found a group of slowly-rotating nitrogen-rich B-type stars in the framework of the VLT-Flames Survey of Massive Stars (Evans et al. 2006). More recently, a similar group of O-type stars has been found in 30 Doradus (aka the Tarantula nebula) by Grin et al. (2017) in the VLT-Flames Tarantula Survey (Evans et al. 2011).

2. Analysis

To investigate the nature of the slowly-rotating nitrogen-rich stars in 30 Doradus we obtained high-quality (R $\sim$ 40000, S/N $>$ 100) spectra of four representative stars using VLT/UVES. These spectra allow the accurate derivation of the key stellar parameters and the surface abundances of helium, carbon, nitrogen, oxygen, and silicon. The spectra were analysed by fitting synthetic spectra from the model atmosphere code FASTWIND (Puls et al. 2005) using a genetic algorithm (GA, Fig. 1). This method allows us to

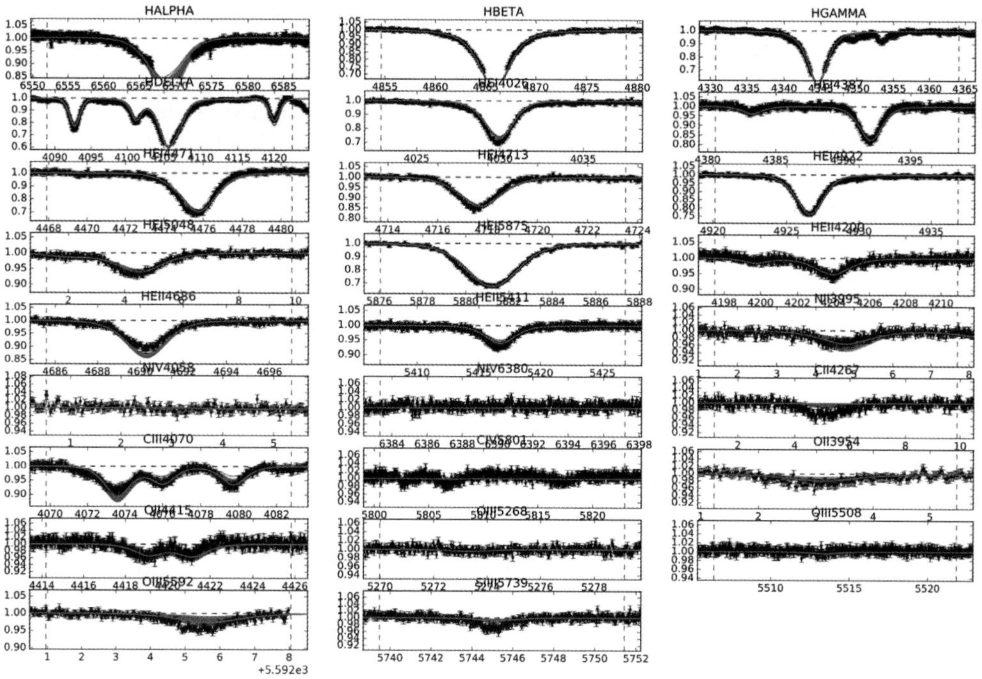

Figure 1. Example of a GA fit to the spectrum of VFTS178.

thoroughly explore the 11-dimensional parameter space (see below) in a reasonable amount of CPU time. Additionaly, the method provides robust error bars on the derived parameters.

The key stellar parameters that are derived from the GA fits are the effective temperature ($T_{\rm eff}$), surface gravity, mass-loss rate, microturbulent velocity, projected rotational velocity, and surface helium abundance. Luminosity and radius are derived based on the absolute magnitude and $T_{\rm eff}$ (see Mokiem et al. 2005). Compared to earlier versions of the GA (Mokiem et al. 2005, Tramper et al. 2011, 2014), we also implemented macroturbulence and surface abundances of C, N, O, and Si as fitting parameters.

3. Future steps

The results of this work will be used to investigate the origin of the anomalous nitrogen abundances. Using the derived carbon and oxygen abundances, the first step will be to see if they are compatible with CNO equilibrium values. The derived parameters will be used to investigate several scenarios which might explain the nature of these stars, e.g.:

- The presence of other mixing processes (e.g., by macroturbulent motions)
- Stripping of the envelope of the stars through prior binary interactions
- Stripping by prior giant mass ejections (e.g., an earlier LBV stage).

Acknowledgements

Based on observations collected at the European Southern Observatory under ESO programme 0100.D-0192(A).

Supplementary material

To view supplementary material for this article, please visit http://dx.doi.org/10.1017/S1743921322002307.

References

Brott, I., de Mink, S. E., Cantiello, M., et al. 2011, *A&A*, 530, A115
Evans, C. J., Lennon, D. J., Smartt, S. J., et al. 2006, *A&A*, 456, 623
Evans, C. J., Taylor, W. D., Hénault-Brunet, V., et al. 2011, *A&A*, 530, A108
Grin, N. J., Ramírez-Agudelo, O. H., de Koter, A., et al. 2017, *A&A*, 600, A82
Hunter, I., Brott, I., Lennon, D. J., et al. 2008, *ApJ* (Letters), 676, L29
Mokiem, M. R., de Koter, A., Puls, J., et al. 2005, *A&A*, 441, 711
Puls, J., Urbaneja, M. A., Venero, R., et al. 2005, *A&A*, 435, 669
Tramper, F., Sana, H., de Koter, A., et al. 2011, *ApJ* (Letters), 741, L8
Tramper, F., Sana, H., de Koter, A., et al. 2014, *A&A*, 572, A36

Massive Stars Near and Far
Proceedings IAU Symposium No. 361, 2024
J. Mackey, J. S. Vink, & N. St-Louis, eds.
doi:10.1017/S1743921322003076

Nitrogen production in population III stars

S. Tsiatsiou, C. Georgy, S. Ekström and G. Meynet

Department of Astronomy, University of Geneva, Chemin Pegasi 51, 1290 Versoix, Switzerland
email: sofia.tsiatsiou@unige.ch

Abstract. The first stars in the Universe have inherited their composition from primordial nucleosynthesis, so they have no metal. These stars, which are also named population III (pop III) stars, began the process of reionization in the Universe and contributed to the metal enrichment with heavy elements. Previous studies showed that they should have been rotating fast due to small or no angular momentum loss, reaching easily the critical velocity since they are massive and have very low stellar winds, thus their mass loss is very low or zero. Our aim is to study how the production of primary nitrogen is affected due to high rotation in the pop III stars. So, we compared grids of pop III stars with zero, average, and high rotation. All these models have been computed using Geneva code (GENEC) in the mass range of $9M_{\odot} \leqslant M_{ini} \leqslant 120M_{\odot}$. Due to the rotational mixing, the carbon produced in the He-burning core is diffused towards the H-burning shell, triggering the CNO cycle and producing primary nitrogen. In some models the transition of the shell from a pp-chain H-burning to a CNO H-burning induces a strong energy release and a complete change of the stellar structure and the nucleosynthesis. The production of nitrogen is boosted for the high rotation models.

Keywords. stars: rotation, stars: interiors, stars: population III, stars: evolution

1. Stellar models

We computed a grid with fast rotating stellar models, $u_{ini} = 0.7 u_{crit}$, using GENEC. Then, we compared grids of stellar models with different initial rotational velocities, zero, average ($u_{ini} = 0.4 u_{crit}$) (Murphy et al. 2021), and high rotation. All the models have been computed with consistency to the papers of Murphy et al. (2021), Ekström et al. (2012), Georgy et al. (2013), and Groh et al. (2019) but with different initial rotations and metallicities. The u_{crit} is the critical limit of rotation and it is defined by $u_{crit} = \sqrt{\frac{2}{3}\frac{GM}{R_p}}$, where R_p is the polar radius at break-up limit. The critical velocity is reached when the gravitational acceleration is counterbalanced by the centrifugal force. Rotational mixing of the chemical elements is mainly due to shear turbulence.

2. Production of nitrogen

1. Shear turbulence at the border of the He-burning core is a key process to bring C into the H-burning shell and produce primary N, left panel of figure 1.
2. The contraction is in part the cause of the Ω gradient that drives the shear turbulence that brings carbon into the H-shell and produce the primary nitrogen.
3. In pop III stars the contraction is less strong than on models with even a small initial content of heavy elements.
4. This is because in pop III stars H-burning occurs at higher temperatures which are near the temperatures that are needed for the 3α-process, right panel in figure 1.
5. Therefore we would expect much less production of primary N in the pop III models than in models with higher metallicities.

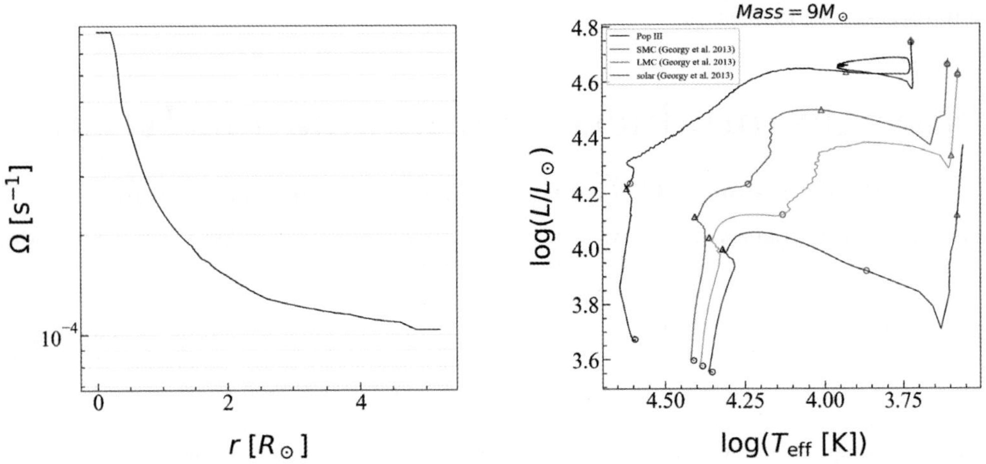

Figure 1. Left panel: The evolution of the angular velocity $\Omega(r)$ at the end of He-burning phase for the $20M_{\odot}$ model. Right panel: HR diagram for $9M_{\odot}$ models with different metallicities (Georgy et al. 2013).

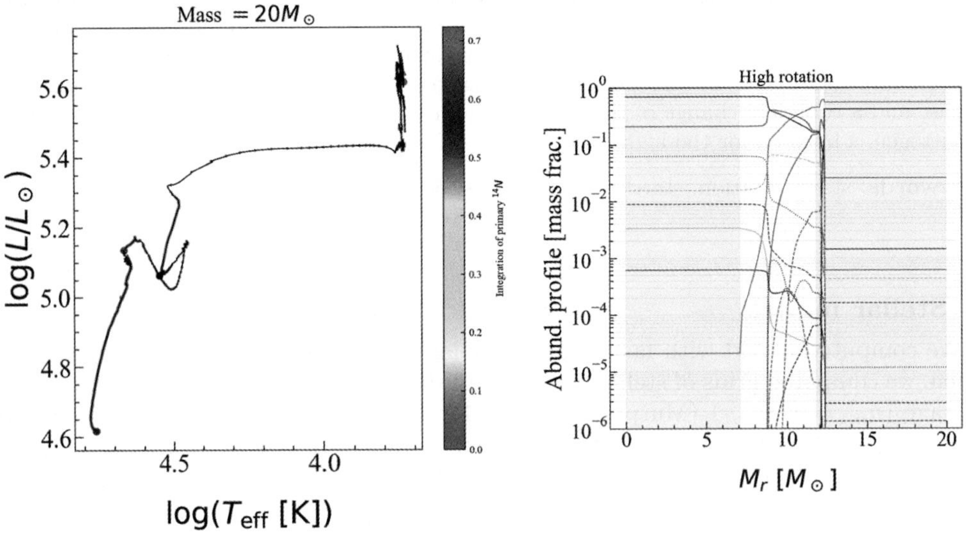

Figure 2. Left panel: HR diagram with the integrated amount of N during its evolution. Right panel: The chemical structure of the $20M_{\odot}$ model with high rotation at the end of He-burning phase from the core to the surface. The grey shaded areas are the convective regions of the star. The green line corresponds to N.

6. However, in the case that the pop III stars are fast rotating models, they may produce a significant amount of primary N as seen in right panel of figure 2.

Supplementary material

To view supplementary material for this article, please visit http://dx.doi.org/10.1017/S1743921322003076.

References

Ekström, S., et al. 2012, A&A, 537, A146
Georgy, C., et al. 2013, A&A, 558, A103
Groh, J. H., et al. 2019, A&A, 627, A24
Murphy, L. J., et al. 2021, MNRAS, 501, 2745

Massive Stars Near and Far
Proceedings IAU Symposium No. 361, 2024
J. Mackey, J. S. Vink, & N. St-Louis, eds.
doi:10.1017/S1743921322002770

Obliquely Rotating Massive Star Magnetospheres

Asif ud-Doula[1] and Stanley P. Owocki[2]

[1]Penn State Scranton, Dunmore, PA, USA
email: auu4@psu.edu

[2]University of Delaware, Newark, DE, USA

Abstract. We present results from 3D MHD simulations of the magnetospheres from massive stars with a dipole magnetic axis that has an arbitrary obliquity angle (β) to the stars rotation axis. As an initial direct application, we examine the global structure of co-rotating disks for tilt angles $\beta = 0$, 45 and 90 degrees using ζ Pup stellar parameters as a prototype. We find that for models with rapid stellar rotation ($\sim$ 0.5 critical rotation), accumulation surfaces closely resemble the form predicted by the analytic Rigidly Rotating Magnetosphere (RRM) model, but with a mass distribution and outer disk termination set by centrifugal breakout processes. However, some significant differences are found including warping resulting from the dynamic nature of the MHD models in contrast to static RRM models. These MHD models can be used to synthesize rotational modulation of photometric absorption and H-alpha emission for a direct comparison with observations.

Keywords. massive stars, magnetohydrodynamics, magnetic fields, rotation

---

Hot, luminous, massive (OB-type) stars lose large amount of mass through radiatively-driven stellar wind (Castor et al. 1975, CAK). In general, winds of these objects are spherically symmetric, but some fraction of them exhibit variability and structure on both small and large scales which can be caused by global magnetic field. Babel & Montmerle (1997) have shown that in Bp stars magnetic field is so strong that it acts like rigid rods that guide material towards the magnetic equator wherein they collide leading to magnetically confined wind shocks (MCWS). Once the material cools, the dense material is a source of strong Hα emission. However, this MCWS was a phenomenological model and did not make any specific predictions about how much material would accumulate within magnetosphere or provide a mechanism for emptying it.

A vast improvement upon MCWS model was made by Townsend et al. (2005) who developed rigidly rotating magnetosphere (RRM) model for Bp star using σ Ori E as a prototype. RRM model predicts the accumulation of circumstellar plasma situated in magnetohydrostatic equilibrium at the intersections between the magnetic and rotational equators. Amount of accumulated material depends on mass feeding rate at the base of rigid field lines. The model also predicts large scale emptying of the magnetosphere via centrifugal breakout (CBO) and has been successfully used to reproduce the periodic modulations observed in the light curve and Hα emission-line profile of σ Ori E.

Key advantage of the RRM model is that it is analytic and does not require expensive numerical computations. However, it lacks self-consistent interactions between magnetic field and the wind that is inherent in magnetohydrodynamic (MHD) simulations. As such, RRM model is unable to predict density structure within the magnetosphere, one of the main goals of this work here.

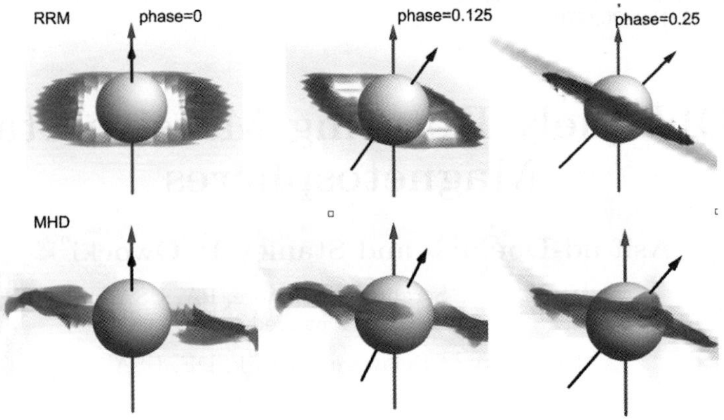

Figure 1. Comparison between analytic Rigidly Rotating Magnetosphere (RRM) vs our fully self-consistent 3D MHD models. They are quite similar but significant differences exist. MHD model shows the sense of rotation with denser part of the disk leading unlike in RRM model. Dragonfly wing-like structure exhibits warping which can have observational implications.

In an extensive series of papers (see ud-Doula & Owocki 2002; ud-Doula et al. 2006, 2008, 2009), we have applied MHD simulations to understand the dynamics of such magnetic wind channeling, the consequent hard X-ray emission and Hα emission. But all these models were either 2D or simplified 3D models where magnetic axis is aligned with rotational axis.

In practice, magnetic massive stars are often oblique rotators, which naturally leads — especially when rotation is rapid — to lateral structures that cannot be modeled by axisymmetric 2-D models. But for rapid rotators with magnetic field tilted with respect to rotational axis, this is not option as axisymmetry is fully broken. Here we present a 3D MHD model where magnetic field is tilted by 45° with respect to the rotational axis.

We use PLUTO code in spherical geometry. We follow the same basic formalism as in ud-Doula & Owocki (2002); ud-Doula et al. (2008). We evolve our model for over 1.5Ms, which allows the model to settle in quasi-steady state. As shown in Fig. 1, fully self consistent MHD model (bottom panel) and the RRM model (top panel) match relatively well. However, there are some notable differences. The MHD model shows the dynamic effects of rotation that leads to warping of the circumstellar structure, making it resemble a 'dragonfly wing'. The overall structure exhibits the sense of rotation with higher density material leading the way. Such MHD models also allow to extract density scaling information that can have observational consequences.

Acknowledgements

We acknowledge NASA ATP grant number 80NSSC22K0628. AuD acknowledges support by NASA through Chandra Award number TM1-22001B and GO2-23003X issued by the Chandra X-ray Observatory 27 Center, which is operated by the Smithsonian Astrophysical Observatory for and on behalf of NASA under contract NAS8-03060.

Supplementary material

To view supplementary material for this article, please visit http://dx.doi.org/10.1017/S1743921322002770.

References

Babel, J., & Montmerle, T. 1997, ApJL, 485, 29
Castor, J. I., Abbott, D. C., & Klein, R. I. 1975, ApJ, 195, 157

Townsend, R. H. D., Owocki, S. P., & Groote, D. 2005, ApJL, 630, L81
ud-Doula, A., & Owocki, S. P. 2002, ApJ, 576, 413
ud-Doula, A., Owocki, S. P., & Townsend, R. H. D. 2008, MNRAS, 385, 97
—. 2009, MNRAS, 392, 1022
ud-Doula, A., Townsend, R. H. D., & Owocki, S. P. 2006, ApJL, 640, L191

Chapter IV:
Stellar multiplicity

Massive Stars Near and Far
Proceedings IAU Symposium No. 361, 2024
J. Mackey, J. S. Vink, & N. St-Louis, eds.
doi:10.1017/S1743921322003180

The Nature of Unseen Companions in Massive Single-Line Spectroscopic Binaries

Hugues Sana[1], Michael Abdul-Masih[2], Gareth Banyard[1], Julia Bodensteiner[3], Dominic M. Bowman[1], Karan Dsilva[1], C. Eldridge[4], Matthias Fabry[1], Abigail J. Frost[1], Calum Hawcroft[1], Soetkin Janssens[1], Laurent Mahy[1,5], Pablo Marchant[1], Norbert Langer[6,7], Timothy Van Reeth[1], Koushik Sen[6,7] and Tomer Shenar[1,8]

[1]Institute of Astronomy, KU Leuven, Celestijnenlaan 200D, 3001 Leuven, Belgium
email: hugues.sana@kuleuven.be

[2]ESO, Alonso de Cordova 3107, Vitacura, Santiago de Chile, Chile

[3]ESO, Karl-Schwarzschild-Strasse 2, 85748 Garching, Germany

[4]Amateur Astronomer, 242 Randell Rd, 6125 Mardella, Western Australia

[5]Royal Observatory of Belgium, Avenue Circulaire/Ringlaan 3, B-1180 Brussels, Belgium

[6]Argelander-Institut für Astronomie, Universität Bonn, Auf dem Hügel 71, 53121 Bonn (D)

[7]Max-Planck-Institut für Radioastronomie, Auf dem Hügel 69, 53121 Bonn, Germany

[8]Anton Pannekoek Institute for Astronomy, Science Park 904, 1098 XH, Amsterdam (NL)

Abstract. Massive stars are predominantly found in binaries and higher order multiples. While the period and eccentricity distributions of OB stars are now well established across different metallicity regimes, the determination of mass-ratios has been mostly limited to double-lined spectroscopic binaries. As a consequence, the mass-ratio distribution remains subject to significant uncertainties. Open questions include the shape and extent of the companion mass-function towards its low-mass end and the nature of undetected companions in single-lined spectroscopic binaries. In this contribution, we present the results of a large and systematic analysis of a sample of over 80 single-lined O-type spectroscopic binaries (SB1s) in the Milky Way and in the Large Magellanic Cloud (LMC). We report on the developed methodology, the constraints obtained on the nature of SB1 companions, the distribution of O star mass-ratios at LMC metallicity and the occurrence of quiescent OB+black hole binaries.

Keywords. binaries: spectroscopic, stars: early-type, stars: evolution, black hole physics, supernovae: general, stars: neutron, gravitational waves

1. Introduction

The majority of massive main-sequence OB stars are found in binary or higher order multiples (Kobulnicky & Fryer 2007; Sana et al. 2014; Moe & Di Stefano 2017). This fact has been established beyond reasonable doubt from observational studies of various clusters and OB associations in our Galaxy (Sana et al. 2012; Kobulnicky et al. 2014; Banyard et al. 2022; Bordier et al. 2022) and in the Tarantula region of the Large Magellanic Cloud (LMC; Sana et al. 2013; Dunstall et al. 2015). Similarly, concrete observational constraints have been obtained on the distributions of orbital periods and eccentricities, again both in the Milky Way (Sana et al. 2012; Barbá et al. 2017;

Banyard et al. 2022), and in the LMC (Almeida et al. 2017; Villaseñor et al. 2021). These studies reveal no strong differences despite the different metallicity environments. The distribution of mass-ratios, however, has remained more difficult to quantify because direct measurements of mass-ratios are mostly obtained from double-lined spectroscopic binaries (SB2s). This requires that the secondary companion is bright enough for its spectral signature to be visible in the spectrum, and that the Doppler separation between the components is large enough for the spectral lines to separate enough. In some cases, e.g., when temperature differences are large enough to result in different ionisation balances, radial velocities (RVs) of the components can be measured from lines unaffected by blending, but this is not the case in the majority of the stars, and even a small degree of contamination can impact the RV measurements (Fabry et al. 2021; Bodensteiner et al. 2021; Banyard et al. 2022).

Focusing on the Galactic population within a few kiloparsecs, about one third of the known O-type spectroscopic binaries show the spectroscopic signature of only one of the binary components (Trigueros Páez et al. 2021; Mahy et al. 2022), i.e. they are single-lined spectroscopic binaries (SB1s). This SB1 fraction increases to over 50% in LMC studies (Almeida et al. 2017), most likely owing to the lower data quality given stars are about 10 times further away. Pending further information, the precise nature of the unseen companions in SB1 systems remains uncertain. One may expect that the majority are 'normal' lower-mass stars whose luminosities are too low for their spectral lines to peak out of the noise in individual-epoch spectroscopy, or whose orbital periods are too large for their spectral lines to sufficiently separate from those of the brighter component. Alternatively, such systems might also harbour binary interaction products where the initially more massive star (mass-donor) has been stripped of (most of) its hydrogen envelope through mass transfer (Paczyński 1967; Pols et al. 1991; Abdul-Masih et al. 2020; Bodensteiner et al. 2020; Shenar et al. 2020; El-Badry & Quataert 2021). SB1s may also be the hiding place of compact objects where the unseen companions is either a black hole (BH) or a neutron star (Langer et al. 2020).

Identifying the nature of unseen companions is important for a number of science cases:

Multiplicity properties: present observational constraints of the mass-ratio distributions of unevolved binaries rely on extrapolation towards the lower-mass-end as low mass-ratios systems ($M_2/M_1 \lesssim 0.4$) are rarely detected as SB2. The initial mass-ratio distribution is, however, critical for binary evolution computations as the mass ratio is a fundamental parameter in defining the outcome of the interaction;

Binary evolution: a frequent outcome for systems that survive a phase of mass-transfer is that the mass-donor is stripped of (most of) its hydrogen envelope while the mass accretor is (probably) a rapidly rotating OB star. Stripped helium stars are candidate hydrogen-poor supernova progenitors (Langer 2012). Furthermore, the distribution of orbital properties and the chemical composition of stars in these systems are fundamental benchmarks to confront the outcome of mass-transfer models, yet these OB+He systems likely appear as SB1 in the optical;

Compact objects: recent binary population synthesis simulations predict that about 2% of O-type stars should be in a binary with a BH companion (Langer et al. 2020). Focusing on SB1 systems, this fraction might rise to 8%. Predicted orbital periods for these OB+BH systems range from about a week to a couple of years. This makes spectroscopy an ideal tool to search for these (most likely) X-ray quiet BH binaries. Alternatively, neutron stars can also be present around lower-mass O stars or around B-type stars. The orbital properties of these systems retain pristine information about the collapse of the BH/NS progenitors (Marchant et al., this volume), information that might be distorted or lost once these systems evolve into X-ray binaries.

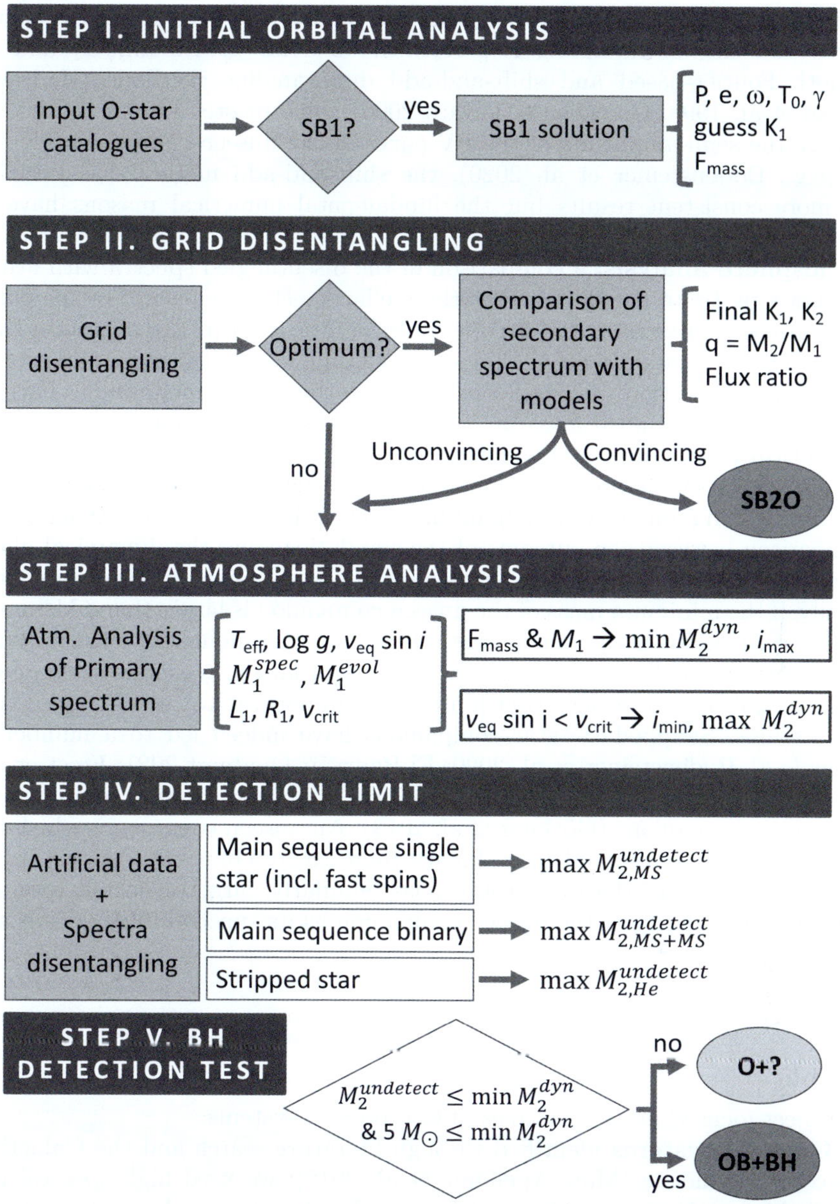

Figure 1. Workflow for identifying the nature of unseen companions in SB1 OB binaries.

2. Method

We worked under the assumption that constraints on the mass, luminosity and spectral properties of the unseen object help us to better constrain its nature and physical properties. We relied on four main steps, described below and summarised in Fig. 1:

• **SB1 orbital analysis:** the SB1 orbital solution of the visible component provides reliable constraints on the orbital periods (P), ephemeris and geometry of the orbit. A first estimate of the semi-amplitude of the RV curve of the visible component (K_1) is obtained but does not need to be trusted, as even a small contamination of its spectral lines with that of a faint unidentified companion may bias the measurements, leading to an underestimation of K_1;

• **Grid-based spectral disentangling:** the well-constrained orbital properties are kept fixed and the disentangling is performed on a 2D grid along K_1 and K_2. We tested both Fourier-based and shift-and-add disentangling methods (Hadrava 1995; Marchenko et al. 1998; González & Levato 2006), with mostly similar results. In some cases where the semi-amplitude of the RV curve of the unseen companion (K_2) is close to null (e.g., Bodensteiner et al. 2020), the shift-and-add method has been yielding slightly more consistent results but the fundamental numerical reasons have not yet been identified;

• **Atmosphere analysis:** a comparison of the disentangled spectra with atmosphere models allows us (i) to decide whether the stellar spectral signature of the companion can be identified, (ii) to estimate the optical light ratio, and (iii) to constrain the stellar properties of the optically bright component, including its spectroscopic and evolutionary masses, projected rotation rate and constraints on the orbital inclination. This provides us with a possible dynamical mass range for the unseen companion;

• **Detection limits:** if no stellar companion is identified by spectral disentangling, we perform simulations to estimate the maximum mass at which the spectral signature of a non-degenerate companion would remain undetected through our disentangling analysis. The comparison between the outcome of the simulations and the dynamical mass range of the unseen companion allows us to put additional constraints on the companion mass. Specifically, if the minimum mass of the unseen companion is larger than 5Msun and non-degenerate companions within the acceptable mass range should have led to a detection with spectral disentangling, then one is left to accept that the system is likely an OB+BH binary. In establishing the detection limits, it is important to consider companions of different natures. Stripped He star companions have indeed led to a number of false detections (e.g., Bodensteiner et al. 2020; El-Badry & Quataert 2021; Frost et al. 2022) because of their significantly different mass-optical brightness relation compared to single main-sequence companions (Götberg et al. 2018). An unseen component which is in fact a very short period near-equal mass binary more easily eludes detection compared to a main sequence star of the same total mass. Similarly, large projected rotation rates decrease the detectability of the unseen companion as its spectral lines are smeared out.

3. Sample

Our project focused on two samples of O-type SB1 systems:

• **32 Galactic systems** identified through literature search and the Galactic O star Spectroscopic Catalogue (Maíz Apellániz et al. 2016): we used high spectral resolving power (R), high-S/N spectra obtained from archival and new observations from ESO, Mercator/HERMES (Raskin et al. 2011) and SALT facilities. Of interest, Cyg X-1, known to host a stellar-mass BH, was included in the sample as it met our selection criteria (O-type SB1 + availability of sufficient spectroscopic data); see Mahy et al. (2022).

• **51 systems in the Tarantula region of the LMC** identified by the VLT-FLAMES Tarantula Survey (VFTS, Evans et al. 2011). These were monitored for 32 spectroscopic epochs by the Tarantula Massive Binary Monitoring (TMBM, Almeida et al. 2017) using the LR02 setup of the VLT/FLAMES multi-object spectrograph and yielding a $R \sim 7000$ from about 3950 to 4550 Å; see Shenar et al. (2022a,b).

Our search for compact objects has also been extended to Galactic and LMC B-type SB1 systems. Preliminary results are presented in Banyard et al. and Villaseñor et al. in this volume; see also Banyard et al. (2022); Villaseñor et al. (2021).

4. Results

Spectral disentangling revealed the signature of non-degenerate stellar companions in over 50% of the systems considered. Future work, including better data, is required to confirm the evolutionary status of these systems, albeit most are likely low mass-ratio binaries but some might be binary interaction products; see Mahy et al. (2022); Shenar et al. (2022a,b) for further details. Our results double the number of TMBM O-type binaries with measured mass-ratios. Combined with the analysis of previously known SB2 systems (Mahy et al. 2020a,a), this gives us direct access to the largest sets of mass-ratio measurements available so far for an O+OB binary population at sub-solar metallicity. Bayesian modelling of the measurements, including correction for observational biases, yields a flat mass-ratio distribution ($f_q \propto q^k$, with $k = 0.2 \pm 0.2$; Shenar et al. 2022a). This value is significantly different than obtained by a previous attempt based on the VFTS data ($k = -1.0 \pm 0.4$), confirming that the modelling of the amplitude of the RV variations offers low quality diagnostics on the mass-ratio distribution (Sana et al. 2013). Of interest, the mass-ratio distribution of O stars in the Tarantula region seem compatible with that derived for O stars in Milky Way clusters and OB associations ($k = -0.1 \pm 0.6$) suggesting no significant impact of the metallicity on the (still uncertain) pairing mechanism of massive binaries.

Among systems in which the signature of the companion cannot be identified, three systems in the Milky Way and three in the TMBM samples meet our requirements for OB+BH systems: (i) the companion mass is larger than 5 $M_{\odot}$ and, (ii) according to our simulations, a non-degenerate component with such high mass should have been detected given our data quality and orbital sampling, and yet it is not. In doing so, we rediscovered the OB+BH nature of Cyg X-1, which builds some confidence in the ability of our method to isolate good OB+BH candidates. Among the identified systems, the two best cases are HD 130298 in the Milky Way ($P = 14.6$ d, $e = 0.47$) and VFTS 243 in the LMC ($P = 10.4$ d, $e < 0.03$). Of interest, both have very different eccentricity properties. The negligible eccentricity in VFTS 243 suggests that the dying primary collapsed into a BH without losing much mass ($\Delta M < 1.1\ M_{\odot}$), thus likely without an associated supernova (SN) explosion. The formation scenario in HD 130298 is different and the system might have lost about 25% of the system total mass prior to SN explosion, i.e. an estimated 8 $M_{\odot}$. This suggests a significant mass loss and a possible BH formation through fallback.

5. Conclusions

We have presented a systematic analysis of a large sample of SB1 systems to identify the nature of their unseen companions. We have shown that grid disentangling along the K_1 and K_2 axes offers significant diagnostic power to extract the optical spectral signature of a faint stellar companion, which is otherwise lost in the noise of individual observational epochs. Simulations of artificial data sets show that signatures of companions representing only 0.3% of the total flux can be retrieved pending sufficiently high-quality time series and that hydrogen lines are better suited for faint objects than metallic lines. We also showed that it is easier to detect the presence a faint stellar companion than to precisely measure its RV-curve semi-amplitude K_2. Simulations of non-idealised data and realistic systems have further shown that companions contributing just 1% are easily retrieved. The lowest mass-ratio new-SB2 system revealed by our analysis has $q \sim 0.15$. The results from the TMBM sample in the LMC point toward a metallicity independent, (nearly) flat mass-ratio distribution for O-type binaries. Further extension towards lower metallicity is desirable, unfortunately current and future surveys do not plan enough multi-epoch coverage to reach the needed sampling and combined S/N. The presented method properly identified Cyg X-1 as a OB+BH system and revealed at least five new OB+BH (strong candidate) systems. We briefly discussed how the orbital properties help constraining the collapse physics.

Acknowledgements

The research leading to these results has received funding from the European Research Council (ERC) under the European Union's Horizon 2020 research and innovation programme (grant agreement number 772225: MULTIPLES), as well as under the Marie Skłodowska-Curie grant agreement No 101024605. The KU Leuven team is grateful to the FWO for multiple support through grants G0F8H6N, 11E1721N, 1286521N, 12ZB620N, 12ZY520N and the Belgian Federal Science Policy Office (BELSPO) in the framework of the PRODEX Programme.

References

Abdul-Masih, M., et al. 2020, Nature, 580, E11
Almeida, L. A., et al. 2017, A&A, 598, A84
Banyard, G., et al. 2022, A&A, 658, A69
Barbá, R. H., Gamen, R., Arias, J. I., & Morrell, N. I. 2017, in IAU Symposium Vol. 329, 89
Bodensteiner, J., et al. 2020, A&A, 641, A43
—. 2021, A&A, 652, A70
Bordier, E., et al. 2022, A&A, 663, A26
Dunstall, P. R., et al. 2015, A&A, 580, A93
El-Badry, K., & Quataert, E. 2021, MNRAS, 502, 3436
Evans, C. J., et al. 2011, A&A, 530, A108
Fabry, M., et al. 2021, A&A, 651, A119
Frost, A. J., et al. 2022, A&A, 659, A3
González, J. F., & Levato, H. 2006, A&A, 448, 283
Götberg, Y., et al. 2018, A&A 615, A78
Hadrava, P. 1995, A&A Supplement, v.114, p.393, 114, 393
Kobulnicky, H. A., & Fryer, C. L. 2007, ApJ, 670, 747
Kobulnicky, H. A., et al. 2014, APJSS, 213, 34
Langer, N. 2012, ARA&A, 50, 107
Langer, N., et al. 2020, A&A, 638, A39
Mahy, L., Sana, H., & Shenar, T. 2022, A&A, 664, 159
Mahy, L., et al. 2020a, A&A, 634, A118
—. 2020b, A&A, 634, A119
Maíz Apellániz, J., et al. 2016, APJSS, 224, 4
Marchenko, S. V., Moffat, A. F. J., & Eenens, P. R. J. 1998, PASP, 110, 1416
Moe, M., & Di Stefano, R. 2017, APJS, 230, 15
Paczyński, B. 1967, Acta Astron., 17, 193
Pols, O. R., Cote, J., Waters, L. B. F. M., & Heise, J. 1991, A&A, 241, 419
Raskin, G., et al. 2011, A&A, 526, A69
Sana, H., et al. 2012, Science, 337, 444
—. 2013, A&A, 550, A107
—. 2014, APJS, 215, 15
Shenar, T., et al. 2020, A&A, 639, L6
—. 2022a, A&A
—. 2022b, Nature Astronomy (https://doi.org/10.1038/s41550-022-01730-y)
Trigueros Páez, E., et al. 2021, A&A, 655, A4
Villaseñor, J. I., et al. 2021, MNRAS, 507, 5348

Massive Stars Near and Far
Proceedings IAU Symposium No. 361, 2024
J. Mackey, J. S. Vink, & N. St-Louis, eds.
doi:10.1017/S1743921322002848

Multiplicity of Galactic Luminous Blue Variable stars

Laurent Mahy[1], Cyprien Lanthermann[2], Eric Gosset[3], Damien Hutsemékers[3], Jacques Kluska[4], Alex Lobel[1], Maddalena Reggiani[4] and Hugues Sana[4]

[1]Royal Observatory of Belgium, Avenue Circulaire/Ringlaan 3, B-1180 Brussels, Belgium
email: laurent.mahy@oma.be

[2]The CHARA Array of Georgia State University, Mount Wilson Observatory, Mount Wilson, CA 91023, USA

[3]Space sciences, Technologies and Astrophysics Research (STAR) Institute, Université de Liège, Allée du 6 Août, 19c, Bât B5c, 4000 Liège, Belgium

[4]Institute of Astronomy, KU Leuven, Celestijnenlaan 200D, 3001, Leuven, Belgium

Abstract. Luminous Blue Variable stars (LBVs) are rare and enigmatic. Often cited as evolutionary stages in the single-star evolution, the idea that binary evolution produces the LBV state was already considered, 30 years ago.

It is now commonly accepted that a significant part of massive stars are born in multiple systems. One aspect that also emerged is that massive stars have on average at least two companions, i.e. they are triples. This immediately implies that a number of LBVs should have evolved as part of multiple systems.

While some LBVs are confirmed as binaries, different methods were used to derive their multiplicity, with different results. We report on a systematic search for multiplicity using spectroscopy, interferometry in a sample of 20 LBVs. Spectroscopy provides us with a bias-corrected binary fraction of $62^{+38}_{-24}\%$, and a percentage of $50-70\%$ is found from interferometry. This has a high impact on the way that these objects might be formed.

Keywords. stars: variables: S Doradus; stars: evolution; binaries: general; stars: massive; Astrophysics - Solar and Stellar Astrophysics

1. Introduction

Luminous Blue Variable stars (LBVs) are enigmatic objects in Astrophysics that are important to understand massive star evolution and their final fate. The classical LBVs experience outbursts, characterised by photometric and spectroscopic variability over periods from years to decades (the so-called S-Doradus cycle). During these cycles, the stars show visual magnitude changes of the order of one or two magnitudes, and they transit from cool to hotter regions in the Hertzsprung-Russell diagram. Their spectra vary from A- or F-type star when they are in a cool phase to B supergiant, O or even WR star when they are in a hot phase. Some LBVs also exhibit giant eruptions with visual magnitude changes higher than $2-3$ magnitudes, mimicking the appearance of type IIn supernovae (SNe). Finally, some stars also present common characteristics with the classical LBVs, but have never been detected to show S-Doradus variability or giant eruption: the dormant or candidate LBVs (Humphreys & Davidson 1994).

LBVs are traditionally thought to be an evolved stage of massive O- or B-type stars, in a transitory phase toward that of the Wolf-Rayet (WR, Conti 1984). Although O and

WR stars have high mass-loss rates, when integrated over the stellar lifetime, these rates are not sufficient to enable a smooth direct evolution. The stars must have gone through a phase of extreme mass loss (the LBV phase) to reveal the bare core that becomes the WR star. This short and extreme mass-loss phase is thought to occur for stars with masses higher than $25M_{\odot}$ stars and ranges in duration from 10^4 to 10^5 yrs.

However, the role of LBVs in massive star evolution appears not so simple. Their identification as progenitors of some high luminosity type IIb and IIn SNe challenged the traditional view of LBVs. Observational and theoretical evidences seem to indeed suggest that LBVs might be formed through binary interactions (e.g. Gal-Yam et al. 2007; Kiewe et al. 2012; Groh et al. 2013, 2020; Justham et al. 2014). While main-sequence O- and B-type stars are mainly found in young open clusters, the relative isolation of LBVs was also pointed out as possible evidence that LBVs are not the product of single-star evolution, but were in fact rejuvenated through mass transfer or merger events (Smith & Tombleson 2015; Smith 2016, 2019; Aghakhanloo et al. 2017, 2022). That claim was however strongly debated in the literature (Humphreys et al. 2016; Davidson et al. 2016; Aadland et al. 2018; Mehner et al. 2021), so that many challenges to understand the role of LBVs in massive star evolution remain.

While the majority (if not all) of massive O- and B-type stars are born in binary or multiple systems (Sana et al. 2012; Bodensteiner et al. 2021; Banyard et al. 2022; Bordier et al. 2022), the multiplicity of LBVs has not been clearly stated. Only a few LBVs in the Milky Way were detected as binaries: η Car (Damineli 1996), HR Car (Boffin et al. 2016, 2022 in prep.), MWC 314 (Lobel et al. 2013), HD 326823 (Richardson et al. 2011), and Pistol star (Martayan et al. 2012) but a general census on the multiplicity fraction was not obtained. Through their X-ray survey, Nazé et al. (2012) detected four objects (η Car, Schulte 12, GAL 026.47+00.02, and Cl* Westerlund 1 W 243) for which the X-ray detections are reminiscent of wind-wind collisions in binary systems, and five objects (P Cyg, AG Car, HD 160529, HD 316285, and Sher 25) in which the X-ray detections reach a strong limit that can be due to binarity, among other explanations. From this study, a binary fraction between 26 and 69% was claimed. Using high-resolution infrared images and adaptive optics, Martayan et al. (2016) investigated the close environments around seven LBVs to search for wide-orbit companions. Such companions were found for two out of seven objects (HD 168625 and MWC 314) but their physical link to the LBVs was not confirmed. From low-number statistics, Martayan et al. (2016) deduced a binary fraction for LBVs of about 30%.

If one accounts for the luminosities and the effective temperatures of LBVs, their radii are expected to be ranged from a few tens of solar radii to several hundreds. At their distances the best observing technique to unveil their multiplicity is the combination of spectroscopy and interferometry (see Fig. 1). We set up an observational campaign to collect new interferometric data of all the LBVs (classical and dormant) brighter than $V = 12.5 - 13.5$ mag and $H = 7.5 - 8$ mag, using VLTI/PIONIER and CHARA/MIRC-X. We combined these data with archival and new spectroscopic observations. Most of these results were presented in Mahy et al. (2022).

2. Data and methods

We built our sample from the LBV catalogues of Clark et al. (2005), Nazé et al. (2012), and Richardson & Mehner (2018). The spectroscopic sample gathers all the Galactic (classical and dormant) LBVs with V magnitudes brighter than 13.5 mag, plus Cl* Westerlund 1 W 243 and the Pistol star, which are fainter, because of the availability of archival data. We removed AG Car because the spectra in the archives are too strongly affected by the S-Doradus cycle to measure the RVs accurately. The interferometric sample gathers all the stars brighter than $H = 7.5$ mag in the northern and $H = 8$ mag in

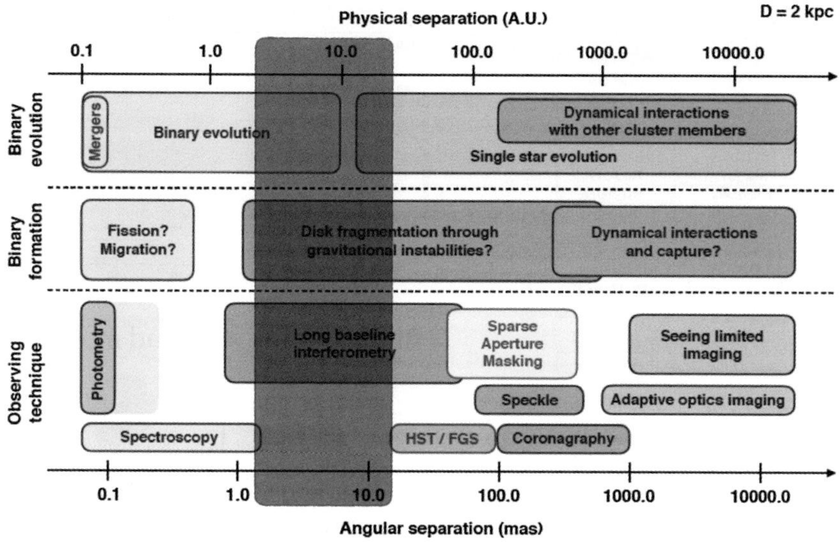

Figure 1. Sketch of the typical physical and angular separations covered by LBVs. A distance of 2 kpc is adopted, as representative of most objects (adapted from Fig. 1 of Sana 2017). Shaded blue area shows the typical separations expected with respect to LBV radii.

the southern hemisphere, because the sensitivities of PIONIER and MIRC-X are different. We collected interferometric snapshots for 20 stars (ESO Programme IDs: 0102.D-0460, 105.20AC, 109.231M, PI: Mahy). To extend this sample, we also used the published interferometric results for η Car (Gravity Collaboration et al. 2018), HR Car (Boffin et al. 2016), and Pistol star (Martayan et al. 2012). The final samples are shown in the left panel of Fig. 2. We stress that more objects are included for the present work than published by Mahy et al. (2022). We also refer the reader to Mahy et al. (2022) for more information about the sample, and the data reduction.

3. Results

3.1. *Spectroscopy*

The detection of secondary components orbiting around LBVs from spectroscopy is difficult because of (1) the difference in luminosity between the components and (2) the strong variability characterising the LBVs (Weis & Bomans 2020). Furthermore, given the LBV radii, their possible companions are expected to be on wide orbits and spectroscopy is not ideal to detect them. We measure the radial velocities (RVs) of the LBVs by focusing on spectra that do not show strong variability due to the S-Doradus cycle (i.e. we focus on spectra taken during the same quiescent/hot phase). We use a cross-correlation technique presented by Zucker (2003).

To classify a star as a likely spectroscopic binary, we use two criteria:

(a) $\frac{|\mathrm{RV}_i - \mathrm{RV}_j|}{\sqrt{\sigma_i^2 + \sigma_j^2}} > 4.0$,
(b) $\Delta\mathrm{RV} = |\mathrm{RV}_i - \mathrm{RV}_j| > \Delta\mathrm{RV}_{\mathrm{min}}$,

where RV_i and RV_j are the individual RV measurements and σ_i and σ_j the respective 1σ errors on the RV measurements at epochs i and j. If and only if at least one pair of RVs measured at different epochs simultaneously satisfies these criteria, the star is reported as spectroscopic binary. The observed binary fraction as a function of the $\Delta\mathrm{RV}_{\mathrm{min}}$ value

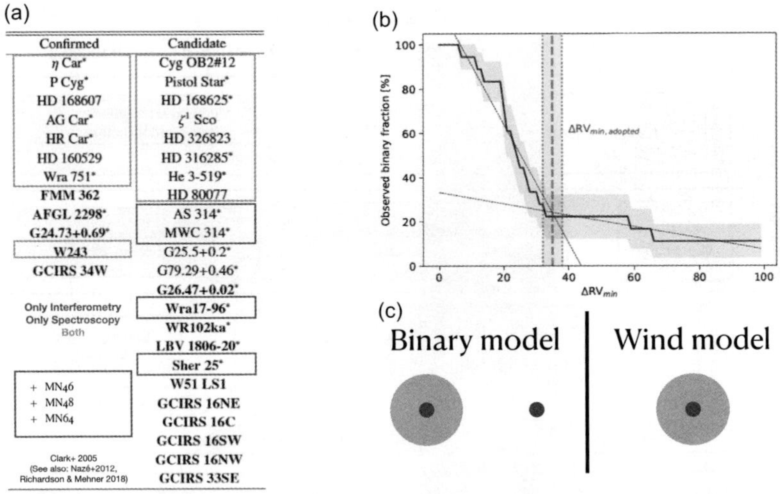

Figure 2. a) List of Galactic classical and dormant LBVs observed for this work, b) Observed binary fraction as a function of the adopted RV-variability threshold ΔRV_{min}. The vertical dotted line gives the adopted threshold value $\Delta RV_{min} = 35 \pm 3$ km s^{-1}, indicated by the red line (the shaded area represents the 1σ error), and c) Interferometric models used for CANDID (left), and LITPRO (right).

is displayed in the top right panel of Fig. 2. A kink is clearly visible at 35 km s^{-1}. We adopted this value as threshold, considering all the variability under that threshold dominated by intrinsic variability, and everything above that threshold dominated by binarity. The observed spectroscopic binary fraction is equal to $f_{obs} = 26^{+16}_{-10}\%$, where binomial statistics have been used to compute the uncertainty.

To obtain the intrinsic spectroscopic binary fraction for the Galactic LBVs, the observed binary fraction needs to be corrected for the observational biases. We run simulations following the approach described in Sana et al. (2013) to compute an estimate of the sensitivity of our observations and this provides us with a correction factor (see more details in Mahy et al. 2022). Once we correct for the observational biases, the intrinsic spectroscopic binary fraction rises to $f_{int} = 62^{+38}_{-24}\%$.

3.2. *Interferometry*

To search for astrometric companions among LBVs, we used the CANDID (Gallenne et al. 2015) and the Lyons Interferometric Tool prototype (LITPRO, Tallon-Bosc et al. 2008) codes. CANDID is a set of python tools that was specifically created to systematically search for high-contrast companions. It assumes a disc for the primary, an unresolved secondary, and additional background flux. However, because LBVs are characterised by strong winds, a wind model, composed of a disc or point-like source and a Gaussian profile to model the winds and centred on the star) also needed to be tested using LITPRO (see the bottom right panel of Fig. 2). For ten stars in our sample (HD 160529, P Cyg, HD 168625, ζ Sco, Schulte 12, Cl* Westerlund 1 W 243, HD 326823, WRAY 15-751, MWC 930 and WRAY 17-96), the fits performed by CANDID provide a lower chi-square, suggesting that these objects are binary systems. For five stars (AG Car, [GKF2010] MN46, and [GKF2010] MN48, WR 31a, and HD 168607), the measured chi-squares with CANDID and LITPRO are equivalent and the difference between the two models become insignificant. If we also consider HR Car, the Pistol star, and η Car (already reported as interferometric binaries), our analysis shows that the binary fraction among the Galactic LBVs is between 50 and 75%.

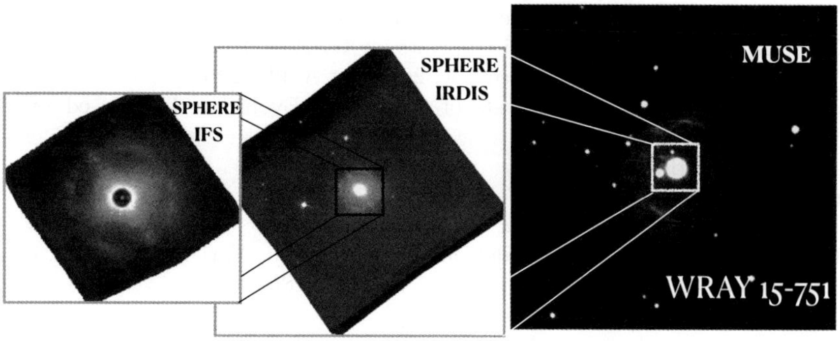

Figure 3. Newly acquired SPHERE/IFS+IRDIS and MUSE/WFM images of WRAY 15-751, with respective field-of-views of 1.73″, 11″, and 1′.

4. Conclusions

Massive stars predominantly live their lives in binary or multiple systems (Sana et al. 2014; Moe & Di Stefano 2017) with on average more than two companions. As a consequence, the presence of a companion severely impacts the evolution of these stars with a significant percentage of stellar mergers ($\sim 24\%$), accretion and spin up or common envelope evolution ($\sim 14\%$), and envelope stripping ($\sim 33\%$). Binary star evolution must therefore be considered as a possible channel to form LBVs (Gallagher 1989). Furthermore, the idea of merger events was proposed to explain the giant eruption observed for η Car (Smith et al. 2018; Hirai et al. 2021).

The search for binarity among the Galactic LBV population could provide a quantitative test to understand how these objects are formed. Our study shows that a large fraction of Galactic LBVs are binary systems, most of them being detected with interferometry. We stress, however, that so far the physical association of these detections with the LBVs (classical or dormant) only relies on statistical arguments and further observations are still needed to derive the orbital parameters of these possible systems. It is also interesting to investigate larger separations to probe their close environments (Fig. 3).

Acknowledgements

L.M. thanks the European Space Agency (ESA) and the Belgian Federal Science Policy Office (BELSPO) for their support in the framework of the PRODEX Programme. A.L. acknowledges funding received in part from the European Union Framework Programme for Research and Innovation Horizon 2020 (20142020) under the Marie Sklodowska-Curie grant Agreement No. 823734 (Physics of Extreme Massive Stars). H.S. acknowledges support from the European Research Council (ERC) under the European Union's Horizon 2020 research and innovation programme (grant agreement numbers 772225: MULTIPLES).

References

Aadland, E., Massey, P., Neugent, K. F., & Drout, M. R. 2018, AJ, 156, 294
Aghakhanloo, M., Murphy, J. W., Smith, N., & Hložek, R. 2017, MNRAS, 472, 591
Aghakhanloo, M., Smith, N., Andrews, J., Olsen, K., Besla, G., & Choi, Y. 2022, arXiv e-prints, arXiv:2202.06887
Banyard, G., Sana, H., Mahy, L., Bodensteiner, J., Villaseñor, J. I., & Evans, C. J. 2022, A&A 658, A69
Bodensteiner, J., et al. 2021, A&A 652, A70
Boffin, H. M. J., et al. 2016, A&A 593, A90

Bordier, E., Frost, A. J., Sana, H., Reggiani, M., Mérand, A., Rainot, A., Ramírez-Tannus, M. C., & de Wit, W. J. 2022, arXiv e-prints, arXiv:2203.05036
Clark, J. S., Larionov, V. M., & Arkharov, A. 2005, A&A 435, 239
Conti, P. S. 1984, in Observational Tests of the Stellar Evolution Theory, ed. A. Maeder & A. Renzini, Vol. 105, 233
Damineli, A. 1996, ApJL, 460, L49
Davidson, K., Humphreys, R. M., & Weis, K. 2016, arXiv e-prints, arXiv:1608.02007
Gal-Yam, A., et al. 2007, ApJ, 656, 372
Gallagher, J. S. 1989, Close Binary Models for Luminous Blue Variable Stars, ed. K. Davidson, A. F. J. Moffat, & H. J. G. L. M. Lamers, Vol. 157, 185
Gallenne, A., et al. 2015, A&A 579, A68
Gravity Collaboration et al. 2018, A&A 618, A125
Groh, J. H., Farrell, E. J., Meynet, G., Smith, N., Murphy, L., Allan, A. P., Georgy, C., & Ekstroem, S. 2020, ApJ, 900, 98
Groh, J. H., Meynet, G., & Ekström, S. 2013, A&A 550, L7
Hirai, R., Podsiadlowski, P., Owocki, S. P., Schneider, F. R. N., & Smith, N. 2021, MNRAS, 503, 4276
Humphreys, R. M., & Davidson, K. 1994, PASP, 106, 1025
Humphreys, R. M., Weis, K., Davidson, K., & Gordon, M. S. 2016, ApJ, 825, 64
Justham, S., Podsiadlowski, P., & Vink, J. S. 2014, ApJ, 796, 121
Kiewe, M., et al. 2012, ApJ, 744, 10
Lobel, A., et al. 2013, A&A 559, A16
Mahy, L., et al. 2022, A&A 657, A4
Martayan, C., et al. 2012, in Astronomical Society of the Pacific Conference Series, Vol. 464, Circumstellar Dynamics at High Resolution, ed. A. C. Carciofi & T. Rivinius, 293
Martayan, C., et al. 2016, A&A 587, A115
Mehner, A., et al. 2021, A&A 655, A33
Moe, M., & Di Stefano, R. 2017, ApJS, 230, 15
Nazé, Y., Rauw, G., & Hutsemékers, D. 2012, A&A 538, A47
Richardson, N. D., Gies, D. R., & Williams, S. J. 2011, AJ, 142, 201
Richardson, N. D., & Mehner, A. 2018, Research Notes of the American Astronomical Society, 2, 121
Sana, H. 2017, in The Lives and Death-Throes of Massive Stars, ed. J. J. Eldridge, J. C. Bray, L. A. S. McClelland, & L. Xiao, Vol. 329, 110–117
Sana, H., et al. 2012, Science, 337, 444
—. 2013, A&A 550, A107
—. 2014, ApJS, 215, 15
Smith, N. 2016, MNRAS, 461, 3353
—. 2019, MNRAS, 489, 4378
Smith, N., & Tombleson, R. 2015, MNRAS, 447, 598
Smith, N., et al. 2018, MNRAS, 480, 1466
Tallon-Bosc, I., et al. 2008, in Society of Photo-Optical Instrumentation Engineers (SPIE) Conference Series, Vol. 7013, Optical and Infrared Interferometry, ed. M. Schöller, W. C. Danchi, & F. Delplancke, 70131J
Weis, K., & Bomans, D. J. 2020, Galaxies, 8, 20
Zucker, S. 2003, MNRAS, 342, 1291

Massive Stars Near and Far
Proceedings IAU Symposium No. 361, 2024
J. Mackey, J. S. Vink, & N. St-Louis, eds.
doi:10.1017/S1743921322003064

Hunting for red supergiant binaries: UVIT photometry of the SMC

Lee R. Patrick[1,2], David Thilker[3], Danny Lennon[4,5], Luciana Bianchi[3], Abel Schootemeijer[6,7], Ricardo Dorda[3,4], Norbert Langer[6,7] and Ignacio Negueruela[1]

[1]Departamento de Física Aplicada, Universidad de Alicante, E-03690 San Vicente del Raspeig, Alicante, Spain
email: lee@lrp-astro.org

[2]School of Physical Sciences, The Open University, Walton Hall, Milton Keynes MK7 6AA, UK

[3]Department of Physics and Astronomy, Johns Hopkins University, 3400 N. Charles Street, Baltimore, MD 21218, USA

[4]Instituto de Astrofísica de Canarias, E-38205 La Laguna, Tenerife, Spain

[5]Universidad de La Laguna, Dpto. Astrofísica, E-38206 La Laguna, Tenerife, Spain

[6]Argelander-Institut für Astronomie, Universität Bonn, Auf dem Hügel 71, 53121, Bonn, Germany

[7]Max-Planck-Institut für Radioastronomie, Auf dem Hügel 69, 53121, Bonn, Germany

Abstract. We present UVIT/Astrosat UV photometry of the RSG population of the Small Cloud galaxy (SMC). As RSGs are extremely faint in the far-UV, these observations directly probe potential companion stars. From a sample of 861 SMC RSGs, we find 88 have detections at far-UV wavelengths: a clear signature of binarity. Stellar parameters are determined for both components, which allows us to study - for the first time - the mass-ratio (q) distribution of RSG binary systems. We find a flat mass-ratio distribution best describes the observations up to $M_{RSG} \sim 15 M_{\odot}$. We account for our main observing bias (i.e. the limiting magnitude of the UVIT survey) to determine the intrinsic RSG binary fraction of $18.8 \pm 1.5\,\%$, for mass-ratios in the range $0.3 < q < 1.0$ and orbital periods approximately in the range $3 < \log P[\mathrm{days}] < 8$.

Keywords. stars: late-type, stars: binaries, galaxies: Magellanic Clouds

1. Introduction

Multiplicity is a key uncertainty in the evolution and eventual fate of massive stars (Langer 2012). In recent times there is a growing realisation that the majority of massive stars are born with a close companion (e.g. Sana et al. 2012). Close companions may interact and merge (Moe & Di Stefano 2017; Langer et al. 2020; Sen et al. 2022), which results in blue straggler stars (Schneider et al. 2014) and exotic stellar systems (in originally higher-order multiple systems). The result of such evolution on the red supergiant star population, which is either one of, or the final, evolutionary stage for the majority of massive stars (between $\sim$8 and 40 $M_{\odot}$), is not well constrained observationally.

Theoretically, when a star within a close binary system evolves off the main sequence towards the RSG phase, an interaction occurs, which either results in a merger event or some form of envelope stripping. The majority of such interactions are expected to result in an evolutionary path that avoids a RSG phase (Eldridge et al. 2008).

Stars that have been regenerated in a merger event return to the main-sequence (Schneider et al. 2019) and subsequently evolve towards the RSG phase where they are expected to appear as red analogues to blue-straggler stars, so called 'red stragglers' (Britavskiy et al. 2019), where they can account for up to 50 % of RSG populations in clusters (Beasor et al. 2019; Britavskiy et al. 2019; Patrick et al. 2020).

Red supergiant stars are very red and luminous, therefore, massive binary systems where one component is a RSG are likely to be dominated by the appearance of the RSG at optical and infrared wavelengths. Such systems are also necessarily longer orbital period systems, where the minimum allowed orbital periods for a RSG are between 100 and 1000 days. As the lifetime of the RSG phase is much shorter than that of the main-sequence, one expects this to place a lower limit on the mass of a potential companion (Neugent et al. 2018), although this lower limit is not well constrained observationally as lower mass, lower mass-ratio systems are typically most difficult to detect (see Section 4 for more discussion).

The impact of binary systems on stellar populations is important to study binary physics and to predict the outcomes of binary evolution. For example, the increased the line-of-sight velocity distribution of young clusters is seen as an indicator of the level of binarity (Gieles et al. 2010), however, this is not observed in older clusters dominated by RSGs (Patrick et al. 2016, 2020), which is an indication that the multiplicity fraction of RSGs is lower than in earlier evolutionary phases.

To identify RSG binary systems there are typically two approaches:

(a) Detect radial velocity variations that cannot be explained as stellar activity from a single star. Given the long orbital periods, long-baseline observations are a requirement for the success of this technique. Burki & Mayor (1983) first performed such a study in the Galaxy, detecting 19 K- and M-type RSG binary systems. Patrick et al. (2019) and Patrick et al. (2020) applied this technique to star clusters in both the Large and Small Cloud galaxies (MCs), however, these authors could not detect individual systems. Dorda & Patrick (2021) calibrated a large baseline set of inhomogeneous observations in the MCs to the Gaia DR2 radial velocity rest frame. These authors were able to identify individual systems, but couldn't correct for the various observational biases to determine intrinsic multiplicity fractions.

(b) Detect a signature of the companion in photometric or spectroscopic observations. Neugent et al. (2018) developed a method that exploits widely-available optical photometry to identify a blue excess that is attributed to the presence of a companion. This method has been refined in a series of studies that identifies RSG binary systems in several Local Group galaxies (Neugent et al. 2020; Neugent 2021). From their observations, these authors determine the intrinsic binary fraction of RSGs using detailed simulations.

Table 1 provides an overview of the results of these various studies.

In this article we summarise the results of a search for RSG binary systems with newly available UV photometry from the UVIT/Astrosat SMC survey (Thilker et al. in prep.). UV photometry has the unique benefit that an early-type companion contributes close to 100% of the flux at such wavelengths and, as such, the companions of a large sample of RSG binary systems can be studied directly, which is the first time that this has been possible.

Table 1. Summary of the determination of the RSG multiplicity fraction in Local Group galaxies.

Galaxy	Multiplicity fraction (%)	Orbital configuration range	Reference
SMC	18.8 ± 1.5	$3 < \log P[\text{days}] < 8$ $0.3 < q < 1.0$	Patrick et al. (2022)
SMC	>15	–	Dorda & Patrick (2021)
NGC 330[1]	30 ± 10	$2.3 < \log P[\text{days}] < 4.3$ $0.1 < q < 1.0$	Patrick et al. (2020)
LMC	$19.5^{+7.6}_{-6.7}$	$0.1 < q < 1.0$[2]	Neugent et al. (2020)
LMC	>15	–	Dorda & Patrick (2021)
30 Dor[3]	30 ± 10	$3.3 < \log P[\text{days}] < 4.3$ $0.3 < q < 1.0$	Patrick et al. (2019)
MW	27 ± 10	$\log P < 2.9$[5] $q > 0.1$	Burki & Mayor (1983)
M31	$33.5^{+8.6}_{-5.0}$	$0.1 < q < 1.0$[2]	Neugent (2021)
M33	15-40[4]	$0.1 < q < 1.0$[2]	Neugent (2021)

Notes:
[1]Young massive cluster in the SMC.
[2]Neugent (priv. comm.)
[3]LMC star-forming region. As part of the VLT-FLAMES Tarantula Survey Evans et al. (2011).
[4]Neugent (2021) find a metallicity dependence of the RSG binary fraction
[5]Burki & Mayor (1983) consider stars down to 5 $M_\odot$, which results in small allowed orbital periods.

2. Methods

As our aim is to identify RSG binary systems we first construct a catalogue of 861 RSGs in the SMC, which is based on the source catalogue of Yang et al. (2020). This is a list of high-probability SMC RSGs, based on the photometric classification at multiple wavelengths and astrometric data provided by Gaia EDR3 (Gaia Collaboration et al. 2021).

Thilker et al. (in prep.) presented the UVIT survey of the SMC in the F172M filter, which has a central wavelength of 1717 Å. This SMC-wide survey† is currently 75 % complete. This survey has a limiting magnitude of $m_{F172M} = 20.3$ ABmag with an astrometric precision of around 0.1" and a 4x better spatial resolution than GALEX.

We identify RSG binary systems by cross-matching the two catalogues. To determine the contribution from chance alignments, we follow the method of Bianchi & Shiao (2020) and use a control catalogue, which is the original RSG catalogue offset by 10" in declination. The results of this analysis show that outside of a cross-match distance (XMD) of 0.5", the probability of chance alignments dominates the matches, which highlights the need for precision astrometry in such studies. To minmise the possibility of chance alignments we select a maximum XMD of 0.4". This results in 88 matches, which we interpret as genuine RSG binary systems.

3. Stellar parameters of both components

We utilise photometry to determine the stellar parameters of both components within each of the 88 RSG binary systems and for the entire RSG catalogue. In the following analysis we assume a single extinction law for all targets, which is characterised with $A_V = 0.35$ and the SMC bar reddening law of Gordon et al. (2003). See Patrick et al. (2022) for further details. Figure 1 displays the stellar parameters determined for the stars within binary systems.

The luminosities of the RSG component are determined from K-band photometry using the SMC calibrations of Davies et al. (2013). The RSG effective temperatures are determined using a calibration to the de-reddened $J - K$ colours (Dorda & Patrick 2021). This calibration is based on the RSG effective temperatures determined for SMC RSGs

† of the SMC ...

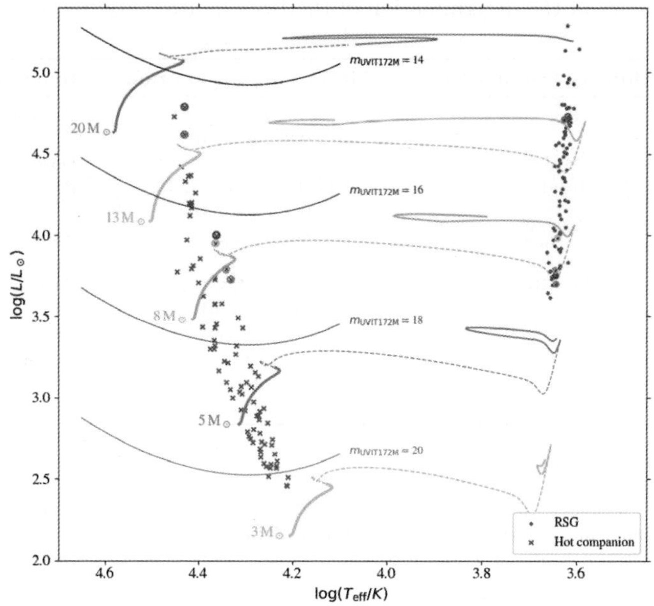

Figure 1. Hertzsprung–Russel diagram showing the RSGs and their companions. The coloured lines show stellar evolutionary tracks that are based on those in Schootemeijer et al. (2019). The solid grey lines highlight lines of constant UVIT magnitudes in the stellar models. Rings of different colours highlight the six systems that have a mass-ratio greater than 1.

in Tabernero et al. (2018). A discussion of SMC RSG effective temperatures is provided in Appendix B of Patrick et al. (2022). We determine the evolutionary masses and ages of the RSG via a comparison with MESA models computed for the SMC by Schootemeijer et al. (2019). These models use a mass-dependent overshooting parameter and a semiconvection parameter of 10. We determine the mass and age of the RSG by comparing the observed luminosity of the RSG with the luminosity of the mid-point of the helium burning lifetime of the models.

To determine the stellar parameters of the companions we make the simplifying assumption that each companion is a single, main-sequence star that is the same age as the RSG. We determine the luminosity of the companion using the UVIT photometry, assuming a bolometric correction taken from the MIST models (Choi et al. 2016; Dotter 2016). At these wavelengths there is effectively no contribution from the RSG. To determine the mass and effective temperature of the companion, we compare the luminosity of the companion with the same MESA models as for the RSG.

4. Key results and discussion

We determine mass-ratios for all of the 88 RSG binary systems detected. Figure 2 shows the mass-ratio distribution as a function of RSG mass and the cumulative distribution function for RSG binary systems for RSGs within the mass range $10 < M/M_\odot < 15$. The left panel highlights the principle observing bias of this survey, which is that the limiting magnitude of the UVIT survey results in a mass-ratio dependent observing bias such that low mass-ratio systems go undetected at low RSG masses.

To minimise the contribution of the observing bias on the mass-ratio distribution, we concentrate on RSG masses in the range $10 < M/M_\odot < 15$ and mass-ratios in the range $0.3 < q < 1.0$. RSG binary systems in this parameter range are best described with a flat

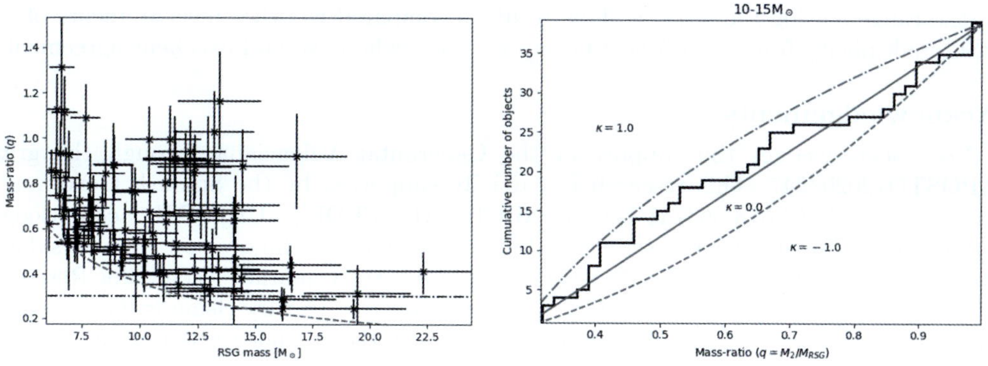

Figure 2. Left: Mass-ratio distribution as a function of RSG mass. The red dashed line shows the mass-ratio dependent observing bias. The black dot-dashed line shows a mass-ratio limit of $q = 0.3$. Right: Mass-ratio cumulative distribution function for RSGs masses between 10 and $15\,M_{\odot}$.

mass-ratio distribution. We also find a lack of high-mass ratio systems at above $15\,M_{\odot}$, which, in principle, should be the easiest systems to detect given our observations.

We account for observational biases by simulating the observed population of RSG binaries assuming a flat mass-ratio distribution with the aim of effectively filling in the area of the left panel of Figure 2, that could not be observed. From 100 000 simulations we determine that we have missed on average 17.7 ± 4.4 systems as the result of the UVIT detection limit. This results in an intrinsic binary fraction of $18.8 \pm 1.5\,\%$, in the range $0.3 < q < 1.0$.

When comparing intrinsic binary fractions of stellar populations, it is fundamental to consider the range of orbital configurations over which the observations are sensitive. We place constraints on the range of orbital periods by considering that the physical size of the RSG places a lower limit on the orbital period of around $\log P[\mathrm{day}] \sim 3$. The upper limit can be estimated using the XMD of 0.4", which at the distance of the SMC corresponds to an orbital period of $\log P[\mathrm{day}] \sim 8$.

Table 1 compares the RSG binary fractions that have been determined for Local Group galaxies. We find excellent agreement with the LMC study of Neugent et al. (2020) and the lower limit placed by Dorda & Patrick (2021). See Patrick et al. (2022) for further comparisons at other evolutionary stages.

5. Conclusions

In this article we present the key results from Patrick et al. (2022), where we aim to identify and study RSG binary systems in the SMC using a newly available UV photometric survey. We have demonstrated that using UV photometry is an extremely effective method to not only identify RSG binary systems, but to characterise their companions, which is something that until now, has only been possible for a very small number of Galactic systems.

The source list of RSGs that we have compiled consists of 861 high-probability RSGs. From a cross-match with UV photometry we identify 88 SMC RSG binary systems and we determine stellar parameters for both components in all systems.

By studying the mass-ratios of SMC RSGs we find a flat distribution best describes the observations over the range $0.3 < q < 1.0$. We use this result to simulate the main observing bias of the UVIT survey to determine the intrinsic RSG multiplicity fraction to be $18.8 \pm 1.5\,\%$, for mass-ratios in the range $0.3 < q < 1.0$ and orbital periods approximately in the range $3 < \log P[\text{days}] < 8$. This result is compared to other measurements of the RSG multiplicity fraction in Local Group galaxies, where we find excellent agreement.

Acknowledgements

LRP acknowledges the support of the Generalitat Valenciana through the grant APOSTD/2020/247. This research is partially supported by the Spanish Government under grant PGC2018-093741-B-C21 (MICIU/AEI/FEDER, UE). DJL acknowledges support from the Spanish Government Ministerio de Ciencia, Innovación y Universidades through grants PGC-2018-091 3741-B-C22 and from the Canarian Agency for Research, Innovation and Information Society (ACIISI), of the Canary Islands Government, and the European Regional Development Fund (ERDF), under grant with reference ProID2017010115.

References

Beasor, E. R., Davies, B., Smith, N., & Bastian, N. 2019, MNRAS, 486, 266
Bianchi, L., & Shiao, B. 2020, ApJS, 250, 36
Britavskiy, N., et al. 2019, A&A, 624, A128
Burki, G., & Mayor, M. 1983, A&A, 124, 256
Choi, J., Dotter, A., Conroy, C., Cantiello, M., Paxton, B., & Johnson, B. D. 2016, ApJ, 823, 102
Davies, B., et al. 2013, ApJ, 767, 3
Dorda, R., & Patrick, L. R. 2021, MNRAS, 502, 4890
Dotter, A. 2016, ApJS, 222, 8
Eldridge, J. J., Izzard, R. G., & Tout, C. A. 2008, MNRAS, 384, 1109
Evans, C. J., et al. 2011, A&A, 527, A50
Gaia Collaboration et al. 2021, A&A, 649, A1
Gieles, M., Sana, H., & Portegies Zwart, S. F. 2010, MNRAS, 402, 1750
Gordon, K. D., Clayton, G. C., Misselt, K. A., Landolt, A. U., & Wolff, M. J. 2003, ApJ, 594, 279
Langer, N. 2012, ARA&A, 50, 107
Langer, N., et al. 2020, A&A, 638, A39
Moe, M., & Di Stefano, R. 2017, ApJS, 230, 15
Neugent, K. F. 2021, ApJ, 908, 87
Neugent, K. F., Levesque, E. M., Massey, P., Morrell, N. I., & Drout, M. R. 2020, ApJ, 900, 118
Neugent, K. F., Massey, P., Morrell, N. I., Skiff, B., & Georgy, C. 2018, AJ, 155, 207
Patrick, L., Bianchi, L. C., Dorda, R., Langer, N., Lennon, D. J., Negueruela, I., & Thilker, D. 2021, Cool stars with hot companions in the Small Magellanic Cloud, HST Proposal. Cycle 29, ID. #16776
Patrick, L. R., Evans, C. J., Davies, B., Kudritzki, R.-P., Hénault-Brunet, V., Bastian, N., Lapenna, E., & Bergemann, M. 2016, MNRAS, 458, 3968
Patrick, L. R., Thilker, D., Lennon, D. J., Bianchi, L., Schootemeijer, A., Dorda, R., Langer, N., & Negueruela, I. 2022, MNRAS, 513, 5847
Patrick, L. R., et al. 2019, A&A, 624, A129
—. 2020, A&A, 635, A29
Sana, H., et al. 2012, Science, 337, 444
Schneider, F. R. N., Ohlmann, S. T., Podsiadlowski, P., Röpke, F. K., Balbus, S. A., Pakmor, R., & Springel, V. 2019, Nature 574, 211
Schneider, F. R. N., et al. 2014, ApJ, 780, 117
Schootemeijer, A., Langer, N., Grin, N. J., & Wang, C. 2019, A&A, 625, A132
Sen, K., et al. 2022, A&A, 659, A98
Tabernero, H. M., Dorda, R., Negueruela, I., & González-Fernández, C. 2018, MNRAS, 476, 3106
Yang, M., et al. 2020, A&A, 639, A116

Discussion

BOSTROEM: How do you deal with line of sight coincides and recognise true binaries?

PATRICK: This is an excellent question and something I think is really important to consider. I didn't have time to elaborate on this much in the talk. We assess the probability of chance alignments using a control catalogue. We find that outside of a cross-match distance of 0.5" false positive detections can account for up to 100% of matches. Within 0.4", the false positive detection rate is effectively negligible. It is only thanks to the high precision of our UV point source catalogue from the UVIT/Astrosat survey that we are able to do this so accurately.

GRAEFENER: Are there signatures of wind emission from the RSG in these detections? Could these detections in the UV be the result of the RSG wind?

PATRICK: As Goetz points out, in the IUE spectra of Galactic RSG binary systems there are signatures of wind emission from the RSG to varying degrees. In the UV photometry that I have shown today, these signatures do not contribute significantly to the flux that we observe and because of that we are confident that the UV flux comes almost exclusively from the hot star companion. We have HST STIS UV spectra from our recent SNAP proposal Patrick et al. (2021) that do show evidence of wind emission features that originate from the RSG. Single RSGs on the other hand, would be far too faint in the UV to be detected with the UV photometry presented here.

STANWAY: I'm not sure I understood how you get both a temp and luminosity from a single UV point. Can you clarify?

PATRICK: We determine the luminosity from the UVIT magnitudes and bolometric correction from the MIST models. We then assume an age (which is taken from the RSG age) to determine the effective temperatures and masses from the MIST models. The grey solid lines in Figure 1 highlight the relationship between effective temperature and luminosity for constant UVIT magnitudes. Because these lines are almost perpendicular to the main sequence, by assuming an age, we can do a good job of determining the stellar parameters of the companions despite the observational limitation of only using a single photometric point. In the future we hope to supplement this with further UV imaging.

Massive Stars Near and Far
Proceedings IAU Symposium No. 361, 2024
J. Mackey, J. S. Vink, & N. St-Louis, eds.
doi:10.1017/S1743921322002836

The impact of binary interaction on the main-sequence morphology of young star clusters

Chen Wang[1,2], Norbert Langer[1,3], Abel Schootemeijer[1], Antonino Milone[4], Ben Hastings[1,3], Xiao-Tian Xu[1,3], Julia Bodensteiner[5,6], Hugues Sana[6], Norberto Castro[7], D. J. Lennon[8,9], Pablo Marchant[6], A. de Koter[10,6] and Selma E. de Mink[2,10,11]

[1]Argelander-Institut für Astronomie, Universität Bonn, Auf dem Hügel 71, 53121 Bonn, Germany
email: cwang@mpa-garching.mpg.de

[2]Max Planck Institute for Astrophysics, Karl-Schwarzschild-Strasse 1, 85748 Garching, Germany

[3] Max-Planck-Institut für Radioastronomie, Auf dem Hügel 69, 53121 Bonn, Germany

[4]Dipartimento di Fisica e Astronomia "Galileo Galilei", Univ. di Padova, Vicolo dell'Osservatorio 3, I-35122 Padova, Italy

[5]European Southern Observatory, Karl-Schwarzschild-Strasse 2, 85748 Garching, Germany

[6]Institute of Astronomy, KU Leuven, Celestijnlaan 200D, 3001 Leuven, Belgium

[7]Leibniz-Institut für Astrophysik Potsdam (AIP), An der Sternwarte 16, 14482 Potsdam, Germany

[8] Instituto de Astrofísica de Canarias, E-38200 La Laguna, Tenerife, Spain

[9] Dpto. Astrofísica, Universidad de La Laguna, E-38205 La Laguna, Tenerife, Spain

[10]Astronomical Institute "Anton Pannekoek", University of Amsterdam, Science Park 904, 1098 XH Amsterdam, The Netherlands

[11]Center for Astrophysics, Harvard-Smithsonian, 60 Garden Street, Cambridge, MA 02138, USA

Abstract. Since massive stars form preferentially as members of close binary systems, we use dense grids of detailed binary evolution models to explore how binary evolution shapes the main-sequence morphology of young star clusters. We propose that binary mergers might be the origin of the blue main sequence stars in young star clusters. Our results imply that stars may either form by accretion, or through a binary merger, and that both paths lead to distinctly different spins, magnetic fields, and stellar mass distributions.

Keywords. stars: early-type, stars: binaries, Hertzsprung-Russell diagram

1. Introduction

Recent high-precision Hubble Space Telescope (HST) photometry reveals that the main sequences (MSs) of young star clusters in the color-magnitude diagram are split into several components (Milone et al. 2009, 2018). This challenges the traditional view of star clusters containing coeval stars born with identical initial conditions. These MS components are characterized by a split MS and an extended MS turn-off (eMSTO).

There is a growing consensus that the split MS is caused by a bi-modal rotation velocity distribution, with the red MS containing fast rotators and the blue MS containing slow rotators. This is supported by the spectroscopic measurements of the rotation velocity of the red and blue MS stars in the Large Magellanic Cloud (LMC) cluster NGC 1818 (Marino et al. 2018). This bi-modal velocity distribution also agrees well with the previous spectroscopic studies of the velocity of B/A type stars (Zorec & Royer 2012; Dufton et al. 2013). Fast rotation is expected from star formation theory, as stars increase their spins during the contraction and accretion phases in their formation, whereas the origin of the slowly rotating stars is still unknown.

About 70% of massive stars are shown to have close companions (Sana et al. 2012). However, the impact of binary interaction in shaping the MS morphology of young star clusters has not been studied in detail so far, due to the lack of proper binary models. We provide such detailed binary models and propose a self-consistent theory to explain all these discrete MS components with co-eval stars.

2. Methods

We use the one-dimensional stellar evolution code MESA (Paxton et al. 2011, 2013; Paxton et al. 2015), version 8845, to compute our binary evolution models. We use the same physics assumptions as the single star models of Brott et al. (2011) and Schootemeijer & Langer (2018). For the details of our binary star models, we refer to Wang et al. (2020).

We consider a Small Magellanic Cloud (SMC) like metallicity ($Z_{\rm SMC} = 0.002179$) for our binary models, such that stellar winds, which are one of the important uncertainties in stellar astrophysics, are weak. We build more than 50 000 binary evolution models, with the primary star mass ranging from $5M_{\odot}$ to $100M_{\odot}$, initial mass ratio ranging from 0.3 to 0.95 and initial orbital periods ranging from 1 day to $\sim$8.6 yr. We consider constant parameter intervals and do not include statistical probabilities, such as the IMF and initial binary parameter distributions, because we only mean to show which parts of the HRD are covered by binary models in different evolutionary stages.

In order to perform isochrone fits to young star clusters, we also compute single star models with MESA version 12115 (Paxton et al. 2019), as this new version allows building star models with rotational velocities as high as 90% of critical rotation. For the details of our single star models, we refer to Wang et al. (2022).

3. Results

3.1. *Binary evolutionary induced main-sequence features*

Figure 1 shows the location of our binary models in the HRD at 30 Myr. Each dot corresponds to the visually brighter component in one binary system. The pre-interaction binaries (blue dots) in this figure are located on a single line, which represents the corresponding single-star isochrone. The binary merger products (yellow dots) are located on the upper left side of the HRD. We assume that the binary merger products have slow rotation according to Schneider et al. (2019). The merger products appear bluer and younger than the other binary systems because of rejuvenation. The post-mass transfer binaries (red dots) form a discrete main sequence on the upper right side of the HRD. They are the accretors that have experienced stable mass transfer and have reached near-critical rotation during accretion. We assume that they will be observed as Be stars, which may have disks and thus have emission lines in the spectrum and IR excess in photometry (Rivinius et al. 2013). The semi-detached binaries (green dots) are those that are

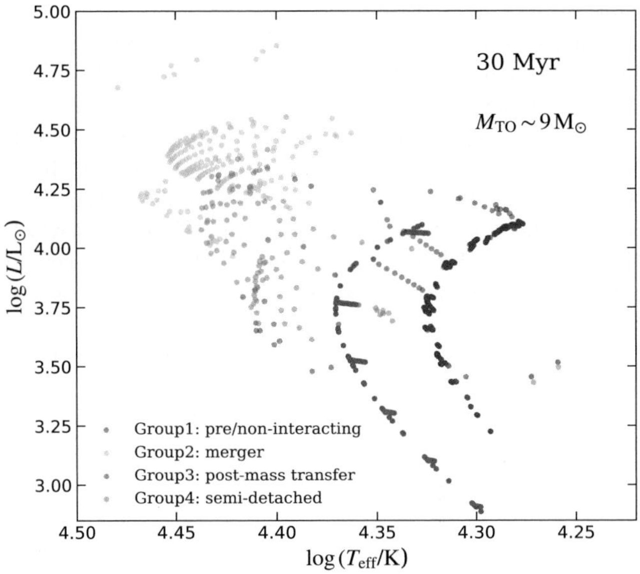

Figure 1. Distribution of our binary models in the Hertzsprung-Russell diagram at 30 Myr. Only the visually brighter component of each binary is plotted. Blue dots are the stellar models that have not yet interacted with a companion. Yellow dots are the binary merger products, while red dots correspond to post-interaction stellar models. Green dots represent the stellar models that are undergoing mass transfer at this time. The MS turnoff mass for the non-interacting stars in a cluster at this age is $\sim 9\,M_\odot$. (Colors can only be seen in the online version.)

undergoing nuclear timescale mass-transfer phase. We stress here that the colors in this figure only mean to indicate different binary evolutionary stages. The red and blue dots do not correspond to the observed red and blue MS stars in young star clusters.

Figure 1 indicates that binary evolution leads to multiple MS components in young star clusters, by producing a population of blue stragglers via binary mergers and a population of red Be stars from stable mass transfer.

3.2. *The origin of the blue main-sequence*

Our detailed binary evolution models show that binary merger products can explain the blue stragglers observed in young star clusters. However, the origin of the blue MS stars below the cluster turn-off, which are also believed to have slow rotation, is yet unknown. We use the LMC cluster NGC 1755 as an example to investigate the origin of the blue MS. In Fig. 2 we show the CMD of the MS stars in NGC 1755 and the isochrone fit. We identify the red and blue MS stars according to the method in Wang et al. (2022). We examine the mass function of the red and blue MS stars in this cluster. The masses of the stars are derived according to the mass-luminosity relation contained in the isochrones in the right panel of Fig. 2.

We plot the cumulative number distribution of the red and blue MS stars in NGC 1755 as a function of mass in Fig. 3. We use a power-law mass function $N(m)dm \propto m^\gamma$ to fit the obtained distribution. For the red MS stars, the derived power law exponent of $\gamma = -2.17 \pm 0.15$ is close to that of a Salpeter law ($\gamma = -2.35$) in the mass range of 2 to 5.5 $M_\odot$. However, in the same mass range, the power law exponent of the blue MS stars is $\gamma = -1.03 \pm 0.32$, representing a much flatter mass distribution. We find similar results in other young Magellanic Cloud clusters (< 100 Myr) that exhibit a split MS feature.

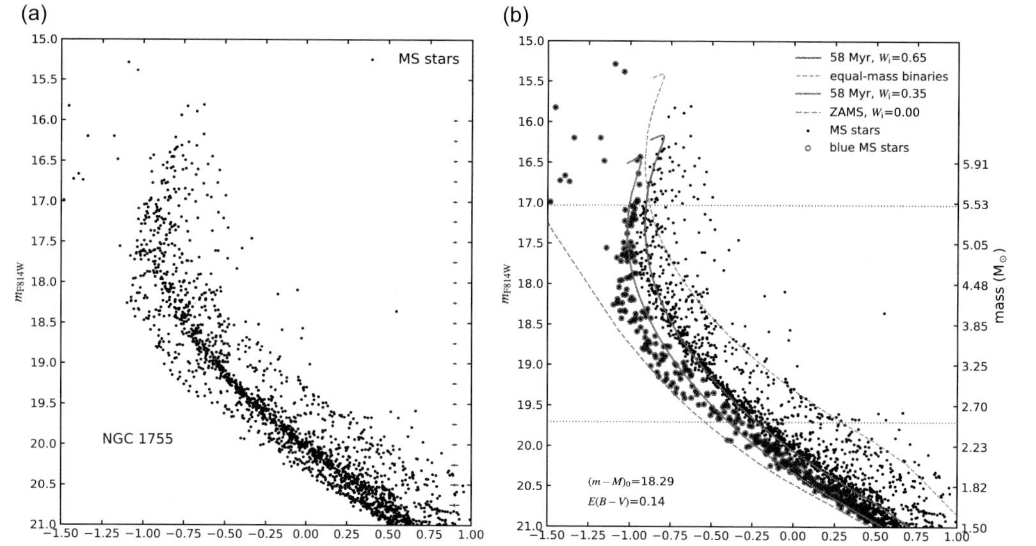

Figure 2. **a**: HST color-magnitude diagram of NGC 1755 (black dots). Red error bars indicate typical errors at corresponding magnitudes. **b**: Isochrone fit for the red and blue MSs in NGC 1755. The red and blue solid lines show the isochrones of 60 Myr single star models whose initial velocities are 65% and 35% of critical rotation. The red dashed line shows the positions of the equal-mass binaries in which both components have 65% of critical rotation initially. The grey dashed line indicates the zero-age MS line of non-rotating stars. The stars between the two horizontal dotted lines are used in our mass function analysis. The right y-axis shows the stellar mass according to the mass-magnitude relation of the fast rotating star models. (Colors can only be seen in the online version.)

Different mass functions may indicate different origin mechanisms of the red and blue MS stars.

We propose that blue MS stars stem from binary mergers. Unlike the blue stragglers that can be explained by binary evolutionary-induced mergers, the progenitors of the blue MS stars below the cluster turn-off do not have time to encounter mass transfer. These blue MS stars may originate from the binary mergers triggered by dynamical processes, e.g. the interaction between the binary systems and their surroundings, including circumbinary disks, ambient gas or a distant tertiary star. This means that binaries are born with large separations, then the two stellar components migrate towards each other due to the above-mentioned interaction. Some binaries merge during this migration phase. The derived flat mass function for the blue MS stars agrees with their merger origin, as it is shown that binary fraction increases with stellar mass (Schneider et al. 2015), and more massive binaries are probably easier to merge (Korntreff et al. 2012). Given the hypothesis of the merger origin of the blue MS stars, we derive an early merger rate peak. This can be seen by the fact that the star density is the highest in the vicinity of the blue isochrone in the right panel of Fig. 2.

There are pieces of observational evidence supporting binary migration. Firstly, a dearth of short period massive binaries is found in young star-forming regions (Sana et al. 2017), meaning that migration within the first several Myrs of the cluster lifetime is required to explain the observed high fraction of close massive binary systems (Sana et al. 2012). Secondly, the observed velocity dispersion of the cluster stars increases with cluster age during their first few million years of evolution (Ramírez-Tannus et al. 2021),

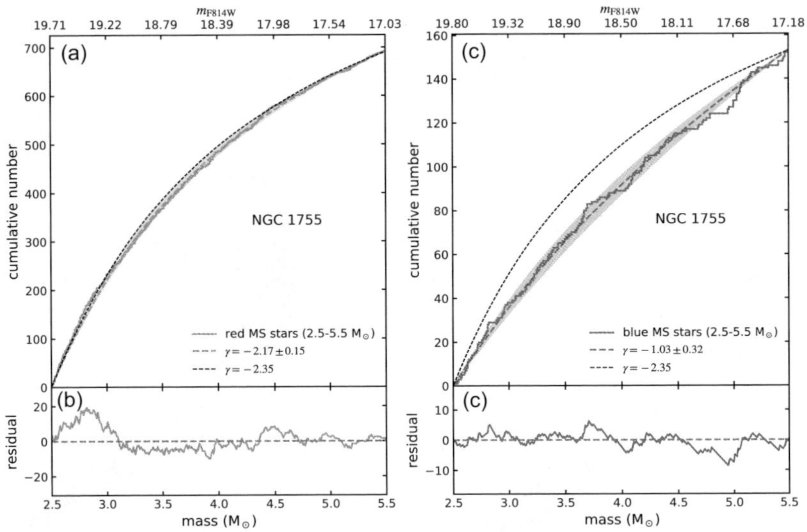

Figure 3. **a** and **c**: Cumulative number distribution of the red MS (red lines) and the blue MS stars (blue lines) in NGC 1755 as a function of their mass. Dashed lines show the best fitting power-law mass function. The shaded areas represent one sigma errors of our power-law fitting. The dashed black lines show the distribution predicted by the Salpeter IMF with a power-law index of $\gamma = -2.35$. The top x-axes show the apparent magnitudes for given masses. **b** and **d**: Residuals, i.e. difference between the colored solid and dashed lines in Panels a and c, as a function of mass. The grey dashed lines indicate zero residual. (Colors can only be seen in the online version.)

which also indicates that the close binary fraction increases with time. From the theoretical side, tidal forces imposed by the circumbinary matter are known to induce a drastic decay of the binary orbit (Korntreff et al. 2012). Recent binary formation models indeed predict ~30% of binary B stars to merge during their pre-MS evolution or early MS evolution (Tokovinin & Moe 2020).

4. Conclusions

We find that binary interaction plays a vital role in shaping the MS morphology of young star clusters. Binary interaction can produce the slowest rotating stars by binary mergers, which will populate the blue MS. Meanwhile, binary interaction can also produce the fastest rotating stars by stable mass transfer, which will populate the red Be sequence. Our results imply that stars might be born in two fundamentally different ways. On the one hand, the majority of stars form by accretion of gas via accretion disks. These stars should have natal spins slightly larger than half of the critical rotation; they should follow the Salpeter IMF and should not be magnetic stars. On the other hand, some of the thus created stars merge and produce blue MS stars, rotating slowly and obeying a flatter mass function than the red MS stars. These stars may have strong magnetic fields. Our work shed new light on the origin of the bi-modal mass, spin, and magnetic field distributions of MS stars. Further spectroscopic observations are encouraged to thoroughly test our theory.

Acknowledgements

The research leading to these results has received funding from the European Research Council (ERC) under the European Union's Horizon 2020 research and innovation programme (grant agreement numbers 772225: MULTIPLES). CW acknowledges funding from CSC scholarship. This work has received funding from the European Research Council (ERC) under the European Union's Horizon 2020 research innovation programme (Grant Agreement ERC-StG 2016, No 716082 'GALFOR', PI: Milone, http://progetti.dfa.unipd.it/GALFOR). APM acknowledges support from MIUR through the FARE project R164RM93XW SEMPLICE (PI: Milone) and the PRIN program 2017Z2HSMF (PI: Bedin). HS and JB acknowledge support from the FWO Odysseus program under project G0F8H6N. NC gratefully acknowledges funding from the Deutsche Forschungsgemeinschaft (DFG) - CA 2551/1-1. PM acknowledges support from the FWO junior postdoctoral fellowship No. 12ZY520N. SdM acknowledges funding by the Netherlands Organization for Scientific Research (NWO) as part of the Vidi research program BinWaves with project number 639.042.728.

References

Brott, I., et al. 2011, A&A, 530, A115
Dufton, P. L., et al. 2013, A&A, 550, A109
Korntreff, C., Kaczmarek, T., & Pfalzner, S. 2012, A&A, 543, A126
Marino, A. F., Przybilla, N., Milone, A. P., Da Costa, G., D'Antona, F., Dotter, A., & Dupree, A. 2018, AJ, 156, 116
Milone, A. P., Bedin, L. R., Piotto, G., & Anderson, J. 2009, A&A, 497, 755
Milone, A. P., et al. 2018, MNRAS, 477, 2640
Paxton, B., Bildsten, L., Dotter, A., Herwig, F., Lesaffre, P., & Timmes, F. 2011, ApJS, 192, 3
Paxton, B., et al. 2013, ApJS, 208, 4
—. 2015, ApJS, 220, 15
—. 2019, ApJS, 243, 10
Ramírez-Tannus, M. C., et al. 2021, A&A, 645, L10
Rivinius, T., Carciofi, A. C., & Martayan, C. 2013, A&A Rev., 21, 69
Sana, H., Ramírez-Tannus, M. C., de Koter, A., Kaper, L., Tramper, F., & Bik, A. 2017, A&A, 599, L9
Sana, H., et al. 2012, Science, 337, 444
Schneider, F. R. N., Izzard, R. G., Langer, N., & de Mink, S. E. 2015, ApJ, 805, 20
Schneider, F. R. N., Ohlmann, S. T., Podsiadlowski, P., Röpke, F. K., Balbus, S. A., Pakmor, R., & Springel, V. 2019, Nature, 574, 211
Schootemeijer, A., & Langer, N. 2018, A&A, 611, A75
Tokovinin, A., & Moe, M. 2020, MNRAS, 491, 5158
Wang, C., Langer, N., Schootemeijer, A., Castro, N., Adscheid, S., Marchant, P., & Hastings, B. 2020, ApJL, 888, L12
Wang, C., et al. 2022, Nature Astronomy, 6, 480
Zorec, J., & Royer, F. 2012, A&A, 537, A120

Questions and Answers

Question: How will your findings affect cluster age estimation?

Answer: Previous studies only use single non-rotating star models to estimate cluster age, however, our findings tell us that we need to consider binary interaction and rotation. But we have not quantified how large the difference will be at this moment. Fabian Schneider suggested a difference of up to a factor of 10 merely from binary interaction.

Question: What is the fraction of the blue MS stars, and what is the fraction of the MS stars that are detected to have magnetic fields?

Answer: The fraction of the blue MS stars is around 10–30%, while the fraction of magnetic upper MS stars is around 10%. This difference may come from magnetic decay or perhaps not all binary mergers can create strong magnetic fields.

Massive Stars Near and Far
Proceedings IAU Symposium No. 361, 2024
J. Mackey, J. S. Vink, & N. St-Louis, eds.
doi:10.1017/S1743921322002599

The spin rate properties of Galactic massive O-type stars

G. Holgado[1], **S. Simón-Díaz**[2,3], **A. Herrero**[2,3] **and R. H. Barbá**[4,†]

[1]Centro de Astrobiologia (CAB), CSIC-INTA, Camino Bajo del Castillo s/n, 28692, Villanueva de la Cañada, Madrid, Spain
email: gholgado@cab.inta-csic.es

[2]Instituto de Astrofísica de Canarias, E-38200 La Laguna, Tenerife, Spain

[3]Departamento de Astrofísica, Universidad de La Laguna, E-38205 La Laguna, Tenerife, Spain

[4]Departamento de Física y Astronomía, Universidad de La Serena, Avenida Juan Cisternas 1200, La Serena, Chile.

Abstract. We present results from a recent study of the spin rate properties of a sample of more than 400 Galactic O-type stars surveyed by the IACOB and OWN projects. We combine vsini, Teff, and logg estimates with information about the spectroscopic binarity status for 285 of the stars in the sample, and provide a renewed overview about how the empirical distribution of projected rotational velocities in the O-star domain depends on mass, evolutionary and binary status. The obtained distributions are then compared with predictions of state-of-the-art population synthesis simulations including binary interaction, and used to provide hints about the initial velocity distribution of stars with masses in the range 15-80 $M_\odot$.

Keywords. Stars: early-type – Stars: rotation – Techniques: spectroscopic – Catalogs – The Galaxy

1. Introduction

Stellar rotation affects the global physical characteristics and the evolution of massive stars (Maeder & Meynet 2000; Langer 2012). As a massive star evolves through the main sequence, in single stars the equatorial velocity usually decreases monotonically, and the net braking rate of the stellar surface depends on the angular momentum losses due to stellar winds and the efficiency of internal transport of angular momentum from the core to the envelope. In binary systems – specially close enough to interact – several effects combine (tides and mass transfer events) which are capable of both decelerating or accelerating each of the individual components (de Mink *et al.* 2013, and references therein). Numerous studies have obtained and analyzed spectroscopic data sets of difference size and quality to advance in our knowledge of rotational properties of massive stars (Conti & Ebbets 1977; Penny 1996; Howarth *et al.* 1997; Penny & Gies 2009; Simón-Díaz and Herrero 2014; Markova *et al.* 2014; Ramírez-Agudelo *et al.* 2013, 2015).

As a continuation of the IACOB series of papers, devoted to the study of more than 400 Galactic O-type stars with IACOB and OWN high-quality, multi-epoch spectra (last described in Simón-Díaz *et al.* 2020; Barbá *et al.* 2017), *we focus on a re-evaluation of the spin-rate properties of the Galactic main sequence stars with masses in the range* $\sim$*15 – 80* $M_\odot$.

† deceased

Table 1. Main characteristics of several reference studies investigating the **rotational** properties of high-mass stars in the Milky Way (MW) and the Large/Small Magellanic Clouds (LMC/SMC) covering the whole O-star domain.

Study	Galaxy	Sample size (O-stars)	$v \sin i/v_{\rm mac}$ separation	Spectroscopic parameters	Specific SB study
This work	MW	285	Yes	Yes	SB1
Conti & Ebbets (1977)	MW	205	No	No	No
Penny (1996)	MW	177	No	No	No
Howarth *et al.* (1997)	MW	225	No	No	No
Simón-Díaz and Herrero (2014)	MW	116	Yes	No	No
Markova *et al.* (2014)	MW	39	Yes	Yes	No
Penny & Gies (2009)	SMC/LMC	39/70	No	No	No
Ramírez-Agudelo *et al.* (2013, 2015)	LMC	326	Yes	No	SB1, SB2

We use the outcome of a homogeneous quantitative spectroscopic analysis performed in the context of the IACOB-project (Simón-Díaz et al. 2011; Simón-Díaz and Herrero 2014) and information on the spectroscopic binary status (Holgado Alijo 2019). We include the global properties of the sample, the interdependence between $v \sin i$ and evolution, an evaluation of these results in the context of the binary population synthesis simulation presented by de Mink *et al.* (2013), and a study on empirical constraints to our theories of massive star formation.

2. Overview and comparison with previous studies

The final sample of targets considered for this work was drawn from an initial sample of 415 Galactic O-type stars included in the IACOB and OWN database of high-resolution ($R = 25000 - 85000$), multi-epoch spectroscopic observations. To avoid too much repetition, we refer the reader to Holgado *et al.* (2020) for a detailed description of how this sample was built. We excluded all those stars clearly identified as double line spectroscopic binaries (SB2, 113 objects), as well as targets presenting features in their spectra which prevent from a successful fitting. They were associated with Oe (6), Wolf-Rayet (4) or magnetic (7) peculiarities. As a result, we ended up with a final sample of 285 stars characterized as either likely-single (LS) or single line spectroscopic binaries (SB1).

The tools and methodology we have followed to perform the quantitative spectroscopic analysis of our working sample of 285 O-type stars can be found in Holgado *et al.* (2018, 2020). In summary, we obtain two line-broadening parameters ($v \sin i$ and $v_{\rm mac}$) using the IACOB-BROAD tool (Simón-Díaz and Herrero 2014) and following the guidelines presented in Holgado *et al.* (2018). In particular, we based our $v \sin i$ determination on the O III λ5592 line in the vast majority of stars.

Our study represents an updated and improved empirical overview of the spin-rate properties of Galactic O-type stars with respect to several commonly referenced studies found in the literature, table 1. It is a clear improvement in size and availability of additional parameters ($T_{\rm eff}$ and $\log g$), and supersede the derived $v \sin i$ with new (more robust) estimates, with an accurate identification of spectroscopic binaries. In addition, it complements the investigation of the rotational velocities of a sample of O-type stars in the 30 Dor region of the LMC performed by Ramírez-Agudelo *et al.* (2013, 2015) in the framework of the VLT-FLAMES Tarantula Survey (VFTS, Evans *et al.* 2011).

3. Bimodality

Figure 1 (left) depicts the resulting $v \sin i$ distribution separated by luminosity class groups. It shows two main features:

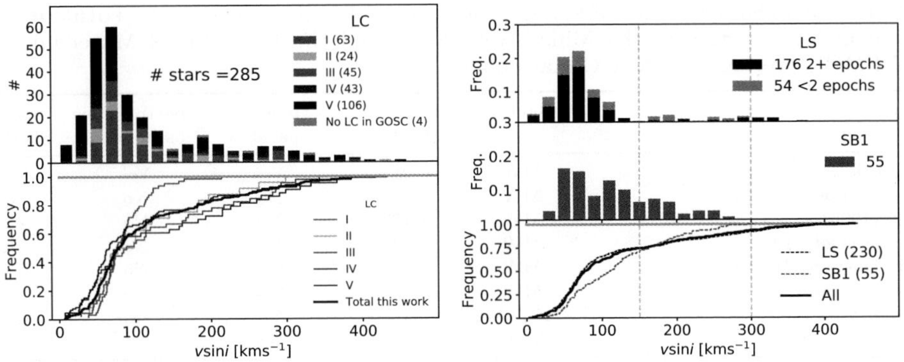

Figure 1. *(Left)* Histogram (*top*) and cumulative distribution (*bottom*) of $v \sin i$ estimates for the 285 O-type stars investigated in this work, separating results for the various luminosity class subgroups. *(Right)* Comparison of the $v \sin i$ distributions of the likely single (LS) and SB1 subsamples, presented as histograms *(top and middle panels, respectively)* and cumulative distributions *(bottom panel)*. Stars in the LS sample for which we only have 1 spectrum are marked in grey.

A clear bimodal character, with a slow velocity peak comprising most of the stars (up to $v \sin i \sim 150$ km s^{-1}), and a tail of fast rotators, reaching $v \sin i$ values up to ~ 450 km s^{-1}.

Different $v \sin i$ range coverage of the various LC groups. The dwarfs (LC V) are distributed all over the whole range and the supergiants (LC I) concentrated within the low velocity peak.

Recent studies have proposed that the bimodal character of the $v \sin i$ distribution is a natural consequence of the effect of binary interaction during massive star evolution (de Mink *et al.* 2013). Our sample include 170 binaries (113 SB2 and 55 clearly detected SB1 stars) and 230 likely single. While the SB2 systems were discarded for this work, we investigate differences between our LS and SB1 sub-samples.

Right panel in Fig. 1 shows the distribution of likely single stars is very similar to the combined one, meaning that they are clearly dominating the global distribution. But the comparison of the LS and SB1 distributions highlight clear differences. The SB1 distribution appears as a single and wider component than the LS stars. In addition, the later is centered at a larger $v \sin i$ (120 vs. 65 km s^{-1}), and there is a lack of SB1 stars with projected rotational velocities larger than 300 km s^{-1}. These differences are likely produced by a combination of effects taking place along the various stages of early evolution of high-mass binaries (such as tides, mass/angular momentum transfer and magnetic braking; see further notes below).

3.1. *Interpreting the effect of binary interactions*

The evolution of the individual spin rates in a binary system may be affected by several effects not present in isolated stars, which could give rise to a noticeable acceleration of the individual stars. We provide interpretation of our $v \sin i$ distribution of O-type stars using as a guideline the predictions from the binary population synthesis simulations performed by de Mink *et al.* (2013). They showed that binary interaction offers a natural explanation for the bimodal characteristic of the distribution.

We note that de Mink *et al.* (2013) simulation includes SB2 stars that have not been considered in our observational sample and that the difference in metallicity –LMC for de Mink's study– could lead to a higher percentage of high spin objects in their simulations. Considering these, the agreement seems to be surprisingly good.

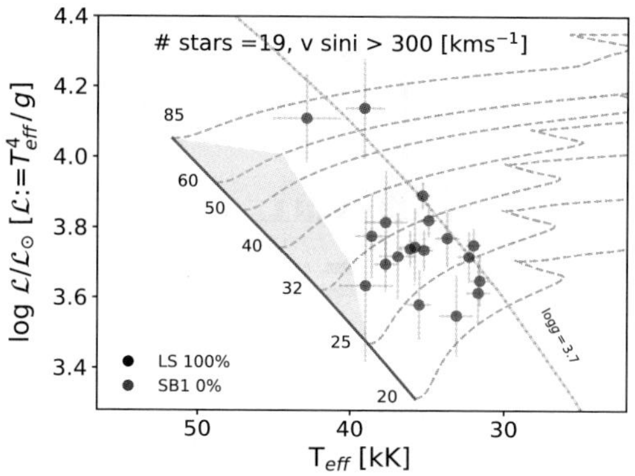

Figure 2. Spectroscopic HR diagram indicating the location of the 19 Galactic O-type stars with $v\sin i$>300km s^{-1}. All of them LS stars. Uncertainties are included as error bars. Evolutionary tracks and position of the ZAMS from the non-rotating, solar metallicity models by Ekström *et al.* (2012), plus the log g=3.7 dex line of constant gravity, are included for reference. The grey shadowed area highlights a region of the diagram which is void of stars (see main text and Holgado *et al.* 2020, for further explanations).

De Mink's simulation is also able to explain the clear dominance of apparently single stars in the tail of fast rotators, with their tail of fast rotators mostly populated by post-interaction products of binaries de Mink, Langer, & Izzard (2011); de Mink *et al.* (2014).

It also explains the upper limit in $v\sin i$ of ∼300 km s^{-1} found in the sample of clearly detected SB1 stars. Following de Mink *et al.* (2014), our SB1 star sample is dominated by pre-interaction objects – i.e. previous to mass exchange –, and tidal interactions produce a tail of accelerated stars that reaches just up to 300 km s^{-1}. These tidal interactions and orbital-rotation synchronization may also be the origin of the difference between the distributions of single and SB1 stars (Fig. 1).

These characteristics are also found in the work by Ramírez-Agudelo *et al.* (2013, 2015), with small differences (e.g., they have higher $v\sin i$ velocities) and without considering stellar masses and effective temperatures. These authors deepened into the empirical existence of the bimodal distribution of $v\sin i$ and the differences between apparent single and binary components (mainly primaries) distributions. They also used the works by de Mink *et al.* (2013, 2014) for the interpretation of their results, finding good agreement between observations and simulations. Our study reproduces these findings.

4. Binary interaction and rejuvenation

We find interesting to remark how the 19 stars with $v\sin i > 300$ km s^{-1}, seen in Fig. 2, are mostly concentrated in the lower mass region of the O-star domain (except for two targets –HD 14442 (O5n(f)p) and HD 229232 (O4 V)– all of them are located below the 40 $M_{\odot}$ evolutionary track). All of them likely single stars. We investigate the specific location in the sHRD of these stars in the context of the simulations including binary interaction in the literature.

As already discussed, the origin of such a high velocity is probably the binary interaction. Mass accretion or merger during the evolution of the binary systems may produce a rejuvenation and spin up of the initial secondary star or the final product (if no additional mechanisms like the generation of magnetic fields play a role, de Mink *et al.* 2013, 2014). They may even appear as blue stragglers in a cluster when compared to the rest of the population (Ahumada & Lapasset 2007; Schneider *et al.* 2014, 2015, 2016; Wang *et al.* 2020). The works of Ahumada & Lapasset (2007) and Schneider *et al.* (2016) find that the

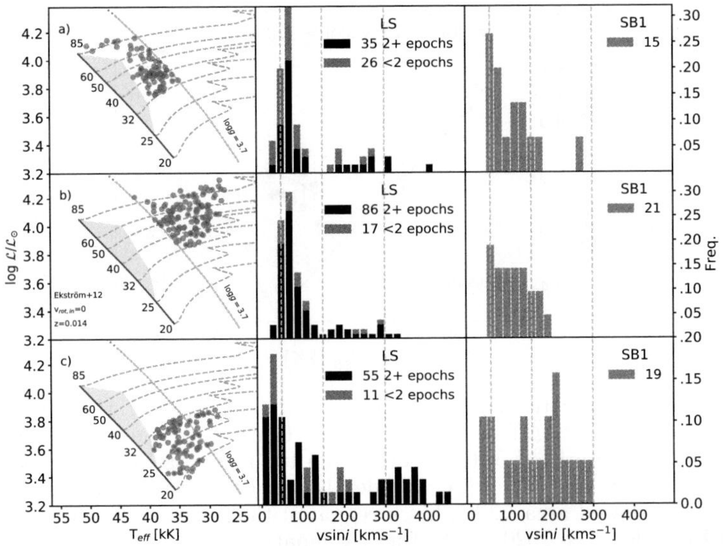

Figure 3. Rotational properties of our working sample of Galactic O-type stars separated in three groups (*from top to bottom*) comprising stars covering different ranges in mass and evolutionary status. *[Left panels]* Location of each subroup in the sHRD. *[Middle and right panels]* Corresponding $v \sin i$ distributions for the LS and SB1 samples, respectively. In each panel, vertical dashed lines mark 50, 150 and 300 $\mathrm{km\,s^{-1}}$.

number of blue-stragglers is larger and their rejuvenation more intense for smaller masses, and more evolved clusters. This, together with a faster evolution at higher masses, could explain the concentration of very fast rotators that we see below 32 $M_{\odot}$.

5. $v \sin i$ and evolution

We investigate how the distribution of projected rotational velocities in the O star domain depends on mass and evolution. To this aim, we split the global sample in three subsamples accordingly to their location in the sHRD. We used the 32 $M_{\odot}$ non-rotating evolutionary track computed by Ekström *et al.* (2012) to roughly separate the sample in two mass ranges. Stars with higher masses are expected to develop winds strong enough to produce significant losses of angular momentum even during the main sequence. Then, we further divided the high mass sample in two, using the $\log g = 3.7$ dex line of constant gravity, which roughly separate the dwarf stars from the more evolved giants and supergiants

We denote the two high-mass subgroups as "a" and "b", while group "c" are below the 32 $M_{\odot}$ evolutionary track. In a single star evolution context, stars and $v \sin i$ distributions in the group "a" are progenitors of those in group "b". We highlight the main characteristics (from a purely empirical point of view) of the $v \sin i$ distributions presented in Fig. 3 and refer the reader to Holgado et al. (2022, accepted) for a further discussion and interpretation of results taking into account our present theoretical knowledge about single and binary star evolution of high-mass stars.

The most important take away message from this study is that we do not detect a strong braking effect on the evolution of massive stars, even above the 32 $M_{\odot}$ evolutionary track. This empirical result has far-reaching implications for our better understanding of the impact that angular momentum losses due to stellar winds, as well as several proposed mechanisms which could transport angular momentum from the stellar core to the envelope, have on the evolution of surface rotational velocities in stars with masses above 32 $M_{\odot}$.

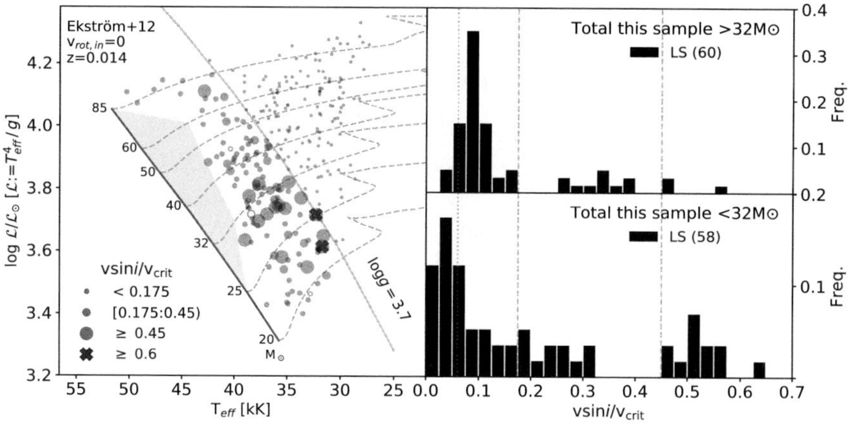

Figure 4. (*Left*) sHRD highlighting a subsample of 118 LS O-type stars with $\log g > 3.7$ dex. The symbol size depends on the value of the ratio $v \sin i/v_{\mathrm{crit}}$, as indicated. *(Right)* $v \sin i/v_{\mathrm{crit}}$ distributions (as normalized histograms) for the indicated subsample separating those stars located above (upper panel) and below (lower panel) the $32 M_{\odot}$ track. The vertical dashed lines mark several values of interest for the $v \sin i/v_{\mathrm{crit}}$ ratio (see text for explanation).

6. Initial velocity

The ratio of initial equatorial velocity (v_{ini}) against the critical velocit(v_{crit}) is an influential input parameter for any evolutionary model of massive stars (see review by Maeder & Meynet 2000, and references therein). To construct an estimated $v \sin i/v_{\mathrm{crit}}$ distribution we select those stars that have evolved as little as possible since birth, having $\log g > 3.7$ dex, using gravity again as a proxy for evolution. We rely on the result that the $v \sin i$ distribution tends to maintain shape during evolution, see above. Lastly, we discard SB1 stars as it is likely that their v_{ini} is affected by the influence of their companions. Left panel of Fig. 4 presents the location of subsample in the sHRD, and the right panel shows the associated $v \sin i/v_{\mathrm{crit}}$ histograms, separating further the investigated subsample in two (considering those stars located above and below the evolutionary track of 32 $M_{\odot}$).

Both histograms show a peak near $v \sin i/v_{\mathrm{crit}} = 0.1$, and two low and high velocity components separated by a gap. We associate this to the effect of binary evolution. The region with $v \sin i/v_{\mathrm{crit}} < 0.175$ would be dominated by single stars or wide binaries which have suffered little evolution and are representative of the initial spin-rate distribution, and the fast rotating components are most likely populated by post-interacting binaries. Therefore, this high velocity components should not be associated with the initial spin-rate conditions of stars at birth.

We conclude that O stars in their early stages have velocities between $v \sin i/v_{\mathrm{crit}} \sim 0.1$-0.175 and are typically born with rotational velocities near $v_{\mathrm{ini}}/v_{\mathrm{crit}} \sim 0.2$.

7. Conclusions

This study is a step forward for rotational properties of O stars, with enough stars to contrast evolutionary subsamples within it. We found significant differences between $v \sin i$ distributions of LS and SB1 stars, possibly related to binary interactions. Our empirical data allow us to conclude that (1) no important braking effect of the stellar surface is detected as O stars evolve along the MS, and (2) single massive O-type stars are likely born with equatorial velocities not larger than 20% of critical velocity.

References

Ahumada, J.A. and Lapasset, E.: 2007, *A&A* **463**, 789.

Barbá, R.H., Gamen, R., Arias, J.I., and Morrell, N.I.: 2017, *The Lives and Death-Throes of Massive Stars* **IAUs 329**, 89.

Conti, P.S. and Ebbets, D.: 1977, *ApJ* **213**, 438.

de Mink, S.E., Langer, N., and Izzard, R.G.: 2011, *Bulletin de la Societe Royale des Sciences de Liege* **80**, 543.

de Mink, S.E., Langer, N., Izzard, R.G., Sana, H., and de Koter, A.: 2013, *ApJ* **764**, 166.

de Mink, S.E., Sana, H., Langer, N., Izzard, R.G., and Schneider, F.R.N.: 2014, *ApJ* **782**, 7.

Ekström, S., Georgy, C., Eggenberger, P., Meynet, G., Mowlavi, N., Wyttenbach, A., et al.,: 2012, *A&A* **537**, A146.

Evans, C.J., Taylor, W.D., Hénault-Brunet, V., Sana, H., de Koter, A., Simón-Díaz, S., et al.,: 2011, *A&A* **530**, A108.

Holgado Alijo, G.: 2019, *Ph.D. Thesis.*

Holgado, G., Simón-Díaz, S., Barbá, R.H., Puls, J., Herrero, A., Castro, N., et al.,: 2018, *A&A* **613**, A65.

Holgado, G., Simón-Díaz, S., Haemmerlé, L., Lennon, D.J., Barbá, R.H., Cerviño, M., et al.,: 2020, *A&A* **638**, A157.

Howarth, I.D., Siebert, K.W., Hussain, G.A.J., and Prinja, R.K.: 1997, *MNRAS* **284**, 265.

Langer, N.: 2012, *ARAA* **50**, 107.

Langer, N. and Kudritzki, R.P.: 2014, *A&A* **564**, A52.

Maeder, A. and Meynet, G.: 2000, *A&A* **361**, 159.

Markova, N., Puls, J., Simón-Díaz, S., Herrero, A., Markov, H., and Langer, N.: 2014, *A&A* **562**, A37.

Penny, L.R.: 1996, *ApJ* **463**, 737.

Penny, L.R. and Gies, D.R.: 2009, *ApJ* **700**, 844.

Ramírez-Agudelo, O.H., Simón-Díaz, S., Sana, H., de Koter, A., Sabín-Sanjulían, C., de Mink, S.E., et al.,: 2013, *A&A* **560**, A29.

Ramírez-Agudelo, O.H., Sana, H., de Mink, S.E., Hénault-Brunet, V., de Koter, A., Langer, N., et al.,: 2015, *A&A* **580**, A92.

Schneider, F.R.N., Izzard, R.G., de Mink, S.E., Langer, N., Stolte, A., de Koter, A., et al.,: 2014, *ApJ* **780**, 117.

Schneider, F.R.N., Izzard, R.G., Langer, N., and de Mink, S.E.: 2015, *ApJ* **805**, 20.

Schneider, F.R.N., Podsiadlowski, P., Langer, N., Castro, N., and Fossati, L.: 2016, *MNRAS* **457**, 2355.

Simón-Díaz, S., Castro, N., Herrero, A., Puls, J., Garcia, M., and Sabín-Sanjulián, C.: 2011, *Journal of Physics Conference Series* **328**, 012021.

Simón-Díaz, S. and Herrero, A.: 2014, *A&A* **562**, A135.

Simón-Díaz, S., Britvaskiy, N., Castro, N., and Holgado, G.: 2020, *XIV.0 Scientific Meeting (virtual) of the Spanish Astronomical Society*, 186.

Wang, C., Langer, N., Schootemeijer, A., Castro, N., Adscheid, S., Marchant, P., et al.,: 2020, *ApJ* **888**, L12.

Discussion

S.E. DE MINK: You say O stars are born with less than 30% of critical rotation. Some speakers have shown stars with lower mass rotating at much higher rotation. How is this difference explained? Is it mass?

G. HOLGADO: I do not know about that. This is a large sample, maybe some peculiar objects that people study are just that, peculiar ones. My summary is that to explain the vast majority of the distribution most (O stars) should be born with 0.2 $v \sin i / v_{\rm crit}$ maximum.

Massive Stars Near and Far
Proceedings IAU Symposium No. 361, 2024
J. Mackey, J. S. Vink, & N. St-Louis, eds.
doi:10.1017/S1743921322002204

Near and far: a hunt for binary-interaction products

J. Bodensteiner[1,2], H. Sana[2], L. Mahy[3], T. Shenar[4], G. Banyard[2] and C. Wang[5]

[1]European Southern Observatory, Karl-Schwarzschild-Str. 2, 85738 Garching bei M/"unchen, Germany
email: julia.bodensteiner@eso.org

[2]Institute for Astronomy, KU Leuven, Celestijnenlaan 200 D, 3001, Leuven, Belgium

[3]Royal Observatory of Belgium, Avenue Circulaire 3, B-1180 Brussel, Belgium

[4]Anton Pannekoek Institute for Astronomy, University of Amsterdam, Postbus 94249, 1090 GE Amsterdam, The Netherlands

[5]Max Planck Institute for Astrophysics, Karl-Schwarzschild-Str. 1, 85748 Garching, Germany

Abstract. Direct observations of the products of binary interactions are sparse, yet they provide important insights on the outcome of the interaction and the physics at play. Young and intermediate-age star clusters are the ideal tool to search for, and characterize such interaction products and allow for a detailed comparison to theoretical predictions. We here report on integral field spectroscopy obtained with MUSE for several such clusters in the Magellanic Clouds.

Keywords. stars: early-type, stars: binaries, galaxies: Magellanic Clouds

1. Theoretical predictions for stellar populations

Massive stars predominantly live their lives in binary systems in which many are close enough to interact (Sana et al. 2012; de Mink et al. 2014). Depending on the initial orbital parameters, different evolutionary pathways are possible (e.g. Podsiadlowski et al. 1992) which can lead to different end products.

In a given stellar population, the number of interaction products should be signicant. Binary population synthesis studies predict observable characteristics for post-interaction populations: a lower binary fraction, merger products above the cluster turnoff, and a population of rapid rotators (e.g., Wang et al. 2020).

2. Star clusters as laboratory to study pre- and post-interaction populations

To assess initial multiplicity properties and binary parameter distributions, B-star populations in young clusters including NGC 6231 (Banyard et al. 2022) in the Milky Way and 30 Doradus (Dufton et al. 2013; Dunstall et al. 2015) in the Large Magellanic Cloud (LMC) were studied. The number of interaction products is, however, predicted to peak in intermediate-age clusters such as the 35 Myr-old cluster NGC 330 in the Small Magellanic Cloud. The different host metallicities make a direct comparison of the observed clusters to theoretical predictions difficult. This will be remedied by a new, homogeneous set of MUSE observations of several clusters at LMC metallicity spanning an age range from 20 to 40 Myr (Fig. 1).

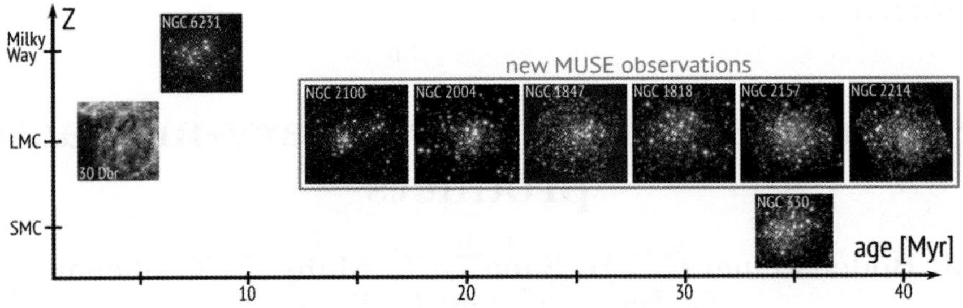

Figure 1. Observations of young massive star clusters in the Milky Way and the Magellanic Clouds.

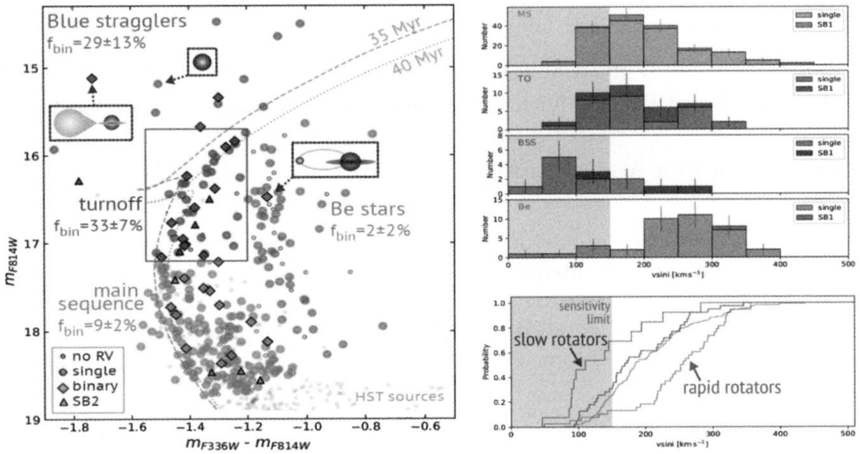

Figure 2. Observed binary fraction and rotational velocities in different CMD regions.

3. NGC 330 - a pilot study

Using six epochs of observation with integral-field spectrograph MUSE at the VLT we extract spectra of $>$ 350 stars (Bodensteiner et al. 2020b), measure radial velocities and compute the bias-corrected binary fraction (Bodensteiner et al. 2021). With additional HST photometry (Milone et al. 2018) and TLUSTY atmospheric models (Lanz & Hubeny 2007) we estimate effective temperatures, surface gravities and rotational velocities for 200 B and Be stars (Bodensteiner et al. in prep.).

The fraction of Be stars (which are often interpreted as mass gainers in binary interactions, e.g. Pols et al. 1991; Bodensteiner et al. 2020a) is $>$30%, and even $>$50% around the cluster turnoff. The overall bias-corrected binary fraction is f_{bin}=34±8%. The color-magnitude-diagram (CMD) shows that the observed binary fraction varies strongly as a function of CMD position, as expected for a population of post-interaction stars. Additional interaction products could be the slowly rotating single blue stragglers, possible merger candidates (see Fig. 2).

4. Conclusions

The new MUSE observations of six intermediate-age LMC clusters will allow us to search for binary interaction products, to map global properties (f_{bin}, rotational velocity distribution, Be star fraction) as a function of cluster age, and to test binary population synthesis predictions.

References

Banyard, G., Sana, H., Mahy, L., Bodensteiner, J., Villaseñor, J. I., & Evans, C. J. 2022, A&A, 658, A69
Bodensteiner, J., Sana, H., Wang, C., & et al. in prep.
Bodensteiner, J., Shenar, T., & Sana, H. 2020a, A&A, 641, A42
Bodensteiner, J., et al. 2020b, A&A, 634, A51
—. 2021, A&A, 652, A70
de Mink, S. E., Sana, H., Langer, N., Izzard, R. G., & Schneider, F. R. N. 2014, ApJ, 782, 7
Dufton, P. L., et al. 2013, A&A, 550, A109
Dunstall, P. R., et al. 2015, A&A, 580, A93
Lanz, T., & Hubeny, I. 2007, ApJS, 169, 83
Milone, A. P., et al. 2018, MNRAS, 477, 2640
Podsiadlowski, P., Joss, P. C., & Hsu, J. J. L. 1992, ApJ, 391, 246
Pols, O. R., Cote, J., Waters, L. B. F. M., & Heise, J. 1991, A&A, 241, 419
Sana, H., et al. 2012, Science, 337, 444
Wang, C., Langer, N., Schootemeijer, A., Castro, N., Adscheid, S., Marchant, P., & Hastings, B. 2020, ApJL, 888, L12

Massive Stars Near and Far
Proceedings IAU Symposium No. 361, 2024
J. Mackey, J. S. Vink, & N. St-Louis, eds.
doi:10.1017/S1743921322002903

Formation of black holes in the pair-instability mass gap: Hydrodynamical simulation of a massive star collision & evolution of a post-collision star

Guglielmo Costa[1,2,3], **Alessandro Ballone**[1,2,3], **Michela Mapelli**[1,2,3], **Alessandro Bressan**[4] and **Morgan MacLeod**[5]

[1]Physics and Astronomy Department Galileo Galilei, University of Padova, Vicolo dell'Osservatorio 3, I-35122 Padova, Italy
email: guglielmo.costa@unipd.it

[2]INAF - Osservatorio Astronomico di Padova, Vicolo dell'Osservatorio 5, I-35122 Padova, Italy

[3]INFN - Padova, Via Marzolo 8, I–35131 Padova, Italy

[4]SISSA, via Bonomea 365, I–34136 Trieste, Italy

[5]Center for Astrophysics | Harvard & Smithsonian 60 Garden Street, MS-16, Cambridge, MA 02138, USA

Abstract. In this work, we study in detail the collision formation scenario of black holes (BHs) which lie in the pair-instability (PI) mass gap. We study the collision scenario of two massive stars by means of a smoothed-particle hydrodynamics (SPH) simulation and the post-collision evolution with detailed stellar evolutionary codes. We find that the stellar collision scenario is a suitable formation channel to populate the BHs' PI mass gap.

Keywords. stars: evolution, hydrodynamics, black hole physics

---

1. Introduction and SPH collision simulation

The detection of GW190521 by the LIGO-Virgo collaboration revealed the existence of black holes (BHs) in the pair-instability (PI) mass gap. Here, we investigate the formation of BHs in the PI mass gap via star-star collisions in young stellar clusters. To avoid PI, the stellar collision product must have a relatively small core and a massive envelope. We investigated this issue by means of the SPH code STARSMASHER (Gaburov et al. 2010) and detailed stellar evolutionary models with PARSEC (Costa et al. 2019) and MESA (Paxton et al. 2019). In this work, we simulated the collision of a core helium burning (CHeB) star of $\sim 58\ M_\odot$ with a main-sequence (MS) star of $\sim 42\ M_\odot$.

In Ballone et al. (2022), we created the initial condition of the two stars using PARSEC. Then, STARSMASHER re-samples these profiles with particles distributed in the 3D space. The CHeB and MS stars are sampled with 8×10^5 and 9×10^4 particles, respectively. Then we put the two stars on a hyperbolic radial orbit with a velocity at infinity of 10 km s^{-1} and an initial separation of 110 $R_\odot$. We simulate a head-on collision to probe the most extreme case in terms of kinetic energy and to obtain an upper limit to the mass lost in the collision. As the MS star plunges into the atmosphere of the CHeB star, its outer layers form a strong shock on the frontal side of the collision, while they lead to a cometary tail on the back side. Then, the MS star is tidally disrupted by the core of

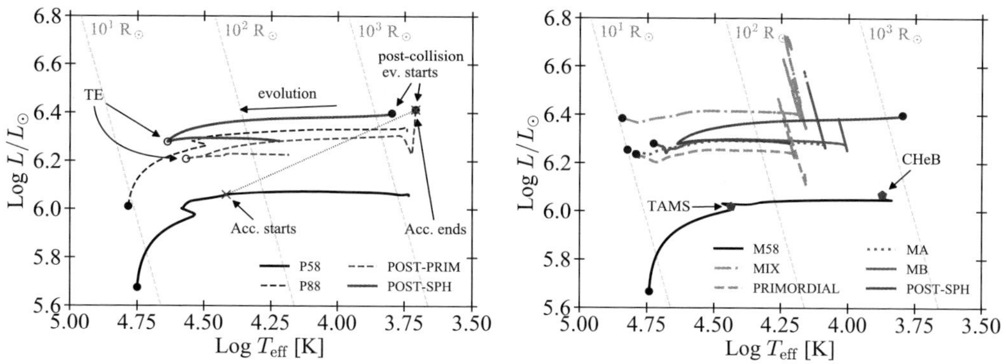

Figure 1. HR diagram of post-collision PARSEC (MESA) tracks in the left (right) panel. Solid black lines indicate the primary star P58 (M58) for PARSEC (MESA). The dashed black line on the left panel shows the 'standard' PARSEC track with 88 $M_\odot$. The red crosses mark the starting and ending points of the accretion. The blue points (empty circles) indicate the post-collision evolution starting (thermal equilibrium, TE) points. Details on the text.

the CHeB star. During collision, up to 12% of the total mass is lost. At the end of the simulation, the post-collision star has a mass of 88 $M_\odot$ and a helium core of 28 $M_\odot$.

2. Post-Collision evolution and Final masses

In Costa et al. (2022), we used the output of the SPH simulation to study the evolution of the collision product with PARSEC and MESA. We build the PARSEC model POST-PRIM by accreting mass on the primary CHeB star until its total mass is 88 $M_\odot$. In this model, we maintain the pristine chemical composition of the envelope. Starting from the POST-PRIM model, we build the POST-SPH model by changing the chemical composition of the envelope to match that of the SPH simulation. Both post-collision stars start their evolution as red supergiants and evolve to become blue supergiants (BSG). After reaching thermal equilibrium (TE), they evolve toward the red part of the HR diagram until they explode (Figure 1). Similarly, we build the post-collision MESA stellar tracks, accreting mass and matching the SPH simulation envelope composition. The model MA is built with the primary in the terminal age main sequence phase (TAMS). The MB model is built with the primary in the CHeB phase. Remarkably, the PARSEC and MESA stellar models evolve in a very similar way after TE, and finish their lives as BSGs (Figure 1). All of them avoid PI and evolve until the final core collapse. We estimate the final BHs masses taking into account the mass ejected during the final collapse, due to shocks induced by neutrino loss; we find that all our models lose less than 0.5 $M_\odot$ during the final collapse because of the relatively high compactness of the stellar envelope. Therefore, we expect that all our models produce BHs with mass ~ 87 $M_\odot$, within the PI mass gap.

Acknowledgements

AB, GC and MM acknowledge financial support from the European Research Council for the ERC Consolidator grant DEMOBLACK, under contract no. 770017. M. MacLeod acknowledges support by the US National Science Foundation under Grant No. 1909203.

Supplementary material

To view supplementary material for this article, please visit http://dx.doi.org/10.1017/S1743921322002903.

References

Ballone, A., Costa, G., Mapelli, M., & MacLeod, M. 2022, arXiv e-prints, arXiv:2204.03493
Costa, G., Ballone, A., Mapelli, M., & Bressan, A. 2022, arXiv e-prints, arXiv:2204.03492
Costa, G., et al. 2019, A&A, 631, A128
Gaburov, E., Lombardi, James C., J., & Portegies Zwart, S. 2010, MNRAS, 402, 105
Paxton, B., et al. 2019, ApJS, 243, 10

Massive Stars Near and Far
Proceedings IAU Symposium No. 361, 2024
J. Mackey, J. S. Vink, & N. St-Louis, eds.
doi:10.1017/S174392132200206X

Constraining the dynamical mass of the massive binary 9 Sagittarii

Matthias Fabry[1], Calum Hawcroft[1], Abigail J. Frost[1], Laurent Mahy[2,1], Pablo Marchant[1], Jean-Baptiste Le Bouquin[3] and Hugues Sana[1]

[1]Institute of Astronomy (IvS), Celestijnenlaan 200D, 3001 Leuven, Belgium
email: matthias.fabry@kuleuven.be

[2]Royal Observatory of Belgium (ROB), Avenue Circulaire 3, 1180 Brussels, Belgium

[3]Institute of Planetology and Astrophysics (IPAG), Grenoble University, Rue de la Piscine 414, 38400 St-Martin d'Hères, France

Abstract. Especially in the upper Hertzsprung-Russell diagram, where stellar physics is least understood, obtaining model independent masses is of great value. Spectroscopic binaries that are also resolved astrometrically are an excellent alternative to eclipsing double-lined spectroscopic binaries where dynamical masses can be measured. 9 Sgr is such a massive binary. However, its characterization is troubled by conflicting conclusions from the spectroscopic analysis on the one hand and the interferometric one on the other hand. In this work, we attempt to resolve this tension by applying a novel approach to spectral disentangling of the spectroscopic data to constrain better the mass of 9 Sgr.

Keywords. stars: early-type, stars: binaries, stars: fundamental parameters

1. Introduction

9 Sgr (HD 164794) is a long-period ($P \approx 9\,\mathrm{yr}$) double-lined spectroscopic binary in the Lagoon nebula. From both the radio synchrotron emission (Abbott et al. 1984) and elevated X-ray emission Rauw et al. (2002a,b); Nazé et al. (2008), it was determined this system contains two massive star components, the winds of which collide and create non-thermal radiation(Pittard & Dougherty 2006; Rauw et al. 2002c; Sana et al. 2004, 2006; Rauw & Nazé 2016). Rauw et al. (2012) determined the first spectroscopic orbit of 9 Sgr and showed that, if the stars are assumed to have masses of similar to their inferred spectral type, the inclination of 9 Sgr should be around 45 degrees. This was later challenged by the interferometric analysis of Le Bouquin et al. (2017), who derived an inclination of 85 degrees, with tight bounds.

In our work (Fabry et al. 2021), we attempt to resolve the existing tension on the mass of 9 Sgr, using the fact that spectral disentangling can reveal whether the radial velocity (RV) semi-amplitudes of Rauw et al. (2012) are representative of the true RV curves.

2. Methods

We rederive the geometrical orbit using archival and new interferometry from PIONIER (Le Bouquin et al. 2011) and GRAVITY (Gravity Collaboration et al. 2017). Similar to Le Bouquin et al. (2017), we find a very high orbital inclination of 86.5 ± 0.5 degrees. Next, we use the technique of spectral disentangling (Simon et al. 1994; Hadrava 1995) to determine the RV semi-amplitudes of the binary, using the Fourier disentangling code from Ilijić (2017). We take a grid-disentangling approach, in which we fix most of the orbital parameters to the ones found from the interferometric analysis, leaving only the semi-amplitudes K_1 and K_2 free. Our method constrains $(K_1, K_2) = (36, 49)$ km s^{-1}, which is significantly different than the values $(K_1, K_2) = (25, 38)$ km s^{-1} of Rauw et al. (2012, 2016). With these RV semi-amplitudes, the dynamical masses inferred are $53^{+7}_{-6} M_\odot$ and $39^{+6}_{-3} M_\odot$ for the primary and secondary respectively, in accordance with their updated spectral type: O3V and O5V.

Subsequent atmospheric analysis with FASTWIND (Santolaya-Rey et al. 1997) and evolutionary modeling by comparing with the stellar evolution tracks of Brott et al. (2011) through BONNSAI (Schneider et al. 2014) shows the components of 9 Sgr are most likely coeval at about 1 Myr and have temperatures and mass loss rates typical for their dynamical mass. This is with the caveat of poorly constrained surface gravities, which is most likely a result of the repeated normalization: first when reducing the raw spectroscopic data and second after the disentangling procedure, which leaves the continuum undetermined.

3. Conclusions

In our work (Fabry et al. 2021), we resolved the tension present in the literature concerning the dynamical mass of the massive binary 9 Sgr. As one of the most massive galactic binaries whose mass is determined model independently, 9 Sgr serves as an important cornerstone when calibrating stellar models in the upper HRD.

Supplementary material

To view supplementary material for this article, please visit http://dx.doi.org/10.1017/S174392132200206X.

References

Abbott, D. C., Bieging, J. H., & Churchwell, E. 1984, ApJ, 280, 671
Brott, I., et al. 2011, A&A, 530, A115
Fabry, M., Hawcroft, C., Frost, A. J., Mahy, L., Marchant, P., Le Bouquin, J.-B., & Sana, H. 2021, A&A, 651, A119
Gravity Collaboration et al. 2017, A&A, 602, A94
Hadrava, P. 1995, A&AS, 114, 393
Ilijić, S. 2017, ASCL, ascl:1705.012
Le Bouquin, J.-B., et al. 2011, A&A, 535, A67
-. 2017, A&A, 601, A34
Nazé, Y., Becker, M. D., Rauw, G., & Barbieri, C. 2008, A&A, 483, 543
Pittard, J. M., & Dougherty, S. M. 2006, MNRAS, 372, 801
Rauw, G., & Nazé, Y. 2016, AdSpR, 58, 761
Rauw, G., Nazé, Y., Gosset, E., Stevens, I. R., Blomme, R., Corcoran, M. F., Pittard, J. M., & Runacres, M. C. 2002a, A&A, 395, 499
Rauw, G., Sana, H., Spano, M., Gosset, E., Mahy, L., De Becker, M., & Eenens, P. 2012, A&A, 542, A95
Rauw, G., Vreux, J.-M., Stevens, I. R., Gosset, E., Sana, H., Jamar, C., & Mason, K. O. 2002b, A&A, 388, 552
Rauw, G., et al. 2002c, A&A, 394, 993
-. 2016, A&A, 589, A121
Sana, H., Rauw, G., Nazé, Y., Gosset, E., & Vreux, J.-M. 2006, MNRAS, 372, 661
Sana, H., Stevens, I. R., Gosset, E., Rauw, G., & Vreux, J.-M. 2004, MNRAS, 350, 809
Santolaya-Rey, A. E., Puls, J., & Herrero, A. 1997, A&A, 323, 488
Schneider, F. R. N., Langer, N., de Koter, A., Brott, I., Izzard, R. G., & Lau, H. H. B. 2014, A&A, 570, A66
Simon, K. P., & Sturm, E. 1994, A&A, 281, 286

Massive Stars Near and Far
Proceedings IAU Symposium No. 361, 2024
J. Mackey, J. S. Vink, & N. St-Louis, eds.
doi:10.1017/S1743921322002216

HR 6819 – a post-interaction binary system originally thought to be a triple system containing a black hole

Abigail J. Frost[1], Julia Bodensteiner[2], Thomas Rivinius[3], Dietrich Baade[2], Antoine Merand[2], Fernando Selman[3], Michael Abdul-Masih[3], Gareth Banyard[1], Emma Bordier[1,3], Karan Dsilva[1], Calum Hawcroft[1], Laurent Mahy[4], Maddalena Reggiani[1], Tomer Shenar[5], Mauricio Cabezas[6], Petr Hadrava[6], Marianne Heida[2], Robert Klement[7] and Hugues Sana[1]

[1]Institute of Astronomy, KU Leuven, Celestijnlaan 200D, 3001 Leuven, Belgium
email: abi.frost@kuleuven.be

[2]European Organisation for Astronomical Research in the Southern Hemisphere (ESO), Karl-Schwarzschild-Str. 2, 85748 Garching b. München, Germany

[3]European Organisation for Astronomical Research in the Southern Hemisphere (ESO), Casilla 19001, Santiago 19, Chile

[4]Royal Observatory of Belgium, Avenue Circulaire 3, B-1180 Brussel, Belgium

[5]Anton Pannekoek Institute for Astronomy, University of Amsterdam, Postbus 94249, 1090 GE Amsterdam, The Netherlands

[6]Astronomical Institute, Academy of Sciences of the Czech Republic, Boční II 1401, 141 31 Praha 4, Czech Republic

[7]The CHARA Array of Georgia State University, Mount Wilson Observatory, Mount Wilson, CA 91023, USA

Abstract. In 2020, HR 6819 was reported to be a triple system containing the closest black hole to Earth. However, these results were contested, with an alternative explanation of a post-interaction binary suggested. Using new integral field spectroscopic and interferometric data, we have been able to determine the true nature of this exotic source.

Keywords. stars: early-type, stars: binaries, stars: fundamental parameters

---

1. Introduction

The majority of main-sequence (MS) massive OB-type stars belong to binary or higher-order multiple systems, among which close binaries ($P < 10\,\mathrm{yr}$) are common (e.g. Sana et al. 2012; Villaseñor et al. 2021) which will most likely interact and leave the involved stars chemically and physically altered (e.g. de Mink et al. 2013). One type of object which potentially shows the effects of such interactions are Be stars. Be stars are non-supergiant B-type stars that exhibit Balmer emission lines such as Hα (see e.g., Zorec & Briot 1997; Rivinius et al. 2013). This emission is associated with a circumstellar decretion disk, of order stellar radii in size, around the equatorial plane of the star (e.g. Struve 1931). While the origin of such Be stars is uncertain, a binary formation channel seems a likely candidate. When the slightly more massive star in the system expands to its giant phase,

Roche Lobe overflow can occur. The subsequent transfer of mass and momentum can spin up the other star in the system, turning it into a Be star. The donor star would appear altered by such an interaction as well, with most of its envelope material removed so that it appears more like the core of a star or 'stripped'.

One system which shows Be star signatures is HR 6819 (d=350pc, V=5.4mag), which made headlines in 2020 when it was claimed to be a triple system containing 'The nearest black hole to Earth' by Rivinius et al. (2020). These authors postulated, based on the analysis of FEROS spectra, that one star in the system (a B star) was on a 40 day orbit with an unseen companion of $4M_\odot$ - a stellar mass black hole (BH) and other emission line signatures present in spectra were coming from a Be star that was on a much wider orbit. However, other authors (Bodensteiner et al. 2020; El-Badry & Quataert 2021) disagreed with these findings. Bodensteiner et al. (2020) reanalysed the FEROS spectra and found that other emission lines associated with the Be star show subtle anti-phase movement with respect to the B star. Therefore, these two stars could be in the 40 day orbit visible in the spectra. The caveat for these findings was that the B star in the system had to have been stripped and of low-mass, in order to explain the orbital velocities of the lines associated with the B star. With these spectral data alone, no definitive distinction between the two scenarios could be made. If the system was a triple, we would expect a bright source at ~100mas separation (the Be star) and one bright source at 1mas scales (B star) which would be orbiting the invisible stellar mass BH. If the system was a binary, we would expect no bright source at ~100mas and two bright sources separated on mas scales, which would be the B and Be star, on the 40 day orbit detected in the spectra. In order to determine the true nature of the source, we (Frost et al. 2022) revisited HR 6819 using MUSE data to trace the 100mas scales of HR 6819 and infrared interferometry to trace the mas scales.

2. Results

In order to probe the system at large, 100mas scales, the first observation we used was the Multi Unit Spectroscopic Explorer (MUSE) situated at the Very Large Telescope (VLT). MUSE is an integral field spectrograph operating over visible wavelengths (480-930nm). The MUSE data were taken using the narrow-field mode (NFM) which covers a 7.5" × 7.5" field of view (or ~2550 AU (or ~0.12 pc) with a spatial sampling of 0.025" × 0.025". When the data from all the IFUs are stacked and combined, we retrieve an image of the source. A single source is seen at 100mas scales, meaning that there is no evidence that a Be star is orbiting the central source of HR 6819 at the large orbit required for the black-hole scenario.

In order to constrain the inner regions of the system, we use GRAVITY/VLTI. GRAVITY is a spectro-interferometer, meaning that it not only provides information on the spatial extent and orientation/asymmetry of the system (from the interferometric observables visibility and phases respectively) but that we can also gain spectral information of the source. These observables were all fit using geometric models from the code PMOIRED (Mérand 2022). From this least-square fitting process, we determined that the system was composed of two point sources at (on average, across the two obtained epochs of data) 1.15±0.03mas. Furthermore, thanks to PMOIRED's ability to assign emission line features to individual model components, we determined that the Brackett Gamma (Brγ) emission seen in the total flux (Figure 1) was associated with the brighter, and therefore likely more massive, star in the system. The Brγ line shows a double-peaked profile, indicative of rotation. We were able to fit a Keplerian disk (Figure 1) model which well fit this profile, as well as the differential phases, within the datasets, with the star hosting the disk being almost twice as bright as the other star. We conclude that this is

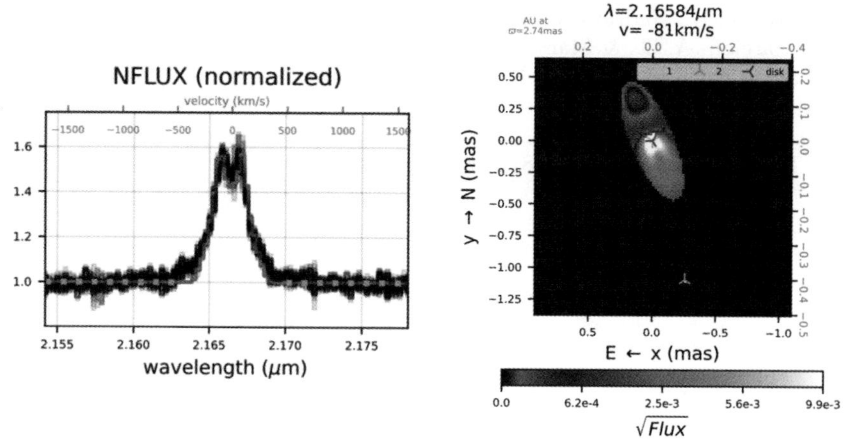

Figure 1. The Brγ line as seen in the normalised flux (NFLUX) from the GRAVITY data (left) and the Keplerian disk model which assisted in its fitting (right).

likely to be the Be star decretion disk. Therefore, the binary scenario appears to explain both our MUSE and GRAVITY data, and HR 6819 is most likely a post-interaction Be+stripped B star binary system, with no black hole.

3. Conclusions

Our new data show that HR 6819 is a binary system and that the presence of a BH on a short-period orbit in this system can be rejected. HR 6819 therefore constitutes a perfect source for investigating the origin of Be stars and their possible formation through a binary channel. In future work, further monitoring of the system with GRAVITY will be crucial. Not only can the orbit be better constrained, but these measurements will provide distance and precise mass estimates of what is likely a newly post-interaction, bloated stripped object and its associated Be star for the first time. Together with higher-resolution spectroscopy (e.g. from UVES), abundances of both stars could be derived. With this information, HR 6819 would constitute a corner-stone object for comparing binary evolution models

Acknowledgements

This research has received funding from the European Research Council (ERC) under the European Union's Horizon 2020 research and innovation programme (grant agreement number 772225: MULTIPLES). JB and MH are supported by an ESO fellowship. TS acknowledges funding received from the European Union's Horizon 2020 under the Marie Skłodowska-Curie grant agreement No. 101024605. We also thank the team behind QFitsView for developing their tool which allowed easy handling of our MUSE data.

References

Bodensteiner, J., et al. 2020, A&A, 641, A43

de Mink, S. E., Langer, N., Izzard, R. G., Sana, H., & de Koter, A. 2013, ApJ, 764, 166

El-Badry, K., & Quataert, E. 2021, MNRAS, 502, 3436

Frost, A. J., et al. 2022, A&A, 659, L3

Mérand, A. 2022, PMOIRED: Parametric Modeling of Optical Interferometric Data, Astrophysics Source Code Library, record ascl:2205.001

Rivinius, T., Baade, D., Hadrava, P., Heida, M., & Klement, R. 2020, A&A, 637, L3
Rivinius, T., Carciofi, A. C., & Martayan, C. 2013, A&A Rev., 21, 69
Sana, H., et al. 2012, Science, 337, 444
Struve, O. 1931, ApJ, 73, 94
Villaseñor, J. I., et al. 2021, MNRAS, 507, 5348
Zorec, J., & Briot, D. 1997, A&A, 318, 443

Massive Stars Near and Far
Proceedings IAU Symposium No. 361, 2024
J. Mackey, J. S. Vink, & N. St-Louis, eds.
doi:10.1017/S1743921322002708

Contact Tracing of Massive Binary Stars

Jan Henneco[1,2], Fabian R. N. Schneider[1,2], Saskia Hekker[1,2] and Eva Laplace[1]

[1]Heidelberg Institute for Theoretical Studies, Schloss-Wolfsbrunnenweg 35, 69118 Heidelberg, Germany
email: jan.henneco@h-its.org

[2]Centre for Astronomy of Heidelberg University, Königstuhl 12, 69117 Heidelberg, Germany

Abstract. Stellar mergers produce more massive, rejuvenated (strongly magnetic) stars, with potentially peculiar properties, and can be detected as luminous red novae. Using a grid of detailed 1D binary evolution models, we aim to determine which binary systems are likely to merge and at what evolutionary stage. This will tell us more about the merger products, and might help us understand some of the trends found in observed single- and multiple-star populations.

Keywords. binaries: general, binaries: evolution

1. Introduction

A significant fraction of massive stars live in binaries (Sana et al. 2012; De Marco & Izzard 2017). When binary components merge as "living" stars they can be observed as luminous red novae (Nandez et al. 2014) and produce more massive, rejuvenated (magnetic) stars (Schneider et al. 2019). Knowing which binaries are likely to merge might provide explanations for some open question regarding multiple main sequences (Wang et al. 2022) and over-massive main sequence (MS) stars in stellar clusters (Schneider et al. 2016). Furthermore, around 10% of early-type massive stars are expected to be merger products (de Mink et al. 2014) and a quarter of massive O-type stars is expected to merge during their lifetime (Sana et al. 2012). The actual merger event between two stars is usually preceded by a contact phase, where both stars overfill their respective Roche lobes and form a "peanut" shaped star. Whether such configurations always lead to an actual merger is a question on its own.

We aim to constrain the progenitor parameter space of stellar mergers, characterise their structure at the point of contact, and describe the mechanisms leading to contact/merging. To this end we compute a grid of ~ 6000 detailed binary evolution models using the stellar structure and evolution code `MESA` (Paxton et al. 2019, and references therein). In this grid, component masses range from 0.5 to $20\,\mathrm{M}_\odot$, initial mass ratios $q_\mathrm{i} = M_2/M_1$ from 0.1 to 0.97, and initial separations a_i from $\sim 1\,\mathrm{R}_\odot$ to $\sim 10^4\,\mathrm{R}_\odot$.

2. Results & Conclusions

In our binary grid, we find a variety of contact/common envelope (CE) configurations that might eventually lead to stellar mergers. Specifically at an initial primary mass of $20\,\mathrm{M}_\odot$, shown in Fig. 1, we see MS contact phases for all initial mass ratios, and a large region of the parameter space at $q_\mathrm{i} > 0.5$ in which stripped stars, that avoid contact, are produced. Furthermore, for this primary mass, case C mass transfer seems to be stable, which is most likely related to strong wind mass loss during the red supergiant phase.

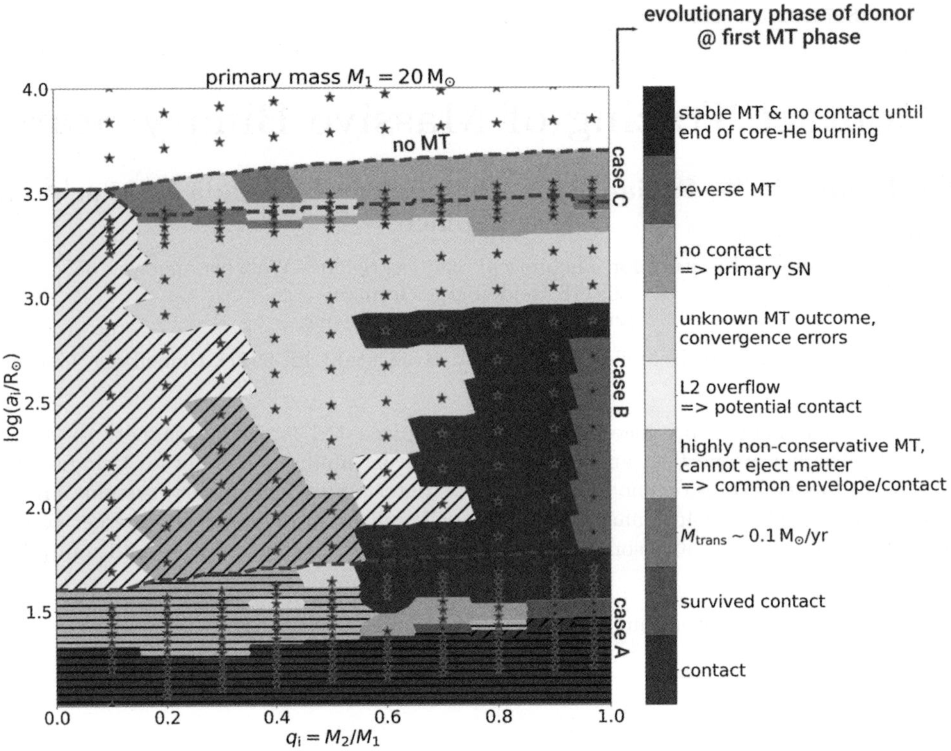

Figure 1. Overview of binary models with initial primary masses of 20 M$_\odot$. Each purple star corresponds to a model. Colours in between models are determined by their nearest neighbour. The evolutionary stages refer to the following; case A: main sequence star, case B: star before core-He exhaustion, and case C: star after core-He exhaustion. Horizontal and diagonal hatching indicate case A + case A and case B + case A contact/common envelope phases respectively.

Lastly, for twin systems ($q_i = 0.97$) with a post-MS primary star, a phase of reverse mass transfer from the secondary occurs, since at mass ratios close to 1, the latter is able to catch up with the primary in terms of evolutionary stage.

Based on the properties of the stars at contact and/or the mechanisms that lead to contact, we can now make predictions for which systems will likely merge. Furthermore we will be able to compare this over a range of initial primary masses.

Lastly, the scope of this grid is not limited to contact tracing alone and can be used to study for example X-ray binaries and Be-type stars.

References

De Marco, O., & Izzard, R. G. 2017, PASA, 34, e001

de Mink, S. E., Sana, H., Langer, N., Izzard, R. G., & Schneider, F. R. N. 2014, ApJ, 782, 7

Nandez, J. L. A., Ivanova, N., & Lombardi, Jr., J. C. 2014, The Astrophysical Journal, 786, 39, aDS Bibcode: 2014ApJ...786...39N

Paxton, B., et al. 2019, ApJS, 243, 10

Sana, H., et al. 2012, Science, 337, 444

Schneider, F. R. N., Ohlmann, S. T., Podsiadlowski, P., Röpke, F. K., Balbus, S. A., Pakmor, R., & Springel, V. 2019, Nature, 574, 211

Schneider, F. R. N., Podsiadlowski, P., Langer, N., Castro, N., & Fossati, L. 2016, MNRAS, 457, 2355

Wang, C., et al. 2022, Nature Astronomy, arXiv: 2202.05552

Massive Stars Near and Far
Proceedings IAU Symposium No. 361, 2024
J. Mackey, J. S. Vink, & N. St-Louis, eds.
doi:10.1017/S1743921322001946

Hunting for massive binaries with a black-hole component using Gaia data

Soetkin Janssens[1], T. Shenar[2], H. Sana[1], S. Faigler[3], N. Langer[4,5], P. Marchant[1], T. Mazeh[3], C. Schürmann[4,5] and S. Shahaf[6]

[1]Institute of Astronomy, KU Leuven, Celestijnenlaan 200D, 3001 Leuven, Belgium
email: soetkin.janssens@kuleuven.be

[2]Anton Pannekoek Institute for Astronomy, Science Park 904, 1098 XH, Amsterdam, The Netherlands

[3]School of Physics and Astronomy, Tel Aviv University, Tel Aviv 6997801, Israel

[4]Argelander-Institut für Astronomie, Universität Bonn, Auf dem Hügel 71, 53121 Bonn, Germany

[5]Max-Planck-Institut für Radioastronomie, Auf dem Hügel 69, 53121 Bonn, Germany

[6]Department of Particle Physics and Astrophysics, Weizmann Institute of Science, Rehovot 7610001, Israel

Abstract. With the upcoming third Gaia data release (DR3), the first Gaia astrometric orbital solutions for binary sources will become available. Potentially, many rarely seen single-degenerate massive binaries with a black hole (OB+BH) will be revealed. Here, we investigate how many OB+BHs are expected to be detected as binaries in Gaia astrometry by using tailored models for the massive star population. We use a method based on the astrometric data to investigate how many OB+BH binaries will be uncovered by Gaia. We estimate that ∼200 OB+BHs are detectable among the sources in the second Alma Luminous Star massive star catalogue, either in DR3 or in upcoming data releases. Moreover, we show that BH-formation scenarios could be constrained from the distributions of parameters such as the orbital periods and eccentricities.

Keywords. black hole physics, binaries:general, astrometry, stars:statistics

---

Binary interactions, such as mass transfer and merger, dominate the evolution of massive binaries (Sana et al. 2012). Single-degenerate massive binaries with a black-hole component (OB+BH binaries) play a crucial rule in the evolution towards double-BH mergers. Accreting OB+BH systems have been detected through X-rays (e.g. Corral-Santana et al. 2016). Non-accreting (quiescent) OB+BH systems are harder to detect. Spectroscopy can in principle help. However, in the past two years, spectroscopically reported OB+BH systems have been challenged by follow-up studies (e.g. LB-1 and HR6819; Abdul-Masih et al. 2020; Bodensteiner et al. 2020; El-Badry & Quataert 2020; Irrgang et al. 2020; Mazeh & Faigler 2020; Shenar et al. 2020; Frost et al. 2022), disfavouring the BH nature of the companion, leaving a handful of good candidates (e.g. Casares et al. 2014; Mahy et al. 2022; Shenar et al. 2022, Sana et al. in these proceedings). However, according to recent simulations, there should be about 1200 OB+BHs in the Milky Way Langer et al. (2020). Here, we present predictions on the capability of finding

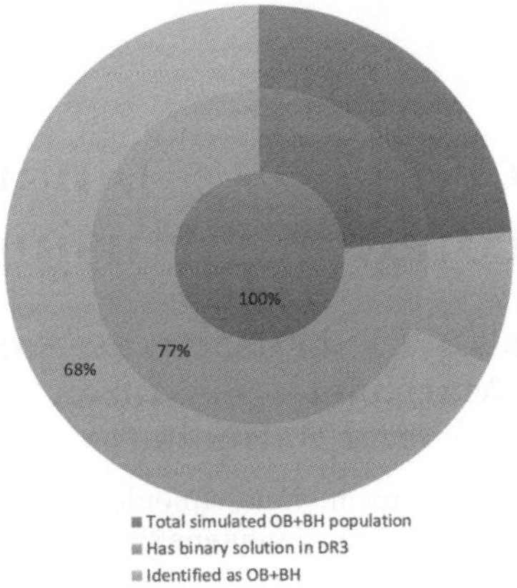

Figure 1. Pie chart of the detectable and identifiable fraction of OB+BHs in the simulated population.

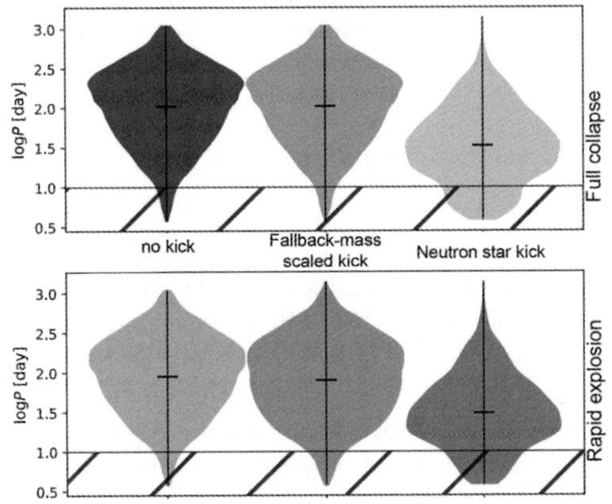

Figure 2. Predicted period distributions of the identifiable OB+BHs for different BH-formation scenarios. The vertical axis indicates different collapse scenarios.

OB+BH binaries with Gaia astrometric orbits using a novel method (Shahaf et al. 2019; Janssens et al. 2022).

We combined tailored simulations for the massive star population (Langer et al. 2020) with the distance distribution of stars in the second Alma Luminous Star massive star catalogue (ALSII; Pantaleoni González et al. 2021) to create a mock population of OB+BH binaries. Assuming that orbits for a system are available if the semi-major axis of the projected photocentre motion α satisfies $\alpha/\sigma_\alpha > 3$, we predicted how many of these OB+BHs are detectable by Gaia and how many are astrometrically identifiable as such with the Astrometric Mass-Ratio Function (AMRF; Shahaf et al. 2019). The method uses the size of the semi-major axis of the photocentre motion of a system to determine the nature of the companion. Almost 70% of our simulated population is identifiable with the AMRF, leading up to around 200 OB+BHs in the ALSII (Fig. 1). Moreover, non-degenerate binaries should not be misidentified as OB+BHs.

The detection of ~200 OB+BHs would increase the sample of known OB+BHs by a factor of around 20. From the distributions of the parameters of the OB+BHs, such as the orbital periods and eccentricities, we can obtain constraints on the kick BHs receive at birth and how much mass is lost in the preceding collapse (Fig. 2).

References

Abdul-Masih, M., et al. 2020, Nature, 580, E11

Bodensteiner, J., et al. 2020, A&A, 641, A43

Casares, J., Negueruela, I., Ribó, M., Ribas, I., Paredes, J. M., Herrero, A., & Simón-Díaz, S. 2014, Nature, 505, 378

Corral-Santana, J. M., Casares, J., Muñoz-Darias, T., Bauer, F. E., Martínez-Pais, I. G., & Russell, D. M. 2016, A&A, 587, A61

El-Badry, K., & Quataert, E. 2020, MNRAS, 493, L22

Frost, A. J., et al. 2022, A&A, 659, L3

Irrgang, A., Geier, S., Kreuzer, S., Pelisoli, I., & Heber, U. 2020, A&A, 633, L5

Janssens, S., et al. 2022, A&A, 658, A129

Langer, N., et al. 2020, A&A, 638, A39

Mahy, L., et al. 2022, arXiv e-prints, arXiv:2207.07752

Mazeh, T., & Faigler, S. 2020, MNRAS, 498, L58

Pantaleoni González, M., Maíz Apellániz, J., Barbá, R. H., & Reed, B. C. 2021, MNRAS, 504, 2968

Sana, H., et al. 2012, Science, 337, 444

Shahaf, S., Mazeh, T., Faigler, S., & Holl, B. 2019, MNRAS, 487, 5610

Shenar, T., et al. 2020, A&A, 639, L6

—. 2022, Nature Astronomy

Massive Stars Near and Far
Proceedings IAU Symposium No. 361, 2024
J. Mackey, J. S. Vink, & N. St-Louis, eds.
doi:10.1017/S1743921322002630

Multiplicity of Northern bright O-type stars with optical long baseline interferometry. Results of the pilot survey

Cyprien Lanthermann

The CHARA Array of Georgia State University, Mount Wilson Observatory, Mount Wilson, CA 91023, USA
email: cyprien@chara-array.org

Abstract. The study of the multiplicity of massive stars gives hints of their formation processes and their evolution path. Optical interferometry is mandatory to fulfill our knowledge of their multiplicity by probing the separation gap between 1 and 50 mas. We demonstrated the capability of the new interferometric instrument MIRC-X, located at the CHARA array, to study a large sample of more than 120 ($H < 7.5$) O-type stars. We observed 29 O-type star systems, including a couple of systems in average atmospheric conditions around a magnitude of $H = 7.5$. Out of these 29 systems, we detected 18 companions in 16 different systems, resulting in a multiplicity fraction $f_m = 16 / 29 = 0.55$, and a companion fraction of $f_c = 18 / 29 = 0.62$. We observed for the first time 11 of these detected companions. This study concludes that a large survey on more than 120 Northern O-type stars is possible with MIRC-X.

Keywords. (stars:) binaries: general, stars: formation, stars: statistics, infrared: stars

1. Introduction

Massive stars are key components of the evolution of their host galaxy. They are the main producers of heavy elements, and the momentum and kinetic energy involved in their death have an influence on a large part of their galaxy (Zinnecker and Yorke (2007)). They are also at the origin of the compact objects that produce the gravitational waves (GW) that we can currently detect (Abbott et al. (2016)).

However, their short lifetime (a few million years) and their rapid formation process (10^5 years) make the observation of their early ages difficult. Indeed, their lifetime makes them rare, and so to observe a large number of massive stars (> 100), one needs to look for them at significant distances, typically 1 to 3 kpc. In consequence, the formation process of massive stars is still actively discussed. The outcomes of the different formation models predict different multiplicity parameters. Hence, the study of the multiplicity of massive stars, after their formation process, should provide relevant constraints on these formation models.

To probe the full range of orbital separations of systems situated at a typical distance of 2 kpc, one needs to use different observational techniques. For separations between 1 and 50 mas, the only technique we can use is Optical Long Baseline Interferometry (OLBI). But until recently, this technique was limited by its sensitivity and could only be applied to a modest sample of massive stars.

In this proceeding, we present the results of the pilot survey performed on 29 systems, with the goal to demonstrate the feasibility of a large survey of more than 100 O-type

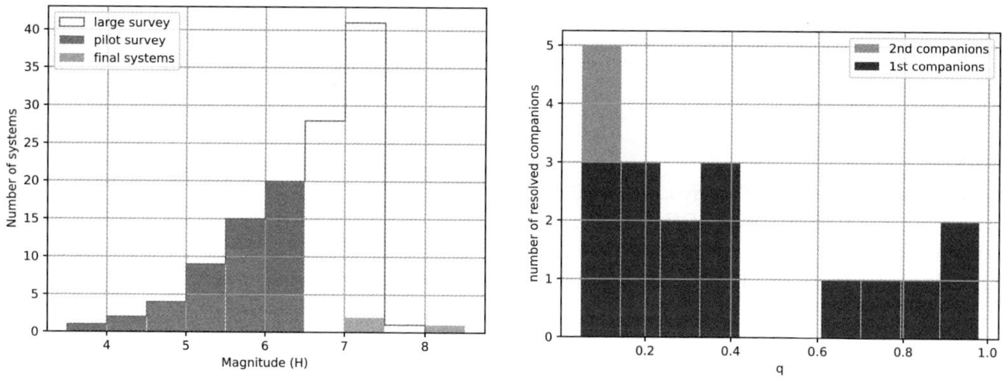

Figure 1. Left: Histogram of the H-band magnitude of our observed sample in green, the original sample planned for this pilot survey in orange and the planned sample for the large survey in blue. Right: Histogram of detected companions as a function of the estimated mass ratio q, when taking into account only the first detected companions (blue), and when adding the second companions detected (orange).

stars with the CHARA/MIRC-X instrument (Kraus et al. 2018), which requires a limiting magnitude of H =7.5.

2. Observations

We built our sample using the Galactic O-star Catalog (GOSC Maíz Apellániz et al. 2013). We selected every O-star observable by CHARA and within the limiting magnitude of MIRC-X, which is H = 6.5 at the time of this pilot study. Our input sample is therefore magnitude limited. During our observation campaign, we could obtain good quality data on 29 O-type star systems, with the histogram of their magnitude displayed in Figure 1 (left). We can see that we observed a couple of systems with magnitude around H = 7.5, in normal seeing conditions, demonstrating the feasibility of observing a large sample of more than 100 O-type stars.

3. Statistics on detected companions

Among the 29 systems observed with good quality data, we confirm the detection of 18 companions for 16 multiple systems. Out of these 18 companions, 11 are detected for the first time. This gives us a multiplicity fraction $f_m = 16/29 = 0.55 \pm 0.09$, and a companion fraction of $f_c = 18/29 = 0.62 \pm 0.13$.

Figure 1 (right) displays the distribution of the estimated mass ratio of the detected companions. In this figure, we can see that the distribution seems to be bi-modal, with a lack of companions between $q = 0.4$ and 0.6, and favoring a lower mass ratio. However, our sample is too small to perform any statistical study accurately enough to conclude on massive star formation models.

4. Conclusion

This pilot survey shows that a large survey of more than 100 O-type stars is possible with CHARA/MIRC-X, and our sample is too small to perform an accurate statistical study, showing that the large survey is required. More detail on this pilot study and the implication for further study can be found in Lanthermann et al. 2022 (in preparation).

Supplementary material

To view supplementary material for this article, please visit http://dx.doi.org/10.1017/S1743921322002630.

References

Abbott, B. P., et al. 2016, Phys. Rev. Lett., 116, 061102
Kraus, S., et al. 2018, Proc.SPIE, 10701, 10701
Maíz Apellániz, J., et al. 2013, in Massive Stars: From alpha to Omega, 198
Zinnecker, H., & Yorke, H. W. 2007, ARA&A, 45, 481

Massive Stars Near and Far
Proceedings IAU Symposium No. 361, 2024
J. Mackey, J. S. Vink, & N. St-Louis, eds.
doi:10.1017/S1743921322003003

Deciphering hard disks

Yaël Nazé[1,2], Gregor Rauw[1], Myron A. Smith[3] and Christian Motch[4]

[1]STAR Institute, Université de Liège, Allée du 6 août 19c, Bât B5c, 4000 Liège, Belgium

[2]FNRS Senior research associate
email: ynaze@uliege.be

[3]NSF OIR Lab, 950 N Cherry Ave, Tucson, AZ 85721, USA

[4]CNRS, Obs. Ast. de Strasbourg, 11 rue de l'Université, F-67000 Strasbourg, France

Abstract. Some Be stars were found to emit very bright and extremely hard thermal X-rays. This so-called γ Cas-star category is rapidly growing, showing that the phenomenon is far from being uncommon and its consequences on Be population feedback should be examined. The origin of this X-ray peculiarity is however much debated. In this contribution, we review the most recent observational clues and derive constraints on the cause of the phenomenon: properties of these stars (multiplicity status, photometric variability), X-ray reaction to disk changes, impact of stripped-star companions on the X-ray emission,...

Keywords. stars: emission-line, Be; stars: early-type; X-rays: stars

1. Introduction

In the 19th century, emission lines were detected in the spectrum of γ Cas, the star at the heart of the Cassiopeia constellation, leading to the definition of the Be category. A century later, the star was however found to be atypical in the X-ray range (for a review, see Smith et al. 2016). In the last two decades, two dozen other Be stars were found to display similar high-energy properties (e.g. Nazé & Motch 2018). Their X-ray emission is bright ($\log(L_X) \sim 31.6 - 33.2$, $\log(L_X/L_{BOL})$ of -6.2 to -4) and hard ($kT > 5$ keV). It also presents short-term "flares" and a fluorescent FeKα line is also detected in the X-ray spectrum near the usual iron complex at 6.7 keV. The origin of such an emission remains debated, with different scenarios proposing to link the emission to a companion (accretion onto a compact object, collision of companion's wind with the Be disk) or to magnetic star-disk interactions.

2. Photometric variability

The variability of most γ Cas stars was studied using SMEI, BRITE, and TESS high-precision photometry (Nazé et al. 2020a,b). The periodograms display red noise, frequency groups, isolated NRP frequencies (including dominant ones and signals at high frequencies). Changes in the frequency content are also detected during/near outbursts. All these characteristics are also found, with a similar incidence, in other Be stars.

3. Are they alone?

While Be stars are now often considered as binary interaction products, only two γ Cas stars were previously known to be binaries (γ Cas and π Aqr). Thanks to a monitoring done with high-resolution spectrographs (UVES, Carmenes, HEROS - Nazé et al. 2022a),

6 new orbital solutions were found (even revealing one quadruple system!) and five other stars showed velocity shifts reminiscent of binary motion although more data are needed to establish an orbital solution. Other objects are too faint or their spectrum too variable to perform a velocity study. For the detected binaries, the small velocity amplitudes and long periods are similar to other Be spectroscopic binary systems.

4. What happens if the disk changes?

A long-term monitoring of the Hα line in several γ Cas stars was undertaken, and X-ray observations were triggered when the disk changed substantially (Rauw et al. 2019 and submitted, Nazé et al. 2019, 2022b). For HD 45314, the hottest γ Cas object, the hard X-rays nearly fully disappeared when the Hα emission was very reduced. In contrast, π Aqr always displayed γ Cas characteristics, even when its disk was extremely small. Its X-ray emission does vary somewhat in brightness and hardness but these changes are not correlated with orbital phase or EW values. In addition, γ Cas has shown a recent strong increase in Hα emission but it didn't lead to much change in the hard X-rays or in V-magnitude. Finally, despite very weak Hα emissions, V767 Cen and HD 119682 always remained γ Cas analogs. Their X-ray properties were unchanged as their disk varied, with a general trend towards smaller Hα emission superimposed on short-term surges in emission.

5. Impact of stripped-star companions in the X-ray range

We have performed the first X-ray survey of all known Be+stripped systems (Nazé et al., submitted). Most systems remain undetected in the X-ray range, with very low X-ray luminosity limits. Amongst the detections, only two systems display γ Cas characteristics while other systems show faint and soft X-ray emissions. The γ Cas incidence appears similar to that of general Be samples. Furthermore, no link was found between X-ray luminosity and orbital period or companion properties, in contrast with expectations from the wind-disk collision scenario.

6. Conclusion

No specific signature of the γ Cas phenomenon was found outside the X-ray range. Many γ Cas stars actually are binaries, but no link is found between the X-ray emission and the orbital phase. Moreover, the γ Cas character of the X-ray spectrum remains even if the Hα emission is very small. No obvious correlation is found between X-ray flux and Hα strength or with orbital phase, but changes are detected in X-rays when both broad-band magnitude and Hα strength vary. All this suggests that the external parts of the disk are of limited importance in the γ Cas phenomenon, thereby favoring the star-disk interaction scenario.

References

Nazé, Y., & Motch, C. 2018, *A&A*, 619, A148
Nazé, Y., Rauw, G., & Smith, M.A. 2019, *A&A*, 632, A23
Nazé, Y., Pigulski, A., Rauw, G., & Smith, M.A. 2020a, *MNRAS*, 494, 958
Nazé, Y., Rauw, G., Pigulski, A. 2020b, *MNRAS*, 498, 3171
Nazé, Y., Rauw, G., Czesla, S., et al. 2022a, *MNRAS*, 510, 2286
Nazé, Y., Rauw, G., Bohlsen, T., et al. 2022b, *MNRAS*, 512, 1648
Rauw, G., Nazé, Y., Smith, M.A., et al. 2019, *A&A*, 615, A44
Smith, M.A., Lopes de Oliveira, R., & Motch, C. 2016, *Adv. in Sp. Res.*, 58, 782

Massive Stars Near and Far
Proceedings IAU Symposium No. 361, 2024
J. Mackey, J. S. Vink, & N. St-Louis, eds.
doi:10.1017/S1743921322002915

Study of the Be/X-ray binary IGR J21343+4738 physical parameters based on the 15-years monitoring

Evgeniia Nikolaeva[1,2,*] **and Ilfan Bikmaev**[1,2]

[1]Kazan (Volga-region) Federal University, Kremlevskaya 18, 420008, Kazan, Russian Federation

[2]Academy of Sciences of Tatarstan, Baumana, 20, 420111, Kazan, Russian Federation
*email: evgeny.nikolaeva@gmail.com

Abstract. We present the results of our long-term RTT-150 photometric and spectroscopic observations Be optical counterpart of the High-mass X-ray binary IGR J21343+4738.

Keywords. X-rays: binaries, stars: emission-line, Be, stars: individual (IGR J21343+4738)

1. Introduction

The hard X-ray source *IGR J*21343 + 4738 was detected by the IBIS telescope (Ubertini et al. 2003) of the INTEGRAL Space Observatory (Winkler et al. 2003) in December 2002 (Krivonos et al. 2007; Bird et al. 2007). In the paper of Bikmaev et al. (2008) for the first time optical counterpart was classified as a B3 star due to the presence of hydrogen HI and helium HeI absorption lines in the spectrum, and the absence of ionized helium HeII lines, and was showed that the optical component is a Be star. Reig & Zezas (2014) were able to determine the spectral class of an optical star more accurately as B1 IVe and find the parameters of its stellar atmosphere ($T_{\rm eff} = 25000K$, log $g = 3.75$, $v \sin i = 380\ km\ s^{-1}$) during the absence of the circumstellar equatorial disk in August 2013.

2. Results

We performed long-term photometric and spectroscopic observations of the source in 2005-2019 using Russian-Turkish 1.5-m optical telescope RTT-150 (TUBITAK National Observatory, Turkey) and TFOSC instrument. The most striking traces of the disk's presence are shown by the H_α line profile, which acquires a complex structure: absorption is framed on both sides by emission peaks. The brightness drop occurs when the equatorial disk shields the star (Fig. 1a). The complex structure of the Ha line is due to the physical structure of a Be star consisting of a fast-rotating B1 star and an equatorial disk surrounding it, formed from photospheric plasma due to the rotation of the star. As a result, emission lines from the equatorial disk are superimposed on the absorption photospheric spectrum of the star. The difference between the wavelengths of the peaks (ΔV) can be interpreted as the outer radius of the emission line formation region, the velocity corresponding to this difference is the rotation speed of the outer parts of the equatorial disk. And since the rotation speed of the disk depends on the radius, from

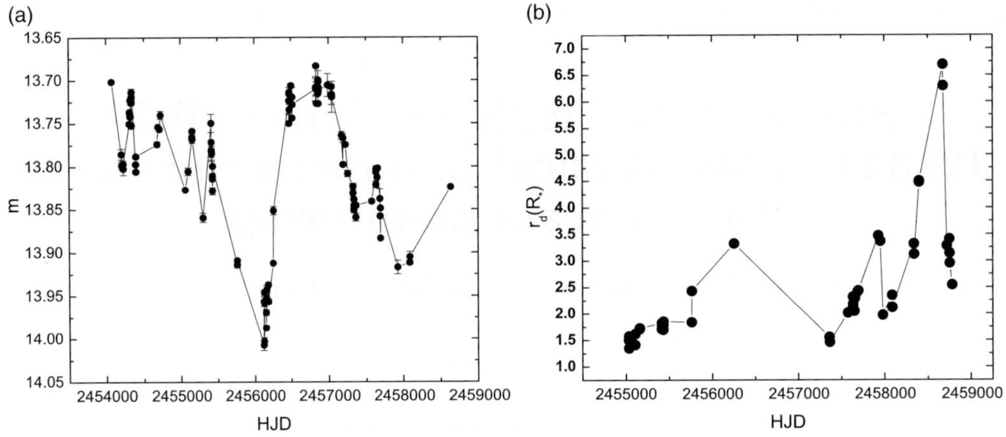

Figure 1. a: Photometric variability in white filter; b: Change in the radius of the equatorial disk with time.

here one can determine the size of the disk:

$$\frac{r_d}{R_*} = \left(\frac{2V_* \sin\, i}{\Delta V}\right)^{1/j},$$

where ΔV is the separation of peaks, defined as the difference between the central wavelengths of the red and violet peaks (Huang 1972). v_* and R_* are the velocity and radius of the star, $j = 1/2$ (for the Keplerian disk), $v \sin i = 380\ km\ s^{-1}$ (Reig & Zezas 2014). The change in the radius of the equatorial disk over time is shown in the Fig. 1b. A correlation is seen between the drop in photometric brightness and the increase in the size of the equatorial disk of the Be star.

This work was supported partially by subsidy no. FZSM-2023-0015 of the Ministry of Education and Science of the Russian Federation allocated to the Kazan Federal University for the state assignment in the sphere of scientific activities. We are grateful to TUBITAK, Space Research Institute, Kazan Federal University, and Academy of Sciences of Tatarstan for their partial support in using RTT-150 (the Russian-Turkish 1.5-m telescope in Antalya).

References

Bikmaev, I.F., Burenin, R.A., Revnivtsev, M.G., Sazonov, S.Yu., Sunyaev, R.A., Pavlinsky, M.N., Sakhibullin, N.A. 2008, *Astron. Lett.*, 34, 653

Bird, A.J., Malizia, A., Bazzano, A., Barlow, E.J., Bassani, L., Hill, A.B., Bélanger, G., Capitanio, F., Clark, D.J., Dean, A.J., Fiocchi, M., Götz, D., Lebrun, F., Molina, M., Produit, N., Renaud, M., Sguera, V., Stephen, J.B., Terrier, R., Ubertini, P., Walter, R., Winkler, C., Zurita, J. 2007, *ApJS*, 170, 175

Huang, S.-S. 1972, *ApJ*, 171, 549

Krivonos, R., Revnivtsev, M., Lutovinov, A., Sazonov, S., Churazov, E., Sunyaev, R. 2007, *A&A*, 475, 775

Reig P. & Zezas A. 2014, *A&A*, 561, A137

Ubertini, P., Lebrun, F., Di Cocco, G., Bazzano, A., Bird, A.J., Broenstad, K., Goldwurm, A., La Rosa, G., Labanti, C., Laurent, P., Mirabel, I.F., Quadrini, E.M., Ramsey, B., Reglero, V., Sabau, L., Sacco, B., Staubert, R., Vigroux, L., Weisskopf, M.C., Zdziarski, A.A. 2003, *A&A*, 411, L131

Winkler, C., Courvoisier, T.J.-L., Di Cocco, G., Gehrels, N., Giménez, A., Grebenev, S., Hermsen, W., Mas-Hesse, J.M., Lebrun, F., Lund, N., Palumbo, G.G.C., Paul, J., Roques, J.-P., Schnopper, H., Schönfelder, V., Sunyaev, R., Teegarden, B., Ubertini, P., Vedrenne, G., Dean, A.J. 2003, *A&A*, 411, L1

Massive Stars Near and Far
Proceedings IAU Symposium No. 361, 2024
J. Mackey, J. S. Vink, & N. St-Louis, eds.
doi:10.1017/S1743921322002721

X-ray emission characteristics of an O+O binary HD 93205

Bharti Arora[1,2] **and Jeewan C. Pandey**[1]

[1]Aryabhatta Research Institute of Observational Sciences, Nainital-263 002, Uttarakhand, India

[2]School of Studies in Physics & Astrophysics, Pt. Ravishankar Shukla University, Raipur−492 010, India
email: bhartiarora612@gmail.com, jeewan@aries.res.in

Abstract. A detailed X-ray study of a massive binary called HD 93205 has been made using long-term XMM-Newton observations. The X-ray spectrum of HD 93205 displays negligible counts above 5 keV. The two thermal plasma emission models with average temperature values as ∼0.20 and ∼0.60 keV are required to explain the X-ray spectra. The X-ray flux variations in the binary are noticed to be in qualitative agreement with the wind-wind collision model but with a few deviations from the expected 1/D trend (D is the binary separation).

Keywords. Stars:early-type; binaries:colliding-winds; X-rays:stars; stars:individual: HD 93205

1. Introduction

The Carina nebula region is an interesting star-forming region of the Galaxy. The open clusters Trumpler 14, Trumpler 16, and Collinder 228 present in this region harbor significant fraction of the known O-type stars in the nearby Milky Way. HD 93205 is a massive binary in Trumpler 16 with binary components as O3.5 V((f))+O8 V (Sota et al. 2014). Conti & Walborn (1976) estimated the first radial velocity solution for the binary orbit of this system and derived an orbital period of 6.0810 ± 0.0007 d, along with minimum masses of primary and secondary stars as 39 and 15 $M_\odot$, respectively. The elliptical orbit of HD 93205 with an eccentricity of ∼0.5 points toward very young evolutionary status of both the components lying near the zero-age main sequence.

HD 93205 is moderately luminous in X-rays with luminosity of $\sim 10^{32}$ erg/s. Using five X-ray observations from XMM-Newton, Antokhin et al. (2003) has suggested that X-ray flux in HD 93205 changes in accordance to the distance between two binary components. In order to deeply investigate colliding winds of HD 93205, we have performed its detailed X-ray analysis using data observed with European Photon Imaging Camera (EPIC) onboard XMM-Newton obtained over a considerably longer time baseline (July 2000 to July 2015). The X-ray data has been reduced following the standard procedure with Science Analysis Software (SAS v.17.0.0).

2. Results: X-ray spectral modeling of HD 93205

The X-ray spectrum of HD 93205 is similar to the hot stars' spectra displaying optically thin thermal plasma emission at high temperature where the soft band is populated by several emission lines. The spectrum of HD 93205 appears much softer with negligible counts above 5 keV (Fig. 1 *left*). Therefore, the spectral fitting was performed

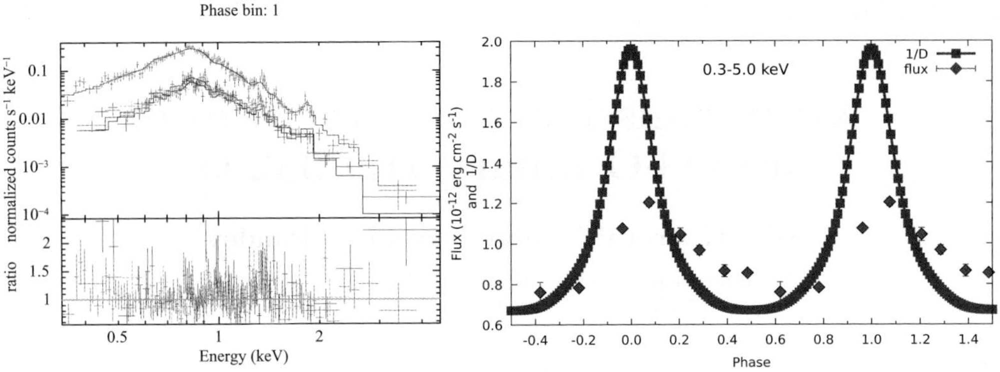

Figure 1. Left: X-ray spectra of HD 93205 fitted with two absorbed components of thermal plasma emission model. Right: Variation of X-ray flux from HD 93205 with orbital phase. The solid line shows the 1/D function, where D is the binary separation in terms of semi-major axis.

in 0.3-5.0 keV energy band using two components of thermal plasma emission model, further corrected for ISM absorption by fixing N_H^{ISM} to 0.24×10^{22} cm^{-2} (Jenkins 2019).

The quality of the EPIC spectrum of HD 93205 was not good enough to properly constrain the spectral parameters from a single observation ID upon modeling. Therefore, all the spectra available within similar orbital phases were fitted jointly (total eight phase bins). Fig. 1 (*left*) presents the joint X-ray fitting of spectra observed near the phase 0.07. The estimated ISM corrected X-ray flux in 0.3-5.0 keV energy band as a function of orbital phase has been shown in Fig. 1 (*right*). Here, the orbital phase is estimated using the ephemeris JD= 2442532.784+6.081E and zero phase represents the time of periastron passage (Conti & Walborn 1976). The average values of two temperatures ($k\mathrm{T}_1$ and $k\mathrm{T}_2$) were estimated to be 0.20±0.07 and 0.60±0.08 keV, respectively.

The phase locked variations in X-ray emission from HD 93205 are clearly displayed being maximum around periastron and minimum near the apastron as expected from the colliding winds of an eccentric massive binary. The X-ray flux is maximum around the orbital phase ~0.07 and minimum at ~0.62 and the maximum to minimum flux ratio is 1.58±0.11. In the case of adiabatic wind collision, the variation of X-ray luminosity is expected to be 1/D, where D is the separation between the two stars. However, theoretical studies suggests that mutual radiative influence of the two components of the binary may significantly alter wind dynamics in the wind collision region leading to deviations from 1/D trend (as observed for WR 25 also by Arora et al. 2019). Such phenomena may also reduce the wind speeds before interacting which further reduces hard X-ray generation from the wind collision. This may be another reason that HD 93205 displays mostly soft X-ray emission additional to fact that O-type stars intrinsically possess relatively lesser wind speeds than their evolved counter parts, Wolf-Rayet stars. To verify these effects, we are working on applying recent wind collision models to our observations and their implications on the X-ray emission will be considered in our upcoming paper.

Supplementary material

To view supplementary material for this article, please visit http://dx.doi.org/10.1017/S1743921322002721.

References

Antokhin, I. I., Rauw, G., Vreux, J.-M., et al. 2003, *Magnetic Fields in O, B and A Stars: Origin and Connection to Pulsation, Rotation and Mass Loss*, 305, 383

Arora, B., Pandey, J. C., & De Becker, M. 2019, *MNRAS* 487, 2624
Conti, P. S. & Walborn, N. R. 1976, *ApJ*, 207, 502
Jenkins, E. B. 2019, *ApJ*, 872, 55
Sota, A., Maíz Apellániz, J., Morrell, N. I., et al. 2014, *ApJS*, 211, 10

Massive Stars Near and Far
Proceedings IAU Symposium No. 361, 2024
J. Mackey, J. S. Vink, & N. St-Louis, eds.
doi:10.1017/S1743921322002745

Massive stars and their large-separation low-mass companions in Sco OB1

T. Pauwels, M. Reggiani, A. Rainot and H. Sana

Institute of Astronomy, KU Leuven, Celestijnenlaan 200D, 3001, Leuven, Belgium
email: `tinne.pauwels@kuleuven.be`

Abstract. Exploring the low-mass end of the companion mass function around massive stars is of crucial importance to constrain massive star formation theories. We present a high-contrast imaging study of 20 O- and early B-type stars in the Scorpius OB1 association. From the analysis of VLT/SPHERE data, we identify a total of 789 sources. The data probe the brown dwarf regime around massive stars, resulting in the discovery of large-separation multiple systems with mass-ratios as low as 0.001 (comparable to Jupiter-Sun mass-ratio).

Keywords. stars: early-type, stars: low-mass, brown dwarfs, binaries: general, stars: formation

1. Introduction

As an important outcome parameter of star formation, the multiplicity characteristics of massive stars put crucial constraints on massive star formation theories. The large variety of mass-ratios and companion separations demand the use of various observational techniques to characterise all companions. Previous multiplicity studies mainly focused on spectroscopy, interferometry and classical imaging, demonstrating that almost all massive stars are found in binary and multiple systems (Sana et al. 2012, 2013, 2014; Sota et al. 2014). Recently, the high-contrast imaging instrument SPHERE on the VLT enabled us to investigate the multiplicity properties of massive stars at relatively large separations (0.15" and 6") and up to much higher contrast magnitudes ($\Delta K \sim 14$ mag) than what previous studies of massive stars have been able to probe (Rainot et al. 2020, 2022; Reggiani et al. 2022). Here, we present SPHERE observations of 13 O-type and 7 early B-type stars in the active star forming region Sco OB1, which is located at a distance of 1.53 kpc (Mel'nik & Dambis 2017) and is 4-8 Myr old (Sung et al. 2013).

2. Detecting low-mass companions around massive stars

We analyse the SPHERE data using the python Vortex Imaging Processing package VIP (Gomez Gonzalez et al. 2017) which allows us to perform angular and spectral differential imaging through a principal component analysis approach.

A total of 789 sources is detected around the 20 stars in our sample. However, the data are highly contaminated by background sources. To distinguish true companions from background contaminants, we calculate the spurious association probability (=probability that a source is not bound to the central star) based on the magnitude-dependent source density in the local region around the central star. We adopt a combination of 2MASS (K<14 mag) (Cutri et al. 2003) and the Besançon Galaxy model (K>14 mag) (Czekaj et al. 2014). Figure 1 shows that the SPHERE data are probing the brown dwarf regime around massive stars, however not yet the planet regime.

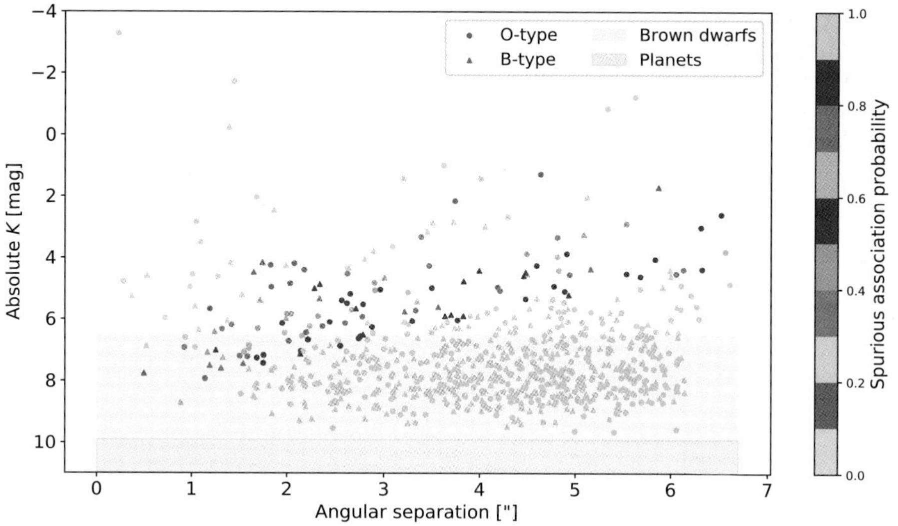

Figure 1. All sources detected in the sample. The brown dwarf and planet regimes are estimated as the average K-band magnitude for a 6 Myr old 0.08 $M_\odot$ and 0.012 $M_\odot$ object respectively.

Taking into account the spurious association probability, the (not yet bias-corrected) high-contrast imaging companion fraction (=average number of companions per star) for O-type stars in our sample is 2.8 ± 0.5. For B-type stars, we find a significantly larger companion fraction of 4.6 ± 0.8, although it could be due to an observational bias, since B-type stars are intrinsically less bright than O-type stars.

3. Conclusions

Since many of the stars in our sample already have known close companions, the addition of the high-contrast imaging companions would make them quadruple or even higher-order multiple systems. Some of the companions are at the stellar/substellar boundary, making them some of the lowest-mass companions ever discovered around OB-type stars.

Acknowledgements

T.P. and M.R. thank the Research Foundation - Flanders (FWO) for the PhD fellowship 1164522N and the postdoctoral fellowship 1280121N respectively.

The research leading to these results has received funding from the European Research Council (ERC) under the European Union's Horizon 2020 research and innovation programme (grant agreement numbers 772225: MULTIPLES).

References

Cutri, R. M., et al. 2003, VizieR Online Data Catalog, II/246
Czekaj, M. A., Robin, A. C., Figueras, F., Luri, X., & Haywood, M. 2014, A&A, 564, A102
Gomez Gonzalez, C. A., et al. 2017, AJ, 154, 7
Mel'nik, A. M., & Dambis, A. K. 2017, MNRAS, 472, 3887
Rainot, A., Reggiani, M., Sana, H., Bodensteiner, J., & Absil, O. 2022, A&A, 658, A198
Rainot, A., et al. 2020, A&A, 640, A15
Reggiani, M., et al. 2022, A&A, 660, A122
Sana, H., et al. 2012, Science, 337, 444

—. 2013, A&A, 550, A107
—. 2014, ApJS, 215, 15
Sota, A., Maíz Apellániz, J., Morrell, N. I., Barbá, R. H., Walborn, N. R., Gamen, R. C., Arias, J. I., & Alfaro, E. J. 2014, ApJS, 211, 10
Sung, H., Sana, H., & Bessell, M. S. 2013, AJ, 145, 37

Massive Stars Near and Far
Proceedings IAU Symposium No. 361, 2024
J. Mackey, J. S. Vink, & N. St-Louis, eds.
doi:10.1017/S1743921322002265

Phase-resolved spectroscopic analysis of the eclipsing black hole X-ray binary M33 X-7

Varsha Ramachandran

Zentrum für Astronomie der Universität Heidelberg, Astronomisches Rechen-Institut, Mönchhofstr. 12-14, 69120 Heidelberg
email: `vramachandran@uni-heidelberg.de`

Abstract. We present a detailed spectroscopic analysis of the only known eclipsing high mass X-ray binary with a black hole companion, M33 X-7. We obtained the first UV spectra of the system accompanied by X-ray observations, taken at three key orbital phases. We performed a detailed analysis of X-Ray, UV, and archival optical spectra using stellar atmosphere models which shed light on the interaction of the stellar wind with the black hole. Our new analysis suggests a large reduction in component masses compared to previous results. Our one-dimensional calculations confirm that the photoionization by the X-ray radiation can significantly change the ionization structure and diminish the wind accelerations. For this system standard wind-fed accretion scenario alone cannot explain the observed X-ray luminosity, indicating an additional mass overflow towards the black hole. Our evolutionary models suggest that the system is transitioning towards a common envelope stage in which both components merge.

Keywords. X-rays: binaries, black hole accretion, stars: fundamental parameters, stars: winds, outflows

---

To understand the complex behaviour of High Mass X-ray binaries (HMXBs) with black hole (BH) companions, detailed knowledge of the massive star donors is essential. However, only a few such systems are known so far. To remedy this situation, we performed a multi-wavelength phase-resolved analysis of the extragalactic HMXB M33 X-7. This eclipsing BH HMXB is reported to contain a very massive O supergiant donor and a massive black hole in a short orbit (Orosz et al. 2007; Pietsch et al. 2006). However, previous spectroscopic analyses were limited to plane-parallel models which are optimized for hot stars with no significant wind.

Using phase-resolved simultaneous *HST*- and *XMM-Newton*-observations, we trace the interaction of the stellar wind with the BH. The UV resonance lines show the Hatchett-McCray effect with a large reduction in absorption strength when the BH is in the foreground due to the strong X-ray ionization (see Fig. 1). Our comprehensive spectroscopic investigation of the donor star (X-ray+UV+optical) yields new stellar and wind parameters for the system that differ significantly from previous estimates. In particular, the masses of the components are considerably reduced to $\approx 38 M_\odot$ for the O-star donor and $\approx 11.4 M_\odot$ for the black hole (see comparison in Fig. 2a). The O giant is overfilling its Roche lobe and shows surface He enrichment.

The derived mass-loss rate of the donor is in good agreement with the Vink et al. (2001) prediction assuming a high depth-dependent microclumping. By incorporating observed X-ray luminosities in models corresponding to different orbital phases, we were able to reproduce the spectral variations at three phases with the same stellar and wind parameters. We investigated the wind driving contributions from different ions and the

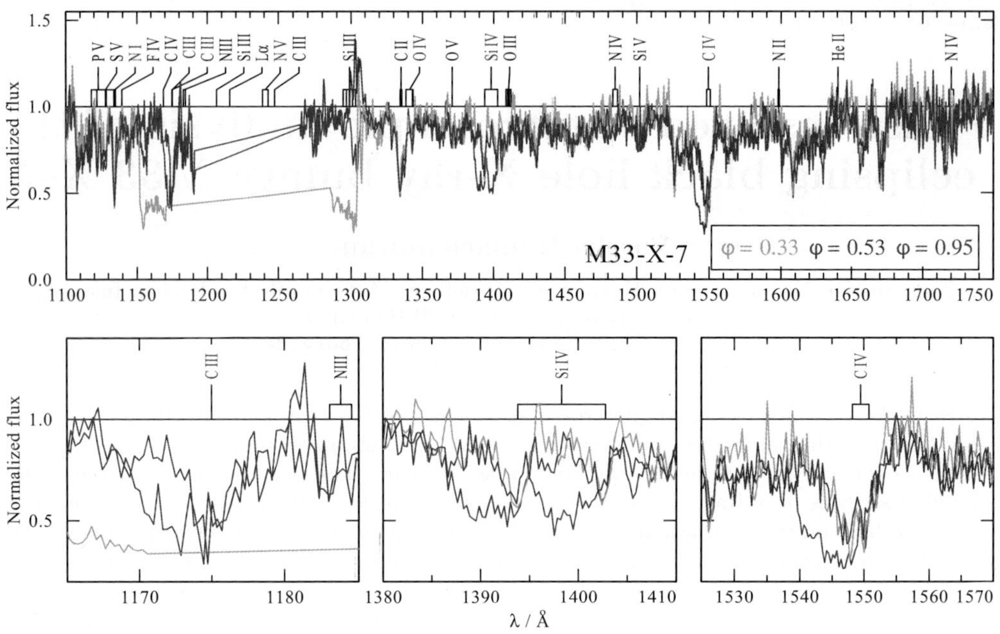

Figure 1. Comparison of HST/COS spectra of M33 X-7 taken during different orbital phases. The observed line variations are due to the Hatchett-McCray effect.

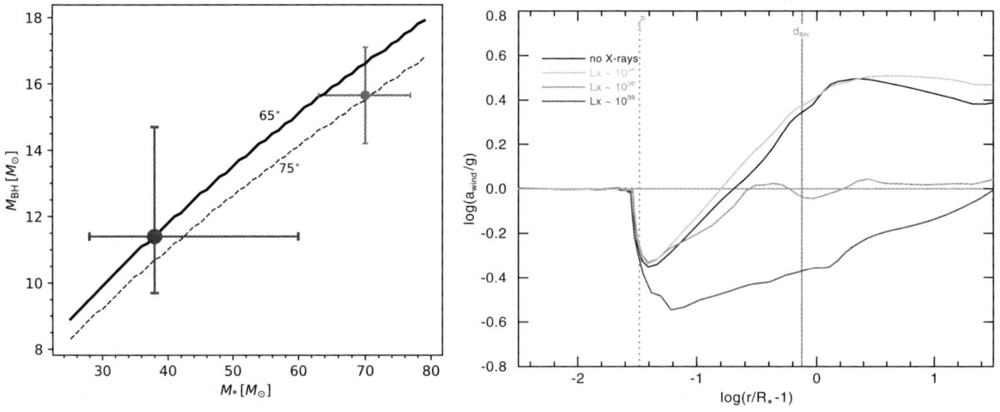

Figure 2. (a) Component masses in the M33 X-7 derived in this work (red) compared to previous values (grey) from Orosz et al. (2007). (b) Comparison of normalized wind acceleration for models including different amounts of X-rays.

changes in the ionization structure due to X-ray illumination. Towards the black hole, the wind is strongly quenched due to strong X-ray illumination (Fig. 2b). For this system, the standard wind-fed accretion scenario alone cannot explain the observed X-ray luminosity, pointing towards an additional mass overflow, in line with our acceleration calculations. The classical distinction between wind-fed and Roche-lobe overflow systems becomes meaningless for our system. Our investigations on wind driving and the impact of X-rays in M33 X-7 can be also applied to other high luminosity HMXB systems in general.

We computed binary evolutionary tracks for the system using MESA. Currently, the system is transitioning towards an unstable mass transfer phase, resulting in a common envelope of the black hole and the O-star donor.

References

Orosz, J. A., McClintock, J. E., Narayan, R., et al. 2007, Nature, 449, 872
Pietsch, W., Haberl, F., Sasaki, M., et al. 2006, ApJ, 646, 420
Vink, J. S., de Koter, A., & Lamers, H. J. G. L. M. 2001, A&A, 369, 574

Massive Stars Near and Far
Proceedings IAU Symposium No. 361, 2024
J. Mackey, J. S. Vink, & N. St-Louis, eds.
doi:10.1017/S1743921322002976

MWC 656: A Be+BH or a Be+sdO?

Th. Rivinius[1], R. Klement[2], S. D. Chojnowski[3], D. Baade[4], K. Shepard[5] and P. Hadrava[6]

[1]ESO - European Organisation for Astronomical Research in the Southern Hemisphere, Chile
email: triviniu@eso.org

[2]The CHARA Array of Georgia State University, Mount Wilson, USA

[3]Department of Physics, Montana State University, USA

[4]ESO - European Organisation for Astron. Research in the Southern Hemisphere, Germany

[5] CHARA and Department of Physics and Astronomy, Georgia State University, USA

[6] Astronomical Institute of the Academy of Sciences of the Czech Republic, Czech Republic

Abstract. MWC 656 has been reported as classical Be star with a black hole companion. Revisited spectral variability properties render this unlikely, with a hot subdwarf more probable.

Keywords. stars: Be, stars: individual (MWC 656), binaries, subdwarfs

1. Introduction

Casares et al. (2014) identified MWC 656 (HD 215227) as a binary consisting of a classical Be star and a black hole (BH). The interest in this star was triggered by a γ-ray detection with 0.6° error circle (Williams et al. 2010). The identification of a BH in MWC 656 hinges on radial velocities (RVs) from optical spectra, most critically on FeII λ4583 emission from a circumstellar disk tracing the motion of the Be primary.

More recently, however, Be stars are being recognized in large numbers as Be+sdOB systems, i.e., with hot subdwarf companions (Wang et al. 2021), some showing striking similarities to MWC 656. For a conclusive distinction between a Be+BH and a Be+sdO system, we re-examined the original data together with new higher-resolution spectra from the past decade, namely the original 32 blue and 64 red spectra of Casares et al. (R=5500), four echelle spectra from amateur observers (R=9 000-12 000) from the BeSS database, three epochs of linear polarimetry, i.e., with very high S/N, from ESPaDOnS (R=48 000) at the CFHT, and twelve spectra from ARCES (R=31 500) at Apache Point Observatory.

2. Results

Re-determination of the orbital period: The combined spectroscopic database proved incompatible with the period of 60.37 ± 0.04 d found by Williams et al. (2010) in suboptimal photometric data and adopted by Casares et al. (2014). Fourier analysis of our new RV measurements from fitting a Gaussian to all the relatively narrow HeII λ4686 emission profiles suggests a similar and highly significant period of 59.12 ± 0.05 d.

Radial velocity amplitudes: The semi-amplitude, K_2, of HeII λ4686 is virtually identical to that obtained by Casares et al. (2014) from the same line, i.e., about 80 km/s (Fig. 1, left). However, the optically thin double-peaked emission lines from the disk surrounding the Be star *do not* simply shift due to orbital motion, but also vary in peak

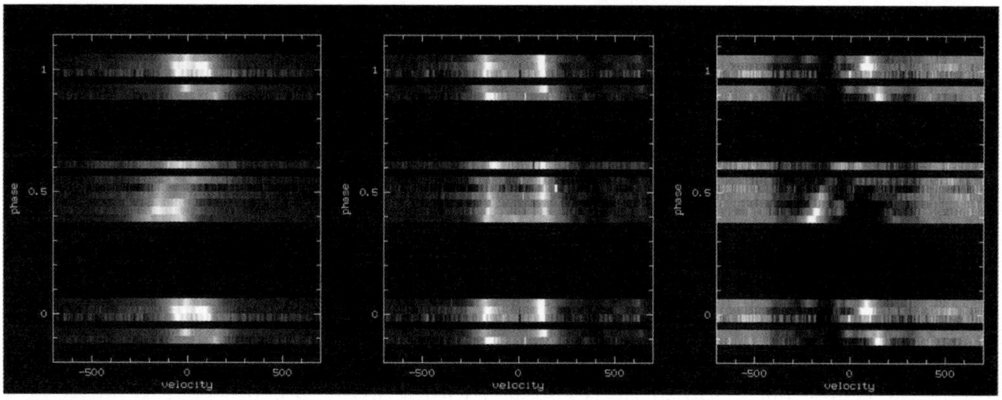

Figure 1. New spectra folded with revised period. ±20% phase extensions are shown for clarity. From left to right, the spectral lines are He II λ4686, Fe II λ5317, and He I λ6678. As phase zero, we adopt the periastron time given by Casares et al. (2014).

separation (Fig. 1, middle). This is due to the tidal distortion of the disk (Panoglou et al. 2016) and makes the line peaks unsuitable for measuring the RV amplitude of the Be star, K_1. The effect is well illustrated also by the base of Hα that changes in width with the orbital period. Since Casares et al. (2014) used such lines to measure K_1, this invalidates their results and thus also the mass ratio which led to the suggestion of a BH.

In the new spectra, we detected photospheric lines from the Be star, in particular He I λ6678 (Fig. 1, right). The outer edges have peak-to-peak RV amplitudes of about 20 km/s (red side) and 30 km/s (blue side), respectively. The implied amplitude of K_1 of 10 to 15 km/s is at least a factor of 2 less than the value of 32 km/s derived by Casares et al. (2014). At $K_1 = 7$ km/s, the RV amplitude of Hα (measured via the bisector of the lower wings) is even lower. This correction reduces the mass of the companion to the Be star by a similar factor and probably removes it from the BH domain.

3. Discussion

The He II λ4686 emission, from which the RV amplitude, K_2, of the companion to the Be star was determined, is double-peaked and arises from an accretion disk around the companion. It is fed by spillover from the disk around the Be star. Such variable emission is also seen in the well-known Be+sdO binaries φ Per, 59 Cyg, and HD 55606 (Chojnowski et al. 2018). In addition, there are circumstellar gas streams, also similar to many sdO stars: Fig. 1 shows the new spectra phased with the orbital period: The He I λ6678 emission traces the orbit of the secondary, just as He II λ4686 does. But there is also an additional, phase locked *absorption* component. Around phase 0.4 it grows into a very deep shell absorption lasting for about 15% of the cycle. Owing to the lack/weakness of X-rays, Casares et al. (2014) posited that the BH is quiescent, i.e., non-accreting. The improved spectroscopic mapping of the gas in the system contradicts this.

Hence, MWC 656 is more likely to host a hot-subdwarf instead of a black-hole, and its proximity no longer serves to indicate a relatively large number of Be+BH binaries.

References

Casares, J., Negueruela, I., Ribó, M., et al. 2014, *Nature*, 505, 378
Chojnowski, S. D., Labadie-Bartz, J., Rivinius, T., et al. 2018, *ApJ*, 865,
Panoglou, D., Carciofi, A. C., Vieira, R. G., et al. 2016, *MNRAS*, 461, 2616.
Wang, L., Gies, D. R., Peters, G. J., et al. 2021, *AJ*, 161, 248.
Williams, S. J., Gies, D. R., Matson, R. A., et al. 2010, *ApJL*, 723, L93

Massive Stars Near and Far
Proceedings IAU Symposium No. 361, 2024
J. Mackey, J. S. Vink, & N. St-Louis, eds.
doi:10.1017/S1743921323003976

Observations south of HD 5980

Andrés Sixtos, Aida Wofford and Antonio Peimbert

Intituto de Astronomía, UNAM
email: jasixtos@astro.unam.mx

Abstract. We study the He II λ4686 emission south of multiple-star system HD 5980, which hosts two WN-type stars. We use optical VLT FORS1 long-slit spectra. The observations are close to the eclipse phase. Broad He II emission with a SNR > 5 is observed as far as 7.6 pc from HD 5980. The He II emission that is closest to HD 5980 is 1.2 pc directly south and has a redshifted component with FWHM 1450 km s-1 originating in the eclipsing WN5-6 star, and a blueshifted component with FWHM 600 km s-1. The second component cannot be explained by either the nearly- eclipsed WN star or the nearby supernova remnant, SNR B0057-724. We suggest that the additional He II emission comes from the colliding winds of the two WN stars.

Keywords. binaries: eclipsing – stars: individual (HD 5980) – stars: Wolf–Rayet – HII regions – Magellanic Clouds

1. Introduction

NGC 346 (the brightest H II region of the Small Magellanic Cloud) hosts a multiple massive-star system, HD5980, that is of great interest for understanding massive-star evolution and binary-black hole formation.

HD 5980 is composed of least three massive stars. Its first component, star A, is a 60 ± 10 $M_\odot$ (Koenigsberger et al. 2014) that went through the eruptive luminous blue variable (LBV) phase during 1993-1994, and that now has a spectrum corresponding to that of a WN5-6 nitrogen-sequence Wolf-Rayet (WR) star. Its second component, star B, is a close eclipsing WN4 companion of similar mass (66 ± 10 $M_\odot$; Koenigsberger et al. 2014). The ultraviolet spectrum of HD 5980 has a third component, star C, which is believed to be itself a binary system containing a late O type supergiant (Hillier et al. 2019).

2. Observations

The VLT FORS1 observations that we use in this work were obtained on September 10, 2002. The observations include a direct V-band image that corresponds to a plate scale of 0.2" / pixel; and long slit (0.51" x 410") optical spectra of three different locations (slit A, slit B and slit C) of NGC 346 that are shown in Figure 2. Spectra were obtained at each location with three grism / filter combinations: low dispersion (3440-7600Å); blue (3560-5970 Å); and red (5330-7485 Å) spectra.

3. Properties of the broad He II emission lines

Tipically, the FWHM of WR stars is $\geqslant$ 1000 km/s, meanwhile the FWHM due to nebular emission is 100 km/s. We divided the slit into 5"-long windows. We find a He II emission maximum in window 22. Figure 1, left panel shows that the He II profile is composed of a redshifted-component with FWHM = 1439 km s^{-1} and a blueshifted-component with FWHM = 603 km s^{-1}. The emission is observed as far as 7.6 pc away from HD 5980. We find that the orbital phase (0.32) of our observations is close to the

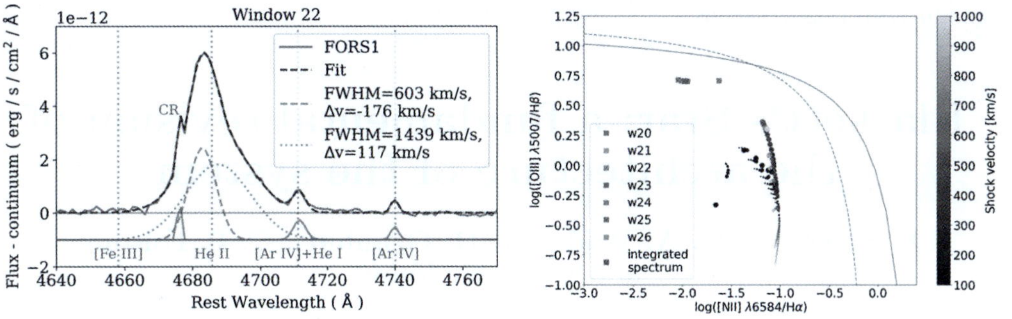

Figure 1. Left-. Continuum-subtracted spectrum corresponding to window 22. He II 4686 line (red curve). The blue curve is a multi-gaussian fit to the observations. The green curves show the individual Gaussian components (we use a dashed curve for He II). The legend gives the FWHM and velocity shift of the He II line components. Right-. [O III] 5007 / Hβ versus [N II] 6584/Hα diagnostic diagram. The dashed and solid lines represent the Kauffmann et al. (2003) and Kewley et al. (2001) curves. The squares represent values measured in windows of slit C with two gaussian components and the red-filled square is the value measured after integrating the spectra of the 7 windows. We also show the shock models of Alarie & Morisset (2019) with SMC metallicity, shock speeds between 200 and 1000 km s-1 (see colour bar) and magnetic field strengths between 0.0001 and 10 μG (given by the size of the filled circle, where the size increases with the strength).

eclipse of the system (0.36), so the second Gaussian component may not correspond to the second WN star of the binary system.

4. X-ray emission around HD 5980

There is X-ray emission around HD5980 due to the presence of the SNR B0057-724. This diffuse emission reaches the region covered by our slit C. Thus, we looked at the position of our observation in optical diagnostic diagrams relative to the radiative shock models of Alarie & Morisset (2019). It is unlikely that the nearby SNR significantly contributes to the He II emission because the observations are far from the location of the shock models.

References

Alarie, A. & Morisset C. 2019, *Rev. Mex. Astron. Astrofís*, 55, 377

Hillier, D. John, Koenigsberger, Gloria, Nazé, Yaël, Morrell, Nidia, Barbá, Rodolfo H. & Gamen, Roberto 2003, *MNRAS*, 486, 725

Kauffmann, Guinevere, Heckman, Timothy M., Tremonti, Christy, Brinchmann, Jarle, Charlot, Stéphane, White, Simon D. M., Ridgway, Susan E., Brinkmann, Jon, Fukugita, Masataka, Hall, Patrick B., Ivezić, Željko, Richards, Gordon T., Schneider & Donald P. 2003, *MNRAS*, 346, 1055

Kewley, L. J., Dopita, M. A., Sutherland, R. S., Heisler, C. A. & Trevena, J. 2001, *ApJ*, 556, 121

Koenigsberger, Gloria, Morrell, Nidia, Hillier, D. John, Gamen, Roberto, Schneider, Fabian R. N., González-Jiménez, Nicolás, Langer, Norbert & Barbá, Rodolfo 2014, *AJ*, 148, 62

Massive Stars Near and Far
Proceedings IAU Symposium No. 361, 2024
J. Mackey, J. S. Vink, & N. St-Louis, eds.
doi:10.1017/S1743921322002198

Plaskett's Star: a fundamental revision of the architecture of the system

E. Stacey[1], G. A. Wade[1], J. H. Grunhut[2], C. P. Folsom[3] and O. Kochukhov[4]

[1]Royal Military College of Canada, Kingston, Canada
email: erik@erikstacey.com

[2]Telus Digital, Toronto, Canada

[3]Tartu Observatory, University of Tartu, Tartu, Estonia

[4]Uppsala University, Uppsala, Sweden

Abstract. A century of study has characterized Plaskett's Star (HD 47129) as an evolved, massive, short-period, equal mass O+O binary system. The discovery of a magnetic field in the broad-line component by Grunhut et al. (2013) renewed interest in the study of this system and led to its establishment as the most rapidly rotating magnetic O-type star. Grunhut et al. (2021) observed the circular polarization signatures of the magnetic star to exhibit no radial velocity variations while the narrow-line star demonstrates radial velocity variations consistent with the established orbital period. This has raised fundamental questions about the architecture of this system and the nature of the magnetic star which have led to a major shift in our understanding of HD 47129.

Keywords. magnetic stars, massive stars, binary evolution

1. Introduction

Plaskett's Star was first discovered by Canadian Astronomer John Stanley Plaskett in 1922 (Plaskett 1922). It has been historically understood as a very massive ($M \approx 100$ $M_\odot$) double-lined, short-period ($P \approx 14.4$ d) spectroscopic binary containing two O-type components (see Linder et al. 2008). The broad-lined component was discovered to contain a magnetic field by Grunhut et al. (2013), establishing it as the most rapidly rotating known magnetic O-type star. Grunhut et al. (2022) have recently presented strong evidence against significant radial velocity variations in the magnetic star, which is supported by spectral disentangling analysis. These results are inconsistent with the near-equal mass binary model for this system and motivate a fundamental revision of our understanding of this system.

2. Observations

Spectroscopy of HD 47129 reveals a narrow-lined O8 component (NLC) and a broad-lined O7.5 component (BLC) (Linder et al. 2008) associated with a clear magnetic signature in Stokes V (Grunhut et al. 2013). The NLC is rich in nitrogen (16 $N_\odot$) and depleted in carbon (0.03 $C_\odot$), whereas the BLC is depleted in nitrogen (0.2 $N_\odot$) and rich in helium (1.76 $He_\odot$) (Linder et al. 2008). Plaskett's Star was recently observed extensively with the high resolution ESPaDOnS and Narval spectropolarimeters as reported by Grunhut et al. (2022). Analyzing these spectra, they found that substantially better fits to the Stokes V profiles were achieved assuming a constant RV for the magnetic star instead of the Linder et al. (2008) model. This is also reflected in the magnetic maps in each case, with the latter yielding an unrealistically complex magnetic geometry. This is further supported by direct measurement of radial velocities from disentangled Stokes I spectra. These results provide strong evidence against significant RV variability of the magnetic star that is required for the historical orbital model. This has motivated a fundamental revision of our model for the composition and structure of Plaskett's Star.

Plaskett's Star has been observed by both the Convection, Rotation, and Planetary Transits (CoRoT) satellite and the Transiting Exoplanet Survey Satellite (TESS) and demonstrates complex photometric variability (see Stacey et al., in prep.). The majority of this is attributable to modulation associated with the rotation of the magnetic star and its magnetosphere. The CoRoT data also present variability at the Linder et al. (2008) orbital period, and stochastic low-frequency variability (Blomme et al. 2011) is present in both datasets and qualitatively similar to the *O bright giant and supergiant* sample of Bowman et al. (2020).

3. Hypothesis 1: An unseen object

We consider a case where the NLC is in a close $P \approx 14.4$ d orbit with a third object, and the BLC is either dynamically unrelated or weakly dynamically bound (e.g. hierarchical trinary). The minimum mass of a companion based on the RV variations of the NLC is approximately 14 $M_\odot$ which is consistent with a potential black hole (BH) companion. There are several analogous examples in literature of O/B+BH systems that support the plausibility of this scenario (e.g. Cyg X-1; Brocksopp et al. 1999, MWC 656; Casares et al. 2014), nevertheless the strong winds expected of an O-type star as difficult to reconcile with the weak X-ray luminosity of HD 47129 ($L_X = 8.34 \times 10^{32}$ erg/s; Linder et al. 2006). Infrared interferometry reveals no targets from 1 to 8000 mas (subtending $\approx 1 - 8000$ AU) that are consistent with the luminosity of the BLC (Sana 2014). Therefore, this solution restricts the BLC to narrow ranges of orbital phase in a hierarchical configuration. Additionally, this solution does not address the rapid rotation of the magnetic star nor the system's chemical peculiarities.

4. Hypothesis 2: A stripped star

A second revised model under consideration for HD 47129 instead enforces the mass ratio as implied by the radial velocity models measured from the disentangled spectra. Assuming the BLC is a "normal" magnetic O star with a mass of (47.3 ± 0.3) $M_\odot$ (Linder et al. 2008), this requires a mass of approximately $1 - 2$ $M_\odot$ for the NLC. This component still demonstrates an O spectral type in the spectroscopic observations and comparable luminosity to the BLC. Therefore, we postulate the NLC may be a stripped sdO star (Götberg et al. 2018). This prescription addresses the radial velocity measurements of both components simultaneously with the anomalously high rotation rate of the magnetic star (through angular momentum transfer in the stripped material). The He/N enhancements and C/N/O depletions associated with these objects (Götberg et al. 2018) is also consistent with the Linder et al. (2008) abundances and apparant lack of oxygen lines in the system. Interpreting the NLC as a stripped sdO star currently appears to best address the outstanding issues with modelling Plaskett's Star while remaining consistent with existing observational results.

Supplementary material

To view supplementary material for this article, please visit http://dx.doi.org/10.1017/S1743921322002198.

References

Plaskett, J. S. 1922, MNRAS, 82, 447

Linder, N.; Rauw, G.; Martins, F.; Sana, H.; De Becker, M. and Gosset, E. 2008, *A&A*, 489.2, 713

Grunhut, J. H.; Wade, G. A.; Leutenegger, M.; Petit, V.; Rauw, G. et al. 2013, *MNRAS*, 428.2, 1686

Grunhut, J. H.; Wade, G. A.; Folsom, C. P.; Neiner, C.; Kochukhov, O. et al. 2022, *MNRAS*, 512.2, 1944

Blomme, R.; Mahy, L.; Catala, C.; Cuypers, J., Gosset, E. et al. 2011, *A&A*, 533, A4

Bowman, D. M.; Burssens, S.; Simón-Díaz, S.; Edelmann, P. V. F. et al. 2020, *A&A*, 640, A36

Brocksopp, C.; Fender, R. P.; Larionov, V.; Lyuty, V.M.; Tarasov, A.E. et al. 1999, *MNRAS*, 309.4, 1063

Casares, J.; Negueruela, I.; Ribó, M.; Ribas, I.; PAredes, J. M. et al. 2014, *Nature*, 505.7483, 378

Linder, N.; Rauw, G.; Pollock, A.M.T. and Stevens, I.R. 2006, *MNRAS*, 370.4, 1623

Sana, H.; Le Bouquin, J.-B.; Lacour, S.; Berger, J.-P.; Duvert, G. et al. 2014, *ApJS*, 215.1, 35

Götberg, Y.; de Mink, S. E.; Groh, J.H; Kupfer, T.; Crowther, P. A. et al. 2018, *A&A*, 615, A78

Massive Stars Near and Far
Proceedings IAU Symposium No. 361, 2024
J. Mackey, J. S. Vink, & N. St-Louis, eds.
doi:10.1017/S1743921322003222

Evolution and final fate of massive post-common-envelope binaries

Dandan Wei[1], Fabian R. N. Schneider[1,2], Philipp Podsiadlowski[3], Eva Laplace[1], Jan Henneco[1], Friedrich K. Röpke[1,2] and Marco Vetter[1]

[1]Heidelberger Institut für Theoretische Studien, Schloss-Wolfsbrunnenweg 35, 69118 Heidelberg, Germany
email: dandan.wei@h-its.org

[2]Zentrum für Astronomie der Universität Heidelberg, Heidelberg, Germany

[3]University of Oxford, St Edmund Hall, Oxford, OX1 4AR, United Kingdom

Abstract. Mergers of neutron stars and black holes are nowadays observed routinely thanks to gravitational-wave astronomy. In the isolated, binary-evolution channel, a common-envelope phase of a red supergiant and a compact object is crucial to sufficiently shrink the orbit and thereby enable a merger via gravitational-wave emission. Here, we use the outcome of three-dimensional hydrodynamic common-envelope simulations of a 9.4 solar mass red supergiant and a 5 solar mass black-hole to explore the further evolution and final fate of the remnant binary. The binary system undergoes another phase of mass transfer during which it is visible as an X-ray binary. We find that the donor star does not explode as an ultra-stripped supernova because of the large remaining envelope mass, but as a Type Ib/c supernova. Supernova kicks are actually required to sufficiently perturb the orbit and thus facilitate a merger within a Hubble time via gravitational-wave emission.

1. Introduction

In the isolated binary-evolution channel, a common-envelope(CE) phase of a red supergiant (RSG) and a compact object is crucial to sufficiently shrink the orbit of the binary, and thereby enable a merger via gravitational-wave emission. Here, we use the outcome of three-dimensional hydrodynamic CE (3D-CE) simulations of a 9.4 $M_\odot$ RSG and a 5 $M_\odot$ black-hole (BH) to explore the further evolution and final fate of the remnant binary.

2. Results and Conclusions

We use the MESA stellar evolution code (Paxton et al. 2018) to let the same RSG star, as employed in a complex 3D-CE simulation from Moreno et al. (2022), lose most of its envelope quickly to get a remnant RSG core, mimicking a 3D-CE event. A BH companion of 5.0 $M_\odot$ orbits this remnant core as a post-CE binary in our MESA model. We simulate the binary with the original parameters from the outcome of the 3D-CE model. However, the remnant RSG core with 2.97 $M_\odot$ in our model expands during the CE envelope removal such that it does not fit into the Roche lobe of the post-CE binary orbit as suggested by the 3D-CE simulation. Further envelope removal ($\sim 0.17\,M_\odot$) is required and was applied. Therefore, we get a remnant RSG core of 2.8 $M_\odot$. The orbital period of this post-CE binary is 3.53 days, determined at the point where the available

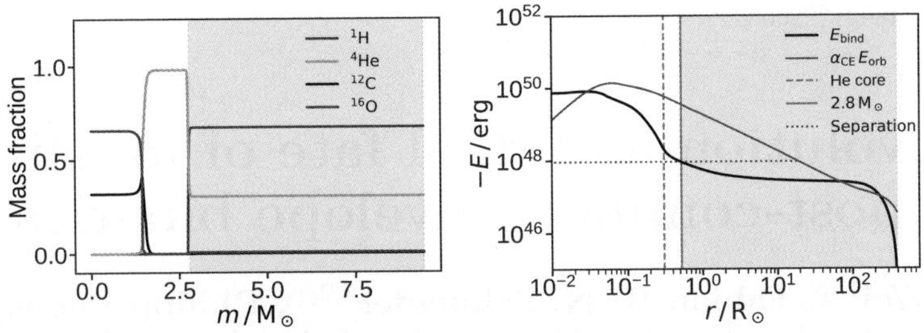

Figure 1. Chemical composition of the RSG donor with $9.4\,\mathrm{M_\odot}$ in the left panel. The gray shaded area indicates the envelope lost in the CE event. The right panel shows binding energy ($E_{\rm bind}$) of the RSG donor and the available orbital energy ($\alpha_{\rm CE}E_{\rm orb}$, $\alpha_{\rm CE} = 0.69$) of the binary with a $5.0\,\mathrm{M_\odot}$ BH companion.

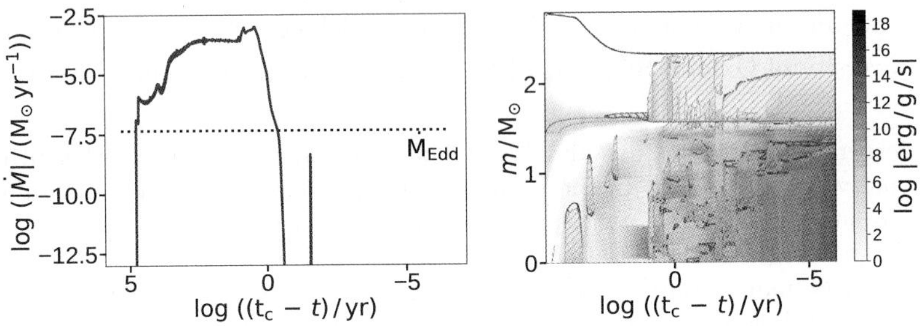

Figure 2. Mass-transfer rate and Kippenhahn diagram of the remnant RSG core, as a function of the remaining stellar age of the remnant RSG core ($t_c - t$) until core collapse.

orbital energy ($\alpha_{\rm CE}\,E_{\rm orb}$) equals the binding energy ($E_{\rm bind}$) of the donor star (see the right panel in Fig. 1). The envelope-ejection efficiency $\alpha_{\rm CE} = 0.69$ (Moreno et al. 2022).

The post-CE binary initiates a mass-transfer (MT) phase, and is fully computed until core-collapse (see Fig. 2). After the MT phase, the mass of the remnant RSG core decreases to $\sim 2.3\,\mathrm{M_\odot}$, but only quite few material ($\sim 0.002\,\mathrm{M_\odot}$) is accreted by the BH companion due to Eddington-limited accretion of $4.42 \times 10^{-8}\,\mathrm{M_\odot\,yr^{-1}}$. The remaining helium-rich envelope of the remnant RSG core is $\sim 0.7\,\mathrm{M_\odot}$. It will not explode as an ultra-stripped SN. The final orbital configuration is too wide (5.16 days) to get a merger within a Hubble time without a SN kick. However, it is possible that roughly 5% of the post-CE binary merge within a Hubble time if the SN kick is favorable (Moreno et al. 2022). The mass of the remnant RSG core is changed during the MT phase, which affects the possibility of a SN explosion. It can also end its life as an electron-capture SN or an oxygen-neon-magnesium white dwarf, in other slightly different binary configurations.

Acknowledgements

This work is supported by the Klaus Tschira Foundation and the Sino-German (CSC-DAAD) Postdoc Scholarship from China Scholarship Council and German Academic Exchange Service. It has received funding from the European Research Council (ERC) under the European Union's Horizon 2020 research and innovation programme (Grant agreement No. 945806).

References

Moreno, M. M., Schneider, F. R. N., Roepke, F. K., Ohlmann, S. T., Pakmor, R., Podsiadlowski, P., & Sand, C. 2022, A&A in print

Paxton, B., et al. 2018, ApJS, 234, 34

Chapter V:
Stellar structure and evolution of single stars

Massive Stars Near and Far
Proceedings IAU Symposium No. 361, 2024
J. Mackey, J. S. Vink, & N. St-Louis, eds.
doi:10.1017/S1743921322003349

Stellar Structure and Evolution of Massive Rotating Single Stars

R. Hirschi[1,2], E. Kaiser[1], P. Eggenberger[3], S. Ekström[3], C. Georgy[3], A. Maeder[3] and G. Meynet[3]

[1]Astrophysics Group, Lennard-Jones Laboratories, Keele University, Keele ST5 5BG, UK
email: r.hirschi@keele.ac.uk

[2]Kavli IPMU (WPI), University of Tokyo, 5-1-5 Kashiwanoha, Kashiwa 277-8583, Japan

[3]Geneva Observatory, Geneva University, CH-1290 Sauverny, Switzerland

Abstract. Rotation plays an important role in the structure and evolution of massive stars. It leads to deviation from spherical symmetry for very fast rotating stars, mixing in otherwise unmixed radiative regions and generally increased mass loss. In addition, magnetic fields interact with rotation and lead to significant transport of angular momentum. In this article, we review the various rotational and magnetic instabilities present in massive stars and their implementation in one-dimensional stellar evolution codes. We then focus on their impact on the evolution of single rotating stars. Finally, we compare rotating models to observations and discuss ways to disentangle between various uncertainties.

Keywords. rotation - hydrodynamics - magnetism - stars: evolution, interiors, massive

1. Structure and Evolution Models of Rotating Stars

Modelling the evolution of stars and populations of stars is still generally limited to one-dimensional (1D) models though the number and range of multi-dimensional hydrodynamic simulations are growing as demonstrated by many contributions in these proceedings (see e.g. contributions by Rieutord and Rizzuti in this Session). 1D models make the assumption of spherical symmetry, which means that physical quantities such as temperature and pressure are considered to be constant on spherical shells and only depend on radius. The mass enclosed by a shell is usually preferred to the radius as the independent variable in the Lagrangian description used in stellar models. The structure of a star can then be obtained by solving the 1D version of the four stellar structure equations (conservation of mass, momentum and energy as well as energy transfer). Modelling the evolution of stars requires additional physics and equations related to nuclear reactions, mass loss, convection, rotation, magnetic fields, binary interactions, equation of state, opacities, neutrino losses, etc. including their dependence on metallicity.

We will focus here on rotation and magnetic fields and their interactions (see other contributions for the other aspects). Rotation has several effects on the structure and evolution of stars. Axial rotation opposes the effects of gravity (centrifugal force) and thus breaks spherical symmetry. It is, however, possible to keep stellar structure equations 1D (sometimes referred to as 1.5D) by making the assumption of shellular rotation (Zahn 1992). This assumes that instabilities can act much faster perpendicular to gravity (i. e. strong horizontal turbulence on isobars) and leads to constant angular velocity on isobars. Other quantities like density and temperature are not constant on isobars and averages must be used in the structure equations (see Meynet & Maeder 1997, for more details). Angular velocity and angular momentum generally vary across the star (unless solid body rotation is imposed, which is usually done at the start of the evolution or in convective regions). The evolution of rotation is modelled using the additional equation of angular momentum transport (the conserved quantity). Other key effects of rotation include enhanced mass loss (see Langer 1998; Maeder & Meynet 2000; Gagnier et al. 2019a, for details and discussions) and internal transport of angular momentum and mixing of the chemical composition in otherwise unmixed radiative zones. Magnetic fields bring further deviations from spherical symmetry and a rotating magnetic star would ideally be modelled using 3D magneto-hydrodynamic simulations (see e.g. Braithwaite & Spruit 2004; Varma & Müller 2021; Yoshida et al. 2021). To simulate the long-term evolution of stars, however, 1D theoretical prescriptions have to be used to model the evolution of magnetic fields and their interactions with rotation.

2. Internal Transport Mechanisms

There are many internal transport mechanisms at play in massive stars: convection, internal gravity waves (e.g. Talon & Charbonnel 2005; Edelmann et al. 2019), rotational (aka hydrodynamic) and magnetic instabilities. Note that in binary stars, tidal effects (Section 4; e. g. Zahn 1977; Ogilvie 2014), are also able to redistribute angular momentum and composition. We limit here the discussion to rotational and magnetic instabilities.

2.1. *Rotational Instabilities*

The two dominant rotational instabilities are meridional (aka Eddington-Sweet) circulation (Talon & Zahn 1997; Zahn 1992; Maeder & Zahn 1998; Maeder 2009) and shear instabilities. Other instabilities have been identified: Solberg-Hoiland, GSF, etc. but they are unlikely to significantly change the evolution of stars (Heger et al. 2000; Hirschi et al. 2004; Caleo et al. 2016). Meridional circulation is driven by the thermal gradients on isobars (polar regions have higher gravity and radiative fluxes and thus are hotter than the equatorial regions). The circulation can transport angular momentum both inward and outward and thus can both build and reduce angular velocity gradients depending on the direction of the circulation (see fig. 1 in Meynet & Maeder 2002). Indeed, angular momentum is transported inward when the circulation goes outward at the pole (where the angular momentum is zero) and inward at the equator and vice versa. Shear instabilities develop when there is differential rotation inside the star since, in this case, adjacent layers move laterally with respect to each other and experience a strain. Shear instabilities can take place on dynamical timescales if the excess kinetic energy is larger than the work against the buoyancy force needed to overturn the shear layer. Dynamical shear is fast but generally limited to very narrow regions at the edge of convective regions and it thus has a very small impact on the structure and evolution of rotating stars (Hirschi et al. 2004; Hirschi & Maeder 2010; Edelmann et al. 2017). When the shear layers are dynamically stable, thermal diffusion can help shear transport angular momentum and composition on a much longer, secular timescale (Zahn 1974; Maeder 1997). Shear, whether dynamical or secular, mixes regions together so can only reduce gradients.

Shear mixing obviously only takes place when there is a gradient of angular velocity. It is important to note that this is not the case with meridional circulation, which can occur even when the star has solid body rotation. Circulation instead will be quite sensitive to the assumptions made concerning the horizontal turbulence. As the evolution proceeds, chemical composition (μ) gradients develop, which will inhibit both instabilities (see Frischknecht et al. 2010, for how the evolution on the MS affects these instabilities).

2.2. *Magnetic Instabilities*

Various magnetic instabilities have been studied in the context of massive stars. Here we will only discuss the Tayler-Spruit dynamo (Spruit 2002; Heger et al. 2005) and its modification by Fuller et al. (2019). Other instabilities include the magneto-rotational instabilitiy (Wheeler et al. 2015; Griffiths et al. 2022) and the $\alpha - \Omega$ dynamo (Potter et al. 2012b). For fossil fields and the connection between magnetic fields and mass loss, see Session 3 (physical processes). In the Tayler-Spruit dynamo (TS dynamo hereinafter; see Tayler 1973; Spruit 2002; Heger et al. 2005; Maeder & Meynet 2005, for details), an initially very small seed field is assumed. Due to differential rotation, the radial component (B_r) of the field is wound up and the azimuthal component (B_ϕ) is thus amplified. The Tayler (pinch-type) instability then generates a stronger radial component (B_r), thus closing the amplification loop. The amplification saturates at the point where the amplification and damping (dominated by magnetic diffusivity) timescales become equal. A minimum amount of differential rotation (denoted $q_{\rm min}$) is required to trigger this instability meaning that the instability is not present in the case of solid body rotation.

Fuller et al. (2019) propose modifications to the Tayler-Spruit dynamo (FTS dynamo hereinafter). The main ones are the component of the magnetic fields (background versus perturbed component) being wound up (perturbed radial field in Spruit 2002) and damped (background azimuthal field damped in Spruit 2002). Both these changes lead to higher saturation levels of the magnetic fields and enhanced transport.

2.3. *Implementation in 1D Stellar Evolution Codes*

Magnetic instabilities (both TS and FTS dynamos) like shear instabilities lead to transport of angular momentum and chemical composition that is suitably represented as a diffusive process (using an effective diffusion). As written above, meridional circulation, on the other hand, can both build and reduce gradients of angular momentum and is thus best described with an advective term in the angular momentum transport equation as done in `GENEC` (Eggenberger et al. 2008; Ekström et al. 2012) and several other codes `FRANEC` (Chieffi & Limongi 2013) and `STARS/ROSE` (Potter et al. 2012a). Taking into account an advective term, however, leads to a fourth-degree equation that is time consuming and difficult to solve. Several other codes `MESA` (e.g. Paxton et al. 2013) `STERN` and `KEPLER` (e.g. Heger et al. 2000; Petrovic et al. 2005), thus treat the meridional (Eddington-Sweet) circulation as a diffusive process. For the transport of chemical elements, all the above instabilities lead to mixing that can be treated as diffusion and for which an effective diffusion coefficient is used for the combined effect of meridional circulation and horizontal turbulence in `GENEC`. In addition to whether meridional circulation is treated as an advective or diffusive process, there are further differences in between codes in the formulae used for specific instabilities as well as parameters modifying the strength of these (e.g. f_c and f_μ, see Heger et al. 2000, for details), which are calibrated to sets of observations. One thus has to be careful when comparing rotating models calculated with different codes, even when they include the same physical processes.

3. Structure and Evolution of Massive Rotating Stars

Massive stars go through six burning phases (H, He, C, Ne, O and Si) after which their iron core collapses into a neutron star or black hole. The evolution of the structure of as star can be summarised in the structure evolution diagram (aka Kippenhahn diagram), an example of which is presented in Fig. 1 for a 15 $M_\odot$ non-rotating model. This figure includes iso-radius contour lines ($\log_{10}(r)$). These nicely show how the core of massive stars contracts (contours moving up in mass coordinate) throughout their evolution. Contraction is modest during a burning stage but then accelerates at the end of each burning phase when nuclear reactions no longer counteracts the gravitational force. The contraction stops (and is sometimes even slightly reversed) when a high enough temperature is reached in the core for the next burning phase to start. The envelope of massive

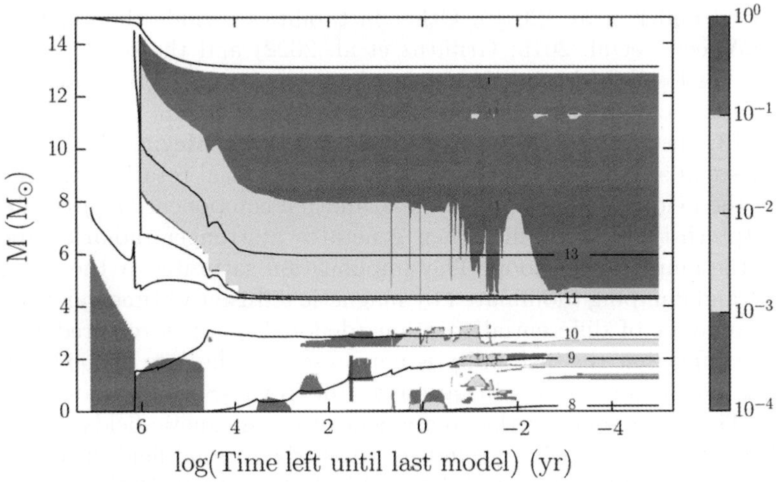

Figure 1. Structure evolution diagram of the 15 $M_\odot$ (non-rotating) solar metallicity `GENEC` stellar model. The horizontal axis is a logarithmic scale of the time left before the predicted collapse of the star in years (the last model in this simulation is before the end of silicon burning but since the time-scale of silicon burning is so short this does not affect the plot for the earlier phases) and the vertical axis is the mass coordinate in solar masses. The total mass and radial contours (in the form $\log_{10}$(r) in cm), are drawn as solid black lines. Shaded areas correspond to convective regions. The colour indicates the value of the Mach number (figure taken from Cristini et al. 2017).

stars, on the other hand, expands through most of their evolution with the largest and fastest expansion taking place at the end of the main sequence (MS) when most massive stars become red or blue supergiants (WR stars do not expand as much). Key effects of rotation on the structure and evolution of massive stars are enhanced mass loss and larger core masses (due to mixing in radiative zones) leading to rotating stars evolving like more massive non-rotating stars (see e. g. Hirschi et al. 2004, for details).

3.1. *Evolution of Rotation*

The evolution of rotation inside a massive stars is predominantly driven by the core contraction and envelope expansion described above and by convective regions (coloured regions in Fig 1) efficiently redistributing angular momentum within these regions. Given that angular momentum is conserved and that the specific angular momentum, j, is proportional to Ωr^2, a contraction (expansion) leads to an increase (decrease) in the angular velocity (Ω). This explains why the angular velocity (Ω) increases in the core and generally decreases in the envelope throughout the evolution. Fig. 2 (*left*) shows the evolution of the angular velocity for a `GENEC` 25 $M_\odot$ rotating stellar model. This model loses its H-rich envelope during He-burning and thus loses its very low Ω tail by C-burning. Stellar evolution models assume that convective regions rotate as solid bodies so have a flat angular velocity profile (though see question at the end). This explains the flat Ω profiles in the core during various burning phases and in the convective envelope (see left panels in Fig. 3). By the end of the evolution, Ω reaches values around 1 s^{-1}.

To determine if angular momentum is transported between the core and envelope or lost by stellar winds, it is easier to consider the evolution of the specific angular momentum (j) profiles (see Fig. 2 *right* panel), which are not modified by contraction or expansion. In `GENEC` models, meridional circulation and shear transport j out of the core mostly during H-burning and to a smaller extent during core He-burning. During the advanced phases, only convection has the time to redistribute angular momentum, creating a saw-tooth

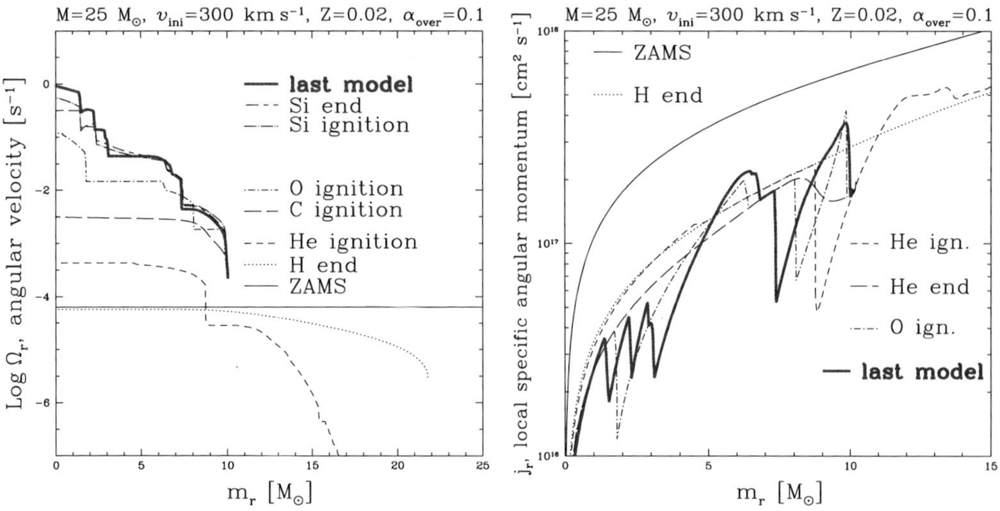

Figure 2. Angular velocity (*Left*) and local specific angular momentum profiles as a function of the Lagrangian mass coordinate, m_r, inside the 25 $M_\odot$ model ($v_{\rm ini} = 300$ km s^{-1}) GENEC solar metallicity model at various evolutionary stages (figure taken from Hirschi et al. 2004).

pattern given the assumption of constant Ω in convective regions. Mass loss removes a significant amount of mass, more than half the initial mass in this 25 M$_\odot$ model. This demonstrates the crucial effect of mass loss for stars above 25 − 30 M$_\odot$. Angular momentum being highest at the surface, stellar winds also remove significant amount of angular momentum from massive stars. Rotation in the core, however, is more affected by internal transport than mass loss at the surface as discussed below.

Before discussing magnetic instabilities, it is instructive to compare non-magnetic rotating models since, as discussed above, different theoretical prescriptions are used in different codes and their technical implementation differ too. The discussion here is limited to one representative `GENEC` model (Fig. 2) taken from Hirschi et al. (2004) (see Frischknecht et al. 2010; Ekström et al. 2012; Chieffi & Limongi 2013, for more detailed discussions) and one representative `MESA` model (*top* row of Fig. 3) taken from Kaiser (2021, see Sect. 2.3 for references to other similar codes). The surface rotation is different because the 25 M$_\odot$ (`GENEC`) model loses its H-rich envelope while the 15 M$_\odot$ (`MESA`) model does not. This is explained by mass loss. The core angular velocity (*left* panels), at first sight, appear to evolve in a similar way with the increase in Ω explained by the contraction of the core. There are key differences, however, in the timing and extent of the angular momentum transport (*right* panels) between the core and envelope. The first one is the absence of j transport between the core and envelope during the MS in the `MESA` model. The second is the smaller overall decrease in the specific angular momentum. These are mainly due to breaks in the transport just above convective cores where composition gradients inhibit instabilities (see Kaiser 2021, for a detailed discussion). Despite these differences, we see that both these `GENEC` and `MESA` non-magnetic models end with a core angular velocity, $\Omega \sim 1\,{\rm s}^{-1}$.

4. Magnetic Instabilities and their Effect on Internal Rotation

Values of $\Omega \sim 1\,{\rm s}^{-1}$ at the end of the evolution of massive stars are too high to explain the observed rotation rate of pulsars (see comparison below). Such models would actually predict neutron stars at critical rotation and this discrepancy between models and observations is one of the main drivers to include magnetic instabilities in stellar models

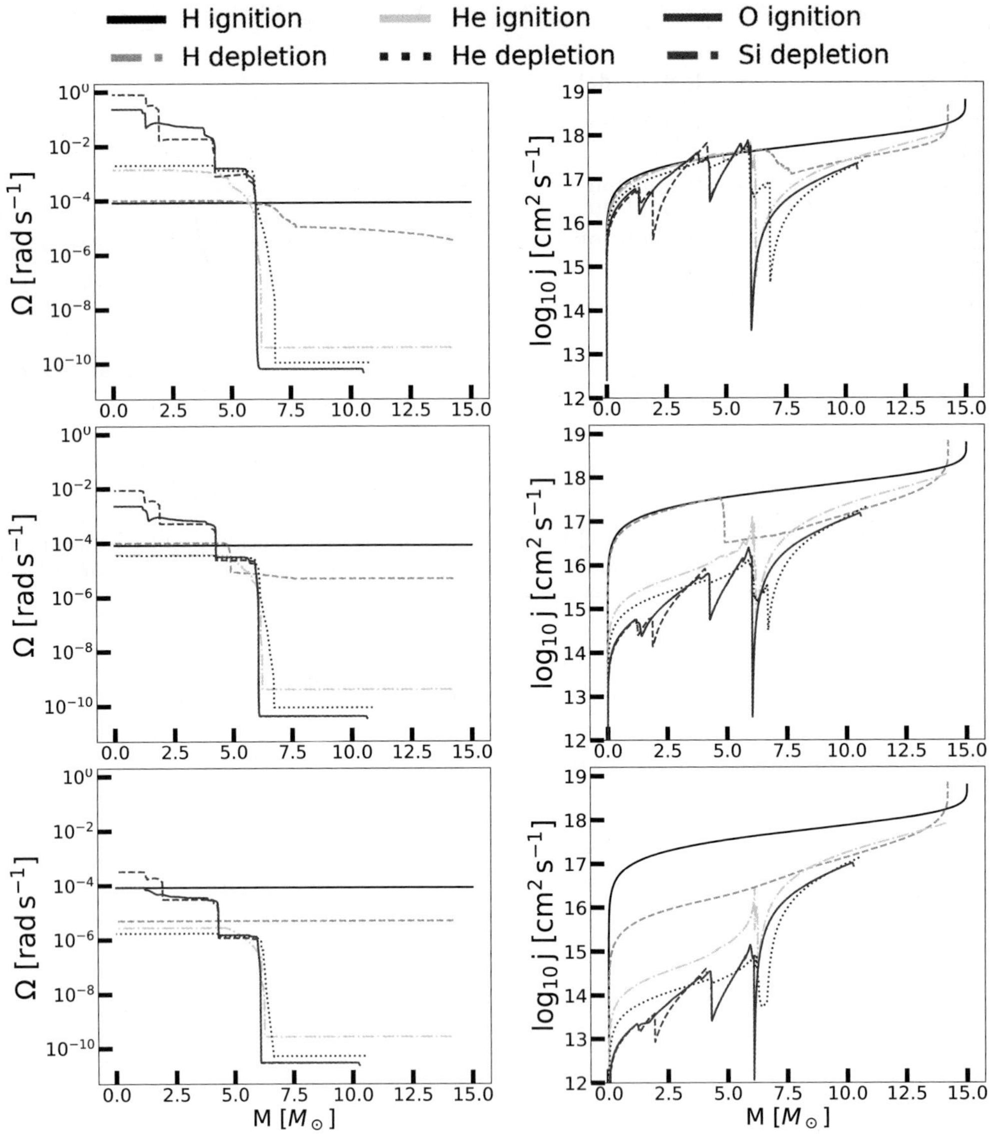

Figure 3. Evolution of rotation profiles for $15\,M_\odot$ `MESA` models including different angular momentum transport mechanisms, *top*: without magnetic fields, *middle*: TS dynamo, *bottom*: FTS dynamo, showing on the left panel the angular rotation velocity, Ω, and on the right panel the specific angular momentum, j. Each line presents the profile at a different evolutionary stage indicated by the colour and linestyle (see legend on top). To enable a better comparison, the axis of each thematic subplot group has been scaled to the same values. These models were computed with the `MESA` stellar evolution code (Paxton et al. 2011, 2013, 2015, 2018), revision 10108 (figure adapted from multiple figures from Kaiser 2021).

to solve this issue (e.g. Heger et al. 2000, 2005; Petrovic et al. 2005; Hirschi et al. 2005; Wheeler et al. 2015; Ma & Fuller 2019).

4.1. *Taylor-Spruit and Fuller-modified Taylor-Spruit Dynamos*

We briefly investigate here the impact of the TS or FTS dynamos on the evolution of rotation in $15\,M_\odot$ `MESA` models (with an initial rotation corresponding to $\Omega/\Omega_{\rm crit} = 0.4$)

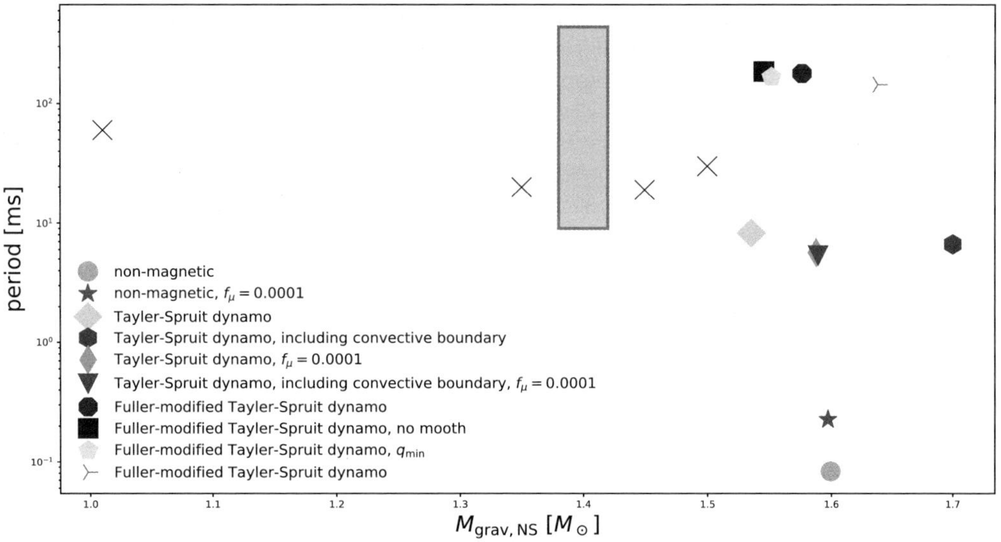

Figure 4. Observed (black crosses) and predicted (coloured symbols) neutron star spin period as a function of the gravitational mass for various 15 $M_\odot$ models including the ones from Fig. 3. The black crosses show observed neutron star spins by Muslimov & Page (1996) and Faucher-Giguère & Kaspi (2006). The grey rectangle depicts an observed range of neutron star spins (Marshall et al. 1998; Kaspi & Helfand 2002; Faucher-Giguère & Kaspi 2006; Popov et al. 2010; Popov & Turolla 2012; Gotthelf et al. 2013) with an assumed neutron star mass of 1.4 $M_\odot$ (figure taken from Kaiser 2021).

and compare these models to a non-magnetic model in Fig. 3. Even though it has a limited impact during the MS (see discussion in Kaiser 2021), the magnetic dynamo viscosity in the TS dynamo model (*middle* panels) is much stronger than the rotational instabilities included in the non-magnetic model (*top* panels). This leads to a strong transport of angular momentum out of the core between the end of core H-burning and the start of core He-burning (the period of strong core contraction and envelope expansion) and a much reduced final angular velocity profile, $\Omega \sim 10^{-2}\,\mathrm{s}^{-1}$. Considering now the FTS dynamo model (*bottom* panels), the even larger magnetic dynamo viscosity already operates during the MS and leads to an even lower core angular velocity, $\Omega \sim 10^{-4}\,\mathrm{s}^{-1}$.

4.2. *Predicted Rotation Rate of Pulsars*

Figure 4 presents the predicted estimated neutron star period as a function of the gravitation mass, $M_{\mathrm{grav,NS}}$, for 15 $M_\odot$ models at solar metallicity with different angular momentum transport assumptions (see legend for key assumptions). The spin period and the gravitational mass are computed following Heger et al. (2005), their Section 3.4. Additionally, Fig. 4 also includes observed spin periods of young neutron stars. A few neutron stars have fast natal spin periods of 11 ms (Marshall et al. 1998) or 20 ms (Muslimov & Page 1996; Kaspi & Helfand 2002) but the majority rotate slower with a spin period of 50 – 100 ms (Muslimov & Page 1996; Faucher-Giguère & Kaspi 2006; Popov et al. 2010; Popov & Turolla 2012). There are also a few very slowly rotating neutron stars with a natal spin of up to 400 ms (Gotthelf et al. 2013). The comparison between the model predictions and observations reveals several points. First, the three different angular momentum transport combinations predict three distinct ranges for the spin period of neutron stars, independently of the various related uncertainties in angular momentum transport (e.g. spatial and temporal smoothing, "q_{min}", f_μ; see Kaiser 2021, for details). However, $M_{\mathrm{grav,NS}}$ does show a dependence on the uncertainty, mainly due to the amount of chemical mixing occurring during the stellar evolution. Second, the non-magnetic models fail to predict the observed rotation rate of neutron stars because of missing angular momentum transport. Third, the two magnetic dynamos predict spin periods that are in the range of observations. However, the estimate of the models with the TS dynamo only covers the very fast rotating neutron stars, whereas the prediction by the models with the FTS dynamo agrees with the slower rotating neutron stars in the sample. None of the models manages to predict the full range of natal neutron star spin periods. Therefore, while the Tayler-Spruit dynamo does not transport enough angular momentum, the FTS version probably slows down the rotation cores too much. Salmon et al. (2022) show, using asteroseismic inference of the internal rotation of the β Cephei star HD 129929, that the FTS dynamo is definitely too strong for this star.

5. Discussion and Conclusion

We briefly investigated the evolution of the structure and rotation in massive stars and the impact of rotational and magnetic instabilities on the evolution of rotation. Importantly, magnetic instabilities bring the predicted spin of neutron stars much closer to the observed ones with the Tayler-Spruit dynamo reaching the fast end of observations and the Fuller-modified Tayler-Spruit dynamo reaching the slow end. Note, however that the comparisons above assume that the angular momentum content of the core at the time of the pre-collapse is the same as the one that will be locked into the neutron star, an hypothesis that may not be always correct. Furthermore, there are still major uncertainties concerning the theories used to describe rotation and magnetic fields in stars and their implementation. In particular, one has to be careful when comparing the predictions of different codes since the implementation of physical processes vary between codes. There are fortunately many other constraints from observations of stars in different mass ranges and evolutionary phases that are and will be able to constrain these processes and disentangle uncertainties. Examples include the key role of rotation and magnetic instabilities in the context of the solar modelling problem (Eggenberger et al. 2022), asteroseismology (see contributions by Burssens and Bowman in these proceedings; e.g. Pamyatnykh et al. 2004; Deheuvels et al. 2014; Aerts et al. 2019; Claret & Torres 2018; Pedersen et al. 2021; Salmon et al. 2022), the luminosity-mass plane (Higgins & Vink 2019), comparison of models to observations of low-mass stars and white dwarfs (Cantiello et al. 2014; Spada et al. 2016; Fuller & Ma 2019; Eggenberger et al. 2019a,b; den Hartogh et al. 2019, 2020). In addition multi-D simulations (e.g. Edelmann et al. 2017; Gagnier et al. 2019b; Varma & Müller 2021; Yoshida et al. 2021; Petitdemange et al. 2022) of stellar interiors will also help guide and constrain theories used in 1D models.

Acknowledgements

RH acknowledges support from the World Premier International Research Centre Initiative (WPI Initiative), MEXT, Japan and the IReNA AccelNet Network of Networks (National Science Foundation, Grant No. OISE-1927130). GM, SE, PE, CG acknowledge funding from the European Research Council (ERC) under the European Union's Horizon 2020 research and innovation program (Grant No. 833925). This article is based upon work from the ChETEC COST Action (CA16117) and the European Union's Horizon 2020 research and innovation programme (ChETEC-INFRA, Grant No. 101008324).

References

Aerts C., Mathis S., Rogers T. M., 2019, ARA&A, 57, 35
Braithwaite J., Spruit H. C., 2004, Nature, 431, 819
Caleo A., Balbus S. A., Tognelli E., 2016, MNRAS, 460, 338

Cantiello M., Mankovich C., Bildsten L., Christensen-Dalsgaard J., Paxton B., 2014, ApJ, 788, 93
Chieffi A., Limongi M., 2013, ApJ, 764, 21
Claret A., Torres G., 2018, ApJ, 859, 100
Cristini A., Meakin C., Hirschi R., Arnett D., Georgy C., Viallet M., Walkington I., 2017, MNRAS, 471, 279
Deheuvels S., et al., 2014, A&A, 564, A27
Edelmann P. V. F., Röpke F. K., Hirschi R., Georgy C., Jones S., 2017, A&A, 604, A25
Edelmann P. V. F., Ratnasingam R. P., Pedersen M. G., Bowman D. M., Prat V., Rogers T. M., 2019, ApJ, 876, 4
Eggenberger P., Meynet G., Maeder A., Hirschi R., Charbonnel C., Talon S., Ekström S., 2008, AP&SS, 316, 43
Eggenberger P., et al., 2019a, A&A, 621, A66
Eggenberger P., den Hartogh J. W., Buldgen G., Meynet G., Salmon S. J. A. J., Deheuvels S., 2019b, A&A, 631, L6
Eggenberger P., Buldgen G., Salmon S. J. A. J., Noels A., Grevesse N., Asplund M., 2022, Nature Astronomy, 6, 788
Ekström S., et al., 2012, A&A, 537, A146
Faucher-Giguère C.-A., Kaspi V. M., 2006, ApJ, 643, 332
Frischknecht U., Hirschi R., Meynet G., Ekström S., Georgy C., Rauscher T., Winteler C., Thielemann F. K., 2010, A&A, 522, A39
Fuller J., Ma L., 2019, ApJL, 881, L1
Fuller J., Piro A. L., Jermyn A. S., 2019, MNRAS, 485, 3661
Gagnier D., Rieutord M., Charbonnel C., Putigny B., Espinosa Lara F., 2019a, A&A, 625, A88
Gagnier D., Rieutord M., Charbonnel C., Putigny B., Espinosa Lara F., 2019b, A&A, 625, A89
Gotthelf E. V., Halpern J. P., Alford J., 2013, ApJ, 765, 58
Griffiths A., Meynet G., Eggenberger P., Moyano F., Aloy M. A., 2022, arXiv e-prints, p. arXiv:2204.00016
Heger A., Langer N., Woosley S. E., 2000, ApJ, 528, 368
Heger A., Woosley S. E., Spruit H. C., 2005, ApJ, 626, 350
Higgins E. R., Vink J. S., 2019, A&A, 622, A50
Hirschi R., Maeder A., 2010, A&A, 519, A16
Hirschi R., Meynet G., Maeder A., 2004, A&A, 425, 649
Hirschi R., Meynet G., Maeder A., 2005, A&A, 443, 581
Kaiser E., 2021, Ph.D. Thesis, Keele University
Kaspi V. M., Helfand D. J., 2002, in Slane P. O., Gaensler B. M., eds, Astronomical Society of the Pacific Conference Series Vol. 271, Neutron Stars in Supernova Remnants. p. 3 (`arXiv:astro-ph/0201183`)
Langer N., 1998, A&A, 329, 551
Ma L., Fuller J., 2019, MNRAS, 488, 4338
Maeder A., 1997, A&A, 321, 134
Maeder A., 2009, Physics, Formation and Evolution of Rotating Stars. Springer, doi:10.1007/978-3-540-76949-1
Maeder A., Meynet G., 2000, A&A, 361, 159
Maeder A., Meynet G., 2005, A&A, 440, 1041
Maeder A., Zahn J.-P., 1998, A&A, 334, 1000
Marshall F. E., Gotthelf E. V., Zhang W., Middleditch J., Wang Q. D., 1998, ApJL, 499, L179
Meynet G., Maeder A., 1997, A&A, 321, 465
Meynet G., Maeder A., 2002, A&A, 390, 561
Muslimov A., Page D., 1996, ApJ, 458, 347
Ogilvie G. I., 2014, ARA&A, 52, 171
Pamyatnykh A. A., Handler G., Dziembowski W. A., 2004, MNRAS, 350, 1022
Paxton B., Bildsten L., Dotter A., Herwig F., Lesaffre P., Timmes F., 2011, ApJS, 192, 3
Paxton B., et al., 2013, ApJS, 208, 4

Paxton B., et al., 2015, ApJS, 220, 15
Paxton B., et al., 2018, ApJS, 234, 34
Pedersen M. G., et al., 2021, Nature Astronomy, 5, 715
Petitdemange L., Marcotte F., Gissinger C., 2022, arXiv e-prints, p. arXiv:2206.13819
Petrovic J., Langer N., Yoon S. C., Heger A., 2005, A&A, 435, 247
Popov S. B., Turolla R., 2012, AP&SS, 341, 457
Popov S. B., Pons J. A., Miralles J. A., Boldin P. A., Posselt B., 2010, MNRAS, 401, 2675
Potter A. T., Tout C. A., Eldridge J. J., 2012a, MNRAS, 419, 748
Potter A. T., Chitre S. M., Tout C. A., 2012b, MNRAS, 424, 2358
Salmon S. J. A. J., Moyano F. D., Eggenberger P., Haemmerlé L., Buldgen G., 2022, A&A, 664, L1
Spada F., Gellert M., Arlt R., Deheuvels S., 2016, A&A, 589, A23
Spruit H. C., 2002, A&A, 381, 923
Talon S., Charbonnel C., 2005, A&A, 440, 981
Talon S., Zahn J. P., 1997, A&A, 317, 749
Tayler R. J., 1973, MNRAS, 161, 365
Varma V., Müller B., 2021, MNRAS, 504, 636
Wheeler J. C., Kagan D., Chatzopoulos E., 2015, ApJ, 799, 85
Yoshida T., Takiwaki T., Aguilera-Dena D. R., Kotake K., Takahashi K., Nakamura K., Umeda H., Langer N., 2021, Monthly Notices of the Royal Astronomical Society: Letters, 506, L20
Zahn J. P., 1974, in Ledoux P., Noels A., Rodgers A. W., eds, IAU Symposium Vol. 59, Stellar Instability and Evolution. p. 185
Zahn J. P., 1977, A&A, 57, 383
Zahn J.-P., 1992, A&A, 265, 115
den Hartogh J. W., Eggenberger P., Hirschi R., 2019, A&A, 622, A187
den Hartogh J. W., Eggenberger P., Deheuvels S., 2020, A&A, 634, L16

Discussion

COMMENT: 1D theoretical models for meridional circulation cells use one spherical harmonic. Recent 2D simulations show a different picture.

ANSWER: Good point. The theory described above is the one used in current stellar models. Multi-D simulations like those of Gagnier et al. (2019b), however, will provide key guidance for future theoretical developments.

QUESTION: What about angular momentum transport in convective region?

ANSWER: As written above, stellar evolution codes assume constant Ω in convective zones. Potter et al. (2012a) investigated the impact of assuming uniform specific angular momentum in convective zones but more multi-D studies similar to those for the Sun would be welcome to better understand the distribution of angular momentum in convective regions in massive stars.

QUESTION: What mass loss rates are being used for rotating models?

ANSWER: A modification factor is generally applied to mass loss rate prescriptions to take into account the effects of rotation (see Langer 1998; Maeder & Meynet 2000; Gagnier et al. 2019a, for details and discussions).

Massive Stars Near and Far
Proceedings IAU Symposium No. 361, 2024
J. Mackey, J. S. Vink, & N. St-Louis, eds.
doi:10.1017/S1743921322002320

Entrainment in 3D hydrodynamics simulations of neon burning

F. Rizzuti[1], R. Hirschi[1,2], C. Georgy[3], W. D. Arnett[4], C. Meakin[5] and A. StJ. Murphy[6]

[1]Astrophysics Group, Lennard-Jones Laboratories, Keele University, Keele ST5 5BG, UK
email: `f.rizzuti@keele.ac.uk`

[2]Kavli IPMU (WPI), University of Tokyo, 5-1-5 Kashiwanoha, Kashiwa 277-8583, Japan

[3]Geneva Observatory, Geneva University, CH-1290 Sauverny, Switzerland

[4]Steward Observatory, University of Arizona, 933 N. Cherry Avenue, Tucson AZ 85721, USA

[5]Pasadena Consulting Group, 1075 N Mar Vista Ave, Pasadena, CA 91104 USA

[6]School of Physics and Astronomy, University of Edinburgh, Edinburgh EH9 3FD, UK

Abstract. Our knowledge of massive star evolution is limited by uncertainties linked with multi-dimensional processes taking place in stellar interiors. Important examples are convective boundary mixing (CBM) and entrainment, which are implemented in 1D stellar evolution models assuming simplified prescriptions. 3D hydrodynamics models can improve these prescriptions by studying realistic multi-D processes for a short timerange (minutes or hours). In these proceedings, we present results coming from a new set of high-resolution hydrodynamics simulations of a neon-burning shell in a massive star, and discuss how the entrainment law can be calibrated from 3D models and then used to improve 1D stellar evolution prescriptions.

Keywords. convection - hydrodynamics - turbulence - stars: evolution - stars: interiors - stars: massive

1. Introduction

Stellar evolution models are one-dimensional (1D) tools employed to simulate the entire lifetime of a star. Their efficiency is gained by making simplifying assumptions, spherical symmetry being the main one. Calibrating the prescriptions often requires comparison of results from specific simulations to observational data. As an example, the mixing length theory (MLT, Böhm-Vitense 1958) is used in 1D models to predict the extension of convective regions in the star.

Multi-dimensional hydrodynamics models provide useful insight in this sense, having several advantages compared to 1D stellar models. In particular, removing the assumption of spherical symmetry, it is possible to model fluid instabilities and reproduce complex processes, e.g. convection, rotation and magnetic activity, without the simplifying assumptions made in 1D models. On the other hand, multi-D models have a high computing cost, much larger than for 1D models, so they are bound to reproduce only short time scales and localized spatial extents. Therefore, an advancement in understanding stellar evolution can be achieved by combining the strengths of 1D and multi-D models, using the latter for constraining prescriptions, and subsequently including them in the former.

In these proceedings, we present a new set of hydrodynamics models simulating convection in a neon-burning shell of a 15 $M_\odot$ star. The simulations have been run with the PROMPI code, successfully used in the past for studying in a similar way other convective regions of massive stars (see Meakin & Arnett 2007; Viallet et al. 2013; Cristini et al. 2019). In particular, we focus here on the evolution of the convective zone in the neon shell, and the modelling of the convective boundary mixing (CBM) employing the entrainment law.

2. Methodology

Initial conditions were taken from a 1D stellar evolution simulation of a 15 $M_\odot$ star, obtained with the GENEC code (Eggenberger et al. 2008). Assumptions of this 1D model include solar metallicity, Schwarzschild criterion for convective boundaries, and penetrative overshoot for hydrogen- and helium-burning cores. For our neon-shell model, we mapped the 1D radial profiles of fundamental variables (e.g. density, temperature, abundances) from the second neon-burning shell onto a cubic box of side 0.64×10^8 cm. The numerical resolution chosen for the model is a grid of 512^3 cells, i.e. each cell is a cube of 1.25×10^5 cm side. All simulations have been run starting from this initial setup.

In our simulations, convection is driven using an explicit nuclear network for neon burning, including the isotopes ^{4}He, ^{16}O, ^{20}Ne, ^{24}Mg and ^{28}Si (the ^{4}He abundance is assumed to be at nuclear equilibrium in this late burning stage). Having all the same burning network, the simulations differ for the so-called "boosting factor", a constant factor that multiplies the nuclear energy generation rates. The factor has been introduced in order to boost convection and reduce the computing cost, but also to study the dependence of key quantities on the driving energy. For this reason, we produced a set of four simulations with boosting factors 1, 10, 100, 1000 times the nominal 1D case (see Rizzuti et al. 2022), in a similar way to what has been done in the work of Cristini et al. (2019) for the carbon shell. The boosting factor has a strong impact on the computing power required to run the simulations. In particular, with a fixed computing budget we are able to simulate a longer time range for models with a larger boosting factor.

3. Results

In Figure 1 we display a vertical cross-section across the domain of the nominal-luminosity simulation, with the velocity magnitude in colour scale. The convective zone is easily recognizable as the central region where the flow has higher speed and forms plumes and eddies. Above and below the convective zone, the stable region is dominated by low-speed internal gravity waves.

In order to study the evolution of this convective zone in our simulations, we need first to identify the convective boundaries. We define them by means of the chemical composition of the layers, using the fact that the convective zone is always homogeneously mixed. In this way, we define the convective boundaries as the location where the horizontally-averaged mean atomic mass is equal to the mid-point between the value in the convective and in the stable zone. With this definition, we are able to follow the time evolution of both the upper and lower convective boundary locations, as shown in Figure 2 for the four simulations with different boosting factors.

At the beginning of the simulations an intermediate phase is present, that we call "initial transient", before convection is fully developed with eddies and plumes. The length of this transient depends on the boosting factor, since a stronger energy generation produces faster convection. After the initial transient the simulation is in a "quasi-steady" state, and it is possible to analyse the results. Figure 2 clearly shows the growth in time of the convective zone, both upwards and downwards. This is due to a phenomenon

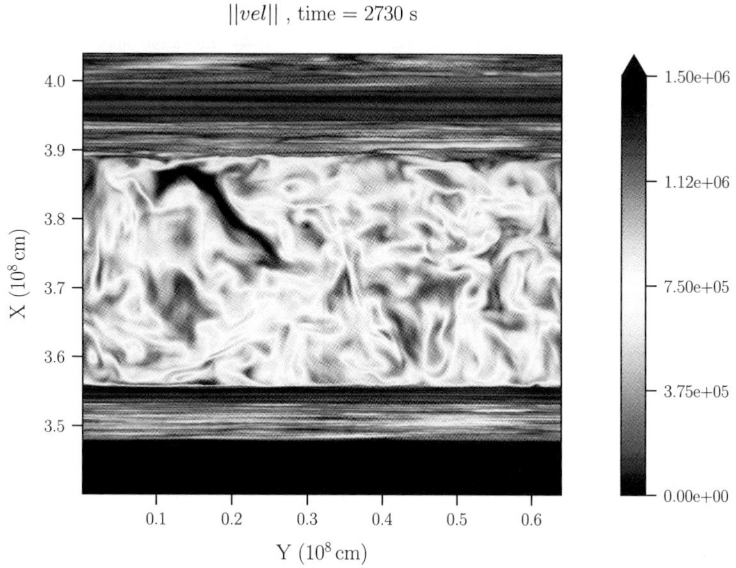

Figure 1. Vertical cross section of the velocity magnitude (values in colour scale, cm s^{-1}) taken towards the end of the nominal-luminosity simulation, at 2730 seconds. The high-velocity eddies identify the convective zone.

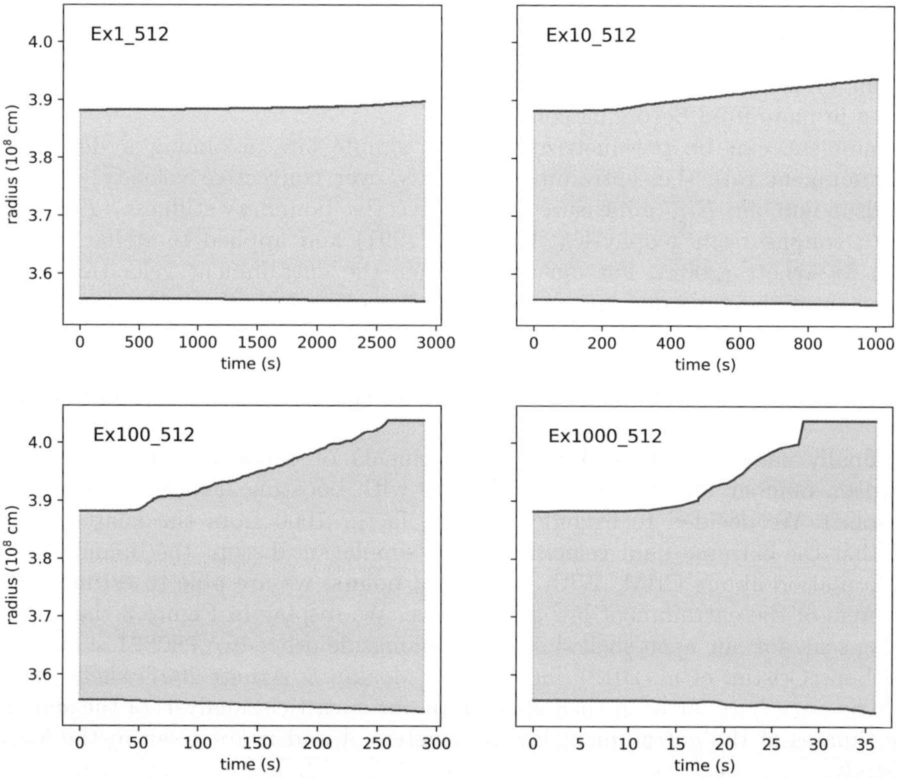

Figure 2. Time evolution of the upper and lower boundaries enclosing the convective zone (in magenta), for the four simulations with boosting factors 1, 10, 100 and 1000. After an initial transient, whose length depends on the boosting, the convective zones grow due to entrainment. It is noticeable how strongly the boosting affects the timescale for evolution.

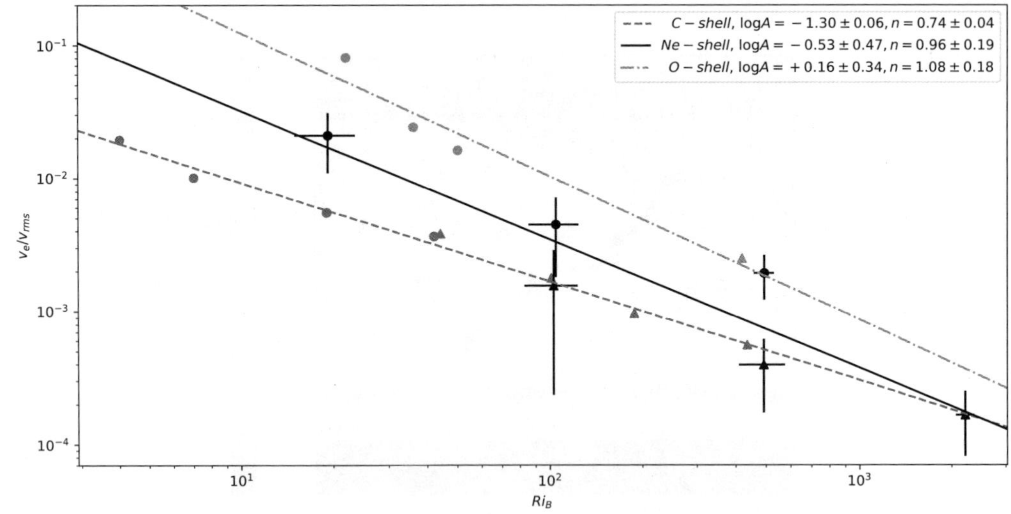

Figure 3. Entrainment rate versus bulk Richardson number in log scale and respective linear regressions: Ne-shell from this study (blue, solid), C-shell from Cristini et al. (2019) (green, dashed), O-shell from Meakin & Arnett (2007) (orange, dot-dashed). Triangles are measurements for the lower convective boundary, circles for the upper boundary. Error bars (only for Ne-shell) are standard deviations. Parameter estimates for the entrainment law $v_e/v_{rms} = A\ Ri_B^{-n}$ are listed in the legend. See Rizzuti et al. (2022).

called "turbulent entrainment", well-known in the literature (see Maeder 1976; Zahn 1991). It results from the shear mixing that takes place near the convective boundaries, where material is continuously entrained from the stable region into the convective zone, resulting in more fluid becoming convective.

Entrainment can be parametrized with a simple law, assuming a dependence of the entrainment rate (i.e. entrainment velocity over convective velocity) on the bulk Richardson number, Ri_B (a measure of the convective boundary stiffness): $E = v_e/v_{rms} = A\ Ri_B^{-n}$, coming from geophysics (Fernando 1991) and applied to stellar interiors by Meakin & Arnett (2007). For our simulations, the entrainment velocities have been obtained from the time derivatives of Figure 2, noticing that a larger boosting factor produces stronger entrainment, and that the entrainment velocity is smaller for the lower boundary, given its larger stiffness. The bulk Richardson number has been computed for each boundary by integrating the Brunt-Väisälä frequency as described in Cristini et al. (2019).

We finally show in Figure 3 the measurements of entrainement rate versus bulk Richardson number for our three simulations with boosting factors 1, 10 and 100 (in blue, solid). We decided to exclude boosting factor 1000 from the analysis, since we found that the extreme fluid velocities in the simulation disrupt the boundaries, losing the information about CBM. With these data points, we are able to estimate the free parameters of the entrainment law given above. We display in Figure 3 the line of best fit in log scale for our neon-shell simulations alongside other two `PROMPI` simulations for the carbon (Cristini et al. 2019) and oxygen (Meakin & Arnett 2007) shells of massive stars. The error bars for neon shell come from our statistical analysis of the simulations. The estimates of the entrainment law parameters A and n are given in the legend for each study.

From the parameter estimates, we can conclude that entrainment takes place in a similar way in multiple burning stages of massive stars. This is also compatible with other hydrodynamics simulations of convective zones in stellar interiors (e.g. Mocák et al.

2009; Gilet et al. 2013; Horst et al. 2021). The best fit for our new neon-shell simulations provides $n = 0.96 \pm 0.19$, which is compatible with the value of 1 expected from geophysics (Fernando 1991). Estimates for A are generally more uncertain, taking into account the large dispersion of measurements and that the fitting has been done in log scale. Our results for the neon shell indicate that values lie between 0.1 and 1.0. The small errors in the estimated parameters may suggest that uncertainties have been underestimated.

4. Conclusions

We present in these proceedings a new set of 3D hydrodynamics simulations of a neon-burning shell in a 15 $M_{\odot}$ star, obtained with the `PROMPI` code. We studied in detail the development and evolution of the convective zone, as well as the mixing which takes place near the convective boundaries. Importantly, we analyzed the entrainment of stable material into the convective zone and parametrized it with a simple law.

We found that significant entrainment is present in our simulations when using the exact same conditions taken from a state-of-the-art 1D stellar model, particularly in the case where the energy generation was not modified (nominal-luminosity case). Estimating the free parameters for the entrainment law in our simulations, we found that results for neon shell are in agreement with previous hydrodynamics studies of other burning phases in massive stars, confirming the occurrence and importance of CBM for the stellar evolution.

There is still a discrepancy between predictions for entrainment in hydrodynamics models versus stellar evolution models. This is well represented by the disagreement on the A parameter for the entrainment law. We refer here to the studies conducted by Staritsin (2013) and Scott et al. (2021), the only ones so far in the literature to have tested the entrainment law in 1D stellar models.

Synergy and convergence between one- and multi-dimensional models are crucial steps towards a better understanding of the complex phenomena in the stellar interiors. On the one hand, hydrodynamics models rely on the initial conditions assumed from 1D models, and they show disagreement on the strength of CBM assumed for 1D stellar evolution. On the other hand, adopting the entrainment law in 1D, especially for late convective phases, will provide a realistic way of modelling CBM, as well as produce more accurate initial conditions for starting new 3D simulations.

Acknowledgements

RH acknowledges support from the World Premier International Research Centre Initiative (WPI Initiative), MEXT, Japan and the IReNA AccelNet Network of Networks (National Science Foundation, Grant No. OISE-1927130). CG has received funding from the European Research Council (ERC) under the European Union's Horizon 2020 research and innovation program (Grant No. 833925). WDA acknowledges support from the Theoretical Astrophysics Program (TAP) at the University of Arizona and Steward Observatory. CG, RH, and CM acknowledge ISSI, Bern, for its support in organising collaboration. This article is based upon work from the ChETEC COST Action (CA16117) and the European Union's Horizon 2020 research and innovation programme (ChETEC-INFRA, Grant No. 101008324). The authors acknowledge PRACE for awarding access to the resource MareNostrum 4 at Barcelona Supercomputing Center, Spain, and the STFC DiRAC HPC Facility at Durham University, UK (Grants ST/P002293/1, ST/R002371/1, ST/R000832/1, ST/K00042X/1, ST/H008519/1, ST/ K00087X/1, ST/K003267/1). The University of Edinburgh is a charitable body, registered in Scotland, with Registration No. SC005336.

References

Böhm-Vitense E., 1958, Zeitschrift für Astrophysik, 46, 108

Cristini A., Hirschi R., Meakin C., Arnett D., Georgy C., Walkington I., 2019, MNRAS, 484, 4645

Eggenberger P., Meynet G., Maeder A., Hirschi R., Charbonnel C., Talon S., Ekström S., 2008, AP&SS, 316, 43

Fernando H. J. S., 1991, Annual Review of Fluid Mechanics, 23, 455

Gilet C., Almgren A. S., Bell J. B., Nonaka A., Woosley S. E., Zingale M., 2013, ApJ, 773, 137

Horst L., Hirschi R., Edelmann P. V. F., Andrássy R., Röpke F. K., 2021, A&A, 653, A55

Maeder A., 1976, A&A, 47, 389

Meakin C. A., Arnett D., 2007, ApJ, 667, 448

Mocák M., Müller E., Weiss A., Kifonidis K., 2009, A&A, 501, 659

Rizzuti F., Hirschi R., Georgy C., Arnett W. D., Meakin C., StJ. Murphy A., 2022, arXiv e-prints, p. arXiv:2207.03223

Scott L. J. A., Hirschi R., Georgy C., Arnett W. D., Meakin C., Kaiser E. A., Ekström S., Yusof N., 2021, MNRAS, 503, 4208

Staritsin E. I., 2013, Astronomy Reports, 57, 380

Viallet M., Meakin C., Arnett D., Mocák M., 2013, ApJ, 769, 1

Zahn J. P., 1991, A&A, 252, 179

Discussion

QUESTION: Do your simulations eventually reach the sub-surface convection? How far from the surface do they get?

ANSWER: The burning shell is deep so it doesn't reach the surface. Simulating convection below the surface would be a different environment. Our simulations are not including the surface.

QUESTION: I was very excited to see that you also see changes at lower boundary as well as the upper boundary, because in 1D stellar models we just use overshooting and not undershooting. I'd like to know if we should try to include undershooting in our models to get closer to 3D results.

ANSWER: Yes, this is something we see happening in 3D, so it should be included. We are expecting it from the fluid motions. Entrainment is much slower at the lower boundary, but for completeness it also should be included.

FOLLOW-UP QUESTION: So why is this? Why do we not see the same effect at the upper and lower boundary?

FOLLOW-UP ANSWER: It depends on the properties of the boundary. The bulk Richardson number is a measurement of the stiffness of boundary. The lower boundary is much stiffer, it has a higher bulk Richardson number, so we have much different properties.

Massive Stars Near and Far
Proceedings IAU Symposium No. 361, 2024
J. Mackey, J. S. Vink, & N. St-Louis, eds.
doi:10.1017/S1743921322002885

Two-dimensional models of fast rotating stars and mixing processes

Michel Rieutord[1] **and Damien Gagnier**[2]

[1]IRAP, Université de Toulouse, CNRS, UPS, CNES, 14, avenue Édouard Belin, F-31400 Toulouse, France
email: mrieutord@irap.om.eu

[2]Institute of Theoretical Physics, Faculty of Mathematics and Physics, Charles University, V Holešovičkách R, Prague 8, 18000, Czech Republic
email: damien.gagnier@mff.cuni.cz

Abstract. In this short review we present the recent progresses in modelling fast rotating stars in two dimensions. We thus give a brief description of the features of the public domain code ESTER that can compute self-consistently the structure and the large-scale flows (differential rotation and meridional circulation) of an axisymmetric stellar model of a (fast) rotating early-type main sequence star. We illustrate these modelling with the recent results obtained on Altair, a nearby extremely fast rotator. We then discuss the various way mixing takes place in the stably stratified radiative envelope of early-type stars, and especially in massive ones where the radiative winds add a new source of large-scale flows, which are shown to be strongly anisotropic and very difficult to represent in one dimension.

Keywords. Rotation, early-type stars, spin-down

1. Introduction

As far as massive stars are concerned, rotation is a concern! Why? Massive stars are intrinsically young objects and their formation fueled them with a lot of angular momentum, so much that most of the time they are members of binary or multiple systems. Making a binary is indeed an efficient way to store angular momentum. Hence, the fact that massive stars are almost never alone may be interpreted as a signature of a formation process that provides the stars with a high amount of angular momentum. Hence, it is no surprise that fast rotation is commonly detected among massive stars.

The foregoing considerations have motivated several extended observational studies to measure the impact of rotation in the population of O-type stars (Ramírez-Agudelo *et al.* 2013, 2015; Holgado *et al.* 2022). From these studies it turns out that the initial rotation velocity of O-type stars is around 20% of the break-up rotational velocity. Binarity and strong winds may be reasons for this not so fast rotation. But if we now consider B-type stars which typically cover the low mass end of massive stars (i.e. from 3 $M_\odot$ to ~ 20 $M_\odot$), the work of Huang *et al.* (2010) shows that $\sim 45\%$ of galactic B-type stars have a $V \sin i \gtrsim 150$ km/s, while still 30% of these stars show a $V \sin i \gtrsim 200$km/s. This latter fraction is even higher in the Large Magellanic Cloud, presumably because of its lower metallicity (e.g. Dufton *et al.* 2013).

Two well studied stars illustrate fast rotation of massive stars:

- ζ Ophiuchi of type O9.5, thus near the upper limit of B-type stars, has a $V \sin i \simeq$ 400 km/s (Kambe *et al.* 1990).
- ϕ Persei of type B2 has a $V \sin i \simeq 450$ km/s (Poeckert 1981).

For both of these stars, centrifugal effect is quite significant: Repolust *et al.* (2004) indicate that the mass of ζ Oph is around 20 $M_\odot$ and that its radius (assumed a polar radius) is $R \simeq 8.9$ $R_\odot$. With a minimum equatorial velocity of 400 km/s, the Roche model gives a flattening

$$\varepsilon = 1 - \frac{R_p}{R_{\rm eq}} = \frac{R_p V_{\rm eq}^2}{2GM} \simeq 0.16$$

where R_p and $R_{\rm eq}$ are the polar and equatorial radii respectively. $V_{\rm eq}$ is the equatorial velocity, G the gravitation constant and M the mass of the star.

The same exercise applied to ϕ Per, with a mass $M \simeq 9.6$ $M_\odot$ (Schootemeijer *et al.* 2018), a polar radius of 5.5 $R_\odot$ and an inclination $i = 80°$ (Gies *et al.* 1998), we get a flattening $\varepsilon \simeq 0.24$.

Such flattened stars obviously require two dimensional models if all the effects of fast rotation are to be taken into account properly.

2. Two-dimensional stellar models

2.1. *The ESTER code*

By 2D stellar models we understand models where the sole symmetries are axisymmetry and symmetry with respect to the equatorial plane. Hence, all quantities describing the stellar structure like pressure or density, or describing large-scale flows, namely differential rotation and meridional circulation, only depend on the spherical coordinates (r, θ). Moreover, they are invariant in the transform $\theta \to \pi - \theta$ (equatorial symmetry).

Any self-consistent 2D model should include the aforementioned large-scale flows. The main flow is the differential rotation, which is driven by the baroclinic torque. Let us recall the origin of these motions. If we consider the momentum equation, written in a galilean frame, it reads:

$$(\vec{v} \cdot \vec{\nabla})\vec{v} = -\frac{1}{\rho}\vec{\nabla}P - \vec{\nabla}\phi_g + \frac{1}{\rho}\vec{F}_{\rm visc} \tag{2.1}$$

for a steady state. In this equation ϕ_g is the gravitational potential and $\vec{F}_{\rm visc}$ the viscous force. Outside boundary layers (or shear layers), the viscous force is negligible and therefore we leave it aside. Since we assume that the model is axisymmetric, the velocity field $\vec{v}$ reads:

$$\vec{v} = r \sin\theta\, \Omega(r, \theta)\vec{e}_\varphi + v_r(r, \theta)\vec{e}_r + v_\theta(r, \theta)\vec{e}_\theta .$$

Here, $\Omega(r, \theta)$ is the differential rotation, while (v_r, v_θ) stand for the components of the meridional circulation. Now, taking the curl of the momentum equation and neglecting the viscous force, yields the so-called thermal wind equation, well-known in atmospheric physics (see Cushman-Roisin 1994, for instance),

$$r \sin\theta \frac{\partial \Omega^2}{\partial z} = \frac{\vec{\nabla}P \times \vec{\nabla}\rho}{\rho^2} \cdot \vec{e}_\varphi \tag{2.2}$$

which shows that the non-alignment of the density and pressure gradients, generates a torque, the so-called baroclinic torque, that forces a gradient of the angular velocity Ω along Oz, the rotation axis direction. The non-alignment of these two gradients results from the fact that pressure and temperature obey two independent equations. Further details on the dynamics of 2D models may be found in Espinosa Lara & Rieutord (2013).

The full problem to be solved for the construction of an axisymmetric model of a fast rotating star is actually given by the four partial differential equations:

$$\begin{cases} \Delta\phi = 4\pi G\rho \\ \rho T\vec{v}\cdot\vec{\nabla}S = -\mathrm{Div}\vec{F} + \varepsilon_* \\ \rho(2\vec{\Omega}_* \wedge \vec{v} + \vec{v}\cdot\vec{\nabla}\vec{v}) = -\vec{\nabla}P - \rho\vec{\nabla}(\phi - \frac{1}{2}\Omega_*^2 s^2) + \vec{F}_{\mathrm{visc}} \\ \mathrm{Div}(\rho\vec{v}) = 0. \end{cases} \tag{2.3}$$

where we recognise Poisson equation, the entropy equation, the momentum equation (here written in a frame rotating at the angular velocity $\vec{\Omega}_*$), and the mass conservation equation. For fluid dynamicists these equation are those of a compressible, self-gravitating fluid harbouring nuclear heating $\varepsilon_*(\rho, T)$. These partial differential equations are completed by the equation of state $P \equiv P(\rho, T)$, the expression of the heat flux $\vec{F}$ and boundary conditions. These latter conditions apply at the surface of the star and at the centre. At the centre, we just demand that the functions be regular, while at the surface the pressure, the velocity and the temperature should meet appropriate conditions depending on where we stop the modelling. A classical one (see Espinosa Lara & Rieutord 2013), is to fix the surface at the optical depth $\tau = 2/3$, where we set the pressure and the heat flux as a black body radiator. In two dimensions we also need to give boundary conditions on the velocity field. At the centre, regularity and axisymmetry impose a zero-velocity while at the surface it is natural to impose stress-free conditions if we neglect any wind (see below). In a steady state, such conditions generate a weak Ekman layer which determines a weak meridional circulation and insures a neat zero-flux of angular momentum across any closed surface inside the star. Of course, if some region of the star is turbulent, $\vec{F}$ and $\vec{F}_{\mathrm{visc}}$ should be completed with the appropriate turbulence model.

The above equations form a rather formidable nonlinear problem, which has been solved for the first time by Espinosa Lara & Rieutord (2013) using a simple turbulence model. The way to reach the numerical solution is detailed in Rieutord *et al.* (2016). Very briefly, it uses the three following items:

- A mapping to follow the star distortion,
- A spectral method (Chebyshev polynomials and spherical harmonics),
- The Newton-Raphson algorithm.

The code where this solution is implemented is the public domain ESTER code that may be obtained on GitHub at `http://ester-project.github.io/ester/`. Presently, this code works well for early-type stars, but still needs developments for late-type stars. The quality of the solutions are gauged with the energy and virial tests, namely two integrals that should be vanishing if the solution is exact. The virial test requires that

$$\int_{(V)} \vec{r}\cdot\left[2\vec{\Omega}\wedge\rho\vec{u} + \rho\vec{u}\cdot\vec{\nabla}\vec{u} + \rho\vec{\nabla}\phi - \rho\Omega^2 s\vec{e}_s - \mathbf{Div}[\sigma]\right] dV = 0$$

where $[\sigma]$ is the viscous stress tensor. The foregoing integral is normalized by the potential energy,

$$W = \frac{1}{2}\int_{(V)} \rho\phi \, d^3\vec{r}.$$

This test usually leads to errors less than 10^{-9} thanks to the spectral discretization. The energy test, which compares the energy produced by nuclear reactions and the energy emitted at the surface, is not as good because of rapid variations (and the tabulated nature) of opacities near the stellar surface.

Table 1. Fundamental parameters of Altair according to Bouchaud *et al.* (2020). The luminosity has been calculated using the same model as Bouchaud *et al.* (2020). The distance is the one given by Hipparcos (van Leeuwen 2007).

Star	Spectral type	Distance pc	Mass $M_\odot$	R_{eq} $R_\odot$	R_{pol} $R_\odot$	T_{eq} K	T_{pol} K	Lum. $L_\odot$
α Aql	A7 IV-V	5.130 ±0.015	1.86 ±0.03	2.008 ±0.006	1.565 ±0.014	6780	8620	10.2
Flattening	Ω_{eq}/Ω_k	*incl.* deg	V_{eq} km/s	P_{eq} day	$X_{env.}$	X_{core}	Z	
0.220 ±0.003	0.744 ±0.01	50.65 ±1.23	313 ±9	0.325 ±0.01	0.733	0.71	0.019	

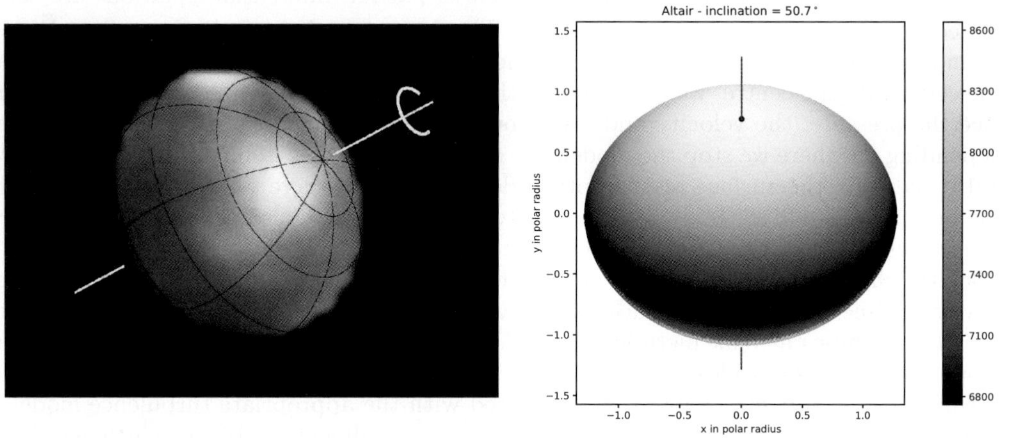

Figure 1. Left: Altair as seen with the CHARA interferometer at Mount Wilson (credit Monnier *et al.* 2007). Right: Distribution of the effective temperature at the surface of Altair's model by Bouchaud *et al.* (2020).

2.2. *Altair's test case*

Altair is a nearby intermediate mass star that is rotating very rapidly (see Tab. 1 for a recap of Altair's fundamental parameters). Because of its proximity, this star has been observed in many ways, notably with interferometers (CHARA at Mount Wilson see Fig. 1, PIONIER and GRAVITY at ESO-Chile) to measure its oblateness. Dedicated spectroscopy determined its $V \sin i$ with constraints on i, the inclination of the rotation axis over the line of sight (see Reiners & Royer 2004), while asteroseismology also brought its contribution to the set of data with a small set of identified acoustic frequencies showing delta-scuti-type oscillations (see Buzasi *et al.* 2005; Bouchaud *et al.* 2020; Le Dizès *et al.* 2021).

This abundant data set motivated Bouchaud *et al.* (2020) to devise a concordance model of Altair that matches all known observables. We show in Fig. 1 the effective temperature distribution that is used to retrieve spectroscopic and interferometric observables.

At the same time the matching between model and data shows us that ESTER models can give a faithful representation of real fast rotating early-type stars. Accompanied with the TOP code (Reese *et al.* 2021), which can compute the eigenmodes of an ESTER model, the main conclusion of Bouchaud *et al.* (2020) was that the mass of Altair is 1.86 ± 0.03 $M_\odot$ and that it is a young star, namely ~100 Myrs old, unlike previous estimates based on 1D models.

3. Rapidly rotating massive stars

3.1. *Introduction*

Altair is by far not a massive star! According to Langer (2012) a massive star is "a star that is massive enough to form a collapsing core at the end of its life and, thus, avoid the white dwarf fate". Hence, the minimum mass is in between 8 and 12 $M_\odot$, depending on metallicity. The final fate of massive stars is a full research field by itself, which is however strongly connected to what happens in the early phases of evolution, especially during the main sequence (see Martins & Palacios 2013). Here, we shall focus on this main sequence stage and review the recent progresses brought by 2D-models.

Compared to Altair and intermediate mass stars, massive stars own a new feature, such as a non-negligible mass-loss. This mass-loss is driven by the strong radiation field at the surface of the star and as far as rotation is concerned, we may anticipate two effects: (i) a lower critical angular velocity due to radiative acceleration, (ii) an intensified mixing due to the spin-down associated with angular momentum losses.

3.2. *Critical angular velocity*

When the radiation field is intense, the radiative acceleration is a non-negligible part of the forces applied to a fluid element. At some height in the atmosphere this acceleration dominates and the wind is launched. However, in the lower parts of the photosphere, where the outflow is largely subsonic, radiative acceleration may help decrease the critical angular velocity over which equatorial surface matter is just orbiting the star.

This question was first investigated by Maeder (1999) and Maeder & Meynet (2000) using 1D models. Since the use of 1D-models gives rather uncertain results when one considers rotation near the critical one, Gagnier *et al.* (2019a) reconsidered this issue in light of 2D-models.

Critical angular velocity is reached when, somewhere at the surface of the star the total acceleration vanishes, namely

$$\vec{g}_{\rm tot} = \vec{g}_{\rm eff} + \vec{g}_{\rm rad} = \vec{0} \tag{3.4}$$

where $\vec{g}_{\rm eff}$ is the effective gravity (the combination of gravity and centrifugal acceleration). Since the radiative acceleration $\vec{g}_{\rm rad}$ is proportional to the radiative flux, namely $\vec{g}_{\rm rad} = \kappa\vec{F}/c$, where κ is a mean opacity weighted by the flux, we immediately see that a vanishing total acceleration is reached if the flux reaches the limit $\vec{F}_{\rm lim} = -c\vec{g}_{\rm eff}/\kappa$. Then, it is usual to introduce the Eddington factor as the ratio

$$\Gamma_\Omega(\theta) = \frac{F}{F_{\rm lim}}$$

and note that

$$\vec{g}_{\rm tot} = \vec{g}_{\rm eff}(1 - \Gamma_\Omega(\theta))$$

where the Eddington factor obviously depends on the rotation rate, because of the centrifugal acceleration, and on the colatitude θ because of the gravity darkening induced by the centrifugal distortion of the star.

Gagnier *et al.* (2019a) used the ω-model of Espinosa Lara & Rieutord (2011), which assumes that the flux in a radiative envelope is just anti-parallel to the effective gravity (but see also Rieutord 2016). This models is very close to complete 2D-models and allows a simple analysis of this problem. The main results, confirmed by the full 2D-models, are that:

- the critical angular velocity is always reached at equator, in line with the results of the first investigations of Maeder & Meynet (2000),

- the critical angular velocity reads

$$\Omega_c = \Omega_k\sqrt{1-\Gamma_{eq}^{3/2}}\ . \tag{3.5}$$

where Ω_k is the Keplerian angular velocity at equator. 1D-models lead to a slightly different law $\Omega_c \propto \sqrt{1-\Gamma_{eq}}$ if $\Gamma \geq 0.639$ and $\Omega_c = \Omega_k$ otherwise, which shows a stronger impact of $\vec{g}_{\rm rad}$ than (3.5).

Actually, the investigations by Gagnier *et al.* (2019a), show that the decrease of opacity with density, still reduces the impact of the radiative acceleration on the critical angular velocity. Gagnier *et al.* (2019a) conclude that below 40 $M_\odot$ Ω_c and Ω_k differ only very slightly.

3.3. *Evolution with mass-loss*

To follow the evolution of a mass-losing star, and especially to follow its rotation rate, one usually peels off the star: some amount of mass is removed with the angular momentum it contains, assuming it comes from the surface layers. Gagnier *et al.* (2019b) used the same technique but applied to 2D-models. To be operative, one needs to know the mass-loss at each colatitude, namely the function $\dot{m}(\theta)$.

In a fast rotator, gravity darkening and centrifugal acceleration control the local radiative flux and the local effective gravity respectively. Hence, with metallicity, to which we may attribute a global value, we need to derive the dependence of the local mass flux $\dot{m}(\theta)$ with respect to the local effective temperature and gravity. Gagnier *et al.* (2019b) devised such a relation based on the works of Castor *et al.* (1975), Pauldrach *et al.* (1986) and Vink *et al.* (1999, 2001), mainly. In particular, the work of Vink *et al.* (1999) noticed the bi-stability jump at $T_{\rm eff} \simeq$25,000 K in radiative winds of early-type supergiants and discovered that it is due to the growing importance of FeIII transitions at effective temperature below 25,000 K. Since the surface of fast rotating stars can support a large range of effective temperatures, thanks to gravity darkening, Gagnier *et al.* (2019b) logically found that such massive stars may be endowed with a two-winds regime: a slow and "cold" wind in the equatorial regions where effective temperatures are below the threshold, and a faster hotter wind near the poles. Not surprisingly, the slow equatorial wind is efficient at removing angular momentum of the star.

The foregoing wind processes lead to the following evolution of rotation in massive stars: As the star evolves (along the main sequence) and inflates, it gets closer to the critical rotation if mass-loss is weak enough. In a few words, this behaviour comes from the fact that as the star evolves, and loses a negligible amount of angular momentum, its rotation rate decreases because of radius expansion, however the Keplerian angular velocity at equator decreases even more, hence the ratio $\omega = \Omega_{\rm eq}/\Omega_k$ increases with time. If the initial angular velocity is large enough the star may reach the critical rotation within the main sequence. However, the wind may modify this evolution. If we consider a star whose initial effective temperature is above 25,000K, it will lose mass in a single wind régime. In addition, mass-loss is stronger at the pole because of gravity darkening. But around the pole angular momentum loss is weak and the star actually follows a route similar to the no-mass-loss case. However, at some time gravity darkening can be strong enough to put the equatorial effective temperature below the threshold and a two winds régime starts. Here the angular momentum loss is much more efficient and the star can move away from criticality. This scenario is illustrated in Fig. 2 taken from Gagnier *et al.* (2019b).

3.4. *Interior flows*

The peeling off process is valid as long as the mass-loss timescale is longer than the timescale of angular momentum transfer inside the star. As shown in Gagnier *et al.* (2019b) (see their Tab. 1), this is quite the case for stars up to 20$M_\odot$.

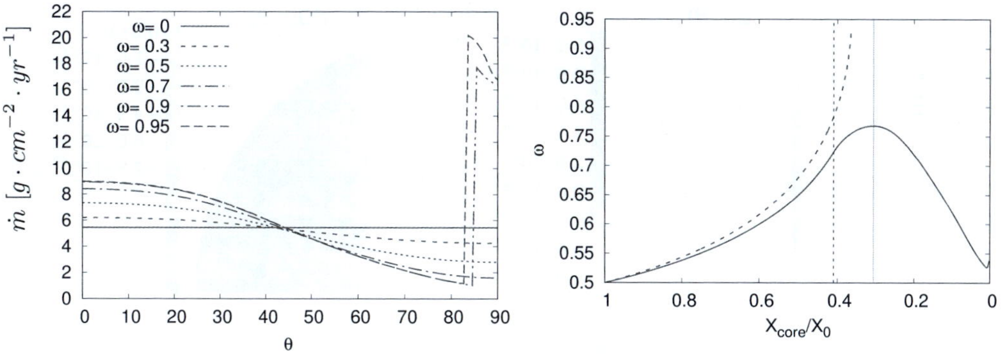

Figure 2. Left: the surface mass flux as a function of colatitude for various rotation rates scaled to the critical one (Gagnier *et al.* 2019a). Right: Evolution of rotation rate scaled by the critical angular velocity according to Gagnier *et al.* (2019b) for a mass-losing model (solid line) and a no-mass-loss model (dashed line); time is represented by the hydrogen mass fraction in the convective core scaled by its initial value. For both figures a 15 $M_\odot$ model is used.

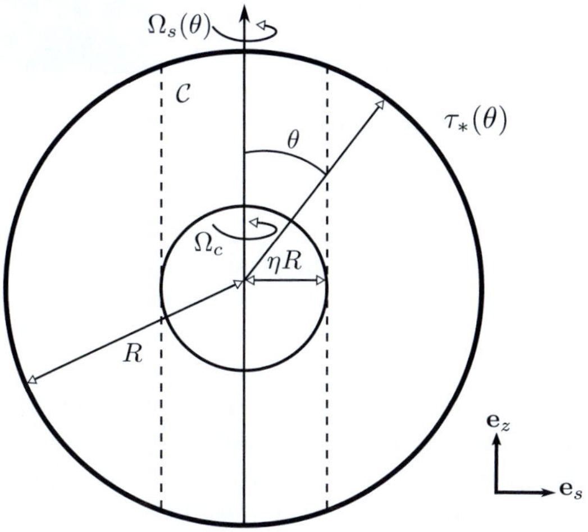

Figure 3. Set-up used to study the internal motion of a fluid in a spinning-down spherical shell braked by a shear stress.

However, this spin-down process also induces some mixing inside the star, which is still not well known. Gagnier & Rieutord (2020) thus investigated the steady flows that results from the spinning down star.

One part of the problem is the determination of the flow that results from the viscous stress generated by the out moving layers. To study the implications of this effect Gagnier & Rieutord (2020) first analysed the flow of a rotating fluid confined into a spherical shell representing the radiative envelope of a massive star. At its surface they imposed a stress of the form:

$$\sigma_{r\varphi} = \tau_*(\theta) = -A\sin\theta$$

which extract angular momentum from the fluid. The set-up is reproduced in Fig. 3.

In this set-up the core is rotating as a solid body and no-slip boundary conditions are used at its interface with the envelope. This is supposed to represent the viscosity jump at the core-envelope boundary. The convective core is endowed with a high viscosity that reflects the turbulent diffusion there (e.g. Rieutord 2006).

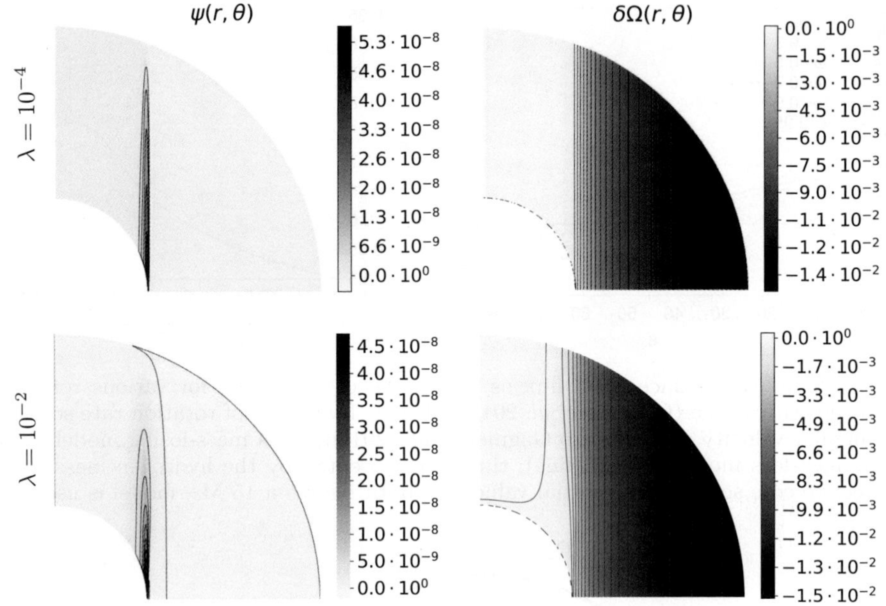

Figure 4. Meridional circulation (left) and differential rotation (right) generated in a spinning-down radially stably stratified spherical shell braked by a shear stress. The Ekman number is $E = 10^{-7}$ and the Boussinesq approximation is used (from Gagnier & Rieutord 2020).

The solution of this spin-down problem is shown in Fig. 4 with a meridional cut. There are two striking features characterizing these solutions. First, we note that the meridional circulation is concentrated in a thin shear layer that is tangent to the core, lying on the tangent cylinder. This layer is well known in the theory of rotating fluids as the Stewartson layer (e.g. Stewartson 1966). In Fig. 4, the fluid in the spherical shell is stably stratified radially, as in the actual radiative envelope. A stable stratification may actually inhibit the raise of a Stewartson layer, but Gagnier & Rieutord (2020) have found that this inhibition is controlled by the parameter

$$\lambda = \mathcal{P}\frac{N^2}{4\Omega^2}$$

where $\mathcal{P}$ is the Prandtl number and N the Brunt-Väisälä. A similar result was also found in previous work (Garaud 2002; Rieutord 2006). In stars, heat diffusion is large and this parameter is small when the star is rapidly rotating. Indeed, $N^2/4\Omega^2$ is less than 100 in fast rotating stars, while the Prandtl number $\mathcal{P}$ is of order of 10^{-6}. Hence, $\lambda \ll 1$ for these stars. In that case the role of stratification is negligible. Of course, if the star rotates slowly λ may be much larger than unity and in that case stratification suppresses the Stewartson layer (e.g. Gagnier & Rieutord 2020).

The Stewartson layer is a feature that is not taken into account in 1D models, while we see that it is linking the core and the surface of the star. Hence, it is able to carry elements from the deep interior to the surface where they may be observed.

The second feature that may be noticed in Fig. 4, is the concentration of the differential rotation in the outer part of the tangent cylinder. As far as the transport properties are concerned, we note that any turbulent transport sustained by the shear associated with the differential rotation is therefore confined outside the tangent cylinder. Moreover, as shown by Fig. 5, when the density stratification is taken into account, this same differential rotation, generated by the spin-down (or the surface stress actually), is now

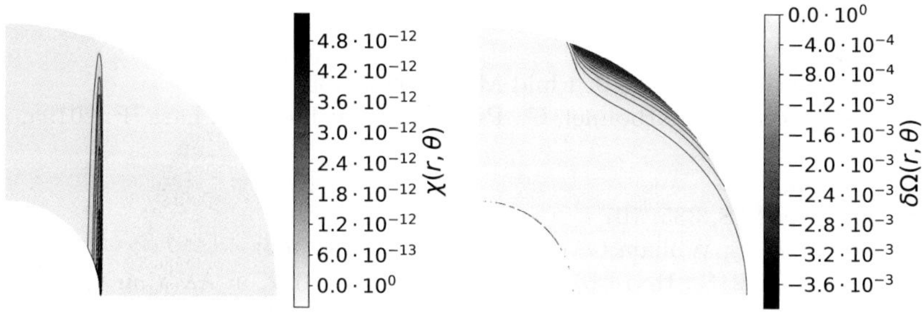

Figure 5. The same as in Fig. 4, but using the anelastic approximation with a density contrast between core and surface of 10^4.

confined in an even narrower region of the radiative envelope. It shows again that the transport properties in a radiative envelope are strongly anisotropic and inhomogeneous.

4. Conclusions and outlook

After this brief overview of the internal dynamics of rapidly rotating massive stars, we certainly feel that much work is ahead of us. Such stars indeed require two-dimensional models, not only because they are flattened by centrifugal effect but also because they own internal flows that strongly break the spherical symmetry. As a consequence, the way chemical elements and angular momentum is transported between the core and the envelope is quite far from what 1D model can predict.

We have seen that baroclinic and spin-down flows span the radiative envelope of these stars, but we should expect that magnetic fields also add their part of anisotropy. They have been detected in many sorts of massive stars like β Cephei or Be stars (Hubrig *et al.* 2011). As is well-known, winds and magnetic fields give rise to magnetic braking which is a strong effect, but of higher complexity since in such a problem the geometry generally loses any symmetry, leaving us with a 3D problem.

But before attacking the foregoing challenging problem, some simpler questions can be addressed with present 2D models. Let us point out three of them:

- The secular time evolution along the main sequence of a 2D-model including mass-loss and its coupling with the interior flows,
- the launch of matter near critical rotation,
- the role of Stewartson layers in the mixing process and the angular momentum transport at the core-envelope interface.

Acknowledgments

MR is very grateful to the organizers for their invitation to participate in this conference and especially to Jonathan Mackay for his kind reading of the manuscript. We also wish to thank the support of the French Agence Nationale de la Recherche (ANR), under grant MASSIF (ANR-21- CE31-0018-02). Numerical simulations have been made with HPC resources from CALMIP supercomputing center (Grant 2022-P0107).

References

Bouchaud, K., Domiciano de Souza, A., Rieutord, M., Reese, D. R., & Kervella, P. 2020, A&A, 633, A78
Buzasi, D. L., Bruntt, H., Bedding, T. R. *et al.* 2005, ApJ, 619, 1072
Castor, J. I., Abbott, D. C., & Klein, R. I. 1975, ApJ, 195, 157
Cushman-Roisin, B. 1994, An introduction to geophysical fluid dynamics (Paris: Prentice-Hall)
Dufton, P. L., Langer, N., Dunstall, P. R. *et al.* 2013, A&A, 550, A109

Espinosa Lara, F., & Rieutord, M. 2011, A&A, 533, A43
— 2013, A&A, 552, A35
Gagnier, D., & Rieutord, M. 2020, J. Fluid Mech., 904, A35
Gagnier, D., Rieutord, M., Charbonnel, C., Putigny, B., & Espinosa Lara, F. 2019a, A&A, 625, A88
— 2019b, A&A, 625, A89
Garaud, P. 2002, MNRAS, 335, 707
Gies, D. R., Bagnuolo, J., William G. *et al.* 1998, ApJ, 493, 440
Holgado, G., Simón-Díaz, S., Herrero, A., & Barbá, R. H. 2022, in A&A, arXiv:2207.12776
Huang, W., Gies, D. R., & McSwain, M. V. 2010, ApJ, 722, 605
Hubrig, S., Ilyin, I., Schöller, M., Briquet, M., Morel, T., & De Cat, P. 2011, ApJ Lett., 726, L5
Kambe, E., Ando, H., & Hirata, R. 1990, Pub. Astron. Soc. Jap., 42, 687
Le Dizès, C., Rieutord, M., & Charpinet, S. 2021, A&A, 653, A26
Maeder, A. 1999, A&A, 347, 185
Maeder, A., & Meynet, G. 2000, A&A, 361, 159
Martins, F., & Palacios, A. 2013, A&A, 560, A16
Monnier, J. D., Zhao, M., Pedretti, E. *et al.* 2007, Science, 317, 342.
Pauldrach, A., Puls, J., & Kudritzki, R. P. 1986, A&A, 164, 86
Poeckert, R. 1981, Pub. Astron. Soc. Pacific, 93, 297
Ramírez-Agudelo, O. H., Sana, H., de Mink, S. E. *et al.* 2015, A&A, 580, A92
Ramírez-Agudelo, O. H., Simón-Díaz, S., Sana, H. *et al.* 2013, A&A, 560, A29
Reese, D. R., Mirouh, G. M., Espinosa Lara, F., Rieutord, M., & Putigny, B. 2021, A&A, 645, A46
Reiners, A., & Royer, F. 2004, A&A, 428, 199
Repolust, T., Puls, J., & Herrero, A. 2004, A&A, 415, 349
Rieutord, M. 2006, A&A, 451, 1025
Rieutord, M. 2016, in Lecture Notes in Physics, vol. 914, 101. 1505.03997
Rieutord, M., Espinosa Lara, F., & Putigny, B. 2016, J. Comp. Phys., 318, 277
Schootemeijer, A., Götberg, Y., de Mink, S. E., Gies, D., & Zapartas, E. 2018, A&A, 615, A30
van Leeuwen, F. 2007, A&A, 474, 653
Vink, J. S., de Koter, A., & Lamers, H. J. G. L. M. 1999, A&A, 350, 181
— 2001, A&A, 369, 574

Massive Stars Near and Far
Proceedings IAU Symposium No. 361, 2024
J. Mackey, J. S. Vink, & N. St-Louis, eds.
doi:10.1017/S1743921322002149

Very Massive Stars: Near and Far

Sébastien Martinet[1], Georges Meynet[1], Sylvia Ekström[1], Cyril Georgy[1], Lionel Haemmerlé[1], Devesh Nandal[1] and Raphael Hirschi[2,3]

[1]Geneva Observatory, University of Geneva, Chemin Pegasi 51, CH-1290 Versoix, Switzerland
email: sebastien.martinet@unige.ch

[2]Astrophysics Group, Keele University, Keele, Staffordshire, ST5 5BG, UK

[3]Institute for Physics and Mathematics of the Universe (WPI), University of Tokyo, 5-1-5 Kashiwanoha, Kashiwa 277-8583, Japan

Abstract. In addition to being spectacular objects, very massive stars (VMS) are suspected to have a tremendous impact on their environment and on the whole cosmic evolution. The nucleosynthesis both during their advanced stages and their final explosion likely contribute greatly to the overall enrichment of the Universe. Their resulting Supernovae are candidates for the most superluminous events and their extreme conditions lead also to very important radiative and mechanical feedback effects, from local to cosmic scale. With the recent implementation of a new equation of state in the GENEC stellar evolution code, appropriate for describing the conditions in the central regions of very massive stars in the advanced phases, we present new results on VMS evolution from Population III to solar metallicity. We explore their evolution and final fate as potential (P)PISNe across the cosmic time. We compare our results to recent spectroscopic observations of VMS in the Large Magellanic Cloud (LMC). We also underline the important radiative feedback of Population III VMS during the reionization epoch and the chemical contribution of these stars at high metallicity, especially for short-lived radionuclei.

Keywords. stars: evolution, stars: supernovae, stars: fundamental parameters

1. Introduction

The Tarantula Nebula in the Large Magellanic Cloud hosts one particular young cluster, known as R136, hosting itself very high mass components. A first spectroscopic study by Cassinelli et al. (1981) showed the presence of very strong P-Cygni features for the center object of this cluster, characteristic of very high ionization flux. From this measurement, they obtained that the central object must either be a few thousands solar masses, or an unresolved cluster of stars. The latter was the solution, found with the observations of Crowther et al. (2006), first with stars ranging up to $150M_\odot$, and then up to $300M_\odot$ after further studies (Crowther et al. 2010). Recent work (Bestenlehner et al. 2020; Brands et al. 2022) confirmed these results with new VLT-FLAMES and HST/STIS optic/spectroscopic measurements and provide interesting stellar characteristics to the most massive stars observed. With the new VMS models computed here, we can confront our results to these new observations and explore the impact of the different physics included.

2. Methods

We present a new grid of VMS models, with initial masses of 180, 250 and $300M_\odot$, including the new EOS for high temperature/density regimes, probing Z=0.000, Z=10^{-5},

Z=0.006 and Z=0.014 metallicities for non-rotating and rotating models at V/V_c=0.4 where V_c is the critical velocity. We adopt an overshooting value of α_{ov}=0.2 following the results of Martinet et al. (2021). We used the Ledoux criterion to determine the position of the convective boundaries, as there are some indications that the Ledoux criterion might be more appropriate (Georgy et al. 2014; Kaiser et al. 2020) for these VMS. The radiative mass-loss rate adopted on the MS is from Vink et al. (2001); the domains not covered by this prescription (see Fig. 1 of Eggenberger et al. 2021) use the de Jager et al. (1988) rates. Gräfener & Hamann (2008) prescriptions are used in their domain of application, while Nugis & Lamers (2000) prescriptions are used everywhere else for the Wolf-Rayet phase†. The radiative mass-loss rate correction factor described in Maeder & Meynet (2000) is applied for rotating models. The dependence on metallicity is taken such that $\dot{M}(Z) = (Z/Z_\odot)^{0.7}\dot{M}(Z)$, except during the red supergiant (RSG) phase, for which no dependence on the metallicity is used. This is done accordingly to van Loon et al. (2005) and Groenewegen (2012a,b) showing that the metallicity dependence for the mass loss rates of these stars do appear to be weak. For the rotation prescription, the models use the shear diffusion coefficient as given by Maeder (1997) and the horizontal diffusion coefficient from Zahn (1992).

3. Comparison with observed VMS in the LMC

Fig. 1 presents the HRD of the Tarantula Nebula, a very active region in the LMC. The compilation of Schneider et al. (2018) is displayed in yellow, while the components of the R136 cluster, obtained by Brands et al. (2022), are displayed in magenta. The tracks used here are the Z=0.006 GENEC tracks from Eggenberger et al. (2021) and the VMS tracks we computed for this work with the new EOS. We can see that this cluster hosts several high-mass components. If we look at the evolution of our VMS models, three stars inside this cluster are above $M_{ini}=200M_\odot$. If we start with the HRD tracks, we can wonder if either the non-rotating or the rotating models fit better their position in the HRD. At first glance, one would favor the non-rotating models, covering a large range of effective temperature on the MS, while the rotating models evolve quasi-chemically homogeneously and stray directly to the bluer part of the HRD, hence incompatible with the observed effective temperature. The picture is however not that easy, as we have indeed seen, the VMS at high metallicity undergo very large mass loss already during the MS. This means, in fact, that the star is embedded inside a "nebula" of very thick wind. When observing such stars, we observe in fact this thick wind, and this blurs the determination of the effective temperature of the surface of the star. Indeed, the wind being much cooler than the surface of the star, we would obtain a lower effective temperature to be observed if we took this effect into account. In fact, if we correct for this process, both non-rotating and rotating models are compatible with the observations.

The evolution of VMS at high metallicity is dominated by mass loss. The prescriptions used in our models are empirical results obtained from lower mass O-type stars, and are a highly controversial topic recently, with new mass loss rates proposed to be both much lower than previously used, or the other way round, much higher, in particularly when getting close to the Eddington limit (see the review of Vink 2021). We confront in Fig. 2 our models to the observed mass loss rates obtained by Brands et al. (2022) and their corresponding fit using the prescription proposed by Bestenlehner et al. (2020). The mass loss rates are plotted as a function of the Eddington parameter (with $\Gamma_{edd,e}$=1 being the Eddington limit). Interestingly, the mass loss we obtain on the MS from the prescriptions

† The Wolf-Rayet phase in GENEC is assumed to begin when the model has an effective temperature larger than 10 000 K and a surface mass fraction of hydrogen at the surface below 0.3.

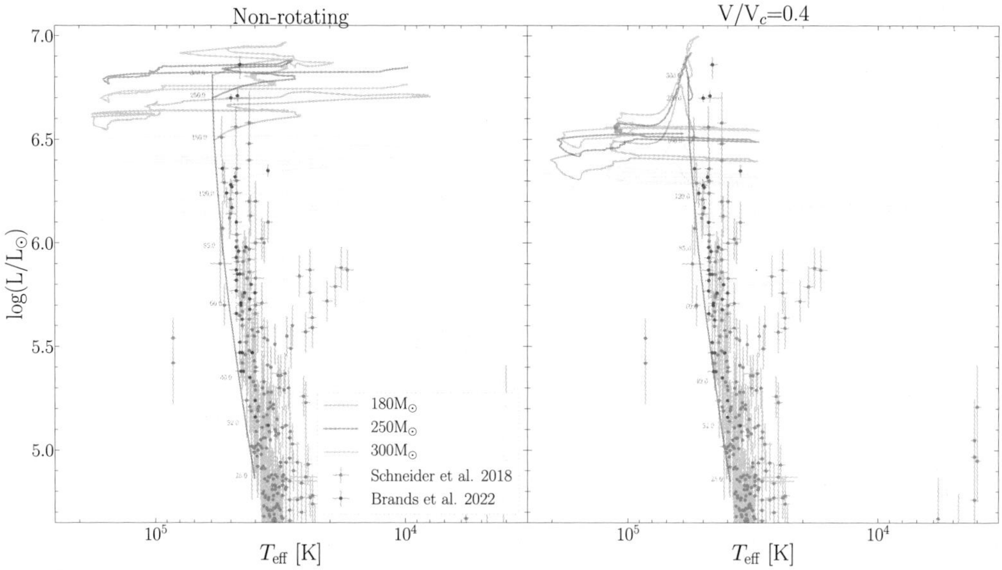

Figure 1. HRD of the Tarantula Nebula from the compilation of Schneider et al. (2018) (regrouping results from Ramírez-Agudelo et al. 2017; Sabín-Sanjulián et al. 2014, 2017; McEvoy et al. 2015; Bestenlehner et al. 2014) and the R136 cluster members from the work of Brands et al. (2022). The GENEC tracks are overplotted in grey, with color emphasis on the VMS models computed here, and the ZAMS is drawn in dark grey. The HRD is plotted for non-rotating (*left*) and rotating at $V/V_c = 0.4$ (*right*) models.

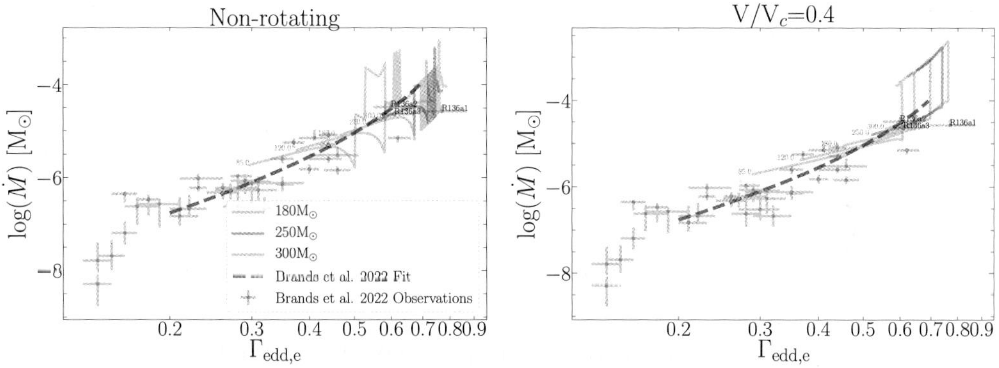

Figure 2. Mass loss rates as a function of the Eddington parameter $\Gamma_{\mathrm{edd,e}}$. The tracks of 85-120$M_\odot$ are from Eggenberger et al. (2021), and the 180-300$M_\odot$ from this work. The observations of Brands et al. (2022) of the R136 components are displayed in gold. The fit they obtained is displayed in purple.

we use are actually in quite good agreement with the observations for the most massive models, and for both the non-rotating and the rotating ones.

As we have spectroscopic observations of these stars, we can obtain surface velocities for the VMS observed in this cluster. The left panel of Fig. 3 shows the values obtained by Bestenlehner et al. (2020) overplotted over the tracks of VMS models. From the spectroscopic studies, the surface velocities observed do not differentiate the surface velocity from the macro-turbulent velocity. This means that the lower limit of the surface velocity would be when the macro-turbulent is at a maximum. We take the maximum macro-turbulent velocities obtained by Simón-Díaz et al. (2017) for stars close to the Eddington limit to obtain the lower uncertainties displayed. The upper limit is obtained from the

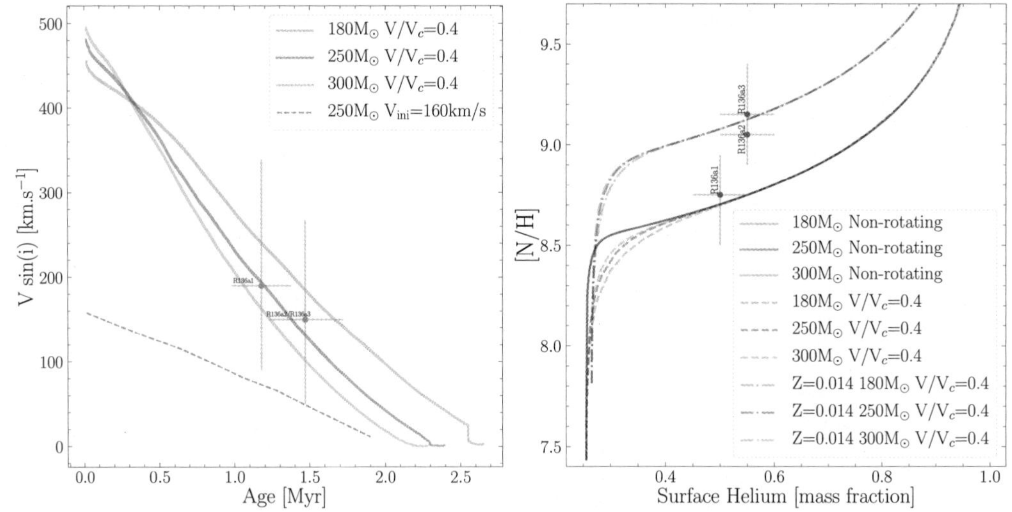

Figure 3. *Left:* Evolution of the surface velocity of VMS models. An additional model with slower initial rotation rate has been computed for this plot. The observed velocities of the three most massive components of R136 have been overplotted (R136a2 and R136a3 have similar velocity obtained). *Right:* Surface enrichment of Nitrogen over Hydrogen as a function of the surface Helium. The observed values of [N/H] are from Brands et al. (2022), and the surface Helium fraction from Bestenlehner et al. (2020).

angle of view effects. We see that the surface velocity is quickly decreasing during the MS due to the large mass loss taking away angular momentum from the envelope, combined to the differential rotation. To obtain the values of the surface velocities observed by Bestenlehner et al. (2020) at this stage of the MS, we need quite high rotation rates already at the ZAMS from the physic we included in these models. Indeed, we computed a 250M$_\odot$ model with lower rotation rate to show that the mass loss still dominates the surface velocity evolution and would not be able to reproduce the high velocities observed for these VMS. To reproduce such high values of surface velocity with lower initial rotation rate, it would be needed to have stronger angular momentum transport inside the star (such as internal magnetic fields) to counteract the large decrease from the mass loss.

The right panel of Fig. 3 shows the observed surface enrichment of Nitrogen as a function of the observed surface Helium respectively from Brands et al. (2022) and Bestenlehner et al. (2020). The non-rotating (in plain lines) and the rotating (in dashed lines) VMS models are overplotted. Interestingly, the evolution of Nitrogen at the surface is very similar between the non-rotating and rotating models. This is because the models quickly reach (before the surface Helium fraction reaches 0.4) the maximum quantity of Nitrogen produced from the CNO cycle in these stars. This quantity depends solely on the initial abundances of CNO. Hence, while the most massive component R136a1 is well reproduced by the models, there is no possibility for these models to explain the very high enrichment value of Nitrogen at this fraction of Helium at the surface of the two other VMS (R136a2/a3). Understanding these discrepancies is difficult, with multiple sources from which they could originate. First, the initial abundances and mixture of CNO could be different from the one used in these Z=0.006 models, leading to not enough nitrogen being able to be produced and brought to the surface. We can see that the Z=0.014 models are able to reproduce such value due to the larger initial abundances of CNO. Second, observational uncertainties cannot be ruled out. Indeed, even if the uncertainties are quantified in these studies, it is not excluded that both unexpected observational and

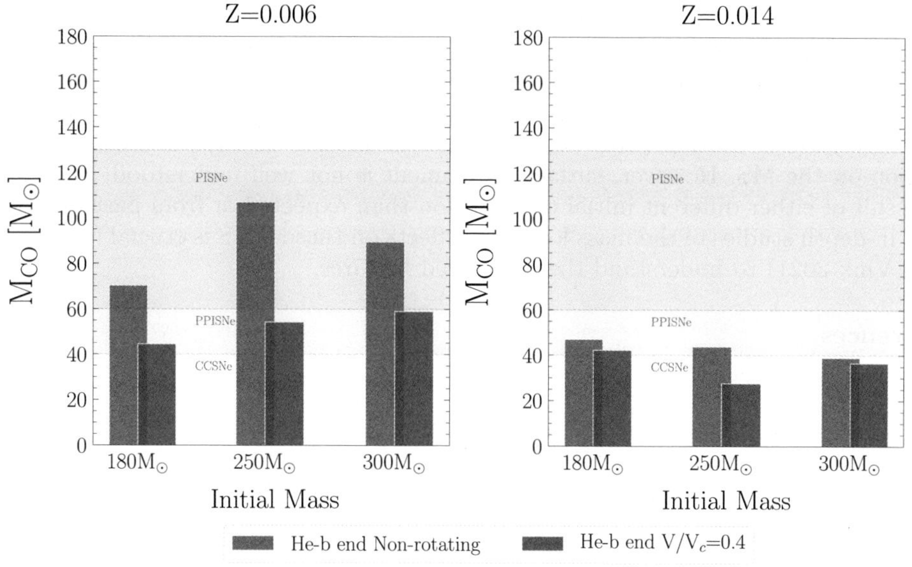

Figure 4. Evolution of the CO core mass of VMS stars at the end of the He-burning phase, for non-rotating and V/V_c=0.4 models at both Z=0.006 and Z=0.014. The(P)PISNe regions are from Farmer et al. (2019).

interpretational effects are present. Third, we can wonder of course if past history of the star, such as binary interaction or early merger could have such effect. A mass transfer from close binary interaction would explain a decrease of the surface Helium by bringing fresh Hydrogen, but would also diminish the [N/H] ratio, and hence cannot explain the nitrogen surface enrichment. An early merger, on the other hand, would simply have at most the maximum quantity from the addition of the initial CNO abundances of both stars, and hence would still leave us with the same problem.

Fig. 4 shows the evolution of the CO core mass of VMS at the end of the He-burning phase, for different rotation rates and metallicities. At the beginning of He-burning, the mass of the core is dominated by the initial mass for the models at Z=0.006 while the stronger line-driven winds occurring for Z=0.014 models result in a large diminution of the core masses, becoming stronger for higher initial masses. During the He-burning phase, the WR winds drive the diminution of the core mass. However, due to their evolution, non-rotating models are undergoing additional mass losses due to their proximity to the Eddington limit. This results for the core masses of models at Z=0.006 to peak at 250$M_\odot$ and then to diminish for higher initial masses. For non-rotating models at Z=0.014, it results to obtain fairly similar core masses at the end of the He-burning phase. All in all, it means that the final fate of this VMS at high metallicity is extremely dependent on both the mass loss rates and the potential extra mixing from rotation. The combination outcome of the two is not trivial, as the resulting evolution can also drive additional mass loss. The final fate can be then extremely different between non-rotating and rotating stars.

4. Conclusions

Very massive stars, while very rare, are extreme objects that can have a very large chemical, mechanical and radiative feedback on a whole stellar population. Indeed, their extreme evolution, combined to very large mass losses could make them important contributors to nucleosynthesis, especially for short-lived radionuclei (see Martinet et al.

2022). Moreover, Population III VMS produce large quantities of ionizing photons, and could make them very important contributors during the reionization epoch (Murphy et al. 2021). We now observe such VMS in the LMC, confirming the large mass loss rates predicted by the models, and their strong ionization flux. These VMS need high initial rotation rates according to our models to explain their current fast rotation on the MS. However, surface enrichment is not well understood, and could be the result of either different initial composition than expected or from past interactions. More in-depth studies of the mass loss rates effects on these VMS is crucial (Higgins et al. 2021; Vink 2021) to understand their observed features.

References

Bestenlehner, J. M., et al. 2014, A&A, 570, A38
—. 2020, MNRAS, 499, 1918
Brands, S. A., et al. 2022, arXiv e-prints, arXiv:2202.11080
Cassinelli, J. P., Mathis, J. S., & Savage, B. D. 1981, Science, 212, 1497
Crowther, P. A., Lennon, D. J., & Walborn, N. R. 2006, A&A, 446, 279
Crowther, P. A., Schnurr, O., Hirschi, R., Yusof, N., Parker, R. J., Goodwin, S. P., & Kassim, H. A. 2010, MNRAS, 408, 731
de Jager, C., Nieuwenhuijzen, H., & van der Hucht, K. A. 1988, A&AS, 72, 259
Eggenberger, P., et al. 2021, A&A, 652, A137
Farmer, R., Renzo, M., de Mink, S. E., Marchant, P., & Justham, S. 2019, ApJ, 887, 53
Georgy, C., Saio, H., & Meynet, G. 2014, MNRAS, 439, L6
Gräfener, G., & Hamann, W. R. 2008, A&A, 482, 945
Groenewegen, M. A. T. 2012a, A&A, 540, A32
—. 2012b, A&A, 541, C3
Higgins, E. R., Sander, A. A. C., Vink, J. S., & Hirschi, R. 2021, MNRAS, 505, 4874
Kaiser, E. A., Hirschi, R., Arnett, W. D., Georgy, C., Scott, L. J. A., & Cristini, A. 2020, MNRAS
Maeder, A. 1997, A&A, 321, 134
Maeder, A., & Meynet, G. 2000, A&A, 361, 159
Martinet, S., et al. 2021, A&A, 648, A126
—. 2022, arXiv e-prints, arXiv:2205.15184
McEvoy, C. M., et al. 2015, A&A, 575, A70
Murphy, L. J., Groh, J. H., Farrell, E., Meynet, G., Ekström, S., Tsiatsiou, S., Hackett, A., & Martinet, S. 2021, MNRAS
Nugis, T., & Lamers, H. J. G. L. M. 2000, A&A, 360, 227
Ramírez-Agudelo, O. H., et al. 2017, A&A, 600, A81
Sabín-Sanjulián, C., et al. 2014, A&A, 564, A39
—. 2017, A&A, 601, A79
Schneider, F. R. N., et al. 2018, A&A, 618, A73
Simón-Díaz, S., Godart, M., Castro, N., Herrero, A., Aerts, C., Puls, J., Telting, J., & Grassitelli, L. 2017, A&A, 597, A22
van Loon, J. T., Cioni, M. R. L., Zijlstra, A. A., & Loup, C. 2005, A&A, 438, 273
Vink, J. S. 2021, arXiv e-prints, arXiv:2109.08164
Vink, J. S., de Koter, A., & Lamers, H. J. G. L. M. 2001, A&A, 369, 574
Zahn, J.-P. 1992, A&A, 265, 115

Discussion

QUESTION: Is there observable allowing us to disantangle between PISNe coming from non-rotating progenitors and from rotating ones ?

MARTINET S.: While we still did not observe any confirmed PISN, we can see from Fig. 4 that rotation can have very different effects at different metallicities, due to the different evolution pathways. Indeed, a rotating model is usually expected to increase the size of the core through rotational mixing. However, the extra mixig also bring the star to evolve in a quasi-chemically homegenous way, reaching much faster the WR phase. This triggers large mass loss that do not occur as early in non-rotating models. In fact, it results in final CO core masses much smaller for rotating models than for non-rotating ones.

Massive Stars Near and Far
Proceedings IAU Symposium No. 361, 2024
J. Mackey, J. S. Vink, & N. St-Louis, eds.
doi:10.1017/S174392132200237X

Massive star interiors revealed by gravity wave asteroseismology and high-resolution spectroscopy

Dominic M. Bowman

Institute of Astronomy, KU Leuven, Celestijnenlaan 200D, 3001 Leuven, Belgium
email: dominic.bowman@kuleuven.be

Abstract. In recent years, it has been discovered that massive stars commonly exhibit a non-coherent form of variability in their light curves referred to as stochastic low frequency (SLF) variability. Various physical mechanisms can produce SLF variability in such stars, including stochastic gravity waves excited at the interface of convective and radiative regions, dynamic turbulence generated in the near-surface layers, and clumpy winds. Gravity waves in particular are a promising candidate for explaining SLF variability as they can be ubiquitously generated in main sequence stars owing to the presence of a convective core, and because they provide the large-scale predominantly tangential velocity field required to explain macroturbulence in spectral line fitting. Here, I provide an overview of the methods and results of studying SLF variability in massive stars from time series photometry and spectroscopy.

Keywords. stars: early-type, stars: oscillations (including pulsations), stars: interiors, waves.

1. Introduction

Modern time-series space photometry, such as from the CoRoT (Auvergne et al. 2009), Kepler/K2 (Borucki et al. 2010; Howell et al. 2014) and TESS (Ricker et al. 2015) missions, have revealed in the last decade that massive stars show a diverse range of variability mechanisms (see Bowman 2020 for a review). In particular, in addition to a high variability fraction above 90% for early-type star light curves, which includes coherent pulsation modes, binarity and rotational modulation (see e.g. Burssens et al. 2020), massive stars almost ubiquitously show stochastic low-frequency (SLF) variability (Bowman et al. 2019a,b, 2020). SLF variability is also common across the different evolutionary stages of massive stars including blue and yellow supergiants, Wolf Rayet stars and luminous blue variables (see e.g., Bowman et al. 2019b; Dorn-Wallenstein et al. 2020; Nazé et al. 2021; Elliott et al. 2022), which show similar amplitude spectra to those of main sequence stars. There are currently a few proposed mechanisms to explain SLF variability in massive stars, but it is unclear which may dominate in particular parameter regimes. Hence the investigation of SLF variability in space photometry has allowed for a novel method to understanding the interiors of massive stars (Bowman 2020).

In their original pilot study, Bowman et al. (2019a) analysed 35 early-type stars from the CoRoT mission to demonstrate how solar granulation scaling laws cannot explain SLF variability in massive stars. Using a much larger sample of over 160 massive stars from the Kepler/K2 and TESS missions, including galactic and Large Magellanic Cloud (LMC) stars, Bowman et al. (2019b) demonstrated that essentially all massive stars exhibit SLF variability, and that its morphology is dependent on the brightness of the star. More recently, Bowman et al. (2020) studied a sample of galactic O and B stars from

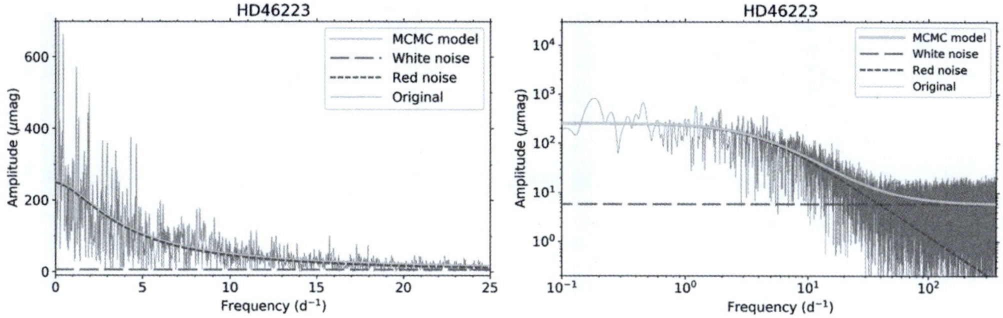

Figure 1. Amplitude spectrum of the 25-d sector 6 TESS light curve of the O4V((f)) star HD 46223. Left and right panels are the same data plotted in a linear-linear and log-log format, respectively. Figures adapted from (Bowman et al. 2020), their figure 1.

the TESS mission and combined it with high resolution spectroscopy (Simón-Díaz et al. 2011; Burssens et al. 2020) to show how the morphology of SLF variability was correlated with mass, age, and amount of macroturbulence needed to fit the spectral line broadening in the massive star regime.

In an approach similar to the forward asteroseismic modelling of coherent pulsation modes (see Aerts 2021), the exploitation of SLF variability in time series photometry offers a route to improving our understanding of stellar structure and evolution (Bowman 2020). In these proceedings, I review the current state-of-the-art in the methods and analysis of SLF variability in massive stars.

2. Stochastic Low Frequency Variability

The SLF variability in massive stars typically covers a broad period (and frequency) range from several days up to minutes. The amplitudes, on the other hand, range between tens of μmag up to tens of mmag in broadband time series photometry (Bowman et al. 2019b). Of course, these metrics depend on the length of the light curve, its cadence, wavelength passband and photometric precision. For example, the wavelength passband of the TESS mission is redder to those of the Kepler/K2 and CoRoT missions, which means TESS is slightly less sensitive to the changes in brightness of blue stars (Bowman et al. 2019a). Similarly, with higher photometric precision and longer duration light curves, smaller changes in flux in a light curve can be measured and an improved frequency resolution is achieved in an amplitude spectrum, respectively. Lastly, a shorter cadence yields a higher Nyquist frequency, which is necessary to unambiguously probe short-period variability in massive stars and avoid instrumental suppression (Bowman 2017). Nevertheless, space mission light curves show ubiquitous SLF variability in massive stars when high enough quality time series data are used (Bowman et al. 2019b).

To characterise the morphology of SLF variability in the Fourier domain, the semi-Lorentzian function:

$$\alpha\left(\nu\right) = \frac{\alpha_0}{1 + \left(\frac{\nu}{\nu_{\mathrm{char}}}\right)^{\gamma}} + C_{\mathrm{w}} , \tag{2.1}$$

has been used to quantify the characteristic frequency, ν_{char}, characteristic amplitude, α_0 and steepness, γ, of the profile, with C_{W} being a white noise term. An example of observed SLF variability in the amplitude spectrum of a massive star, using the TESS data of the O4V((f)) star HD 46223, is shown in Fig. 1. In the left and right panels, the linear and logarithmic amplitude spectra are shown, each with the best-fit profile using

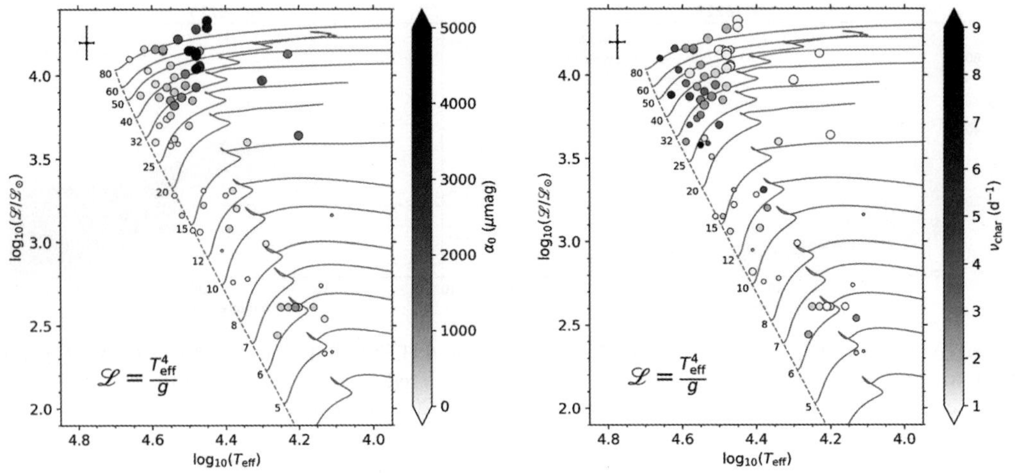

Figure 2. Spectroscopic HR diagram of the massive star sample of (Bowman et al. 2020) in which the morphology of SLF variability in the amplitude spectra of TESS light curves was quantified using Eqn. (2.1). The left and right panels contain the same sample, but each star, shown as a circle, is colour-coded by its α_0 and $\nu_{\rm char}$ parameter value, respectively. Figures adapted from (Bowman et al. 2020), their figure 2.

Eqn. (2.1) as determined by Bowman et al. (2020). HD 46223 was also observed by the CoRoT mission and the characteristic amplitude of its SLF variability was somewhat smaller in the TESS data, as expected (Bowman et al. 2019a).

3. Trends in Mass and Age

Beyond the discovery and characterisation of SLF variability in massive stars, the study by Bowman et al. (2020) combined high-photometric and short-cadence TESS light curves of massive stars with high-resolution spectroscopy from the IACOB and OWN surveys (Simón-Díaz et al. 2011; Burssens et al. 2020). For the first time, it was shown how the morphology of SLF variability, and specifically the characteristic amplitudes and frequencies, α_0 and $\nu_{\rm char}$, directly probe a star's location in the Hertzsprung–Russell (HR) diagram. More massive and more evolved stars have larger amplitudes (i.e. α_0) and lower frequencies (i.e. $\nu_{\rm char}$) in the morphology of their SLF variability, on average. Consequently, the characterisation of SLF variability provides constraints on a massive star's mass and age without the need for spectroscopic constraints. This is demonstrated in Fig. 2, in which the left and right panels denote each star's location in the spectroscopic HR diagram colour-coded by α_0 and $\nu_{\rm char}$, respectively.

An explanation for these trends is that more massive and more evolved stars have larger radii. This means that for larger radii, SLF variability exhibits longer periods, and vice versa. In terms of amplitudes, more evolved stars are observed to have larger amplitudes in their SLF variability, which means that the excitation mechanism(s) responsible exist from the zero-age main sequence (ZAMS) through to the terminal-age main sequence (TAMS), but are more efficient for more evolved stars. It is important to note that the sample of massive stars in Fig. 2 is far from complete and suffers from biases. For example, the B-star regime is severely under sampled, as is the post-TAMS phase of evolution. Therefore, the observed trends in SLF variability with mass and age are only evident for main-sequence O stars, and does include some astrophysical scatter.

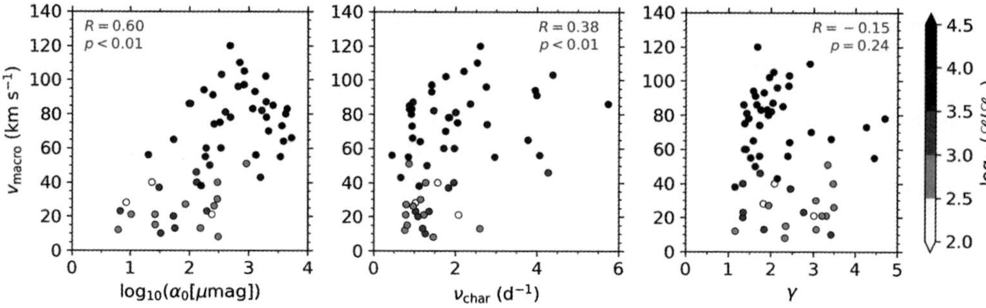

Figure 3. Pairwise correlation statistics between the amount of spectroscopic macroturbulence, $v_{\rm macro}$, and the morphology parameters of SLF variability determined by (Bowman et al. 2020). Figure adapted from (Bowman et al. 2020), their figure 3.

4. Correlation with Macroturbulence

In addition to placing stars in the HR diagram, Bowman et al. (2020) found a strong correlation between the characteristic amplitudes and frequencies of SLF variability and macroturbulence in massive stars. Macroturbulence is a non-rotational broadening mechanism needed to fit spectral lines of massive stars, which has both radial and tangential components and amplitudes of order 100 km s^{-1} among O stars (Simón-Díaz et al. 2014, 2017). Macroturbulence is typically the dominant form of spectral line broadening in massive stars, so various studies have placed specific emphasis on investigating its physical cause (see e.g. Aerts et al. 2009; Grassitelli et al. 2015).

From their sample of massive stars, Bowman et al. (2020) found a statistically significant relationship between the amount of macroturbulence needed to fit the spectral line profiles and the characteristic amplitudes and frequencies of SLF variability in TESS photometry. Specifically, stars with larger macroturbulence have large values of α_0 and smaller values of $\nu_{\rm char}$. This correlation is demonstrated in Fig. 3, in which the correlation coefficient, R, and the resulting p-value from a hypothesis test is shown in the top corner of each panel. On the other hand, there is no significant correlation among the steepness, γ, of the SLF variability morphology, the amount of macroturbulence, nor the location of the star in the HR diagram. This demonstrates that a typical γ value of $1 \lesssim \gamma \lesssim 3$ exists for all massive stars, regardless of their mass, age or the physical mechanism(s) that cause SLF variability.

Due to the limited sample size, it was not possible for Bowman et al. (2020) to investigate further correlations with, for example, rotation rate or metallicity. However, such an investigation with a much larger sample of massive stars will undoubtedly shine further light on the physical relationship between macroturbulence in spectroscopy and SLF variability in time-series photometry. For example, rotation is expected to significantly modify the structure of a star and also shift the frequencies of pulsation modes, which may explain the scatter in Figs. 2 and 3.

It was first Lucy (1976) who postulated that macroturbulence was caused by pulsations. Later, Aerts et al. (2009) showed using detailed simulations that non-radial gravity modes provide the necessary pre-dominantly tangential velocity field to reproduce spectral line profile broadening caused by macroturbulence. More recently, Aerts & Rogers (2015) extended this and demonstrated how an ensemble of stochastically excited gravity waves provides the necessary tangential velocity field to explain macroturbulence as a pulsational broadening mechanism. Together, this makes pulsations, specifically gravity waves, a plausible mechanism for macroturbulence and SLF variability in massive stars.

5. Discussion and Conclusions

The discovery of SLF variability in massive stars has inspired and energised various theoretical studies dedicated to providing possible explanations of this new type of observational signal. Motivated by there always being a convective core for main-sequence stars, Aerts & Rogers (2015) postulated that the SLF variability detected at the surface of three O stars was caused by stochastically excited gravity waves generated at the interface of the convective core and radiative envelope. This was based on the qualitatively similar amplitude spectra of these three O stars with those predicted from 2D hydrodynamical simulations (Rogers et al. 2013; Rogers 2015; Rogers & McElwaine 2017). More recently, further 2D and 3D hydrodynamical simulations of stellar interiors using different numerical setups have shown that gravity waves excited at the boundaries of convective cores produce detectable surface velocity and flux perturbations (Edelmann et al. 2019; Horst et al. 2020). Such predictions are indeed quantitatively comparable to the observed amplitude spectra of massive stars (Bowman et al. 2019b, 2020).

On the other hand, the presence of a sub-surface convection zone and how its properties depend on a star's location in the HR diagram has led other studies to postulate that SLF variability and/or macroturbulence originates much closer to the surface (Cantiello et al. 2009, 2021; Grassitelli et al. 2015). Moreover, numerical calculations of wave transfer functions for gravity waves generated near the convective cores of main-sequence stars predict them to appear as a series of equally-spaced peaks in the amplitude spectrum, which correspond to the stellar eigenfrequencies, rather than as SLF variability (Lecoanet et al. 2019, 2021). However, the effects of non-linearity and rotation should be studied in detail in the context of wave propagation, because the Coriolis force is a significant restoring force for gravity waves in massive stars (Aerts et al. 2019). Furthermore, the dynamics of sub-surface convection zones in massive stars are expected to be quite turbulent. This is demonstrated by hydrodynamical simulations (Jiang et al. 2015, 2018; Schultz et al. 2022), which as a consequence also predict SLF variability in time-series photometry. However, the efficiency of sub-surface convection zones is strongly metallicity dependent, such that some stars may have no subsurface convection zones at all (Jermyn et al. 2022). Finally, for more evolved stars near or beyond the TAMS, their radiatively driven and optically thick winds are predicted to produce SLF variability in time-series observations (e.g. Krtička & Feldmeier 2018, 2021).

All of the aforementioned mechanisms that predict SLF variability in massive stars are not mutually exclusive. This means that the challenge ahead is in disentangling which mechanism dominates within specific parameter regimes in the HR diagram. This remains an open question because the properties of (sub-surface) convection, winds, and the driving and propagation of gravity waves depend on the mass, radius, metallicity, and rotation rate of a massive star. What is clear is that any mechanism that aims to unanimously explain SLF variability in main-sequence stars must do so for stars that span a broad range of masses (i.e. $3 \lesssim M \lesssim 100\ \mathrm{M}_{\odot}$), ages from the ZAMS to the TAMS and beyond, rotation rates between zero and large fractions of the critical breakup velocity, and for both metal-poor (i.e. $Z \leq 0.002$) and metal-rich stars (i.e. $Z \geq 0.014$).

Since the work by Bowman et al. (2019a) and Bowman et al. (2020), much larger samples of massive stars with longer light curves are available thanks to the ongoing TESS mission. In the future, new ground-based spectroscopic monitoring campaigns and space missions, such as the CubeSpec mission (Bowman et al. 2022), will provide invaluable constraints on the time-dependent nature of macroturbulence and its link to pulsations in massive stars. From renewed theoretical work on understanding the interiors and atmospheres of massive stars, the constraints on the physical mechanisms that govern their structure and evolution are ever improving. Asteroseismology of SLF variability not only has the potential to provide independent masses and ages of massive stars, but also provide insight of their interior mixing and rotation properties to calibrate stellar structure and evolution theory.

Acknowledgements

DMB gratefully acknowledges funding from the Research Foundation Flanders (FWO) by means of a senior postdoctoral fellowship (grant agreement No. 1286521N). DMB thanks the TESS science team for the excellent data, without which massive star asteroseismology and much of this work would not have been possible.

References

Aerts et al. 2019, *ARAA*, 57, 35-78
Aerts et al. 2009, *A&A*, 508, 409-419
Aerts & Rogers 2015, *ApJL*, 806, L33
Aerts 2021, *Reviews of Modern Physics*, 93, 015001
Auvergne et al. 2009, *A&A*, 506, 411-424
Borucki et al. 2010, *Science*, 327, 977
Bowman 2017, *Springer*, Amplitude modulation of pulsation modes in delta Scuti stars
Bowman et al. 2019, *A&A*, 621, A135
Bowman et al. 2019, *Nature Astronomy*, 3, 760-765
Bowman et al. 2020, *A&A*, 640, A36
Bowman et al. 2022, *A&A*, 658, A96
Bowman 2020, *Fron. Astron. Space Sci.*, 7, 70
Burssens et al. 2020, *A&A*, 639, A81
Cantiello et al. 2009, *A&A*, 499, 279-290
Cantiello et al. 2021, *ApJ*, 915, 112
Dorn-Wallenstein et al. 2020, *ApJ*, 878, 155
Edelmann et al. 2019, *ApJ*, 876, 4
Elliott et al. 2022, *MNRAS*, 509, 4246-4255
Grassitelli et al. 2015, *ApJL*, 808, L31
Horst et al. 2020, *A&A*, 641, A18
Howell et al. 2014, *PASP*, 126, 398-408
Jermyn et al. 2022, *ApJ*, 926, 221
Jiang et al. 2015, *ApJ*, 813, 74
Jiang et al. 2018, *Nat*, 561, 498-501
Krtička & Feldmeier et al. 2018, *A&A*, 617, A121
Krtička & Feldmeier et al. 2021, *A&A*, 648, A79
Lecoanet et al. 2019 *ApJL*, 886, L15
Lecoanet et al. 2021 *MNRAS*, 508, 132-143
Lucy 1976 *ApJ*, 206, 499
Nazé et al. 2021, *MNRAS*, 502, 5038-5048
Ricker et al. 2015, *J. Astron. Telesc., Instrum., Sys.*, 1, 014003
Rogers et al. 2013, *ApJ*, 772, 21
Rogers 2015, *ApJL*, 815, L30
Rogers & McElwaine 2017, *ApJL*, 848, L1
Schultz et al. 2022, *ApJL*, 924, L11
Simón-Díaz et al. 2011, *Bulletin de la Societe Royale des Sciences de Liege*, 80, 514-518
Simón-Díaz et al. 2014, *A&A*, 562, A135
Simón-Díaz et al. 2017, *A&A*, 597, A22

Massive Stars Near and Far
Proceedings IAU Symposium No. 361, 2024
J. Mackey, J. S. Vink, & N. St-Louis, eds.
doi:10.1017/S1743921322002642

Fundamental Properties of O and B Stars with Optical Interferometry

Kathryn D. Gordon[1], **Douglas R. Gies**[2] **and Gail H. Schaefer**[3]

[1]University of Tampa
401 W. Kennedy Blvd., Tampa, FL, USA
email: kgordon@ut.edu

[2]Georgia State University
CHARA, Department of Physics and Astronomy, Atlanta, GA, USA
email: dgies@gsu.edu

[3]The CHARA Array
Mount Wilson, CA, USA
email: gschaefer@gsu.edu

Abstract. We used interferometric observations made with the CHARA Array of 25 B-type stars and 6 O-type stars to obtain precise measurements of angular size, radius, and effective temperature to test stellar atmospheric models for massive stars. Our measured angular diameters range from 1.09 milli-arcseconds (mas) for β Tau down to 0.11 mas for 10 Lac, the smallest star yet resolved with the CHARA Array. The rotational oblateness of the rapidly rotating star ζ Oph is directly measured for the first time. We collected ultraviolet to infrared spectrophotometry for all sample stars and derived temperatures, angular diameters, and reddening estimates that best fit the spectra. There is generally good agreement between the observed and spectral fit angular diameters for the O and B stars, indicating that the fluxes predicted from model atmospheres are reliable. The derived and model temperatures for the O stars are also in fair agreement, however the sample size is small and several of the O stars results we consider to be preliminary. On the other hand, the temperatures derived from angular diameters and fluxes tend to be larger (by $\approx 4\%$) for the B stars than those from published results based on analysis of the line spectrum (Gordon et al. 2018, 2019).

Keywords. stars: early-type, stars: fundamental parameters (classification, colors, luminosities, masses, radii, temperatures, etc.), techniques: interferometric

1. Introduction

Massive stars have a critical impact on their surroundings due to their intense radiation, strong stellar winds and explosive ends as supernovae. They are also the progenitors of many interesting astrophysical objects, from black holes to Wolf-Rayet stars and Luminous Blue Variables. It is therefore important to have an accurate understanding of the parameters of these stars and how these parameters relate to their evolutionary state. Fundamental parameters such as radius and effective temperature, for individual O and B-type stars are generally determined through continuum flux measurements or by detailed analysis of the line spectrum. The determination of fundamental parameters of individual O-type stars is mostly based on analysis of the line spectrum, in which the temperature and gravity are derived from the He line ionization balance and the pressure broadening of the hydrogen lines. For the hot and intermediate-mass B-type stars an estimation of effective temperature is generally found by continuum flux measurements or by detailed analysis of the line spectrum. Both of these methods rely on comparisons with predictions from stellar model atmospheres. It is therefore of critical importance to test these models through independent means.

With long-baseline optical interferometry, we are able to resolve the disks of these stars and directly measure their angular diameters. This can be combined with a known distance, from *Gaia* parallax measurements (Gaia Collaboration et al. 2018), to obtain an independent determination of the radius of the star. The effective temperature is derived from the relationship between the extinction corrected bolometric flux, angular diameter θ, and $T_{\rm eff}$:$f_{\rm bol} = \frac{1}{4}\theta^2\sigma T_{\rm eff}^4$. We can then compare our observationally determined angular diameters and effective temperatures to those predicted by stellar atmosphere models to test their validity. Once physical radii and effective temperatures are determined, stars can be plotted on an observational HR diagram and age estimates can be obtained from model isochrones. We have utilized the TLUSTY OSTAR 2002 (Hubeny & Lanz 1995; Lanz & Hubeny 2003), the TLUSTY BSTAR2006 (Lanz & Hubeny 2007), and the ATLAS9 (Kurucz 1992; Castelli & Kurucz 2004) models as the temperature of each star dictated. The TLUSTY models are non-LTE (non-local thermal equilibrium), which becomes more important in hotter atmospheres, and the ATLAS9 models are LTE. All models assumed solar abundances.

In these proceedings, we present the results of our angular diameter measurements of a sample of 25 B stars and 6 O stars. We discuss the fits of the SED as a function of temperature, angular size, and reddening for each star, and the comparison of observed and derived parameters to model predictions. More detailed discussions of these studies can be found in their respective papers, with the O star results in Gordon et al. (2018) and the B star results in Gordon et al. (2019).

2. Observations

The stars in our observational sample consisted of six O spectral type stars ranging from O7.5 to O9 and 25 B-type stars with spectral types ranging from B2 to B9. All luminosity classes are represented with 3 supergiants, 16 giants, one sub-giant, and 11 dwarfs, or main sequence stars. All target stars had a V magnitude of 5 or less and had a predicted angular diameter range of 0.21.0 mas. Choosing stars with these constraints on brightness, size, and with a declination of greater than $-14°$ ensured that they would be observable and resolvable with the CHARA Array. All O star observations were made using the Precision Astronomical Visible Observations (PAVO) beam combiner (Ireland et al. 2008) at the CHARA Array (ten Brummelaar et al. 2005), located at Mount Wilson Observatory in California. B stars were observed with PAVO or with the CLassic Interferometry with Multiple Baselines (CLIMB) beam combiner (ten Brummelaar et al. 2013) as their predicted angular sizes dictated.

The CHARA Array is an optical interferometer composed of six 1m telescopes arranged in a Y-shaped configuration. Combinations of the telescopes allow for 15 different usable baselines ranging in length from 34 to 331 m. Combining the longest usable baseline currently available in the world and the operating wavelength range of the PAVO beam combiner (650800 nm), we were able to achieve an extremely high angular resolution for our targets of about 0.2 mas. The CLIMB beam combiner was operated in the H-band (1.67 μm) and has a resolution limit of about 0.5 mas.

Because of their shorter lifetimes, it is common for massive stars to have a close companion or be a member of a multiple star system. The extra flux from these companions can affect the fitting of our interferometric data and spectophotometric modeling and must be accounted for. A literature search was done for each target star to uncover known companions. Targets with companions close enough to contribute incoherent flux in the

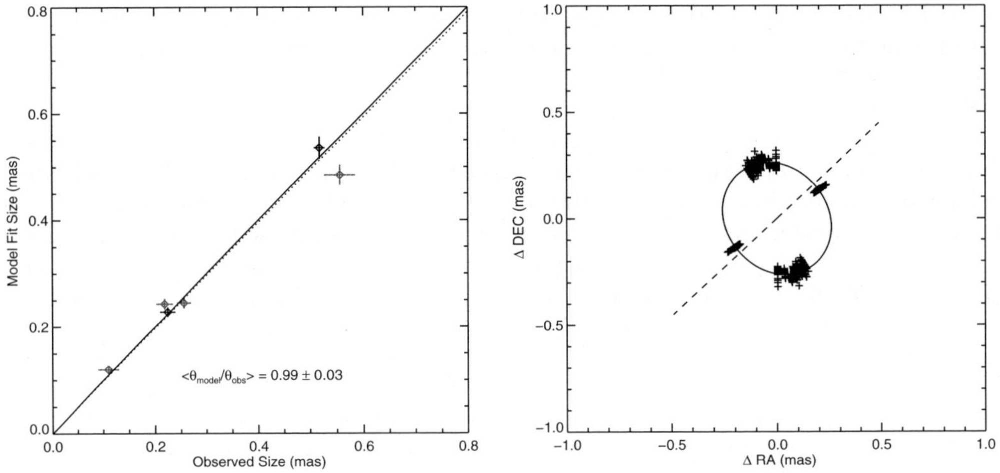

Figure 1. (a) (Left) Observed angular size θ_{LD} compared to the angular size $\theta_{LD}(T_{\rm eff})$ derived from the published $T_{\rm eff}$ and fit to the SED for the O star sample. The solid line shows a line with a slope of unity for reference, and the dashed line shows the trend for the mean ratio of these diameters. Blue points indicate diameter estimates based upon only a single data bracket or an extremely small angular size in the case of 10 Lac. From Gordon et al. (2018).
Figure 1. (b) (Right) A simple ellipse fitted to our interferometric data for ζ Oph. Each symbol shows the derived angular size of a limb-darkened star whose visibility equals the observed value, and each is plotted at a position angle derived from the (u, v) spatial frequencies of the observation. The fit was made of the major and minor axes with the position angle of the minor axis set by published polarimetry. The dashed line shows the adopted rotational axis of the star at a position angle of 132.°5. From Gordon et al. (2018).

field of view of the detector were corrected with the method described in Boyajian et al. (2008) for the interferometric data and by adding the calculated flux of the companion to the spectrum for the spectrophotometric modeling.

3. Results

3.1. *O Stars*

Our initial survey of O stars finds that the measured diameters are generally in good agreement with the expected diameters based upon the published temperatures and model fits of the SED. A comparison of observed $\theta_{\rm LD}$ (limb-darkened angular diameter) and the predicted angular size $\theta_{\rm LD}(T_{\rm eff})$ is shown in Figure 1.(a) with the average ratio of $\theta_{\rm LD}(T_{\rm eff})$ to $\theta_{\rm LD}$ as approximately 0.99 ± 0.03. Two evolved stars in our sample, ξ Per and ζ Ori A, do not fit with model predictions as well but have limited data coverage and we consider these results preliminary. Additional interferometric observations are needed to confirm the diameters found in these cases and the O star sample needs to be expanded to give greater statistical significance to our results and allow us to investigate effective temperature trends.

ζ Oph is a particularly interesting O star with an extremely rapid rotational velocity. It has an estimated $v \sin i$ of 311 km/s (Simón-Díaz et al. 2014), or about 90% of its critical velocity (Howarth & Smith 2001). We observed ζ Oph on two different baselines with different position angles to measure its oblateness due to its rapid rotation. By setting the position angle from a past spectropolarimetric study (Poeckert et al. 1979), we made an elliptical fit to the diameter as a function of position angle in the sky determined from our observations. This fit is shown in Figure 1.(b) and constitutes the first direct measurement of the shape of ζ Oph.

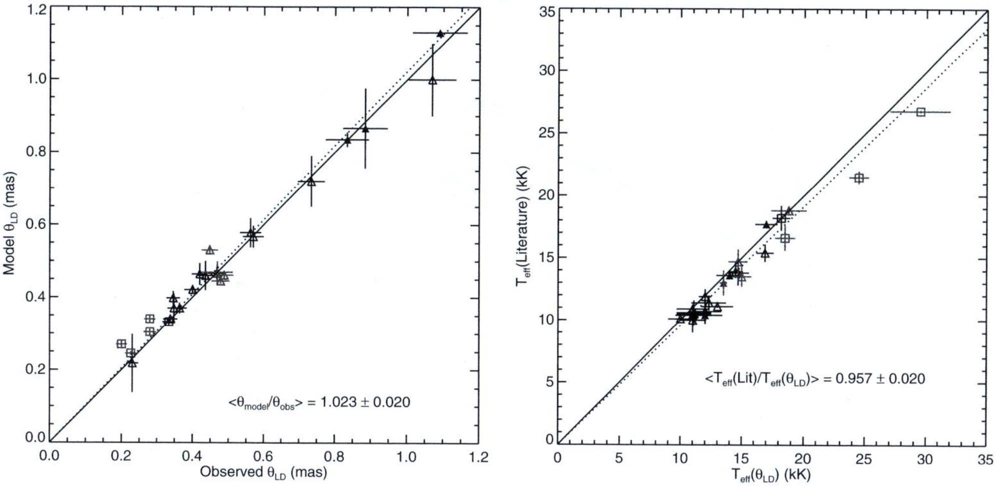

Figure 2. **(a)**(Left) A comparison of our observed angular diameters to the SED-derived model size for the B star sample. The solid line is a unit slope line for reference, and the dotted line indicates the weighted mean of $\theta_{model}/\theta_{obs}$. Square symbols indicate stars fit with the TLUSTY BSTAR2006 models, triangles indicate stars fit with ATLAS9 models, and filled triangles indicate stars with CLIMB data. Red points are measurements considered preliminary and are not included in the weighted mean ratio shown. From Gordon et al. (2019).
Figure 2.(b) (Right) A comparison of our derived temperatures to the average literature temperature $T_{\rm eff}$(Lit) for the B star sample. The solid line is a unit line for reference and the dotted line indicates the weighted mean of the ratio of variables being compared. Square symbols indicate stars fit with the TLUSTY BSTAR2006 models, triangles indicate stars fit with ATLAS9 models, and filled triangles indicate stars with CLIMB data. Red points are measurements considered preliminary and are not included in the weighted mean ratios shown. From Gordon et al. (2019).

3.2. *B Stars*

The results of our angular diameter survey of 25 B stars found that our measured angular diameters are in good agreement with model predicted angular sizes. This result does set aside some observed stars with data we consider preliminary due to complications discussed in detail in our B star paper (Gordon et al. 2019). A comparison of measured and model angular diameters is shown in Figure 2.(a). Our derived interferometric effective temperatures are also in good agreement with the model predicted effective temperatures. This agreement confirms the validity of the application of the TLUSTY (Lanz & Hubeny 2007) and ATLAS9 (Kurucz 1992; Castelli & Kurucz 2004) stellar atmosphere models for most main-sequence B-type stars. However, there is some discrepancy between our derived effective temperatures and those determined primarily from spectral line studies. A comparison of these two $T_{\rm eff}$ measurements is shown in Figure 2.(b), and the error weighted average ratio is $< T_{\rm eff}(\mathrm{Lit})/T_{\rm eff}(\theta_{LD}) >= 0.957 \pm 0.022$ after removal of the preliminary results. This represents a 2σ smaller estimate for $T_{\rm eff}$ from line studies compared to that from the continuum SED fit at the observed angular size.

Another notable discrepancy in our B star sample comes from our observations of 55 Cyg, the only supergiant in our sample. The observed angular size, at 0.448 ± 0.023 mas, is 15% smaller than the model prediction, 0.530 ± 0.013 mas, which leads to a best fit temperature of 18.8 ± 1.4 kK, which is 14% higher than the model fit value of 16.5 ± 1.0 kK. However, a previous spectral study of 55 Cyg by Kraus et al. (2015) using the FASTWIND stellar atmosphere code (Santolaya-Rey et al. 1997) found an effective

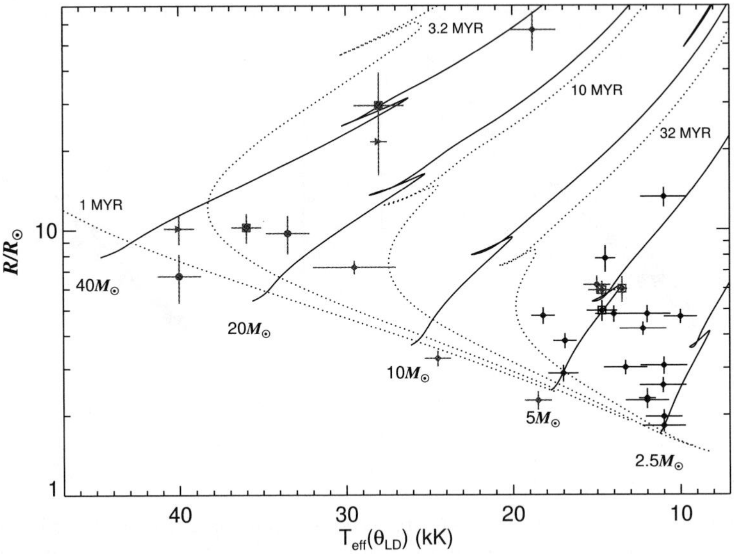

Figure 3. Observational HR diagram for our targets based on our observed diameters and derived effective temperatures. The black diamonds indicate B stars, blue circles O stars, and triangles are O stars with only upper or lower effective temperature estimates. Red points indicate results we consider to be preliminary. Symbols enclosed by squares indicate stars that are cluster members. Evolutionary tracks and isochrones are from the MESA grid. From Gordon et al. (2019).

temperature of 18.8 kK, which is in excellent agreement with our derived temperature. Kraus et al. (2015) also gives a stellar radius for 55 Cyg of $57R_\odot$, which when combined with the distance from the recent *Gaia* DR2 parallax (Gaia Collaboration et al. 2018), gives an angular diameter of 0.45 ± 0.07 mas. This is in good agreement with our measured angular diameter. We suspect that this discrepancy is due to the fact that the TLUSTY BSTAR2006 and ATLAS9 models do not incorporate the effects of stellar winds, which can be quite significant for massive supergiant stars. It will be important to compare interferometric results and fluxes of the supergiants to models like CMFGEN (Hillier & Miller 1998), and we are working on studying this effect with a sample of B supergiants.

We combined our observed angular diameters with known distances to calculate stellar radii. *Gaia* DR2 parallaxes were used to calculate distance where available. If no *Gaia* DR2 parallax was available, then the Hipparcos (van Leeuwen 2007) value was adopted. With our derived estimates for stellar radius and effective temperature, we placed our sample stars on an observational HR diagram. This is shown in Figure 3, and includes the O star and B star sample. Circles indicate the O stars, while the triangles mark the two O stars for which we only have upper or lower limits on $T_{\rm eff}$. Overplotted are evolutionary tracks and isochrones for the MESA grid (Paxton et al. 2013; Choi et al. 2016; Dotter 2016). This allows for stellar age and mass estimates that are observationally defined and less model dependent than previous methods.

References

Boyajian, T. S. et al. 2008, ApJ, 683, 424
Castelli, F., & Kurucz, R. L. 2004, ArXiv Astrophysics e-prints
Choi, J., Dotter, A., Conroy, C., Cantiello, M., Paxton, B., & Johnson, B. D. 2016, ApJ, 823,

102
Dotter, A. 2016, ApJS, 222, 8
Gaia Collaboration et al. 2018, A&A, 616, A1
Gordon, K. D., Gies, D. R., Schaefer, G. H., Huber, D., & Ireland, M. 2019, ApJ, 873, 91
Gordon, K. D., Gies, D. R., Schaefer, G. H., Huber, D., Ireland, M., & Hillier, D. J. 2018, ApJ, 869, 37
Hillier, D. J., & Miller, D. L. 1998, ApJ, 496, 407
Howarth, I. D., & Smith, K. C. 2001, MNRAS, 327, 353
Hubeny, I., & Lanz, T. 1995, ApJ, 439, 875
Ireland, M. J. et al. 2008, in Proc. SPIE, Vol. 7013, Optical and Infrared Interferometry, 701324
Kraus, M. et al. 2015, A&A, 581, A75
Kurucz, R. L. 1992, in IAU Symposium, Vol. 149, The Stellar Populations of Galaxies, ed. B. Barbuy & A. Renzini, 225
Lanz, T., & Hubeny, I. 2003, ApJS, 146, 417
——. 2007, ApJS, 169, 83
Paxton, B. et al. 2013, ApJS, 208, 4
Poeckert, R., Bastien, P., & Landstreet, J. D. 1979, AJ, 84, 812
Santolaya-Rey, A. E., Puls, J., & Herrero, A. 1997, A&A, 323, 488
Simón-Díaz, S., Herrero, A., Sabín-Sanjulián, C., Najarro, F., Garcia, M., Puls, J., Castro, N., & Evans, C. J. 2014, A&A, 570, L6
ten Brummelaar, T. A. et al. 2005, ApJ, 628, 453
——. 2013, Journal of Astronomical Instrumentation, 2, 1340004
van Leeuwen, F. 2007, A&A, 474, 653

Discussion

GARCIA: Have you made some tests with the R value, total to selective extinction?

GORDON: No, we have used the standard 3.1 for that and we didn't change that at all.

GARCIA: Is it possible that would effect your effective temperature determination?

GORDON: Early on we did do a few tests to see how sensitive the temperature was to changing that. So we did change it change it a little bit and we saw that the effective temperature was not very sensitive to that value.

Q2: I was wondering, for zeta Oph where you have a measurement of oblateness and a measurement of vsini, if you could use these together, assuming some oblateness of the star due to its rotation, to back out an inclination angle?

GORDON: That is certainly possible, something I could look into. Not my area of expertise. There is a position angle already from a spectropolarimetry study but it would be interesting to compare to that what we got from our observations.

Massive Stars Near and Far
Proceedings IAU Symposium No. 361, 2024
J. Mackey, J. S. Vink, & N. St-Louis, eds.
doi:10.1017/S1743921322001983

On the elusive detection of O-type stars close to the ZAMS: the empirical birthline of massive stars

G. Holgado[1], S. Simón-Díaz[2,3] and L. Haemmerlé[4]

[1]Centro de Astrobiologia (CAB), CSIC-INTA, Camino Bajo del Castillo s/n, 28692, Villanueva de la Cañada, Madrid, Spain
email: gholgado@cab.inta-csic.es

[2]Instituto de Astrofísica de Canarias, E-38200 La Laguna, Tenerife, Spain

[3]Departamento de Astrofísica, Universidad de La Laguna, E-38205 La Laguna, Tenerife, Spain

[4]Département d'Astronomie, Université de Genève, chemin des Maillettes 51, CH-1290 Versoix, Switzerland

Abstract. In this contribution we present the results from a † homogeneous quantitative spectroscopic analysis of ∼400 Galactic O-type stars targeted by the IACOB and OWN surveys. The ultimate objective is to perform a modern reassessment of one of the long-standing problems in the field of massive stars: the elusive detection of mid O-type stars close to the "canonical" theoretical ZAMS. We first provide statistically significant evidence of the existence of a clear lack of stars in our sample populating the region of the spectroscopic HR diagram approximately delimited by the theoretical ZAMS, the ∼40 and ∼70 $M_\odot$ single evolutionary tracks and the 2 Myr isochrone. We then evaluate if this empirical result could be a result of possible limitations of our analysis strategy and/or the existence of potential observational biases affecting the compiled sample. Once both explanations are investigated, we evaluate the possibility that a modification of the efficiency of mass accretion during the star formation process could lead to a new (corrected) theoretical ZAMS in better accordance with our empirical results.

Keywords. Stars: early-type – Stars: massive – Techniques: spectroscopic – Hertzsprung-Russell and C-M diagramas – Stars : evolution

---

The apparent lack of massive O-type stars near the zero-age main sequence, or ZAMS (at ages < 2 Myr), is a topic that has been widely discussed in the past 40 years. Different explanations for the elusive detection of these young massive stars have been proposed from the observational and theoretical side, but no firm conclusions have been reached yet.

We reassess this empirical result here, benefiting from the high-quality spectroscopic observations of (more than 400) Galactic O-type stars gathered by the IACOB and OWN surveys (Holgado *et al.* 2018, 2020).

We used effective temperatures and surface gravities resulting from a homogeneous semi-automatized IACOB-GBAT/FASTWIND spectroscopic analysis to locate our sample of stars in the Kiel and spectroscopic Hertzsprung-Russell (sHR) diagrams. We evaluated the completeness of our magnitude-limited sample of stars as well as potential observational biases affecting the compiled sample using information from the Galactic O star

† Link to the talk: Here

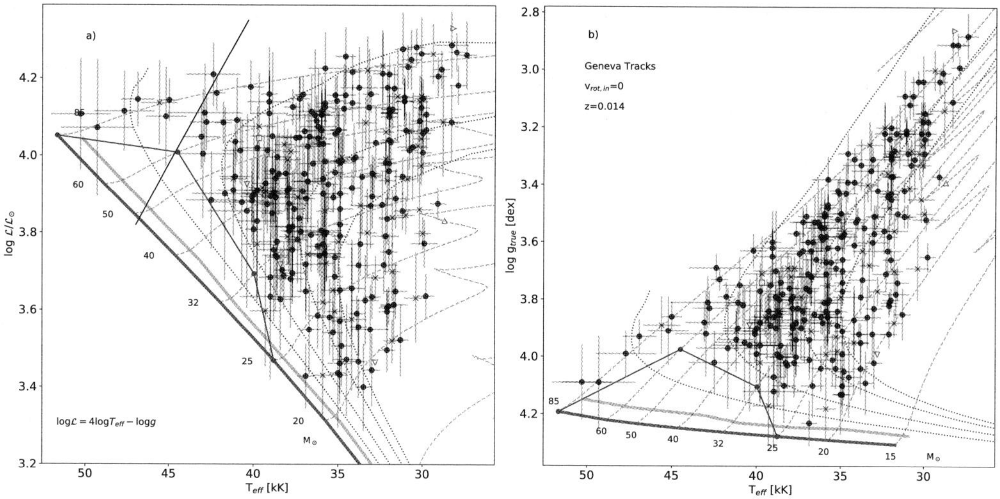

Figure 1. 285 presumably single and SB1 Galactic O-type stars in the Kiel (*right*) and spectroscopic HR (*left*) diagrams. Crosses are SB1. Open symbols, triangles, and squares mark stars with limits. Evolutionary tracks, isochrones and position of the ZAMS (thick solid line) from the nonrotating solar metallicity models by Ekström et al. (2012) are included for reference. The thick solid blue line represents the ZAMS for similar models, but with an initial rotation speed of 40% the critical speed. The diagonal line separates the region where no He I lines are available. We delineate in red the region close to the ZAMS where no stars are found.

catalog (GOSC). We discuss limitations and possible systematics of our analysis method, and compare our results with other recent studies using smaller samples of Galactic O-type stars. We mainly base our discussion on the distribution of stars in the sHR diagram in order to avoid the use of still uncertain distances to most of the stars in our sample. However, we also performed a more detailed study of the young cluster Trumpler-14 as an illustrative example of how *Gaia* cluster distances can help to construct the associated classical HR diagram (see full discussion in Holgado *et al.* 2020).

We find that the apparent lack of massive O-type stars near the ZAMS with initial evolutionary masses in the range between ∼40 and 70 $M_{\odot}$ still persist even when spectroscopic results from a large non-biased sample of stars are used. We do not find any correlation between the dearth of stars close to the ZAMS and obvious observational biases, limitations of our analysis method, and/or the use of one example spectroscopic HR diagram instead of the classical HR diagram. Finally, by investigating the effect of the efficiency of mass accretion during the formation process of massive stars, we conclude that an adjustment of the mass accretion rate towards lower values than canonically assumed might reconcile the hotter boundary of the empirical distribution of optically detected O-type stars in the spectroscopic HR diagram and the theoretical birthline for stars with masses above ≈30 $M_{\odot}$. Last, we also discuss how the presence of a small sample of O2O3.5 stars found much closer to the ZAMS than the main distribution of Galactic O-type star might be explained in the context of this scenario when the effect of nonstandard star evolution (e.g. binary interaction, mergers, and/or homogeneous evolution) is taken into account.

References

Ekström, S., Georgy, C., Eggenberger, P., Meynet, G., Mowlavi, N., Wyttenbach, A., et al.,: 2012, *A&A* **537**, A146.

Holgado, G., Simón-Díaz, S., Barbá, R.H., Puls, J., Herrero, A., Castro, N., et al.,: 2018, *A&A* **613**, A65.

Holgado, G., Simón-Díaz, S., Haemmerlé, L., Lennon, D.J., Barbá, R.H., Cerviño, M., et al.,: 2020, *A&A* **638**, A157.

Massive Stars Near and Far
Proceedings IAU Symposium No. 361, 2024
J. Mackey, J. S. Vink, & N. St-Louis, eds.
doi:10.1017/S1743921322003441

How to make an 85 Solar Mass Black Hole

Ethan Winch[1,2], Jorick S. Vink[1], Erin Higgins[1] and Gautham Sabhahit[1,2]

[1]Armagh Observatory and Planetarium, College Hill, Armagh, BT61 9 Northern Ireland
email: ethan.winch@armagh.ac.uk

[2]Queen's University Belfast, University Road, Belfast, BT7 1NN Northern Ireland

Abstract. We present in-progress resolution test and parameter space studies for very massive stars using MESA, showcasing current MESA version convergence studies.

Keywords. stars: evolution, stars: fundamental parameters, black hole physics

---

1. Introduction

The Pair Instability (PI) mass gap is predicted from approximately 60 to 130 $M_\odot$ (Fowler & Hoyle (1964)), and defines a region where no black hole from direct core collapse should be found. However, with the gravitational wave detection, GW190521, two black holes of 66 (+17, -18) and 85(+21, -14) solar masses merged to form a 146 solar mass black hole (Abbott *et al.* (2020)). This prompted the theory that the two component black holes were themselves the products of mergers from black holes below the PI mass gap. However, Vink *et al.* (2021) showed that a black hole on the order of 85 solar masses could be formed from single star evolution, so long as the star in question maintained a low core mass. Extending on that research, this work aims to investigate stellar evolution inside of the PI mass gap and provide a likelihood of black hole formation, across a range of initial conditions. To do this, I use the Modules for Experiments in Stellar Astrophysics (MESA) (Paxton *et al.* (2011)) code to evolve stars throughout their full evolution until core collapse. The aim of this work is to critically and systematically assess black hole formation in the PI mass gap, provide insight into stellar evolution at high stellar mass regimes, but primarily it is to establish the total likelihood of black hole production in this regime and attempt to provide a clearer answer on the origin of large stellar mass black holes, such as those seen in GW190521. By exploring a range of parameters which is outlined in Section 2, it will become clear which regions of the parameter space are more prone towards making black holes, and the limits of these parameters in this context.

2. Methods

Using the MESA code we have a large grid of models spanning the Pair Instability mass gap across full evolution, with multiple varying parameters. Firstly, the resolution of the models should be to converged to ensure that future results are consistent. Secondly, the grid is set up and run to provide insight into the final fates of single stars who are in the PI mass gap.

The parameters varied are:

- Initial mass, M_i, between 60 and 150 solar masses,
- Initial metallicity, Z_i, between 1×10^{-5} and 1×10^{-1},

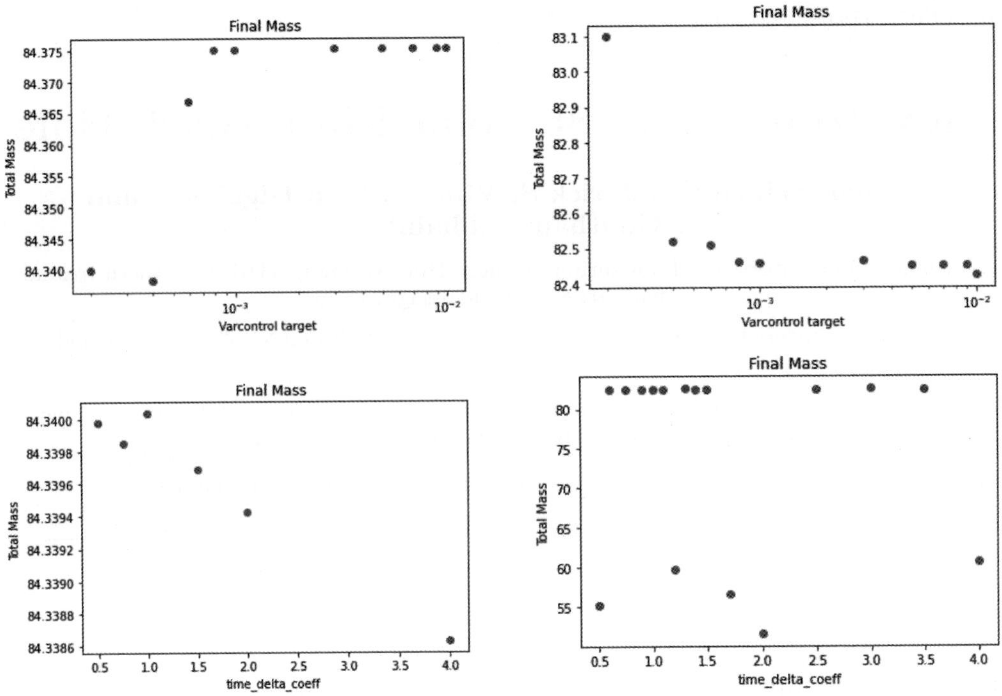

Figure 1. Final mass of stellar evolution models. The left panels are for main sequence evolution, while the right panels represent stars which have completed core Helium burning. The varying resolution parameter in the top two panels is *mesh_delta_coefficient*, while the bottom two are for *time_delta_coeff*.

- Initial rotation, $\Omega/\Omega_{critical}$, from 0, so non-rotating, to 0.6,
- Overshooting, f_0, from 0.1 to 0.5,
- and the Spruit-Taylor Dynamo which is either *on* or *off* at 1 or 0.

In order to be sure that the results of the aforementioned grid are converged, studies on the resolution parameters in MESA need to be conducted. The MESA version used, r15140, has two main resolution parameters - *mesh_delta_coefficient* and *time_delta_coefficient*. However, it should be noted that for versions before this, the parameter *varcontrol_target* was used instead. This variable was phased out in r15140 in favour of *time_delta_coefficient* and a limit is now placed on *varcontrol_target*, though this can be overridden.

3. Conclusions

Due to the size of the stars in this grid, the resolution parameters for MESA have shown to produce inconsistent results for small variations, with no clear convergence for the given spatial or temporal parameters. As shown in Figure 1, stars during Helium burning (right panels) did not provide a clear trend of the resolution parameters to final mass when varying *time_delta_coefficient* (bottom right panel), unlike the top right panel which shows a varying *varcontrol_target*. The mass discrepancy seen in the bottom right panel of Figure 1 is an unphysical numerical result at the boundary of the core and the shell wherein the model runs away and becomes hot and luminous, with a large convective region. If using a metallicity dependent wind, then this will lead to increased mass loss.

Acknowledgements

Thanks to the MESA developers for the public release of their software, and colleagues for their help in editing.

References

Abbott, R., et al. 2020, *ApJL*, 900, L13
Fowler, W. A., & Hoyle, F. 1964, *ApJS*, 9, 201
Paxton, B., Bildsten, L., Dotter, A., Herwig, F., Lesaffre, P., & Timmes, F. 2011, *ApJS*, 192, 3
Vink, J. S., Higgins, E. R., Sander, A. A. C., & Sabhahit, G. N. 2021, *MNRAS*, 504, 146

Chapter VI:
Cool supergiants

Massive Stars Near and Far
Proceedings IAU Symposium No. 361, 2024
J. Mackey, J. S. Vink, & N. St-Louis, eds.
doi:10.1017/S1743921322002563

Variable mass loss on 100 yr timescales from massive yellow hypergiants

René D. Oudmaijer[1] **and Evgenia Koumpia**[2]

[1]School of Physics & Astronomy, University of Leeds, Woodhouse Lane, LS2 9JT Leeds, UK
email: R.D.Oudmaijer@leeds.ac.uk

[2]ESO Vitacura, Alonso de Córdova 3107 Vitacura, Casilla 19001 Santiago de Chile, Chile

Abstract. This contribution presents new results on two members of the class of post-Red Supergiants, IRAS 17163-3907, the central star of the Fried Egg nebula and IRC +10420. New optical spectra in the blue spectral range confirm their spectral type to be of A-supergiant class. Our VLTI/GRAVITY *K*-band interferometry reveals that the neutral Na I 2.2 μm line emitting region is smaller than that of the hydrogen Brγ emission. This can be explained with the hydrogen emission the result of collisional excitation populating the higher levels in a neutral region instead them being populated through recombination in an ionised environment as mostly inferred in stellar winds. Finally, the central star of the Fried Egg nebula, has undergone 3 distinct mass loss episodes over the last hundreds of years. As it is likely that at least the last mass loss event occurred when the star was already a Yellow Hypergiant and not a Red Supergiant, we put forward the bi-stability mechanism as explanation for the mass loss.

Keywords. stars: evolution, stars: mass-loss, stars: AGB and post-AGB, stars: individual: IRAS 17163-3907; IRC+10420, circumstellar matter

1. Introduction

The post-Red Supergiant (post-RSG) stars are evolved massive stars that are moving to the left in the HR diagram. Although occupying similar locations in the HR diagram, they are different from the Yellow Hypergiants (YHG) by having massive dust shells betraying a previous mass losing phase which has been identified with the RSG stage. At present their ultimate fate is not overly clear. They may evolve to higher temperatures to become Luminous Blue Variables, B[e] supergiants or Wolf-Rayet stars, or even "bounce back" from what has been coined to be the Yellow Void or White Wall and move back to cooler temperatures (see reviews by de Jager 1998; Oudmaijer et al. 2009; Gordon and Humphreys 2019). Not many such objects are known, and in this contribution we highlight our recent work on the objects IRAS 17163-3907 and IRC +10420 presented in Koumpia et al. (2020) and Koumpia et al. (2022) respectively. IRC +10420 had been known for a long time to be a massive evolved object (Humphreys et al. 1973; Jones et al. 1993; Oudmaijer et al. 1996), while despite its IR brightness, IRAS 17163-3907 (or Hen 3-1379), the central star of the Fried Egg Nebula was only recognised relatively late to be of massive nature (Lagadec et al. 2011). Below we will focus on the optical spectroscopy and near-infrared spectro-interferometry we have obtained.

2. Stellar Temperatures

The optical spectra of both objects are swamped with emission lines in the red part of the spectrum. Wallström et al. (2015) compared the emission line spectra of both objects

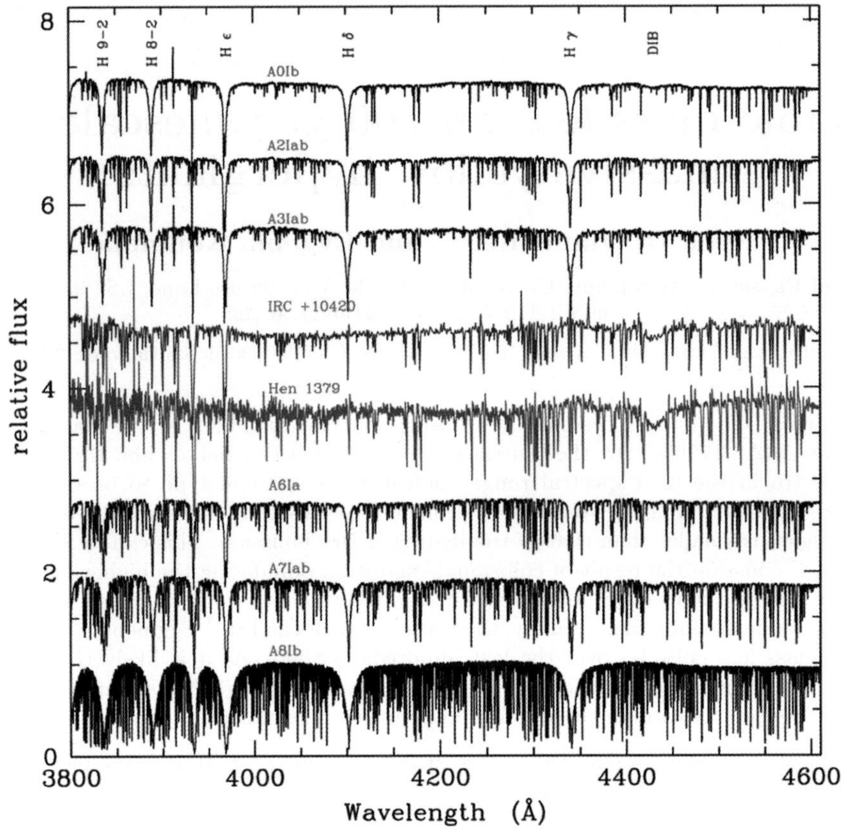

Figure 1. Blue optical spectra of of Hen 3-1379 and IRC+10420 obtained with X-Shooter and as presented in Koumpia et al. (2020) and Koumpia et al. (2022) respectively. Data have been rebinned to improve the signal-to-noise, are continuum normalised and offset by increases of 1. For reference spectra of spectral standard stars from Bagnulo et al. (2003) are plotted. Both objects have a spectrum consistent with a mid-A supergiant spectral type.

and find a striking similarity. Given the relative faintness and red colours of the objects, not many blue spectra have been published. It is in this spectral region that the emission lines are less prominent or absent, allowing spectral typing to be done. For IRC +10420, using the only blue spectrum published up until Koumpia et al. (2022), Oudmaijer (1998) was able to assign an A-supergiant type, while, based on the same data, in a dedicated study Klochkova et al. (1997) computed a temperature of 8500 K. For the Fried Egg, Koumpia et al. (2020) determined a similar spectral type based on the only blue spectrum so far. In Fig. 1 we show X-Shooter data of both objects and compare them with those of several spectral standard stars. The spectra of both objects are similar and can be assigned a similar spectral class. The IRC +10420 spectrum was obtained in 2009, and is presented in Koumpia et al. (2022). It does not show any significant variations with respect to its spectrum taken 15 years earlier, signifying that its earlier observed rapid increase in temperature (2000 K in 20 years - Oudmaijer 1998) has halted.

3. VLTI/GRAVITY data around Brγ and Na I

We observed both objects with near-infrared interferometry in the *K*-band spectral region using VLTI/GRAVITY. In Fig. 2, we show the spectrum as well as their visibilities at the longest baseline ($\sim$120 m). To recap, a visibility of 1 would indicate an

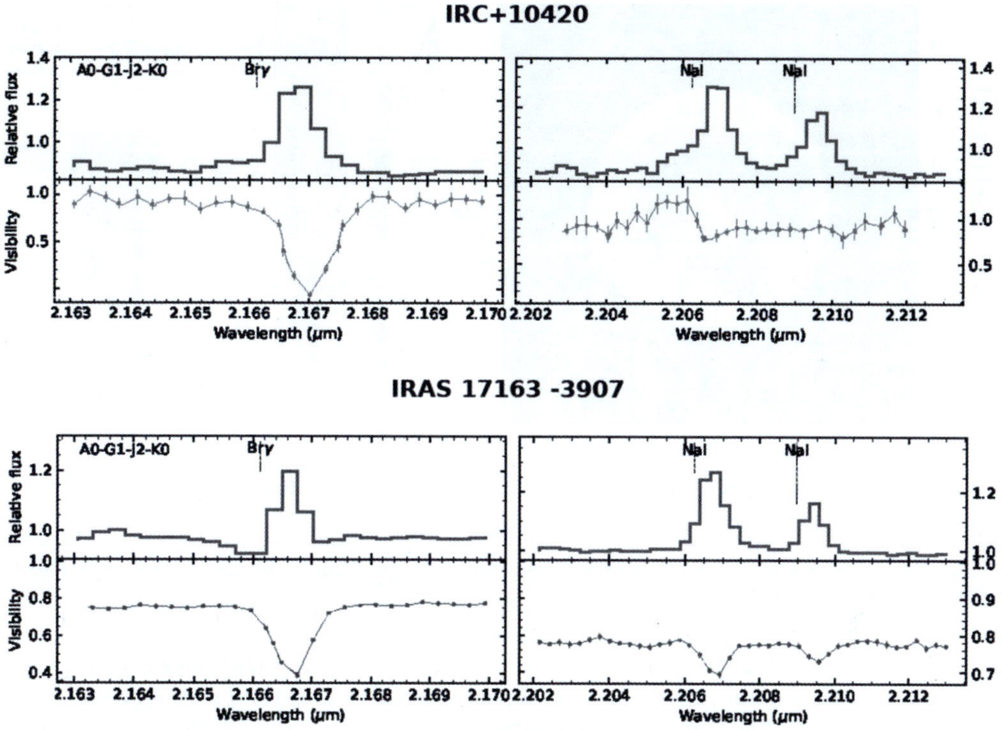

Figure 2. The GRAVITY spectra and visibilities at the longest baselines (~120 m) which are sensitive to the smallest scales for the IRC+10420 (top) and Fried Egg (bottom). Note that in both cases the central source is barely resolved and is identified with the star, while the Na I doublet at 2.2 μm appears to come from a smaller emitting regions than Brγ at 2.16 μm.

unresolved point source (roughly 2 milli-arcsec in the present data) within the field-of-view of the interferometer, while the lower the visibility, the more extended the source is. In both cases, the continuum appears unresolved and traces the stellar continuum, while also in both cases, the Na I emission region is smaller than that of the hydrogen emission. It may be interesting to note that although rare, Na I 2.2 μm doublet emission appears to be present in the spectrum of many YHGs (IRC+10420, HD 179821, HR 8752, and ρ Cas (Lambert et al. 1981; Hrivnak et al. 1994; Hanson et al. 1996; Oudmaijer and de Wit 2013). Various explanations for the origin of the Na I 2.2 μm doublet emission are discussed in literature (for an overview see e.g. Oudmaijer and de Wit 2013), the spatial information here helps us to constrain the various models.

We would like to draw attention to the fact that the neutral sodium's emission region is smaller than that of the hydrogen, Brγ, emission. The puzzling issue here is that we have very strong hydrogen emission from a relatively cool star, while in addition, neutral sodium originates from inside the hydrogen emitting region. The latter is counter-intuitive, generally it is assumed that hydrogen emission is due to recombination. This of course implies that the hydrogen has to be ionised to begin with, yet the ionisation potential of sodium is much lower than that of hydrogen (5.1 eV versus 13.6 eV). So the question arises why Na would be neutral while it is located closer to the star than the ionised hydrogen.

A solution could be that the hydrogen emission is not due to recombination, but due to collisional excitation; in regions of high density, collisions can excite electrons to higher energy levels, while their de-excitation can give rise to line emission without the need

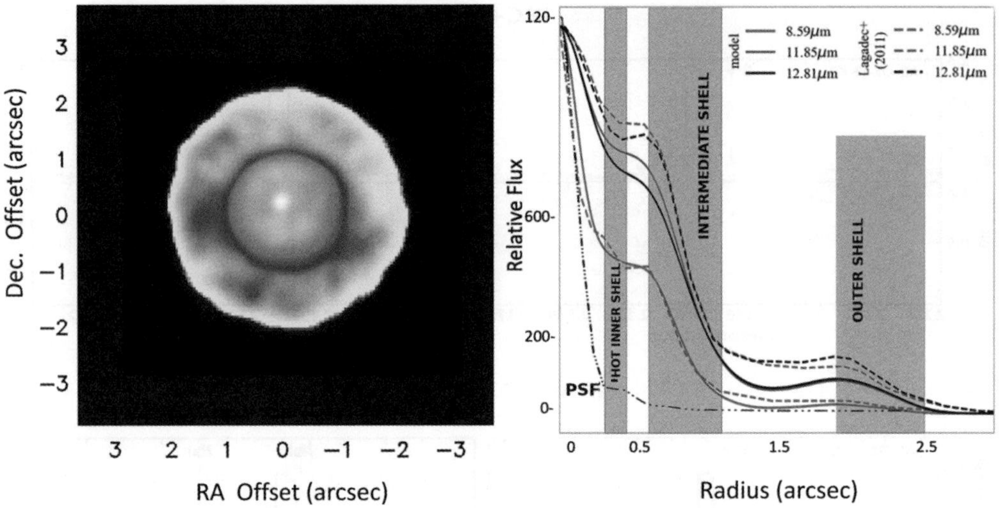

Figure 3. Left: 10 μm VISIR image of the Frieg Egg nebula adapted from Lagadec et al. (2011). The presence of two separate shells is readily visible. Right: the radial profile reveals the presence of 3 resolved shells.

for ionisation, and thus occur in a neutral region. We explored this notion using simple LTE models and are able to explain the 2 μm observations of the Fried Egg for the size scales and emission line strengths (Koumpia et al. 2020). The next step is to approach this with kinematic modelling and ultimately to arrive at a description of the immediate circumstellar environment of the stars.

4. Variable mass loss

The 10μm diffraction limited images that led Lagadec et al. (2011) to introduce the name "Fried Egg" (see Fig. 3) already reveals a discontinuity in the dust density distribution. In fact, the radial profile of the emission (also in the Figure) reveals the presence of 3 distinct shells, which all have kinematic ages less than hundreds of years. The simultaneous modelling of the object's spectral energy distribution and the spatial information, revealed that these shells can be understood in terms of short periods of enhanced mass loss (Koumpia et al. 2020). These mass loss periods were preceded by a much earlier period of mass loss which is traced on larger scales by Herschel observations at longer wavelengths (Hutsemékers et al. 2013).

An outstanding question is what could lead to these short bursts of mass loss. Traditionally, in the case of Yellow Hypergiants, pulsational instabilities have been proposed to do so (e.g. Lobel et al. 1994), however not much is known about any photometric or spectroscopic variability of the central star of the Fried Egg nebula that could reveal information about its pulsational properties. Instead, it could be argued that the object has remained fairly stable in its spectral appearance and photometry since Lebertre et al. (1989). As the innermost shell has a kinematic age that is comparable to the time that has lapsed since Lebertre et al. (1989)'s observations, it is reasonable to assume it was ejected when the star was of A-spectral type.

This inference led us to consider an alternative mass loss mechanism which is due to the so-called "bi-stability jump". This mechanism is known to occur around 20,000 K. At this temperature, the ionisation stage of iron changes from Fe IV to Fe III. The line opacity of iron far exceeds that of the more abundant H and He, and is the main contribution to the opacity responsible for driving winds of hotter stars. As the opacity depends on the ionisation levels, any change in temperature around this ionisation boundary will result in different wind properties. For example, a slight decrease in temperature across the "bi-stability" jump, will lead to recombination of Fe, leading to a larger wind density and smaller wind velocities by a factor of 2. This has been well-studied for the "first" bi-stability jump at T $\sim$ 20,000 K, however, a "second" bi-stability jump, which thus far has not received much attention, is expected to occur at T $\sim$ 8800 K.

The A-supergiants under consideration here straddle the temperature range of the second bi-stability jump (the temperature at which the dominant ionisation stage of iron changes from triply to doubly ionised, Vink et al. 2000, 2001; Petrov et al. 2016). Koumpia et al. (2020) pointed out that a similar change in mass loss could arise as for the, much better known, first bi-stability jump (Vink et al. 1999) that occurs at temperatures of order 20,000 K.

As even slight changes in temperature can lead to comparatively large changes in mass loss rates in the Yellow Hypergiant spectral type range, the observations of very recent mass loss episodes for the central star of the Fried Egg nebula means that this relatively unexplored "second" bi-stability jump warrants further study.

5. Concluding remarks

New spectral and near-infrared interferometric data on key post-Red Supergiants allowed us to reach several conclusions:

• The optical spectra of both the central star of the Fried Egg nebula and IRC +10420 are remarkably similar in both their emission line spectrum in the red, as are their absorption line spectra in the blue spectral range. Both have spectral types corresponding to A-type supergiants. IRC +10420 appears not to have changed its spectral type since 1994 signifying it has halted its previously observed blueward evolution in the HR diagram.

• The VLTI/GRAVITY *K*-band interferometry reveal an unresolved continuum for both stars which we identify with the stellar photospheres. Any extended *K*-band emission due to thermal emission from dust falls outside the interferometer's field-of-view. The neutral Na I at 2.2 μm line emitting region is smaller than that of the hydrogen Brγ emission. This can be explained with the hydrogen emission being due to collisional excitation populating the higher levels, followed by de-excitations instead of the usually assumed recombination in an ionised gas. Simple LTE model approaches can reproduce the line strengths and the respective size scales.

• Hen 3-1379, the central star of the Fried Egg nebula, has undergone 3 distinct mass loss episodes over the last hundreds of years. As it is likely that at least the last mass loss event occurred when the star was already a Yellow Hypergiant and not a Red Supergiant, we put forward the bi-stability mechanism as explanation for the mass loss.

Acknowledgements

EK was funded by the STFC (ST/P00041X/1) This project has received funding from the European Union's Framework Programme for Research and Innovation Horizon 2020 (2014-2020) under the Marie Skłodowska-Curie Grant Agreement No. 823734.

References

Bagnulo, S., Jehin, E., Ledoux, C., Cabanac, R., Melo, C., Gilmozzi, R., & ESO Paranal Science Operations Team 2003, The UVES Paranal Observatory Project: A Library of High- Resolution Spectra of Stars across the Hertzsprung-Russell Diagram. *The Messenger*, 114, 10–14.

de Jager, C. 1998, The yellow hypergiants. *A&A Rev.*, 8(3), 145–180.

Gordon, M. S. & Humphreys, R. M. 2019, Red Supergiants, Yellow Hypergiants, and Post-RSG Evolution. *Galaxies*, 7(4), 92.

Hanson, M. M., Conti, P. S., & Rieke, M. J. 1996, A Spectral Atlas of Hot, Luminous Stars at 2 Microns. *ApJS*, 107, 281.

Hrivnak, B. J., Kwok, S., & Geballe, T. R. 1994, Near-infrared spectroscopy of proto-planetary nebulae. *ApJ*, 420, 783–796.

Humphreys, R. M., Strecker, D. W., Murdock, T. L., & Low, F. J. 1973, IRC+10420 - Another Eta Carinae? *ApJL*, 179, L49.

Hutsemékers, D., Cox, N. L. J., & Vamvatira-Nakou, C. 2013, A massive parsec-scale dust ring nebula around the yellow hypergiant Hen 3-1379. *A&A*, 552, L6.

Jones, T. J., Humphreys, R. M., Gehrz, R. D., Lawrence, G. F., Zickgraf, F.-J., Moseley, H., Casey, S., Glaccum, W. J., Koch, C. J., Pina, R., Jones, B., Venn, K., Stahl, O., & Starrfield, S. G. 1993, IRC +10420: A Cool Hypergiant near the Top of the HRD. *ApJ*, 411, 323.

Klochkova, V. G., Chentsov, E. L., & Panchuk, V. E. 1997, Optical spectrum of the IR source IRC+10420 in 1992-1996. *MNRAS*, 292, 19.

Koumpia, E., Oudmaijer, R. D., de Wit, W. J., Mérand, A., Black, J. H., & Ababakr, K. M. 2022, Tracing a decade of activity towards a yellow hypergiant. The spectral and spatial morphology of IRC+10420 at au scales. *MNRAS* in press,, arXiv:2207.05812.

Koumpia, E., Oudmaijer, R. D., Graham, V., Banyard, G., Black, J. H., Wichittanakom, C., Ababakr, K. M., de Wit, W. J., Millour, F., Lagadec, E., Muller, S., Cox, N. L. J., Zijlstra, A., van Winckel, H., Hillen, M., Szczerba, R., Vink, J. S., & Wallström, S. H. J. 2020, Optical and near-infrared observations of the Fried Egg Nebula. Multiple shell ejections on a 100 yr timescale from a massive yellow hypergiant. *A&A*, 635, A183.

Lagadec, E., Zijlstra, A. A., Oudmaijer, R. D., Verhoelst, T., Cox, N. L. J., Szczerba, R., Mékarnia, D., & van Winckel, H. 2011, A double detached shell around a post-red supergiant: IRAS 17163-3907, the Fried Egg nebula. *A&A*, 534, L10.

Lambert, D. L., Hinkle, K. H., & Hall, D. N. B. 1981, Circumstellar shells of luminous supergiants. I - Carbon monoxide in Rho Cassiopeiae and HR 8752. *ApJ*, 248, 638–650.

Lebertre, T., Epchtein, N., Gouiffes, C., Heydari-Malayeri, M., & Perrier, C. 1989, Optical and infrared observations of four suspected proto-planetary objects. *A&A*, 225, 417–431.

Lobel, A., de Jager, C., Nieuwenhuijzen, H., Smolinski, J., & Gesicki, K. 1994, Pulsation of the yellow hypergiant ρ Cassiopeiae in 1970. *A&A*, 291, 226–238.

Oudmaijer, R. D. 1998, High resolution spectroscopy of the post-red supergiant IRC+10420. I. The data. *A&AS*, 129, 541–552.

Oudmaijer, R. D., Davies, B., de Wit, W.-J., & Patel, M. Post-Red Supergiants. In Luttermoser, D. G., Smith, B. J., & Stencel, R. E., editors, *The Biggest, Baddest, Coolest Stars* 2009,, volume 412 of *Astronomical Society of the Pacific Conference Series*, 17.

Oudmaijer, R. D. & de Wit, W. J. 2013, Neutral and ionised gas around the post-red supergiant IRC +10 420 at AU size scales. *A&A*, 551, A69.

Oudmaijer, R. D., Groenewegen, M. A. T., Matthews, H. E., Blommaert, J. A. D. L., & Sahu, K. C. 1996, The spectral energy distribution and mass-loss history of IRC+10420. *MNRAS*, 280, 1062–1070.

Petrov, B., Vink, J. S., & Gräfener, G. 2016, Two bi-stability jumps in theoretical wind models for massive stars and the implications for luminous blue variable supernovae. *MNRAS*, 458(2), 1999–2011.

Vink, J. S., de Koter, A., & Lamers, H. J. G. L. M. 1999, On the nature of the bi-stability jump in the winds of early-type supergiants. *A&A*, 350, 181–196.

Vink, J. S., de Koter, A., & Lamers, H. J. G. L. M. 2000, New theoretical mass-loss rates of O and B stars. *A&A*, 362, 295–309.

Vink, J. S., de Koter, A., & Lamers, H. J. G. L. M. 2001, Mass-loss predictions for O and B stars as a function of metallicity. *A&A*, 369, 574–588.

Wallström, S. H. J., Muller, S., Lagadec, E., Black, J. H., Oudmaijer, R. D., Justtanont, K., van Winckel, H., & Zijlstra, A. A. 2015, Investigating the nature of the Fried Egg nebula. CO mm-line and optical spectroscopy of IRAS 17163-3907. *A&A*, 574, A139.

Discussion

Jesús Maíz Appelániz: I checked Gaia EDR3 and find that the distance to IRAS 17163-3907 is 5 kpc, but with a large errorbar.

René Oudmaijer: It is indeed. The pre-Gaia distance to the object was mainly based on the kinematic distance of both object and interstellar absorption lines and the large interstellar extinction value. This led to a distance of 1 kpc, which happens to be consistent with the Gaia DR2 distance. 1 kpc already results in a luminosity that is half a million times solar, so 5 kpc would really give an extremely large value for the brightness.

Sally Oey: Do you have kinematic info on the Brγ and Na I?

René Oudmaijer: Yes, however, Brγ might be affected by possible P Cygni absorption. We have LSR velocities for Na (25 km/s), CO mm (18 km/s), H30α (21 km/s) and interstellar KI (-32, -10, 6), many pointing at large distances indeed.

Sally Oey: Thanks, was wondering whether there was an offset between them, but apparently not interesting.

Alexis Quintana: Following Maíz Appelániz' question, I remember that Bailer-Jones's last catalogue gives two distances typically, including photogeometric distances. If I remember correctly these set of distances take extinction into account. Since this object has a very high extinction ($A_V = 11$ mag), could the photogeometric distance be more reliable, or are the uncertainties too high as well?

René Oudmaijer: That is a good point - however the distances that take into account reddening etc may work well for the 99.999% of normal stars in the Galaxy, but I have learned the hard way that they do not work well for objects that are a-typical. It will be interesting to see how DR3 and later will deal with such objects.

Massive Stars Near and Far
Proceedings IAU Symposium No. 361, 2024
J. Mackey, J. S. Vink, & N. St-Louis, eds.
doi:10.1017/S174392132200240X

Analytic, Turbulent Pressure Driven Mass Loss from Red Supergiants

N. Dylan Kee[1] and the MAESTRO Project[2]

[1]National Solar Observatory, 22 Ohi'a Ku St, Makawao, HI 96768, USA
email: dkee@nso.edu

[2]Institute of Astronomy, KU Leuven, Celestijnenlaan 200D, B-3001 Leuven, Belgium

Abstract. Despite the important role mass loss in the red supergiant phase plays in controlling stellar evolution and massive stars final supernova fates, a theoretical explanation of the mechanism driving this mass loss has been elusive. In this contribution we present a recent breakthrough (Kee et al. 2021) showing that turbulent pressure alone is sufficient to markedly extend the atmospheres of red supergiants and allow a wind to be launched. The resulting theory provides a fully analytic prescription for red supergiant mass-loss rates. Moreover, the theoretical mass-loss rates computed from observationally inferred turbulent velocities are in overall good agreement with observationally inferred red supergiant mass loss. A particularly interesting aspect of this theory is that it is not sensitive to metallicity, providing important implications for stellar evolution and the so-called "red-supergiant problem" for supernova progenitors in various environments.

Keywords. supergiants, stars: mass loss, stars: winds, convection, turbulence, stars: evolution, supernovae: general

1. Introduction

Although red supergiants (RSGs) are observationally inferred to be undergoing vigorous mass loss ($\dot{M} \gtrsim 10^{-7} M_{\odot}\ \mathrm{yr}^{-1}$), the physical mechanism driving these stars' winds has remained unknown (see, e.g., Levesque 2017). For lower mass evolved stars, specifically asymptotic giant branch stars, atmospheric pulsations are thought to levitate material up to distances at which dust can form, before radiation pressure on dust opacity drives material out of the stellar gravitational potential (see, e.g., Höfner and Olofsson 2018, for a review). However, due to the higher effective temperatures and lower pulsational amplitudes of RSGs, studies of this mechanism find that an insufficient amount of material can be levitated to distances away from the star where dust can form to be able to explain the winds of RSGs (e.g. Arroyo-Torres et al. 2015).

An alternate explanation is that the deep seated convection cells present in the atmospheres of RSGs (e.g. Freytag et al. 2012) may seed sufficient atmospheric turbulence to augment or replace pulsations in extending the atmospheric scale height (Gustafsson and Plez 1992; Josselin and Plez 2007). Spurred by recent observations revealing just such atmospheric turbulence (e.g. Josselin and Plez 2007; Ohnaka et al. 2017), we build upon the work of Gustafsson and Plez (1992) and Josselin and Plez (2007) to develop an analytic theory leveraging atmospheric turbulence as the primary driving mechanism of the winds of RSGs (Kee et al. 2021).

2. Model

The following section discusses the derivation of the analytic mass-loss rate. As this discussion is abridged, we direct the interested reader to Kee et al. (2021) for the detailed derivation.

As our goal is to obtain a steady state mass loss rate $\dot{M} \equiv 4\pi\rho v r^2$, we need to know velocity v and density ρ at a single radius r. To this end, we begin from the isothermal, steady-state equation of motion

$$v\left(1 - \frac{c_s^2 + v_{\rm turb}^2}{v^2}\right)\frac{\partial v}{\partial r} = \frac{2\left(c_s^2 + v_{\rm turb}^2\right)}{r} - \frac{GM_*\left(1-\Gamma\right)}{r^2}, \tag{2.1}$$

expressed in terms of the isothermal sound speed c_s, turbulent velocity $v_{\rm turb} = \sqrt{P_{\rm turb}/\rho}$, and Eddington factor $\Gamma \equiv \kappa L_*/(4\pi G M_* c)$. Turbulent velocity and the Eddington factor are further functions of turbulent pressure $P_{\rm turb}$ and opacity κ, stellar luminosity L_*, and stellar mass M_* respectively. Examining Equation 2.1, a convenient radius to consider is the location where $v(r) = \sqrt{c_s^2 + v_{\rm turb}^2}$, as the left side of Equation 2.1 is zero at this radius. As this is basically the method used by Parker (1958) to derive the gas pressure driven solar wind, here with a modified, "effective" sound speed $c_{\rm s,eff} = \sqrt{c_s^2 + v_{\rm turb}^2}$, we refer to this radius as the modified Parker radius

$$R_{\rm p,mod} = \frac{GM_*\left(1-\Gamma\right)}{2\left(c_s^2 + v_{\rm turb}^2\right)}. \tag{2.2}$$

With velocity at a specific radius in hand, what remains is to find the density at the same position. Here we assume that the portion of the wind below the modified Parker radius is nearly in hydrostatic equilibrium, such that

$$\rho(r) = \rho_* \exp\left[-\frac{2R_{\rm p,mod}}{R_*}\left(1 - \frac{R_*}{r}\right) - \frac{1}{2}\right], \tag{2.3}$$

where the extra factor of $e^{-1/2}$ accounts for the velocity stratification of the region below $R_{\rm p,mod}$. The normalization density ρ_* is obtained to be

$$\rho_* = \frac{4R_{\rm p,mod}}{3\,\kappa\,R_*^2}\left[1 - \exp\left(-\frac{2R_{\rm p,mod}}{R_*}\right)\right]^{-1}. \tag{2.4}$$

by taking the stellar radius to be at spherically modified optical depth 2/3 (Lucy 1971). Taken all together this gives an analytic mass-loss rate

$$\dot{M}_{\rm an} = 4\pi\rho(R_{\rm p,mod})\sqrt{c_s^2 + v_{\rm turb}^2}\,R_{\rm p,mod}^2, \tag{2.5}$$

for

$$\rho(R_{\rm p,mod}) = \frac{4R_{\rm p,mod}e^{3/2}}{3\kappa R_*^2\left(e^{2R_{\rm p,mod}/R_*} - 1\right)}. \tag{2.6}$$

Relaxing the assumption of isothermality allows us to iteratively converge on a final mass rate for our model. To replace the isothermal temperature structure Kee et al. (2021) took the temperature structure from the Lucy (1971) spherically-modified grey model atmosphere. Comparing the associated numerically determined mass-loss rate to the analytic expression provides us a final mass-loss rate

$$\dot{M} = \dot{M}_{\rm an}\left(\frac{v_{\rm turb}/(17\,{\rm km\,s^{-1}})}{v_{\rm esc}/(60\,{\rm km\,s^{-1}})}\right)^{1.30}, \tag{2.7}$$

for $v_{\rm esc}$ the escape speed from the stellar surface.

3. Analysis and implications

Comparing the mass-loss rates provided by our model to the mass-loss rates observationally inferred by Josselin and Plez (2007) and Ohnaka et al. (2017) returns a required turbulent velocity for our model of, on average, $v_{\rm turb} = 18.2 \pm 3.4$ km s^{-1} (Kee et al. 2021). This agrees to better than one standard deviation with the average observationally inferred turbulent velocity inferred by these same authors of $v_{\rm turb} = 20.3 \pm 3.3$ km s^{-1}. Further comparing our mass-loss rate using this average turbulent velocity to the current state-of-the-art in empirical mass loss prescriptions for RSGs used in stellar evolution calculations (e.g. de Jager et al. 1988; Nieuwenhuijzen and de Jager 1990; Vanbeveren et al. 1998; van Loon et al. 1999; Beasor et al. 2020) again shows generally good agreement between our theory and what is observationally inferred. However, it is important to note that there is substantial scatter between the stellar properties and mass-loss rates of individual RSGs, and that this scatter carries over to the differences between different empirical mass-loss rate prescriptions, which in some places in the parameter space grow to multiple orders of magnitude. Therefore, as is the case for the empirically inferred mass-loss rate prescriptions, these theoretical mass-loss rates are robust in describing the average behavior of the population of RSGs, while some care should be taken in describing the mass-loss rates of individual RSGs.

To explain the efficacy of this model, we note that a key parameter is the argument of the exponential stratification in density in Equation 2.3

$$\frac{R_{\rm p,mod}}{R_*} \propto \frac{v_{\rm esc,eff}^2}{c_{\rm s}^2 + v_{\rm turb}^2}, \tag{3.1}$$

for effective escape speed $v_{\rm esc,eff} = \sqrt{2GM(1-\Gamma)/R_*}$. For cool evolved stars, like RSGs, the surface escape escape speed is much lower ($\sim$ 60 km s^{-1}) than main-sequence stars ($\sim$ 600 km s^{-1}) such that a relatively modest amount of turbulent velocity $\sim 4\ c_{\rm s}$ is sufficient to make the modified Parker radius an order unity factor times the stellar radius. Meanwhile, for the solar wind, the 10^6 K corona and attendant factor ~ 15 increase in sound speed is required. Beyond this, RSGs are intrinsically quite large compared to main-sequence stars with $R_* \sim 1000 R_\odot$, which only amplifies the effects of looser atmospheric binding as mass-loss rate scales like radius squared.

Given the success of this model, an interesting first step in applying these new theoretical rates is to examine their effects on stellar evolution. To do so, we run two sets of `MESA` models that are identical except in their treatment of the "cool" mass-loss prescription for stars with effective temperature ≤ 8 kK. For the first set, we apply the standard de Jager et al. (1988) mass-loss rates in this regime, while for the second we preferentially use our new Kee et al. (2021) mass-loss rates. As this new theoretical description is intended to apply to the mass loss of RSGs, we take the maximum of the de Jager et al. (1988) and Kee et al. (2021) mass-loss rates for the region where effective temperature is between 5 and 8 kK, and always apply the Kee et al. (2021) rates for stars with effective temperature ≤ 5 kK. The implementation of this "Leuven-modified Dutch mass loss scheme" in the stellar evolution code `MESA` (Paxton et al. 2011, 2013) is available at https://doi.org/10.5281/zenodo.4333564. The inlists describing the remaining input parameters for `MESA` are also available at https://doi.org/10.5281/zenodo.4333564.

The resulting stellar evolution tracks are shown in Figure 1. All stars shown are evolved until carbon core depletion. The difference in evolution between the stars evolved with the de Jager et al. (1988) and Kee et al. (2021) rates is immediately evident, as all stars shown using the de Jager et al. (1988) die on the RSG branch, while stars of initial mass $\geq 17 M_\odot$ evolve back to the blue with the Kee et al. (2021) mass-loss rates. This upper limit to the initial mass of stars that end their lives in core collapse on the RSG branch

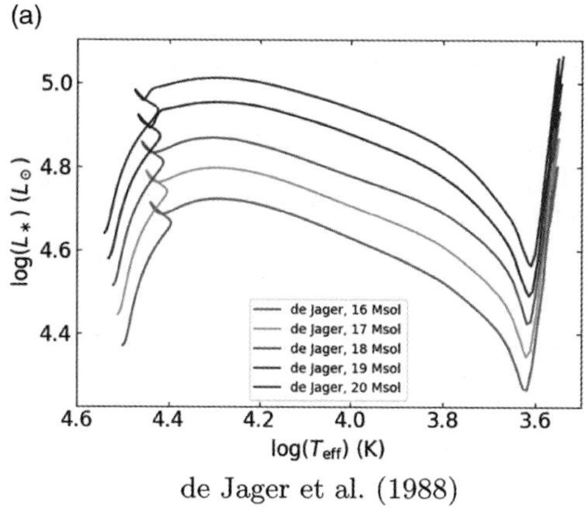

Figure 1. Comparison of stellar evolution tracks using the de Jager et al. (1988) mass-loss rates (left) and the Kee et al. (2021) ('Leuven') mass-loss rates (right) during the RSG phase. The different tracks in each panel correspond to stars with initial masses from 16 to 20 $M_\odot$ in steps of 1 $M_\odot$.

is in broad agreement with what is observationally inferred for the progenitors of Type II-P/L supernovae ($16 \sim 23 M_\odot$, Smartt et al. 2009). On the other hand, this differs from the recent findings of Beasor et al. (2021) (discussed by E. Beasor in these proceedings) where stars of initial masses up to 30 $M_\odot$ die on the RSG branch.

This difference is at its core related again to the aforementioned sensitivity of this mass-loss prescription to the surface gravity of the star. While the Kee et al. (2021) mass-loss rates are in general agreement with the de Jager et al. (1988) and Beasor et al. (2020) rates over the majority of a star's RSG lifetime (e.g. Kee et al. 2021, B. Davies priv. comm.), as a star evolves toward the tip of the RSG branch, its surface gravity decreases and as such the Kee et al. (2021) mass-loss rates increase. This initial increase is still in general agreement with other empirical rates which also increase with increasing stellar luminosity, but has additional implications for the Kee et al. (2021) mass-loss rates. Specifically, as a star loses mass near the tip of the RSG branch its surface gravity further decreases allowing the star to enter a positive feedback loop. This loop drives the star into a short-lived phase (3-14% of the star's RSG lifetime for stars born with 17-20 $M_\odot$) with stellar mass-loss exceeding $10^{-4} M_\odot$ yr^{-1} during which the star sheds its hydrogen envelope and then evolves toward the blue.

Both the initial stellar mass leading to a star entering this feedback loop and the duration a star will spend in this phase, as well as the overall scale of stellar mass loss throughout the remainder of the RSG phase, are of course all dependent on the value of $v_{\rm turb}$ and any dependencies it may have on stellar parameters. Therefore, it is important to disentangle the origins of the atmospheric turbulence to make direct comparisons with the recent studies by Davies and Plez (2021) (discussed in these proceedings by B. Davies) and Beasor and Smith (2022) placing upper limits on the amount of time an RSG could spend in this type of phase of high mass-loss, as well as studies of the mass and luminosity extent of the population of RSGs that are progenitors of Type II-P/L supernovae (e.g. Smartt et al. 2009).

4. Potential origins of atmospheric turbulence

As mentioned in the introduction, a likely origin of the turbulent motions we employ for our model is the deep seated and vigorous convection present in much of the volume of an RSG. Recent radiation-hydrodynamic simulations from Goldberg et al. (2022) (discussed in these proceedings by J. Goldberg) do indeed show characteristic velocities in the outer layers of RSGs in line with the velocities that are observationally inferred and those that are required for our model. Analysing those three-dimensional radiation-hydrodynamic models shows that in part the strong convection and resulting atmospheric turbulence are driven by the stellar atmosphere locally breaching the Eddington limit ($\Gamma > 1$ locally) in the regions of enhanced opacity corresponding to hydrogen and helium opacity. Indeed, this is a property that is shared by the total (radiative+convective) luminosity of the 1D stellar evolution models above. This has two immediately evident important implications for the Kee et al. (2021) mass-loss rates. First, assuming otherwise unchanged stellar structure, the hydrogen and helium opacity peaks are metallicity independent, and as such the turbulent velocities and thus also RSG mass-loss as predicted by Kee et al. (2021) are at most weakly dependent on metallicity for otherwise fixed stellar structures (see, however Sabhahit et al. 2021, discussed in these proceedings by G. Sabhahit, for an example of how altered stellar evolution with *decreased* metallicity causes a star to evolve to an altered structure on the RSG branch such that the strength of these opacity peaks is *increased*). Second, the degree to which an RSG exceeds the Eddington limit is linearly dependent on luminosity, such that $v_{\rm turb}$ may be expected to increase with evolutionary phase as a star climbs the RSG branch. This would imply that the feedback loop is even stronger and the duration of a high mass-loss phase shorter than shown in the prior section.

References

Arroyo-Torres, B., Wittkowski, M., Chiavassa, A., Scholz, M., Freytag, B., Marcaide, J. M., Hauschildt, P. H., Wood, P. R., & Abellan, F. J. 2015, What causes the large extensions of red supergiant atmospheres?. Comparisons of interferometric observations with 1D hydrostatic, 3D convection, and 1D pulsating model atmospheres. *A&A*, 575, A50.

Beasor, E. R., Davies, B., & Smith, N. 2021, The Impact of Realistic Red Supergiant Mass Loss on Stellar Evolution. *ApJ*, 922(1), 55.

Beasor, E. R., Davies, B., Smith, N., van Loon, J. T., Gehrz, R. D., & Figer, D. F. 2020, A new mass-loss rate prescription for red supergiants. *MNRAS*, 492(4), 5994–6006.

Beasor, E. R. & Smith, N. 2022, The extreme scarcity of dust-enshrouded red supergiants: consequences for producing stripped stars via winds. *arXiv e-prints*,, arXiv:2205.02207.

Davies, B. & Plez, B. 2021, The impact of winds on the spectral appearance of red supergiants. *MNRAS*, 508(4), 5757–5765.

de Jager, C., Nieuwenhuijzen, H., & van der Hucht, K. A. 1988, Mass loss rates in the Hertzsprung-Russell diagram. *A&AS*, 72, 259–289.

Freytag, B., Steffen, M., Ludwig, H. G., Wedemeyer-Böhm, S., Schaffenberger, W., & Steiner, O. 2012, Simulations of stellar convection with CO5BOLD. *Journal of Computational Physics*, 231(3), 919–959.

Goldberg, J. A., Jiang, Y.-F., & Bildsten, L. 2022, Numerical Simulations of Convective Three-dimensional Red Supergiant Envelopes. *ApJ*, 929(2), 156.

Gustafsson, B. & Plez, B. Can classical model atmospheres be of any use for the study of hypergiants. In de Jager, C. & Nieuwenhuijzen, H., editors, *Instabilities in Evolved Super- and Hypergiants* 1992,, 86.

Höfner, S. & Olofsson, H. 2018, Mass loss of stars on the asymptotic giant branch. Mechanisms, models and measurements. *A&AR*, 26(1), 1.

Josselin, E. & Plez, B. 2007, Atmospheric dynamics and the mass loss process in red supergiant stars. *A&A*, 469(2), 671–680.

Kee, N. D., Sundqvist, J. O., Decin, L., de Koter, A., & Sana, H. 2021, Analytic, dust-independent mass-loss rates for red supergiant winds initiated by turbulent pressure. *A&A*, 646, A180.

Levesque, E. M. 2017,. *Astrophysics of Red Supergiants*. IOP Publishing.

Lucy, L. B. 1971, The Formation of Resonance Lines in Extended and Expanding Atmospheres. *ApJ*, 163, 95.

Nieuwenhuijzen, H. & de Jager, C. 1990, Parametrization of stellar rates of mass loss as functions of the fundamental stellar parameters M, L, and R. *A&A*, 231, 134–136.

Ohnaka, K., Weigelt, G., & Hofmann, K. H. 2017, Vigorous atmospheric motion in the red supergiant star Antares. *Nature*, 548(7667), 310–312.

Parker, E. N. 1958, Dynamics of the Interplanetary Gas and Magnetic Fields. *ApJ*, 128, 664.

Paxton, B., Bildsten, L., Dotter, A., Herwig, F., Lesaffre, P., & Timmes, F. 2011, Modules for Experiments in Stellar Astrophysics (MESA). *ApJS*, 192(1), 3.

Paxton, B., Cantiello, M., Arras, P., Bildsten, L., Brown, E. F., Dotter, A., Mankovich, C., Montgomery, M. H., Stello, D., Timmes, F. X., & Townsend, R. 2013, Modules for Experiments in Stellar Astrophysics (MESA): Planets, Oscillations, Rotation, and Massive Stars. *ApJS*, 208(1), 4.

Sabhahit, G. N., Vink, J. S., Higgins, E. R., & Sander, A. A. C. 2021, Superadiabaticity and the metallicity independence of the Humphreys-Davidson limit. *MNRAS*, 506(3), 4473–4487.

Smartt, S. J., Eldridge, J. J., Crockett, R. M., & Maund, J. R. 2009, The death of massive stars - I. Observational constraints on the progenitors of Type II-P supernovae. *MNRAS*, 395(3), 1409–1437.

van Loon, J. T., Groenewegen, M. A. T., de Koter, A., Trams, N. R., Waters, L. B. F. M., Zijlstra, A. A., Whitelock, P. A., & Loup, C. 1999, Mass-loss rates and luminosity functions of dust-enshrouded AGB stars and red supergiants in the LMC. *A&A*, 351, 559–572.

Vanbeveren, D., De Donder, E., Van Bever, J., Van Rensbergen, W., & De Loore, C. 1998, The WR and O-type star population predicted by massive star evolutionary synthesis. *New Astron.*, 3(7), 443–492.

Massive Stars Near and Far
Proceedings IAU Symposium No. 361, 2024
J. Mackey, J. S. Vink, & N. St-Louis, eds.
doi:10.1017/S1743921322003040

Strong lithium lines in red supergiants at different metallicities

Ignacio Negueruela[1], Javier Alonso-Santiago[2], Ricardo Dorda[3] and Lee R. Patrick[1]

[1]Departamento de Física Aplicada, Facultad de Ciencias, Universidad de Alicante, Carretera de San Vicente s/n, E03690, San Vicente del Raspeig, Spain
email: ignacio.negueruela@ua.es

[2]INAF–Osservatorio Astrofisico di Catania, via S. Sofia 78, 95123 Catania, Italy

[3]Instituto de Astrofísica de Canarias, Vía Láctea s/n, E38200, La Laguna, Tenerife, Spain

Abstract. Current models of stellar evolution predict that stars more massive than $\sim 6\,\mathrm{M}_\odot$ should have completely depleted all lithium (Li) in their atmospheres by the time when they reach the He core burning phase. Against this, a non-negligible number of red giants with masses $\gtrsim 6\,\mathrm{M}_\odot$ presenting strong Li lines have recently been reported. Motivated by this finding, we have carried out a spectroscopic survey of red supergiants (RSGs) in the Perseus Arm and a selection of young open clusters in the Magellanic Clouds to assess the presence of the Li I 6708Å doublet line. Based on a sample of > 70 objects, close to one third of RSGs in the Perseus Arm display noticeable Li lines, with perhaps a trend towards a lower fraction among more luminous stars. The samples in the Magellanic Clouds are not as large, but hint at a metallicity dependence. Twenty one RSGs in 5 LMC clusters show a very high fraction of Li detection, around 40%. Conversely, 17 RSGs in 5 SMC clusters give only one secure detection. The interpretation of these observational results is not straightforward, but a mechanism for Li production seems most likely. Further characterisation work is ongoing, while theoretical studies into this matter are urgently needed.

Keywords. stars: evolution, stars: supergiants, stars: fundamental parameters, – open clusters and associations – Hertzsprung-Russell and colour-magnitude diagrams

---

1. Introduction

Lithium (Li) is the heaviest element created during primordial nucleosynthesis. In the Milky Way, low-metallicity stars consistently display Li abundances around $A(\mathrm{Li}) = 2.2$ (Spite plateau), while stars with metallicity (Z) more similar to the Sun's show higher values, around $A(\mathrm{Li}) = 3.2$ (e.g. Bensby & Lind 2018). The lower value has been interpreted as the primordial abundance – although it is very significantly lower than predictions based on models of Big Bang nucleosynthesis –, while the higher value in younger stars suggests the existence of a production mechanism through the aeons (Randich & Magrini 2021, and references therein).

Lithium is easily destroyed at temperatures slightly above $T_{\mathrm{eff}} = 2 \times 10^6$ K. It captures a proton and disintegrates into two He nuclei. In stars with convective envelopes, Li is transported from the outer layers to higher temperature zones and depleted on a timescale of a few to tens of Ma (e.g. Jeffries 2006). On the other hand, intermediate-mass stars with radiative envelopes can keep their original Li abundance in a thin outer layer. As the star leaves the main sequence and starts to swell, Li is diluted. Later, during the first dredge-up, strong Li depletion further reduces its abundance. This evolution is nicely illustrated in clusters containing stars with masses $\lesssim 2\,M_\odot$ (e.g. Twarog et al. 2020), where Li is severely depleted in subgiants and falls below detectability in most red giants, including clump stars (e.g. Anthony-Twarog et al. 2021). More massive stars are expected to behave similarly, although Li lines are not seen in stars with $T_{\rm eff} > 8\,500$ K (Lyubimkov 2016). Studies of low-luminosity supergiants (Luck 1977) and stars in moderately young clusters (Luck 1994) did not hint at significant differences with respect to less massive stars with radiative envelopes.

Lithium is not only destroyed in stars. Under some circumstances, it can also be created. A small fraction ($\sim 1\%$) of K-type giants show high Li abundances, in some cases well above those observed in main-sequence stars, strongly hinting at a production mechanism. Different authors have argued for a transient phenomenon, connected to the star's evolution (e.g. de la Reza 2020). Recent studies based on large samples of field stars suggest that this phenomenology may be related to the He flash (Zhang et al. 2021). Many Li-rich stars seem to be in the clump, and several authors have suggested that more than one mechanism must be at play (e.g. Martell et al. 2021; Zhou et al. 2022). High Li abundances are also seen in massive AGB stars. These are very luminous giants of late-M type, believed to represent the last phase in the life of stars with initial masses $4-9\,M_\odot$. Evolved AGB stars, understood to be close to the end of the thermal-pulse phase, show high Rb abundances and a very wide range of A(Li). In the Magellanic Clouds, such objects display $-7 \lesssim M_{\rm bol} \lesssim -6$ (García-Hernández et al. 2007). Other similar objects, assumed to be in a slightly earlier phase, have been found to display very high $A({\rm Li}) > 3$, while not showing high abundances of s-process elements (García-Hernández et al. 2013). This is taken as support for models that predict strong Li production at the start of the thermal-pulse phase, when Hot Bottom Burning is activated, as the base of the envelope of the star reaches $T \gtrsim 4 \times 10^7$ K, and the Cameron-Fowler mechanism (Cameron & Fowler 1971) can proceed (e.g. van Raai et al. 2012). In the Cameron-Fowler mechanism, enhanced production of beryllium (^{7}Be) will occur at the base of the envelope. Convection during the third dredge-up will move it to regions of lower $T_{\rm eff}$ in the outer layers, where it may decay into lithium (Mazzitelli et al. 1999).

The behaviour of Li in massive stars has not been studied in depth. Lyubimkov et al. (2012) observed a sample of Galactic F and G-type supergiants, determining stellar parameters and Li abundances. They found that stars with initial masses $\lesssim 6\,M_\odot$ may show a wide range of Li abundances, going from non-detectable to the same value as main-sequence stars, a spread that they interpret as a consequence of different initial rotations – suggesting that stars with very low rotation may keep most of their original surface Li until the first dredge-up, and a reduced abundance $A({\rm Li}) \lesssim 1.3$ after it has happened. Conversely, Lyubimkov et al. (2012) barely detect any lithium in any star with estimated mass $\gtrsim 6\,M_\odot$, a fact that they explain resorting to stellar models that include the effect of rotation (e.g. Brott et al. 2011; Ekström et al. 2012). These models predict that, even for relatively low initial rotational velocities, Li is depleted below detectability by the end of the main sequence, due to rotational mixing. In models without rotation, a precipitous drop in Li abundances happens during the B-type giant phase for stars of 7 to $15\,M_\odot$. Such models are also supported by limited observations of the behaviour of Be, another light element (Proffitt et al. 2016). Nevertheless, recent observations of red supergiants in open clusters, with estimated masses $7-9\,M_\odot$, have shown a non-negligible fraction of objects with strong Li lines (Negueruela et al. 2020), in open contradiction with models and previous results. In an attempt to check if this behaviour extends to higher masses, we have performed two observational campaigns targeting large populations of RSGs at different metallicities.

Table 1. Clusters containing RSGs observed with *X-shooter*. An estimate of the cluster age and the number of RSGs observed in each cluster are given.

SMC			LMC		
Name	**Age (Ma)**	N	**Name**	**Age (Ma)**	N
NGC 299	20	3	NGC 2100	18	7
NGC 376	25	4	NGC 1772	30	4
Bruck 71	35	3	NGC 1818	30	2
NGC 241/2	60	5	NGC 1711	40	4
NGC 256	90	2	NGC 1755	60	4

2. Observations

We conducted an exploratory survey with the 80 cm Telescopi Joan Oró at the Montsec Observatory, which operates in robotic mode (Vilardell et al. 2013). We used the ARES spectrograph with the Red VPH, which provides a resolving power $R \approx 10\,000$ over the 630 to 673 nm range. Observations, taken between December 2019 and June 2020, targeted a small sample of RSGs from the Perseus arm, selected from the lists of Dorda et al. (2018) and the references included there. Since these spectra indicated that a sizable fraction of the RSGs present Li lines, we proceeded to observe large samples of stars.

A sample of Perseus arm RSGs was observed with the high-resolution FIbre-fed Echelle Spectrograph (FIES) attached to the 2.56 m Nordic Optical Telescope (NOT; La Palma, Spain) during a run on 2020, October 2 – 5. FIES is a cross-dispersed high-resolution echelle spectrograph that covers the 370 – 830 nm range. We used the large aperture, which provides a resolving power $R \approx 25\,000$. The spectra were homogeneously reduced using the FIEStool software in advanced mode. We observed > 50 RSGs, distributed over the Northern sky, most of them from the Cassiopeia region of the Perseus Arm. When combined with over 20 RSGs from the region of Per OB1 in de Burgos et al. (2020), observed with the same instrument, this results in a sample of over 70 Perseus Arm RSGs.

Samples of RSGs at different metallicities were obtained by observing RSGs in Magellanic Cloud clusters. These objects were observed with *X-shooter*, mounted on the VLT UT3 (Melipal). Most of the observations used the UVB arm with an $1.3''$ slit ($R \approx 4\,000$) and the VIS arm with a $0.9''$ slit ($R \approx 9\,000$). Observations were obtained in service mode between 2020 October 13 and December 29. The targets observed were selected to belong to open clusters of different ages, in order to cover a wide range of masses. The clusters observed are listed in Table 1.

3. Results

Analysis of the spectra obtained is ongoing, in order to provide a full characterisation of all the target stars that will serve as a context for the detection of Li lines. Meanwhile, visual inspection can inform us about the basic properties of the stars.

Among the Perseus Arm sample, we find a high fraction of stars with Li lines. About 25% of the stars of all spectral types (stars observed fall mainly in the K5 – M4 range, typical of Milky Way RSGs – see Dorda et al. 2018) display strong Li lines. A few others have weak Li lines that appear blended with neighbouring weak metallic features. This result is in agreement with a recent report by Fanelli et al. (2022), based on a smaller sample.

Our RSG sample in the Perseus Arm consists mostly of field stars, which cannot be assigned ages, and thus masses, to compare to evolutionary tracks. To address this issue, all our targets in the Magellanic Clouds are members of open clusters, which can provide an evolutionary context for them. In Table 1, we list the clusters targeted, together

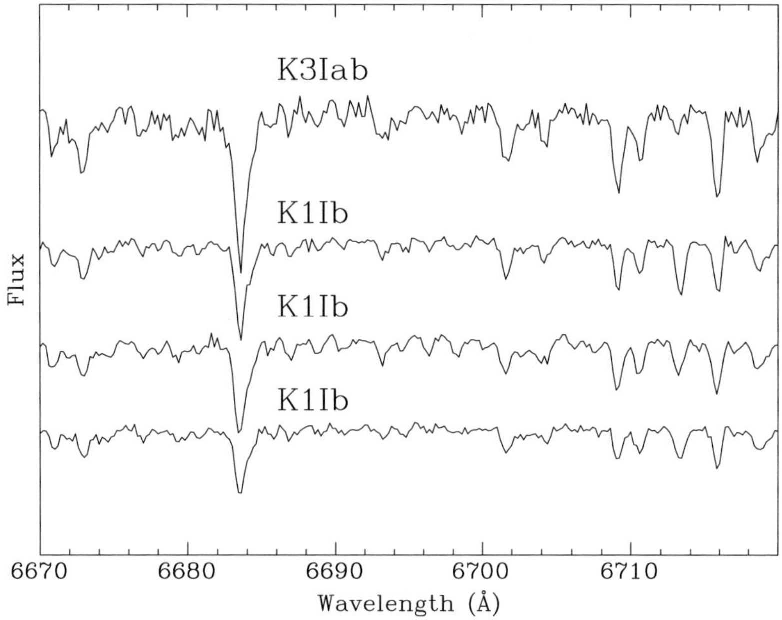

Figure 1. A small section of the *X-shooter* spectra of 4 RSGs in the LMC cluster NGC 1711. The Li I 6708Å line, displaced to close to 6714Å because of the systemic velocity of the LMC, is seen in the three bottom stars.

with an estimate of their ages (based on isochrones without binary interaction and the assumption that the *bulk* of the RSGs determines the age).

Our Magellanic samples are smaller than that from the Milky Way, but likely large enough to be statistically representative. The LMC sample includes one of the most well-populated RSG clusters, NGC 2100, which is likely younger than 20 Ma, and a range of clusters extending to the well studied NGC 1755, which is believed to contain RSGs of $\sim 7\,M_{\odot}$. In total, we have observed 21 RSGs. Stars with Li lines are found in all clusters, except in NGC 1818, where only 3 could be observed and one turns out to have an early (G) spectral type. As an example, Fig. 1 shows a small section of the spectrum of the 4 RSGs observed in NGC 1711, three of which display a clearly noticeable Li line. In total, about 40% of the stars observed have lithium. Given the small sample size, this value in consistent with that found in the Milky Way.

In the SMC, the number of large young open clusters is much smaller than in the LMC, and we could not find many suitable targets. As a consequence, our clusters cover a wider range in ages, reaching NGC 256, where the RSGs are expected to have $\sim 6\,M_{\odot}$. Among 17 stars observed, there is only one with a moderately strong Li line, in Bruck 71. Two stars in NGC 241/2 may have weak Li present. Although the sample is small, this result very strongly suggests a different behaviour from that seen at higher Z.

4. Discussion

The most massive intermediate-mass stars ($M_* \gtrsim 6\,M_{\odot}$) and most high-mass stars (except for the most massive ones) appear as RSGs (i.e. are classified as luminosity class I objects) when evolved. Thus, RSGs below a threshold mass – that likely depends on Z and probably also the evolutionary history of the star – will behave as intermediate-mass stars, entering the AGB and leaving behind a white dwarf as final remnant (see, e.g. discussion in Negueruela et al. 2017). Modern models (e.g. Poelarends et al. 2008;

Doherty et al. 2015) indicate quite high values (approaching $10\,M_\odot$) for this transition at $Z_\odot$, although binary interaction may have a significant effect on the final outcome. At Z_{SMC}, the minimum mass to explode as a supernova should be closer to $8\,M_\odot$. The reliability of all these models, however, depends on our ability to understand the complex physics governing the transport of mass and angular momentum between the core and the outer layers and the behaviour of convection at different boundary layers.

Given our current understanding, Li should not be present in the atmospheres of evolved massive stars – especially if they have started their lives as moderate or fast rotators. Despite this, we find strong Li lines – much stronger than seen in any normal red giant – in a substantial fraction of RSGs in the Milky Way and the LMC. How can these observations be interpreted?

At present, there is no obvious explanation for these results. A number of scenarios can be invoked, but they all suffer from severe shortcomings. Among these scenarios, we can list:

• **The lithium observed is preserved throughout the life of the star.** This would require that no rotational mixing and no mass loss has occurred. Given that only a fraction of the RSGs have Li and this fraction is smaller in the SMC, it would be tempting to identify the RSGs with Li with the original slow rotators. However, it must be noted that this preservation is not seen in lower mass stars (which generally rotate slowly). Moreover, it is difficult to envisage how stars of $> 10\,M_\odot$ may have reached the RSG phase without substantial mass loss.

• **RSGs with Li are in fact early-AGB stars producing Li via the Cameron-Fowler mechanism.** In recent years, the use of isochrones based on tracks including binary interaction has resulted in older age determinations for young open clusters, under the assumption that stars close to the main-sequence turn-off are mostly mass gainers (and, hence, blue stragglers). We could thus assume that most of the RSGs that we have observed, even those in Per OB1, are less massive than $\lesssim 12\,M_\odot$, and behave as AGB stars. Although this interpretation can provide us with an already-tested physical basis for the presence of Li, it does not hold on evolutionary terms. A massive star spends $\sim 10\%$ of its lifetime as a He-core burning RSG, but an intermediate-mass star spends $< 1\%$ of its lifetime as an E-AGB star. There are simply too many RSGs with Li (and specifically, too many RSGs with Li in open clusters) for this interpretation to be correct.

• **The RSGs with Li have engulfed brown dwarf companions.** Engulfment of giant planets has been proposed as an explanation for the existence of some Li-rich giants Aguilera-Gómez et al. (2016). For a massive star, a very hefty planet would be needed to provide enough Li to be detected, and brown dwarfs are likely necessary. The viability of such a process has not been studied, but the much higher fraction of RSGs with Li as compared to solar-mass giants renders this scenario suspect.

• **A mechanism to take Li to the surface works in RSGs**. The Cameron-Fowler mechanism requires third dredge-up convection to penetrate below the H-shell burning layer and retrieve newly created Be to the surface, where it can decay to lithium. A mechanism for penetrating the active H layer in RSGs is not obvious, but it might provide an explanation to the observed high frequency. The lower fraction at Z_{SMC} will serve as a consistency check on any mechanism devised.

In summary, the detection of Li in a high fraction of RSGs is extremely surprising under the light of our current stellar models. Further theoretical developments are certainly needed to provide an understanding of this phenomenon.

Acknowledgements.

Partially based on observations collected at the European Organisation for Astronomical Research in the Southern Hemisphere under ESO programme 106.212J.001 and on observations made with the Nordic Optical Telescope, operated by the Nordic Optical Telescope Scientific Association at the Observatorio del Roque de los Muchachos (La Palma, Spain) of the Instituto de Astrofísica de Canarias. We thank the director of the Montsec Observatory for DDT time on the TJO. This research is partially supported by the Spanish Government under grants PGC2018-093741-B-C21/C22 (MICIU/AEI/FEDER, UE), and by the Generalitat Valenciana under grant PROMETEO/2019/041.

References

Aguilera-Gómez, C., Chanamé, J., Pinsonneault, M. H., et al. 2016, *ApJ*, 829, 127
Anthony-Twarog, B. J., Deliyannis, C. P., & Twarog, B. A. 2021, *AJ*, 161, 159
Bensby, T. & Lind, K. 2018, *A&A*, 615, A151
Brott, I., de Mink, S. E., Cantiello, M., et al. 2011, *A&A*, 530, A115
de Burgos, A., Simon-Díaz, S., Lennon, D. J., et al. 2020, *A&A*, 643, A116
Cameron, A. G. W., & Fowler, W. A. 1971, *ApJ*, 164, 111
Doherty, C. L., Gil-Pons, P., Siess, L., Lattanzio, J. C., & Lau, H. H. B. 2015, *MNRAS*, 446, 2599
Dorda, R., Negueruela, I., & González-Fernández, C. 2018, *MNRAS*, 475, 2003
Ekström, S., Georgy, C., Eggenberger, P., et al. 2012, *A&A*, 537, A146
Fanelli, C., Origlia, L., Mucciarelli, A., et al. 2022, *ApJ*, 931, 61
García-Hernández, D. A., García-Lario, P., Plez, B., et al. 2007, *A&A*, 462, 711
García-Hernández, D. A., Zamora, O., Yagüe, A., et al. 2013, *A&A*, 555, L3
Jeffries, R. D. 2006, in *Chemical Abundances and Mixing in Stars in the Milky Way and its Satellites*, p163.
Lyubimkov, L. S. 2016, Astrophysics, 59, 411
Lyubimkov, L. S., Lambert, D. L., Kaminsky, B. M., et al. 2012, *MNRAS*, 427, 11
Luck, R. E. 1977, *ApJ*, 218, 752
Luck, R. E. 1994, *ApJS*, 91, 309
Mazzitelli, I., D'Antona, F., & Ventura, P. 1999, *A&A*, 348, 846
Martell, S. L., Simpson, J. D., Balasubramaniam, A. G., et al. 2021, *MNRAS*, 505, 5340
Negueruela, I., Alonso-Santiago, J., Tabernero, H. M., et al. 2017, *MemSAIt*, 88, 368
Negueruela, I., Alonso-Santiago, J., Tabernero, H. M., et al. 2020, *MemSAIt*, 91, 114
Poelarends, A. J. T., Herwig, F., Langer, N., & Heger, A. 2008, *ApJ*, 675, 614-625
Proffitt, C. R., Lennon, D. J., Langer, N., et al. 2016, *ApJ*, 824, 3
van Raai, M. A., Lugaro, M., Karakas, A. I., et al. 2012, *A&A*, 540, A44
Randich S., & Magrini L., 2021, *FrASS*, 8, 6
de la Reza, R. 2020, *MemSAIt*, 91, 63
Twarog, B. A., Anthony-Twarog, B. J., Deliyannis, C. P., et al. 2020, *MemSAIt*, 91, 74
Vilardell, F., Colomé, J., Sanz, J., et al. 2013, Highlights of Spanish Astrophysics VII, 958
Zhang, J., Shi, J.-R., Yan, H.-L., et al. 2021, *ApJL*, 919, L3
Zhou, Y., Wang, C., Yan, H., et al. 2022, *ApJ*, 931, 136

Massive Stars Near and Far
Proceedings IAU Symposium No. 361, 2024
J. Mackey, J. S. Vink, & N. St-Louis, eds.
doi:10.1017/S1743921322003428

On the Z-(in)dependence of the Humphreys-Davidson Limit

Gautham N. Sabhahit[1,2], **Jorick S. Vink**[1], **Erin R. Higgins**[1] and **Andreas A. C. Sander**[1,3]

[1]Armagh Observatory and Planetarium, College Hill, Armagh, BT61 9 Northern Ireland
email: gautham.sabhahit@armagh.ac.uk

[2]Queen's University Belfast, University Road, Belfast, BT7 1NN Northern Ireland

[3]Zentrum für Astronomie der Universität Heidelberg, Astronomisches Rechen-Institut, Mönchhofstr. 12-14, 69120 Heidelberg, Germany

Abstract. The temperature independent part of the Humphreys-Davidson (HD) limit sets the boundary for evolutionary channels of massive stars that either end their lives as red supergiants (RSGs) or as the hotter blue supergiants (BSGs) and Wolf-Rayet stars. Recent downward revision of most luminous RSGs the Galaxy below $\log(L/L_{\odot}) \approx 5.5$, more in line with the Magellanic Clouds, might hint towards a metallicity (Z)-independent HD limit. We present MESA single star models in the 15-40 $M_{\odot}$ range and study the different Z-dependent processes that could potentially affect the location of the upper luminosity limit of RSGs.

Keywords. Hertzsprung-Russell diagram, stars: evolution, stars: late-type, stars: fundamental parameters

1. Introduction

The observed absence of supergiants in the Galaxy and the Magellanic Clouds above a certain luminosity threshold has been known for decades (Humphreys & Davidson 1979). The upper luminosity boundary for the supergiants in these galaxies consists of two parts: a temperature dependent part characterised by decreasing luminosity with decreasing temperature for the hottest supergiants which merges with a temperature independent part (at $\sim$ 10 kK) for Red supergiants (RSGs). Humphreys & Davidson (1979) found an upper magnitude of $M_{\rm bol} \approx -9.5$ to 10 mag in both the Galaxy and the Large Magellanic Cloud (LMC) which translates to a luminosity of $\log(L/L_{\odot}) \approx 5.8$.

Ever since supergiants were not detected above this luminosity threshold, the general absence of these stars above this limit was assumed to be due to stellar wind mass loss in close proximity to the Eddington limit. The shape of the temperature dependent part of the HD limit when reproduced in the temperature-gravity diagram closely resembles the Eddington limit (Lamers & Fitzpatrick 1988), and the reversal of the redward expansion of stars is due to high mass loss as deeper layers of the stars get exposed. Lamers & Fitzpatrick (1988) also explain the 'plateau'ing of the observed limit as below a certain critical luminosity the stars are stable against radiation pressure even at the location of the maximum opacity and the Eddington limit is not reached as the star expands redwards.

Line-driven mass loss inherently scales with the metal content: the lower the metal mass fraction Z, the smaller the content of line-driving ions such as iron and hence lower the mass loss rates. If the upper luminosity limit of supergiants is indeed set by mass loss, the critical luminosity would also scale with Z shifting it to higher luminosities at lower Z.

So it was quite a surprise when Davies et al. (2018) re-evaluated the luminosity limit of RSGs in the Magellanic Clouds and found no evidence for the limit to be higher in the Small Magellanic Cloud (SMC, $Z \approx 0.2Z_{\odot}$) compared to the Large Magellanic Cloud (LMC, $Z \approx 0.5Z_{\odot}$). In both these sub-solar metal content environments, Davies et al. (2018) report a RSG luminosity limit of $\log(L/L_{\odot}) \approx 5.5$. More recently, Davies & Beasor (2020) re-analysed some of the brightest known RSGs in the Galaxy and suggest a Galactic RSG limit also at $\log(L/L_{\odot}) \approx 5.5$, hinting towards a Z-independent upper luminosity limit. If this finding is confirmed this would imply that the physics setting the HD limit remains elusive, and mass loss is just part of the answer.

The luminosities of massive stars steeply increase with initial stellar mass resulting in large L/M ratios that could bring such stars on the brink of radiative instability where one could potentially have the outward radiative force exceed the inward pointing gravitational force locally. The ratio of radiative force to Newtonian gravity, given by the so-called Eddington parameter $\Gamma_{\rm rad}$, could potentially approach unity (and even cross it) in layers associated with the opacity bumps of hydrogen (H), helium (He) and metals (mostly iron) (Sanyal et al. 2015). The metallicity dependence of the Eddington limit being reached in the iron bump clearly predicts a growing luminosity limit with decreasing Z (Sanyal et al. 2017). This is because the iron bump opacities increase with the content of metals, allowing the stability of more massive stars, hence higher maximum luminosity at lower Z.

The uncertain single-star physics of stellar envelopes in close proximity to the Eddington limit is of key relevance to understanding both the HD limit as well as luminous blue variable (LBV) S Doradus variations (Grassitelli et al. 2021). Furthermore, it has direct implications on binary evolution involving the interaction probability of close binary systems and gravitational wave (GW) events (e.g. Klencki et al. 2021). A better understanding of supergiant envelopes is also critical for understanding the maximum black hole (BH) mass of blue supergiant (BSG) models up to approximately 85 $M_{\odot}$ (Vink et al. 2021).

2. Methods

The one-dimensional stellar evolution code Modules for Experiments in Stellar Astrophysics (MESA version 12115) (Paxton et al. 2011, 2013, 2015) is used to compute our grid of stellar models. The initial mass in our grid ranges from 15 $M_{\odot}$ to 40 $M_{\odot}$ at intervals of 2.5 $M_{\odot}$, and three different initial metallicities corresponding to the Galaxy ($Z = 0.017$), LMC ($Z = 0.008$) and SMC ($Z = 0.004$). We also have over-shooting mixing immediately above (or below) convective zones with two different values for their efficiencies: low ($f_{\rm ov} = 0.01$) and high ($f_{\rm ov} = 0.05$), implemented using a diffusive approach with an exponential profile as described in Herwig (2000). We use the Ledoux criteria for convection, with a high semiconvective efficiency of $\alpha_{\rm sc} = 100$. Further models are run to test the effects of certain processes such as mass loss, rotation and semi-convection on our results. All the models are evolved until core helium exhaustion when the central helium mass fraction ($Y_{\rm c}$) drops below 0.01.

Convection is treated using the standard mixing length theory (MLT) developed by Cox & Giuli (1968) with the free parameter $\alpha_{\rm MLT} = 1.82$ (Choi et al. 2016). The exact value of $\alpha_{\rm MLT}$ can change the RSG temperatures by a few hundred Kelvin, but has negligible effects on the stars that evolve bluewards and on the RSG limit. While the existence of such density inversions and inflated envelopes is heavily debated, we can still study the Z-dependence of the opacity bumps by using standard MLT and understand the stability of massive stars based on the local $\Gamma_{\rm rad}$ in these bumps.

Mass loss in our models is implemented similar to the "Dutch wind scheme" in MESA comprising of three different recipes. Hot O stars ($T_{\rm eff} > 10\,$kK) with surface hydrogen mass fraction $X_s > 0.4$ use the metallicity-dependent mass loss recipe from Vink et al. (2001). The mass-loss rates used in our models scale with the surface iron abundance ($Z_{\rm Fe}$) instead of the total metal fraction at the surface. Cooler stars with $T_{\rm eff} < 10\,$kK use the dust driven mass loss recipe from de Jager et al. (1988). For temperatures $10\,{\rm kK} < T_{\rm eff} < 11\,{\rm kK}$, we interpolate between the rates derived from Vink et al. (2001) and de Jager et al. (1988). Post-main sequence stars that lose their envelope ($X_s < 0.4$) and evolve bluewards use the recipe adopted from Nugis & Lamers (2000), which only become relevant for stars towards the upper end of the initial mass range considered in our study.

3. Z dependent processes

Over the last few decades, theoretical efforts have been made to predict the radiation-driven mass loss as a function of fundamental stellar parameters such as luminosity, mass and metallicity (Lucy & Solomon 1970; Castor et al. 1975). Simulations such as Monte Carlo calculations in Vink et al. (2001) show the mass loss rate to scale with the surface iron abundance, $\dot{M} \sim Z_{\rm Fe}^{0.7}$. The mass loss-$Z$ scaling has also been verified empirically where the modified wind-momentum luminosity relation (WLR) $D_{\rm Mom}(L) = \dot{M} v_{\rm inf} R^{0.5}$ for the LMC stars lies in between the WLRs of Galactic and SMC stars. The Z-dependence of the mass loss causes the critical luminosity of the RSGs to grow at lower Z due to lower total mass being lost.

Internal mixing processes can potentially affect the surface temperatures of supergiants and set a limit on their luminosity. Gilkis et al. (2021) find the over-shooting above the convective core during MS to significantly affect the fraction of post-MS lifetime (core helium burning) spent as RSGs. Semi-covective mixing, that arises in layers with a chemical gradient that can potentially stabilize the convective instability, can also cause blueward evolution (Langer et al. 1985; Schootemeijer et al. 2019). Metallicity can indirectly affect the formation of semi-convective layers in the envelopes of massive stars through mass loss (Higgins & Vink 2020). Semi-convection only becomes relevant at low enough initial Z where massive stars can retain most of their envelopes thus allowing semi-convective regions to form. This lowered the maximum luminosity of RSGs in LMC and SMC models which agreed well with the then empirical limits in the respective galaxies reported by Davies et al. (2018).

As mentioned in the Introduction, radiative instability in opacity bumps inside massive stars also influences the maximum luminosity of supergiants, and depends on the host Z. Unlike pure He stars and hot O supergiants where the envelope density stratification follows the topology of the iron bump (Gräfener et al. 2012), it is the H opacity bump that dominates the physics of red supergiant envelopes. In Fig 1, we show the variation of Rosseland mean opacity with temperature inside a red supergiant (profile selected at $Y_{\rm c} = 0.5$) for the range of Z considered in our model grid. In stark contrast to the Z-dependence of the iron bump, the hydrogen bump gets stronger at lower Z. This can be understood by realizing that stars at lower Z have lost less mass during their evolution and are also more compact. This two-fold effect results in higher densities in their envelope layers and consequently a stronger H bump opacity at lower Z.

The reversal of the H bump opacity with the initial metallicity due to an indirect effect of mass loss means the radiative instability favours the maximum luminosity of red supergiants to grow at higher Z. This dependence on metallicity is a qualitatively different behavior compared to the previously discussed direct effect of mass loss and its dependence on Z.

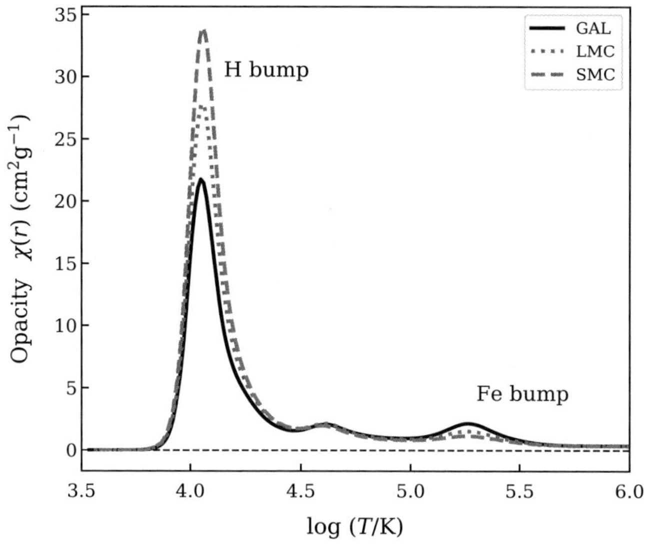

Figure 1. Rosseland mean opacity as a function of temperature showing the H and Fe opacity bumps at $\log(T_{\rm eff}) \approx 4$ and 5.3 respectively. The red supergiant profiles are taken halfway through core helium burning for Galactic (solid), LMC (dotted) and SMC (dashed) metallicity with $M_{\rm init} = 30 M_\odot$

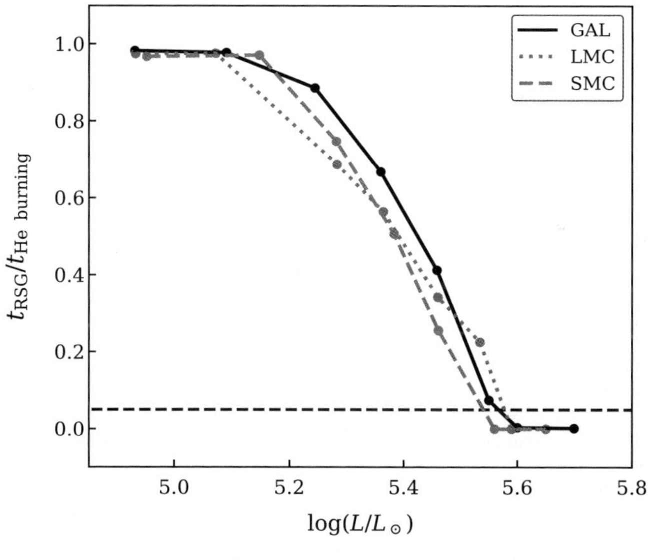

Figure 2. Distribution of fraction of He burning time spent below the threshold effective temperature $\log(T_{\rm eff}/{\rm K}) = 3.68$ as a function of the luminosity of the RSG. The solid, dotted and dashed lines correspond to Galactic, LMC and SMC-like metallicity respectively. The horizontal line marks the 5% limit that we assume to set the HD limit.

Any process that might destabilize and remove density inversions on top of inflated envelopes, be it turbulent convection or enhanced mass loss, will depend on the Eddington parameter in these zones given by

$$\Gamma(r) = \frac{g_{\rm rad}}{g_{\rm newt}} = \frac{\chi(r) l_{\rm rad}(r)}{4\pi G c m(r)} \tag{3.1}$$

which depends on the opacity. In Figure 2 we show the fraction of helium burning lifetime spent as RSG for models where the convective efficiencies in the H bump are scaled with the local Eddington parameter (for full grid see Sabhahit et al. (2021)). Models with sufficiently high over-shooting that spend a significant fraction of their post-MS with access to the H bump all have a very similar luminosity cutoff at $\log(L/L_{\odot}) \approx 5.5$ regardless of the initial Z. The effects of Z-dependent mass loss and the H bump radiative instability balances each other out and could potentially explain the Z-independence of the observed maximum luminosity of red supergiants.

4. Conclusions

In Sabhahit et al. (2021) we present a possible qualitative explanation for the observed Z-independence of the HD limit according to recent findings. Line-driven mass loss scales with the metal content: the lower the metallicity, the lower the mass loss rates. The direct effect of mass loss is thus to shift the luminosity boundary upwards at low Z.

On the other hand we have radiative instability in opacity bumps which could also set the HD limit. In contrast to pure helium stars and hot O stars whose envelope physics is dictated by the iron bump, the physics of red supergiant envelopes considered in this study is primarily dominated by the conditions in the H bump. In realistic stellar evolution models, we find the H bump to be stronger at lower Z due to higher densities - an indirect effect of mass loss. This can have a qualitatively different trend with Z on the maximum RSG luminosity compared to the direct of mass loss, and if balanced could potentially explain the Z-independence of the observed HD limit.

Finally, if future observational constraints further confirms the Z-independence of the limit or strongly favours a limit that varies with metallicity, it informs us about the physical processes inside the star and which effects dominate over the other.

References

Castor, J. I., Abbott, D. C., & Klein, R. I. 1975, ApJ, 195, 157
Choi, J., Dotter, A., Conroy, C., Cantiello, M., Paxton, B., & Johnson, B. D. 2016, ApJ, 823, 102
Cox, J. P., & Giuli, R. T. 1968, Principles of stellar structure
Davies, B., & Beasor, E. R. 2020, MNRAS, 493, 468
Davies, B., Crowther, P. A., & Beasor, E. R. 2018, MNRAS, 478, 3138
de Jager, C., Nieuwenhuijzen, H., & van der Hucht, K. A. 1988, A&AS, 72, 259
Gilkis, A., Shenar, T., Ramachandran, V., Jermyn, A. S., Mahy, L., Oskinova, L. M., Arcavi, I., & Sana, H. 2021, MNRAS
Gräfener, G., Owocki, S. P., & Vink, J. S. 2012, A&A, 538, A40
Grassitelli, L., Langer, N., Mackey, J., Gräfener, G., Grin, N. J., Sander, A. A. C., & Vink, J. S. 2021, A&A, 647, A99
Herwig, F. 2000, A&A, 360, 952
Higgins, E., & Vink, J. S. 2020, Astronomy and Astrophysics, 635, 1937374
Humphreys, R. M., & Davidson, K. 1979, ApJ, 232, 409
Klencki, J., Nelemans, G., Istrate, A. G., & Chruslinska, M. 2021, A&A, 645, A54
Lamers, H. J. G. L. M., & Fitzpatrick, E. L. 1988, ApJ, 324, 279
Langer, N., El Eid, M. F., & Fricke, K. J. 1985, A&A, 145, 179
Lucy, L. B., & Solomon, P. M. 1970, ApJ, 159, 879
Nugis, T., & Lamers, H. J. G. L. M. 2000, A&A, 360, 227
Paxton, B., Bildsten, L., Dotter, A., Herwig, F., Lesaffre, P., & Timmes, F. 2011, ApJS, 192, 3
Paxton, B., et al. 2013, ApJS, 208, 4
—. 2015, ApJS, 220, 15
Sabhahit, G. N., Vink, J. S., Higgins, E. R., & Sander, A. A. C. 2021, MNRAS, 506, 4473
Sanyal, D., Grassitelli, L., Langer, N., & Bestenlehner, J. M. 2015, A&A, 580, A20
Sanyal, D., Langer, N., Szécsi, D., -C Yoon, S., & Grassitelli, L. 2017, A&A, 597, A71

Schootemeijer, A., Langer, N., Grin, N. J., & Wang, C. 2019, A&A, 625, A132
Vink, J. S., de Koter, A., & Lamers, H. J. G. L. M. 2001, A&A, 369, 574
Vink, J. S., Higgins, E. R., Sander, A. A. C., & Sabhahit, G. N. 2021, MNRAS, 504, 146

Discussion

TOMER SHENAR: It is interesting to speculate whether the seeming Z-independence of the HD limit is caused by various effects that are cancelling each other out. What happens when the metallicity approaches zero

GAUTHAM SABHAHIT: With no metal lines the line driven mass loss would be negligible, but the star could still lose substantial amount of mass through continuum driven winds or other processes such as rotation and pulsations, which comes with its own uncertainties. Unless properly tested with all these relevant physics included, it would be hard to speculate what happens in these primordial conditions.

Massive Stars Near and Far
Proceedings IAU Symposium No. 361, 2024
J. Mackey, J. S. Vink, & N. St-Louis, eds.
doi:10.1017/S1743921322003210

Red Supergiants in M31: The Humphreys-Davidson limit at high metallicity

Sarah L. E. McDonald[1], **Ben Davies**[1] and **Emma R. Beasor**[2]

[1]Astrophysics Research Institute, Liverpool John Moores University, Liverpool Science Park ic2,146 Brownlow Hill, Liverpool, L3 5RF, UK
email: s.e.mcdonald@2015.ljmu.ac.uk

[2]NSF's National Optical-Infrared Astronomy Research Laboratory, 950 N. Cherry Ave., Tucson, AZ 85719, USA

Abstract. The empirical upper limit to Red Supergiant (RSG) luminosity, known as the Humphreys-Davidson (HD) limit, has been commonly explained as being caused by the stripping of stellar envelopes by metallicity-dependent, line-driven winds. As such, the theoretical expectation is that the HD limit should be higher at lower metallicity, where weaker mass-loss rates mean that higher initial masses are required for an envelope to be stripped. In this work, we test this prediction by measuring the luminosity function of RSGs in M31 and comparing to those in the LMC and SMC. We find that $\log(L_{\rm max}/L_{\odot}) = 5.53 \pm 0.03$ in M31 (Z $\gtrsim$ Z$_{\odot}$), consistent with the limit found for both the LMC (Z $\sim$ 0.5 Z$_{\odot}$) and SMC (Z $\sim$ 0.25 Z$_{\odot}$), while the RSG luminosity distributions in these 3 galaxies are consistent to within 1σ. We therefore find no evidence for a metallicity dependence on both the HD limit and the RSG luminosity function, and conclude that line-driven winds on the main sequence are not the cause of the HD limit.

Keywords. stars: late-type, stars: fundamental parameters, Hertzsprung-Russell diagram, stars: mass loss

1. Introduction

It is well established that there is an empirical upper limit to Red Supergiant (RSG) luminosity (Stothers 1969; Sandage & Tammann 1974), often referred to as the 'Humphreys-Davidson (HD) Limit' (Humphreys & Davidson 1979). The HD limit is often explained as being a manifestation of mass loss (e.g. Humphreys & Davidson 1979) during the lifetime of the star, caused by strong stellar winds or episodic periods of mass-loss, where the fraction of mass lost from the stellar envelope is dependent on the initial mass of the star. Under this explanation, lower initial mass supergiants ($\sim 8M_{\odot} - 15M_{\odot}$) experience winds which are not strong enough to remove the entire hydrogen envelope on the main sequence (MS, Maeder 1981; Maeder & Meynet 2003) so the star is able to evolve to the RSG phase, where it resides before dying as a core-collapse supernova. Higher initial mass stars ($\sim 15M_{\odot} - 30M_{\odot}$) can lose a considerable fraction of their envelope, causing the star to undergo only a brief RSG phase before evolving to a Wolf Rayet (WR) star (Stothers & Chin 1979). At even higher masses ($\gtrsim 30M_{\odot}$), the entire envelope can be lost by the time hydrogen in the core is exhausted, preventing evolution to the cool red side of the Hertzsprung-Russell (HR) diagram. These stars instead evolve directly from the

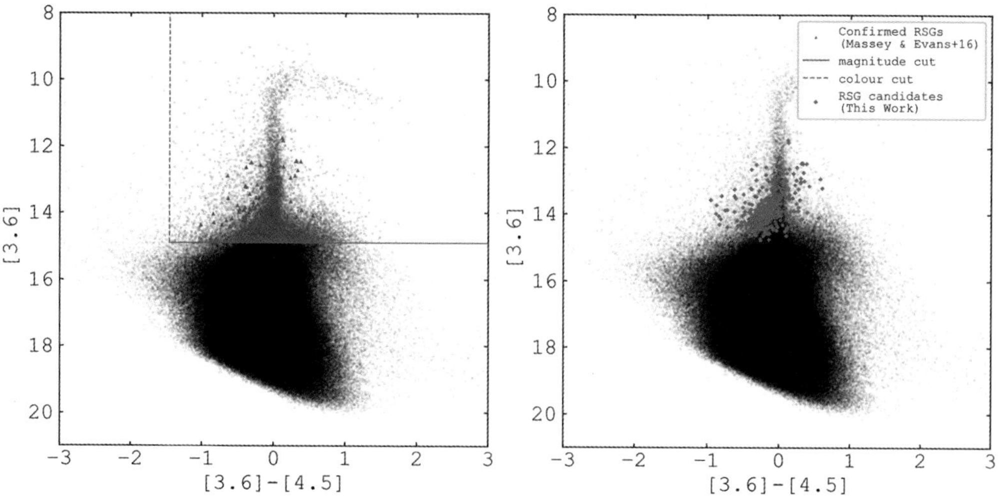

Figure 1. (a) A colour magnitude diagram, where the black points in both panels indicate all the M31 point sources detected by Spitzer IRAC/MIPS (Khan 2017). The grey points show the sources which fit the criteria to be a likely RSG candidate based on the colour (dashed grey line) and magnitude (solid grey line) cuts applied, to find the first constraint towards establishing a sample of RSG candidates. The red triangles indicate known M31 RSGs with determined spectral classifications by Massey & Evans (2016) from which we have based our colour and magnitude cuts around. (b) The magenta points indicate all the RSG candidates we find and use in the present work after all photometric and astrometric cuts have been applied.

MS to a WR star, completely bypassing the RSG phase (Stothers & Chin 1978). Under this scenario, the HD limit therefore represents the luminosity which corresponds to the most massive star that may still experience a RSG phase.

Evolutionary models predict that lower metallicity environments should produce more luminous supergiants due to this dependency of mass loss on metallicity (Kudritzki et al. 1987; Maeder & Meynet 2003). This means the HD limit should therefore also be metallicity dependent.

The HD limit has been measured previously in the literature, the first being a hard upper limit of $\log(L/L_{\odot}) = 5.8 \pm 0.1$ inferred by Humphreys & Davidson (1979), using a sample of cool supergiants in the Milky Way and the Large Magellanic cloud (LMC) (later revised to $\log(L/L_{\odot}) = 5.66$ in Humphreys (1983)). Davies, Crowther & Beasor (2018, hereafter, DCB18) revisited the HD limit in the Magellanic Clouds, with higher precision multi-wavelength photometry, finding an upper limit of $\log(L/L_{\odot}) = 5.5$ for both the Small Magellanic Cloud (SMC) and the LMC.

In this work, we complement the study of DCB18 with focus on the high luminosity end of the M31 RSG luminosity function as well as make quantitative comparisons with RSG populations in lower metallicity galaxies.

2. Method

2.1. *Compiling the sample of RSGs*

To locate our target stars, we constructed colour-magnitude diagrams (CMDs) using the Spitzer photometry (Khan 2017), see Figure 1, and overplotted a sample of known RSGs from Massey & Evans (2016) to define the location of RSGs in mid-IR colour-magnitude space. The total number of stars rejected from our sample are discussed further in Sections 2.1 and 3.3 in McDonald et al. (2022), where reasons for rejection are outlined in detail. These stars were then cross-matched to Local Group Galaxy Survey

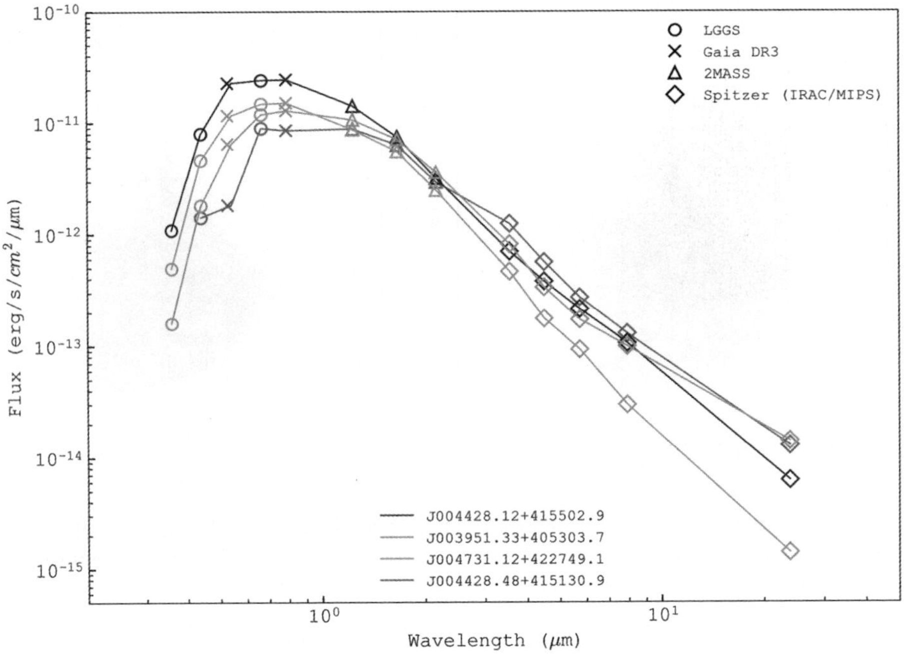

Figure 2. Spectral energy distributions of the most luminous Red Supergiant (RSG) candidates. These have $\log(L/L_{\odot}) > 5.4$ with complete de-reddened photometry ranging from the optical through to the mid-infrared. The symbols in the upper legend indicate the catalogue source of the photometry and the lower legend provides the LGGS star name for each candidate.

(LGGS) UBVRI photometry Massey et al. (2006), Gaia EDR3 photometry (BP and RP bands) and astrometry (proper motion and parallax) Gaia Collaboration (2020) and Two Micron All Sky Survey (2MASS) JHK photometry Cutri et al. (2003).

2.2. *Foreground Extinction*

To correct for foreground extinction, we use an extinction map of M31 from Dalcanton et al. (2015), surveyed by The Panchromatic Hubble Andromeda Treasury project (PHAT, Dalcanton et al. 2012). Each RSG candidate was then de-reddened according to the Cardelli et al. (1989) reddening law for the optical photometry, and Rieke & Lebofsky (1985) for the near-IR.

3. Luminosity Distributions and L_{max}

3.1. *Determining bolometric luminosity*

We converted the de-reddened photometry into fluxes using Vega calibrated zero point fluxes for each filter from the SVO Filter Profile Service (Rodrigo & Solano 2020). Using these fluxes, we plot spectral energy distributions (SEDs) for each RSG candidate and integrate under the SED to determine bolometric luminosity. Figure 2 shows the SEDs of the most luminous candidates.

The observational luminosity function of M31 RSGs is seen in Figure 3. The light grey distribution shows the number of RSG candidates per log luminosity bin for M31. The two darker grey distributions show the number of RSG candidates we use in this study which are also found in previous M31 RSG studies. Each of these luminous RSGs are discussed in more detail in McDonald et al. (2022).

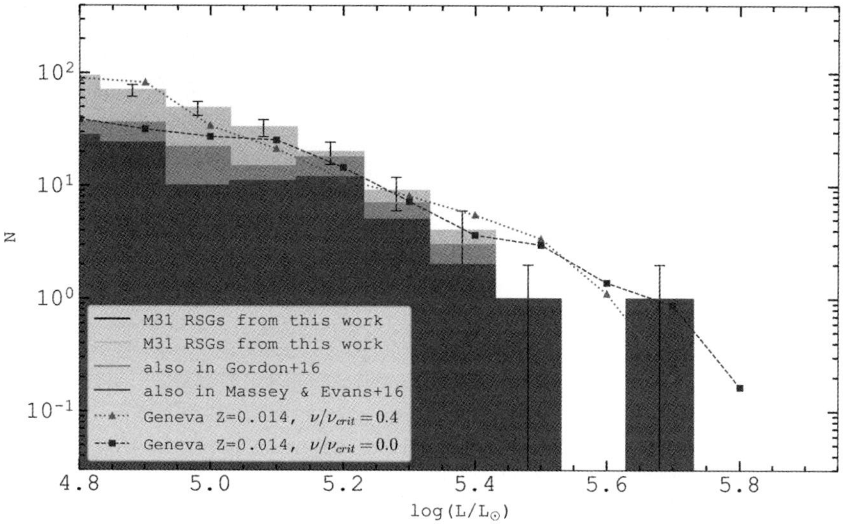

Figure 3. The Red Supergiant luminosity distribution for M31. The observed luminosity distribution from this work is shown in light grey, with the two darker grey distributions showing the number of RSG candidates we use in this study that are also found in previous M31 RSG studies. Over-plotted are the rotating ($\nu/\nu_{crit} = 0.0$) and non-rotating ($\nu/\nu_{crit} = 0.4$) model predicted distributions from the GENEVA models at solar metallicity (Z=0.014) from Ekström et al. (2012). N.b. The brightest star at $\log(L/L_{\odot}) = 5.71$ cannot be definitively ruled out, but is a borderline M31 candidate due to its proper motion.

4. Statistical Analysis

To make a broader test of the metallicity dependence of $L_{\rm max}$ and the luminosity function, we perform two comparisons. Firstly, we compare the empirical luminosity functions of the LMC and SMC with M31. Secondly, we compare the M31 luminosity function and $L_{\rm max}$ to theoretical expectations of lower metallicities using population synthesis.

4.1. *Observational comparisons between the LMC and SMC*

We look at the cumulative RSG luminosity function for M31 and compare with the empirical SMC and LMC distributions from (Davies, Crowther & Beasor 2018), looking at all RSGs with $\log(L/L_{\odot}) > 5$, where our sample is considered to be complete. The left panel of Figure 4 shows the similarities of the observed cumulative luminosity functions for M31, SMC and LMC. We perform a Kolmogorov-Smirnov (KS) test to evaluate these similarities by measuring the differences between the cumulative distribution functions. We find for the empirical M31 distribution compared with the SMC and LMC, a 60% and 44% probability, respectively, that they are drawn from the same parent distribution. Hence, the probability that the RSG luminosity function in the three galaxies are consistent with one another is within 1σ. Furthermore, each galaxy has the same $L_{\rm max}$ to within 0.1dex, at $\log(L/L_{\odot}) \sim 5.5$.

We do the same for the model cumulative luminosity function of RSGs at SMC-like (Z=0.002) metallicity (SMC-like tracks are from Georgy et al. 2013) when compared with M31, seen in the right panel of Figure 4. Here we find a probability of 5% (rotating) and 0.1% (non-rotating) for the M31 models compared with observations and a 0.02% (rotating) and 10^{-6}% (non-rotating) probability for the SMC models compared with observations. These low probabilities lead us to conclude that there is little similarity

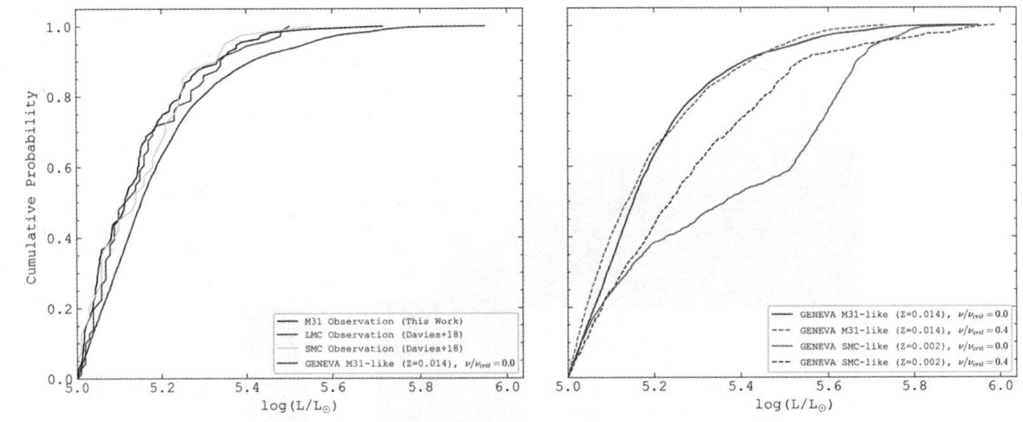

Figure 4. Left Panel: The cumulative luminosity distribution of all the Red Supergiants with an observational luminosity $\log(L/L_{\odot}) > 5$ in M31 from this work, as well as for the Large and Small Magellanic Clouds from Davies, Crowther & Beasor (2018). Right Panel: Cumulative luminosity distribution of the cool supergiants with a luminosity $\log(L/L_{\odot}) > 5$ from the model luminosity functions predicted by GENEVA, at both solar (from Ekström et al. 2012) and SMC-like metallicities (from Georgy et al. 2013) for both the rotating and non-rotating models. We include the 'M31-like' non-rotating model predicted distribution in the left panel for comparison.

between the model distributions in the two galaxies and they are unlikely to be drawn from the same parent distribution.

4.2. *Comparisons to theoretical predictions of* $L_{\rm max}$

To investigate the effects of sample size on $L_{\rm max}$, we perform another Monte Carlo experiment to find the average $L_{\rm max}$ for each sample size of N cool supergiants with $\log(L/L_{\odot}) > 5$. The results of this Monte Carlo are shown in Figure 5 where the empirical $L_{\rm max}$ for the sample size we observe for that galaxy is denoted by the black star. Although M31 shows agreement within 3σ, the SMC shows a disagreement beyond the 99.7% confidence limit. In summary, we find no significant difference in $L_{\rm max}$ within the errors across a metallicity baseline of ($0.25Z_{\odot}$ to $\gtrsim Z_{\odot}$). This is in clear disagreement with theoretical expectations because $L_{\rm max}$ predictions from the models are simply too high compared to observational measurements and this effect is predicted to only increase with decreasing metallicity.

5. Conclusions

We compiled a sample of mid-IR selected cool supergiants to measure the luminosity function of the red supergiant (RSG) population in M31 to investigate the Humphreys-Davidson limit ($L_{\rm max}$).

• We find that the luminosity function of RSGs is independent of metallicity, based on the range of metallicities studied here (from SMC-like to M31-like).

• $L_{\rm max}$ is also independent of metallicity, where we find the HD limit for M31 is $\log(L/L_{\odot}) = 5.53 \pm 0.03$, within 0.1dex of the SMC and LMC. We are in agreement with Davies, Crowther & Beasor (2018) who find a lack of evidence for a metallicity dependent $L_{\rm max}$. This suggests that mass loss from line-driven winds are **not** the cause of the HD limit.

• A population synthesis analysis shows that the single star Geneva evolutionary models not only over-predict the number of luminous cool supergiants at the high luminosity end, but also over-predict $L_{\rm max}$, particularly at lower metallicities.

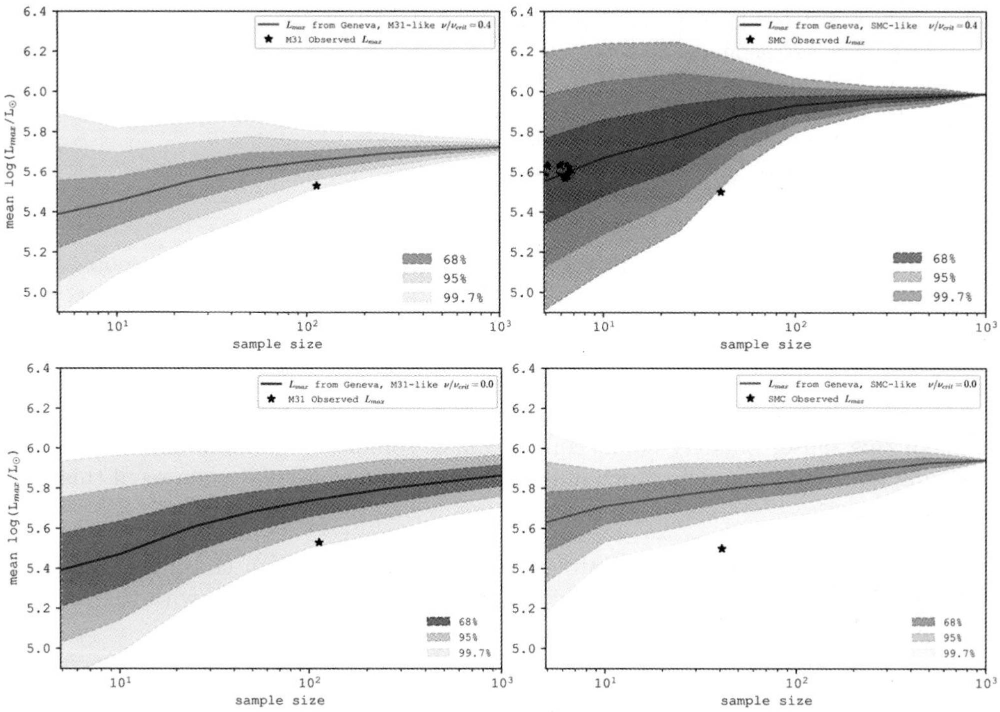

Figure 5. The expected $L_{\max}$ for a range of sample sizes as predicted by the Geneva rotating models for both solar (Z=0.014) and SMC-like (Z=0.002) metallicities. The shaded regions indicate the confidence limits on $L_{\max}$ as shown in the legend and the black stars indicate the observed $L_{\max}$ and sample size for M31 from this work and the same for the SMC from Davies, Crowther & Beasor (2018).

References

Cardelli, J. A., Clayton, G. C., & Mathis, J. S. 1989, ApJ, 345, 245

Cutri, R. M., et al. 2003, VizieR Online Data Catalog, II/246

Dalcanton, J. J., et al. 2012, ApJS, 200, 18

—. 2015, ApJ, 814, 3

Davies, Crowther & Beasor. 2018, MNRAS, 478, 3138

Ekström, S., et al. 2012, A&A, 537, A146

Gaia Collaboration. 2020, VizieR Online Data Catalog, I/350

Georgy, C., et al. 2013, A&A, 558, A103

Humphreys, R. M. 1983, ApJ, 265, 176

Humphreys, R. M., & Davidson, K. 1979, ApJ, 232, 409

Khan, R. 2017, ApJS, 228, 5

Kudritzki, R. P., Pauldrach, A., & Puls, J. 1987, A&A, 173, 293

Maeder, A. 1981, A&A, 102, 401

Maeder, A., & Meynet, G. 2003, in IAU Symposium, Vol. 212, A Massive Star Odyssey: From Main Sequence to Supernova, ed. K. van der Hucht, A. Herrero, & C. Esteban, 267

Massey, P., & Evans, K. A. 2016, ApJ, 826, 224

Massey, P., Olsen, K. A., Hodge, P. W., Jacoby, G. H., McNeill, R. T., Smith, R. C., & Strong, S. B. 2006, in American Astronomical Society Meeting Abstracts, Vol. 209, American Astronomical Society Meeting Abstracts, 27.01

McDonald, S. L. E., Davies, B., & Beasor, E. R. 2022, MNRAS, 510, 3132

Rieke, G. H., & Lebofsky, M. J. 1985, ApJ, 288, 618

Rodrigo, C., & Solano, E. 2020, in Contributions to the XIV.0 Scientific Meeting (virtual) of the Spanish Astronomical Society, 182
Sandage, A., & Tammann, G. A. 1974, ApJ, 191, 603
Stothers, R. 1969, ApJ, 155, 935
Stothers, R., & Chin, C. W. 1978, ApJ, 226, 231
—. 1979, ApJ, 233, 267

Discussion

LEE PATRICK: How do you think the idea that most of the RSGs that we observe are merger products fits into this picture of of the metallicity independent HD limit?

SARAH MCDONALD: Given the binary fraction of the most massive RSGs is expected to be high, the contribution to mass loss from a companion star would mean that the most massive stars are losing more mass than would be expected from winds in single stars alone. In this sense, binarity could control Lmax somewhat. We are currently looking into using the BPASS models which implement binarity, so we may soon see if this solves the discrepancy.

PHILLIP MASSEY: Did you look into what would happen if you considered subsets of your spatial sampling to see how your comparison of the observed LF with the predicted LF holds up?

SARAH MCDONALD: The effects of variable SFH is somewhat similar to the stochastic effects which we try to quantify with the Monte Carlo analysis of sample size effects. To some degree, the stochasticity is dominated by the relatively low number of RSGs in the SMC, and so admittedly we havent given too much attention to the SFH. That being said, Philip you are correct that Harris & Zaritsky 2009 (fig 19) implies we are close to the average SF level across all epochs, at the current epoch. i.e. we are not in a particular rut of star formation at current times, and so the HD limit measured in our work should be close to the typical value over all epochs.

Massive Stars Near and Far
Proceedings IAU Symposium No. 361, 2024
J. Mackey, J. S. Vink, & N. St-Louis, eds.
doi:10.1017/S1743921322003106

Luminous red supergiants in the Magellanic Clouds

S. De Wit[1,2], A. Z. Bonanos[1], F. Tramper[4], M. Yang[1,5], G. Maravelias[1,3], K. Boutsia[6], N. Britavskiy[7] and M. Zapartas[1]

[1]IAASARS, National Observatory of Athens, Athens, Greece

[2]National and Kapodistrian University of Athens, Athens, Greece
email: sdewit@noa.gr

[3]Institute of Astronomy, KU Leuven, Belgium

[4]Key Laboratory of Space Astronomy and Technology, National Astronomical Observatories, Chinese Academy of Sciences, Beijing, People's Republic of China

[5]Institute of Astrophysics, FORTH, Heraklion, Greece

[6]Carnegie Observatories, Las Campanas Observatory, La Serena, Chile

[7]University of Liège, Liège, Belgium

Abstract. There is evidence that some red supergiants (RSGs) experience phases of episodic mass-loss. These episodes yield more extreme mass-loss rates, further stripping the envelope of the RSG, significantly affecting the further evolution towards the final collapse of the star. Mass lost through RSG outbursts/superwinds will flow outwards and form dust further out from the surface and this dust may be detected and modelled. Here, we aim to derive the surface properties and estimate the global properties of Mid-IR bright RSGs in the Magellanic Clouds. These properties will then be compared to evolutionary predictions and used for future spectral energy distribution fitting studies to measure the mass-loss rates from present circumstellar dust.

Keywords. massive stars, mass-loss, stellar parameters

1. Observing plan and modelling

The presence of circumstellar dust reveals past mass-loss events of RSGs. We aim to model such dusty RSGs and learn about their properties. For this, we selected previously unstudied, bright and dusty supergiant candidates using colour-magnitude criteria from the Near-IR, Mid-IR and optical colour-magnitude diagrams (see Fig. 1, top panel). For eleven objects satisfying the criteria, we obtained Magellan/MagE spectra ($R = \frac{\lambda}{\delta\lambda} \sim 4000$). Eight objects were classified as RSGs. We modelled these spectra using MARCS atmospheric models (Gustafsson et al. 2008), to obtain the effective temperature, the optical extinction factor and the surface gravity through a minimal χ^2 fitting routine (see Fig. 1, bottom panel for an example). More specifically, we have used the Ca II triplet to constrain the surface gravity, the molecular TiO bands to constrain the effective temperature and the slope of the spectrum to derive the extinction. Subsequently, using the K-band magnitude, we have estimated global properties such as the stellar luminosity and radius. The full analysis is presented in de Wit et al. (submitted).

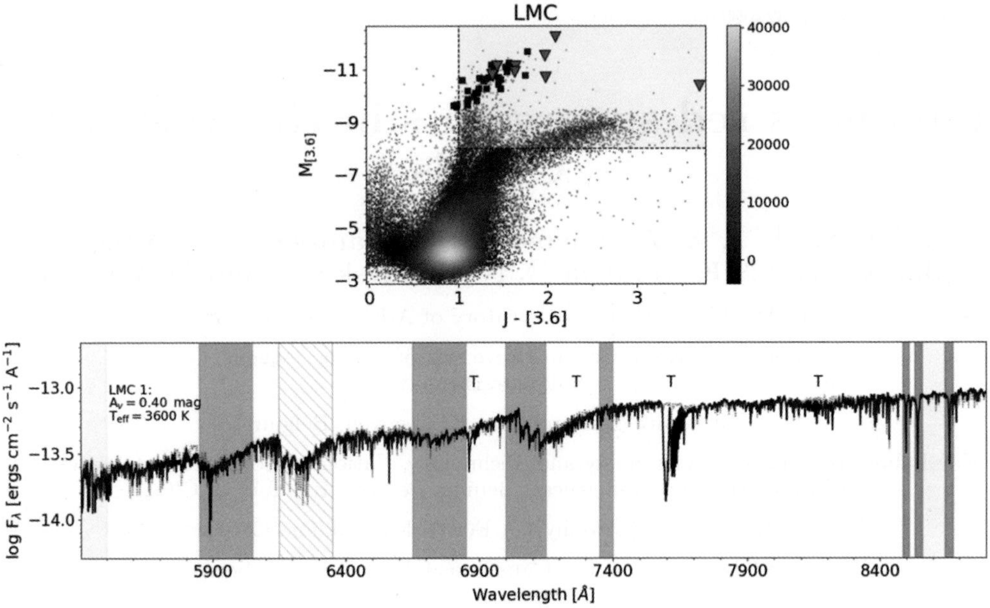

Figure 1. Top: Colour-magnitude diagram of targets in the LMC. The blue shade highlights the applied selection criteria. The red triangles indicate the observed red supergiants in the LMC, whereas the black boxes show red supergiants of similar studies. The SMC diagram looks similar but is not shown here. Bottom: Example of a best-fit spectrum. The model spectrum is shown in red and the observed spectrum is shown in black. Shaded regions were included in the analysis to calculate the effective temperature (blue), optical extinction (yellow) and surface gravity (green). The best-fit properties are indicated in the top left.

2. Main results

In de Wit et al. (submitted), we will present the full table containing the spectral types and derived surface properties of the observed RSGs. Two other evolved massive stars have been identified in this work. One is a new yellow supergiant (F8I), and we re-classified a known SgB[e] from B8I[e] to A0I[e]. In this work, we derived typical RSG effective temperatures in the range of 3500-3750 K, with extinction factors ranging from 0.3-1.0 mag, which is low for RSGs due to the contribution of circumstellar dust to the optical extinction. However, the reason for this has been reported frequently and attributed to the temperature structure used in the MARCS models, such that using the models to fit the optical TiO bands results in unreasonable extinction factors (Davies et al. 2013).

One of the stars studied in the LMC, [W60]B90, lies at the bright end of the RSG luminosity function (Davies et al. 2018) at $\log(L_*/L_\odot) \sim 5.5$ and has a stellar radius around 1400 $R_\odot$. These extreme properties put this object in the range of the largest star known in the LMC, WOH G64. WOH G64 is a well studied dust enshrouded RSG with a current estimate on the radius around 1550 $R_\odot$ (Levesque et al. 2009; Beasor & Smith 2022). A follow up to our study is to employ SED fitting techniques (i.e. DUSTY; Ivezic & Elitzur 1997) to derive the mass-loss rate of [W60]B90 and see whether this object is just as extreme. In a separate future work we aim to extend the analysis here, by modelling ~ 200 newly obtained FORS2 spectra of RSGs. These RSGs are located in various nearby galaxies targeted by the ERC ASSESS project on episodic mass loss in evolved massive stars (Bonanos et al. 2022). The aim is to model and study their surface and mass-loss properties and compare them as a function of metallicity.

To view supplementary material for this article, please visit http://dx.doi.org/10.1017/S1743921322003106.

References

Beasor, E. R., & Smith, N. 2022, ApJ, 933, 41
Bonanos, A. Z., et al. 2022
Davies, B., Crowther, P. A., & Beasor, E. R. 2018, MNRAS, 478, 3138
Davies, B., et al. 2013, ApJ, 767, 3
Gustafsson, B., Edvardsson, B., Eriksson, K., Jørgensen, U. G., Nordlund, Å., & Plez, B. 2008, A&A, 486, 951
Ivezic, Z., & Elitzur, M. 1997, MNRAS, 287, 799
Levesque, E. M., Massey, P., Plez, B., & Olsen, K. A. G. 2009, AJ, 137, 4744

Massive Stars Near and Far
Proceedings IAU Symposium No. 361, 2024
J. Mackey, J. S. Vink, & N. St-Louis, eds.
doi:10.1017/S1743921322003155

ASSESSing evolved massive stars in NGC 6822 and IC 10

G. Munoz-Sanchez[1,2], G. Maravelias[1,3], A. Z. Bonanos[1], F. Tramper[4], S. de Wit[1,2] and M. Yang[1,5]

[1]IASSARS, National Observatory of Athens, Athens, Greece

[2]National and Kapodistrian University of Athens, Athens, Greece
email: gonzalom@noa.gr

[3]Institute of Astrophysics, FORTH, Heraklion, Greece

[4]Institute of Astronomy, KU Leuven, Belgium

[5]Key Laboratory of Space Astronomy and Technology, National Astronomical Observatories, Chinese Academy of Sciences, Beijing, People's Republic of China

Abstract. The role of mass loss from massive stars, especially episodic mass loss, is one of the outstanding open questions facing stellar evolution theory. Multiple lines of evidence are pointing to violent, episodic mass-loss events being responsible for removing a large part of the massive stellar envelope, especially in low-metallicity galaxies. The ERC ASSESS project aims to determine whether episodic mass loss is a dominant process in the evolution of the most massive stars by conducting the first extensive, multi-wavelength survey of evolved massive stars in the nearby Universe. The project hinges on the fact that mass-losing stars form dust and are bright in the mid-infrared. We aim to investigate the properties of evolved targets in nearby galaxies and estimate the amount of ejected mass, which will constrain evolutionary models. In this work we present some of our first observational results from the galaxies NGC 6822 and IC 10 obtained with OSIRIS (GTC).

Keywords. stars: mass loss, stars: winds and outflows, circumstellar matter

---

1. Introduction

The episodic mAss loSS in Evolved maSsive Stars (ASSESS†) is one of the outstanding open questions facing stellar evolution theory. In this project we propose a study of nearby galaxies within a wide range of metallicities ($1/15 - 2\,Z_{\odot}$) to disentangle the effect of mass loss in massive stars by observing luminous-reddened objects.

We focus on reddened stars in the mid-infrared because it indicates the presence of dust in their surroundings. Specifically, we have used the 3.6μm and 4.5μm bands from Spitzer photometry to create a catalog of reddened stars in the galaxies selected. Given the amount of possible targets, we created a priority system of objects to observe (Table 1). VIS/NIR indicates whether photometry exists in the optical/near-infrared. The catalog was filtered with Gaia DR2 to discard contamination of foreground stars.

2. Observations

The observations of NGC 6822 and IC 10 took place on August 2020 with OSIRIS‡ (GTC) in the multi-slit mode using the grism 1000R, providing a spectral coverage between 5200-9200 Å and a 2.6 Å/px dispersion. We observed the galaxies with two and

† http://assess.astro.noa.gr/
‡ http://www.gtc.iac.es/instruments/osiris/

Table 1. Priority system of ASSESS. The number x/y means x objects detected in the observations out of y objects selected.

	Prio 1	**Prio 2**	**Prio 3**	**Prio 4**	**Prio 5**	**Prio 6**
$m_{3.6} - m_{4.5}$ (mag)	≥ 0.5	≥ 0.25	≥ 0.5	≥ 0.25	≥ 0.1	≥ 0.1
$M_{3.6}$ (mag)	≤ -9.75	≤ -9.75	≤ -9.75	≤ -9.75	≤ 9.0	≤ -9.0
VIS/NIR	Yes	Yes	No	No	Yes	No
NGC 6822	5/7	3/3	0/1	–	22/28	1/1
IC 10	1/1	2/2	1/2	–	11/14	7/15

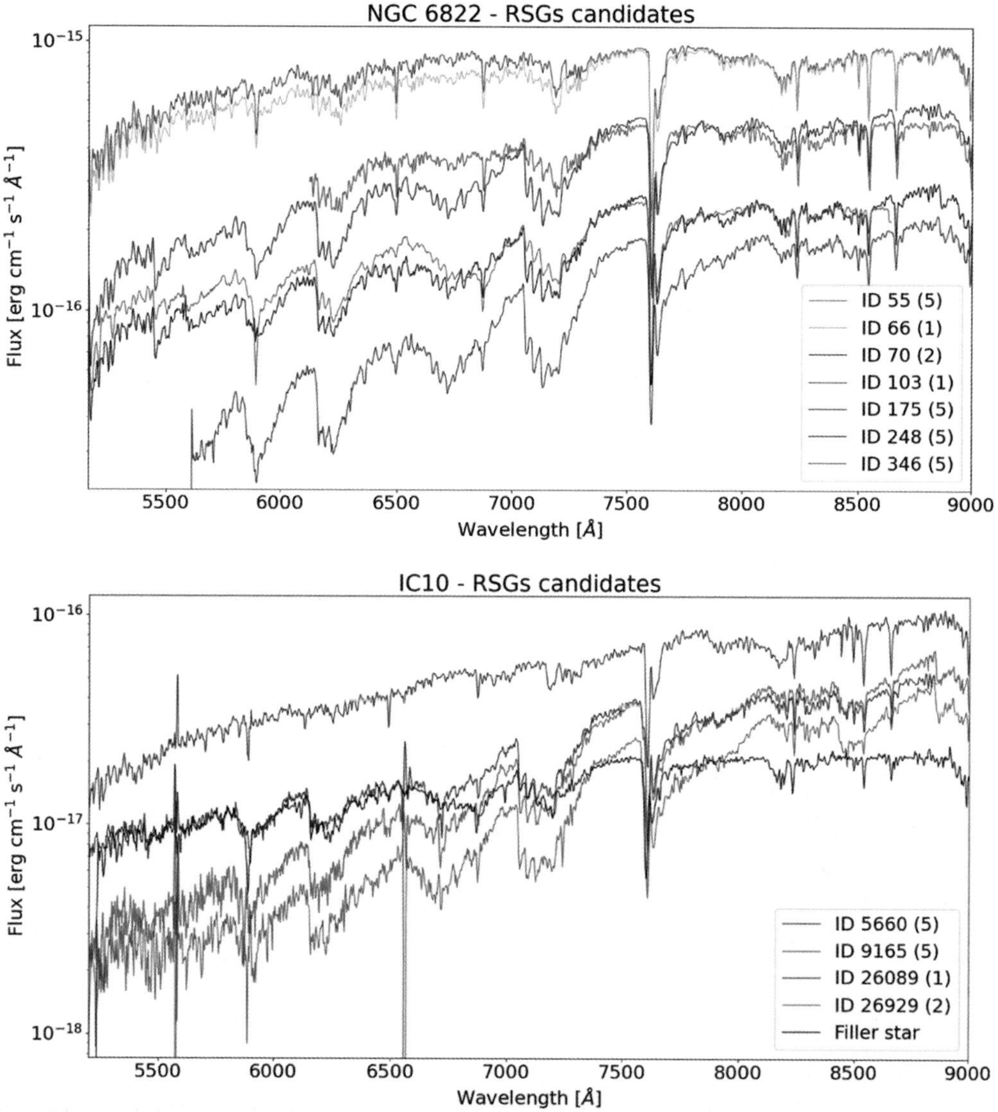

Figure 1. *Top*: RSG candidates in NGC 6822. *Bottom*: RSG candidates in IC 10. The number in parentheses corresponds to the priority.

one pointing respectively, for a total of 90 minutes (1350s x 4 exposures) per field. We used IRAF for the data reduction and we treated each slit individually guaranteeing the correct processing of the data. After a first spectral classification by visual inspection, we found different types of stars such as RSGs, two OB-stars within a HII region, several carbon stars and a WR star. As the RSG was the most common type in both galaxies, we decided to present the most luminous ones in this work (Fig. 1). The spectra show the CaII triplet at 8498 Å, 8542 Å and 8662 Å and most of them have the TiO molecular bands at 6150 Å and 7054 Å. We will use Gaia DR3 to confirm with parallaxes and proper motions if these stars belong to the host galaxy candidate or if they are foreground stars.

3. Future work

In this contribution we have presented some results for the galaxies NGC 6822 and IC 10. However, the campaign with OSIRIS (GTC) is not completed yet, as we are waiting for the galaxies M 81, NGC 2403, NGC 4214 and NGC 4736 to be observed (proposal submitted). Once the sample is complete, we will obtain the parameters of these stars by modelling them and we will evaluate their properties with evolutionary tracks and their SED fitting.

Supplementary material

To view supplementary material for this article, please visit http://dx.doi.org/10.1017/S1743921322003155.

Chapter VII:
Unsteady mass loss

Massive Stars Near and Far
Proceedings IAU Symposium No. 361, 2024
J. Mackey, J. S. Vink, & N. St-Louis, eds.
doi:10.1017/S1743921322001934

Luminous blue variables and B[e] supergiants

Andrea Mehner

ESO - European Organisation for Astronomical Research in the Southern Hemisphere, Alonso de Cordova 3107, Vitacura, Santiago de Chile, Chile
email: amehner@eso.org

Abstract. Luminous blue variables (LBVs) and B[e] supergiants (B[e]SGs) are some of the most massive stars that display extreme and puzzling behavior. Their rarity indicates that they belong to short evolutionary phases or short-lived phenomena in the post-main sequence evolution of massive stars. However, their strong mass loss and episodic mass eruptions may be crucially impacting massive star evolution. LBVs are a group of evolved massive stars that exhibit irregular variability and eruptive mass loss. Various subtypes, including S Doradus variables, giant eruptions, and pre-supernova outbursts, exist. The physical cause of the LBV phenomenon remains heavily debated. B[e]SGs have strong forbidden line emission and infrared excess from dust that are thought to arise in a circumstellar disk or torus. The formation mechanism of their disk-like structures is yet to be settled. The evolutionary phases of LBVs and B[e]SGs and their connection to other evolved massive stars are important unanswered questions in massive star evolution.

Keywords. stars: emission-line, Be, stars: evolution, (stars:) circumstellar matter, stars: winds, outflows, stars: variables: other

1. Introduction

The upper Hertzsprung-Russell (HR) diagram is populated by evolved, luminous, and variable massive stars. Many of them show evidence for strong stellar winds and periods of enhanced or eruptive mass loss. Prominent stellar groups are the luminous blue variables (LBVs) and the B[e] supergiants (B[e]SGs). Extensive observing campaigns and theoretical studies have been conducted over the past decades to investigate these massive evolved stars. LBV studies centered on identifying the instability mechanism, potential triggers of the instability, and the mass-loss drivers of their enhanced mass-loss episodes. B[e]SG studies focused on understanding the formation mechanism of their dusty circumstellar environments. The low-number statistics and the complexity of the physics involved, such as the impact of initial mass, metallicity, mass-loss history, rotation, and binarity, are major challenges to the understanding of their place in massive star evolution and eventual fate.

The term LBV was introduced to describe the diverse group of unstable evolved luminous hot stars in the upper HR diagram (Conti 1984). Because of the observed diversity of the LBV phenomenon and lack of understanding of the underlying physics, we distinguish three LBV subtypes in this review; classical LBVs, giant eruption LBVs, and LBV-like pre-supernova eruptions. The B[e] phenomenon describes B-type stars with strong Balmer emission, low excitation permitted emission, forbidden emission, and a strong near- or mid-infrared excess due to hot circumstellar material (Kraus 2019). Stars of different mass and different evolutionary stages can exhibit the B[e] phenomenon (Lamers et al. 1998). We only discuss the B[e]SGs, because we restrict this review to evolved massive stars.

1.1. *Classical and giant eruption LBVs*

Classical LBVs, often called S Doradus variables, experience "outbursts" and episodes of enhanced mass loss on timescales of years to decades. During an outburst, they transit in the HR diagram from a hot quiescent state ($T_{\rm eff} \sim 12\,000 - 30\,000$ K) to lower temperatures ($T_{\rm eff} \sim 7\,000 - 8\,000$ K), i.e., their spectral appearance changes from a hot supergiant to a much cooler A or F-type supergiant. Their visual magnitudes change by $1-2$ mag at almost constant bolometric luminosity. Classical LBVs are very rare with only about $40-50$ known (e.g., HR Car, AG Car, RMC 71; Richardson & Mehner 2018). They are generally divided into high-luminous LBVs ($log\ L/L_\odot > 5.8$), which evolve from the most massive stars with $M > 50\ M_\odot$, and less-luminous LBVs, which have initial masses of the order $M \sim 25 - 40\ M_\odot$ and which have presumably lost more than half of their initial mass during a red supergiant (RSG) phase.

Giant eruption LBVs, represented by the famous Galactic stars η Car and P Cygni, exhibit eruptions with visual magnitude changes of $2-3$ mag or more, during which their bolometric luminosity increases. Supernova (SN) surveys are finding increasingly such transients in other galaxies, which are referred to as "SN imposters," as they mimic the appearance of SN IIn.

The terminal velocities of LBV winds are in the range of $100 - 250$ km s^{-1}, significantly lower than those of normal OB supergiants. LBV mass-loss rates are up to $10^{-5} - 10^{-4}\ M_\odot$ yr^{-1} a factor of $10-100$ larger than those of normal supergiants. During quiescent phases, LBVs lose mass most likely via ordinary line-driving and the bistability mechanism is a good candidate to account for their transitional mass-loss variations (Vink 2021). The large mass-loss rates and giant eruptions observed in some LBVs suggest that also continuum-driven winds and/or explosions could play a role (Shaviv 1998; 2001, Owocki et al. 2004, Smith & Owocki 2006, Rest et al. 2012, Davidson & Humphreys 2012, Owocki & Shaviv 2016, González et al. 2022). As the different mechanisms can produce the same observables, the question remains whether the LBV giant eruptions are better understood in terms of a continuum-driven wind, an explosive model, or a combination of both.

Classical and giant eruption LBVs show evidence for extensive circumstellar matter. The morphology, kinematics, and chemical composition of these nebulae trace the star's mass-loss history and evolution. Typically, LBV nebulae have sizes of $0.2-5$ pc, ionized masses of $1-4\ M_\odot$, and expansion velocities of several tenths to a few hundreds km s^{-1} (Weis 2003). Bordiu et al. (2021) found that molecular gas may account for $> 30\%$ of the total mass lost around LBVs. Most LBV nebulae exhibit axisymmetric morphologies. At least 50% of the LBV nebulae are bipolar and several nebulae show in addition equatorial ring-like structures. To explain the observed nebular morphologies, a range of models have been proposed (see, e.g., Nota et al. 1995, Frank 1999, Frank et al. 1998, Owocki & Gayley 1997, Dwarkadas & Owocki 2002, Maeder & Desjacques 2001, Soker 2004).

Chemical abundances are consistent with material processed in the CNO cycle, and were the first proof that LBVs are post-main-sequence objects. Several abundance studies have attempted to infer the evolutionary phase in which the material was ejected (Smith et al. 1997; 1998, Lamers et al. 2001), but results range from potential ejections during a previous RSG phase, a yellow supergiant (YSG) phase, or a blue supergiant (BSG) phase. The dust grains in LBV nebulae are unusually large and have an unusual composition compared to grains in the interstellar medium and the asymptotic giant branch stars. They may trace the cool outbursts or giant eruptions (Kochanek 2011) or they could have formed in a prior RSG phase (Waters et al. 1997). Agliozzo et al. (2021)

found that the dust mass of individual LBVs does not correlate with their stellar parameters, which suggests that the dust production mechanism is independent of the initial stellar mass or that the stars have different evolutionary paths and mass-loss histories.

1.2. *LBV-like pre-SN eruptions*

Several SNe have displayed precursor eruptive events (Smith 2017). The responsible mechanism could be the same as those of the classical or giant eruption LBVs or may be triggered by core instabilities in the very final burning stages of massive stars. The most dramatic and well-studied case so far has been SN 2009ip. Similar transients have been observed, which are characterized by a long-lasting phase of erratic variability that ends with two luminous outbursts a few weeks apart. SN 2009ip descended from a very luminous progenitor consistent with a $50 - 80\ M_{\odot}$ star (Smith et al. 2010, Foley et al. 2011). On the other hand, the progenitor of AT 2016jbu, which has been classified as a SN 2009ip-like transient, was identified as a YHG of only $\sim 20\ M_{\odot}$ (Brennan et al. 2022). There is much debate about the nature of these transients and their final state. If they are true SNe and not impostors (Smith et al. 2022), the progenitors could be LBVs or YHGs. To match the ejecta mass and energy would require giant LBV eruptions. However, such events are very infrequent and many LBVs and YHGs have massive shells that are thousands of years old. It seems thus more likely that the pre-SN eruptions are related to the imminent core collapse.

There is increasing observational and theoretical evidence that LBVs could be direct progenitors of some high-luminosity type IIb and IIn SNe as signatures in the SN spectra indicate dense circumstellar material, which requires eruptive pre-SN mass-loss (Groh et al. 2013, Justham et al. 2014, Smith 2014, Groh 2017, Ustamujic et al. 2021). This poses a serious challenge for the traditional single-star evolutionary picture of LBVs as the transitional phase leading to Wolf-Rayet (WR) stars.

1.3. *B[e] supergiants*

LBVs and B[e]SGs are spectroscopically similar, but can be distinguished based on spectral lines and spectral energy distribution (Humphreys et al. 2017b, Agliozzo et al. 2021). B[e]SGs appear to originate from a lower initial mass population with $M \sim 15 - 40\ M_{\odot}$, compared to LBVs. They show two components in their circumstellar environment with a hot, fast (~ 1000 km s^{-1}) line-driven polar wind and a cool, slow (~ 100 km s^{-1}), dense equatorial wind (Zickgraf et al. 1985). The fast wind in the polar region is similar to the winds of normal BSGs. Strong near- and mid-infrared excess emission is indicative of hot circumstellar dust.

Imaging and interferometry have spatially resolved the environment of the closest and brightest Galactic objects, providing precise measurements of the disk inclinations and disk sizes (e.g., Domiciano de Souza et al. 2007; 2008, Wheelwright et al. 2012). The kinematics of the equatorial gaseous material favor Keplerian rotation (Kraus et al. 2007; 2010, Millour et al. 2011, Wheelwright et al. 2012), likely accumulated in multiple, partial rings and possible spiral arm-like structures and not a smooth disk (Aret et al. 2012, Kraus et al. 2016, Maravelias et al. 2018). Several B[e]SGs have been discovered to be in short-period binary systems and the dynamical and spatial information indicates that the prevalent geometry is a circumbinary ring or disk (Kraus et al. 2013, Porter et al. 2022). Large dust shell structures and nebulae are detected around some objects, hinting at major mass-loss events (Liimets et al. 2022).

2. LBV instability mechanisms

It is yet unclear what causes the transitions of LBVs across the HR diagram, the giant LBV eruptions, and the LBV-like pre-SN eruptions. Arguments in favor of instabilities in the outer layers use the facts that (1) LBVs are close to the Eddington limit (Sanyal et al. 2015), (2) the distinction between less- and high-luminous LBVs with potentially different evolutionary phases, and (3) the timescales, mass, and energy of ejecta of classical LBV outbursts. Arguments in favor of core instabilities are (1) the total mass and energy of giant LBV eruptions and (2) the timescales of LBV-like pre-SN eruptions. A single mechanism may be operating over a wide range of energy and mass, different mechanisms may produce the same observables, or several mechanisms may cooperate. Apart from this, rotation, binary interaction, and stellar mergers may also play a crucial role in shaping the LBV phenomenon.

Radiation and turbulent pressure instabilities. One of the first suggestions to explain the LBV phenomenon was that stars become unstable near the Humphreys-Davidson limit due to radiation pressure (Humphreys & Davidson 1984, Lamers 1986). As massive stars ($M > 40\ M_{\odot}$) evolve, their photospheres become increasingly unstable against radiation pressure and eventually reach a critical point where the radiation pressure and gravity are balanced, leading to mass loss, which ends their redward evolution. Turbulent pressure may ease the requirement for radiative instability (Stothers 2003). These two mechanisms may be working together with the opacity-modified Eddington limit (Lamers & Fitzpatrick 1988, Ulmer & Fitzpatrick 1998).

Ionization-induced dynamical instabilities. There is a large family of proposed ionization-induced instabilities related to the iron, helium, and hydrogen opacity peaks at different temperatures in the stellar layers. However, several are unable to explain the LBV outbursts. A geyser-like model (Maeder 1992) is not applicable for the highest temperature LBVs. Dynamical instabilities (Stothers & Chin 1993; 1994; 1997) and strange modes (Glatzel & Kiriakidis 1993, Kiriakidis et al. 1993) can likely only account for some of the observed photometric microvariations (Glatzel & Kiriakidis 1998, Lovekin & Guzik 2014).

Jiang et al. (2018) found with physically realistic three-dimensional global radiation hydrodynamic simulations that the physical mechanism responsible for the LBV phenomenon may be the interaction of the large radiative flux with the opacity peaks in the optically thick envelope as these stars expand and cool. Grassitelli et al. (2021) found unstable configurations in time-dependent hydrodynamic stellar evolution models due to feedback between the wind mass loss and the stellar envelope structure, which reproduce the observed LBV variability. Inflated envelopes (Gräfener et al. 2012) and the bistability mechanism (Vink et al. 1999, Petrov et al. 2016) are key ingredients in their models (cf., Björklund et al. 2022).

Pulsational pair instability (PPI). Woosley et al. (2007) noted the potential of PPI in the stellar core for powering pre-SN and giant LBV eruptions. PPI appears only in the final years to decades before core collapse. However, the ages of LBV nebulae is $10^2 - 10^4$ yr, indicating that the stars survived longer than expected if PPI was at play. In addition, the PPI is predicted to occur only for extremely massive stars at low metallicity. Thus, PPI is not a viable explanation for LBV giant eruptions, but it could be an option for the pre-SN eruptions.

Deep rooted energy release or addition. A large amount of extra energy can be deposited deep in a massive star's envelope by a number of mechanisms, including unsteady burning (Smith & Arnett 2014), explosive shell burning instabilities (Dessart et al. 2010, Smith & Arnett 2014), wave-driven mass loss (Meakin & Arnett 2007, Quataert & Shiode 2012, Shiode & Quataert 2014), and stellar collisions or mergers in a binary system (Podsiadlowski et al. 2010, Smith & Arnett 2014). For energy releases or additions that are on the order of the envelope binding energy, partial envelope ejection can result, reminiscent of LBV giant eruptions (Dessart et al. 2010, Owocki et al. 2019). In case of the even more energetic pre-SN LBV-like eruptions, late nuclear burning instabilities may be applicable.

Stellar rotation. Rotation modifies drastically the evolution of stars and some LBVs exhibit very rapid rotation, near their break-up rates (Groh et al. 2006; 2009). LBV outbursts might be associated with luminous stars encountering the critical rotation limit (Langer 1998), but it is not clear how reaching this limit could induce a sudden mass loss eruption. Nevertheless, rotation may play a key role as an instability trigger and for the resulting asymmetric mass-loss geometry. Recent studies found rotationally driven instabilities that could potentially trigger LBV outbursts, such as centrifugal instabilities (Zhao & Fuller 2020) and viscous instabilities (Shi & Fuller 2022).

Binarity. Most massive stars reside in multiple systems and the processes of mass loss and mass transfer in binary systems impact the evolution and fate (De Marco & Izzard 2017, Sana et al. 2012, de Mink et al. 2014; 2013). LBVs were long considered as single stars, but this picture is subject to discussion. Eta Car and HR Car are confirmed binary systems (e.g., Corcoran et al. 1997, Boffin et al. 2016) and Mahy et al. (2022) found that the spectroscopic and interferometric binary fraction of Galactic LBVs is similar to other massive stars, albeit with wide orbits.

Interacting binaries and stellar mergers offer many free parameters to explain the diversity in LBV eruptions. Proposed binary scenarios include common envelope evolutionary phases (Kenyon & Gallagher 1985, Gallagher 1989), violent periastron non-merger collisions (Smith 2011, Owocki et al. 2019, Quataert et al. 2016), mass accretion by the secondary star (Soker 2001; 2004; 2007, Kashi & Soker 2009), and tidal influence of the companion star near periastron (Smith et al. 2003, Koenigsberger 2004). The scenario of a stellar merger in a triple system has gained attention to explain the giant LBV eruptions (Iben 1999, Portegies Zwart & van den Heuvel 2016, Hirai et al. 2021).

Binary evolution for LBVs? The topic of a single versus binary evolutionary path for LBVs has been developing into a lively scientific discussion in the past years. The crucial question is whether binary interactions are the key to the LBV phenomenon, or whether they are only modifying or triggering a single-star instability. Smith & Tombleson (2015), Smith (2016), Aghakhanloo et al. (2017), Smith (2019), Aghakhanloo et al. (2022) claimed that LBVs reside outside young massive clusters with early O-type stars and inferred that LBVs may thus be primarily the product of binary evolution. In this scenario, LBVs are rejuvenated mass gainers that get enriched, spun up, and sometimes kicked far from their birth sites by a SN of their companion. Humphreys et al. (2016), Davidson et al. (2016), Aadland et al. (2018), Mehner et al. (2021) claimed the contrary, that indeed LBVs are associated with luminous young stars and supergiants appropriate for their luminosities and position in the HR diagram.

3. B[e]SG disk formation

The origin of the B[e] phenomenon is still uncertain. Generally, dust formation in the winds of single hot stars is inhibited by the low particle densities and the harsh ultraviolet environment. Thus, the formation of dust and its continuous presence close to the hot B[e]SGs are intriguing.

The formation of a slowly expanding, rotating disk (or ring) has been linked to near-critical rotation combined with bistable radiation-driven winds (Zickgraf et al. 1986, Lamers & Pauldrach 1991, Oudmaijer et al. 1998, Pelupessy et al. 2000, Curé et al. 2005, Kraus 2006, Madura et al. 2007, Kraus et al. 2010). Optical and infrared spectroscopic observations showed that the circumstellar material of B[e]SGs is concentrated in multiple, clumpy rings on stable Keplerian orbits and pulsations might thus play a role (Kraus 2017). Evidence between the occurrence of the B[e] phenomenon and binarity in B-type supergiants is growing (Podsiadlowski et al. 2006, Wang et al. 2012, Maravelias et al. 2018, Miroshnichenko et al. 2020) and circumbinary Keplerian rings are the most probable geometry for the equatorial outflows. Merger products, maybe in triple systems, are good candidates for B[e]SGs (Podsiadlowski et al. 2006, Pasquali et al. 2000, Wu et al. 2020). Several different scenarios may lead to the B[e] phenomenon, with binarity and mergers appearing to be preferable for several objects.

4. The role of LBVs and B[e]SGs in massive star evolution

Initial mass, metallicity, mass-loss, rotation, and binarity may all crucially affect the outcome of massive star evolution. In particular, the effects of mass loss complicate the model predictions (Chiosi & Maeder 1986, Langer 2012). Observations and theory find that mass-loss is too weak to expel the hydrogen-envelopes to form WR stars. As a result, an important open question is whether massive star evolution is facilitated mainly by mass loss through stellar winds or whether episodic mass loss is required, e.g., during an eruptive LBV phase.

Stars of different masses and different evolutionary stages can exhibit the B[e] phenomenon. In the case that LBV outbursts depend also only on the physical conditions within their stellar envelopes, the LBV phenomenon could also appear in different evolutionary phases. Thus, for a discussion on the evolutionary stages and the relationships of LBVs and B[e]SGs to other massive stars, it is critical to note the possibility that they are spectroscopic and not evolutionary phases.

LBVs. LBVs are generally assumed to be post-main-sequence stars. Their hydrogen surface abundances and luminosities indicate properties between O-type and WN stars. The LBV phenomenon appears to be intermittent and part of the population might be dormant (e.g., the Ofpe/WN stars). In the traditional view, LBVs have been interpreted as a transitional phase of evolution when a massive star moves from core-hydrogen burning to core-helium burning (Conti 1978, Chiosi & Maeder 1986, Humphreys & Davidson 1994). The upper luminosity boundary to the HR diagram provides a valuable clue and constraint on the understanding of these stars.

Above an initial mass of $40 - 50\ M_{\odot}$, stars do not evolve across the HR diagram to become RSGs. Groh et al. (2014) found that LBVs arise naturally in their single-star evolution models at the end of the main sequence when the mass-loss rate increases as a consequence of crossing the bistability limit. Lower mass supergiants evolve across the HR diagram through BSG and YSG phases, before reaching the RSG phase. High mass loss during the RSG stage causes the star to evolve again bluewards, through a second YSG phase, and possibly becoming a post-RSG warm hypergiant, an LBV, or WR star (Maeder 1982, Lamers et al. 1983, Chiosi & Maeder 1986, Stothers & Chin 1996, Groh et al. 2013).

B[e]SGs. Most of the B[e]SGs lie below the LBV instability strip in the HR diagram, which suggests that the phenomena are not related in an evolutionary sense and that B[e]SGs may be more common at lower stellar masses. Several studies support a post-RSG state (Kastner et al. 2010, Aret et al. 2012, Humphreys et al. 2017a), while others concluded that the C12/C13 ratios support a post-main-sequence state in a likely pre-RSG stage (Oksala et al. 2013). This analysis depends strongly on stellar evolution models and on the rotational speed of the stars.

In summary, LBVs and B[e]SGs are likely post-main-sequence objects. There are sufficient differences in their spectra, their luminosities, spatial distribution, and the presence of circumstellar dust, to conclude that they are not related in an evolutionary sense. However, overlap between these two classifications may occur since the B[e]SGs are a spectroscopic phase (and the LBVs could be as well). The less-luminous LBVs are post-RSGs, maybe post-warm hypergiants. There may be more than one path to become a B[e]SG. Some B[e]SGs may be pre-RSGs and some may be post-RSGs. Hot supergiants, warm and cool hypergiants, B[e]SGs, and Of/late-WN stars may be evolutionary neighbors or even quiescent LBVs as they are neighboring in the HR diagram and show all evidence for high mass loss and instabilities.

5. Outlook

The diversity and rarity of massive stars and their descendants challenge our understanding of their evolution and fate. Despite considerable advances in the past decades, many questions remain. On a positive outlook, due to their brightness and extended nebulosities, massive stars are excellent targets for almost any observing technique and wavelength range, enabling a multi-wavelength approach that provides different diagnostics and consistency checks. Upcoming large survey missions (e.g., Euclid, Roman, Rubin) will provide an improved census of evolved massive stars, massive star transients, and SN precursors. High-multiplex spectroscopic surveys (e.g., MOONS and 4MOST at the VLT, WEAVE at the WHT) will allow classification and follow-up of the candidates identified in those surveys. Increasingly sophisticated computational methods and connecting observations to theory will be a major task, but with statistically meaningful observational samples it may be possible to unveil the nature and fate of these fascinating objects.

References

Aadland, E., Massey, P., Neugent, K. F., & Drout, M. R. 2018, AJ, 156, 294
Aghakhanloo, M., Murphy, J. W., Smith, N., & Hložek, R. 2017, MNRAS, 472, 591
Aghakhanloo, M., Smith, N., Andrews, J., et al. 2022, arXiv e-prints, arXiv:2202.06887
Agliozzo, C., Phillips, N., Mehner, A., et al. 2021, A&A, 655, A98
Aret, A., Kraus, M., Muratore, M. F., & Borges Fernandes, M. 2012, MNRAS, 423, 284
Björklund, R., Sundqvist, J. O., Singh, S. M., Puls, J., & Najarro, F. 2022, arXiv e-prints, arXiv:2203.08218
Boffin, H. M. J., Rivinius, T., Mérand, A., et al. 2016, A&A, 593, A90
Bordiu, C., Bufano, F., Cerrigone, L., et al. 2021, MNRAS, 500, 5500
Brennan, S. J., Fraser, M., Johansson, J., et al. 2022, MNRAS, 513, 5666
Chiosi, C., & Maeder, A. 1986, ARA&A, 24, 329
Conti, P. S. 1978, ARA&A, 16, 371
Conti, P. S. 1984, in Observational Tests of the Stellar Evolution Theory, ed. A. Maeder & A. Renzini, Vol. 105, 233
Corcoran, M. F., Ishibashi, K., Swank, J. H., et al. 1997, Nature, 390, 587
Curé, M., Rial, D. F., & Cidale, L. 2005, A&A, 437, 929
Davidson, K., & Humphreys, R. M. 2012, Nature, 486, E1
Davidson, K., Humphreys, R. M., & Weis, K. 2016, arXiv e-prints, arXiv:1608.02007
De Marco, O., & Izzard, R. G. 2017, PASA, 34, e001
de Mink, S. E., Langer, N., Izzard, R. G., Sana, H., & de Koter, A. 2013, ApJ, 764, 166
de Mink, S. E., Sana, H., Langer, N., Izzard, R. G., & Schneider, F. R. N. 2014, ApJ, 782, 7
Dessart, L., Livne, E., & Waldman, R. 2010, MNRAS, 405, 2113
Domiciano de Souza, A., Kervella, P., Bendjoya, P., & Niccolini, G. 2008, A&A, 480, L29
Domiciano de Souza, A., Driebe, T., Chesneau, O., et al. 2007, A&A, 464, 81
Dwarkadas, V. V., & Owocki, S. P. 2002, ApJ, 581, 1337
Foley, R. J., Berger, E., Fox, O., et al. 2011, ApJ, 732, 32
Frank, A. 1999, New A Rev, 43, 31

Frank, A., Ryu, D., & Davidson, K. 1998, ApJ, 500, 291
Gallagher, J. S. 1989, ASSL, Vol. 157, IAU Colloq. 113: Physics of Luminous Blue Variables, ed. K. Davidson, A. F. J. Moffat, & H. J. G. L. M. Lamers, 185
Glatzel, W., & Kiriakidis, M. 1993, MNRAS, 263, 375
Glatzel, W., & Kiriakidis, M. 1998, MNRAS, 295, 251
González, R. F., Zapata, L. A., Raga, A. C., et al. 2022, A&A, 659, A168
Gräfener, G., Owocki, S. P., & Vink, J. S. 2012, A&A, 538, A40
Grassitelli, L., Langer, N., Mackey, J., et al. 2021, A&A, 647, A99
Groh, J. H. 2017, Philos. Trans. R. Soc. A, 375, 20170219
Groh, J. H., Hillier, D. J., & Damineli, A. 2006, ApJL, 638, L33
Groh, J. H., Meynet, G., & Ekström, S. 2013, A&A, 550, L7
Groh, J. H., Meynet, G., Ekström, S., & Georgy, C. 2014, A&A, 564, A30
Groh, J. H., Damineli, A., Hillier, D. J., et al. 2009, ApJL, 705, L25
Hirai, R., Podsiadlowski, P., Owocki, S. P., Schneider, F. R. N., & Smith, N. 2021, MNRAS, 503, 4276
Humphreys, R. M., & Davidson, K. 1984, Science, 223, 243
Humphreys, R. M., & Davidson, K. 1994, PASP, 106, 1025
Humphreys, R. M., Davidson, K., Hahn, D., Martin, J. C., & Weis, K. 2017a, ApJ, 844, 40
Humphreys, R. M., Gordon, M. S., Martin, J. C., Weis, K., & Hahn, D. 2017b, ApJ, 836, 64
Humphreys, R. M., Weis, K., Davidson, K., & Gordon, M. S. 2016, ApJ, 825, 64
Iben, Jr., I. 1999, in ASP Conf. Ser., Vol. 179, Eta Carinae at The Millennium, ed. J. A. Morse, R. M. Humphreys, & A. Damineli, 367
Jiang, Y.-F., Cantiello, M., Bildsten, L., et al. 2018, Nature, 561, 498
Justham, S., Podsiadlowski, P., & Vink, J. S. 2014, ApJ, 796, 121
Kashi, A., & Soker, N. 2009, NewA, 14, 11
Kastner, J. H., Buchanan, C., Sahai, R., Forrest, W. J., & Sargent, B. A. 2010, AJ, 139, 1993
Kenyon, S. J., & Gallagher, J. S., I. 1985, ApJ, 290, 542
Kiriakidis, M., Fricke, K. J., & Glatzel, W. 1993, MNRAS, 264, 50
Kochanek, C. S. 2011, ApJ, 743, 73
Koenigsberger, G. 2004, RMxAA, 40, 107
Kraus, M. 2006, A&A, 456, 151
Kraus, M. 2017, ASPCS, Vol. 508, The B[e] Phenomenon: Forty Years of Studies, ed. A. Miroshnichenko, S. Zharikov, D. Korčáková, & M. Wolf, 219
Kraus, M. 2019, Galaxies, 7, 83
Kraus, M., Borges Fernandes, M., & de Araújo, F. X. 2007, A&A, 463, 627
Kraus, M., Borges Fernandes, M., & de Araújo, F. X. 2010, A&A, 517, A30
Kraus, M., Oksala, M. E., Nickeler, D. H., et al. 2013, A&A, 549, A28
Kraus, M., Cidale, L. S., Arias, M. L., et al. 2016, A&A, 593, A112
Lamers, H. J. G., & Pauldrach, A. W. A. 1991, A&A, 244, L5
Lamers, H. J. G. L. M. 1986, A&A, 159, 90
Lamers, H. J. G. L. M., de Groot, M., & Cassatella, A. 1983, A&A, 123, L8
Lamers, H. J. G. L. M., & Fitzpatrick, E. L. 1988, ApJ, 324, 279
Lamers, H. J. G. L. M., Nota, A., Panagia, N., Smith, L. J., & Langer, N. 2001, ApJ, 551, 764
Lamers, H. J. G. L. M., Zickgraf, F.-J., de Winter, D., Houziaux, L., & Zorec, J. 1998, A&A, 340, 117
Langer, N. 1998, A&A, 329, 551
Langer, N. 2012, ARA&A, 50, 107
Liimets, T., Kraus, M., Moiseev, A., et al. 2022, Galaxies, 10, 41
Lovekin, C. C., & Guzik, J. A. 2014, MNRAS, 445, 1766
Madura, T. I., Owocki, S. P., & Feldmeier, A. 2007, ApJ, 660, 687
Maeder, A. 1982, A&A, 105, 149
Maeder, A. 1992, in Instabilities in Evolved Super- and Hypergiants, ed. C. de Jager & H. Nieuwenhuijzen, 138
Maeder, A., & Desjacques, V. 2001, A&A, 372, L9

Mahy, L., Lanthermann, C., Hutsemékers, D., et al. 2022, A&A, 657, A4
Maravelias, G., Kraus, M., Cidale, L. S., et al. 2018, MNRAS, 480, 320
Meakin, C. A., & Arnett, D. 2007, ApJ, 667, 448
Mehner, A., Janssens, S., Agliozzo, C., et al. 2021, A&A, 655, A33
Millour, F., Meilland, A., Chesneau, O., et al. 2011, A&A, 526, A107
Miroshnichenko, A. S., Zharikov, S. V., Korčaková, D., et al. 2020, Contributions of the Astronomical Observatory Skalnate Pleso, 50, 513
Nota, A., Livio, M., Clampin, M., & Schulte-Ladbeck, R. 1995, ApJ, 448, 788
Oksala, M. E., Kraus, M., Cidale, L. S., Muratore, M. F., & Borges Fernandes, M. 2013, A&A, 558, A17
Oudmaijer, R. D., Proga, D., Drew, J. E., & de Winter, D. 1998, MNRAS, 300, 170
Owocki, S. P., & Gayley, K. G. 1997, ASPCS, Vol. 120, Luminous Blue Variables: Massive Stars in Transition, ed. A. Nota & H. Lamers, 121
Owocki, S. P., Gayley, K. G., & Shaviv, N. J. 2004, ApJ, 616, 525
Owocki, S. P., Hirai, R., Podsiadlowski, P., & Schneider, F. R. N. 2019, MNRAS, 485, 988
Owocki, S. P., & Shaviv, N. J. 2016, MNRAS, 462, 345
Pasquali, A., Nota, A., Langer, N., Schulte-Ladbeck, R. E., & Clampin, M. 2000, AJ, 119, 1352
Pelupessy, I., Lamers, H. J. G. L. M., & Vink, J. S. 2000, A&A, 359, 695
Petrov, B., Vink, J. S., & Gräfener, G. 2016, MNRAS, 458, 1999
Podsiadlowski, P., Ivanova, N., Justham, S., & Rappaport, S. 2010, MNRAS, 406, 840
Podsiadlowski, P., Morris, T. S., & Ivanova, N. 2006, ASPCS, Vol. 355, Stars with the B[e] Phenomenon, ed. M. Kraus & A. S. Miroshnichenko, 259
Portegies Zwart, S. F., & van den Heuvel, E. P. J. 2016, MNRAS, 456, 3401
Porter, A., Blundell, K., & Lee, S. 2022, MNRAS, 509, 1720
Quataert, E., Fernández, R., Kasen, D., Klion, H., & Paxton, B. 2016, MNRAS, 458, 1214
Quataert, E., & Shiode, J. 2012, MNRAS, 423, L92
Rest, A., Prieto, J. L., Walborn, N. R., et al. 2012, Nature, 482, 375
Richardson, N. D., & Mehner, A. 2018, RNAAS, 2, 121
Sana, H., de Mink, S. E., de Koter, A., et al. 2012, Science, 337, 444
Sanyal, D., Grassitelli, L., Langer, N., & Bestenlehner, J. M. 2015, A&A, 580, A20
Shaviv, N. J. 1998, ApJL, 494, L193
Shaviv, N. J. 2001, MNRAS, 326, 126
Shi, Y., & Fuller, J. 2022, MNRAS, 513, 1115
Shiode, J. H., & Quataert, E. 2014, ApJ, 780, 96
Smith, L. J., Nota, A., Pasquali, A., et al. 1998, ApJ, 503, 278
Smith, L. J., Stroud, M. P., Esteban, C., & Vilchez, J. M. 1997, MNRAS, 290, 265
Smith, N. 2011, MNRAS, 415, 2020
Smith, N. 2014, ARA&A, 52, 487
Smith, N. 2016, MNRAS, 461, 3353
Smith, N. 2017, Philos. Trans. R. Soc. A, 375, 20160268
Smith, N. 2019, MNRAS, 489, 4378
Smith, N., Andrews, J. E., Filippenko, A. V., et al. 2022, arXiv e-prints, arXiv:2205.02896
Smith, N., & Arnett, W. D. 2014, ApJ, 785, 82
Smith, N., Davidson, K., Gull, T. R., Ishibashi, K., & Hillier, D. J. 2003, ApJ, 586, 432
Smith, N., & Owocki, S. P. 2006, ApJL, 645, L45
Smith, N., & Tombleson, R. 2015, MNRAS, 447, 598
Smith, N., Miller, A., Li, W., et al. 2010, AJ, 139, 1451
Soker, N. 2001, MNRAS, 325, 584
Soker, N. 2004, ApJ, 612, 1060
Soker, N. 2007, ApJ, 661, 482
Stothers, R. B. 2003, ApJ, 589, 960
Stothers, R. B., & Chin, C.-W. 1993, ApJL, 408, L85
Stothers, R. B., & Chin, C.-W. 1994, ApJL, 426, L43
Stothers, R. B., & Chin, C.-W. 1996, ApJ, 468, 842

Stothers, R. B., & Chin, C.-W. 1997, ApJ, 489, 319
Ulmer, A., & Fitzpatrick, E. L. 1998, ApJ, 504, 200
Ustamujic, S., Orlando, S., Miceli, M., et al. 2021, A&A, 654, A167
Vink, J. S. 2021, arXiv e-prints, arXiv:2109.08164
Vink, J. S., de Koter, A., & Lamers, H. J. G. L. M. 1999, A&A, 350, 181
Wang, Y., Weigelt, G., Kreplin, A., et al. 2012, A&A, 545, L10
Waters, L. B. F. M., Morris, P. W., Voors, R. H. M., & Lamers, H. J. G. L. M. 1997, ASPCS, Vol. 120, Luminous Blue Variables: Massive Stars in Transition, ed. A. Nota & H. Lamers, 326
Weis, K. 2003, A&A, 408, 205
Wheelwright, H. E., de Wit, W. J., Weigelt, G., Oudmaijer, R. D., & Ilee, J. D. 2012, A&A, 543, A77
Woosley, S. E., Blinnikov, S., & Heger, A. 2007, Nature, 450, 390
Wu, S., Everson, R. W., Schneider, F. R. N., Podsiadlowski, P., & Ramirez-Ruiz, E. 2020, ApJ, 901, 44
Zhao, X., & Fuller, J. 2020, MNRAS, 495, 249
Zickgraf, F. J., Wolf, B., Stahl, O., Leitherer, C., & Appenzeller, I. 1986, A&A, 163, 119
Zickgraf, F. J., Wolf, B., Stahl, O., Leitherer, C., & Klare, G. 1985, A&A, 143, 421

Massive Stars Near and Far
Proceedings IAU Symposium No. 361, 2024
J. Mackey, J. S. Vink, & N. St-Louis, eds.
doi:10.1017/S1743921322002782

Introducing the ASSESS project: Episodic Mass Loss in Evolved Massive Stars - Key to Understanding the Explosive Early Universe

A. Z. Bonanos[1], G. Maravelias[1,3], M. Yang[1,5], F. Tramper[4], S. de Wit[1,2], M. Zapartas[1], K. Antoniadis[1,2], E. Christodoulou[1,2] and G. Munoz-Sanchez[1,2]

[1]IASSARS, National Observatory of Athens, Athens, Greece
email: bonanos@astro.noa.gr

[2]National and Kapodistrian University of Athens, Athens, Greece

[3]Institute of Astrophysics, FORTH, Heraklion, Greece

[4]Institute of Astronomy, KU Leuven, Belgium

[5]Key Laboratory of Space Astronomy and Technology, National Astronomical Observatories, Chinese Academy of Sciences, Beijing, People's Republic of China

Abstract. Episodic mass loss is not understood theoretically, neither accounted for in state-of-the-art models of stellar evolution, which has far-reaching consequences for many areas of astronomy. We introduce the ERC-funded ASSESS project (2018-2024), which aims to determine whether episodic mass loss is a dominant process in the evolution of the most massive stars, by conducting the first extensive, multi-wavelength survey of evolved massive stars in the nearby Universe. It hinges on the fact that mass-losing stars form dust and are bright in the mid-infrared. We aim to derive physical parameters of ~1000 dusty, evolved massive stars in ~25 nearby galaxies and estimate the amount of ejected mass, which will constrain evolutionary models, and quantify the duration and frequency of episodic mass loss as a function of metallicity. The approach involves applying machine-learning algorithms to select dusty, luminous targets from existing multi-band photometry of nearby galaxies. We present the first results of the project, including the machine-learning methodology for target selection and results from our spectroscopic observations so far. The emerging trend for the ubiquity of episodic mass loss, if confirmed, will be key to understanding the explosive early Universe and will have profound consequences for low-metallicity stars, reionization, and the chemical evolution of galaxies.

Keywords. Stars: massive – Stars: mass loss – Stars: evolution – Stars: fundamental parameters – Supergiants – Surveys

1. Introduction

The role of mass loss from massive stars, especially episodic mass loss in evolved massive stars, is one of the outstanding open questions facing stellar evolution theory (Smith 2014). While the upper limit to the masses of stars is thought to be 150 $M_\odot$(Figer 2005, Oey & Clarke 2005), and was even claimed to exceed 300 $M_\odot$(Crowther et al. 2010, Banerjee et al. 2012), the masses of hydrogen-deficient Wolf-Rayet (WR) stars do not exceed 20 $M_\odot$(Crowther 2007). Classical line-driven wind theory (Kudritzki & Puls 2000), once thought to be responsible for removing the envelopes of massive stars, has been shown inadequate, both on theoretical grounds (due to clumping, Owocki & Puls

1999) and estimations based on spectral lines (Bouret et al. 2005, Fullerton et al. 2006, Cohen et al. 2014), which demand reductions in the mass-loss rates by a factor of ~2-3. So how do massive stars shed their envelopes? Binary interactions via Roche-Lobe overflow (RLOF) are predicted to occur in 70% of massive stars and strip the envelopes in ~ 30% of O stars, given the high binarity fraction (~ 70%) of massive stars (Sana et al. 2012). Episodic mass loss is possibly the dominant process that operates in single stars, however, the physical mechanism responsible remains a mystery(Smith 2014).

The importance of episodic mass loss has come to the forefront in both the massive star and supernova (SN) communities. *Spitzer* images have revealed numerous circumstellar shells surrounding massive, evolved stars in our Galaxy (Gvaramadze et al. 2010, Wachter et al. 2010). Episodes of enhanced mass loss have been recorded not only in luminous blue variables (LBVs), but also in extreme red supergiants (RSGs, e.g. VY CMa, Decin et al. 2006). Moreover, untargeted supernova surveys have found dusty circumstellar material around superluminous supernovae (SLSN, Gal-Yam 2012), and mysterious optical transients with luminosities intermediate between novae and supernovae. The presence of circumstellar material implies a central role of episodic mass loss in the evolution of massive stars and this proposal aims to confirm this hypothesis. Tantalizing evidence suggests that SLSN occur in low-metallicity host galaxies (Neill et al. 2011), implying that such supernovae dominated the metal-poor early Universe. The overluminous Type IIn SN 2010jl is a well-studied example of a SLSN, with a massive progenitor star (30 $M_\odot$) surrounded by a dense circumstellar shell (Smith & Frew 2011, Zhang et al. 2012), which exploded in a low-metallicity galaxy (Stoll et al. 2011). SN2008S, a well-studied example of the class of intermediate-luminosity optical transients, was found to have a dust-enshrouded progenitor (8–10 $M_\odot$, Prieto et al. 2008) in pre-explosion *Spitzer* images of the host galaxy NGC 300. Finally, the remarkable SN2009ip involves a 50–80 $M_\odot$ progenitor that underwent a series of episodic mass loss events. Its spectacular finale included a series of eruptions in 2009 and 2010 until its final explosion in 2012 as a Type IIn supernova (Mauerhan et al. 2013), although this was contested (Pastorello et al. 2013, Fraser et al. 2013). These examples strongly suggest that episodic mass loss in massive stars is central to their evolution and therefore has profound consequences for the enrichment of the interstellar medium and the chemical evolution of the early Universe.

The physics of LBV eruptions, pre-SN eruptions and extreme RSG mass-loss is still in its infancy and, as stated in the review by Smith (2014), "is a major unsolved problem in astrophysics". Models of single-star evolution adopt empirical, constant mass-loss prescriptions, which highly influence the outcome (Meynet et al. 2015). The ASSESS project tackles the role of episodic mass loss in massive stars by using the fact that mass-losing stars form dust and are bright in the mid-infrared (mid-IR). Physically, there are a number of ways a massive star can become a source of significant mid-IR emission. First, dust can form in a dense, but relatively steady stellar wind. In the most extreme cases, such as in the progenitors of the SN 2008S and the NGC300-OT 2008 transient (Bond et al. 2009), the wind is optically thick even in the near-IR and the source star is seen only in the mid-IR (Prieto et al. 2008). Second, a very massive star can have an impulsive mass ejection or eruption with dust forming in the ejected shell of material. Initially the optical depth and dust temperatures are high, but then drop as the shell expands. The most famous example is the "great eruption" of η Carinae in the 19th century (Humphreys & Davidson 1994, Davidson & Humphreys 1997, Smith & Frew 2011), which ejected several solar masses of material. Third, the dust can be located in a circumstellar disk and emit over a broad range of temperatures, as is seen in supergiant B[e] stars (sgB[e]) stars (Zickgraf 2006).

While stars with significant mid-IR emission are intrinsically rare, many of the most interesting massive "superstars", such as η Car or "Object X" in M33 (Khan et al. 2011), belong to this class. It is clear that searching for analogs of these interesting stars using mid-IR photometry of nearby galaxies is the way to go. The existing mid-IR "roadmaps" for interpreting luminous massive stars (Bonanos et al. 2009; 2010) are based on known massive stars in the LMC and the SMC. They have identified LBVs, sgB[e], and RSGs among the brightest mid-IR sources, due to their intrinsic brightness and due to being surrounded by their own dust. What is new about ASSESS is the idea of conducting - for the first time - a systematic study of mass loss in massive stars, by selecting targets using mid-IR photometry of nearby galaxies obtained with *Spitzer*.

2. Methodology

We have collected recently published mid-IR photometric catalogs from *Spitzer* of galaxies with high star-formation rates within 5 Mpc: (a) seven dwarf galaxies within 1.5 Mpc from the DUSTiNGS project (Boyer et al. 2015): IC 10, IC 1613, Phoenix, Pegasus, Sextans A, Sextans B, and WLM, (b) 13 galaxies within 5 Mpc (Khan et al. 2015, Khan 2017): M31, M33, NGC 247, NGC 300, NGC 1313, NGC 2403, M81, M83, NGC 3077, NGC 4736, NGC 4826, NGC 6822, and NGC 7793, and (c) five galaxies within 4 Mpc (Williams & Bonanos 2016): NGC 55, NGC 253, NGC 2366, NGC 4214, and NGC 5253. The mid-IR photometry made available by the SAGE surveys of the LMC (Meixner et al. 2006) and SMC (Gordon et al. 2011) has been also searched for undetected, dust-obscured targets in our nearest neighbor galaxies. These catalogs contain mid-IR photometry of over 5 million point sources in 27 nearby galaxies, 19 of which have Pan-STARRS1 coverage (Chambers et al. 2016), providing an ideal dataset for a systematic study of luminous, dusty, evolved massive stars. We have compiled mid-IR photometric catalogs for these galaxies, including their counterparts in Pan-STARRS1 ($g, r, i, z, y-$bands), 2MASS (Cutri et al. 2003), VISTA Science Archive, WISE (Cutri & et al. 2012) and other archival surveys of particular galaxies to construct their spectral energy distributions (SEDs) out to 24 μm. The single epoch 5σ depth of Pan-STARRS1 ranges from 22nd magnitude in $g-$band to 20th magnitude in $y-$band, corresponding to absolute magnitudes brighter than -6 in g and -8 in y at 3.5 Mpc, respectively, which include the most luminous, evolved targets.

Based on these catalogs, we have selected over 1000 luminous and red sources (selected by their colors in $[3.6] - [4.5]$) in these 27 galaxies and are conducting follow-up low-resolution spectroscopy of these sources, mainly with FORS2 on VLT and OSIRIS on GTC. The spectra yield stellar types, luminosity classes, effective temperatures and an estimate of the reddening. High-resolution spectra will be obtained for particularly interesting targets for further analysis.

SED modeling with DUSTY (Ivezic & Elitzur 1997) will provide radii and age estimates of the circumstellar shell, as well as the dust temperature, ejected mass, and bolometric luminosity. SED shapes will be quantified to estimate the timescales of episodic mass loss and lifetimes of the various evolved stages as a function of spectral type and metallicity. Evidence of binarity (from spectra, SEDs, light curves) will provide an estimation of the relative contribution of RLOF to the observed dusty evolved stages of massive stars. Armed with all these parameters for a sample of $\sim$ 1000 dusty, evolved stars, spanning a range of metallicity ($\sim$1/15 $-$ 2 $Z_{\odot}$), we will perform a comparison with state-of-the-art stellar evolutionary models (Brott et al. 2011, Ekström et al. 2012, Georgy et al. 2013, Meynet et al. 2015) to evaluate the input mass-loss rates and predicted outcomes. We plan to reverse-engineer the target stars to quantify and confirm the amount of "input" episodic mass loss needed to match the measurements.

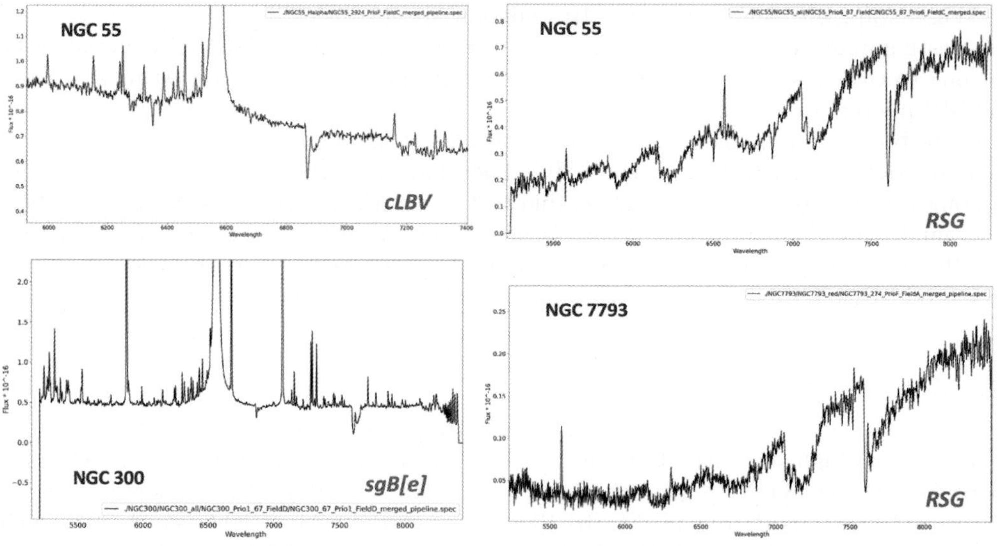

Figure 1. Newly identified spectra of evolved massive stars in our FORS2 data from the VLT, including a candidate LBV and RSG in NGC 55 (top row), a sgB[e] in NGC 300 and a RSG in NGC 7793 (bottom row).

3. Results

3.1. *Photometric Classifier*

We have employed state-of-the-art machine-learning algorithms to automatically classify and select types of mass-losing stars, thereby accelerating and systematizing the investigation of multi-wavelength photometry. We developed a classifier for evolved massive stars based on known massive stars in M31 and M33 and using color indices as features to classify evolved massive stars into the following categories: blue, yellow, red supergiants, LBVs, classical Wolf-Rayet stars, sgB[e]. We also included a class for outliers (e.g. background galaxies, AGNs). The classifier is found to be on average 83% accurate (Maravelias et al. 2022a). We are currently applying this classifier to classify over one million sources in 25 nearby galaxies (see Maravelias et al. 2022b; this volume).

3.2. *Observational survey*

We have prioritized our targets based on their luminosity and IR excess, specifically, giving highest priority to targets with $m_{3.6} - m_{4.5} \geq 0.5$ mag and $M_{3.6} \leq -9.75$ mag. We have obtained multi-object spectroscopy with both the VLT and GTC starting in 2020, giving priority to the galaxies that had enough high-priority targets to justify multi-object spectroscopy. We used the FORS2 spectrograph (Tramper et al., in prep.) and obtained spectra of over 400 high-priority and over 500 "filler" stars in M83, NGC 55, NGC 247, NGC 253, NGC 300, NGC 7793, Sextans A and WLM over 43h. The spectra have a resolving power of $R = 1000$ and a wavelength coverage around $5400 - 8200$ Å, which is suitable for classification and parameter estimation. Figure 1 shows examples of dusty, evolved massive stars identified in NGC 55, NGC 300 and NGC 7793.

We also used the GTC OSIRIS spectrograph (for details see Munoz-Sanchez et al. 2022; this volume) and have so far obtained 48 high-priority stars in NGC 6822 and 33 in IC 10. The GTC spectra have a resolving power of $R \sim 500 - 700$ and a wavelength coverage around $5200 - 9200$ Å, which are being used to classify the sources and obtain their parameters.

In the Magellanic Clouds, we have similarly selected dusty, evolved sources and obtained spectra with the MagE spectrograph on Magellan and identified 8 new RSGs. Among them is a luminous, extreme RSG, with similar properties to WOH G64. Our results are presented by de Wit et al. (2022; this volume).

3.3. *Mass loss rates*

We have set out to determine the mass loss rates (MLR) of red supergiants in the Small Magellanic Cloud, based on the catalogs of Yang et al. (2020) and Ren et al. (2021). Comprehensive photometry in over 50 bands (from the UV to $24\mu m$) for over 2000 RSG has been compiled and a grid of DUSTY models (Ivezic & Elitzur 1997) was created for both silicate and carbon dust. This grid was used to perform a χ^2 fit of the SEDs and determine the dust parameters, optical depth and the mass loss rate for each supergiant.

From the distribution of MLR, we find a typical value of $\sim 10^{-6}\,\mathrm{M}_{\odot}\,yr^{-1}$, with a few outliers at around $\sim 10^{-4}$ and $10^{-3}\,\mathrm{M}_{\odot}\,yr^{-1}$. We determine a new MLR vs. L relation based on an unbiased sample of RSG in the SMC, which shows an upturn at around $\log(L/\mathrm{L}_{\odot}) = 4.7$, with enhanced mass loss occurring at higher L. Compared to previously determined relations in the SMC, our result (Yang et al. 2022, in prep.) is most similar to the relations of Feast (1992) and van Loon et al. (2005). We plan to apply the procedure to all our program galaxies and determine the MLR at a range of metallicities.

4. Conclusions

We have presented the first results of an ambitious systematic study of episodic mass loss in ~ 1000 evolved massive stars. This survey is timely, given the recent availability of mid-IR catalogs and ambitious, as it plans to increase the number of evolved massive stars in nearby galaxies by a factor of 5. The *James Webb Space Telescope* is operating concurrently with this project. The enormous boost in sensitivity and angular resolution will revolutionize our understanding of these nearby objects. However, to fully exploit this we need to be able to tie the JWST results into the more general population. This project provides this anchor.

The results of this study will not only provide the first quantitative inventory and characterization of dusty massive stars in 27 galaxies in the nearby Universe at a range of metallicities, but may also reveal new classes of enshrouded stars and rare transitional objects. A byproduct of the survey will be the release of multi-wavelength photometric catalogs of luminous sources in 27 galaxies, including their classifications, which will be valuable for various scientific projects.

Acknowledgments

We acknowledge funding support from the European Research Council (ERC) under the European Union's Horizon 2020 research and innovation programme (Grant agreement No. 772086).

References

Banerjee, S., Kroupa, P., & Oh, S. 2012, MNRAS, 426, 1416
Bonanos, A. Z., Lennon, D. J., Köhlinger, F., et al. 2010, AJ, 140, 416
Bonanos, A. Z., Massa, D. L., Sewilo, M., et al. 2009, AJ, 138, 1003
Bond, H. E., Bedin, L. R., Bonanos, A. Z., et al. 2009, ApJL, 695, L154
Bouret, J. C., Lanz, T., & Hillier, D. J. 2005, A&A, 438, 301
Boyer, M. L., McQuinn, K. B. W., Barmby, P., et al. 2015, ApJS, 216, 10
Brott, I., de Mink, S. E., Cantiello, M., et al. 2011, A&A, 530, A115

Chambers, K. C., Magnier, E. A., Metcalfe, N., et al. 2016, arXiv e-prints, arXiv:1612.05560
Cohen, D. H., Wollman, E. E., Leutenegger, M. A., et al. 2014, MNRAS, 439, 908
Crowther, P. A. 2007, ARA&A, 45, 177
Crowther, P. A., Schnurr, O., Hirschi, R., et al. 2010, MNRAS, 408, 731
Cutri, R. M., & et al. 2012, VizieR Online Data Catalog, II/311
Cutri, R. M., Skrutskie, M. F., van Dyk, S., et al. 2003, 2MASS All Sky Catalog of point sources.
Davidson, K., & Humphreys, R. M. 1997, ARA&A, 35, 1
de Wit, S., Bonanos, A. Z., Tramper, F., et al. 2022
Decin, L., Hony, S., de Koter, A., et al. 2006, A&A, 456, 549
Ekström, S., Georgy, C., Eggenberger, P., et al. 2012, A&A, 537, A146
Feast, M. W. 1992, in Instabilities in Evolved Super- and Hypergiants, ed. C. de Jager & H. Nieuwenhuijzen, 18
Figer, D. F. 2005, Nature, 434, 192
Fraser, M., Inserra, C., Jerkstrand, A., et al. 2013, MNRAS, 433, 1312
Fullerton, A. W., Massa, D. L., & Prinja, R. K. 2006, ApJ, 637, 1025
Gal-Yam, A. 2012, Science, 337, 927
Georgy, C., Ekström, S., Eggenberger, P., et al. 2013, A&A, 558, A103
Gordon, K. D., Meixner, M., Meade, M. R., et al. 2011, AJ, 142, 102
Gvaramadze, V. V., Kniazev, A. Y., & Fabrika, S. 2010, MNRAS, 405, 1047
Humphreys, R. M., & Davidson, K. 1994, PASP, 106, 1025
Ivezic, Z., & Elitzur, M. 1997, MNRAS, 287, 799
Khan, R. 2017, ApJS, 228, 5
Khan, R., Stanek, K. Z., Kochanek, C. S., & Bonanos, A. Z. 2011, ApJ, 732, 43
Khan, R., Stanek, K. Z., Kochanek, C. S., & Sonneborn, G. 2015, ApJS, 219, 42
Kudritzki, R.-P., & Puls, J. 2000, ARA&A, 38, 613
Maravelias, G., Bonanos, A. Z., Tramper, F., et al. 2022a, A&A, in press, arXiv:2203.08125
Maravelias, G., Bonanos, A. Z., Tramper, F., et al. 2022b
Mauerhan, J. C., Smith, N., Filippenko, A. V., et al. 2013, MNRAS, 430, 1801
Meixner, M., Gordon, K. D., Indebetouw, R., et al. 2006, AJ, 132, 2268
Meynet, G., Chomienne, V., Ekström, S., et al. 2015, A&A, 575, A60
Munoz-Sanchez, G., Maravelias, G., Bonanos, A. Z., et al. 2022
Neill, J. D., et al. 2011, ApJ, 727, 15
Oey, M. S., & Clarke, C. J. 2005, ApJL, 620, L43
Owocki, S. P., & Puls, J. 1999, ApJ, 510, 355
Pastorello, A., Cappellaro, E., Inserra, C., et al. 2013, ApJ, 767, 1
Prieto, J. L., Kistler, M. D., Thompson, T. A., et al. 2008, ApJL, 681, L9
Ren, Y., Jiang, B., Yang, M., et al. 2021, ApJ, 923, 232
Sana, H., et al. 2012, Science, 337, 444
Smith, N. 2014, ARA&A, 52, 487
Smith, N., & Frew, D. J. 2011, MNRAS, 415, 2009
Stoll, R., Prieto, J. L., Stanek, K. Z., et al. 2011, ApJ, 730, 34
van Loon, J. T., Cioni, M. R. L., Zijlstra, A. A., & Loup, C. 2005, A&A, 438, 273
Wachter, S., Mauerhan, J. C., Van Dyk, S. D., et al. 2010, AJ, 139, 2330
Williams, S. J., & Bonanos, A. Z. 2016, A&A, 587, A121
Yang, M., Bonanos, A. Z., Jiang, B.-W., et al. 2020, A&A, 639, A116
Zhang, T., Wang, X., Wu, C., et al. 2012, AJ, 144, 131
Zickgraf, F. J. 2006, in Astronomical Society of the Pacific Conference Series, Vol. 355, Stars with the B[e] Phenomenon, ed. M. Kraus & A. S. Miroshnichenko, 135

Discussion

PAUL CROWTHER: Could some of the low-luminosity sources be AGB stars?

ALCESTE BONANOS: We do not think we have much AGB contamination, as we have used conservative selection criteria based on the work of Yang et al. (2020).

ALEX DE KOTER: We expect RSGs to predominantly have Oxygen (i.e. Si) dust. What are the luminosities of the sources that needed carbon dust to be fitted?

ALCESTE BONANOS: I have to check this. However, we first modeled all sources with Si dust and only proceeded with a carbon dust fit for those that did not fit well with Si dust. We don't expect to have much AGB contamination.

FLOOR BROEKGARDEN: Why were the stars in your survey quite different compared to Emily Levesque? Especially in the HR diagram Levesque seemed to find more low T systems. Is this just because of the difference in surveys?

ALCESTE BONANOS: We used mid-IR selection criteria, so we expect to find more extreme RSGs.

PAT MORRIS: Following the question by Alex de Koter about finding silicate dust in the SMC RSG fits to be a little suspicious, more likely in lower mass (luminosity) AGB stars. Might this be an effect of more than one thermal component? I did not catch whether this was considered in the fits, might also help with the poorer fits you showed.

ALCESTE BONANOS: We have not modeled more than one thermal component, we should try this.

Massive Stars Near and Far
Proceedings IAU Symposium No. 361, 2024
J. Mackey, J. S. Vink, & N. St-Louis, eds.
doi:10.1017/S1743921322003337

Using machine learning to investigate the populations of dusty evolved stars in various metallicities

Grigoris Maravelias[1,2]†, Alceste Z. Bonanos[1], Frank Tramper[3], Stephan de Wit[1,4], Ming Yang[1,5], Paolo Bonfini[6], Emmanuel Zapartas[1], Konstantinos Antoniadis[1,4], Evangelia Christodoulou[1,4] and Gonzalo Muñoz-Sanchez[1,4]

[1]IAASARS, National Observatory of Athens, Greece

[2]Institute of Astrophysics, FORTH, Greece
email: `gmaravel@physics.uoc.gr`

[3]Institute of Astronomy, KU Leuven, Belgium

[4]Department of Physics, National and Kapodistrian University of Athens, Greece

[5]Key Laboratory of Space Astronomy and Technology, National Astronomical Observatories, Chinese Academy of Sciences, People's Republic of China

[6]Ballista Technology Group, Florida, USA

Abstract. Mass loss is a key property to understand stellar evolution and in particular for low-metallicity environments. Our knowledge has improved dramatically over the last decades both for single and binary evolutionary models. However, episodic mass loss although definitely present observationally, is not included in the models, while its role is currently undetermined. A major hindrance is the lack of large enough samples of classified stars. We attempted to address this by applying an ensemble machine-learning approach using color indices (from IR/*Spitzer* and optical/Pan-STARRS photometry) as features and combining the probabilities from three different algorithms. We trained on M31 and M33 sources with known spectral classification, which we grouped into Blue/Yellow/Red/B[e] Supergiants, Luminous Blue Variables, classical Wolf-Rayet and background galaxies/AGNs. We then applied the classifier to about one million *Spitzer* point sources from 25 nearby galaxies, spanning a range of metallicites ($1/15-\sim 3\ Z_{\odot}$). Equipped with spectral classifications we investigated the occurrence of these populations with metallicity.

Keywords. Stars: massive – Stars: mass-loss – Stars: evolution – Galaxies: individual: WLM, M31, IC 1613, M33, Sextans A – Methods: statistical

1. Introduction

Although our knowledge on the modeling of stellar evolution for massive stars has significantly advanced over the last couple of decades, we still struggle to understand the influence of some key parameters. One such example is mass-loss, for which the theoretical predictions are at odds with the values derived from observations. This becomes more evident as we move away from the main-sequence (e.g. Smith 2014). Certain phases of evolved massive stars display - additionally to the stellar winds (e.g. Vink 2021) - episodic mass-loss activity, such as in Luminous Blue Variables (LBVs; e.g. Davidson

† contact: maravelias@noa.gr

2020), Wolf-Rayet stars (WRs; e.g. Crowther & Walborn 2011), B[e] supergiants (BeBRs; e.g. Kraus 2019), Yellow Supergiants (YSGs; e.g. de Jager 1998), Red Supergiants (RSGs; e.g. Levesque 2017). We already know that the majority of the massive stars are born into multiple systems (e.g. Kobulnicky et al. 2014, Sana et al. 2014) which results in a significant fraction of binary interactions (e.g. de Mink et al. 2014). However, binarity may not be the only channel for episodic mass-loss. Moreover, it is only now that we start to explore these populations in low metallicity environments, which can help us understand better the underlying mechanisms (e.g. Lorenzo et al. 2022). Simultaneously, it is those environments that seem to host superluminous supernovae, which may be the result of the interaction of ejecta with a complex dusty circumstellar environment formed by possible multiple mass-loss events prior to the explosion (Neill et al. 2011, Gal-Yam 2012).

The ASSESS project† (P.I. Alceste Bonanos, see also Bonanos et al. 2022 this volume) aims to understand the importance of episodic mass-loss by studying a large number of evolved stars in nearby galaxies. One of the approaches is to use machine-learning methods to classify sources without any spectral classification, and to perform follow-up observations to verify their nature. Given this tool we can also perform a study of those populations across different metallicity environments. The current paper summarizes our first results towards this goal.

2. The classifier

Since the classifier is presented in detail in Maravelias et al. (2022), we provide only a brief overview in this section, so that the reader understands the post-processing steps and the final results.

After careful search in literature we compiled a *Gaia* cleaned sample of 932 M31 and M33 sources with known spectral types. We grouped these into seven classes of interest, i.e. in BSG, YSG, RSG, BeBR, LBV, (classic) WR, and background galaxies (GAL; as a collective class of outliers). We used photometry, and in particular consecutive color indices, from the *Spitzer* ([3.6] and [4.5]) and Pan-STARRS (r, i, z, and y) bands (not corrected for extinction as there is not enough data to perform this and the influence in those selected bands is minimal). As this initial sample is highly imbalanced we employed synthetic data approaches to overcome the limitations imposed by the underrepresented classes. Although this does not completely alleviate the problem with LBV it helps with other classes (e.g. WR). Our classifier is an ensemble approach of three supervised machine-learning algorithms, that is Support Vector Classification, Random Forests, and Multilayer-Perceptron (a type of neural network). Their output results probabilities per class and these are combined in order to obtain the final probability distribution, while the final class is derived by using the class that corresponds to the maximum probability.

The results from the application of this approach are fairly good. We managed to correctly identified from the initial training sample $\sim 60\%$ to $\sim 80\%$ for BSG, GAL, and YSG, $\sim 73-80\%$ for BeBR, and $\sim 45\%$ for WR, with the best results obtained for the RSG ($\sim 94\%$) and the worst for LBV ($\sim 28\%$ - only from the Support Vector Classification). These results are comparable or slightly better when compared to the results from Dorn-Wallenstein et al. (2021), who worked with a much less homogeneous (with respect to the labels) but more populated Galactic sample. Our overall weighted balanced accuracy is $\sim 83\%$.

Our test sample consisted of a (limited) number of sources collected for IC 1613, WLM, and Sextans A galaxies. Due to the fact that $\sim 14\%$ of these sources have missing values (in the selected feature space, i.e. not having measurements across all selected *Spitzer* and

† `http://assess.astro.noa.gr/`

Pan-STARRS bands), we performed data imputation by replacing the features' values using iterative imputer (which models, iteratively, each feature with missing values as a function of other features, and uses that estimate for imputation). Although overall the missing data imputation does not affect much the results on this particular test dataset, our results showed that we can retain a relative good prediction score even when missing up to three features (out of the total five available, i.e. the consecutive color indices). The accuracy obtained for this test sample (at $\sim 70\%$) was lower than that for the M31 and M33 sample. We attribute this discrepancy partly to photometric issues and to the total effect of metallicity, which might affect the intrinsic colors of the sources and extinction due to the different galactic environments. However, we still lack a statistically significant sample to study and understand the influence of the metallicity, a target towards which the ASSESS project is actively working to address, both through the machine-learning applications such this classifier as well as dedicated observational campaigns (see Munoz-Sanchez et al. 2022, de Wit et al. 2022 in this volume).

3. Application and Results

After completing the training of the model using the whole available sample of sources in M31 and M33, we proceeded with its application. We used source lists for 25 nearby galaxies, spanning a range of 1/15 to $\sim 3\ Z_{\odot}$, that have been compiled from *Spitzer* photometric catalogs, and after *Gaia* screening. The classifier returns spectral types for 1156925 sources in total, including information on band completeness (the fraction of color indices without missing data) and probability. From all these sources only a tiny $\sim 0.5\%$ have spectroscopic classifications.

We applied a number of selection cuts in order to first examine secure candidate sources. For this we employed a limit for band completeness so that we keep sources with measurements across all bands (i.e. band completeness equal to 1), which means that we excluded all sources with missing values. Then we considered only sources with a classification probability of at least 0.51. This value corresponds to the peak of the incorrect classifications for the M31 and M33 sources (see Fig. 6 Maravelias et al. 2022).

Given the spectral classifications and photometric information we can proceed and plot indicative Color-Magnitude Diagrams (CMDs) to examine our results. In Figure 1 we plot the optical $g - i$ vs. i and the IR [3.6]-[4.5] vs. [4.5] CMDs for 708 WLM sources. We can easily notice the distributions of sources as expected in the optical with the BSG, YSG, and RSG occupying different distinct locations. There are numerous WR though which seem to overlap with other populations. This is also evident in the IR CMD, and it is a possible effect of a higher rate of false positives for this class. In the IR CMD we sse the clustering of RSG above the BSG populations as expected (c.f. Bonanos et al. 2009). Interestingly, we identify one LBV and two BeBR candidate sources, which shows that the classifier is able to highlight the most promising targets.

Having obtained the predictions for all galaxies, we can examine the predictions in a more global framework. We considered the fractions of the number of predicted sources (per class) over the total number of sources in each galaxy. In Figure 2 we present these fractions with metallicity for a subsample of our galaxies. It is interesting to point out the increasing (decreasing) trend of BSG (RSG) with lower metallicities consistent with theoretical expectations that more (less) time is spend in these phases (e.g. Ren et al. 2021, Massey 2002). Also, WR display decreasing trend because of the drop in the strength of the stellar winds with metallicity. YSG display a rather flatter behavior (with the exception of IC 1613 which deserves a revisit), while we notice far fewer BeBR than LBVs in low metallicity galaxies.

However, we need to be careful with the interpretation of these plots, as currently we have not taken into account the errors with respect to the predictions that should

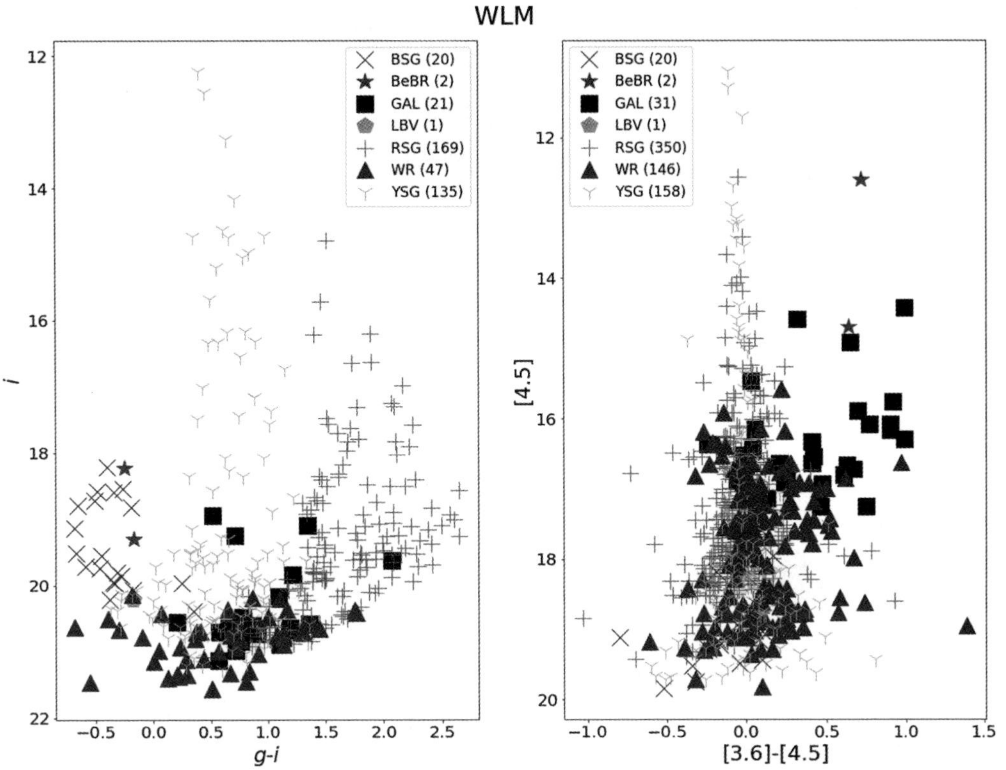

Figure 1. An optical (left) and IR (right) CMD for WLM. We can easily notice the distributions of BSG (blue 'X'), YSG (yellow slim triangles), RSG (red crosses), WR (purple full triangles), GAL (black boxes), BeBR (dark violet stars), LBV (pink pentagons). Note that the number of sources in the optical CMD is smaller than the IR one (708 sources in total) because not all sources have g magnitudes.

propagate through the end fractions, as well as completeness issues. As our samples are based on *Spitzer* catalogs we are looking for (and predicting) stars with substantial amounts of dust. Although this can be true for the majority of the RSG and the BeBR it is not necessary for BSG (which includes main-sequence massive stars) or YSGs. We also need to quantify the contamination. For example, the BSG class may get mixed with HII regions, while bright AGB stars may still be present among the RSGs. However, at the distances of the galaxies we are looking at we are actually probing the upper part of the Hertzsprung-Russell diagram, which minimizes the contamination by low-mass stars. Last but not least, we need to further study in detail the influences of our selection cuts. For example Fig. 8 in (Maravelias et al. 2022) shows that the peak probability of incorrect classification shifts to higher probability values when examining the test galaxies (lower-Z environments).

4. Conclusions

We developed a machine-learning classifier based on optical (Pan-STARRS) and IR (*Spitzer*) photometry, using M31 and M33 sources as training sample, with an overall weighted balanced accuracy of $\sim 83\%$. Then we applied this classifier to 25 nearby galaxies (with $1/15 - \sim 3\ Z_{\odot}$) to get spectral classification (including Blue/Yellow/Red Supergiants, B[e] Supergiants, Wolf-Rayet, Luminous Blue Variables, and background galaxies) for about 1.1 million sources. Selecting some initial quality cuts for the output

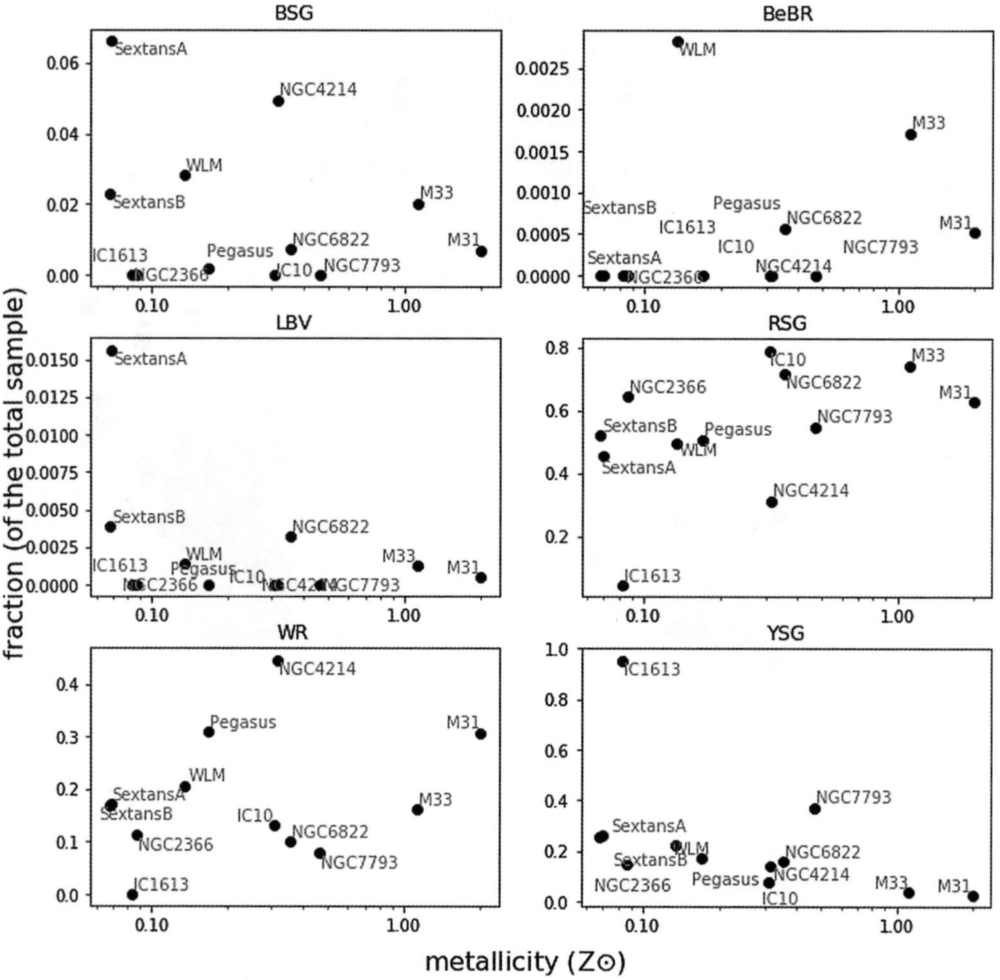

Figure 2. The fractions, of the predicted class members over the total sample size for each galaxy, with metallicity.

class probability and data availability we explore the population statistics with respect to the metallicity. We identify potentially interesting trends, such as the increase/decrease of Blue/Red supergiants and drop of WR with decreasing metallicity. However, to fully exploit these results further processing is needed to correctly account for predictions errors, completeness issues, and influences of quality cuts. Nevertheless, this catalog is an excellent tool for identifying interesting objects and prioritizing targets for observations, especially for *JWST*.

Acknowledgments

We acknowledge funding support from the European Research Council (ERC) under the European Union's Horizon 2020 research and innovation programme (Grant agreement No. 772086).

References

Bonanos, A. Z., et al. 2009, AJ, 138, 1003
Bonanos, A. Z., et al. 2022
Crowther, P. A., & Walborn, N. R. 2011, MNRAS, 416, 1311

Davidson, K. 2020, Galaxies, 8, 10
de Jager, C. 1998, A&A Rev., 8, 145
de Mink, S. E., Sana, H., Langer, N., Izzard, R. G., & Schneider, F. R. N. 2014, ApJ, 782, 7
de Wit, S., Bonanos, A. Z., Tramper, F., Yang, M., Maravelias, G., Boutsia, K., Britavskiy, N., & Zapartas, M. 2022
Dorn-Wallenstein, T. Z., Davenport, J. R. A., Huppenkothen, D., & Levesque, E. M. 2021, ApJ, 913, 32
Gal-Yam, A. 2012, Science, 337, 927
Kobulnicky, H. A., et al. 2014, ApJS, 213, 34
Kraus, M. 2019, Galaxies, 7, 83
Levesque, E. M. 2017, Astrophysics of Red Supergiants
Lorenzo, M., Garcia, M., Najarro, F., Herrero, A., Cerviño, M., & Castro, N. 2022, arXiv e-prints, arXiv:2207.09700
Maravelias, G., Bonanos, A. Z., Tramper, F., de Wit, S., Yang, M., & Bonfini, P. 2022, arXiv e-prints, arXiv:2203.08125
Massey, P. 2002, ApJS, 141, 81
Munoz-Sanchez, G., Maravelias, G., Bonanos, A. Z., Tramper, F., de Wit, S., & Yang, M. 2022
Neill, J. D., et al. 2011, ApJ, 727, 15
Ren, Y., Jiang, B., Yang, M., Wang, T., Jian, M., & Ren, T. 2021, ApJ, 907, 18
Sana, H., et al. 2014, ApJS, 215, 15
Smith, N. 2014, ARA&A, 52, 487
Vink, J. S. 2021, arXiv e-prints, arXiv:2109.08164

Discussion

MIRIAM GARCIA: I have two questions: a. What about using spectroscopic data in the machine-learning method? and b. In IC 1613 we have detected an LBV-candidate and a WO star, which I do not see in your results.

GRIGORIS MARAVELIAS: As far as the first question, we want to maximize the number of targets for which we want to have predicted spectral types. The largest lists of stars can come from photometry. In supervised machine learning you need the input data and their corresponding classes for proper training. When you apply the classifier you have to do it in on a dataset which has the same input (in our case photometry in the form of color indices) as your training. That's why we tried to get as many spectral types as possible from literature and link them to photometry. If we would like to use other input data to train (e.g. spectra, spectral parameters) we should have the exact same features to apply to. And unfortunately there is not such a set (and if there was, e.g. spectra, the sample size would be much more limited in contrast to photometry).
For the second question, I actually presented only a part of our results, based on some quality cuts including those without any missing data. However, this would open the discussion about how we handle cases with missing values and I would redirect you to the paper as this demands much more time for discussion. It is most possible that these sources can be found in the other parts.

GRIFFIN HOSSEINZADEH: Is it possible to train the algorithm with lower metallicity samples?

GRIGORIS MARAVELIAS: With the current use of *Spitzer* and Pan-STARRS photometry only M31 and M33 have significant samples. The Magellanic Clouds are not covered by Pan-STARRS and the inclusion of other surveys would include more systematics. However, this will change as we do have a large observing campaign to obtain spectra in many of these galaxies (along with others of course, e.g. see Marta Lorenzo's work on Sextans A), so that we can re-train the classifier in the future with more data.

Massive Stars Near and Far
Proceedings IAU Symposium No. 361, 2024
J. Mackey, J. S. Vink, & N. St-Louis, eds.
doi:10.1017/S1743921322002733

Wray 15-906 low mass Luminous Blue Variable in a pre-supernova stage

O. Maryeva[1], V. Gvaramadze[2], A. Kniazev[3,4,2] and L. Berdnikov[2]

[1]Astronomical Institute, Czech Acad. Sci., Fričova 298, 25165 Ondřejov, Czech Republic
email: olga.maryeva@gmail.com

[2]Sternberg Astronomical Institute, MSU, Universitetsky pr., 13, Moscow, 119992, Russia

[3]South African Astronomical Observatory, 7935 Observatory, Cape Town, South Africa

[4]Southern African Large Telescope Foundation, 7935 Observatory, Cape Town, South Africa

Abstract. The evolutionary link between Red Supergiants and Luminous Blue variables is interesting, but still poorly understood. We present the results of a study of the Galactic candidate luminous blue variable Wray 15-906, revealed via the detection of its infrared circumstellar shell (of ≈2 pc in diameter) with the Wide-field Infrared Survey Explorer (WISE) and the Herschel Space Observatory. Using the stellar atmosphere code CMFGEN and the Gaia parallax, we found that Wray 15-906 is a relatively low-luminosity, $log(L/L_\odot) \approx 5.4$, star with a temperature of 25 ± 2 kK. In the framework of single star evolution, the obtained results suggest that Wray 15-906 is a post-red supergiant star with an initial mass of $\approx 25\,M_\odot$ and that before exploding as a supernova it could transform for a short time into a WN11h star. The presence of a shell with a mass $2.9 \pm 0.5\,M_\odot$ indicates that Wray 15-906 has suffered substantial mass loss in the recent past.

Keywords. stars: evolution, stars: atmospheres, circumstellar matter

---

1. Wray15-906

This note is devoted to the evolutionary status of the new Galactic candidate Luminous Blue Variable (cLBV) Wray 15-906†. The star, also known as ALS 2533 or Hen 3-729, attracted our attention after we discovered a circular shell around it using *WISE* data (Kniazev & Gvaramadze 2015). In the *WISE* 22 μm image the nebula appears as a circular limb-brightened shell with a diameter ≈ 2 pc, which is typical of circumstellar nebulae around LBVs (e.g. (Weis 2001)). The star shows spectral line variability (Figure 1) and low amplitude photometric variability (Maryeva et al. 2020).

The basic parameters of Wray 15-906 were determined using the CMFGEN atmospheric modeling code (Hillier & Miller 1998). For fitting, we used the medium resolution spectrum of Wray 15-906 obtained with the SALT telescope. The derived parameters of Wray 15-906 allow to find the position of the star in the Hertzsprung–Russell (HR) diagram (Figure 1).

If Wray 15-906 was born single, then its initial mass could be estimated to be $25\,M_\odot$, which is close to the minimum mass for single (non-rotating) stars to become a WR star (e.g. Georgy et al. (2012)). The surface He abundance (65 per cent by mass) derived for Wray 15-906 is typical of LBVs (e.g. Crowther (1997)) and is intermediate between that of O and WR stars. This suggests that Wray 15-906 has already gone through the red supergiant phase and is currently near the end of the He-burning stage, and implies an

† The object has coordinates $\alpha = 11^h54^m43^s54$, $\delta = -63^\circ13'31''.1$ at J2000 epoch.

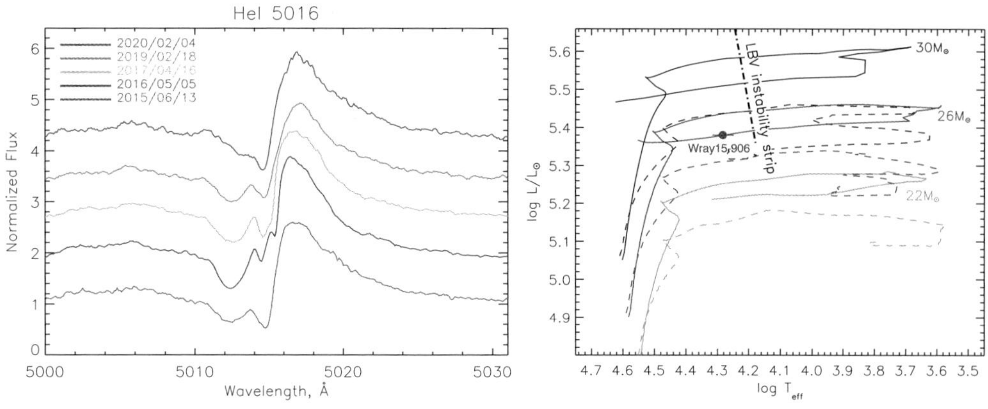

Figure 1. *Left:* Variability of the absorption components of the He I λ5016 P Cygni profile in the medium resolution spectra taken in 2015–2020 with the SALT telescope. *Right:* Position of Wray 15-906 in the Hertzsprung–Russell diagram. The horizontal lines represent the evolutionary tracks, taken from the Geneva library (Ekström et al. 2012).

evolutionary age of this star of ≈7-8 Myr. If Wray 15-906 was born with a low rotational velocity then it would need only several 1000 yr to finish its life with a supernova explosion (Georgy et al. 2012). The possibility that Wray 15-906 may soon explode is indirectly supported by the curious finding by Groh et al. (2013a) and Groh et al. (2013b) that single stars with $M_{\rm init} = 20 - 25\, M_\odot$ have LBV-like spectra before exploding as a supernova. Although unexpected, this finding conforms with previous claims (e.g. Kotak & Vink (2006)) that LBVs could be the immediate precursors of supernovae.

More details may be found in our work: Maryeva et al. (2020).

Acknowledgements

We dedicate this publication to the memory of our dear colleague Vasilii Gvaramadze, who passed away on September 2nd, 2021. The search for evolved stellar objects by means of detection of their circumstellar nebulae in modern sky surveys has been the topic that Vasilii successfully developed for last ten years, and he assembled and inspired our diverse team with his intellectual and organizational talents. This work has not been possible without his efforts.

O.M. is grateful to IAU grant for financial support for her participation in the IAU S361.

References

Crowther, P. A. 1997, in Astronomical Society of the Pacific Conference Series, Vol. 120, Luminous Blue Variables: Massive Stars in Transition, ed. A. Nota & H. Lamers, 51

Ekström, S., et al. 2012, A&A, 537, A146

Georgy, C., Ekström, S., Meynet, G., Massey, P., Levesque, E. M., Hirschi, R., Eggenberger, P., & Maeder, A. 2012, A&A, 542, A29

Groh, J. H., Meynet, G., & Ekström, S. 2013a, A&A, 550, L7

Groh, J. H., Meynet, G., Georgy, C., & Ekström, S. 2013b, A&A, 558, A131

Hillier, D. J., & Miller, D. L. 1998, ApJ, 496, 407

Kniazev, A., & Gvaramadze, V. 2015, in SALT Science Conference 2015 (SSC2015), 49

Kotak, R., & Vink, J. S. 2006, A&A, 460, L5

Maryeva, O. V., Gvaramadze, V. V., Kniazev, A. Y., & Berdnikov, L. N. 2020, MNRAS, 498, 5093

Weis, K. 2001, Reviews in Modern Astronomy, 14, 261

Chapter VIII:
WR stars and Stellar Feedback

Massive Stars Near and Far
Proceedings IAU Symposium No. 361, 2024
J. Mackey, J. S. Vink, & N. St-Louis, eds.
doi:10.1017/S1743921322003015

Wolf-Rayet stars: recent advances and persisting problems

Tomer Shenar

Anton Pannekoek Institute for Astronomy, Science Park 904, 1098 XH, Amsterdam, The Netherlands
email: T.Shenar@uva.nl

Abstract. Wolf-Rayet (WR) stars comprise a class of stars whose spectra are dominated by strong, broad emission lines that are associated with copious mass loss. In the massive-star regime, roughly 90% of the known WR stars are thought to have evolved off the main sequence. Dubbed classical WR (cWR) stars, these hydrogen-depleted objects represent a crucial evolutionary phase preceding core collapse into black holes, and offer a unique window into hot-star wind physics. Their formation is thought to be rooted in either intrinsic mass-loss or binary interactions. Results obtained from analyses using contemporary model atmospheres still fail to reconcile the derived properties of WR stars with predictions from stellar evolution. Importantly, stellar evolution models cannot reproduce the the bulk of cWR stars, a problem that becomes especially severe at subsolar metallicity. Next-generation model atmospheres and upcoming observational campaigns to hunt for undetected companions promise a venue for progress.

Keywords. Wolf-Rayet stars, binaries:general

1. Introduction

Wolf-Rayet (WR) stars are first and foremost a spectral class. Per definition, WR stars exhibit broad emission lines in lines that typically belong to helium, nitrogen for N-rich WR stars (WN), and carbon and oxygen for C- and O-rich WR stars (WC/WO). The current census of WR stars in our Milky Way Galaxy lists 667 objects (van der Hucht 2001, Crowther 2007, Rosslowe & Crowther 2015), but many thousands are known in resolved (e.g., Breysacher et al. 1999, Neugent et al. 2018, Neugent & Massey 2019) and unresolved (e.g. Hadfield & Crowther 2006, Kehrig et al. 2013, Gómez-González et al. 2021) extragalactic populations such as those of the Small and Large Magellanic Clouds (SMC, LMC, Fig. 1).

The winds that characterise massive WR stars arise due to their proximity to the Eddington limit (i.e., high luminosity-to-mass, or L/M ratio). On the one hand, proximity to the Eddington limit is reached by very massive stars ($M_{\rm ini} \gtrsim 80 - 100\,M_\odot$) already on the main sequence due to their extreme luminosities (e.g., de Koter et al. 1997, Rauw et al. 2004, Shenar et al. 2021). On the other hand, massive stars that have evolved off the main sequence and have shed much of their hydrogen-rich layers (Conti 1976) launch powerful winds due to the mass reduction. About 90% of the known WR stars belong to the latter class of classical WR stars (cWR): hot ($T_{\rm eff} \gtrsim 30\,$kK), evolved (typically core He-burning), hydrogen-depleted massive stars. As the putative immediate progenitors of black holes, the properties and formation of cWR stars has enormous implications on the production of stellar-mass black holes, which are the seeds that produce the gravitational waves we are now regularly observing. These proceedings focus on cWR stars.

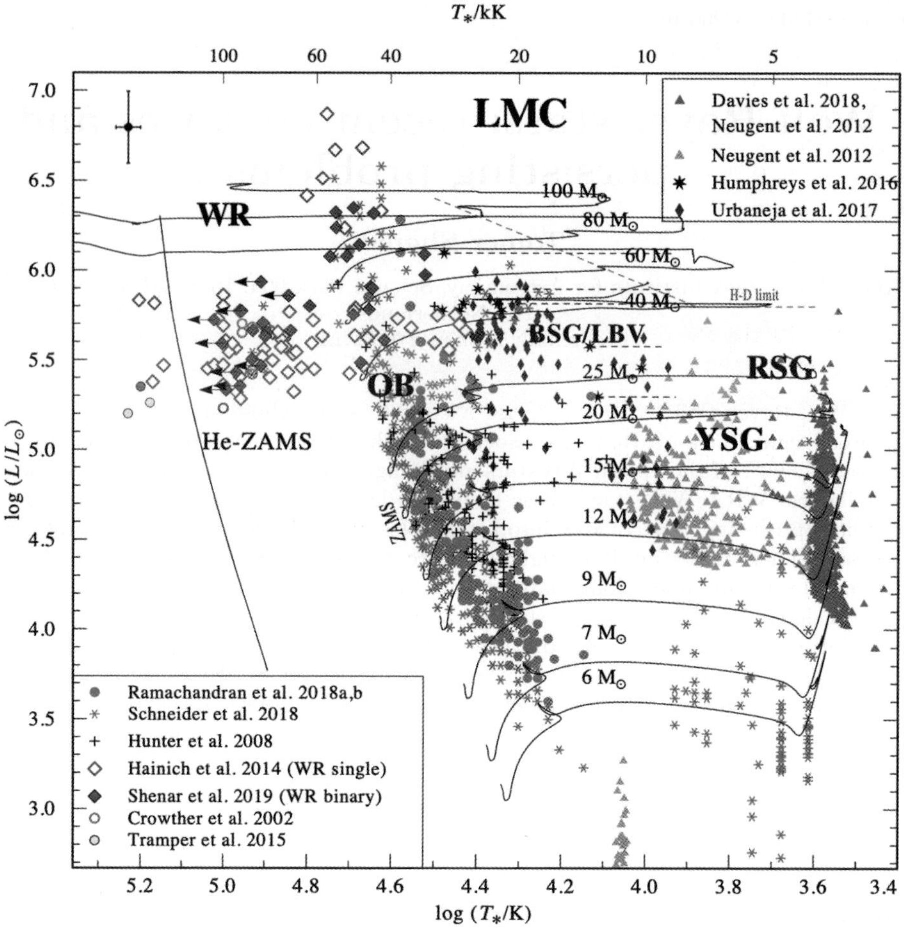

Figure 1. Various stellar populations (see legend) shown on a Hertzsprung-Russell diagramm (HRD) of the LMC, adapted from Ramachandran et al. (2019) and: Davies et al. (2018), Neugent et al. (2012), Humphreys et al. (2016), Urbaneja et al. (2017), Schneider et al. (2018), Shenar et al. (2019), Hainich et al. (2014), Crowther et al. (2002), Tramper et al. (2015), and Hunter et al. (2008). Tracks are single-star evolution tracks computed with the Modules for Experiments in Stellar Astrophysics (MESA) code (Paxton et al. 2011), taken from Gilkis et al. (2021). Tracks are single-star evolution tracks computed with the MESA code, adopted from Gilkis et al. (2021). The WN population (which comprises roughly 80% of the total LMC WR population) is shown on the left part, with filled symbols depicting known binaries, and empty symbols depicting apparently-single WN stars. The WR population is not reproduced, notably the apparently-single stars. A similar picture is observed with other widely-used tracks, such as Geneva (Eggenberger et al. 2021) and BPASS (Eldridge & Stanway 2009, Eldridge et al. 2017). Highlighted is the overlap between cool supergiants and the WR populations, which includes only the most luminous cool supergiants. Figure courtesy of Varsha Ramachandran and Avishai Gilkis.

2. Physical properties of cWR stars

The analysis of WR spectra relies on sophisticated models that solve the radiative transfer problem in expanding atmospheres while relaxing the assumption of local thermodynamic equillibrium (non-LTE; Hillier 1991, Hamann & Gräfener 2004). Along with the mass-loss rate $\dot{M}$ and luminosity L, a central goal of such a fitting process is to derive the base effective temperature of the star T_*, which typically refers to the radial layer

$r = R_*$ at which the mean Rosseland optical depth $\tau_{\rm Ross}$ reaches 10 or 20 (the difference is usually negligible). Hundreds of WR stars in our Galaxy and neighbouring galaxies have been thus analysed (Hainich et al. 2014; 2015, Shenar et al. 2016; 2019, Hamann et al. 2019, Sander et al. 2020). However, the resulting base temperatures are on average a factor of 2-3 lower than what is predicted from evolution models, a problem that became known as the WR "temperature problem" or, equivalently, the "radius problem".

What is "observed" in the spectrum are photons emitted from regions well above the stellar surface, which are closer to the photosphere ($\tau_{\rm Ross} \approx 2/3$, at $r = R_{2/3}$). While the derivation of $T_{\rm eff}(\tau = 2/3) = T_{2/3}$ is therefore relatively robust, the derivation of the base temperature T_* fully depends on the adopted velocity law (see Lefever et al. in these proceedings). In the vast majority of analyses thus published, the velocity fields of the winds are fixed to the so-called β-law (Castor et al. 1975), typically adopting $\beta = 1$.

Early investigations of the motion of clumps in the spectra of WR stars by Lépine & Moffat (1999) already suggested that $\beta = 1$ laws do not yield a consistent solution to the outer velocity field (assuming the clumps trace the velocity field). Moreover, 1D hydrodynamically-consistent models computed by Gräfener & Hamann (2005) and, recently, by Sander, Vink, & Hamann (2020) show that the β-law is often a bad approximation. This is also seen in semi-analytical models provided by Ro & Matzner (2016) and Poniatowski et al. (2021), often revealing a "double-peaked" profile for the velocity field, where the first peak is associated with a dynamical inflation (see also Petrovic et al. 2006, Grassitelli et al. 2016, Gräfener et al. 2012). This dynamical inflation may be the origin for the "temperature problem" of WR stars. However, it remains to be seen that such models can replicate the observed spectra of WR stars.

And yet, 3D effects may also play an important role. First attempts at 3D modelling of the winds of WR stars have been recently published by Moens et al. (2022). Among the plethora of rich results, an interesting finding is that the average velocity field in the 3D model no longer shows the "double-peaked" profile obtained in 1D models. However, this needs to be confirmed by future studies across the parameter regime. Surely, much progress and many insights will flow from this direction over the coming years.

Luckily, the measured luminosities of WR stars are much more robust. This is owing to the Stephan-Boltzmann equation, according to which $L \propto R_*^2 T_*^4 \approx R_{2/3}^2 T_{2/3}^4$ (neglecting the energy loss due to wind-driving). As the photospheric properties are generally well constrained, the luminosities are as well. As shown below, it is the luminosities of WR stars that are difficult to reconcile with evolution models.

3. Formation and evolution

Why should stars get rid of their hydrogen-rich envelopes? It was Paczyński (1967) who proposed that stripping by a binary companion could explain the formation of WR stars (see also Vanbeveren et al. 1998). However, soon thereafter it was realised that massive stars can lose copious mass throughout various phases of their evolution, potentially involving eruptions during a Luminous Blue Variable (LBV) phase. Hence, stars exceeding a certain threshold mass may naturally evolve into the cWR phase, without any help from a companion. This became known as the Conti scenario, after Peter Conti who originally proposed this (Conti 1976).

The question has not been settled ever since. In the late 20$^{\rm th}$ century, intrinsic mass-loss generally took precedence, since mass-loss prescriptions published at the time, in combination with rotation, could broadly explain the census and observed ratios of the WR populations (e.g., WC/WN, WN/O, Meynet & Maeder 2003). However, in the past two decades, two major realisations emerged: 1. Binary interactions are highly common among massive stars (Sana et al. 2012, Kobulnicky et al. 2014), and 2. Mass-loss rates at various stages (Fullerton et al. 2006, Puls et al. 2006, Krtička & Kubát 2017, Björklund et al. 2020, Beasor et al. 2020) phases appear to be lower than previously thought. These two facts combined have led to a resurgence of the binary channel in forming WR stars.

4. The Wolf-Rayet binary fraction as a function of metallicity

A fundamental parameter in this context is the WR binary fraction, which, for the Galaxy, is long-known to be $\approx 40\%$ (Vanbeveren et al. 1998, van der Hucht 2001). However, this fraction is not bias-corrected. A modern radial velocity survey for Galactic WR stars has been performed by Dsilva et al. (2020; 2022), and in prep. The observed binary fraction of WN stars remains comparable ($f_{\rm obs,WN} = 0.41 \pm 0.09$), and a bias correction of the WN population in the range $0 < \log P < 5$ [d] implies that roughly half the Galactic WR stars are in binaries ($f_{\rm int,WN} = 0.52 \pm 0.13$). However, for the putative decendents of WN stars – WC stars – an observed binary fraction of $f_{\rm obs,WC} = 0.58 \pm 0.14$ and an intrinsic fraction of $f_{\rm obs,WC} > 0.74$ are derived. It is important to note that these fractions assume an underlying mass distribution for the companions: it is possible that all "apparently-single" WR stars have low mass companions ($M_2 \lesssim 5\,M_\odot$), if such companions are over-represented in nature. Moreover, the period distributions of the WN and WC populations are found to be distinct: Galactic WN binaries tend to exhibit periods of the order of days to weeks, while WC binaries tend to exhibits periods of months to many years (see Dsilva et al. in these proceedings). These striking differences force us to reconsider the evolutionary connection between WN and WC stars, potentially implying that short-period WN binaries end up merging (Dsilva et al. 2022).

Binaries are thought to be especially critical at low metallicity (Z), where the winds generally expected and observed to be even weaker. To investigate this, Bartzakos et al. (2001), Foellmi et al. (2003a;b), and Schnurr et al. (2008) established the binary fraction of WR stars in the SMC and LMC, which probe metallicity contents of $\approx 1/5\,Z_\odot$ and $\approx 1/2\,Z_\odot$, respectively. The results appeared rather surprising, since the the observed binary fraction were found to be relatively Z-independent ($\approx 30 - 40\%$). This is also observed in other galaxies such as M31 and M33 (Neugent & Massey 2014).

A potential explanation for this result arises when one considers that fact that WR stars comprise a spectral class of stars with strong winds. But at low metallicity, not only is it harder to strip stars via winds, but it is also harder for the stripped stars to drive a wind. Hence, a rough threshold initial mass exists below which the stripped product would not appear as a WR star. Owing to comprehensive analyses of apparently-single and binary WR stars by Hamann et al. (2006; 2019), Hainich et al. (2014; 2015), and Shenar et al. (2016; 2019) in the Milky Way, LMC, and SMC, these mass thresholds can be estimated empirically (Fig. 2). Hence, Shenar et al. (2020a) estimate that in the Milky Way, stars initially less massive than $\approx 20\,M_\odot$ would not appear as WR stars after stripping, while in the SMC, this limit grows almost to $\approx 40\,M_\odot$ (see also Aguilera-Dena et al. 2022). This means that, at low Z, neither of the WR formation channel is effective, and it is not obvious that a metallicity trend should be present.

5. "Single" WR stars: truly single, or binary-interaction products?

In spite of the many advances, a major problem remains: many cWR stars do not have any identified companions. In fact, for the LMC population (Fig. 1), this population of apparently-single WR stars comprises over 70% of the entire cWR population.

One possibility is that these stars are not truly single, but rather binary. The companions are unlikely to be very massive or bright (give or take a few exceptions), because

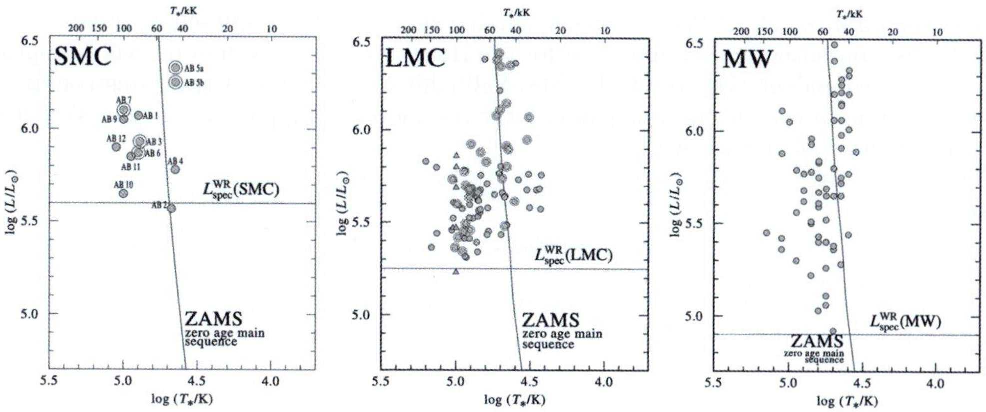

Figure 2. The positions of the WN populations of the SMC, LMC, and Galaxy, adapted from Shenar et al. (2020b). Red circled symbols are binaries. The Wolf-Rayet phenomenon stops below a Z-dependent luminosity threshold: stripped stars below this threshold would not exhibit the necessary winds to be classified as WR stars. The thresholds correspond to initial masses of $\approx 20\,M_\odot$, $\approx 25\,M_\odot$ and $\approx 40\,M_\odot$ for the Galaxy, LMC, and SMC, respectively.

they would have been identified in previous surveys, unless they are so far to have anyhow avoided interaction with the WR component ($\log P \gtrsim 3.5$ [d]). Rather, if they are there, they are likely low-mass and faint (or even completely dark: black holes). While stable mass transfer does not seem to be feasible in this case, stripping via common-envelope evolution might be. This was proposed by Schootemeijer & Langer (2018), who are now actively searching for hidden companions among the SMC WR population (PI: Schootemeijer, ESO program ID: 108.22M1). A different campaign aims to hunt for companions among the LMC WC population (PI: Shenar, ESO program ID: 105.207A).

Another scenario related to binary interaction, which is however harder to probe, is that the apparently-single WR stars were affected by binary interaction, but are now truly single. Mergers is one option, though why merging should lead to the formation of a WR star is not clear. Another option is that they are mass gainers in distrupted binaries (e.g., due to a supernova kick of the mass donor), and the resulting rapid rotation pushes them to evolve homogeneously (Maeder 1987). Yet another option is that they were disrupted from the binary after stripping via third-body interactions (Toonen et al. 2020). These ideas need to be tested with modeling (e.g., population syntheses) and observations (e.g., proper motion and isolation of the targets) to investigate whether they can truly explain such a large part of the WR population.

However, it is also possible that single stars in the relevant mass regime ($M_{\rm ini} \gtrsim 25\,M_\odot$ in the LMC, Shenar et al. 2020b) do actually evolve into the WR phase naturally. One important mass-loss agent is represented by LBVs and their mass eruptions, which are still heavily debated (Justham et al. 2014, Smith & Tombleson 2015, Humphreys et al. 2016, Mahy et al. 2022). It is also important to realise the mass-loss of evolved stars with initial masses above $\approx 30\,M_\odot$ is not well constrained, since they generally do not become cool supergiants (the Humphreys-Davidson limit, Humphreys 1978). Hence, red supergiant mass-loss has very little to do with WR stars, except perhaps for the few most luminuos red supergiants. In fact, some of the measured mass-loss rates derived for the most luminous red supergiants do reach the levels necessary for self-striping (van Loon et al. 2005, Beasor & Smith 2022), notably WOH G64.

Finally, increased mixing spurs bluewards evolution, whether induced by rotation (Ekström et al. 2012) or processes like convection (Higgins & Vink 2019,

Ramachandran et al. 2019). Enhanced mixing was recently proposed by Gilkis et al. (2021) as a potential underlying agent for the Hupmhreys-Davidson limit, which appears to be Z-independent (Davies et al. 2018, Sabhahit et al. 2021). If true, this could also open a potential resolution for explaining the production of apparently-single cWR stars from a single-star prespective.

References

Aguilera-Dena, D. R., Langer, N., Antoniadis, J., Pauli, D., Dessart, L., Vigna-Gómez, A., Gräfener, G., & Yoon, S.-C. 2022, A&A, 661, A60
Bartzakos, P., Moffat, A. F. J., & Niemela, V. S. 2001, MNRAS, 324, 18
Beasor, E. R., Davies, B., Smith, N., van Loon, J. T., Gehrz, R. D., & Figer, D. F. 2020, MNRAS, 492, 5994
Beasor, E. R., & Smith, N. 2022, arXiv e-prints, arXiv:2205.02207
Björklund, R., Sundqvist, J. O., Puls, J., & Najarro, F. 2020, arXiv e-prints, arXiv:2008.06066
Breysacher, J., Azzopardi, M., & Testor, G. 1999, A&AS, 137, 117
Castor, J. I., Abbott, D. C., & Klein, R. I. 1975, ApJ, 195, 157
Conti, P. 1976, in Proc. 20th Colloq. Int. Astrophys., Univ. Liege, 193
Crowther, P. A. 2007, ARA&A, 45, 177
Crowther, P. A., Hillier, D. J., Evans, C. J., Fullerton, A. W., De Marco, O., & Willis, A. J. 2002, ApJ, 579, 774
Davies, B., Crowther, P. A., & Beasor, E. R. 2018, MNRAS, 478, 3138
de Koter, A., Heap, S. R., & Hubeny, I. 1997, ApJ, 477, 792
Dsilva, K., Shenar, T., Sana, H., & Marchant, P. 2020, arXiv e-prints, arXiv:2006.13957
—. 2022, arXiv e-prints, arXiv:2204.12518
Eggenberger, P., et al. 2021, A&A, 652, A137
Ekström, S., et al. 2012, A&A, 537, A146
Eldridge, J. J., & Stanway, E. R. 2009, MNRAS, 400, 1019
Eldridge, J. J., Stanway, E. R., Xiao, L., McClelland, L. A. S., Taylor, G., Ng, M., Greis, S. M. L., & Bray, J. C. 2017, PASA, 34, e058
Foellmi, C., Moffat, A. F. J., & Guerrero, M. A. 2003a, MNRAS, 338, 360
—. 2003b, MNRAS, 338, 1025
Fullerton, A. W., Massa, D. L., & Prinja, R. K. 2006, ApJ, 637, 1025
Gilkis, A., Shenar, T., Ramachandran, V., Jermyn, A. S., Mahy, L., Oskinova, L. M., Arcavi, I., & Sana, H. 2021, arXiv e-prints, arXiv:2102.03102
Gómez-González, V. M. A., Mayya, Y. D., Toalá, J. A., Arthur, S. J., Zaragoza-Cardiel, J., & Guerrero, M. A. 2021, MNRAS, 500, 2076
Gräfener, G., & Hamann, W. R. 2005, A&A, 432, 633
Gräfener, G., Owocki, S. P., & Vink, J. S. 2012, A&A, 538, A40
Grassitelli, L., Chené, A. N., Sanyal, D., Langer, N., St-Louis, N., Bestenlehner, J. M., & Fossati, L. 2016, A&A, 590, A12
Hadfield, L. J., & Crowther, P. A. 2006, MNRAS, 368, 1822
Hainich, R., Pasemann, D., Todt, H., Shenar, T., Sander, A., & Hamann, W. R. 2015, A&A, 581, A21
Hainich, R., et al. 2014, A&A, 565, A27
Hamann, W. R., & Gräfener, G. 2004, A&A, 427, 697
Hamann, W. R., Gräfener, G., & Liermann, A. 2006, A&A, 457, 1015
Hamann, W. R., et al. 2019, A&A, 625, A57
Higgins, E. R., & Vink, J. S. 2019, A&A, 622, A50
Hillier, D. J. 1991, A&A, 247, 455
Humphreys, R. M. 1978, ApJS, 38, 309
Humphreys, R. M., Weis, K., Davidson, K., & Gordon, M. S. 2016, ApJ, 825, 64
Hunter, I., et al. 2008, ApJL, 676, L29
Justham, S., Podsiadlowski, P., & Vink, J. S. 2014, ApJ, 796, 121
Kehrig, C., et al. 2013, MNRAS, 432, 2731

Kobulnicky, H. A., et al. 2014, ApJS, 213, 34
Krtička, J., & Kubát, J. 2017, A&A, 606, A31
Lépine, S., & Moffat, A. F. J. 1999, ApJ, 514, 909
Maeder, A. 1987, A&A, 178, 159
Mahy, L., et al. 2022, A&A, 657, A4
Meynet, G., & Maeder, A. 2003, A&A, 404, 975
Moens, N., Poniatowski, L. G., Hennicker, L., Sundqvist, J. O., El Mellah, I., & Kee, N. D. 2022, arXiv e-prints, arXiv:2203.01108
Neugent, K., & Massey, P. 2019, Galaxies, 7, 74
Neugent, K. F., & Massey, P. 2014, ApJ, 789, 10
Neugent, K. F., Massey, P., & Morrell, N. 2018, ApJ, 863, 181
Neugent, K. F., Massey, P., Skiff, B., & Meynet, G. 2012, ApJ, 749, 177
Paczyński, B. 1967, Acta Astronomica, 17, 355
Paxton, B., Bildsten, L., Dotter, A., Herwig, F., Lesaffre, P., & Timmes, F. 2011, ApJS, 192, 3
Petrovic, J., Pols, O., & Langer, N. 2006, A&A, 450, 219
Poniatowski, L. G., et al. 2021, A&A, 647, A151
Puls, J., Markova, N., Scuderi, S., Stanghellini, C., Taranova, O. G., Burnley, A. W., & Howarth, I. D. 2006, A&A, 454, 625
Ramachandran, V., et al. 2019, A&A, 625, A104
Rauw, G., et al. 2004, A&A, 420, L9
Ro, S., & Matzner, C. D. 2016, ApJ, 821, 109
Rosslowe, C. K., & Crowther, P. A. 2015, MNRAS, 447, 2322
Sabhahit, G. N., Vink, J. S., Higgins, E. R., & Sander, A. A. C. 2021, MNRAS, 506, 4473
Sana, H., et al. 2012, Science, 337, 444
Sander, A. A. C., Vink, J. S., & Hamann, W. R. 2020, MNRAS, 491, 4406
Schneider, F. R. N., et al. 2018, A&A, 618, A73
Schnurr, O., Casoli, J., Chené, A. N., Moffat, A. F. J., & St-Louis, N. 2008, MNRAS, 389, L38
Schootemeijer, A., & Langer, N. 2018, A&A, 611, A75
Shenar, T., Gilkis, A., Vink, J. S., Sana, H., & Sander, A. A. C. 2020a, A&A, 634, A79
Shenar, T., et al. 2016, A&A, 591, A22
—. 2019, A&A, 627, A151
—. 2020b, A&A, 639, L6
—. 2021, A&A, 650, A147
Smith, N., & Tombleson, R. 2015, MNRAS, 447, 598
Toonen, S., Portegies Zwart, S., Hamers, A. S., & Bandopadhyay, D. 2020, A&A, 640, A16
Tramper, F., et al. 2015, A&A, 581, A110
Urbaneja, M. A., Kudritzki, R. P., Gieren, W., Pietrzyński, G., Bresolin, F., & Przybilla, N. 2017, AJ, 154, 102
van der Hucht, K. A. 2001, New A, 45, 135
van Loon, J. T., Cioni, M. R. L., Zijlstra, A. A., & Loup, C. 2005, A&A, 438, 273
Vanbeveren, D., De Loore, C., & Van Rensbergen, W. 1998, A&A Rev., 9, 63

6. Q&A

DANY VANBEVEREN: Can it be that we are still missing WR binaries in surveys? Are there any biases involved?

TOMER SHENAR: Surveys of WR stars in the Magellanic Clouds are claimed to be virtually complete. However, I do not think that this accounts for WR stars with lines that are strongly diluted by a bright companion. It is hard to estimate the corresponding correction, since the frequency of such binaries depends on uncertainties in binary evolution.

However, since WR stars per definition exhibit substantial emission in their spectra, I would be surprised if we are missing a large population of WR binaries.

SALLY OEY: One clue to the binary nature of Wolf-Rayet stars is whether they are field stars. From what I recall from studies by Nathan Smith, Wolf-Rayet stars are not very "fieldy", unlike for example Oe/Be stars. This may suggest that single-star channels for some of the WR stars are relevant.

RAPHAEL HIRSCHI: I would also provide support the idea of underestimated mixing. Most of the studies adopt a constant mixing, but models show that mixing processes such as overshooting may be stronger for high-mass stars.

TOMER SHENAR: I agree with both comments

KOUSHIK SEN: What would be the mass ratio need to be for you to find the O-type companion in a WR binary?

TOMER SHENAR: This depends on the method. If you are relying on the radial velocity measurements of the Wolf-Rayet component, then in principle the mass ratio does not matter, just the amplitude of the Wolf-Rayet component. But if you want to identify the O-type star in the spectrum, then it would depend on a lot of parameters, and naturally the data quality. However, from experience, I would argue that it would generally be difficult to hide an O-type star in the spectrum.

Massive Stars Near and Far
Proceedings IAU Symposium No. 361, 2024
J. Mackey, J. S. Vink, & N. St-Louis, eds.
doi:10.1017/S1743921322002071

The enigmatic winds of Wolf-Rayet stars: Results from dynamically consistent atmosphere modelling

Andreas A. C. Sander

Zentrum für Astronomie der Universität Heidelberg, Astronomisches Rechen-Institut, Mönchhofstr. 12-14, 69120 Heidelberg, Germany
email: andreas.sander@uni-heidelberg.de

Abstract. Line-driven stellar winds are ubiquitous among hot massive stars. In some cases they can become so strong, that the whole star is cloaked by an optically thick wind. The strong outflow gives rise to large emission lines, defining the class of so-called Wolf-Rayet (WR) stars. While being major players in the evolution of massive stars, the formation of heavy black holes, and the distribution of elements, the occurrence and nature of WR winds is still quite enigmatic.

A promising instrument towards a better theoretical understanding are stellar atmospheres allowing for a consistent inclusion of the hydrodynamics. By coupling stellar and wind parameters and the inclusion of a detailed non-LTE radiative transfer, they allow us to go beneath the observable layers and study the onset of WR-type winds. Establishing larger sets of models, we were able to make ground-breaking progress by identifying trends with mass and metallicity that deviate significantly from present empirical descriptions. Our modelling efforts reveal a complex picture for WR-type winds with strong, non-linear dependencies. Besides covering metallicity and mass, we further identify surface hydrogen as an important ingredient to retain WR-type mass loss at lower metallicity. Here, we present a summary of recent insights on the nature and onset of WR-type winds in massive stars including the consequences for stellar evolution, remaining open questions, and current efforts to overcome them.

Keywords. stars: atmospheres, stars: mass loss, stars: massive, stars: winds, outflows, stars: Wolf-Rayet, stars: evolution, stars: black holes, galaxies: stellar content

1. Introduction

The Wolf-Rayet (WR) phenomenon indicates the presence of strong winds that can partially or completely hide the underlying stellar surface. Consequently, the connection of stellar and wind parameters has been a challenge for these objects due to inherent degeneracies in the spectrum formation (e.g., Hamann & Gräfener 2004, Lefever et al. 2022, these proceedings). Motivated by theoretical considerations of optically thick winds launched in the deeper atmospheres (e.g., Nugis & Lamers 2002), the introduction of hydrodynamically consistent stellar atmosphere models for WR stars Gräfener & Hamann (2005), Sander et al. (2020) has opened a new window to overcome these issues. Already the first model by Gräfener & Hamann (2005) reprising a "classical", i.e. helium-burning, WC star confirmed the suggestion of Nugis & Lamers (2002) that the opacities of the iron M-shell ions are fundamental to launch and drive the winds of classical WR stars. At the same time, independent Monte Carlo simulations at a lower temperature regime by Vink & de Koter (2005) also yielded a fundamental Fe-dependence for WR-type winds.

For the cooler, very massive WNh stars, Gräfener & Hamann (2008) calculated grids of models. Together with the work of Vink et al. (2011), their study demonstrated that

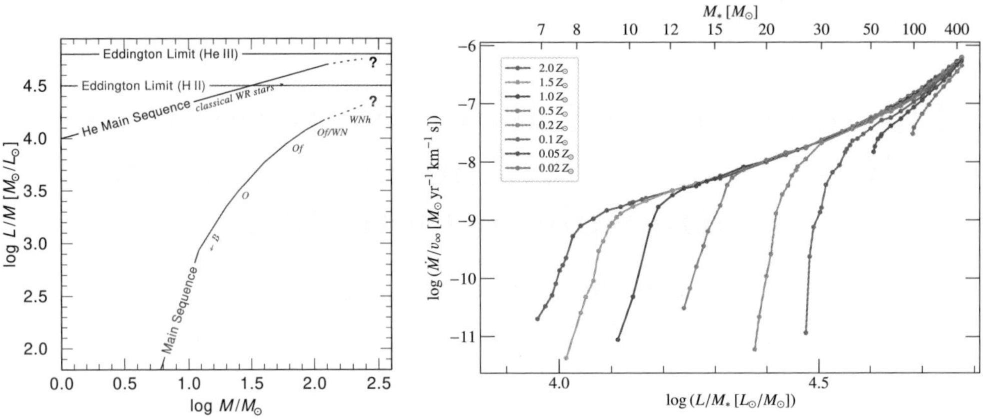

Figure 1. Left panel: Schematic occurrence of WR-type spectra in the $\log L/M$ vs. $\log M$ plane. When approaching the Eddington limit for hydrogen-containing stars at the upper end of the main sequence, we see a transition from Of to WN spectral type. Classical WR stars located on the He main sequence are close to the Eddington limit for hydrogen-free stars and thus also show strong stellar winds. Right panel: Ratio of the mass-loss rate $\dot{M}$ and the terminal velocity v_∞ as a function of L/M for the models from Sander & Vink (2020).

the increased mass loss of WR stars is a natural consequence when stars approach the Eddington limit of $\Gamma_\mathrm{e} \to 1$. The corresponding limit in L/M differs between hydrogen-free and hydrogen-rich stars due to $\Gamma_\mathrm{e} \propto q_\mathrm{ion} L/M$ (see, e.g., Sander et al. 2015) for detailed formulae) as hydrogen changes the value of q_ion. This is schematically illustrated in the left panel of Fig. 1, where we see the trends in L/M versus M for the main sequence and the He main sequence including their corresponding Eddington limits in L/M.

2. Mass-loss rates for classical WN stars

For the hotter, classical WR stars on the He main sequence, Sander & Vink (2020) calculated the first large set of hydrodynamically-consistent models. Employing the mass-luminosity relation from Gräfener et al. (2011) to reduce the parameter space and fixing the inner boundary constrain $T_*(\tau_\mathrm{Ross} = 20) = 141\,\mathrm{kK}$, they provide a description of $\dot{M}$ from sequences in L/M at multiple metallicities Z. A summary of the Sander & Vink (2020) model results is depicted in the right panel of Fig. 1, where the ratio of $\dot{M}$ and the terminal wind velocity v_∞ is shown as a function of L/M. Surprisingly and very much in contrast to OB-type winds, the curves for different metallicities start to align in the limit of dense winds, which is not the case when looking only at $\dot{M}$. The slope of these aligned curves can be brought into a linear dependence when plotting against $\log\left[-\log\left(1-\Gamma_\mathrm{e}\right)\right]$. Another important result from the models is a Z-dependent breakdown of $\dot{M}$ when the sonic point of the wind approaches the optically thin regime. At higher Z, strong mass-loss can be maintained down to lower L/M ratios as more line opacities are available. The finding of this rapid, Z-dependent breakdown qualitatively explains the empirical lower cutoffs found in observed WN populations of different Z (cf. Shenar et al. 2020).

As illustrated later on in the WN-WC comparison (cf. right panel of Fig. 3), the breakdown of $\dot{M}$ is accompanied by a change in the trend for v_∞, which reaches a minimum before eventually taking a steep turn towards higher values once the winds become optically thin. This result is tied to the choice of T_* (and thus the flux at the inner boundary). With our high value of 141 kK, the obtained thin-wind solutions are still launched by the opacities of the high Fe bump (as illustrated in Sander et al. 2020). The conditions in the resulting hot and thin winds are sufficient to keep Fe (and other elements) at higher ionization stages for thousands of stellar radii. In such cases, the

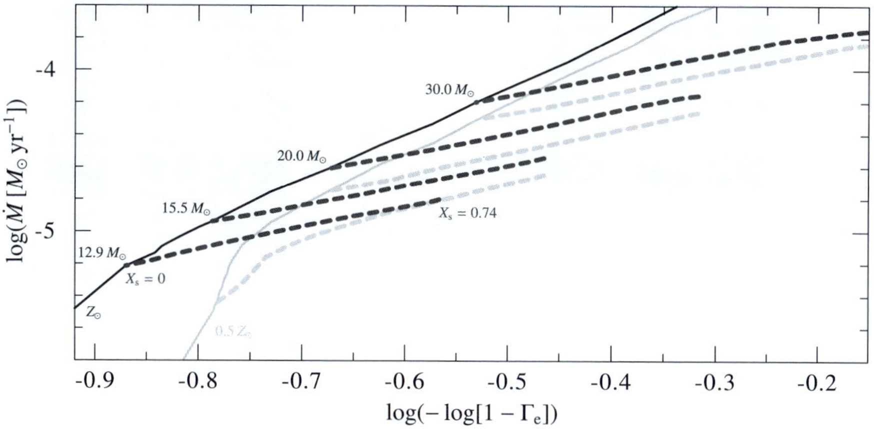

Figure 2. Influence of remaining surface hydrogen on the mass-loss rate $\dot{M}$: The blue curves show our sequence of hydrogen-free WN-type models with increasing L/M. For selected (L,M)-sets, the trend of including hydrogen in the atmospheric calculations (at the cost of helium) is illustrated by red, dashed curves. The opaque lines denote models at $Z_\odot$, while the semi-transparent lines refer to the same set of calculations as $0.5\,Z_\odot$.

acceleration of the thin, fast wind can be maintained and there is no lower limit in $\dot{M}$, in contrast to the results obtained in stellar structure calculations (e.g. Grassitelli et al. 2018, Ro 2019). It is presently unclear whether such massive, luminous, and hot stars do exist in nature or if their formation is prohibited by structural constraints. The detection of these objects, especially in unresolved low-Z galaxies, would be non-trivial as their predicted spectra neither show strong optical emission lines, nor pronounced P Cygni profiles in the UV.

3. Influence of leftover hydrogen at the surface

In addition to the two-dimensional parameter study along L/M and Z published in Sander & Vink (2020), we present first results from our work accounting for remaining hydrogen on the surface of WN-type stars. For the results depicted in Fig. 2, the models assume that the total hydrogen content is negligible in regard to the stellar mass, meaning that L and M are identical to the hydrogen-free calculations. In nature this would correspond to stars having only a thin, hydrogen-containing layer on the surface.

We compare the general trend of $\dot{M}$ for increasing L/M with the trends arising when increasing the surface hydrogen mass fraction X_H for two different metallicities applying models with fixed L, T_*, and M in Fig. 2. In the limit of optically thick winds, we always obtain the linear scaling of $\log \dot{M}$ with $\log\left[-\left(1-\Gamma_\mathrm{e}\right)\right]$ found in Sander & Vink (2020). Interestingly, the slope of the curves where we vary X_H is significantly shallower than the curve denoting the trend with varying L/M. The $20\,M_\odot$ model at $Z_\odot$ with $X_\mathrm{H}=0.2$ has approximately the same Γ_e as the $30\,M_\odot$ model with $X_\mathrm{H}=0$, but its mass-loss rate is a factor of two lower. The terminal velocity of the hydrogen-containing model is also about $250\,\mathrm{km\,s^{-1}}$ lower than the hydrogen-free $30\,M_\odot$ model, which is contrary to what happens when carbon is increased at the cost of helium as we will discuss further below.

Independent of our calculations representing classical WR stars, Sabhahit et al. (2022) recently found qualitatively similar results from stellar evolution calculations for very massive (WNh-type) stars. In their calculations, they needed the scaling of $\dot{M}$ with X_H to be weaker than the scaling with L/M in order to ensure that the resulting evolutionary tracks were covering only the observed region of WNh stars in the HRD. Otherwise, the mass loss would be too weak to prevent stars from expanding to cooler, unobserved

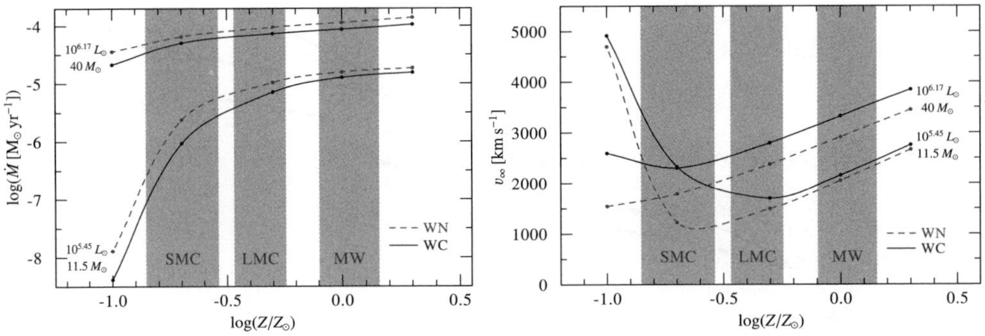

Figure 3. Mass loss rates (left panel) and terminal velocities (right panel) as a function of metallicity Z for four different model sequences. For each of the two different sets of basic stellar parameters (L, M), two different sets of surface abundances representing WN (green, dashed) and WC (blue, solid) stars were calculated.

temperatures. We are presently following up these findings with dynamically consistent calculations in the very massive star regime in order to check whether our insights on $\dot{M}(X_{\rm H})$ for classical WR are also applicable in this different temperature and mass regime or if the scaling diverges otherwise.

4. Similarities between WN and WC stars

As discussed above, the winds of WR stars are mainly launched by Fe M-shell opacities in the calculated temperature regime of $T_* \approx 140\,\mathrm{kK}$. In the outer winds, Fe ions usually remain the leading opacity for further accelerating the wind, but are supported by a plethora of further elements including carbon and oxygen. An immediate consequence of this situation is a strong similarity in the $\dot{M}$-dependences of WN and WC stars.

In Fig. 3, we illustrate this similarity by showing WN and WC model sequences for two different sets of basic stellar parameters (both assuming $T_* = 141\,\mathrm{kK}$). For the mass-loss rate, the WC model curves closely resemble the WN model curves, but with slightly *lower* absolute values for $\dot{M}$. The latter difference mainly roots in the slightly lower amount of free electron scattering available (and thus lower $\Gamma_{\rm e}$) in the WC models. In the WN models, helium has a fraction of 0.95 to 0.99 of the atmospheric mass (with detailed values depending on the metallicity of the model). These values are reduced by 0.45 ($40\,M_\odot$ sequence) and 0.65 ($11.5\,M_\odot$ sequence) in the WC models where we assume $X_{\rm C} = 0.4$ or 0.6 and $X_{\rm O} = 0.05$ for carbon and oxygen respectively. Although both C and O are in higher ionization stages at the launch of the wind and thus do not provide considerable line-driving opacities, both of these elements are not fully ionized. Hence, the budget of free electrons is reduced in comparison to the WN models and consequently there is less total acceleration available when adding up line and continuum opacities in the deeper layers. This leads to a slightly lower $\dot{M}$ for the same L and M_*, in contrast to often applied empirical recipes such as Nugis & Lamers (2000). However, when comparing WN and WC models with the same $\Gamma_{\rm e}$, the WC model yields the higher $\dot{M}$ (cf. Fig. 15 in Sander et al. 2020).

A second effect of changing the surface abundances from a WN to a WC composition is shown in the right panel of Fig. 3: As the higher $X_{\rm C}$ and $X_{\rm O}$ provide more line opacities in the wind, the WC models show a higher v_∞ than the WN models. This difference gets stronger for higher L/M, i.e. when getting closer to the Eddington limit. The examples further illustrate that the minimum of the v_∞-curves is different for all four depicted sequences. As this minimum is related to the transition to optically thin winds, it underlines that this transition is not only affected by changing basic parameters such as L and M, but also by (significant) abundance changes. Due to the generally lower $\dot{M}$ in the WC models, the transition to higher v_∞ occurs already at higher metallicities.

5. Summary & Conclusions

By calculating sequences of hydro-dynamically consistent atmosphere models along multiple dimensions, we obtained the first locally consistent $\dot{M}$-description for classical WR stars based only on first principles. The models further reveal that WR-type winds differ in various aspects substantially from OB-type winds. With increasing wind strength, we reach an optically thick ("pure WR") regime where the sonic point of WR-type wind located very deep in the atmosphere where the conditions do no longer deviate significantly from LTE (Sander & Vink 2020). However, the outer wind regime, which determines v_∞, remains in strong non-LTE, preventing any straight-forward simplification of the wind calculations. The precise dependencies of $\dot{M}$ as a function of stellar parameters are complex, although clear patterns emerge from calculating series of models along the L/M- and Z-dimension. In the limit of the pure WR regime, we find $\dot{M} \propto 1/\log(1-\Gamma_e)$ and a surprisingly shallow metallicity-dependence of $\dot{M} \propto Z^{0.3}$. In Higgins et al. (2021) we performed a first test of implementing the new $\dot{M}$-description into stellar evolution models of He stars, generally yielding lower total mass loss and potentially allowing for pair-instability supernovae up to $0.5\,Z_\odot$. Significant open questions remain, including the importance of further dependencies (e.g. T_*, X_H), where our calculations hint that additional terms need to be incorporated into future $\dot{M}$-descriptions. Eventually, a coherent picture also has to take into account the multi-dimensional WR wind structure, for which intriguing new insights are coming from recent 3D calculations (e.g. Moens et al. 2022). Together with colleagues, we will test and extend current approximation techniques (e.g. Poniatowski et al. 2021) to bring 1D and 3D modelling closer together and go beyond LTE in 3D calculations as well as implement 3D-insights into 1D atmosphere modelling.

References

Gräfener, G. & Hamann, W.-R. 2005, *A&A*, 432, 633
Gräfener, G. & Hamann, W.-R. 2008, *A&A*, 482, 945
Gräfener, G., Vink, J. S., de Koter, A., et al. 2011, *A&A*, 535, A56
Grassitelli, L., Langer, N., Grin, N. J., et al. 2018, *A&A*, 614, A86
Hamann, W.-R. & Gräfener, G. 2004, *A&A*, 427, 697
Higgins, E. R., Sander, A. A. C., Vink, J. S., et al. 2021, *MNRAS*, 505, 4874
Lefever, R. R., Shenar, T., Sander, A. A. C., et al. 2022, *these proceedings*
Moens, N., Poniatowski, L. G., Hennicker, L., et al. 2022, arXiv:2203.01108
Nugis, T. & Lamers, H. J. G. L. M. 2000, *A&A*, 360, 227
Nugis, T. & Lamers, H. J. G. L. M. 2002, *A&A*, 389, 162
Poniatowski, L. G., Sundqvist, J. O., Kee, N. D., et al. 2021, *A&A*, 647, A151
Ro, S. 2019, *ApJ*, 873, 76
Sabhahit, G. N., Vink, J. S., Higgins, E. R., et al. 2022, *MNRAS*, accepted, stac1410
Sander, A., Shenar, T., Hainich, R., et al. 2015, *A&A*, 577, A13
Sander, A. A. C., Hamann, W.-R., Todt, H., et al. 2017, *A&A*, 603, A86
Sander, A. A. C., Vink, J. S., & Hamann, W.-R. 2020, *MNRAS*, 491, 4406
Sander, A. A. C. & Vink, J. S. 2020, *MNRAS*, 499, 873
Shenar, T., Gilkis, A., Vink, J. S., et al. 2020, *A&A*, 634, A79
Vink, J. S. & de Koter, A. 2005, *A&A*, 442, 587
Vink, J. S., Muijres, L. E., Anthonisse, B., et al. 2011, *A&A*, 531, A132

Discussion

STEVANCE: I have a question about your statement that some surface hydrogen might increase the mass loss: What do you mean by a negligible mass fraction?

SANDER: By negligible I mean that it does not change the L/M-ratio when you compare it to a hydrogen-free star. This is what we have done in the presented models. Of course, if you put a considerable amount of hydrogen-rich mass on top of a helium star, the effect of the higher gravity will take over and your mass-loss rate should go down. We will also work on models with non-negligible mass additions, but this adds a further dimension and thus will take time. Generally you can expect that for a significant additional mass your mass-loss rate to be lower than the result coming out of the helium star model.

LAPLACE: I'm also very excited about the remaining hydrogen. I was wondering that if you have this increased mass loss in case of surface hydrogen, does it mean you will get rid of the hydrogen? This would really matter for the further evolution and end stage of the star including its spectra.

SANDER: As long as the additional hydrogen-rich mass is significant, I would expect that the resulting mass loss is decreased. But indeed the last bit of hydrogen should be very easy to remove because of this enhanced mass-loss effect.

LANGER: Great work, very much needed. With respect to your talk, I also liked that you highlight the importance of the Eddington limit. From the stellar structure considerations, if stars go there you find inflation and I was surprised that this did not come up in your talk. In the old days we had problems to find consistent WR mass-loss rates together with small radii. Are your new helium star models inflated or are they compact?

SANDER: Thank you very much for this excellent question that allows me to cover stuff that I could otherwise not fit in a 12-minute talk. The models that I talked about all have winds launched by the hot iron bump and their sonic points are on the order of 1 R_*. So they are compact, but they do not appear compact because they have this optically thick inner wind region. The winds will only get optically thin at much larger radii. Thus, from a "naive" β-law analysis, one would conclude that they have inflated radii. What we need to figure out now, also by comparing with observed spectra, is whether the solution of the current models is always the correct one or whether an inflated (hydrostatic) radius is an actual alternative for at least some observed stars. Moreover, there can also be deceleration regions in the wind. We did not see them in the models presented here, but we are now working on further models where we need to do some mitigations in order to treat them in the code. So, I would say the question is not settled yet, but my current feeling is that the winds are actually launched deep, but then sometimes meet a deceleration region. This would mean that we have a moving atmosphere rather than a hydrostatic inflation, but this is just my surmise at the moment.

Massive Stars Near and Far
Proceedings IAU Symposium No. 361, 2024
J. Mackey, J. S. Vink, & N. St-Louis, eds.
doi:10.1017/S1743921322003404

X-raying massive stars and their feedback near and far

Lidia Oskinova

Institute for Physics and Astronomy, University Potsdam, D-14476 Potsdam, Germany
email: lida@astro.physik.uni-potsdam.de

Abstract. Massive stars emit X-rays. Despite modest X-ray luminosities of single hot massive stars, the ongoing large observing campaigns combining X-ray and UV spectroscopy provide a tomographic view of stellar winds. It is now established that X-ray radiation is modulated with stellar rotation and shows the same period as discrete absorption components (DACs) in the UV resonance lines. The latter are associated with corotating interaction regions (CIRs) in stellar winds, therefore the mechanisms responsible for generation of X-rays and CIRs appear to be physically linked. Binary massive stars with accreting compact companions – high-mass X-ray binaries (HMXBs) – are routinely observed by modern X-ray observatories at Mpc distances. Joint observations in X-ray and UV allow to determine realistic properties of these systems. The brightest sources among HMXBs are ultraluminous X-ray sources (ULXs). Their powerful radiation is an important source of stellar feedback. HMXBs are the products of massive binary evolution and are typically found in the vicinity of young massive star clusters. The superstar clusters blow hot superbubbles which fill large areas in star-forming dwarf galaxies. Recent models show that X-ray emission from superbubbles is likely the dominant source of He II ionization in metal-poor star-forming dwarf galaxies. To conclude, X-ray observations provide an important window for studying massive stars and their feedback near and far.

Keywords. X-rays: stars, X-rays: binaries, X-rays: ISM, galaxies: star clusters

---

1. Introduction

Since the first discovery of X-rays from sources outside of the Solar System exactly sixty years ago (Giacconi et al. 1962), X-ray observations remain at the frontiers of space observational astronomy. Among currently operating X-ray telescopes, two observatories are of special importance for studies of massive stars. The NASA's *Chandra* X-ray observatory provides an unprecedented sub-arcsecond angular resolution, as well as possibilities for low- and high-resolution X-ray spectroscopy. The ESA's X-ray Multi-Mirror Mission (*XMM-Newton*) is the largest X-ray telescope operating simultaneously all its instruments, including two high-resolution X-ray spectrographs. *Chandra* and *XMM-Newton* have similar passband ($1.2 - 120$ Å). Data obtained by these telescopes during more than 20 yr of operations are accumulated in the most recent *XMM-Newton* and *Chandra* catalogs – 4XMM-DR11 and CSC 2.0. The median X-ray flux of objects in these catalogs is $F_X \sim 10^{-14}\,\mathrm{erg\,s^{-1}\,cm^{-2}}$ in the $0.3 - 7.0$ keV band (for comparison, the most sensitive *Chandra* observations reach a point source sensitivity $\sim 10^{-18}\,\mathrm{erg\,s^{-1}\,cm^{-2}}$).

Distances at which telescopes detect stars chiefly depend on the stellar X-ray luminosity and the intervening absorbing column. The typical X-ray luminosity of an O-type supergiant is $\sim 10^{33}\,\mathrm{erg\,s^{-1}}$ (Nebot Gómez-Morán & Oskinova 2018) – therefore these objects could be detected with median length exposures within 30 kpc. Colliding wind binaries have an order of magnitude higher X-ray luminosities (Rauw & Nazé 2016), and a higher than usual X-ray luminosity of a massive star is a reliable indicator of its binarity (Oskinova 2016). On the other hand, normal or low X-ray luminosity does not rule out that a massive star is a binary – enhanced X-ray luminosity is not a necessary attribute of massive binaries (Oskinova 2005, Nazé 2009). By now we know that all types of single hot massive stars, except of WC stars and LBVs, are X-ray sources. The X-ray loud WC stars and LBVs are colliding wind binaries (Oskinova et al. 2003, Nazé et al. 2012).

Massive stars which have a relativistic companion, a neutron star or a black hole, accreting stellar matter are called high-mass X-ray binaries (HMXBs). Typically, X-ray luminosities of persistent HMXBs are $> 10^{35}\,\mathrm{erg\,s^{-1}}$ while their X-ray spectra are hard. This allows easy detection of HMXBs at distances up to $\sim 300\,\mathrm{kpc}$. The hardness of spectra drastically reduces the influence of the interstellar absorption and allows to observe HMXBs across the Galaxy. In fact, the census of persistent HMXBs is nearly complete in the Galaxy and in the Magellanic Clouds (Liu et al. 2006, Krivonos et al. 2012, Haberl & Sturm 2016). The most luminous sources among the HMXB population belong to the class of ultraluminous X-ray sources (ULXs) with $L_{\mathrm{X}} > 10^{39}\,\mathrm{erg\,s^{-1}}$ (Walton et al. 2022). These enigmatic objects are easily detectable within $\sim 3\,\mathrm{Mpc}$.

2. X-rays from single massive stars are modulated with stellar rotation

X-ray spectroscopic diagnostics revealed that in single massive stars X-rays are generated close to photosphere in hot plasma expanding with velocities similar to the general cool wind velocity (Waldron & Cassinelli 2007). X-rays suffer absorption in the cool wind but wind clumping allows more radiation to escape than would be possible if winds were smooth (Oskinova et al. 2006). Observations of HMXBs allow to directly detect spectroscopic signatures of clumps which are optically thick for X-rays (Goldstein et al. 2004, Oskinova et al. 2012, Martínez-Chicharro et al. 2021). At the same time, dedicated searches using excellent data ruled out the presence of stochastic variability which could be linked to strong shocks in stellar winds (Nazé et al. 2013, Nichols et al. 2015).

There is firm evidence that X-rays from single O and WR stars are variable at $\sim 10-20\%$ level on the rotational time scale (Oskinova et al. 2001, Ignace et al. 2013, Massa et al. 2014). Nichols et al. (2021) found that the X-ray flux of the prototypical O-type supergiant ζ Puppis (O4I) is modulated with $P = 1.78\,\mathrm{d}$, i. e. the same period as observed in the broad-band optical photometric light-curve albeit in anti-phase with it.

The association of X-rays with the large scale structures in rotating star winds – corotating interaction regions (CIRs) – has been long predicted (Mullan 1984). The signatures of the CIRs are observed as discrete absorption components (DACs) in the UV resonance lines which originate in stellar winds (Prinja & Howarth 1986, Hamann et al. 2001). To obtain a tomographic view of stellar winds and uncover links between the CIRs and X-ray emission we initiated observing campaigns using the *HST* STIS UV spectrograph and the *XMM-Newton* X-ray telescope. We selected two fast rotating single O-type stars: ξ Per (O7.5III) and ζ Oph (O9.2IV). These objects have different wind strengths, with ξ Per's wind being more dense than ζ Oph's wind, and also different pulsation and rotation periods.

Massa et al. (2019) presented the results of joint *XMM-Newton* and *HST* observations of ξ Per. The detailed study of the UV spectral variability revealed that it remains consistent for at least 22 years elapsed since the previous UV spectral study of this object. The DACs recurrence time, 2.1 d, remained stable. The X-ray variability period is the same as the DACs period, however we established that there is a phase shift between normalized X-ray and the UV wind line fluxes.

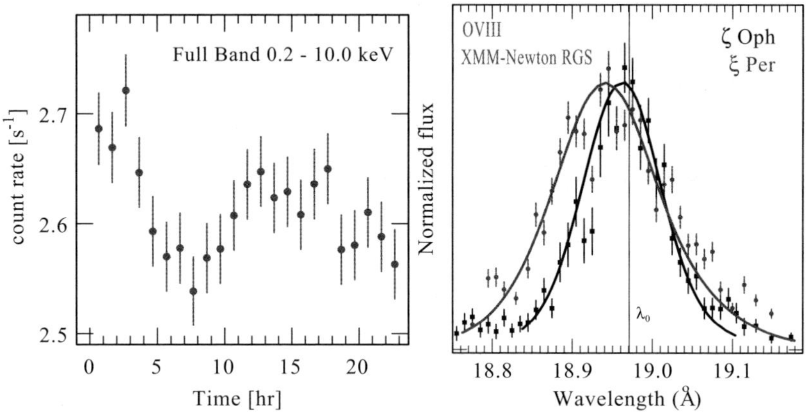

Figure 1. *Left: XMM-Newton* EPIC pn background-subtracted X-ray light curve of ζ Oph in the $0.2 - 10.0$ keV band. The data are binned to 30 min. The horizontal axis denotes the time after the beginning of the observation in hours. The vertical axis shows the count rate as measured by the EPIC pn camera. The error bars (1σ) correspond to the combination of the error in the source counts and the background counts. *Right: XMM-Newton* RGS spectra of the resonance line O VIII λ18.97 Å obtained during the observations of ξ Per (red error bars) and ζ Oph (black error bars). The solid lines show corresponding line profiles predicted by our clumped wind models (Oskinova et al. 2006).

Similar results are found from the analysis of the recent *XMM-Newton* observations of ζ Oph which covered the full rotational period of the star, $P_{\rm rot} \approx 15.28$ hr, as well as its DACs recurrence period, $P_{\rm UV} \approx 20.8$ hr (Howarth et al. 1993). As illustrated in Fig. 1, the X-ray light curve is modulated with rotational period. The analysis of *HST* spectroscopic observations which are nearly simultaneous with the X-ray observations is ongoing.

Thus, the current data point out that the generation of X-rays in massive stars is linked with rotation and is associated with large scale wind structures. Furthermore, the presence of the CIRs was invoked by Bozzo et al. (2017) to explain the observed super-orbital modulation of the X-ray flux in supergiant HMXBs.

3. Hot winds as an explanation of the weak-wind problem

Figure 1 shows the O VIII λ18.97 Å resonance line in the spectra of ξ Per and ζ Oph. The strong absorption of X-rays in the dense wind of ξ Per leads to the line blue-shift (Oskinova et al. 2006). In contrast, the wind of ζ Oph is essentially transparent for emergent X-rays in agreement with the low mass-loss rates inferred from the analyses of UV spectra (Martins et al. 2005). As a result, the observed X-ray emission line is not blue-shifted.

X-rays associated with hot plasma distributed in stellar winds have long been invoked as a promising solution of the so-called *weak wind problem* (Cassinelli et al. 1994, Martins et al. 2005, Lagae et al. 2021). Lucy (2012) suggested that in the late-type O-dwarfs the bulk of the wind is hot, with temperatures of a few MK. The cool radiatively driven gas is confined to dense clumps. Further out in the wind, these cool clumps are destroyed by heat conduction, and the outflow is dominated by a hot thermal wind which reaches a supersonic terminal velocities.

Independently, similar solution to the weak-wind problem was proposed by Huenemoerder et al. (2012) based on the empiric study of the high-resolution X-ray spectra of late O-type dwarfs, μ Col and σ Ori AB. It was shown that the emission measure of the X-ray emitting plasma is much larger than emission measure of the cool wind. Huenemoerder et al. (2012) proposed that the weak-wind problem is resolved when the hot dominant component of the wind is accounted for in the total mass-loss rate budget. It was suggested that "the wind is not weak, but it is hot and its bulk is only detectable in X-rays." The preliminary results of our ongoing study of another O-dwarf, ζ Oph, corroborate this conclusion.

4. High-mass X-ray binaries and ULXs

Traditionally, the focus in studies of HMXBs lies on the analysis of their X-ray spectra and light-curves. However, recent work shows that detailed modeling of donor star spectra in optical and UV is needed to obtain realistic physical description of these systems.

The HMXBs with OB supergiant donor stars are classified in two groups according to their X-ray properties: persistent sources and supergiant fast X-ray transients (SFXTs, Martínez-Núñez et al. 2017). It has been suggested that this dichotomy depends on the characteristics of stellar winds, such as enhanced clumping or weaker winds in SFXTs compared to the persistent sources. Hainich et al. (2020) analyzed stellar and wind parameters of six HMXBs using UV spectra obtained with the *HST*. This thorough study showed that winds of donor stars in SFXTs and persistent HMXBs are similar. Recently, (Sidoli et al. 2021) detected intrinsic X-ray emission from the O-type donor in a quiescent SFXT, further confirming that winds of donors in SFXTs are not special. The SFXT phenomenon could be explained by interactions between the neutron star's magnetic field, its spin, and the dynamics of donor star wind (Bozzo et al. 2008, Shakura et al. 2014, Bozzo et al. 2016). Grids of spectral models for OB stars at different metallicities can be used for analyzes of donor stars in HMXBs (Hainich et al. 2019).

HMXBs and ULXs are important sources of feedback (Lebouteiller et al. 2017, Sazonov & Khabibullin 2018, Schaerer et al. 2019). Oskinova et al. (2019) discovered three ULXs in the metal-poor galaxy ESO 338 and suggested that, in order to adequately model stellar feedback in this and similar low-metallicity dwarf galaxies, X-rays have to be taken into account.

5. X-ray emission from hot superbubbles explain nebular He II emission in metal-poor dwarf galaxies

Star-forming galaxies are filled by diffuse X-rays produced by a few MK hot gas (e.g. Martin 1996, Tenorio-Tagle et al. 2006, Danehkar et al. 2021). Observations establish that its X-ray luminosity is proportional to the star-formation rate (SFR) of the host galaxy (Smith et al. 2019, Mineo et al. 2012, Lehmer et al. 2022). Oskinova & Schaerer (2022) suggest that the nebular He II emission observed in dwarf galaxies can be explained by X-rays from superbubbles blown up by cluster winds of superstar clusters with $M_{\rm cl} > 10^5\,M_\odot$. The kinetic energy feedback of massive stars increases as star clusters evolve, and reaches its maximum in a few Myr old clusters (Oskinova 2005). Initially, the energy input is dominated by stellar winds, while later on it gradually increases due to the input from core-collapse supernovae.

The temperature of superbubbles scales with input kinetic energy. Because at lower metallicities stellar winds are weaker, superbubble temperatures are also lower. Therefore, at low metallicity, the bulk of the X-rays emerging from superbubbles has energies close to the He II ionization edge.

In order to estimate the He II ionizing power of the whole ensemble of young star clusters in a star-forming galaxy, Oskinova & Schaerer (2022) employed an empirical correlation between the cluster formation rate and SFR in galaxies. It was shown that X-ray luminosity of hot diffuse gas in a galaxy is proportional to its SFR $[M_\odot\,{\rm yr}^{-1}]$ as $\log L_{\rm X} \approx \log {\rm SFR} + 39.6\,[{\rm erg\,s^{-1}}]$, i.e. similar to observations (Lehmer et al. 2022). Hence, according to this model, the observed prevalence of He II nebulae in low-Z galaxies is

Table 1. He II ionizing photon rate in WR stars and hot superbubbles (SBB) with $L_X(\mathrm{SBB}) = 10^{40}\,\mathrm{erg\,s^{-1}}$. The metallicity is $Z = 0.07Z_\odot$.

Object	Model*	Temperature	$\log Q(\mathrm{He}^+)$ [s^{-1}]
WNE star	WNE-2014 model 16-14	$T_* = 141\,\mathrm{kK}$	48.2
WNL star	WNL-H60-20	$T_* = 110\,\mathrm{kK}$	47.8
WC star	WC-19-18	$T_* = 200\,\mathrm{kK}$	48.5
WNE star	WNE-18-13	$T_* = 178\,\mathrm{kK}$	48.4
WNL star	WNL-H60-14-06	$T_* = 112\,\mathrm{kK}$	47.7
SBB	*apec*	$kT = 0.5\,\mathrm{keV}$	49.7
SBB	*apec*	$kT = 0.2\,\mathrm{keV}$	50.2
SBB	*apec*	$kT = 0.1\,\mathrm{keV}$	51.0

(*) For WR stars, the model number is according to the PoWR model grid accessible at `www.astro.physik.uni-potsdam.de/PoWR`.

due to the combination of (1) lower temperature of superbubbles, (2) the linear scaling between the X-ray output of star clusters and the SFR of their host galaxy.

Besides superbubbles, WR stars also contribute to the He II ionizing flux (Table 1). The Oskinova & Schaerer (2022) model links star and cluster formation with stellar evolution and feedback. Consistent modeling of galaxies exhibiting He II nebula emission is the next required step in understanding stellar feedback at low metallicity.

Acknowledgements

This work would have been impossible without my co-authors and collaborators. Special thanks and appreciation of their valuable work are to the current PhD students – Matthew Rickard, Timothy Parsons, Sabela Reyero, and Daniel Pauli.

References

Bozzo, E., Falanga, M., & Stella, L. 2008, ApJ, 683, 1031
Bozzo, E., Oskinova, L., Feldmeier, A., & Falanga, M. 2016, A&A, 589, A102
Bozzo, E., Oskinova, L., Lobel, A., & Hamann, W. R. 2017, A&A, 606, L10
Cassinelli, J. P., Cohen, D. H., Macfarlane, J. J., Sanders, W. T., & Welsh, B. Y. 1994, ApJ, 421, 705
Danehkar, A., Oey, M. S., & Gray, W. J. 2021, ApJ, 921, 91
Giacconi, R., Gursky, H., Paolini, F. R., & Rossi, B. B. 1962, Physical Review Letters, 9, 439
Goldstein, G., Huenemoerder, D. P., & Blank, D. 2004, AJ, 127, 2310
Haberl, F., & Sturm, R. 2016, A&A, 586, A81
Hainich, R., Ramachandran, V., Shenar, T., Sander, A. A. C., Todt, H., Gruner, D., Oskinova, L. M., & Hamann, W. R. 2019, A&A, 621, A85
Hainich, R., et al. 2020, A&A, 634, A49
Hamann, W. R., Brown, J. C., Feldmeier, A., & Oskinova, L. M. 2001, A&A, 378, 946
Howarth, I. D., et al. 1993, ApJ, 417, 338
Huenemoerder, D. P., Oskinova, L. M., Ignace, R., Waldron, W. L., Todt, H., Hamaguchi, K., & Kitamoto, S. 2012, ApJL, 756, L34
Ignace, R., Gayley, K. G., Hamann, W. R., Huenemoerder, D. P., Oskinova, L. M., Pollock, A. M. T., & McFall, M. 2013, ApJ, 775, 29
Krivonos, R., Tsygankov, S., Lutovinov, A., Revnivtsev, M., Churazov, E., & Sunyaev, R. 2012, A&A, 545, A27
Lagae, C., Driessen, F. A., Hennicker, L., Kee, N. D., & Sundqvist, J. O. 2021, A&A, 648, A94
Lebouteiller, V., et al. 2017, A&A, 602, A45
Lehmer, B. D., Eufrasio, R. T., Basu-Zych, A., Garofali, K., Gilbertson, W., Mesinger, A., & Yukita, M. 2022, ApJ, 930, 135
Liu, Q. Z., van Paradijs, J., & van den Heuvel, E. P. J. 2006, A&A, 455, 1165
Lucy, L. B. 2012, A&A, 544, A120

Martin, C. L. 1996, ApJ, 465, 680
Martínez-Chicharro, M., et al. 2021, MNRAS, 501, 5646
Martínez-Núñez, S., et al. 2017, Space Science Reviews, 212, 59
Martins, F., Schaerer, D., Hillier, D. J., Meynadier, F., Heydari-Malayeri, M., & Walborn, N. R. 2005, A&A, 441, 735
Massa, D., Oskinova, L., Fullerton, A. W., Prinja, R. K., Bohlender, D. A., Morrison, N. D., Blake, M., & Pych, W. 2014, MNRAS, 441, 2173
Massa, D., Oskinova, L., Prinja, R., & Ignace, R. 2019, ApJ, 873, 81
Mineo, S., Gilfanov, M., & Sunyaev, R. 2012, MNRAS, 426, 1870
Mullan, D. J. 1984, ApJ, 283, 303
Nazé, Y. 2009, A&A, 506, 1055
Nazé, Y., Oskinova, L. M., & Gosset, E. 2013, ApJ, 763, 143
Nazé, Y., Rauw, G., & Hutsemékers, D. 2012, A&A, 538, A47
Nebot Gómez-Morán, A., & Oskinova, L. M. 2018, A&A, 620, A89
Nichols, J., et al. 2015, ApJ, 809, 133
Nichols, J. S., et al. 2021, ApJ, 906, 89
Oskinova, L. M. 2005, MNRAS, 361, 679
—. 2016, Advances in Space Research, 58, 739
Oskinova, L. M., Bik, A., Mas-Hesse, J. M., Hayes, M., Adamo, A., Östlin, G., Fürst, F., & Otí-Floranes, H. 2019, A&A, 627, A63
Oskinova, L. M., Clarke, D., & Pollock, A. M. T. 2001, A&A, 378, L21
Oskinova, L. M., Feldmeier, A., & Hamann, W. R. 2006, MNRAS, 372, 313
Oskinova, L. M., Feldmeier, A., & Kretschmar, P. 2012, MNRAS, 421, 2820
Oskinova, L. M., Ignace, R., Hamann, W. R., Pollock, A. M. T., & Brown, J. C. 2003, A&A, 402, 755
Oskinova, L. M., & Schaerer, D. 2022, A&A, 661, A67
Prinja, R. K., & Howarth, I. D. 1986, ApJS, 61, 357
Rauw, G., & Nazé, Y. 2016, Advances in Space Research, 58, 761
Sazonov, S., & Khabibullin, I. 2018, MNRAS, 476, 2530
Schaerer, D., Fragos, T., & Izotov, Y. I. 2019, A&A, 622, L10
Shakura, N., Postnov, K., Sidoli, L., & Paizis, A. 2014, MNRAS, 442, 2325
Sidoli, L., Postnov, K., Oskinova, L., Esposito, P., De Luca, A., Marelli, M., & Salvaterra, R. 2021, A&A, 654, A131
Smith, B. J., Wagstaff, P., Struck, C., Soria, R., Dunn, B., Swartz, D., & Giroux, M. L. 2019, AJ, 158, 169
Tenorio-Tagle, G., Muñoz-Tuñón, C., Pérez, E., Silich, S., & Telles, E. 2006, ApJ, 643, 186
Waldron, W. L., & Cassinelli, J. P. 2007, ApJ, 668, 456
Walton, D. J., Mackenzie, A. D. A., Gully, H., Patel, N. R., Roberts, T. P., Earnshaw, H. P., & Mateos, S. 2022, MNRAS, 509, 1587

Massive Stars Near and Far
Proceedings IAU Symposium No. 361, 2024
J. Mackey, J. S. Vink, & N. St-Louis, eds.
doi:10.1017/S1743921322003246

Properties of Primitive Galaxies

Sara R. Heap[1], **I. Hubeny**[2], **J.-C. Bouret**[3], **T. Lanz**[4] **and J. Brinchmann**[5]

[1]NASA Goddard Space Flight Center, Greenbelt, MD,
email: sara.heap@Gmail.com

[2]Steward Observatory, University of Arizona, Tucson, AZ,

[3]Aix-Marseille Univ. CNRS, CNES, LAM, Marseille, France,

[4]Observatoire de Côte d'Azur, Nice, France,

[5]University of Porto, Porto, Portugal

Abstract. We report on a study of 9 nearby primitive galaxies observed by Hubble's COS far-UV spectrograph that can serve as templates of high-z galaxies to be observed by JWST. By "primitive galaxies," we mean galaxies having a low stellar mass, $\log(M_*/M_\odot) \leqslant 8$ and low gas metallicity, $\log O/H + 12 \leqslant 8$, whether they are local or at high redshift. We find that far-UV spectra of these galaxies show evidence of hard radiation, including X-rays. Following Thuan et al. (2004), we identify these galaxies as massive X-ray binaries containing a massive accreting stellar black hole. We further find that the lower the metallicity, the higher the probability of extremely strong X-radiation. Following Heger et al. (2003), we suggest that the accreting black hold is produced by direct collapse of stars having initial masses greater than $50\,M_\odot$. The X-radiation produced by black hole disk directly affects the surrounding interstellar medium, and many of these effects are observable in far-UV spectra.

Keywords. techniques: spectroscopic, ultraviolet: galaxies, X-rays: binaries, galaxies: dwarf

1. Introduction

We report on a study of far-UV spectra of nearby primitive galaxies that could give insights into processes operating in the high-z galaxies thought to be responsible for the reionization of the universe. By "primitive galaxies," we mean galaxies having a low stellar mass, $\log(M_*/M_\odot) \leqslant 8$ and low gas metallicity, $\log O/H + 12 \leqslant 8$ and an even lower iron abundance, whether they are local or at high redshift. How does a low metallicity affect the physical properties and evolutionary path of low-metallicity massive stars? One way is through its role in the generation of high-velocity winds of hot, massive stars. Another way is through the presence of hard radiation. X-radiation in extremely deficient stars was first pointed out by Thuan et al. (2004) who found that the three most metal-deficient galaxies are ultra-luminous X-ray emitters (ULX). A ULX is defined by having an X-ray luminosity, $L_x > 1 \times 10^{39}$ erg s^{-1}. They identified these X-ray sources as massive X-ray binaries (MXRB's), each containing a massive, accreting stellar black hole. The optical spectral region also shows evidence of hard radiation. Shirazi & Brinchmann (2012) found that spectra of primitive galaxies observed by the SDSS show *nebular* He II λ4686 in emission, the result of ionization to He III and subsequent recombination. Recently, spectroscopic surveys by Izotov et al. (2021) added six compact star-forming galaxies with $12 + log\mathrm{O/H} = 7.46 - 7.88$ all showing evidence of hard radiation in the form of high-ionization emission lines such as [Ne V].

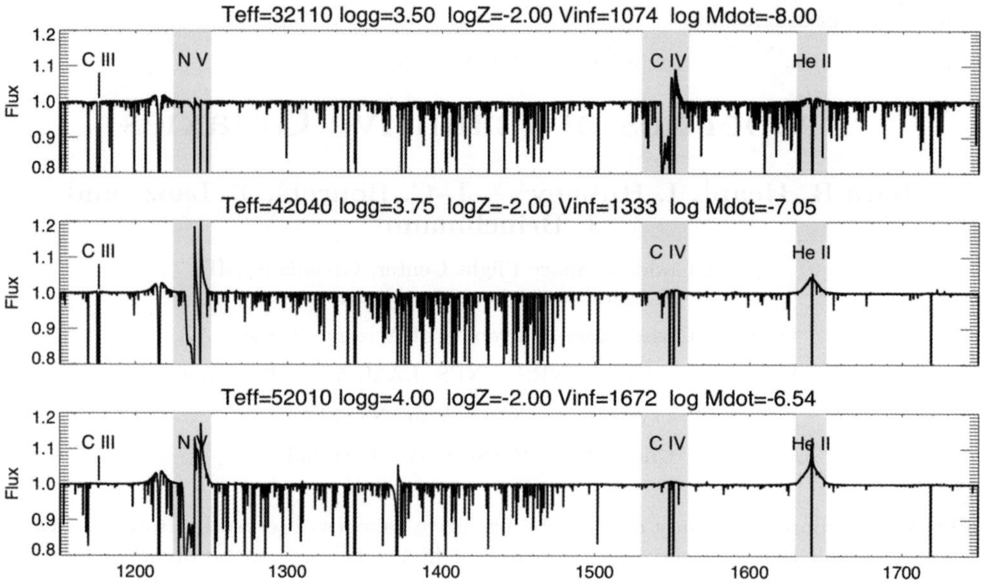

Figure 1. CMFGEN models of the far-UV spectra of main-sequence O-type stars. The stellar parameters are given above each model. They are: effective temperature, log g, log Z, terminal velocity of the wind in km/s, log mass-loss rate in solar masses per year. In a single star, the N V doublet and C IV doublet cannot both be strong, whereas in a star cluster or galaxy, both features can be strong. As He II emission is formed in the stellar wind, It is a broad feature, easily distinguishable from nebular emission which is narrow.

2. Observations of primitive galaxies

The far-UV spectral region is the most informative region for studying stellar winds of primitive stars as it has resonance lines of abundant elements in several ionization stages. There is empirical evidence from far-UV spectra of hot, massive stars in the galaxy and Magellanic Clouds that the strength of the stellar wind increases with increasing metallicity as $Z^{0.7}$ (Vink et al. 2001), but this relation has not been tested below $\log Z \sim 8.0$. We have made predictions of the far-UV spectra of stars based on the grid of non-LTE photospheric models by Lanz & Hubeny (2003; 2007) and unified spectra (photosphere + wind) calculated by Bouret using the CMFGEN code (Hillier & Miller 1998). Figure 1 shows predicted spectra of massive stars showing that even at $\log Z/Z_\odot = -2$, the N V doublet and C IV doublet are still detectable stellar-wind features. When compared to the spectrum of the most metal-deficient galaxy in our sample, I Zw 18-NW, there is qualitative agreement between theory and observation, given that we are comparing the spectra of a galaxy to that of a single star.

Far-UV spectra are similarly informative about the circumstellar, interstellar, and circumgalactic medium of primitive galaxies. Figures 2–3 show the far-UV spectra of 9 galaxies having a nebular oxygen abundances, $\log \mathrm{O/H} + 12 = 7.0 - 8.3$. Spectral regions encompassing the C IV $\lambda\lambda 1548, 1550$ doublet, He II $\lambda 1640$, and O III] $\lambda\lambda 1661, 1666$ are highlighted in yellow. The spectra are arranged in order of increasing oxygen abundance. In Figure 2, covering $\log \mathrm{Z} = 7.0 - 7.8$, not only nebular He II $\lambda 1640$ (IP = 54 eV) is in emission but also nebular C IV doublet (IP=64 eV) in emission superimposed on the stellar spectrum, in which C IV doublet has a P Cygni profile characteristic of a radiatively driven, high-velocity stellar wind. In Figure 3, covering $\log \mathrm{Z} = 7.8 - 8.2$, nebular He II $\lambda 1640$ emission weakens with increasing metallicity to the point that it is

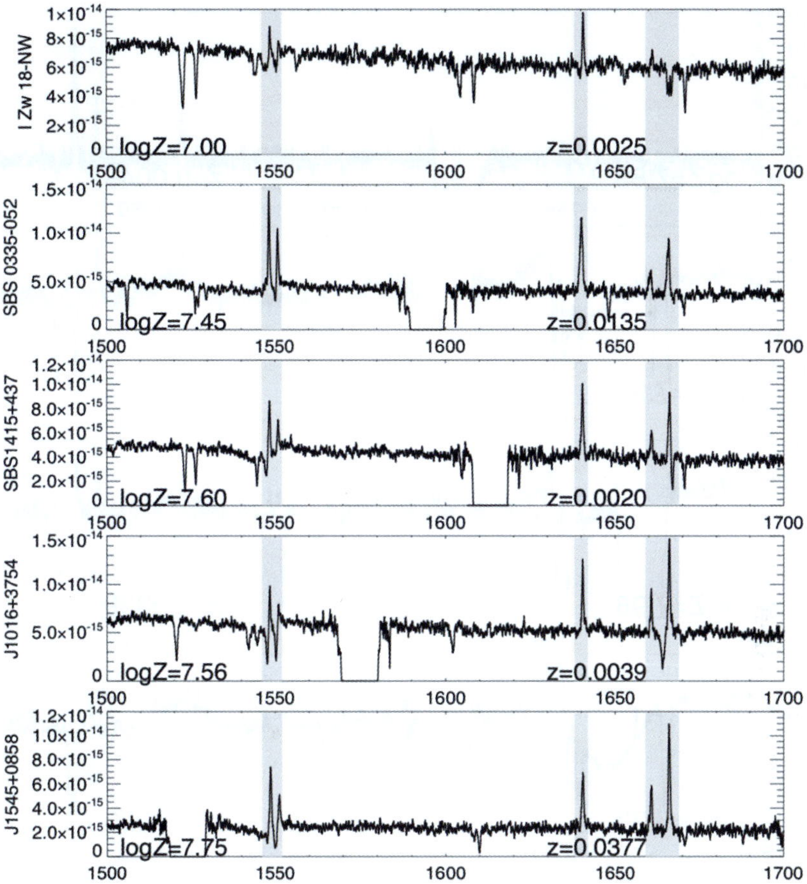

Figure 2. Hubble/COS far-UV spectra of 9 low-Z galaxies. Spectral regions encompassing the C IV $\lambda\lambda1548, 1550$ doublet, He II $\lambda1640$, and O III] $\lambda\lambda1661, 1666$ are highlighted. The galaxy ID is given on the Y-axis of the plot, and its redshift, at the bottom right of the plot. The nebular abundance of oxygen, $\log \mathrm{O/H} + 12$, is listed at the bottom left, denoted as log Z. The spectra are arranged in in order of increasing metal abundance. The spectra of I Zw 18-NWwere obtained directly in programs ID = 11523, 12028 (PI:Green), and most of the others in program ID=14120 (PI:Brinchmann). The two exceptions are SBS 0335-053 and SBS 1415+437, which were obtained from the Hubble archive (ID 15193, PI: Aloisi).

undetectable at $\log \mathrm{Z} = 8.2$. At the same time, broad stellar emission, characteristic of Wolf-Rayet stars, becomes strong in two of the galaxies.

3. Discussion and Conclusions

How are we to understand this metallicity sequence of spectra? To find out, we used CLOUDSPEC, a combination of CLOUDY (Ferland et al. 2017) and SYNSPEC (Hubeny & Lanz 1995) to model the far-UV spectra of galaxies for 3 scenarios. In each scenario, there is a 5 Myr-old primitive stellar population (called "SPop" in Fig. 4) like I Zw 18-NW. In the first scenario (red), there is a dormant black-hole companion or no black hole at all. In the second (green), there is ia black-hole disk radiating X-rays at a rate of 10^{39} erg/s, and in the third (blue), a X-ray source ten times more luminous. Figure 4 shows the results. Nebular emission by He II $\lambda1640$ is not present when there is no black hole, it is detectable in the case of $L_x = 10^{39}$ erg/s, and it quite strong in the $L_x = 10^{40}$ erg/s case. The C IV doublet is seen in absorption when there is no black hole,

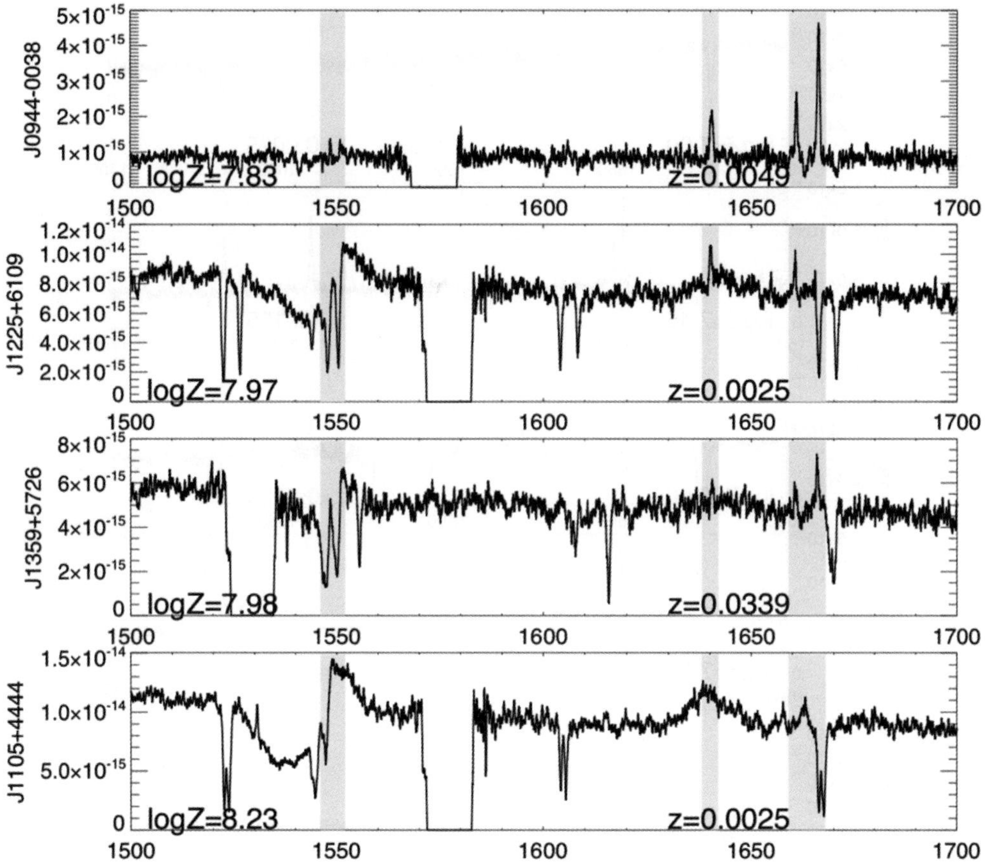

Figure 3. The same as Fig. 2, showing far-UV spectra of higher-z galaxies of our sample.

the absorption lines are partially filled in for the case of $L_x = 10^{39}$ erg/s, and strongly in emission at $L_x = 10^{40}$ erg/s. Although the statistics are limited, we conclude that the presence of C IV emission is associated with galaxies whose metallicity,log $Z \leqslant 7.83$, while nebular He II emission is still present in galaxies until log $Z > 8.0$.

Why should the presence of a black hole be correlated with metallicity? One answer comes from a paper by Heger et al. (2003), titled: "How Massive Single Stars End Their Life". It suggests that stars having a final mass greater than about 40 $M_{\odot}$ collapse directly into a black hole with little to no mass lost in an explosion. Since very low metallicity stars lose very little mass during their lifetime because of such weak stellar winds, their final mass is 80 % of their initial mass (e.g. Groh et al. 2019), so stars formed with an initial mass of 50 $M_{\odot}$ or more will eventually collapse to a massive stellar black hole.

In this interpretation, we are viewing the former secondary orbiting a black hole, which formed from the collapsed primary, and the former secondary is now feeding the black hole via a black hole disk. The black hole disk is hot, particularly at its inner edge where it is 1-2 million degrees, so its emission is centered on the X-ray spectral region. The reason why the disk is hot is because a stellar-mass black hole, as opposed to a supermassive black hole, forms a very deep potential well, so that an accreted material is heated to such high temperatures.

There are still issues to be cleared up, mainly the evolution of massive stars in binary systems, since most hot, massive stars are in binary systems (Sana et al. 2012). Other factors such as the rotation and the C/O ratio must be explored.

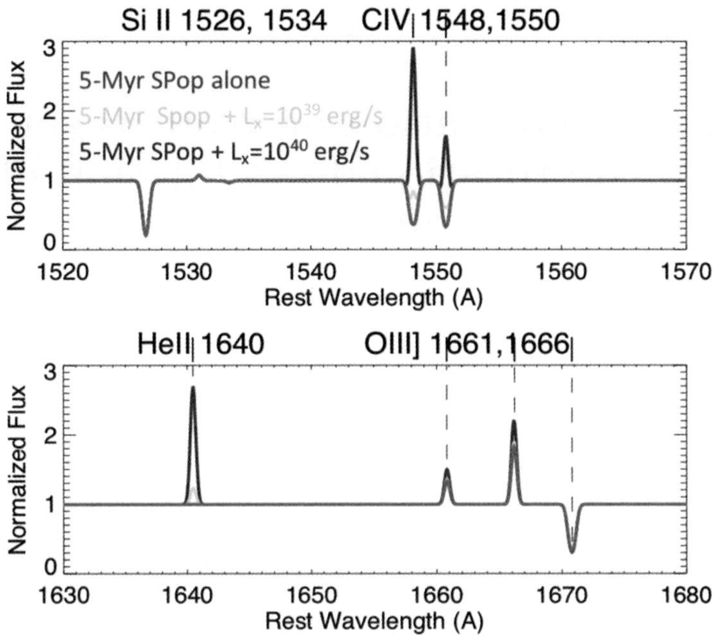

Figure 4. Sample CLOUDSPEC model (Hubeny et al. 2000) of a primitive galaxy like I Zw 18-NW showing the effects of an embedded stellar black hole. Inputs to CLOUDSPEC include: full SED of a stellar cluster at age=5 Myr, log Z = 7.2; BH parameters: Lx as shown, temperature of accretion disk, $T = 10^6$ K; nebular parameters: election density, Ne=10 cm^{-3}, H I column density =2×10^{21} cm^{-2}, covering fraction=0.4, no dust.

We conclude that the properties of massive, extremely low metallicity are extensions of those of low metallicity, e.g. spectral lines are generally weaker, and stellar winds are weaker. However, the final fate and afterlives of primitive massive stars is qualitatively different from those of more metal-rich stars in showing evidence of hard radiation that can best be explained by a hot black hole disk.

References

Ferland, G. 2017, *RMxAA*, 53, 385

Groh, J., Ekström, S., Georgy, C. 2019, *A&A*, 627, 24

Heger, A., Fryer, C., Woosley, S. et al. 2003, *ApJ*, 591, 288

Hillier D. J. & Miller, D. 1998, *ApJ*, 496, 407

Hubeny, I., Heap, S., Lanz, T. 2000, *AAS*, 196, 291

Hubeny, I. & Lanz, T. 1995, *ApJ*, 439, 875

Izotov, Y., Thuan, T.X., Guseva, N. G. 2021, *MNRAS*, 508, Issue 3, p. 2556

Lanz, T. & Hubeny, I. 2003, *ApJS*, 146, 417

Lanz, T. & Hubeny, I. 2007, *ApJS*, 169, 83

Lebouteiller, V., Heap, S., Hubeny, I. Kunth, D. 2013, *A&A*, 553, 16

Lebouteiller, V., Péquignot, D., Cormier, D., et al. 2017, *A&A*, 602, 45

Sana, H., de Mink, S. E., de Koter, A. et al. 2012, *arXiv*:1211.4740

Shirazi, M. & Brinchmann, J. 2012, *MNRAS*, 421, 1043

Thuan, T., Bauer, F. E., Papaderos, P. et al. 2004, *ApJ*, 606, 213

Vink, J., de Koter, A., Lamers, H. 2001, *A&A*, 369, 574

Massive Stars Near and Far
Proceedings IAU Symposium No. 361, 2024
J. Mackey, J. S. Vink, & N. St-Louis, eds.
doi:10.1017/S1743921322002241

Intrinsic polarization of Wolf-Rayet stars due to the rotational modulation of the stellar wind

Slah Abdellaoui[1], **Jiří Krtička**[1] **and Petr Kurfürst**[1,2]

[1]Department of Theoretical Physics and Astrophysics, Masaryk University, Kotlářská 2, 611 37 Brno, Czech Republic
email: slah@physics.muni.cz

[2]Institute of Theoretical Physics, Charles University, V Holešovičkách 2, 180 00 Praha 8, Czech Republic

Abstract. Wolf-Rayet stars are regarded as candidates for progenitors of core-collapse supernovae, and they are expected to be progenitors of long gamma-ray bursts. These types of stars are considered to be fast rotators. Their high rotation speed breaks the sphericity of the star and leads to an axisymmetric wind density structure. In such a case, the electron scattering takes place in a nonspherical environment, and as a result, we might expect an intrinsic polarization. We present a 2.5D radiation hydrodynamic stellar wind model of these stars. The model simulations account for the deformation of the stellar surface due to rotation, gravity darkening, and nonradial forces. We computed the polarization from the density variable of the hydrodynamic model, derived the upper limit of rotational velocities, and found no conflict with the previous studies of Wolf-Rayet stars.

Keywords. stars: rotation, stars: Wolf-Rayet, stars: mass loss, techniques: polarimetric

---

1. Introduction

Collapse of WR stars to a black hole could lead to a long gamma-ray burst (Woosley 1993). From spectropolarimetry observations, Stevance et al. (2018) derived the upper limit of rotational velocities of two WR stars. We used 2.5D hydrodynamic simulation (Owocki et al. 1994) to obtain the wind density structure of these stars (Fig. 1), from which we computed the polarization (Fig. 2).

2. Intrinsic polarization

We intend to compute the polarization due to electron scattering in continuum (Thomson scattering). The analytic expression of the polarization was developed by (Brown & McLean 1977), which depends on the inclination angle, optical depth, and geometry factor, as follows:

$$P_R \approx \bar{\tau}\,(1-3\gamma)\sin^2 i, \tag{2.1}$$

where $\bar{\tau}$ is the averaged Thomson scattering optical depth defined as the integral of the number density over the radius and the colatitudinal angle θ and γ is determined by the density distribution. To take the depolarization effect of finite stellar geometry into account, Cassinelli et al. (1987) introduced the correction factor $D(r) = \sqrt{1 - R_\star^2/r^2}$.

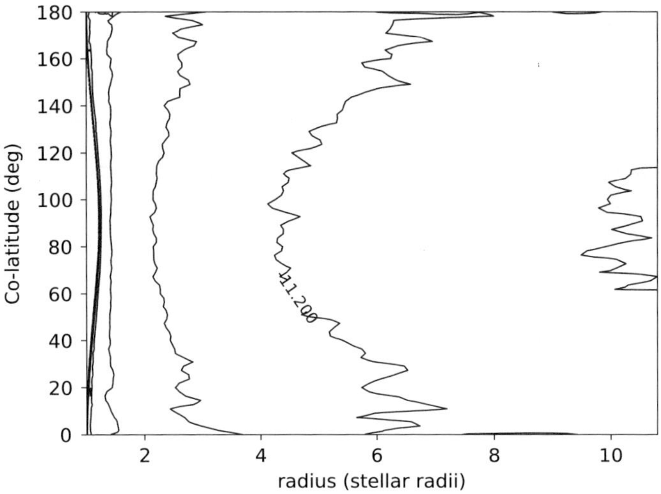

Figure 1. Density (in $\mathrm{g\,cm^{-3}}$) contours of stellar wind of the star WR93b versus radius and colatitude θ, in log scale spaced by 0.8 dex; denoted contours correspond to $\log\rho = -11.2$, for rotation $V_{\mathrm{rot}} = 1100\,\mathrm{km\,s^{-1}}$, with nonradial forces and gravity darkening.

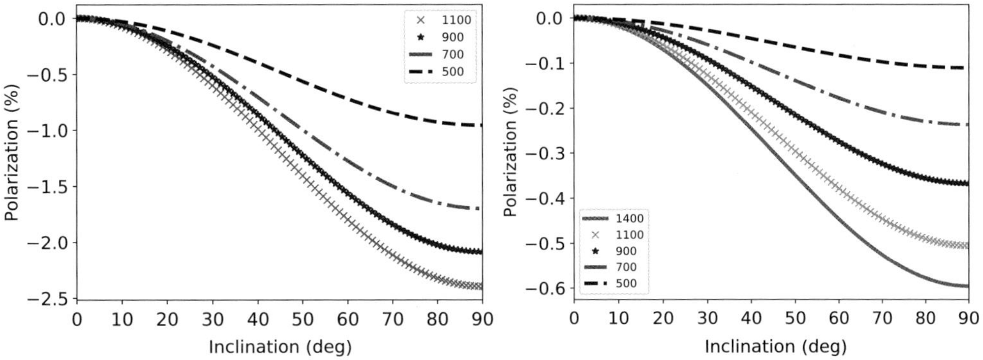

Figure 2. Polarization of WR93b (left), and WR102 (right), as a function of the inclination for different rotational velocities (denoted in the graph in units of $\mathrm{km\,s^{-1}}$) determined from the models, including nonradial forces and gravity darkening.

3. Results

The obtained limits of angular momentum (Abdellaoui et al. 2022) exceed the threshold of MacFadyen & Woosley (1999); as a result, the two stars could be progenitors of a long gamma-ray burst.

4. Conclusions

Fast rotation of WR stars leads to prolate density distribution, which produces a weak polarization signature that is smaller than the derived upper limit of polarization from observation (Stevance et al. 2018). The derived limit of rotational velocities is not in conflict with the model of a long gamma-ray burst.

Acknowledgements

This work was supported from Operational Programme Research, Development and Education - "Project Internal Grant Agency of Masaryk University" (No. CZ.02.2.69/0.0/0.0/19_073/0016943)

References

Abdellaoui, S., Krtička, J., & Kurfürst, P. 2022, A&A, 658, A46
Brown, J. C. & McLean, I. S. 1977, A&A, 57, 141
Cassinelli, J. P., Nordsieck, K. H., & Murison, M. A. 1987, ApJ, 317, 290
MacFadyen, A. I. & Woosley, S. E. 1999, ApJ, 524, 262
Owocki, S. P., Cranmer, S. R., & Blondin, J. M. 1994, ApJ, 424, 887
Stevance, H. F., Ignace, R., Crowther, P. A., et al. 2018, MNRAS, 479, 4535
Woosley, S. E. 1993, ApJ, 405, 273

Massive Stars Near and Far
Proceedings IAU Symposium No. 361, 2024
J. Mackey, J. S. Vink, & N. St-Louis, eds.
doi:10.1017/S1743921322003118

Dependence of Wolf-Rayet wind clumping on the surface temperature

André-Nicolas Chené[1], Nicole St-Louis[2], Guillaume Lenoir-Craig[2], Anthony F. J. Moffat[2] and Kenneth Gayley[3]

[1]Gemini Observatory/NSF's NOIRLab, 670 N. A'ohoku Place, Hilo, Hawai'i, 96720, USA
email: andrenicolas.chene@gmail.com

[2]Centre de Recherche en Astrophysique du Québec, Département de physique, Université de Montréal, Complexe des Sciences, Montréal, QC H2V 0B3, Canada

[3]Department of Physics and Astronomy, University of Iowa, Iowa City, IA 52242, USA

Abstract. We present our measurements of the amplitude of photometric and spectroscopic variability due to clumping in the wind of Wolf-Rayet (WR) stars. Photometric variability was assessed using *TESS* light-curves, while spectroscopic variations were obtained from almost 20 years of monitoring of nearly 100 classical (presumably single) stars. Our results show an apparent dependence of the variability amplitude with the stars' surface temperature and/or terminal velocity. Our interpretation is that it supports the idea that the dominating driver of the clumps in WR winds is a sub-surface convection region.

Keywords. stars: Wolf-Rayet, stars: mass loss, techniques: photometric, techniques: spectroscopic

1. Introduction

The discovery more than 35 years ago of stochastic clumps in the winds of a few Wolf-Rayet (WR) stars led to the current evidence that most if not all hot-star winds are pervaded by hierarchies of thousands of small, randomly varying clumps. This fundamental discovery has had a profound effect, such as the necessary reduction in the estimated mass-loss rates by a factor 2-5, which in turn affects the whole evolutionary history of massive stars. But the question of the origin of these structures remains uncertain and controversial.

The presence of clumps in WR winds leads to photometric, spectroscopic and polarimetric variability. In this work, we measured the amplitude of that variability for a large number of WR stars using *TESS* light-curves and spectra from various ground-based observatories.

2. Methods and results

From the *TESS* light-curves, we measure the amplitude of variability, α_0. It is obtained from a fit of the "amplitude spectrum" (equivalent to a periodogram, Lenoir-Craig et al. 2022). In spectroscopy, we measure $\sigma(\%)$, the amplitude of line profile variations expressed as a percent of the line intensity (Chené et al. 2020). It is obtained from a modified Temporal Variance Spectrum. The fundamental parameters of the WR stars, such as T_{eff} and $v_{\rm inf}$, were taken directly from the work of Hamann et al. (2019) and

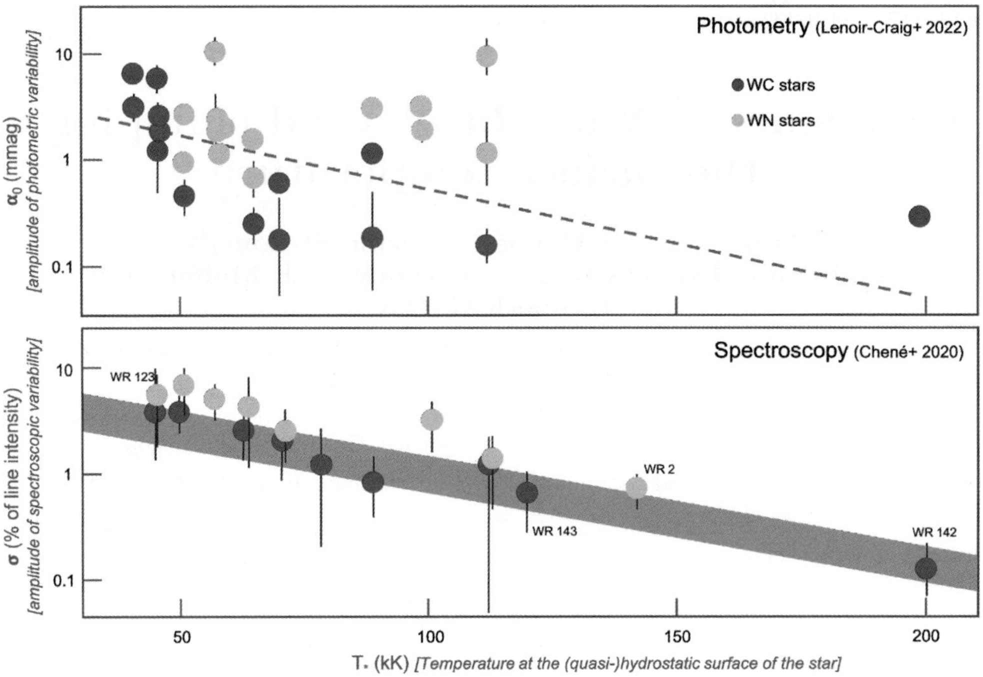

Figure 1. Amplitude of the photometric (*top*) and spectroscopic (*bottom*) wind variability for around 100 WR stars. The spectral types WN and WC are plotted in yellow and blue, respectively. In the bottom panel, stars with the same T_{eff} were grouped a single point.

Sander et al. (2019). Figure 1 shows our results. Both the photometric and spectroscopic amplitudes show a decline with increasing surface temperature (T_{eff}). Interestingly, that same decline is observed when the terminal velocity is used instead of T_{eff}. The same behaviour was also observed in polarimetry by Robert & Moffat (1989). That is in contradiction with the prediction that the inherently unstable nature of radiative line-driving should increase in hotter, faster winds.

3. Conclusions

Our results support the idea that a subsurface convection region is present in hot, massive stars, and is the main origin of wind clumps (Cantiello et al. 2009). It could also have many implications in our understanding of wind clumping, and on future works in atmosphere modeling. This poster opened the discussion with Nicolas Moens and Luka Poniatowski from KU Leuven who published their first 3D radiation-hydrodynamic simulations of WR winds (Moens et al. 2022). The plan is to expand the model to wider radii and extract realistic observables.

Acknowledgements

ANC gratefully acknowledges support by the international Gemini Observatory. AFJM and NSL thank NSERC (Canada) for financial aid.

References

Cantiello, M., et al. 2009, A&A, 499, 279
Chené, A.-N., St-Louis, N., Moffat, A. F. J., & Gayley, K. G. 2020, ApJ, 903, 113
Hamann, W. R., et al. 2019, A&A, 625, A57

Lenoir-Craig, G., et al. 2022, ApJ, 925, 79
Moens, N., Poniatowski, L. G., Hennicker, L., Sundqvist, J. O., El Mellah, I., & Kee, N. D. 2022, arXiv e-prints, arXiv:2203.01108
Robert, C., & Moffat, A. F. J. 1989, ApJ, 343, 902
Sander, A. A. C., Hamann, W. R., Todt, H., Hainich, R., Shenar, T., Ramachandran, V., & Oskinova, L. M. 2019, A&A, 621, A92

Massive Stars Near and Far
Proceedings IAU Symposium No. 361, 2024
J. Mackey, J. S. Vink, & N. St-Louis, eds.
doi:10.1017/S1743921322002393

The multiplicity of Galactic Wolf-Rayet stars

Karan Dsilva[1], **Tomer Shenar**[1,2], **Hugues Sana**[1] **and Pablo Marchant**[1]

[1]Institute of Astronomy, KU Leuven, Celestijnenlaan 200D, 3001 Leuven, Belgium
email: karandsilva413@gmail.com

[2]Anton Pannekoek Institute for Astronomy, University of Amsterdam, Postbus 94249, 1090 GE Amsterdam, The Netherlands

Abstract. We present the results of a magnitude-limited spectroscopic survey of Galactic Wolf-Rayet stars with the HERMES spectrograph mounted on the Mercator telescope. Using cross-correlation to measure radial velocities, we measured the observed binary fractions of the Galactic carbon- (WC) and nitrogen-rich (WN) Wolf-Rayet stars to be $f_{\rm obs}^{\rm WC} = 0.58 \pm 0.14$ and $f_{\rm obs}^{\rm WN} = 0.41 \pm 0.09$. We used Monte-Carlo simulations with a Bayesian framework to derive the intrinsic multiplicity properties and found $f_{\rm int}^{\rm WN} = 0.52^{+0.14}_{-0.12}$ and $f_{\rm int}^{\rm WC} = 0.96^{+0.04}_{-0.22}$. We find that the majority of WN binaries reside in short-period systems, similar to O stars. However, the orbital period distribution of the Galactic WC population peaks at 5000 d, a discrepancy that challenges our current understanding of binary evolution in Wolf-Rayet stars.

Keywords. stars: Wolf-Rayet, stars: binaries, stars: evolution

1. Introduction

Wolf-Rayet (WR) stars are a spectroscopic class of objects characterised by strong, broad emission-line spectra. They are the descendants of O-type stars with initial masses $M_i \gtrsim 20\text{-}30\, M_\odot$ (for a review, see Crowther 2007). Based on the chemical composition of their atmospheres, they are classified as carbon- (WC), nitrogen- (WN) or oxygen-rich (WO). Originally, it was thought that all WR stars are formed in binaries through envelope stripping by a close companion (Paczyński 1967). However, Conti (1976) hypothesised that an O star could strip its own envelope through mass loss via stellar winds (the Conti scenario). Knowledge of the multiplicity properties of WR stars is an important ingredient to understand the dominant formation channel of WR stars (see, e.g. Shenar et al. 2020).

In this contribution, we present the results of a homogeneous, magnitude-limited, spectroscopic survey of northern Galactic WR stars using the HERMES spectrograph (Raskin et al. 2011) on the Mercator telescope. Our sample consists of 27 WN and 12 WC stars. We collected at least six epochs for each object over four years. In Dsilva et al. (2020), we analysed the WC sample and found an intrinsic multiplicity fraction of at least 0.72 at the 10% significance limit. In Dsilva et al. (2022), we used a Bayesian framework to simulate the measured RVs and found the intrinsic fraction of early-type WN (WNE) stars to be $f_{\rm int}^{\rm WNE} = 0.56^{+0.20}_{-0.15}$. After analysing the late-type WN (WNL) stars, we found similar parameters to the WNE sample and hence performed a combined Bayesian analysis for the parent WN population.

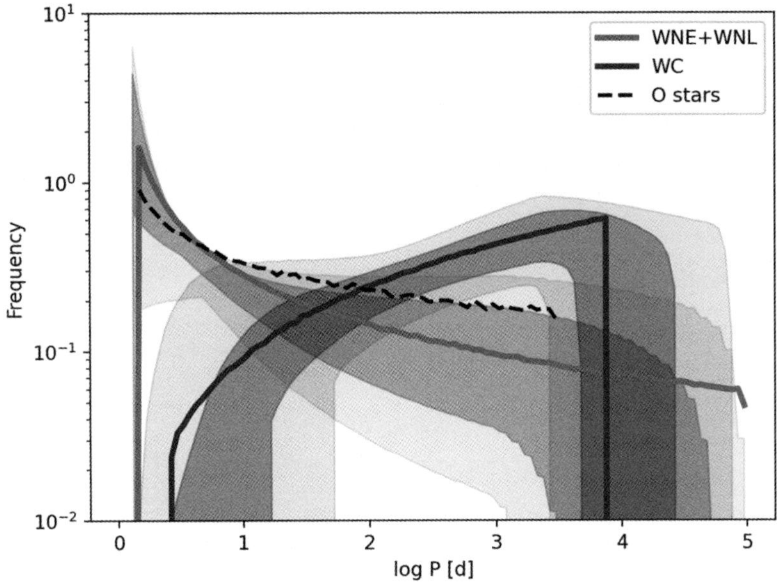

Figure 1. Visualisation of 10^7 period distributions for the WC (blue) and WN (red) populations. The dark and light shaded areas represent the parameter space covered by 68% and 95% of the simulated distributions.

2. Bayesian analysis of the intrinsic multiplicity properties

After measuring radial velocities (RVs) using cross-correlation, we calculated the RV amplitude for each object (ΔRV). To correct for observational biases, we divided the sample into various ΔRV bins and simulated our observations over the power-law index (π), minimum ($\log P_{\rm min}$) and maximum bounds ($\log P_{\rm max}$) of the underlying period distribution. The defined ΔRV bins for the WC and WN populations can be found in Dsilva et al. (2022). We assumed random eccentricity, and orientation in three dimensional space and simulated 10 000 populations of WN and WC subtypes. For the combined WN population, we calculated the highest probability intervals to be $f_{\rm int}^{\rm WN} = 0.52^{+0.14}_{-0.12}$, $\pi^{\rm WN} = -0.99^{+0.57}_{-0.50}$ and $\log P_{\rm max}^{\rm WN} = 4.99^{+0.00}_{-1.11}$ and assumed $\log P_{\rm min}^{\rm WN} = 0.15$ [d]. We sampled these posteriors and those for the WC (Dsilva et al. 2022) 10^7 times and simulated their respective period distributions. This can be seen in Fig. 1, where the red and blue solid lines show the distributions for the WN and WC respectively constructed using the best-fit posteriors.

3. Conclusions

The period distribution of the WN population is similar to that of the O stars. Our simulations indicate a shift in the period distribution of the WC population, which cannot easily be reconciled with our current theory of binary evolution. It is possible that we do not detect long-period companions to WN stars due to wind variability, but short-period WN binaries do not seem to become WC binaries, potentially due to mergers.

References

Conti, P. 1976, in Proc. 20th Colloq. Int. Astrophys., Univ. Liege, 193
Crowther, P. A. 2007, Annual Review of Astronomy & Astrophysics, vol. 45, Issue 1, pp.177-219, 45, 177

Dsilva, K., Shenar, T., Sana, H., & Marchant, P. 2020, Astronomy & Astrophysics, Volume 641, id.A26, 16 pp., 641, A26
—. 2022, Astronomy & Astrophysics, in press (arXiv: 2204.12518)
Paczyński, B. 1967, Acta Astronomica, 17, 355
Raskin, G., et al. 2011, Astronomy and Astrophysics, Volume 526, id.A69, 12 pp., 526, A69
Shenar, T., Gilkis, A., Vink, J. S., Sana, H., & Sander, A. a. C. 2020, Astronomy & Astrophysics, Volume 634, id.A79, 12 pp., 634, A79

Massive Stars Near and Far
Proceedings IAU Symposium No. 361, 2024
J. Mackey, J. S. Vink, & N. St-Louis, eds.
doi:10.1017/S1743921322002927

M 1-67 and RCW 58: nebulae around WN8h stars formed through CE evolution

P. Jiménez-Hernández[1], S. J. Arthur[1], J. A. Toalá[1] and A. P. Marston[2]

[1]Instituto de Radioastronomía y Astrofísica, UNAM, Antigua Carretera a Pátzcuaro # 8701, Ex-Hda. San José de la Huerta, Morelia, Michoacán, México C.P. 58089
email: p.jimenez@irya.unam.mx

[2]European Space Astronomy Centre, ESA, PO Box 78, 28691 Villanueva de la Cañada, Madrid, Spain

Abstract. We present the analysis of the dust properties of the Wolf-Rayet nebulae M 1-67 and RCW 58 around the WN8h stars WR 124 and WR 40, respectively. Modeling with the photoionization code Cloudy shows that in both nebulae the IR spectral energy distributions and ionized gas properties can be reproduced by a dust shell consisting of two populations of dust grains. Furthermore, taking into account the initial mass, the morphology and the kinematics of the nebulae we propose M 1-67 and RCW 58, together with their progenitor stars, as the first observational evidences of post-common envelope evolution in nebulae around massive stars.

Keywords. stars: circumstellar matter, stars: evolution, stars: individual (WR 40, WR 124)

1. Introduction

Some late-type WNh Wolf-Rayet (WR) stars are surrounded by clumpy or irregular ejecta nebulae, suggesting a violent mass-loss episode as their origin. The study of WR nebulae properties can provide information of the mass-loss history of the progenitor massive stars in the late stages of their evolution. In this work we look at the dust properties in WR nebulae around WN8h stars in order to understand the post-main sequence evolution of these objects.

2. Methods

Cloudy (Ferland et al. 2017) allows us to treat the interaction between the UV flux from the central star and the nebular gas and dust simultaneously. For the spectrum of the central stars we used PoWR stellar atmosphere models (Hamann et al. 2006).

To constrain our models we used the shape and flux of the IR spectral energy distribution (SED) to fit dust emission. The IR SED was built from public data corresponding to *Spitzer*, *WISE* and *Herschel* (and ATCA to the RCW 58 case) observations. We estimated the quantity of ionized gas from the Hα and Hβ emission line fluxes and radio observations, when available.

Multi-layer models were required to reproduce the observed SEDs of M 1-67 and RCW 58. A global model of RCW 58 is hampered by the extended background emission and so representative regions RC and LC were modeled to infer the properties of the ring nebula. Fig. 1 shows color-composite images of these two nebulae. Schematic views of the model density distributions are shown in Fig. 2. From our models (see Jiménez-Hernández et al. 2020; 2021; for more detail): we need an inner layer of pure gas and

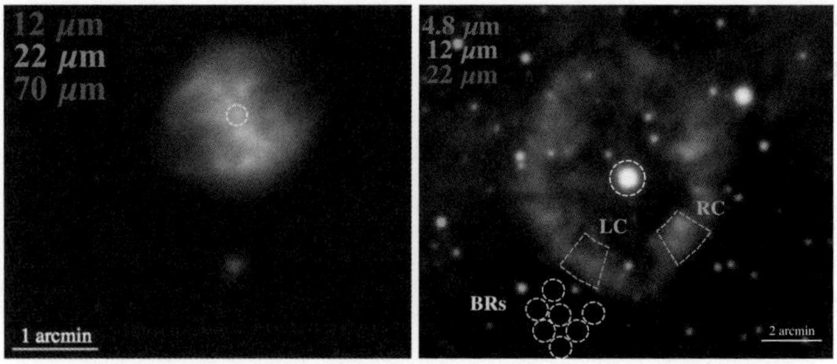

Figure 1. Left: Color-composite IR image of M 1-67. Right: Color-composite IR image of RCW 58. Cyan polygonal shapes indicate the selected regions for study. North is up, east is left.

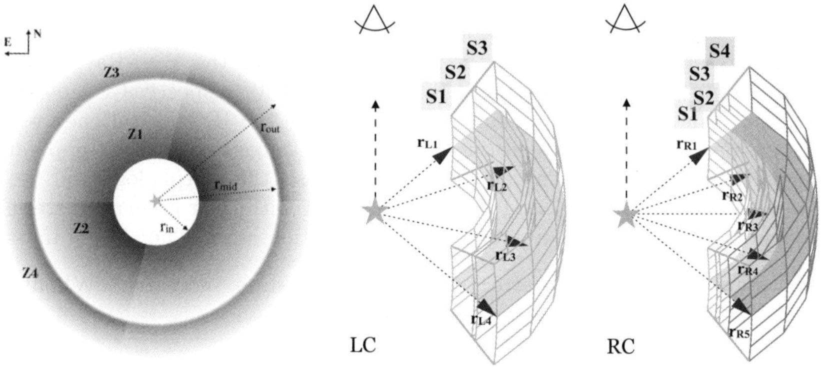

Figure 2. Left: Schematic view of the distribution of gas (Z1,Z2,Z4) and dust (Z3) in our model of M 1-67. Right: Schematic view of the gas and dust distribution in LC and RC clumps, on which our models of RCW 58 are based. See Jiménez-Hernández et al. (2020; 2021)

a dusty outer layer; and the dusty layer has two populations of dust, small ($0.002\ \mu$m $\leq a \leq 0.05\ \mu$m) and big ($0.6\ \mu$m $\leq a \leq 0.9\ \mu$m) grains. In our models we include only spherical silicate grains, in particular, olivine.

3. Single or binary origin?

The maximum grain size of $0.9\ \mu$m supports an eruptive formation for both M 1-67 and RCW 58. This is because mass-loss rates above $\dot{M} \geq 10^{-3}$ $M_{\odot}$ yr^{-1} are required to shield the dust-formation region from stellar UV photons (Kochanek 2011).

We estimated an initial massess for WR 124 and WR 40 (nebular mass calculated plus current mass) around 40 $M_{\odot}$, this rules out the possibility that the WR star had a LBV phase. In addition, the morphology (bipolar and torus pole-on) and the dynamics of each nebula allows us to suggest dust formation though a common envelope (CE) ejection.

CE evolution can lead to the ejection of the common envelope and tighter binary (Ivanova et al. 2013). Furthermore, the presence of a compact object of WR 124 and WR 40 can not be rule out because their dense stellar wind.

Acknowledgements

This research was supported by DGAPA-PAPIIT projects IA100720 and IN107019.

Supplementary material

To view supplementary material for this article, please visit http://dx.doi.org/10.1017/S1743921322002927.

References

Ferland, G. J., et al. 2017, RMxAA, 53, 385
Hamann, W. R., Gräfener, G., & Liermann, A. 2006, A&A, 457, 1015
Ivanova, N., et al. 2013, A&A Rev., 21, 59
Jiménez-Hernández, P., Arthur, S. J., & Toalá, J. A. 2020, MNRAS, 497, 4128
Jiménez-Hernández, P., Arthur, S. J., Toalá, J. A., & Marston, A. P. 2021, MNRAS, 507, 3030
Kochanek, C. S. 2011, ApJ, 743, 73

Massive Stars Near and Far
Proceedings IAU Symposium No. 361, 2024
J. Mackey, J. S. Vink, & N. St-Louis, eds.
doi:10.1017/S1743921322005014

Double Vision: Combining X-ray and Spectropolarimetric Observations of WR Binaries

R. A. Johnson[1], Y. Nazé[2], T. Panzera[1], A. G. Fullard[3], J. R. Lomax[4], J. L. Hoffman[1] and K. H. Nordsieck[5]

[1]Department of Physics & Astronomy, University of Denver, Denver, CO, USA,
email: rachel.johnson387@du.edu

[2]FNRS/Université de Liège, Liège, Belgium

[3]Department of Physics & Astronomy, Michigan State University, Lansing, MI, USA

[4]Department of Physics, United States Naval Academy, Annapolis, MD, USA

[5]Department of Astronomy, University of Wisconsin-Madison, Madison, WI, USA

Abstract. We present X-ray and spectropolarimetric observations of the WN+O binaries WR71 and WR97, which are analogs of the well-studied V444 Cygni. The combined results have the potential to constrain the locations and properties of wind interaction regions in these binaries, give clues to their subsequent evolution, and address the commonalities among WR+O systems.

Keywords. Spectropolarimetry, X-rays, WR stars, binary systems

1. Introduction

Massive, evolved Wolf-Rayet (WR) stars are often found in binary systems. WR+O systems are of particular interest due to their theorized capacity to evolve into gamma ray burst (GRB)-producing supernovae upon their deaths. Binary interactions may provide a mechanism to create the rapidly rotating WR stars necessary for GRB formation in the collapsar model (Woosley et al. 1993; Vink & de Koter 2005). Studying the strong wind collisions between stars in WR+O binaries allow us to investigate their mass loss and mass transfer properties, identify signatures of rapid rotation, and predict the nature of their explosive deaths.

By combining X-ray and spectropolarimetric observations of these systems, we can determine the location and characteristics of colliding wind regions. Optical spectropolarimetry constrains the shapes of the regions that emit and scatter continuum and line photons. X-ray observations reveal the locations and other properties of high-temperature wind collisions. Lomax et al. (2015) found a correlation between the X-ray and line polarization phase behavior in the WN+O binary system V444 Cygni. Analysis of these data pinpointed the location and structure of the colliding winds, revealing a shock cone with a wide opening angle and an off-center orientation due to Coriolis deflection, detected in V444 Cyg for the first time.

We present preliminary analysis of X-ray and optical spectropolarimetric variations in WR71 and WR97, short-period WN+O binaries chosen for their spectral similarity to V444 Cyg. We seek to understand how structurally comparable the three systems are.

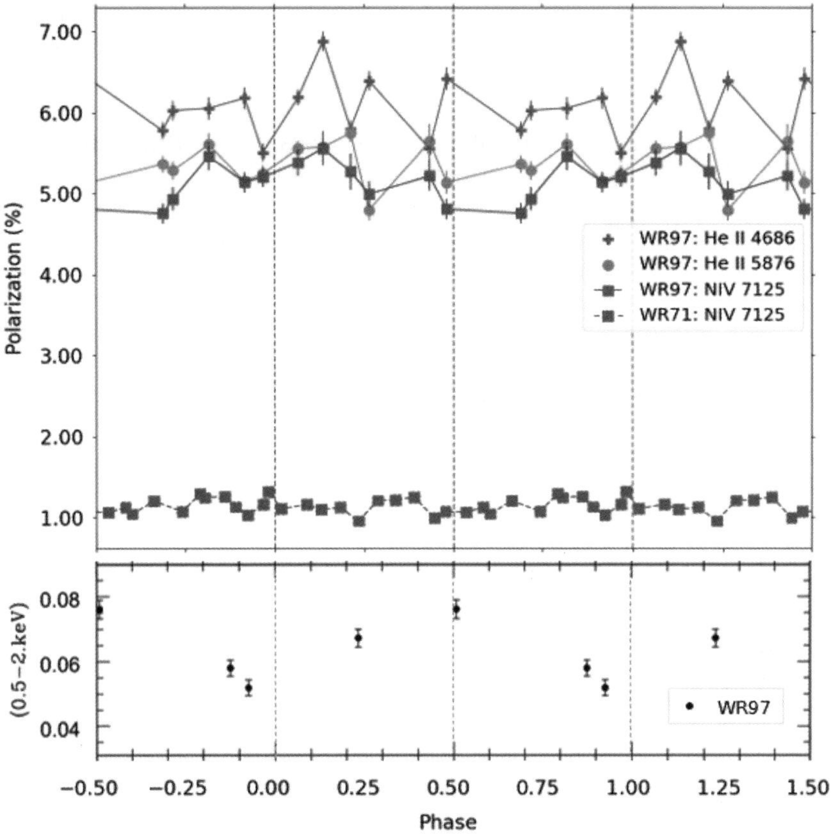

Figure 1. RSS line polarization phase variations for 3 emission lines in WR97 (*top, upper solid lines*), compared with its *Chandra* X-ray variations in the 0.5 − 2 keV band (*bottom*). Polarization of one representative line of WR71 is also shown for comparison (*top, lower dashed line*), with error bars smaller than the points.

2. Methodology

We collected phase-sampled optical spectropolarimetry for WR71 and WR97 between 2018 and 2022 using the RSS instrument on the 11-m Southern African Large Telescope (Kobulnicky et al. 2003). We extracted narrow-band continuum (Massey 1984) and integrated polarization in the strong He and N emission lines for both targets. Lomax et al. (2015) found that these lines were significantly polarized and correlated with X-ray behavior in V444 Cyg. We also obtained 2 observations of WR71 with *XMM-Newton* and 4 observations of WR97 with *Chandra*, and combined these with archival data from *XMM-Newton*, *Chandra*, and *Swift*.

3. Results

WR97 shows both X-ray emission and line polarization variations with phase. There is some possible correlation between the two as in V444 Cyg, particularly in He II λ4686, which has intrinsic polarization changes as large as 1.5% (Fig. 1). However, it is difficult to explain the X-ray behavior with a colliding wind model. We are working to update the ephemeris and construct models that reproduce both the X-ray and polarization data.

By contrast, WR71 does not emit bright X-rays, and its continuum polarization variations suggest it may be a single rotating WR star rather than a binary. Several emission lines show low-amplitude polarization variability similar to that of the continuum (Fig. 1); these may be attributable to corotating interaction regions within the wind.

References

Kobulnicky H. A., Nordsieck K. H., Burgh E. B. et al. 2003, *Proc. SPIE*, 4841, 1634
Lomax, J.R., Nazé, Y., Hoffman, J.L. et al. 2015, *A&A*, 573, A43
Massey, P. 1981, *ApJ*, 281, 789
Vink, J., & de Koter, A. 2005, *A&A*, 442, 587
Woosley, S., Langer, N., & Weaver, T.N. 1993, *ApJ*, 411, 823

Massive Stars Near and Far
Proceedings IAU Symposium No. 361, 2024
J. Mackey, J. S. Vink, & N. St-Louis, eds.
doi:10.1017/S1743921322002083

Investigating the impact of different velocity fields on the spectral appearance of Wolf-Rayet stars

R. R. Lefever[1,2], T. Shenar[2,3], A. A. C. Sander[1], L. Poniatowski[2], K. Dsilva[1] and H. Todt[4]

[1]Zentrum für Astronomie der Universität Heidelberg, Astronomisches Rechen-Institut, Mönchhofstr. 12-14, 69120 Heidelberg, Germany
email: roel.lefever@uni-heidelberg.de

[2]Institute for Astronomy (IvS), KU Leuven, Celestijnenlaan 200D, 3000 Leuven, Belgium

[3]Anton Pannekoek Institute for Astronomy, Science Park 904, 1098 XH, Amsterdam, The Netherlands

[4]Institut für Physik und Astronomie, Universität Potsdam, Karl-Liebknecht-Str. 24/25, D-14476 Potsdam, Germany

Abstract. The emission line spectra of WR stars are often formed completely in the optically thick stellar wind. Hence, any assumption on the wind velocity law in a spectral analysis has a profound impact on the determination of the stellar parameters. By comparing Potsdam Wolf-Rayet (PoWR) model spectra calculated with different β laws, we show that the velocity field heavily influences the spectra: by using the appropriate β laws, the entire range of late and early types can be covered with the same stellar model.

Keywords. Spectropolarimetry, X-rays, WR stars, binary systems

---

Wolf-Rayet (WR) stars play a pivotal role in the energetics of their host galaxies, in the enrichment and ionisation of their local environments and as supernova progenitors. Their powerful, line-driven winds form an optically thick regime around the star, causing WR spectra to be dominated by wind features. For the inference of stellar parameters of WR stars, we rely on model atmosphere codes such as PoWR (Gräfener *et al.* 1999, Hamann *et al.* 2003, Sander *et al.* 2015). To describe the wind velocity field, typically a pre-described velocity law is utilised, the β-velocity law (e.g. Hamann *et al.* 2006), which is described as $v\,(r) = v_\infty\,\left(1 - \frac{r_0}{r}\right)^\beta$, with v_∞, where for WR stars often $\beta = 1$ is adopted for large samples.

When comparing the observed temperatures $T_{\rm eff}(\tau_{\rm ross} = 20) = T_*$ of WR stars from atmosphere modelling to the $T_{\rm eff}$ of evolutionary models yielding hydrogen-depleted, core-helium burning stars (e.g. Ekström *et al.* 2012), there is a clear discrepancy: the observed temperatures are generally considerably lower than the temperatures from evolutionary models due to the extended, optically thick wind. Several solutions to solve this problem have been proposed, e.g. via static (e.g. Petrovic *et al.* 2006) or dynamical (e.g. Grassitelli *et al.* 2018) inflation. Another factor to address here are the $\beta = 1$ velocity laws: studies investigating line profile variations (e.g. Lepine *et al.* 1999) show values of $\beta \sim 5 - 20$, significantly different from a $\beta = 1$ assumption for WR stars.

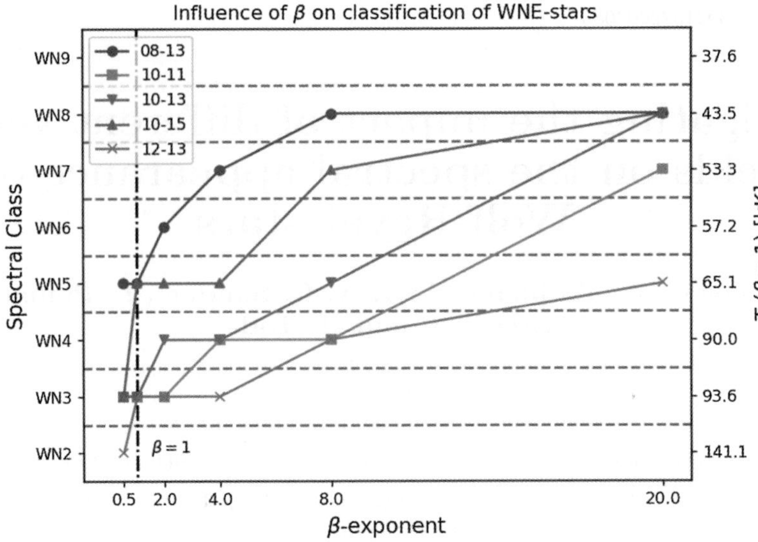

Figure 1. Spectral (sub-)classification for the WN-star models as a function of β. For labelling we use the same indices as the grids on the PoWR website.

To investigate how this β-parameter influences the spectra, we calculated and compared sets of PoWR model spectra by using different β laws. By investigating diagnostic spectral lines, we see that the β parameter has a strong effect on the line spectrum. This is illustrated by classifying the spectra of several models using the scheme of Smith *et al.* (1996). The results for H-depleted WN-star models are shown in Fig. 1, with similar results being obtained for WNh and WC stars. As a general trend, the model spectra are systematically classified as later-type WR stars for higher β. In fact, by only changing the β law, we could cover almost the entire range of spectral types for models with the same T_* if we chose β values in the range of 0.5 to 20. As spectral types are typically associated with an apparent temperature, models with an increased β parameter would be interpreted as cooler stars. This is also shown with the average T_* per spectral type inferred with $\beta = 1$ models on the right axis in Fig. 1. This effect of spectra appearing cooler by using higher-β velocity laws causes a degeneracy in the model spectra: by using different β laws, models with different T_*, $\dot{M}$ and v_∞ can appear similar in the optical spectrum. However, this degeneracy can at least in part be solved by P-Cygni lines in the UV as the absorption trough can be used constrain v_∞ (Prinja *et al.* 1990).

References

Ekström, Sylvia and Georgy, Cyril and Eggenberger, Patrick and Meynet, Georges and Mowlavi, Nami and Wyttenbach, Aurélien and Granada, Anahí and Decressin, Thibaut and Hirschi, Raphael and Frischknecht, Urs 2006, *A&A*, 537, A146

Gräfener, G. and Koesterke, L. and Hamann, W-R 2002, *A&A*, 387, 244–257

Grassitelli, L. and Langer, N. and Grin, N. J. and Mackey, J. and Bestenlehner, J. M. and Gräfener, G. 2018, *A&A*, 614, A86

Hamann, W-R and Gräfener, G. 2003, *A&A*, 410, 993–1000

Hamann, W-R and Gräfener, G. and Liermann, Adriane 2006, *A&A*, 457, 1015–1031

Lépine, Sébastien and Moffat, Anthony F.J. 1999, *ApJ*, 514, 909

Petrovic, J. and Pols, O. and Langer, N. 2006, *A&A*, 450, 219–225

Prinja, Raman K. and Barlow, M. J. and Howarth, Ian D. 1990, *ApJ*, 361, 607

Sander, Andreas and Shenar, Tomer and Hainich, Rainer and Gímenez-García, Angel and Todt, Helge and Hamann, W-R 2015, *A&A*, 577, A13

Smith, Lindsey F. and Michael M., Shara and Moffat, Anthony F.J. 1996, *MNRAS*, 281, 163-191

Massive Stars Near and Far
Proceedings IAU Symposium No. 361, 2024
J. Mackey, J. S. Vink, & N. St-Louis, eds.
doi:10.1017/S1743921322003131

A search for correlations between Wolf-Rayet stochastic wind variability and stellar parameters

Guillaume Lenoir-Craig[1], Nicole St-Louis[1], Anthony F. J. Moffat[1] and Herbert Pablo[2]

[1]Centre de Recherche en Astrophysique du Québec, Département de physique, Université de Montréal, Complexe des Sciences, Montréal, QC H2V 0B3, Canada
email: guillaume.lenoir-craig@umontreal.ca

[2]American Association of Variable Star Observers, 49 Bay State Road, Cambridge, MA 02138, USA

Abstract. We present the results of our analysis of 122 light-curves from 50 Wolf-Rayet (WR) stars using a red+white noise analysis, where we compare the fitted red noise features with stellar parameters to assess the presence of correlations with stellar parameters. A significant correlation between the amplitude of variability α_0 and v_∞ was found for the whole sample, along with several other correlations satisfying the Spearman-Rank $p < 0.001$ criterion for both He-burning and WNh stars. Our results are compatible with several plausible processes that can have an influence on the level of variability in the winds of these stars, including a subsurface convection zone and core-generated internal gravity waves.

Keywords. stars: Wolf-Rayet, stars: fundamental parameters, stars: variables, stars: interiors

1. Introduction

Like the winds of their O-star progenitors, Wolf-Rayet (WR) winds are radiatively driven and susceptible to various instabilities that ultimately lead to the formation of small-scale regions of higher density and lower velocity referred to as "clumps". While moving away from the star, these clumps create stochastic photometric, spectroscopic and polarimetric variability. In this work, we analyzed the amplitude of the temporal variability observed in 122 photometric datasets from 50 WR stars, including 36 classical He-burning stars (cWR) and 14 WNh stars from the MOST (Walker et al. 2003), BRITE-Constellation (Weiss et al. 2014) and TESS (Ricker et al. 2015) satellites, and search for the presence of correlations with various stellar parameters.

2. Methods and results

We characterize the amplitude of variability by fitting a semi-Lorentzian model with both a red and white noise component, following the same methodology presented in Bowman et al. (2019). The fitted red noise parameters are α_0, representing the amplitude of the frequency-dependent red noise in the limit of zero frequency, ν_{char} the characteristic frequency of the variability and γ corresponding to the logarithmic amplitude gradient.

The wind and stellar parameters were taken from Hamann et al. (2019) for WN stars and Sander et al. (2019) for WC and WO stars. Using an F-test comparing the χ^2 of the trend with a linear flat-model, we found a statistically significant negative trend between α_0 and v_∞ for both cWR and WNh stars as shown in Figure 1.

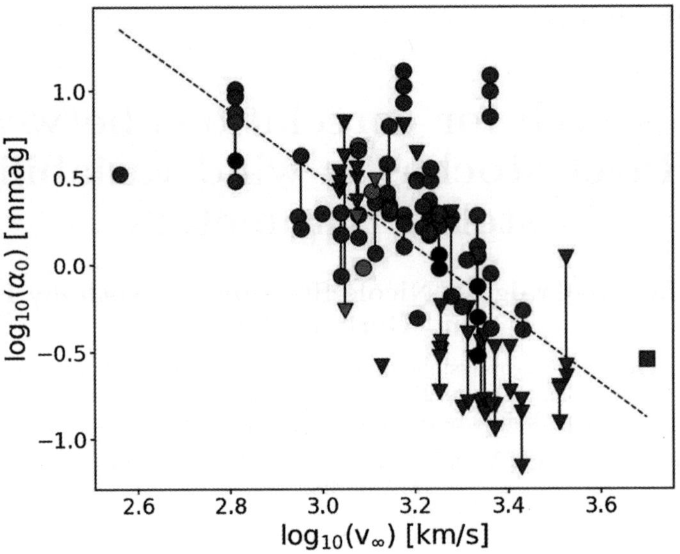

Figure 1. Amplitude of the photometric wind variability α_0 plotted against the terminal speed v_∞ of the wind for all the WR stars of our sample. Vertical black lines link the different datasets of individual stars observed by the same telescope, with TESS data shown in blue, BRITE-Constellation in black and MOST in red. Inverted triangles represent WC stars and circles are for WN stars.

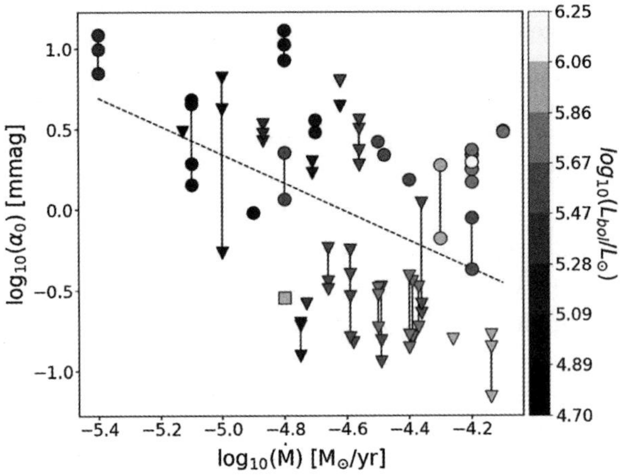

Figure 2. Amplitude of the photometric wind variability α_0 plotted against the mass-loss rate $\dot{M}$ for cWR stars, with a color-coding indicating the bolometric luminosity of each star.

For cWR stars only, correlations satisfying the Spearman rank (R_s) $p < 0.001$ criterion (but not the F-test) were observed between α_0 and both $\dot{M}$ and L_{bol}, as shown in Figure 2. Correlations also satisfying the R_s criterion where found for WNh stars between α_0 and T_* (evaluated in the references mentioned above through the Stefan-Boltzmann relation), along with correlations between ν_{char} and both v_∞ and T_*.

3. Conclusions

Our results show that hotter WR stars with faster winds tend to exhibit lower levels of stochastic variability in their photometry, which is in agreement with both the spectroscopic observations of Chené et al. (2020) and the linear polarimetric variability observed by Robert et al. (1989) and could be an indication that a subsurface convection region is present in these stars (Cantiello et al. 2009). Other plausible processes that are compatible with our results (and non-mutually exclusive) are the presence of core-generated gravity waves (e.g. Bowman et al. (2019)) and the action of the line de-shadowing instability (e.g. Sundqvist et al. (2018)).

References

Bowman, D. M., et al. 2019, Nature Astronomy, 3, 760

Cantiello, M., et al. 2009, Astronomy and Astrophysics, 499, 279

Chené, A.-N., St-Louis, N., Moffat, A. F. J., & Gayley, K. G. 2020, ApJ, 903, 113

Hamann, W. R., et al. 2019, Astronomy and Astrophysics, 625, A57

Ricker, G. R., et al. 2015, Journal of Astronomical Telescopes, Instruments, and Systems, 1, 014003

Robert, C., Moffat, A. F. J., Bastien, P., Drissen, L., & St. -Louis, N. 1989, ApJ, 347, 1034

Sander, A. A. C., Hamann, W. R., Todt, H., Hainich, R., Shenar, T., Ramachandran, V., & Oskinova, L. M. 2019, A&A, 621, A92

Sundqvist, J. O., Owocki, S. P., & Puls, J. 2018, Astronomy and Astrophysics, 611, A17

Walker, G., et al. 2003, PASP, 115, 1023

Weiss, W. W., et al. 2014, PASP, 126, 573

Massive Stars Near and Far
Proceedings IAU Symposium No. 361, 2024
J. Mackey, J. S. Vink, & N. St-Louis, eds.
doi:10.1017/S174392132200254X

PION: Simulating bow shocks and circumstellar nebulae around massive stars

Jonathan Mackey[1], Samuel Green[1], Maria Moutzouri[1], Thomas J. Haworth[2], Robert D. Kavanagh[3], Maggie Celeste[1,4], Robert Brose[1] and Davit Zargaryan[1]

[1]Dublin Institute for Advanced Studies, Astronomy & Astrophysics Section, DIAS Dunsink Observatory, Dublin, D15 XR2R, Ireland
email: jmackey@cp.dias.ie

[2]Astronomy Unit, School of Physics and Astronomy, Queen Mary University of London, London E1 4NS, UK

[3]Leiden Observatory, Leiden University, PO Box 9513, 2300 RA, Leiden, The Netherlands

[4]School of Physics, Trinity College Dublin, The University of Dublin, Dublin 2, Ireland

Abstract. Expanding nebulae are produced by mass loss from stars, especially during late stages of evolution. We describe the algorithms and methods implemented in the radiation-magnetohydrodynamics (MHD) code PION for highly scalable simulations using static mesh-refinement. We present results from 3D MHD simulations of bow shocks around runaway massive stars, and of the expansion of a fast wind from a Wolf-Rayet star into the slow wind from a previous red supergiant phase of evolution. PION is free software that can be downloaded from `https://www.pion.ie/`

Keywords. Hydrodynamics - methods: numerical - ISM: bubbles - Stars: winds, outflows

1. Introduction

Massive stars have a strong effect on their surroundings through their intense radiation (especially extreme-UV, ionizing radiation), strong winds, eruptive explosions and supernova explosions at the end of their lives. Understanding these effects is important for understanding the dynamical and chemical evolution of galaxies, and also the structure and dynamics of the interstellar gas in our own Galaxy. In some cases it is also possible to infer the evolutionary history of a specific star by studying and modelling the nebula surrounding it.

In Mackey et al. (2021) we presented a public release of PION, a software package for simulating nebulae around massive stars. The source code and tutorials for PION v2.0 are available at https://www.pion.ie. PION is a grid-based code that solves the Euler or ideal-magnetohydrodynamics (MHD) equations on a uniform or statically refined mesh. The default integration scheme is second-order accurate in time and space, a finite-volume scheme using a Riemann solver to calculate fluxes between grid cells. Green et al. (2022) used PION to make 3D MHD simulations of the bow shock around the nearby O star, ζ Ophiuchi, and more details of the algorithms and post-processing methods can be found there.

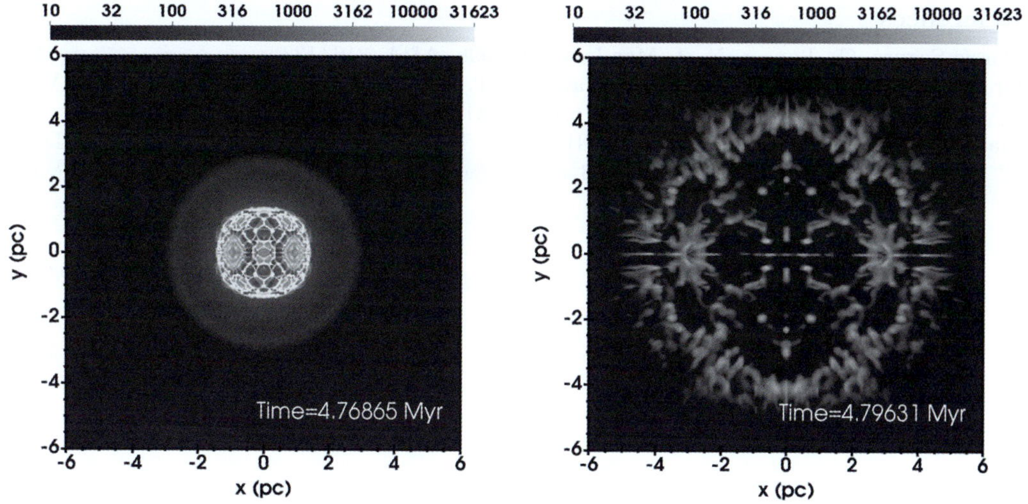

Figure 1. Emission measure (EM) from the expanding nebula 14 000 years (left) and 42 000 years (right) after the RSG→WR transition. The EM is shown with a logarithmic scale (units cm^{-6} pc). Time is shown since the beginning of the stellar evolution calculation; the RSG phase ended at $t \approx 4.755$ Myr.

2. Wolf-Rayet Nebulae

Nebulae around Wolf-Rayet (WR) stars are bright and short-lived ($\sim 10^5$ years) expanding nebulae, as the fast wind and hot radiation from the stellar core sweeps through the extended stellar envelope. The envelope was ejected either through a slow stellar wind from a previous evolutionary phase as red supergiant (RSG), or through binary stripping. A beautiful example is the nebula M1-67 around WR 124 (see poster by Jiménez Hernández in this Volume). Using a stellar evolution calculation for a 35 $M_\odot$ star from García-Segura et al. (1996), we made a 3D radiation-hydrodynamical calculation of the expanding Wolf-Rayet nebula with resolution 256^3 cells on each level and with 4 levels of refinement (see Mackey et al. 2021 for more details).

Fig. 1 shows synthetic observations of the simulation, calculating the projected Emission Measure (EM) and using a raytracing method that neglects internal absorption. In the left panel, about 14 000 years after the transition from RSG to WR, the cool and dense RSG wind is being slowly eroded from the inside by the hot and fast WR wind. The bright inner sphere is the swept-up material in a thin shell. The right panel shows the situation 28 000 years later, when the shock driven by the WR wind has broken out of the RSG wind into the relic hot-wind bubble from the even earlier main-sequence phase. In future work we will develop these simulations and compare quantitatively with observed nebulae.

Acknowledgements

The authors acknowledge funding from a Royal Society-Science Foundation Ireland University Research Fellowship (JM), an Irish Research Council (IRC) Starting Laureate Award (DZ,RB), a Royal Society Research Fellows Enhancement Award (MM), and a Royal Society Dorothy Hodgkin Fellowship (TJH), and the DJEI/DES/SFI/HEA Irish Centre for High-End Computing (ICHEC) for computational facilities and support.

Supplementary material

To view supplementary material for this article, please visit http://dx.doi.org/10.1017/S174392132200254X.

References

García-Segura, G., Langer, N., & Mac Low, M. 1996, A&A, 316, 133
Green, S., Mackey, J., Kavanagh, P., Haworth, T. J., Moutzouri, M., & Gvaramdaze, V. V. 2022, Astronom & Astrophysics, in press, arXiv:2203.06331
Mackey, J., Green, S., Moutzouri, M., Haworth, T. J., Kavanagh, R. D., Zargaryan, D., & Celeste, M. 2021, MNRAS, 504, 983

Massive Stars Near and Far
Proceedings IAU Symposium No. 361, 2024
J. Mackey, J. S. Vink, & N. St-Louis, eds.
doi:10.1017/S1743921322002460

Metal-poor Wolf-Rayet star enriching the ISM by nitrogen in the galaxy NGC 4068

Anastasiya Yarovova[1], Oleg Egorov[1,2], Alexei Moiseev[1,3] and Olga Maryeva[3,4]

[1]Lomonosov Moscow State University, Sternberg Astronomical Institute, Universitetsky pr. 13, Moscow 119234, Russia
email: yaan.ph@gmail.com

[2]Astronomisches Rechen-Institut, Zentrum für Astronomie der Universität Heidelberg, Mönchhofstraße 12-14, D-69120 Heidelberg, Germany

[3]Special Astrophysical Observatory, Russian Academy of Sciences, Nizhnii Arkhyz 369167, Russia

[4]Astronomical Institute of the Czech Academy of Sciences, Fričova 298, 25165 Ondřejov, Czech Republic

Abstract. We discovered a nebula in the low-metallicity (Z=0.1 $Z_\odot$) nearby ($D = 4.4$ Mpc) dwarf galaxy NGC 4068, which reveals broad H_α line profile and unusual emission line fluxes in its spectrum. The object also shows significant nitrogen overabundance, not typical for metal-poor environment. We assumed that the nebula could be ionized by an evolved massive star with $M \approx 80\ M_\odot$, Wolf-Rayet or Blue Supergiant, and built models of the nebula using Cloudy and CMFGEN codes. Our models successfully reproduce the optical emission spectrum of the object, including the peculiar [SII]/[NII] ratio and the presence of the HeII λ4686 line.

Keywords. stars: Wolf-Rayet, stars: mass loss, ISM: abundances

1. Observations

Analysing the small-scale kinematics of the ionized gas in the galaxy NGC 4068 with a scanning Fabry-Perot interferometer at the SAO RAS 6-m telescope BTA, we found compact nebulae (Object #A) with both high surface brightness and line-of-sight velocity dispersion in the Hα line. We obtained four long-slit spectra of it. The spectra exhibit several peculiarities: unusually low [SII]/[NII] ratio, presence of HeII λ4686 line. To estimate the flux in [OIII] λ5007 line, we performed a photometry at the BTA with the narrow-band filter and obtained $m_{5007} = -5.42^{+0.29}_{-0.39}$ mags thus excluding its apearance as a planetary nebula. Hubble Space Telescope (HST) images of NGC 4068 clearly show that the Object #A is a nebula ionized by a single star instead of a small cluster.

Taking into account the high luminosity of the object ($M_{bol} = -10.97$ from HST photometry), its position in the colour-magnitude diagram, strong [OIII] λ5007 line and significant nitrogen overabundance (log(N/O)≈ -0.2), we put forward a hypothesis about the ionization of the nebula by a massive star ($70 - 80\ M_\odot$) at a late stage of evolution (Blue Supergiant star (BSG) or Wolf-Rayet star – WR).

2. Modelling

We modeled the spectrum of the Object #A assuming it is a nebula ionized by a massive single BSG or WR star ($M \sim 80\ M_\odot$) with high mass loss rate. At the first step we created a model of the ionizing star using the CMFGEN code

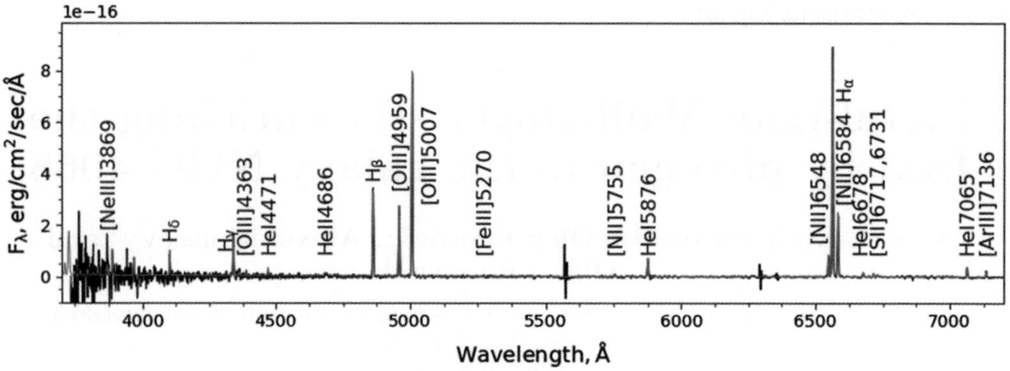

Figure 1. Comparison of the nebula model, ionized by WR star of $M = 80\ M_\odot$ with the spectrum observed with SCORPIO-2 (Afanasiev & Moiseev 2011) at the 6-m telescope BTA.

(Hillier & Miller 1998). The output spectrum was used as an ionization source for the next step, where we estimated parameters of the surrounding nebula using Cloudy photoionization code (Ferland et al. 2017).

Initially, we used calculated evolutionary tracks for massive stars at low metallicity from the work of Georgy et al. (2013). According to their calculations, metal-poor stars become colder and move along the track to the right side of the diagram after the end of hydrogen burning in the core and do not return to the left side of the diagram. Therefore, the most suitable point is the beginning of the BSG stage.

Changing the abundances scaling can dramatically change the massive star evolution. Grasha et al. (2021) calculated the grid of evolutionary tracks for massive stars in the wide range of metallicities implementing the Galactic Concordance abundances to the stellar evolution models, allowing stars with the initial mass of $M \geq 70\ M_\odot$ and rotation $V/V_{crit} = 0.2$ move to the right after the main sequence and then return to the left and to create high-mass WR stars for a low metallicity.

Our best-fit model fits well to the observations (Fig. 1). Our WR models describe the observed properties better than BSG. The in-details analysis of this object will be presented in Yarovova et al. (in prep.).

3. Conclusions

We describe and model the peculiar low-metallicity nebula identified in the nearby galaxy NGC 4068. We argue that it is a WR star of mass $\sim 80\ M_\odot$ in the transitional evolutionary stage, ionizing the nebula and enriching ISM with nitrogen. Such objects are not well-studied yet. High resolution 3D spectroscopy can help to identify such objects thanks to their high luminosity and velocity dispersion in the H_α line.

Acknowledgements

This study was supported under the Ministry of Science and Higher Education of the Russian Federation grant 075-15-2022-262 (13.MNPMU.21.0003). The authors acknowledge the support from IAU for providing a travel grant.

References

Afanasiev, V. L., & Moiseev, A. V. 2011, Baltic Astronomy, 20, 363
Ferland, G. J., et al. 2017, RMxAA, 53, 385
Georgy, C., et al. 2013, A&A, 558, A103
Grasha, K., Roy, A., Sutherland, R. S., & Kewley, L. J. 2021, arXiv e-prints, arXiv:2101.01197
Hillier, D. J., & Miller, D. L. 1998, ApJ, 496, 407

Massive Stars Near and Far
Proceedings IAU Symposium No. 361, 2024
J. Mackey, J. S. Vink, & N. St-Louis, eds.
doi:10.1017/S1743921322003179

A SITELLE view of the Galactic Wolf-Rayet Nebulae NGC6888 and NGC2359

N. St-Louis[1], C. Dumontier[1], M. Ruest[2] and L. Drissen[2]

[1]Département de physique, Université de Montréal, Campus MIL, Montréal (Qc), Canada
email: nicole.st-louis@umontreal.ca

[2]Départ. de physique, de génie physique et d'optique Université Laval, Québec (Qc), Canada

Abstract. In this paper, we present a glimpse of our observations of two Wolf-Rayet (WR) nebulae, NGC2359 and NGC6888 obtained with the SITELLE imaging Fourier transform spectrograph. The data are of unprecedented spatial coverage and cover a broad wavelength range.

Keywords. ISM: bubbles, stars: Wolf-Rayet, HII regions, ISM: NGC2359, ISM: NGC6888

1. Introduction

The winds of massive stars sculpt their surrounding interstellar medium (ISM) leaving traces of the geometry, abundances and strength of the mass-loss from the star during its life. These powerful outflows create cavities with diverse morphologies and ionization structures, with which the supernova ejecta will interact. The comparison of the physical properties of the surrounding ISM with evolutionary models helps to shed light on some unanswered questions of massive-star evolution. The WR phase is critical as it is thought to immediately precede the supernova and follow either a Red Supergiant or Luminous Blue Variable phase, depending on its initial mass. Here we present preliminary results for two such nebula, NGC2359 and NGC6888 from data of unprecedented spatial coverage using the SITELLE instrument on the Canada-France-Hawaii Telescope (CFHT).

2. The observations

We obtained several data cubes using different filters of NGC2359 and NGC6888 with SITELLE. With a field-of-view of $11'' \times 11''$, this instruments provides spectra with a spectral resolution selected by the observer for each $0.32''$ pixel of the image. The many filters available cover all main transitions that are used to determine the physical conditions of the gas (extinction, electronic densities and temperatures, abundances).

3. NGC2359

NGC2359 surrounds the very hot (T_*=112kK; Hamann et al. 2019) WN4-s star HD 56925. This nebula presents strikingly different views, depending on which line is used, revealing the ionization state of the gas that results from varying physical parameters (gas abundances, distance of the gas from the ionizing source, gas density). Fig. 1 shows a composite view with Hα in red, [NII]λ6548,6584 in green and [SII]λ6716,6731 in blue. It is clear that the bubble is nearly invisible in the two latter doublets but strong in Hα. It also emits strongly in [OIII]λ5007 presented in Fig. 2 in blue together with [OII]λ3729,3729 in yellow. This figure clearly illustrates the striking ionization contrast between the bubble and the arc feature. In Fig. 3 we show a section of a spectrum of NGC2359 encompassing

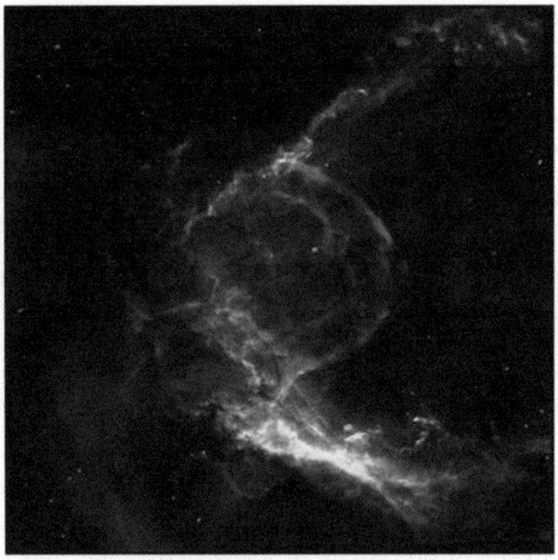

Figure 1. A composite SITELLE image of NGC2359. Hα is in red, [NII] in green and [SII] in blue.

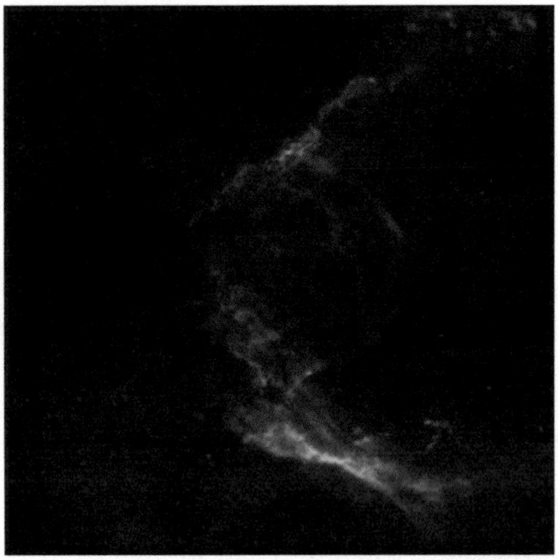

Figure 2. Change in the ionization state of the gas in NGC2359 illustrated as [OIII] in blue and [OII] in yellow.

the resolved components of the [OII] doublet, which ratio will enable the determination of the electronic temperature of the oxygen gas. The faint H9 line is also detected.

4. NGC6888

Figs. 4 and 5 show SITELLE observations of the south-western half of NGC6888 that is ionized by the WN6(h) star HD192163 in Hα and [OIII]λ5007 respectively. It is considered a prototypical wind-blown bubble with its main shell expanding at $\sim$70 km/s (Esteban & Vilchez 1992). Reyes-Pérez et al. (2015) simultaneously modelled the star and the nebula and were able to reproduce both the stellar and nebular emission-line strengths using a temperature for the star of T_*=70000K and a uniform abundance pattern throughout the gas. They found that the star and the nebula were overabundant in N and C

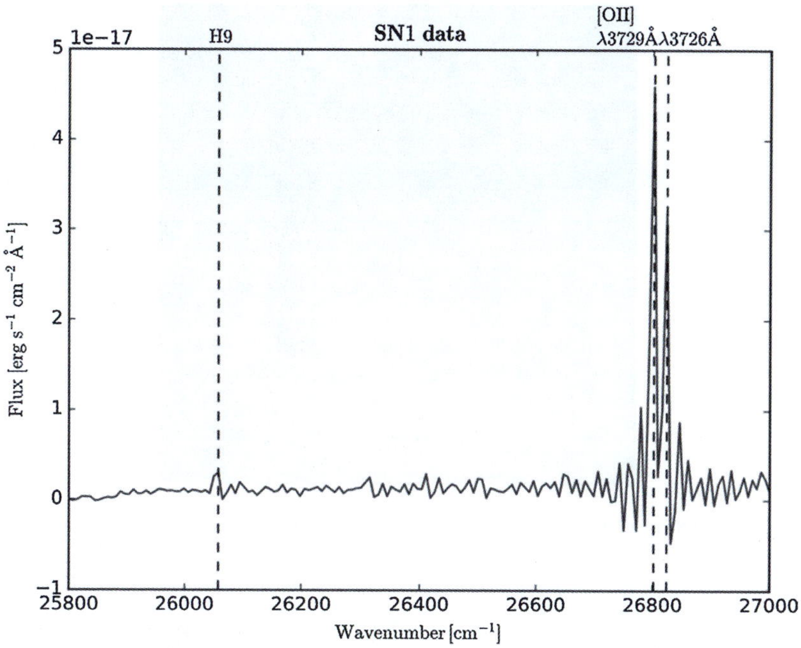

Figure 3. Part of a spectrum of NGC2359 from a section of the arc showing the resolved [O II] doublet.

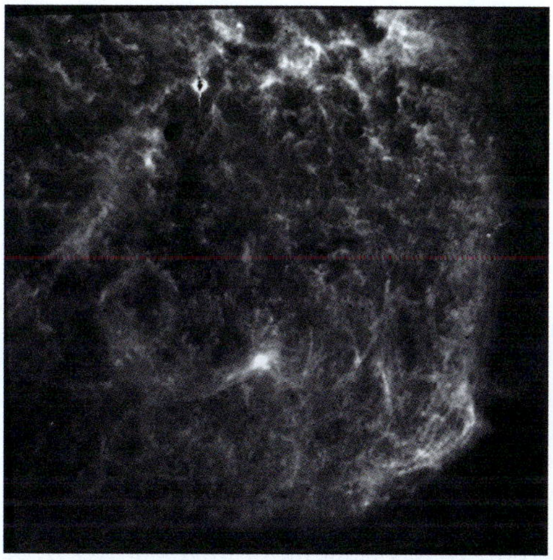

Figure 4. SITELLE image of the filamentary nature of the gas in the NGC6888 nebula as seen in Hα.

but depleted in O. The kinematics of the gas can be studied by fits to various nebular lines. In Fig. 6 we present a histogram of velocities from a two-component fit to the Hα line.

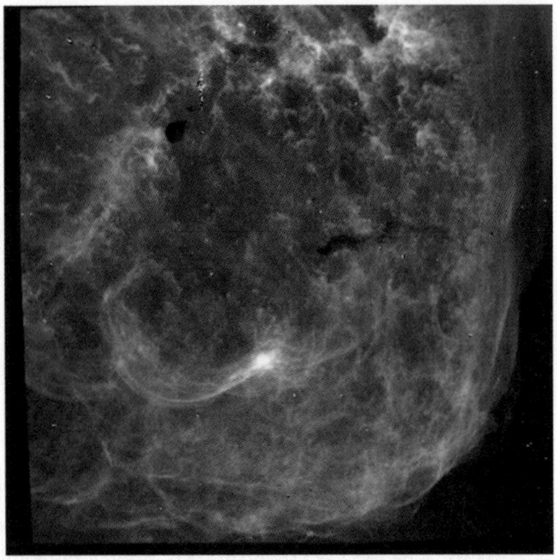

Figure 5. NGC6888 in [O III]λ5007 highlighting the thin outer shock interacting with the ISM.

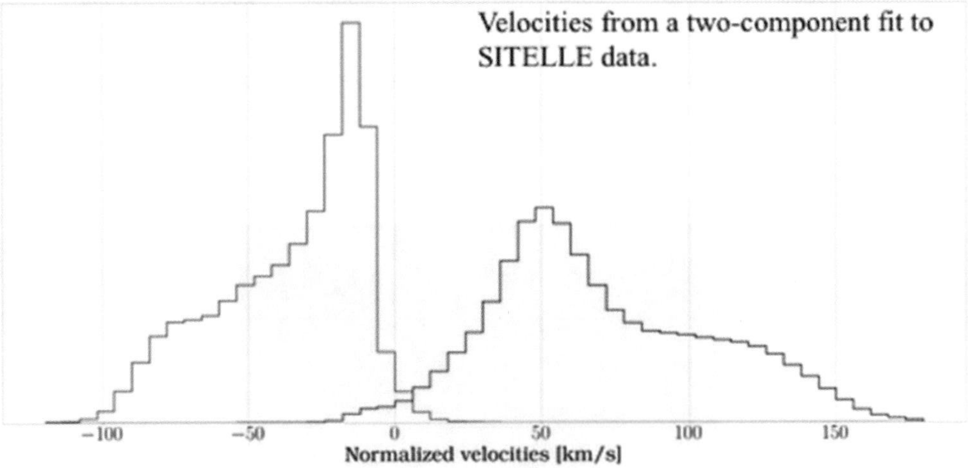

Figure 6. Histogram of velocities of the gas of NGC6888 from a two component fit of the Hα line.

References

Hamann, W.-R., Gräfener, G., Liermann, A., Hainich, R. Sander, A. A. C., Shenar, T., Ramachandran, V., Todt, H. & Oskinova, L. M. 1995, *A& A*, 625, A57

Esteban, C., Vilchez, J.M. 1992, *ApJ*, 390, 536

Reyes-Pérez, J., Morisset, C., Peña, M. and Mesa-Delgado, A. 2015, *MNRAS*, 452, 1764

Chapter IX:
Massive star formation near and far

Massive Stars Near and Far
Proceedings IAU Symposium No. 361, 2024
J. Mackey, J. S. Vink, & N. St-Louis, eds.
doi:10.1017/S1743921322002988

Formation of Very Low-metallicity Stars

Kazuyuki Omukai

Astronomical Institute, Graduate School of Science, Tohoku University
Aoba, Sendai 980-8578, Japan
email: omukai@astr.tohoku.ac.jp

Abstract. I describe (i) our recent updates on first star formation, with particular emphasis on their binaries, (ii) formation of low-metallicity stars and the transition of their initial mass functions with metal enrichment, and finally (iii) formation of supermassive stars from slightly metal-enriched gas by the newly found super-competitive accretion channel.

Keywords. stars: formation, stars: Population II

1. Some Updates on Formation of the First Stars

First stars in the universe are formed from the primordial pristine gas composed only of hydrogen, helium and a trace amount of lithium. They are thought to be very massive with a few tens to even a few hundred $M_\odot$ (see contribution by Ralf Klessen, this volume). This theoretical expectation is based on the following reasoning: i) the host dense core, inside of which a protostar is eventually formed, is massive $\sim 1000 M_\odot$ reflecting the high temperature and thus the Jeans mass. ii) after its formation, the protostar grows in mass by accretion, but the accretion is hard to stop by stellar feedback because the accretion rate is high and the opacity is low. We can naturally expect that the continued accretion from the large mass reservoir results in the formation of massive stars. How massive they are quantitatively is the question to be addressed below.

Some years ago, Hosokawa et al. (2011) studied the accretion growth of a first protostar starting from the cosmological initial condition of a minihalo taken from Yoshida et al. (2008) by way of 2D radiation hydrodynamics simulation assuming the axisymmetry and found that UV feedback from the protostar finally shuts off the accretion by photoevaporating the protostellar disk and sets the final mass of the star. By repeating similar simulations for hundreds of first-star forming sites identified in a larger cosmological simulation box, Hirano et al. (2014; 2015) statistically derived mass distribution of forming first stars, which has a wide mass range form a few tens to hundreds of the Sun. Those calculations were, however, idealistic 2D simulation and did not allow the formation of multiple stars by construction. In fact, in 3D simulations of first star formation with its radiative feedback included, stellar masses are predicted to be smaller by a factor of a few as the mass reservoir is shared among multiple forming stars (Susa et al. 2014). So far, 3D simulations of this sort only treat the central star as a UV source (e.g., Hosokawa et al. 2016). To predict the final mass of stars, however, in particular in case that a binary/multiple stellar system forms, we need to consider radiation from each forming protostar.

Recently, we carried out such 3D simulations considering multiple radiation sources by using AMR hydrodynamics code SFUMATO (Matsumoto 2007), which is combined with a newly developed radiative transfer module using ART method (Abel & Wandelt

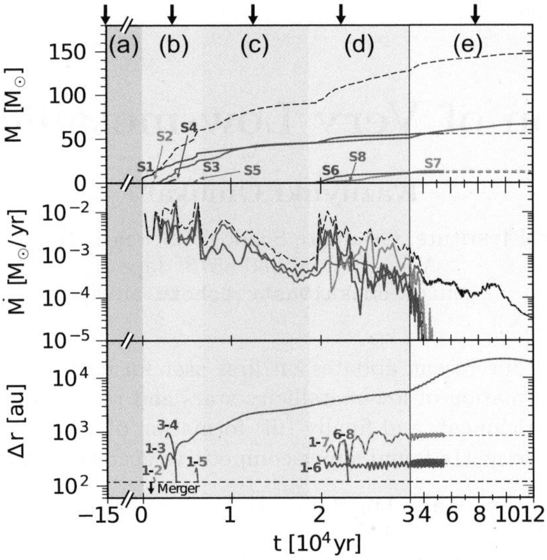

Figure 1. The time evolution of a forming first-star system; the masses of protostars, their accretion rates, and separations among them. Reproduced from Sugimura et al. (2020).

2002). Radiative transfer is solved for ionizing and dissociating photons, and a usual set of the primordial-gas microphysics is included.

As initial conditions, three minihalos (cases A, C, and D in Hosokawa et al. 2016) are picked up from the cosmological simulation of Hirano et al. (2015). Results of one case (case C) has been published in Sugimura et al. (2020), and is reproduced in Fig. 1. We find that a binary system composed of 70 and 60 $M_\odot$ massive stars is forming, along with a few other smaller stars. Such system is very interesting as an origin of merging binary black holes observed by gravitational waves if its separation is close enough ($<$ 0.1au, e.g., Kinugawa et al. 2014). Unfortunately, the separation of this system becomes wider in time due to accretion of high angular momentum matter initially locating at outer part and reaches as wide as $\sim 10^4$ au. In the other two cases studied, we also find massive binary systems with similar masses ($370 + 130 M_\odot$ and $60 + 30 M_\odot$) with very wide separations of $10^3 - 10^4$ au (Sugimura et al. in prep.). Although our numerical resolution prohibits to follow systems with the separations below ~ 100 au, their binary separations are much wider. This suggests that formation of close first-star binary systems needs some other tricks, e.g., angular momentum transport by magnetic field, hierarchical system etc. The bottom line of our calculation is that *massive but wide binaries seem to be common among the first stars.*

2. Formation of Pop II Stars and Transition of IMF with Metallicity

As we have just seen, the first stars are typically massive $\sim 100 M_\odot$ according to theoretical studies. On the other hand, we know that low-mass very-metal-poor Pop II stars are present in Milky Way and other galaxies, and of course Pop I stars in the solar neighborhood are typically low-mass, $0.1 - 1 M_\odot$, following a Salpeter-like IMF. This means that there has been a transition in the characteristic stellar mass from very massive to low-mass ones in the cosmic history (*"Pop III-II transition"*).

The exact physical mechanism is still not very clear, but probably triggered by the accumulation of metals and dust grains in the ISM, which alters the thermal properties of star-forming gases (Omukai 2000, Schneider et al. 2001, Omukai et al. 2005; see Fig. 2).

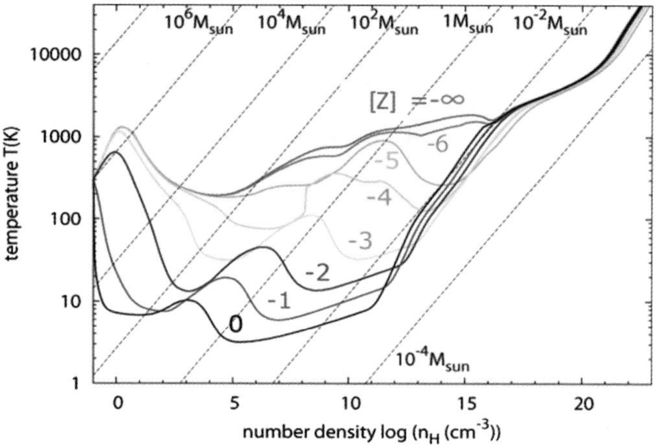

Figure 2. The temperature in star-forming clouds with different metallicities. The Jeans masses are indicated by dotted lines. Reproduced from Omukai et al. (2010).

With respect to the temperature in the primordial gas, which is controlled mostly by the H_2 cooling, with metallicity of $\log Z/Z_\odot \gtrsim -5$, a temperature dip begins to develop due to dust cooling at high densities ($\simeq 10^{12}\mathrm{cm}^{-3}$), where the Jeans mass is sub-solar. Because a rapid temperature drop likely causes fragmentation of the cloud, the dust cooling produces low-mass fragments and perhaps low-mass stars. With more metallicity, another temperature drop appears at lower densities due mainly to the metal (fine-structure) lines of C and O, but the Jeans mass at those densities is rather high $\gtrsim 100 M_\odot$ and resultant fragmentation does not produce low-mass stars.

This theoretical expectation has also been confirmed by more detailed numerical simulations (Tsuribe & Omukai 2006, Clark et al. 2008, Dopcke et al. 2013). Recently, we performed hydrodynamics simulations for star cluster formation from low-metallicity star-forming gas (Chon, Omukai & Schneider 2021) until a relatively late phase of protostellar accretion to obtain the mass distribution of forming stars. These are SPH simulations using Gadget3, with simplified chemistry and radiative processes of a low-metallicity gas. As an initial condition, we use a Bonner-Ebert sphere with transonic turbulence as is often used in studies on the present-day star formation.

The results can be summarized as follows: at relatively high metallicity, say $\gtrsim 10^{-3} Z_\odot$, the temperature drops soon due to efficient cooling, supersonic turbulence makes the cloud filamentary in shape, while at lower metallicities, turbulent effect is not obvious due to inefficient cooling and the cloud morphology is more centrally concentrated with a massive core at the center. The difference in the cloud morphology naturally affects the fragmentation property, and roughly speaking the fragmentation is more vigorous at higher metallicities.

What is more interesting to us might be the mass distribution of stars formed. As we can see in Fig. 3, the IMF is bimodal at low metallicities, consisting of low-mass Salpeter-like and massive log-normal components. At lowest metallicity, only the massive component is present. At $\log Z/Z_\odot \gtrsim -5$, in addition to the massive component, a low-mass component appears and grows in fraction with increasing metallicities. With metallicity as high as $\log Z/Z_\odot \sim -1$, only the low-mass component remains and the massive component disappears. *The IMF transition proceeds rather gradually with metallicity* in this way, rather than abrupt change from top-heavy Pop-III to low-mass Pop-I IMFs at some single metallicity.

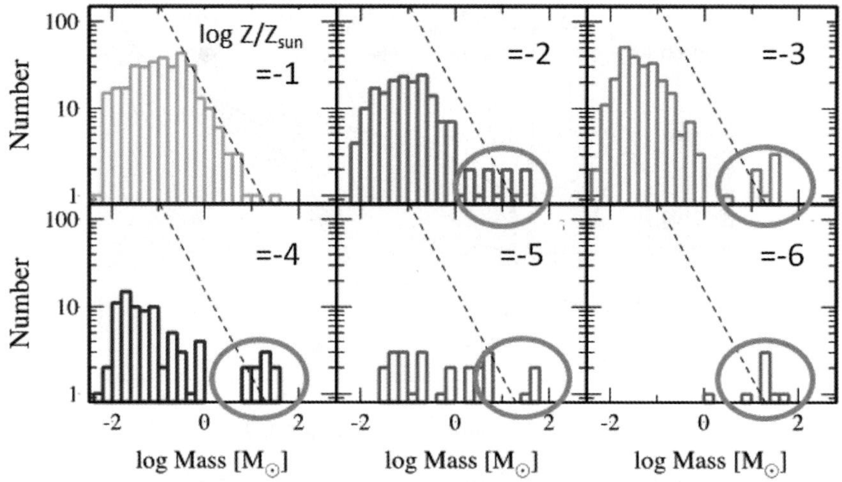

Figure 3. The mass distribution of stars, or the IMF, formed from clouds with different metallicities. The IMF consists of two components, low-mass and massive ones. The latter is indicated by circles in the panels. Plotted is at our final snapshot when the total stellar mass is $150M_{\odot}$. Although the mass distribution changes with time, it has become almost stationary by this time. Reproduced from Chon, Omukai & Schneider (2021).

3. Formation of Super-Massive Stars

Finally, we consider super-massive stars (SMSs), with mass $\gtrsim 10^5 M_{\odot}$. Originally in 1960's, those monsters are proposed as the engine of active galactic nuclei, which are now commonly believed to be accreting super-massive BHs (SMBHs). Interestingly, fifty years later, in the last decade or so, interests on SMSs revived as an origin (i.e., seeds) of SMBHs. SMSs begin at some point in their evolution to ungergo gravitational collapse due to the general relativistic instabilily and form similar mass BHs (Umeda et al. 2016, Woods et al. 2017, Haemmerlé et al. 2018, see contribution by Tyrone Woods this volume).

But, how such massive objects are formed? A current popular scenario supposes the collapse of a massive primordial gas cloud irradiated by a strong far-UV field, which disables H_2 cooling. Without H_2, the cloud cools solely by atomic cooling and collapses isothermally at 8000K. This causes monolithic collapse without fragmentation and the rapid accretion at a rate as high as $1M_{\odot}/\text{yr}$ on to a protostar formed inside (Regan & Haehnelt 2009, Inayoshi et al. 2014, see contribution by John Regan in this volume). Also protostellar evolution becomes very different in cases with such rapid accretion: the stellar envelope inflates enormously to ~ 10 au, 10^3 times larger than the main-sequence stellar radius. As a result, the protostar looks like a red giant with the effective temperature ~ 6000K and thus emit a negligible amount of UV photons. With no radiative feedback to stop accretion, the star is expected to eventually become supermassive by continued accretion.

In conventional views, it is supposed that the above SMS formation channel is operative only for a primordial pristine cloud. This is because even a slight amount of dust causes steep temperature decline in the cloud and fragmentation into smaller pieces of $< 1M_{\odot}$ (Omukai et al. 2008, Latif et al. 2016). The actual outcome of the collapse of a slightly metal-enriched cloud irradiated by strong FUV radiation field has not been clear. Recently, we studied this problem by way of SPH simulations (Chon & Omukai 2020). The initial condition is taken from a cloud in a halo that is strongly irradiated in

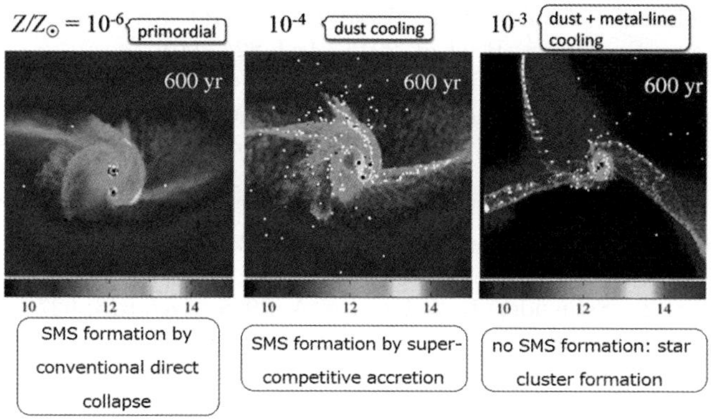

Figure 4. The outcome of the gravitational collapse of strongly irradiated clouds with different metallicities 10^{-6} (left), 10^{-4} (middle) and $10^{-3}Z_\odot$ (right). On the density distribution, protostars are overplotted by dots (black: massive ($> 10M_\odot$), white: less massive). Reproduced from Chon & Omukai (2020).

the cosmological simulation of Chon et al. (2016) and we put different metallicity values on it.

The result is presented in Figure 4. In the lowest metallicity case of $10^{-6}Z_\odot$ (left panel) the cloud collapse proceeds as in the conventional direct collapse and (binary) SMSs form at the center with few small fragments. With some more metals ($> 10^{-5}Z_\odot$, middle panel), the cloud in fact violently fragments by the dust cooling and numerous small protostars are formed. This, however, does not immediately mean a SMS does not form in the cloud. Fragmentation induced by the dust cooling occurs at very high densities $\gtrsim 10^{12}\mathrm{cm}^{-3}$, while global flow patterns set at lower densities effectively feed the central objects the material in spite of the fragmentation during inflow. By this way, as shown for the case of $10^{-4}Z_\odot$, the central objects become super-massive in a similar but more scaled-up way as in the so-called competitive accretion in the context of present-day star formation (*super-competitive accretion*). Once the metallicity becomes as high as $10^{-3}Z_\odot$, the metal-line cooling becomes important at lower densities. As a result, the global flow pattern is significantly changed and no clear central object appears. Also the accretion rates onto the stars are largely reduced due to lower temperature. In such a case, a star cluster appears without SMS formation. In conclusion, *SMSs can be formed not only from the primordial gas but also from slightly metal-enriched gas* as long as the metal-line cooling is not effective yet. This new SMS formation pathway by the super-competitive accretion would enhance the number of formed direct collapse BHs by some factor (see Sassano et al. 2021) and might be able to explain the universal origin of SMBHs in galaxies.

References

Abel, T & Wandelt, B. D. 2002, *MNRAS*, 330, L53

Clark, P. C., Glover, S. C. O. & Klessen, R. S. 2008, *ApJ*, 672, 757

Chon, S., Hirano, S., Hosokawa, T. & Yoshida, N. 2016, *ApJ*, 832, 134

Chon, S., & Omukai, K. 2020, *MNRAS*, 494, 2851

Chon, S., Omukai, K. & Schneider, R. 2021, *MNRAS*, 508, 4175

Dopcke, G., Glover, S. C. O., Clark, P. C., & Klessen, R. S. 2013 *ApJ*, 766, 103

Haemmerlé, L., Woods, T. E., Klessen, R. S. et al. 2018, *MNRAS*, 474, 2757

Hirano, S., Hosokawa, T., Yoshida, N. et al. 2014, *ApJ*, 781, 60

Hirano, S., Hosokawa, T., Yoshida, N. et al. 2015, *MNRAS*, 448, 568

Hosokawa, T., Omukai, K., Yoshida, N. & Yorke, H. W. 2011, *Science*, 334, 1250
Hosokawa, T., Hirano, S., Kuiper, R. et al. 2016, *ApJ*, 824, 119
Inayoshi, K., Omukai, K., & Tasker, E. 2014, *MNRAS*, 445, L109
Kinugawa, T., Inayoshi, K., Hotokezaka, K. et al. 2014, *MNRAS*, 442, 2963
Latif, M. A., Omukai, K., & Habouzit, M. et al. 2016, *ApJ*, 823, 40
Matsumoto, T. 2007, *PASJ*, 59, 905
Omukai, K. 2000, *ApJ*, 534, 809
Omukai, K., Hosokawa, T. & Yoshida, N. 2010, *ApJ*, 722, 1793
Omukai, K., Schneider, R, & Haiman, Z. 2008, *ApJ*, 686, 801
Omukai, K., Tsuribe, T., Schneider, R, & Ferrara, A. 2005, *ApJ*, 626, 627
Regan, J. A. & Haehnelt, M. G. 2009, *MNRAS*, 393, 858
Sassano, F., Schneider, R., Valiante, R. et al. 2021, *MNRAS*, 506, 613
Schneider, R., Ferrara, A., Natarajan, P. & Omukai, K. 2001 *ApJ*, 571, 30
Sugimura, K., Matsumoto, T., Hosokawa, T. et al. 2020, *ApJ* (Letters), 892, L14
Susa, H., Hasegawa, K., & Tominaga, N. 2014, *ApJ*, 792, 32
Tsuribe, T. & Omukai, K. 2006, *ApJ*, 642, L61
Umeda, H., Hosokawa, T. Omukai, K. & Yoshida, N. 2016, *ApJ*, 830, L34
Woods, T. E., Hegar, A., Whalen, D. J., et al. 2017, *ApJ*, 842, L6
Yoshida, N., Omukai, K. & Hernquist, L. 2008, *Science*, 321, 669

Discussion

ANONYMOUS: If SMSs are so big and luminous, are they bright enough to be observed by current or future telescopes?

OMUKAI: Yes, they are above the JWST sensitivity limit. But if those sources are rare, we cannot find them without a clue to where to look for. Still, there have been some claims of candidate detection.

FROST: In the first-star binary formation simulation you showed, you highlighted a merger happening before the bipolar HII regions formed for each of the main binary stars. It seemed like there were discs surrounding these stars and one of these also fragmented after the bipolar flows were generated - can you comment on whether those companions would also merge or whether there is any possibility of them surviving and creating higher order multiple systems?

OMUKAI: In that case, fragments are formed after HII region starts developing. They survived until the end of our simulation without merger. Although an $N > 2$ system tends to be dynamically unstable and is easily subject to ejection, it is possible that a hierarchical higher-order system is formed.

QUINTANA: I was wondering: now that we can see the evolution of the mass distribution for population III to population I stats, with population III having more massive stars, have there been extrapolations/predictions to know what will be the mass distribution of future generation of stars?

OMUKAI: We can carry out similar simulation for future high-metallicity, low CMB temperature environment and predict stellar mass distribution. But the greatest uncertainty is external radiation other than CMB: Since the temperature will be very low, even its small contribution would change the result.

OEY: So intriguing that you find fragmentation in filaments at solar and core concentration below. Does this mean that SF is only hierarchical when filamentary SF occurs? This should affect the IMF.

OMUKAI: We first expected to be so, but the result indicates the power-law mass functional form of the low mass component do not change regardless of filament or disk fragmentation. The reason is not yet very clear due to stochasticity of fragmentation and ejection process.

OEY: It sounds maybe that both modes are hierarchical then? I ask because the Salpeter slope can be obtained from a simplistic hierarchical scenario.

OMUKAI: If hierarchical fragmentation predicts the Salpeter like slope, mixture of hierarchical fragmentation and growth of protostars by competitive accretion may be setting the slope. Both of them give a similar slope. The slope may remain the same for this reason although fragmentation appears more hierarchical at higher metallicities and not so at lower metalicities.

Massive Stars Near and Far
Proceedings IAU Symposium No. 361, 2024
J. Mackey, J. S. Vink, & N. St-Louis, eds.
doi:10.1017/S1743921322002939

The origin of massive stellar systems via disk fragmentation

G. André Oliva[1] and Rolf Kuiper[2]

[1]Institute for Astronomy and Astrophysics, University of Tübingen, Auf der Morgenstelle 10, D-72076, Tübingen, Germany
email: astro@gandreoliva.org

[2]Faculty of Physics, University of Duisburg–Essen, Lotharstraße 1, D-47057, Duisburg, Germany

Abstract. In this contribution, we explore the question on the formation of multiple massive stellar systems via disk fragmentation with the help of the highest-resolution simulations to date of a fragmenting disk in the context of massive star formation. The simulations start from a collapsing cloud of 200 solar masses, followed by the formation of an accretion disk that develops spiral arms and fragments. Due to the high resolution of our grid, we are able to self-consistently form the fragments without the need for a subgrid module such as sink particles. We track the formed fragments into the first stages of companion formation, which allows us to give an estimate of the multiplicity of the final system due to disk fragmentation. We find in total around ~ 6 fragments, some at orbits of ~ 1000 au, and some close (possibly spectroscopic) companions.

Keywords. stars: massive – stars: formation – accretion, accretion disks – stars: binaries: close – methods: numerical

1. Introduction

Massive stars are observed in clusters with high multiplicity (Zinnecker & Yorke 2007). With recent increasing evidence that massive stars form in accretion disks like their low-mass siblings (see, e.g., Ilee et al. 2018; Beuther et al. 2017; Johnston et al. 2020; Maud et al. 2019), we explore the question of whether massive multiple stellar systems can be formed via disk fragmentation during the star formation process.

We base this contribution in the findings reported in Oliva & Kuiper (2020). In that study, we performed high-resolution simulations of the formation of a massive star surrounded by a fragmenting accretion disk. An overview of the evolution of the system can be summarized in the following way: after an initial epoch dominated by gravitational collapse, lasting around 4.5 kyr, a Keplerian-like accretion disk is formed. It becomes Toomre-unstable, forms spiral arms and fragments. There is a fragmentation-dominated epoch lasting around 11 kyr, where fragments form, interact and get destroyed or have the potential to evolve further into companions. Beyond the fragmentation epoch, we observe a quiescent epoch characterized by the stabilization of the inner disk.

2. Methods

We performed a set of five simulations with increasing spatial resolution of the formation of a massive star from the collapse of a slowly-rotating cloud core of 0.1 pc in radius and $200\,M_\odot$ in mass. The gas hydrodynamics is solved using the `Pluto` code (Mignone et al. 2007), with additional modules for self-gravity (Kuiper et al. 2011) and radiation transport. Specifically, we used `Makemake` (Kuiper et al. 2020) to model frequency-dependent irradiation of the (proto)star, and the thermal (re-)emission from the gas and dust with the gray flux-limited diffusion approximation. Stellar evolution is accounted for with the evolutionary tracks of Hosokawa & Omukai (2009). For the dust opacities, we use the tabulated values from Laor & Draine (1993), and we calculate the local evaporation of the dust grains by means of the formula of Isella & Natta (2005).

The mass of the cloud core is initially distributed with a density profile $\rho \propto r^{-1.5}$ (with r being the spherical radius). The initial radial velocity is zero, and the initial angular velocity is distributed such that $\Omega \propto R^{-0.75}$ (R being the cylindrical radius), with a normalization constant determined by the ratio of kinetic energy to gravitational potential energy, which is set to 5%. The exponent of the density profile power law and the ratio of kinetic to gravitational energy are within the expected values measured in star-forming cloud cores (see, e.g., Gieser et al. 2022; Beuther et al. 2002; Goodman et al. 1993). The choice for the exponent of the initial angular velocity profile keeps the initial ratio of kinetic to gravitational energy independent of radius, and allows for the formation of a more massive (and thus more fragmentation-prone) accretion disk earlier in time.

The simulations were run in a time-independent three-dimensional grid in spherical coordinates, where the radial position increases logarithmically, and the polar angle is spaced with the cosine function in such a way that the maximum resolution is achieved in the equatorial plane. The azimuthal angle is discretized uniformly. Our simulation domain has an inner boundary of 30 au, which we call the *sink cell*, and inside it, we model the formation of the central massive protostar. On purpose, we do not use any sub-grid sink particle model for the disk fragments. Thanks to the high resolution of the grid, the disk fragments are self-consistently described within the methods of hydrodynamics. In this contribution we focus on the highest-resolution simulation of the set, that corresponds to a grid of $536 \times 161 \times 512$ cells, of which ~ 26 million are used to model the disk physics.

3. From disk fragments to companion stars

When the accretion disk becomes Toomre-unstable at around $t = 4.5$ kyr, two initial spiral arms form which in turn quickly give rise to fragments. The data reveal that fragments always form in association to a spiral arm, for example due to growing overdensities in the spiral structure or because of collisions between spiral arms. The fragments continuously interact among themselves and with the disk and spiral arms: they merge, migrate and get sheared. Fragments are observed to develop secondary accretion disks of their own, with their own spiral arms that give rise to further fragments in a process that supports the hierarchical fragmentation scenario. An example of the disk fragmentation dynamics is captured in Fig. 1.

We developed an algorithm for the identification and tracking of fragments within the three-dimensional density and temperature data of the disk. Thanks to this tool, we were able to follow the evolution of individual fragments in their journey to become companions, up to the first stages of star formation. In the innermost ~ 150 au of the disk, the grid has cells of a sufficiently small size to resolve the size of a first Larson core according to the calculations by Bhandare et al. (2018) (a few astronomical units). Thus we are able to observe the formation of first Larson cores, that is, the first gravitational collapse of a disk fragment which is halted by thermal pressure. In the data, they are distinguished as a maximum in density and temperature, such that the enclosed mass within a region of the order of 1 au is of the order of $10^{-2}\,M_\odot$, consistent with the calculations by Bhandare et al. (2018). However, the total available mass around the fragments (including their secondary accretion disks and surrounding material up to a

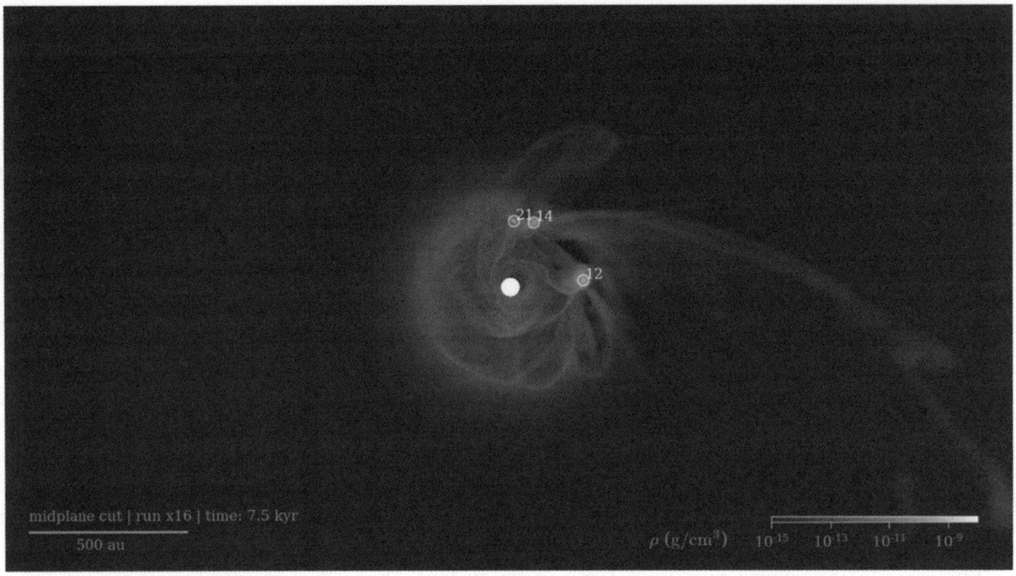

Figure 1. Midplane cut through the density field of the fragmenting accretion disk. Fragments labeled as 21 and 14 are interacting gravitationally. Fragment 12 is surrounded by a secondary accretion disk, and is also featured in Fig. 2. The numbers are only identification tags generated by the fragment-tracking algorithm and do not necessarily give information on the total number of fragments produced.

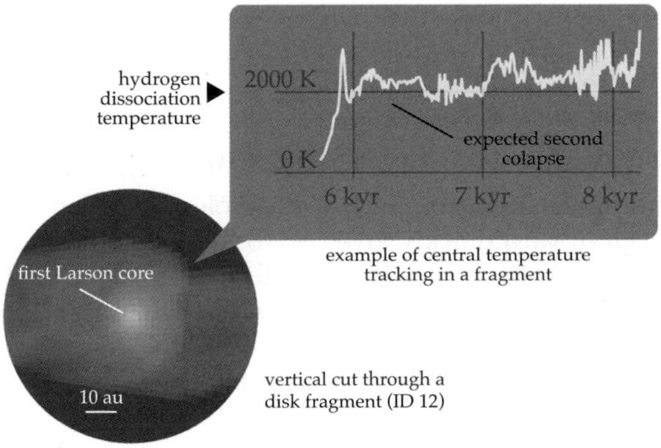

Figure 2. Evolution of a disk fragment into the first stages of stellar evolution.

radius of 40 au) is of the order of a few solar masses. Fragments formed later in time tend to be more massive.

During this stage, the orbits of the fragments are eccentric and chaotic because of their interactions. We observe that the orbits of some fragments become too close to the central massive protostar and get accreted by it. In the process, gravitational energy is converted into radiation energy in the form of an accretion burst.

When the central temperature of a first core increases consistently beyond 2000 K, the gravitational energy is used to dissociate the chemical bonds of molecular hydrogen instead of increasing the thermal pressure, and the gravitational collapse can continue. As a result of this process, a second Larson core of a few solar radii in size is expected

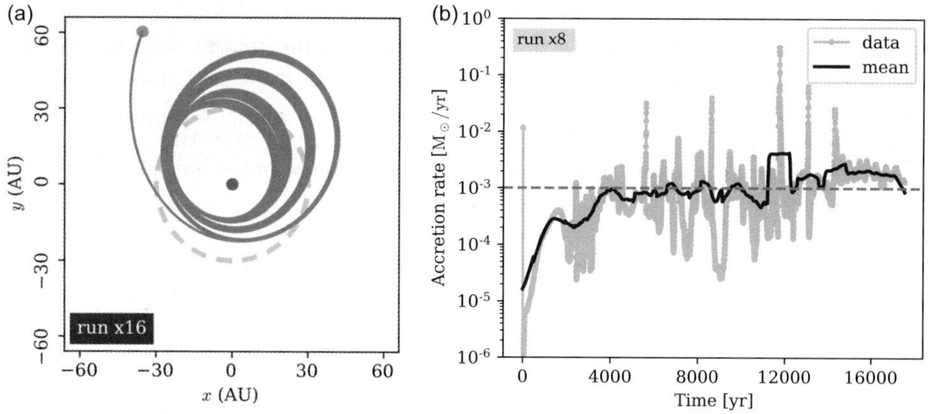

Figure 3. a) Integration of an orbit of a fragment that moves into the sink cell region in the simulation, showing a mechanism for the formation of spectroscopic multiples. b) Accretion rate onto the central massive protostar as a function of time.

to form (Bhandare et al. 2020). Even though our simulation methods did not include hydrogen dissociation, the fragment tracking algorithm enables us to predict when the hydrogen dissociation temperature is reached in the center of the fragment. The high degree of compactness of a second core object leads us to conjecture that the orbits would stabilize: even though second cores would feel the same gravitational interactions as first cores, the pressure gradient force arising from the gas in the disk would decrease. An example of the process first core formation and further evolution into a second core is illustrated with Fig. 2.

4. Multiplicity from disk fragmentation

Because of the interactions experienced by the fragments, only a fraction of them are expected to survive and evolve further into companions. From the analysis in this study, we expect first Larson cores to produce accretion bursts when they go inside of the region belonging to our numerical sink cell. However, second cores have higher probabilities of becoming companion stars, forming close companions if they move to the region close to the central massive protostar and wide companions if they remain in the outer disk.

In the highest resolution simulation of the set, we track 6 mergers and around 5 accretion bursts (very low mass and short-lived fragments are not tracked). At the end of the simulation ($t = 13.5$ kyr), 10 objects are expected to form second cores, 3 of them would become close companions and 3 continue to be located in the disk. The rest of the fragments that should form second cores merge with other fragments (but those mergers must be confirmed with the inclusion of hydrogen dissociation within the methods). Comparing with lower resolution simulations that we were able to run for a longer time, we estimate that the multiplicity produced by disk fragmentation is of the order of six, among close and wide companions. In contrast, previous studies that use sink particles and Cartesian adaptive mesh refinement grids predict numbers of companions that range between none and the order of hundreds (see, e.g., Girichidis et al. 2012; Rosen et al. 2016; Klassen et al. 2016).

From the outcomes of the full simulation set, we estimate that the number of wide orbit companions is of the order of one, with a semi-major axis ~ 1200 au and a mass of 5 $M_\odot$. Close companions would have orbits of semi-major axes smaller than 50 au and masses of the order of a few solar masses. However, because of the secondary accretion disks around forming companions, we cannot rule out the possibility that the final masses of the close companions might be higher than a few solar masses. Additionally, we integrated as an exercise the orbit of a second core object under the assumption that it is affected by the gravity of the central massive protostar and the average gravity of the disk. The results are displayed in Fig. 3a: we obtained an orbit that reduces over time due to the gain in mass of the central massive protostar. This constitutes a potential mechanism for the formation of spectroscopic multiple systems.

References

Beuther, H., Schilke, P., Menten, K. M., Motte, F., Sridharan, T. K., & Wyrowski, F. 2002, ApJ, 566, 945

Beuther, H., Walsh, A. J., Johnston, K. G., Henning, T., Kuiper, R., Longmore, S. N., & Walmsley, C. M. 2017, A&A, 603, A10

Bhandare, A., Kuiper, R., Henning, T., Fendt, C., Flock, M., & Marleau, G.-D. 2020, A&A, 638, A86

Bhandare, A., Kuiper, R., Henning, T., Fendt, C., Marleau, G.-D., & Kölligan, A. 2018, A&A, 618, A95

Commerçon, B., González, M., Mignon-Risse, R., Hennebelle, P., & Vaytet, N. 2022, A&A, 658, A52

Gieser, C., et al. 2022, A&A, 657, A3

Girichidis, P., Federrath, C., Banerjee, R., & Klessen, R. S. 2012, MNRAS, 420, 613

Goodman, A. A., Benson, P. J., Fuller, G. A., & Myers, P. C. 1993, ApJ, 406, 528

Hosokawa, T., & Omukai, K. 2009, ApJ, 691, 823

Ilee, J. D., Cyganowski, C. J., Brogan, C. L., Hunter, T. R., Forgan, D. H., Haworth, T. J., Clarke, C. J., & Harries, T. J. 2018, ApJ, 869, L24

Isella, A., & Natta, A. 2005, A&A, 438, 899

Johnston, K. G., et al. 2020, A&A, 634, L11

Klassen, M., Pudritz, R. E., Kuiper, R., Peters, T., & Banerjee, R. 2016, ApJ, 823, 28

Kuiper, R., Klahr, H., Beuther, H., & Henning, T. 2011, ApJ, 732, 20

Kuiper, R., Yorke, H. W., & Mignone, A. 2020, ApJS, 250, 13

Laor, A., & Draine, B. T. 1993, ApJ, 402, 441

Maud, L. T., et al. 2019, A&A, 627, L6

Mignon-Risse, R., González, M., Commerçon, B., & Rosdahl, J. 2021, A&A, 652, A69

Mignone, A., Bodo, G., Massaglia, S., Matsakos, T., Tesileanu, O., Zanni, C., & Ferrari, A. 2007, ApJS, 170, 228

Oliva, G. A., & Kuiper, R. 2020, A&A, 644, A41

Rosen, A. L., Krumholz, M. R., McKee, C. F., & Klein, R. I. 2016, MNRAS, 463, 2553

Zinnecker, H., & Yorke, H. W. 2007, ARA&A, 45, 481

Discussion

Audience member: What is the role of stellar feedback and magnetic fields on disk fragmentation?

Oliva: Our simulations include stellar irradiation, which stabilizes the innermost regions of the disk at late times. These simulations do not include magnetic fields, although other authors have found that it reduces fragmentation up to a certain degree (see, e.g., Commerçon et al. 2022; Mignon-Risse et al. 2021).

Audience member: How much of the mass of the central massive protostar is gained by accreting fragments?

Oliva: From the curve of accretion rate as a function of time (Fig. 3), we see that the accretion process is not smooth, but rather dominated by discrete accretion events.

However, the average accretion rate is still well in line with the expected accretion rate assuming smooth accretion.

RALF KLESSEN: Are fragments ever ejected from the disk?

OLIVA: This is not observed in the highest-resolution simulation presented in this contribution, but in one of the lower resolution runs, a fragment that is in a wide orbit could actually be ejected. This requires further confirmation by running the simulation for a longer time.

Massive Stars Near and Far
Proceedings IAU Symposium No. 361, 2024
J. Mackey, J. S. Vink, & N. St-Louis, eds.
doi:10.1017/S1743921322002526

Metallicity Dependences of Massive Star Formation from Theoretical and Observational Perspectives

K. E. I. Tanaka[1,2], Y. Zhang[3], T. Shimonishi[4] and R. Matsukoba[5]

[1]University of Colorado Boulder, Boulder, CO 80309, USA
email: kei.tanaka@colorado.edu

[2]National Astronomical Observatory of Japan, 2-21-1 Osawa, Mitaka, Tokyo 181-8588, Japan

[3]University of Virginia, Charlottesville, VA 22904-4325, USA

[4]Niigata University, Ikarashi-ninocho 8050, Nishi-ku, Niigata, 950-2181, Japan

[5]Kyoto University, Sakyo, Kyoto 606-8502, Japan

Abstract. Massive stars play crucial roles in astrophysical settings across cosmic history, and thus it is a fundamental problem to understand whether their formation processes are universal or diverse in various galactic environments. In particular, metallicity is the essential characteristic of cosmic evolution. Our theoretical studies have suggested some degrees of metallicity dependence of massive star formation. In the extremely metal-poor case of $< 10^{-2}\,Z_{\odot}$, protostellar disks are significantly unstable, and the photoionization feedback is more efficient. We also execute an ALMA survey targeting massive protostars in the Large Magellanic Clouds (LMC) with $0.5\,Z_{\odot}$. We found that the outflow properties of LMC protostars (mass, momentum, energy) are consistent with those of Galactic protostars, suggesting the universality of massive star formation at least in the range of ~ 0.5–$1\,Z_{\odot}$.

Keywords. stars: formation, (stars:) binaries (including multiple): close, (galaxies:) Magellanic Clouds

1. Introduction

Massive stars play pivotal roles in various fields of astrophysics. They are the sources of UV radiation, turbulent energy, and heavy elements in galaxies. Close binaries of massive stars are likely progenitors of gravitational wave binaries. Moreover, from a theoretical perspective, the first stars formed just after the big bang, known as Population III stars, are predicted to be very massive as ~ 10–$100\,M_{\odot}$. Because of the importance of massive stars across cosmic history, It is fundamental to understand whether their formation processes are universal or diverse in various environments across galactic evolution. In particular, metallicity is the key environmental parameter as it increases with cosmic time. We conduct theoretical and observational studies to investigate the potential metallicity dependences in massive star formation near and far.

2. Theoretical modeling

Various dynamical processes control the star formation efficiency (SFE) from the initial cloud core with ~ 0.1 pc (Rosen et al. 2020). Protostellar feedback and disk fragmentation are particularly important to determine the initial mass of each newborn star.

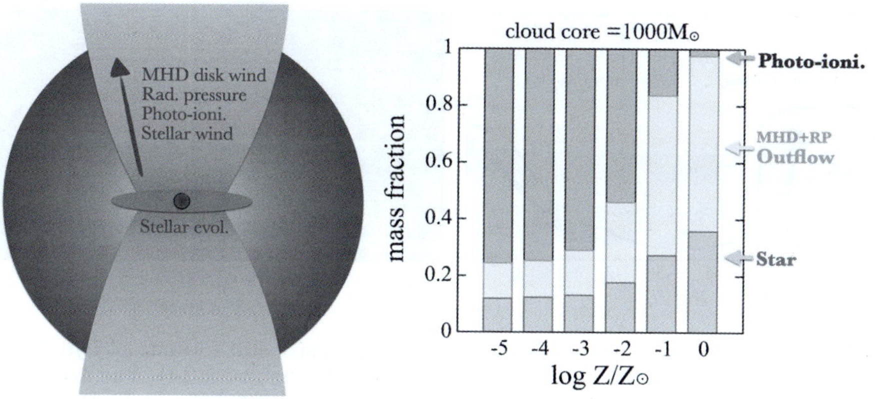

Figure 1. The semi-analytic model of multiple feedback in massive star formation (Tanaka et al. 2017; 2018). *left*: The schematic picture of the feedback model. We take into account MHD disk winds, radiation pressure, and photo-ionization, and investigate their metallicity dependences. *right*: A typical result from the model calculations, showing the mass fractions of the stars (pink), outflows (yellow), and photo-ionized gas (blue) in the case started from massive cloud cores of 1000 $M_{\odot}$. The photo-ionization feedback becomes dominant feedback at the lower metallicity of $\lesssim 10^{-2}\,Z_{\odot}$.

2.1. *Multiple feedback processes*

In the formation of Galactic low-mass protostars, it is well-known that the magnetically-driven disk winds is the dominant feedback (Machida & Hosokawa 2013). Radiation pressure acting on dusty gas was regarded as the crucial feedback in present-day massive star formation for a long time (Krumholz et al. 2009). In the formation of Population III stars from primordial gas, photo-ionization is suggested to be the main feedback (McKee & Tan 2008). Classically, these multiple feedback processes were investigated independently. We developed a semi-analytic model of massive star formation including multiple feedback processes, i.e., MHD disk winds, radiation pressure, and photo-ionization, over the wide range of metallicity (left panel of Figure 1, Tanaka et al. 2017; 2018). Our model suggested that, MHD disk wind is the dominant feedback at solar metallicity, rather than radiation pressure even in the formation of very-massive stars over 100 $M_{\odot}$. In this sense, in the Galactic environment, massive star formation can be considered as the scaled-up version of low-mass star formation. On the other hand, photo-ionization becomes more significant as the metallicity decreases, reducing the SFE compared to the solar-metallicity case, because photo-ionization is more efficient in the absence of dust grain (right panel of Figure 1). The same metallicity dependence of radiative feedback was also found in radiative hydrodynamical simulations by Fukushima et al. (2020). In recent years, multi-feedback have been investigated with radiative MHD simulations by various groups (e.g., Commerçon et al. 2022, Rosen 2022), but the number of theoretical works that take into account the metallicity is still limitted.

2.2. *Disk stability and fragmentation*

In our Galaxy, most of massive stars have close companions (Sana et al. 2012). Low-metallicity massive close-binaries are likely the progenitors of gravitational wave binaries (Abbott et al. 2019). Such massive close-binaries would form by gravitational fragmentation of unstable disks. We developed an analytic disk model in massive star formation, and investigated the metallicity dependences of disk stability (left panel of Figure 2, Tanaka & Omukai 2014). The massive protostellar disk is typically (marginally) unstable

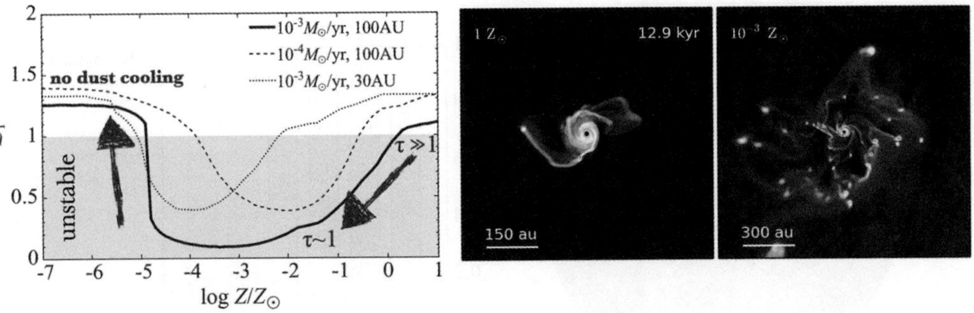

Figure 2. Metallicity dependence of disk stability in massive star formation. *left*: The analytic model of disk stability with various metallicity (Tanaka & Omukai 2014). The disk is predicted to be significantly unstable at the metallicity range of $\sim 10^{-2}$–10^{-5} $Z_\odot$. *right*: Radiative hydrodynamical simulations of protostellar disks at metallicity of 1 and 10^{-3} $Z_\odot$ (Appendix C of Matsukoba et al. 2022). As predicted by the analytic model (the right panel), the disk is more unstable at the low-metallicity case.

with the Toomre's parameter $Q \sim 1$ at the solar metallicity. The disk becomes even more unstable as the metallicity decreases from $Z_\odot$, because the disk optical structure changes from a very optically thick to a thinner (easy-to-cool) regime. The Toomre parameter Q becomes as small as ~ 0.1 at $\sim 10^{-2}$–10^{-4} $Z_\odot$. Such an unstable disk cannot hold its disk structure, and catastrophic fragmentation could occur and form a small stellar cluster. The disk is more stable at the extremely metal-poor case of $\lesssim 10^{-5}$ $Z_\odot$ since the dust cooling does not work effectively. We also conduct radiative hydrodynamical simulations to investigate dynamical evolution of protostellar disks at various metallicities, which agree well with the analytic prediction, i.e., massive protostellar disks are more unstable at 10^{-3} $Z_\odot$ than $Z_\odot$ (right panel of Figure 2 Matsukoba et al. 2022). This metallicity dependence of disk stability may explain the higher close-binary fraction at lower metallicity reported by Moe et al. (2019).

3. ALMA observation

Observational studies of massive star formation have made significant progress in recent years due to improvements in observational techniques. So far, observational studies of massive star formation are mostly limited in the Galactic environments of $\sim Z_\odot$. However, thanks to the high-resolution and high-sensitivity of ALMA, people start to detect protostellar outflows with sub-pc scales in the the Large Magellanic Cloud (e.g., Fukui et al. 2015, Shimonishi et al. 2016). The Large and Small Magellanic Clouds (LMC and SMC) are the excellent laboratory to investigate low-metallicity star formation with ~ 0.5 and $0.2\,Z_\odot$, respectively. Here we conducted a new ALMA survey project, called "MAGellanic Outflow and chemistry Survey (MAGOS)", in which we target 40 massive protostars in the LMC and SMC (K. Tanaka et al., in prep.). As a first science case with the MAGOS data, we investigated the physical properties of molecular outflows of 30 LMC protostars (left panel of Figure 3). We found that the masses, momenta, and energies of LMC outflows are correlated with their protostellar luminosities. These correlations are basically consistent with the well-known Galactic outflow properties (e.g., Beuther et al. 2002, Maud et al. 2015) (see right panel of Figure 3). This similarity could suggest protostellar outflows, i.e., the guideposts of MHD disk accretion at the small scale, have relatively universal properties at the metallicity range of ~ 0.5–$1\,Z_\odot$. Further ALMA studies will observationally examine the universality of massive star formation at various metallicities.

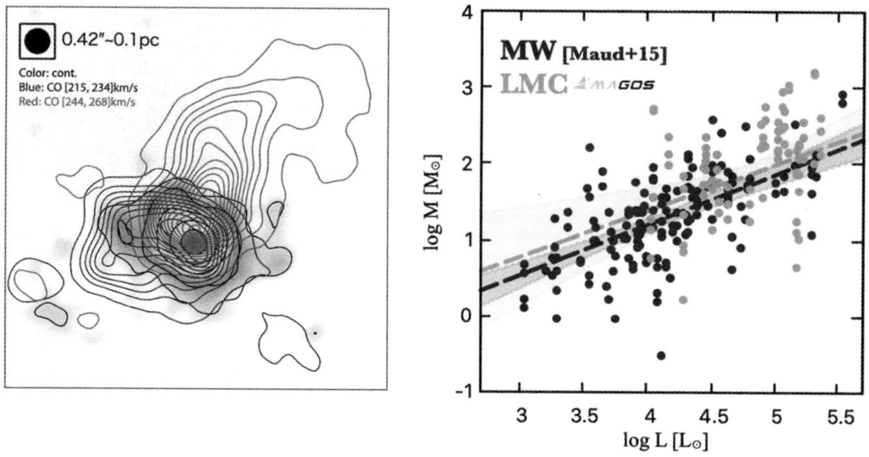

Figure 3. Protostellar outflows in the LMC from our MAGOS project (K. Tanaka et al., in prep.). *left*: An example of bipolar CO outflows detected. *right*: The outflow masses versus the protostellar luminosities of the LMC outflows (orange, MAGOS) and the Galactic outflows (blue, Maud et al. 2015). The LMC outflow properties are reasonably consistent with the Galactic counterparts.

4. Summary

Massive stars are important objects in cosmic time, and thus we investigate their formation at various metallicity from theoretical and observational perspective. Our theoretical models show that star formation dynamics could be qualitatively different in very low-metallicity environments of $\lesssim 10^{-2}\, Z_{\odot}$. At such a low-metallicity, the protostellar feedback is dominated by photo-ionization, and the protostellar disk becomes significantly unstable forming a small stellar cluster. We also conduct a new ALMA survey project called MAGOS, targeting low-metallicity protostars in the LMC and SMC. The LMC outflow properties (mass, momentum, and energy) are similar to those of the Galactic counterparts, possibly suggesting the universality of massive star formation dynamics at the metallicity range of ~ 0.5–$1\, Z_{\odot}$. Further studies from theoretical and observational perspectives are required to understand the universality and diversity of massive star formation.

References

Abbott, B. P., Abbott, R., Abbott, T. D., et al. 2019, ApJL, 882, L24
Beuther, H., Schilke, P., Menten, K. M., et al. 2002, Hot Star Workshop III: The Earliest Phases of Massive Star Birth, 267, 341
Commerçon, B., González, M., Mignon-Risse, R., et al. 2022, A&A, 658, A52
Fukui, Y., Harada, R., Tokuda, K., et al. 2015, ApJL, 807, L4
Fukushima, H., Hosokawa, T., Chiaki, G., et al. 2020, MNRAS, 497, 829
Krumholz, M. R., Klein, R. I., McKee, C. F., et al. 2009, *Science*, 323, 754
Machida, M. N. & Hosokawa, T. 2013, MNRAS, 431, 1719
Matsukoba, R., Tanaka, K. E. I., Omukai, K., et al. 2022, arXiv:2206.03497
Maud, L. T., Lumsden, S. L., Moore, T. J. T., et al. 2015, MNRAS, 452, 637
McKee, C. F. & Tan, J. C. 2008, ApJ, 681, 771
Moe, M., Kratter, K. M., & Badenes, C. 2019, ApJ, 875, 61
Rosen, A. L., Offner, S. S. R., Sadavoy, S. I., et al. 2020, SSRv., 216, 62
Rosen, A. L. 2022, arXiv:2204.09700
Sana, H., de Mink, S. E., de Koter, A., et al. 2012, *Science*, 337, 444

Shimonishi, T., Onaka, T., Kawamura, A., et al. 2016, ApJ, 827, 72
Tanaka, K. E. I. & Omukai, K. 2014, MNRAS, 439
Tanaka, K. E. I., Tan, J. C., & Zhang, Y. 2017, ApJ
Tanaka, K. E. I., Tan, J. C., Zhang, Y., et al. 2018, ApJ, 861, 68

Massive Stars Near and Far
Proceedings IAU Symposium No. 361, 2024
J. Mackey, J. S. Vink, & N. St-Louis, eds.
doi:10.1017/S1743921322002551

Formation of Supermassive Stars and the Direct Collapse to Black Holes

John A. Regan

Department of Theoretical Physics, Maynooth University Maynooth, Ireland
email: john.regan@mu.ie

Abstract. Supermassive stars represent a promising avenue for seeding the (super-)massive black holes observed in the centres of massive galaxies. In these proceedings I review the motivation on the need for supermassive stars as a progenitor pathway for seeding massive black holes. I discuss the currently understood limitations of seeds produced by less massive stars (i.e. remnants from the first generation of stars) and advocate that more massive stars - with masses up to $M_* \sim 10^5 M_\odot$ - formed under the conditions of hierarchical structure formation, in rare haloes, are the favoured pathway. Finally, I discuss some recent high resolution simulations demonstrating the formation of supermassive stars in early galaxies.

Keywords. Massive Black Holes, Early Universe, Supermassive Stars

1. Introduction

Observations point to an occupation fraction of very close to 100% for massive black holes (MBHs) in massive galaxies (Faber et al. 1997, Fan et al. 2006, Kormendy and Ho 2013). However, the origin of MBHs remains unknown. Over the past four decades many theoretical proposals have been put forward to address the origin of MBHs in galaxy centres beginning with Martin Rees in the late 1970's (Rees 1978; 1984). In Figure 1 we illustrate the most up-to-date theoretical proposals for the formation pathways of MBH seeds.

One of the main challenges presented by the detection of MBHs in galactic nuclei is their existence at very early times - less than a billion years after the Big Bang. Almost 200 high-z quasars have now been discovered at $z \geqslant 6$. This means that MBHs with masses close to, or in excess of, $M_{\rm MBH} \sim 10^9 M_\odot$ are already in place well before the end of reionisation. The presence of MBHs at the centres of massive galaxies in the early Universe poses a challenge to our understanding of both the formation and growth of MBHs.

The main avenues to forming and growing MBHs can be broken down into *light* and *heavy* seeds. In these proceedings I will flesh out the details of each pathway making the case that *heavy* seeds - forming out of (super-)massive stars are currently favoured both theoretically and empirically based on the evidence at hand.

2. The Massive Black Holes Mass Spectrum

As noted in the Introduction MBHs display a strong correlation with their host galaxy. This relationship, originally discovered more than two decades ago, is most easily seen using the so-called M-σ relation (Magorrian et al. 1998). Initially the relationship was confined (empirically) to black hole masses in excess of $10^6 M_\odot$ and to galaxy masses above the dwarf galaxy threshold ($M_{\rm gal} \sim 10^{11} M_\odot$). However, in more recent years MBHs in

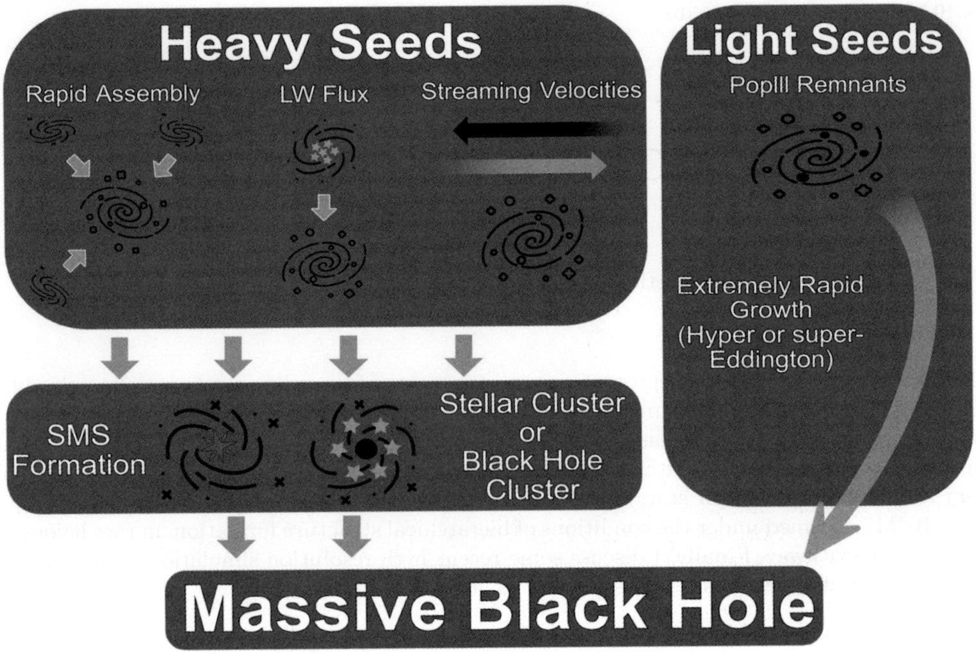

Figure 1. The formation pathways leading to both the creation of heavy and light seeds. Heavy seed pathways can lead to either the realisation of a monolithic collapse and a super-massive star or in the event of significant fragmentation the realisation of a dense stellar cluster or subsequently a black hole cluster. Light seeds are born from the remnants of the first generation of stars and are expected to have typical masses in the range $M_{BH} \sim 10^1 - 10^3 M_{\odot}$. Heavy seeds are expected to cover the range of $M_{BH} \sim 10^3 - 10^6 M_{\odot}$.

dwarf galaxies have been detected and so are beginning to populate the same relation. In Figure 2 we show a recent plot of the M-σ relation from Baldassare et al. (2020) where they were able to identify MBHs below the $10^6 M_{\odot}$ threshold.

Probing MBH masses between $M_{MBH} = 10^{4-6} M_{\odot}$ is now the next frontier in the relationship between MBHs and their host galaxies. This is the regime in which we expect *heavy* seeds to be born or for *light* seeds to reach masses of $M_{MBH} \sim 10^5 M_{\odot}$ before settling into the (dwarf) galactic centre. Identifying MBHs in dwarf galaxies at the masses of what we expect the seed masses to be does not however by itself identify the seeding mechanism. This is because as black holes accrete and grow, memory of their initial formation (i.e. seeding) is erased. Instead we must rely on the demographics of MBHs found in dwarf galaxies to try to piece together the evolutionary histories of the MBH population as a whole. However, this will be no easy task since the number of theorised pathways to forming MBHs is large with some predicting large seed masses (e.g. Regan and Downes 2018) while others predict 'intermediate' (e.g. Lupi et al. 2016) seed masses which grow into the *heavy* seed regime within certain environments (e.g. Stone et al. 2017, Natarajan 2021).

3. The Formation Pathways to Massive Black Holes

MBHs are expected to be seeded at high-z ($z \gtrsim 10$) in either metal-free or metal-poor environments. The demographics of MBHs inside early galaxies is currently unknown with quantitative estimates of the number densities of MBHs as a function of redshift varying by several orders of magnitude (Agarwal et al. 2012; 2014, Dijkstra et al. 2014, Habouzit et al. 2016, Wise et al. 2019). The variation can be partly explained by our lack

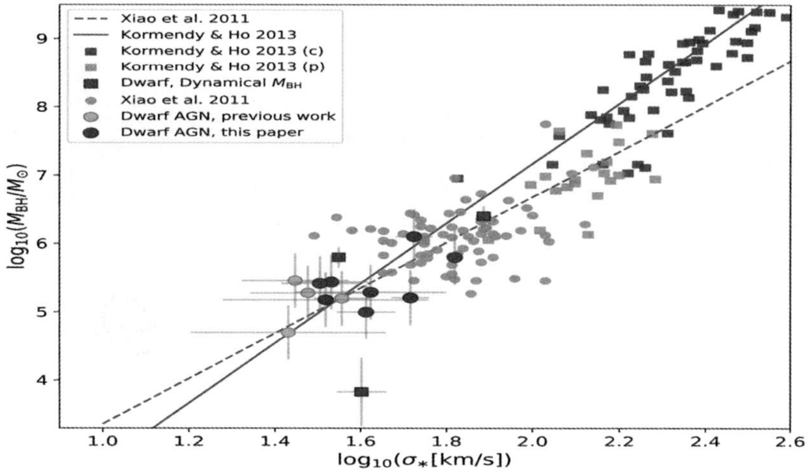

Figure 2. Massive Black hole mass vs. stellar velocity dispersion. The gray squares are from the compilation of Kormendy and Ho (2013); dark gray squares show galaxies with classical bulges and light gray squares show galaxies with pseudo-bulges. Light blue circles show data for low-mass AGN from Xiao et al. (2011). Active dwarf galaxies with stellar velocity measurements (from the Keck II Echellette Spectrograph and Imager) analysed by Baldassare et al. (2020) are shown as red circles with active dwarf galaxies with analysed by Baldassare et al. (2020) from the existing literature are shown in orange. Dwarf galaxies with dynamical BH mass estimates are shown as dark blue squares. We also show the fits to MBH-σ from Xiao et al. (2011) (dashed line) and Kormendy and Ho (2013) (solid line). Reproduced with modification from Baldassare et al. (2020).

of understanding of both the exact processes responsible and the exact environmental conditions required to seed MBHs.

The (astrophysical) seeds from which MBH grow can be broken down into *light* and *heavy* seeds. Light seeds are born from the remnants of the first generation of stars and are expected to have typical masses in the range $M_{BH} \sim 10^1 - 10^3 M_\odot$. Heavy seeds are expected to cover the range of $M_{BH} \sim 10^3 - 10^6 M_\odot$. In the following sections we will describe in detail both the formation pathways and growth prospects of *light* and *heavy* seeds. We will break the *heavy* seed formation scenario into two complementary parts. The first section will deal with the dynamical formation of a MBH seed within a dense, gaseous or stellar, environment (e.g. a dense stellar cluster, a nuclear cluster or within a dense black hole star cluster) and the subsequent formation of a MBH(s) at the cluster centre. The second section will explore the related idea of (super-)massive star formation in a similar environment and the subsequent transition of the system with one or more MBHs at the centre. Both scenarios may overlap potentially distinguished only by the degree of fragmentation at the outset.

3.1. *Light Seed Growth*

In Figure 1 we show a schematic of the pathways of both *light* and *heavy* seeds and their possible evolutionary avenues to MBH realisation. In the right hand panel we illustrate the formation of *light* seeds. These are seeds born out of the remnants of the very first stars. The mass spectrum of the first stars is unknown but there are strong reasons to suspect that the initial mass function of early stars will be top-heavy (Hirano et al. 2017). Resulting from the formation of massive stars will be a population of stellar mass black holes with masses in the range $M_{BH} \sim 10^{1-3} M_\odot$. If these seeds (or a small subset of them) can grow efficiently (Madau and Rees 2001) then they can explain both the high-z

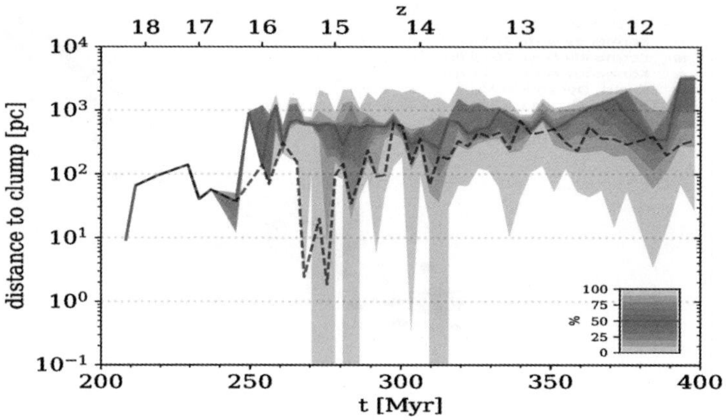

Figure 3. Distribution of distances between a black hole and a high accretability gas clump (greater than $10^{-3}M_{\odot}^{-1}$) for all progenitor haloes of the halo hosting the most black holes (shown in Figure 7.) Where the shaded region extends to the bottom of the figure, some black holes exist within high accretion rate clumps - but crucially this is short lived. Values shown in red correspond to all black holes within a halo's virial radius. The black, dashed line denotes the median separation for black holes within 0.25 of the virial radius. Reproduced with modification from Smith et al. (2018)

quasars as well as the population of MBH observed across cosmic time. However, both semi-analytic models and detailed numerical calculations have shown that growing light seeds is exceedingly difficult (Alvarez et al. 2009, Milosavljević et al. 2009). In a recent investigation by Smith et al. (2018) we found that when analysing the high resolution Renaissance simulations containing over 20,000 PopIII remnant black holes that none grew by more than 10%. In Figure 3 we show the growth of the entire population of black holes from the Renaissance simulations. What we show in Figure 3 is the mean distance of all stellar mass black holes from the nearest clump from which it could accrete (and grow). On average these PopIII remnant black holes are over 100 pc from the nearest gas clump. The black holes are too small to sink to the dense inner parts of their host galaxy and as a result wander the outer regions of the core where they cannot grow.

This characteristic of black hole growth was further investigated by Pfister et al. (2019) who found that stellar mass black holes do not sink and that black hole must reach a mass of $M_{MBH} \sim 10^5 M_{\odot}$ before they become sufficiently massive and can sink to the centre of their host galaxy.

3.2. *Heavy Seed Formation & Growth*

Given the challenges outlined above in relation to the growth of light seeds numerous research studies over the last few decades have postulated that it may be possible (or likely indeed) that more massive seeds could have driven the formation of MBHs. Initial investigations of the collapse of primordial gas clouds and the formation of heavy seeds date back to early papers by Loeb and Rasio (1994) and Eisenstein and Loeb (1995) who looked at the collapse of the central regions of haloes (albeit under highly idealised conditions) and found that seed masses up to approximately $10^6 M_{\odot}$ were possible under favourable conditions. Using a cosmological setup Regan and Haehnelt (2009) used the adaptive mesh refinement code Enzo (Bryan et al. 2014, Brummel-Smith et al. 2019) to model the collapse of gas within a dark matter halo at high redshift. They found that centrifugally supported disks formed with masses capable of supporting the formation of an intermediate stage en-route to MBH formation. Over the last decade rapid progress in understanding the formation pathways to MBHs via a heavy seed route has ensued.

In Figure 1 the potential pathways to heavy seed formation are outlined in the left hand panel. We begin by examining the Lyman-Werner (LW) and Baryonic Streaming Velocity channels to forming MBHs.

Emission in the LW band can readily dissociate H_2 cutting off a critical cooling route to forming PopIII stars. If this cooling channel is either cutoff or even disrupted then normal PopIII star formation is prevented and this can lead to (much more massive) star formation in more massive haloes where cooling is instead achieved via atomic hydrogen emission lines. Numerous authors have investigated this pathway over the last decade with the synchronised halo pair scenario emerging as a leading contender to produce heavy seeds (Shang et al. 2010, Dijkstra et al. 2014, Regan et al. 2017). In this scenario two haloes evolve closely separated in space and time with one halo beginning star formation first and irradiating the nearby halo and through LW irradiatiion leaving the neighbouring halo free of H_2. The neighbouring halo's gas then cools via atomic cooling (rather than molecular) and this leads to the formation of (super-)massive stars. The crucial parameter in the LW channel scenario is the level of flux required written as $J_{21,crit}$ where $J_{21,crit}$ is the critical level of flux in units of J_{21}†, required to dissociate H_2 and allow collapse via atomic hydrogen. Regan et al. (2017) found that neighbouring haloes must be separated by less than 300 pc with the receiving flux then being approximately $J_{crit} \sim 10^2$ J_{21}. Such high levels of flux are only possible due to the very tight separation of the haloes. The fine-tuned conditions for this pathway impacts on the number density of haloes that can support this particular pathway and hence it may be that this scenario cannot explain the full MBH population but maybe important for a subset of haloes hosting MBHs.

Baryonic streaming velocities originate following recombination and refers to the offset in velocities between baryons and dark matter. The mean offset is zero but fluctuations on either side of the mean, result in offsets between the baryons and the dark matter. This pathway has been extensively explored by many authors (e.g. Tanaka and Li 2014, Schauer et al. 2017, Hirano et al. 2017) in the context of heavy seed formation. Hirano et al. (2017) probed the collapse of haloes in rare regions of the Universe subject to 3σ fluctuations in the root mean squared offset between baryonic and dark matter velocities. They found that in these regions baryonic collapse inside dark matter haloes is delayed as the baryons cannot initially settle into the dark matter potential wells due to their excess kinetic energy. As a result larger inflow fluxes develop leading the formation of very masses stars with masses up to $M_* \sim 10^5 \mathrm{M}_\odot$. Similarly to the LW pathway it is unlikely that this particular pathway can explain all MBHs but it could also be, that it can work in tandem with the LW pathway to drive the formation of very massive stars which can be the seeds for the most MBHs (Schauer et al. 2017).

Finally, the rapid assembly process relies on the hierarchical structure formation pathway as the main driver either through dynamical heating offsetting cooling (Yoshida et al. 2003, Fernandez et al. 2014, Wise et al. 2019) or through turbulent cold flows acting to pressure support the gas against collapse (Latif et al. 2022). In many ways this is the most natural pathway with simulations now focusing on the initial mass function of stars produced with this pathway. If the rapid assembly process (or a derivative of it) can produce heavy seeds then a key finding would be the continuum of heavy seed masses supported and of how the seeds produced populate galaxies as they grow and evolve over cosmic time. Regan et al. (2020b) used high resolution adaptive mesh refinement simulations to follow the collapse of single halo in an overdense region which had previously been identified as a rapidly growing halo Regan et al. (2020a). In Figure 4 we show visualisation from the high resolution simulations of Regan et al. (2020b). Using a sophisticated subgrid model for star formation they showed the formation of massive

† J_{21} is defined as 10^{-21} erg cm^{-2} s^{-1} Hz^{-1} sr^{-1}

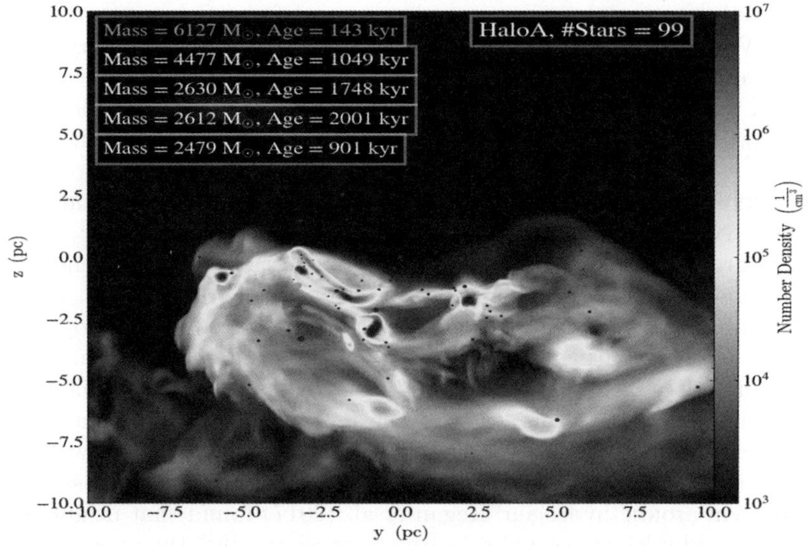

Figure 4. Visualisation of the stars in the high resolution runs of Regan et al. (2020b). Stars are marked with filled circles. The most massive star is coloured orange. Other stars are coloured blue. The stars are plotted on top of the baryonic gas density (in projection weighted by density). The most massive stars are labelled in the top left. Reproduced from Regan et al. (2020b)

stars with stellar masses $M_* \gtrsim 6000\mathrm{M}_\odot$. This (super-)massive star will directly collapse into a MBH of approximately the same mass as significant mass loss is not expected for metal free stars (Woods et al. 2017). It should be noted however that these stars are likely at the Eddington limit (e.g. Haemmerlé 2021) and detailed analysis is still lacking in understanding the exact mass loss rates and stellar winds produced by such objects. More simulations and greater statistics are now required to probe this and other related heavy seed formation channels (e.g. Latif et al. 2022) which may ultimately both predict and explain the observed number density of MBHs.

4. Observational Guidance and Outlook

Definitive observational signatures of the seeding process responsible for the formation of MBHs is still beyond the range of current (and indeed near future) telescopes. With the seeding process existing at small mass scales at very high redshift direct, clear, evidence is likely to be very difficult to achieve in practice. Nonetheless, there are a number of avenues that can be explored that together can build up sufficient circumstantial evidence to point towards certain avenues. In this section we explore how high-z observations of quasars, low-z observations of fossil dwarf galaxies and future gravitational wave measurements can all contribute to understanding the origin of MBHs.

The best direct method to detect the seeding process of MBHs may have to await the launch of a next generation X-ray telescope. As discussed in detail by Haiman et al. (2019) they argue that to detect seed black hole massed with masses in the $10^5\mathrm{M}_\odot$ range at $z \gtrsim 10$ requires a telescope with capabilities similar to that of the proposed Lynx mission † with flux sensitivities down to 10^{-19} $\mathrm{erg s^{-1} cm^{-2}}$. While Lynx will be sensitive enough to detect these embryonic black holes a combination of JWST ‡ and the Roman telescope § will be required for investigation of the host galaxy environment through

† https://www.lynxobservatory.com/
‡ https://www.jwst.nasa.gov/
§ https://roman.gsfc.nasa.gov/

their Optical and Near Infrared emission signatures. In additional to probing the seeding mechanisms of MBH at high-z it is also possible (and indeed a requirement) to probe the fossil remnants of heavy seeds at low-z. Dwarf galaxies may hold the key in this regard as dwarf galaxies resemble, in terms of the mass, the earliest galaxies. For dwarf galaxies that remain today some have had quiescent histories meaning that they retain information on the early Universe and of their foundation in from very early times. As a result the centres of dwarf galaxies today may host MBH that were seeded at high-z and have undergone only mild evolution since that epoch (Volonteri et al. 2008, van Wassenhove et al. 2010). More over recent detections of AGN in dwarf galaxies have resulted in the detection of tens of MBH candidates in the nuclei, as well as off-nuclear candidates, of dwarf galaxies (Mezcua 2017, Mezcua and Domínguez Sánchez 2020, Baldassare et al. 2020). Theoretically the existence of MBHs in dwarf galaxies is well founded (e.g. Dunn et al. 2018, Sharma et al. 2019) and the key now lies in understanding the demographics inside dwarf galaxies (Pacucci et al. 2021) and their connection to their high-z progenitors.

The advent of the detection of gravitational waves via the LIGO and Virgo detectors has opened up a new window into how we observe black holes (e.g. Abbott et al. 2016). While the current generation of GW observatories cannot observe MBH in the range of interest here the planned next generation of space based GW observatories † will have the capabilities to observe the mergers of MBHs out to redshifts up to $z = 20$. LISA, due for launch, in the second half of the next decade will be the premier observatory for detecting mergers of MBHs but by itself will not have the capability to directly distinguish between seeding pathways (Bellovary et al. 2019, Valiante et al. 2021). This is due to the fact the mergers will carry no information on the birth of the seeds. Instead a multi-messenger approach will ultimately be required with statistics being required from multiple different observations to piece together the full picture.

5. Conclusions

MBHs populate the centres of massive galaxies all the way down from the most massive central cluster galaxies into the regime of dwarf galaxies. Understanding both the formation pathways of MBHs and also their demographics inside galaxies remains an open question in modern astrophysics. While observations in both the electro-magnetic (EM) spectrum and gravitational wave spectrum have discovered hundreds of stellar mass black holes, and observations in the EM have detected MBH with masses in excess of $M_{MBH} \gtrsim 10^6 M_\odot$ - MBH detections in the window between $M_{MBH} = 10^{3-6}$ remain elusive.

Here we discuss the avenues to MBH formation in the early Universe - focusing on the formation of supermassive stars as a progenitor pathway. This pathway requires more "exotic" scenarios although we argue here that such environments do arise naturally through hierarchical structure formation and look promising in terms of producing number densities that match current observational constraints. In particular for supermassive stars formed through the rapid assembly of galaxies - be that through dynamical heating induced from minor mergers or through supersonic inflows - seeds with masses in the range $10^4 M_\odot$ appear quite plausible. Further work on building up the statistics of these formation pathways as well as (multi-messenger) observational signatures are now required.

References

B. P. Abbott, R. Abbott, T. D. Abbott, M. R. Abernathy, F. Acernese, K. Ackley, C. Adams, T. Adams, P. Addesso, R. X. Adhikari, and et al. Observation of Gravitational Waves from a Binary Black Hole Merger. *Physical Review Letters*, 116(6):061102, February 2016. doi: 10.1103/PhysRevLett.116.061102.

† `https://www.elisascience.org/`

B. Agarwal, S. Khochfar, J. L. Johnson, E. Neistein, C. Dalla Vecchia, and M. Livio. Ubiquitous seeding of supermassive black holes by direct collapse. *MNRAS*, 425:2854–2871, October 2012. doi: 10.1111/j.1365-2966.2012.21651.x.

B. Agarwal, C. Dalla Vecchia, J. L. Johnson, S. Khochfar, and J.-P. Paardekooper. The First Billion Years project: birthplaces of direct collapse black holes. *MNRAS*, 443:648–657, September 2014. doi: 10.1093/mnras/stu1112.

M. A. Alvarez, J. H. Wise, and T. Abel. Accretion onto the First Stellar-Mass Black Holes. *ApJL*, 701:L133–L137, August 2009. doi: 10.1088/0004-637X/701/2/L133.

Vivienne F. Baldassare, Claire Dickey, Marla Geha, and Amy E. Reines. Populating the Low-mass End of the M_{BH}-sigma Relation. *ApJL*, 898(1):L3, July 2020. doi: 10.3847/2041-8213/aba0c1.

Jillian M. Bellovary, Colleen E. Cleary, Ferah Munshi, Michael Tremmel, Charlotte R. Christensen, Alyson Brooks, and Thomas R. Quinn. Multimessenger signatures of massive black holes in dwarf galaxies. *MNRAS*, 482(3):2913–2923, January 2019. doi: 10.1093/mnras/sty2842.

Corey Brummel-Smith, Greg Bryan, Iryna Butsky, Lauren Corlies, Andrew Emerick, John Forbes, Yusuke Fujimoto, Nathan Goldbaum, Philipp Grete, Cameron Hummels, Ji-hoon Kim, Daegene Koh, Miao Li, Yuan Li, Xinyu Li, Brian OShea, Molly Peeples, John Regan, Munier Salem, Wolfram Schmidt, Christine Simpson, Britton Smith, Jason Tumlinson, Matthew Turk, John Wise, Tom Abel, James Bordner, Renyue Cen, David Collins, Brian Crosby, Philipp Edelmann, Oliver Hahn, Robert Harkness, Elizabeth Harper-Clark, Shuo Kong, Alexei Kritsuk, Michael Kuhlen, James Larrue, Eve Lee, Greg Meece, Michael Norman, Jeffrey Oishi, Pascal Paschos, Carolyn Peruta, Alex Razoumov, Daniel Reynolds, Devin Silvia, Samuel Skillman, Stephen Skory, Geoffrey So, Elizabeth Tasker, Rick Wagner, Peng Wang, Hao Xu, and Fen Zhao. ENZO: An Adaptive Mesh Refinement Code for Astrophysics (Version 2.6). *The Journal of Open Source Software*, 4(42):1636, Oct 2019. doi: 10.21105/joss.01636.

G. L. Bryan, M. L. Norman, B. W. O'Shea, T. Abel, J. H. Wise, M. J. Turk, and The Enzo Collaboration. ENZO: An Adaptive Mesh Refinement Code for Astrophysics. *ApJS*, 211: 19, April 2014. doi: 10.1088/0067-0049/211/2/19.

Mark Dijkstra, Andrea Ferrara, and Andrei Mesinger. Feedback-regulated supermassive black hole seed formation. *MNRAS*, 442(3):2036–2047, August 2014.

Glenna Dunn, Jillian Bellovary, Kelly Holley-Bockelmann, Charlotte Christensen, and Thomas Quinn. Sowing Black Hole Seeds: Direct Collapse Black Hole Formation with Realistic Lyman-Werner Radiation in Cosmological Simulations. *ApJ*, 861(1):39, July 2018. doi: 10.3847/1538-4357/aac7c2.

D. J. Eisenstein and A. Loeb. Origin of quasar progenitors from the collapse of low-spin cosmological perturbations. *ApJ*, 443:11–17, April 1995. doi: 10.1086/175498.

S. M. Faber, Scott Tremaine, Edward A. Ajhar, Yong-Ik Byun, Alan Dressler, Karl Gebhardt, Carl Grillmair, John Kormendy, Tod R. Lauer, and Douglas Richstone. The Centers of Early-Type Galaxies with HST. IV. Central Parameter Relations. *AJ*, 114:1771, November 1997. doi: 10.1086/118606.

X. Fan, M. A. Strauss, R. H. Becker, R. L. White, J. E. Gunn, G. R. Knapp, G. T. Richards, D. P. Schneider, J. Brinkmann, and M. Fukugita. Constraining the Evolution of the Ionizing Background and the Epoch of Reionization with z˜6 Quasars. II. A Sample of 19 Quasars. *AJ*, 132:117–136, July 2006. doi: 10.1086/504836.

R. Fernandez, G. L. Bryan, Z. Haiman, and M. Li. H_2 suppression with shocking inflows: testing a pathway for supermassive black hole formation. *MNRAS*, 439:3798–3807, April 2014. doi: 10.1093/mnras/stu230.

Mélanie Habouzit, Marta Volonteri, Muhammad Latif, Yohan Dubois, and Sébastien Peirani. On the number density of 'direct collapse' black hole seeds. *MNRAS*, 463(1):529–540, November 2016.

L. Haemmerlé. General-relativistic instability in rapidly accreting supermassive stars: The impact of rotation. *A&A*, 650:A204, June 2021. doi: 10.1051/0004-6361/202140893.

Zoltan Haiman, William N. Brandt, Alexey Vikhlinin, Jillian Bellovary, Elena Gallo, Jenny Greene, Kohei Inayoshi, Joseph Lazio, Bret Lehmer, Bin Luo, Piero Madau, Priya Natarajan, Feryal Ozel, Fabio Pacucci, Alberto Sesana, Daniel Stern, Cristian Vignali, Eli Visbal, Fabio Vito, Marta Volonteri, and Joan Wrobel. Electromagnetic Window into the Dawn of Black Holes. BAAS, 51(3):557, May 2019.

Shingo Hirano, Takashi Hosokawa, Naoki Yoshida, and Rolf Kuiper. Supersonic gas streams enhance the formation of massive black holes in the early universe. *Science*, 357(6358): 1375–1378, September 2017. doi: 10.1126/science.aai9119.

J. Kormendy and L. C. Ho. Coevolution (Or Not) of Supermassive Black Holes and Host Galaxies. *ARA&A*, 51:511–653, August 2013. doi: 10.1146/annurev-astro-082708-101811.

Muhammad A. Latif, Daniel J. Whalen, Sadegh Khochfar, Nicholas P. Herrington, and Tyrone E. Woods. Turbulent Cold Flows Gave Birth to the First Quasars. *arXiv e-prints*, art. arXiv:2207.05093, July 2022.

Abraham Loeb and Frederic A. Rasio. Collapse of Primordial Gas Clouds and the Formation of Quasar Black Holes. *ApJ*, 432:52, September 1994. doi: 10.1086/174548.

A. Lupi, F. Haardt, M. Dotti, D. Fiacconi, L. Mayer, and P. Madau. Growing massive black holes through supercritical accretion of stellar-mass seeds. *MNRAS*, 456(3):2993–3003, March 2016. doi: 10.1093/mnras/stv2877.

P. Madau and M. J. Rees. Massive Black Holes as Population III Remnants. *ApJ*, 551:L27–L30, April 2001. doi: 10.1086/319848.

J. Magorrian, S. Tremaine, D. Richstone, R. Bender, G. Bower, A. Dressler, S. M. Faber, K. Gebhardt, R. Green, C. Grillmair, J. Kormendy, and T. Lauer. The Demography of Massive Dark Objects in Galaxy Centers. *AJ*, 115:2285–2305, June 1998. doi: 10.1086/300353.

Mar Mezcua. Observational evidence for intermediate-mass black holes. *International Journal of Modern Physics D*, 26(11):1730021, Jan 2017. doi: 10.1142/S021827181730021X.

Mar Mezcua and Helena Domínguez Sánchez. Hidden AGNs in Dwarf Galaxies Revealed by MaNGA: Light Echoes, Off-nuclear Wanderers, and a New Broad-line AGN. *ApJL*, 898 (2):L30, August 2020. doi: 10.3847/2041-8213/aba199.

M. Milosavljević, S. M. Couch, and V. Bromm. Accretion Onto Intermediate-Mass Black Holes in Dense Protogalactic Clouds. *ApJ*, 696:L146–L149, May 2009. doi: 10.1088/0004-637X/696/2/L146.

Priyamvada Natarajan. A new channel to form IMBHs throughout cosmic time. *MNRAS*, 501 (1):1413–1425, February 2021. doi: 10.1093/mnras/staa3724.

Fabio Pacucci, Mar Mezcua, and John A. Regan. The Active Fraction of Massive Black Holes in Dwarf Galaxies. *ApJ*, 920(2):134, October 2021. doi: 10.3847/1538-4357/ac1595.

Hugo Pfister, Marta Volonteri, Yohan Dubois, Massimo Dotti, and Monica Colpi. The erratic dynamical life of black hole seeds in high-redshift galaxies. *MNRAS*, 486(1):101–111, June 2019. doi: 10.1093/mnras/stz822.

M. J. Rees. Accretion and the quasar phenomenon. *Physica Scripta*, 17:193–200, March 1978.

M. J. Rees. Black Hole Models for Active Galactic Nuclei. *ARA&A*, 22:471–506, 1984. doi: 10.1146/annurev.aa.22.090184.002351.

J. A. Regan and M. G. Haehnelt. The formation of compact massive self-gravitating discs in metal-free haloes with virial temperatures of ~13000-30000K. *MNRAS*, 393:858–871, March 2009. doi: 10.1111/j.1365-2966.2008.14088.x.

J. A Regan, E. Visbal, J. H. Wise, Z Haiman, P. H. Johansson, and G. L Bryan. Rapid formation of massive black holes in close proximity to embryonic protogalaxies. *Nature Astronomy*, 1:0075, April 2017. doi: 10.1038/s41550-017-0075.

John A. Regan and Turlough P. Downes. Rise of the first supermassive stars. *MNRAS*, 478(4): 5037–5049, August 2018. doi: 10.1093/mnras/sty1289.

John A. Regan, John H. Wise, Brian W. O'Shea, and Michael L. Norman. The emergence of the first star-free atomic cooling haloes in the Universe. *MNRAS*, 492(2):3021–3031, February 2020a. doi: 10.1093/mnras/staa035.

John A. Regan, John H. Wise, Tyrone E. Woods, Turlough P. Downes, Brian W. O'Shea, and Michael L. Norman. The Formation of Very Massive Stars in Early Galaxies and Implications for Intermediate Mass Black Holes. *The Open Journal of Astrophysics*, 3(1): 15, December 2020b. doi: 10.21105/astro.2008.08090.

Anna T. P. Schauer, John Regan, Simon C. O. Glover, and Ralf S. Klessen. The formation of direct collapse black holes under the influence of streaming velocities. *MNRAS*, 471(4): 4878–4884, November 2017. doi: 10.1093/mnras/stx1915.

C. Shang, G. L. Bryan, and Z. Haiman. Supermassive black hole formation by direct collapse: keeping protogalactic gas H_2 free in dark matter haloes with virial temperatures T_{vir} ¿ rsim 10^4 K. *MNRAS*, 402:1249–1262, February 2010. doi: 10.1111/j.1365-2966.2009.15960.x.

Ray Sharma, Alyson Brooks, Rachel S. Somerville, Michael Tremmel, Jillian Bellovary, Anna Wright, and Thomas Quinn. Black Hole Growth and Feedback in Isolated Romulus25 Dwarf Galaxies. *arXiv e-prints*, art. arXiv:1912.06646, Dec 2019.

Britton D. Smith, John A. Regan, Turlough P. Downes, Michael L. Norman, Brian W. O'Shea, and John H. Wise. The growth of black holes from Population III remnants in the Renaissance simulations. *MNRAS*, 480:3762–3773, November 2018. doi: 10.1093/mnras/sty2103.

Nicholas C. Stone, Andreas H. W. Küpper, and Jeremiah P. Ostriker. Formation of massive black holes in galactic nuclei: runaway tidal encounters. *MNRAS*, 467(4):4180–4199, June 2017. doi: 10.1093/mnras/stx097.

Takamitsu L. Tanaka and Miao Li. The formation of massive black holes in z $\sim$ 30 dark matter haloes with large baryonic streaming velocities. *MNRAS*, 439(1):1092–1100, March 2014. doi: 10.1093/mnras/stu042.

Rosa Valiante, Monica Colpi, Raffaella Schneider, Alberto Mangiagli, Matteo Bonetti, Giulia Cerini, Stephen Fairhurst, Francesco Haardt, Cameron Mills, and Alberto Sesana. Unveiling early black hole growth with multifrequency gravitational wave observations. *MNRAS*, 500 (3):4095–4109, January 2021. doi: 10.1093/mnras/staa3395.

S. van Wassenhove, M. Volonteri, M. G. Walker, and J. R. Gair. Massive black holes lurking in Milky Way satellites. *MNRAS*, 408(2):1139–1146, October 2010. doi: 10.1111/j.1365-2966.2010.17189.x.

M. Volonteri, G. Lodato, and P. Natarajan. The evolution of massive black hole seeds. *MNRAS*, 383:1079–1088, January 2008. doi: 10.1111/j.1365-2966.2007.12589.x.

John H. Wise, John A. Regan, Brian W. O'Shea, Michael L. Norman, Turlough P. Downes, and Hao Xu. Formation of massive black holes in rapidly growing pre-galactic gas clouds. *Nature*, 566(7742):85–88, January 2019. doi: 10.1038/s41586-019-0873-4.

T. E. Woods, A. Heger, D. J. Whalen, L. Haemmerlé, and R. S. Klessen. On the Maximum Mass of Accreting Primordial Supermassive Stars. *ApJ*, 842:L6, June 2017. doi: 10.3847/2041-8213/aa7412.

Ting Xiao, Aaron J. Barth, Jenny E. Greene, Luis C. Ho, Misty C. Bentz, Randi R. Ludwig, and Yanfei Jiang. Exploring the Low-mass End of the M $_{BH}$-σ_* Relation with Active Galaxies. *ApJ*, 739(1):28, September 2011. doi: 10.1088/0004-637X/739/1/28.

N. Yoshida, T. Abel, L. Hernquist, and N. Sugiyama. Simulations of Early Structure Formation: Primordial Gas Clouds. *ApJ*, 592:645–663, August 2003. doi: 10.1086/375810.

Discussion

Ralf Klessen: The pathways to *heavy* seeds could be via dense stellar clusters or SMS star formation. Which pathway do you think is more likely

Regan: I think some fragmentation is nearly inevitable in most systems and very difficult to prevent. However, it could be that after initial fragmentation some of the stars (fragments) merge with the central object as shown by Chon et al.

Kaz Omukai: Have you looked at SMS formation in metal enriched haloes

REGAN: Yes we have. We looked in the Renaissance simulations to see the fraction of rapidly growing haloes that occur with some prior metal enrichment. We found that including slightly enriched haloes increases the number of candidates haloes by a factor of approximately 3. Further, study of heavy seed formation in these metal-poor, rapidly assembling galaxies would be very useful.

Massive Stars Near and Far
Proceedings IAU Symposium No. 361, 2024
J. Mackey, J. S. Vink, & N. St-Louis, eds.
doi:10.1017/S1743921322002952

Constraints on the stellar upper mass limit from simulations of UV disk ablation

N. Dylan Kee[1] and Rolf Kuiper[2]

[1]National Solar Observatory, 22 Ohi'a Ku St, Makawao, HI 96768, USA
email: dkee@nso.edu

[2]Faculty of Physics, University of Duisburg-Essen, Lotharstraße 1, D-47057 Duisburg, Germany

Abstract. This contribution presents recent advances in identifying the stellar upper mass limit using simulations of UV radiative feedback during the star formation process. Generally, due to computational costs and a focus on au to parsec scales, simulations of massive star formation do not trace the flow of material to distances closer than a few au from the forming star. However, UV line-acceleration acts directly on accreting material in the sub-au circumstellar region, thereby efficiently ablating the surface layers off the protostellar disk. For stars on the order of a few hundred solar masses, this disk destruction rate exceeds the accretion rate, destroying the disk faster than it is replenished, and setting a maximum stellar mass as a function of metallicity that can be attained by single star formation channels.

Keywords. stars: formation, stars: mass function, stars: outflows, stars: early-type, stars: circumstellar matter

1. Introduction

One of the current major questions in the study of luminous massive stars concerns the stellar upper mass limit. While observations strongly suggest that such a limit exists, theory has thus far failed to conclusively identify either its physical origin or the actual value of this limit. However, it is clear that understanding this limit is crucially important due to the extreme impact of the most massive stars on their environments, the role these stars play as tracers of the structure and evolution of distant galaxies, and their place as the evolutionary precursors to exotic supernovae and high mass black holes.

In this contribution, we discuss recent work that constrains the upper mass limit of stars by examining the interplay of ultraviolet (UV) radiation with the circumstellar accretion disk at sub-au distances from forming massive stars (Kee et al. 2018, Kee and Kuiper 2019). By zooming in on this near-star region, we directly probe a volume that is normally omitted in studies of massive star formation due to the high computational cost of tracing accreting gas from the parsec-scale distances characteristic of a pre-stellar molecular cloud down to length scales on order the stellar radius of a zero-age main-sequence massive star. By examining this region, we capture critically important physics for setting the stellar upper mass limit also omitted by the current state-of-the-art simulations of massive star formation (e.g. Klassen et al. 2016, Kuiper and Hosokawa 2018, Tanaka et al. 2018, Mignon-Risse et al. 2021, Rosen 2022), namely the direct acceleration of material off of the accretion disk surface by stellar UV photons, and the associated disk mass loss. In the simulations presented here, we show that this disk mass-loss rate can exceed the accretion rate for stars with masses a few hundred times the mass of the Sun, thereby dispersing the inner disk faster than it can be replenished and preventing the star from growing to higher mass.

2. Description of the simulation suite

For this study, we set up a grid of simulations considering the innermost 20 stellar radii (R_*) around forming massive stars of a range of masses. As the crucial physics is the interaction of stellar UV photons with circumstellar disk material, we select stars with main-sequence-like radii for their mass and luminosity, namely stars which are hot enough to emit substantial UV photon fluxes. Based on computations of stellar structure in the presence of disk accretion by Hosokawa et al. (2010), the Kelvin-Helmholtz timescale becomes shorter than the accretion timescale around 20-30 $M_\odot$, fairly independent of the accretion rate. As such, we simulate stars of 25, 50, 75, and 100 $M_\odot$. For each stellar mass, we also consider a range of disk masses. By comparison with observations (see e.g. Beltrán and de Wit 2016; for a review) and prior simulations (e.g. Klassen et al. 2016, Tanaka et al. 2018, Oliva and Kuiper 2020, Rosen 2022), we extrapolate to the near-star conditions appropriate to disks of 10, 20, and 30 $M_\odot$ in 1000 au.

To describe the radiation force imparted on disk material through the interaction of UV photons with spectral lines of ionized metal species, we follow the Castor et al. (1975) formalism extended to be fully three-dimensional as discussed, for instance, in Kee et al. (2016). This extension to 3D is crucially important to the problem at hand, as the Castor et al. (1975) formalism focuses on the interaction of radially propagating photons with radially accelerating plasma. However, the finite angular size of the star allows photons to interact with non-radial acceleration, here the Keplerian shear of the circumstellar accretion disk, and thereby directly accelerate disk material away from the forming star.

3. Disk destruction rate

The most important property of the simulations carried out here is the rate at which material is stripped, or ablated, off the circumstellar disk, $\dot{M}_{ab}$. To quantify this, we measure the rate at which material leaves the simulation volume through the outer boundary and subtract off the spherically symmetric wind mass-loss rate predicted for each stellar mass. The resulting disk destruction rate is plotted for all models in Figure 1.

Considering each panel alone, we see that the disk mass does not play an important role in setting the ablation rate. Specifically, the variation of ablation rate between simulations with the same star and different disks can be attributed to a time delay caused by the more massive disks taking longer to evacuate than the less massive disks. As such, more massive disks can be taken as a proxy for accretion more efficiently replenishing the disk and more effectively keeping the disk and star connected. Meanwhile, the tendency for ablation rate to increase at later times in simulations with less massive disks can be attributed to the radial falloff in disk density allowing less dense material to be exposed to the direct irradation from the star at later times in the simulation once the inner disk is eaten away. The one exception to this is the set of simulations with a 25 $M_\odot$ star, where the ablation rate is notably the lowest, and as such none of the simulations with a 25 $M_\odot$ star evacuate a substantial fraction of the circumstellar disk in one year.

This brings us to a second important take away from these simulations, namely that comparing the panels of Figure 1 shows an almost two order-of-magnitude spread in ablation rate from simulations with a 25 $M_\odot$ star and those with a 100 $M_\odot$ star. In order to examine this trend, it is natural to scale the ablation rates with the predicted spherically symmetric wind mass-loss rate predicted for each individual star. As we are interested in the case where the ablation rate is most akin to what it would be in the

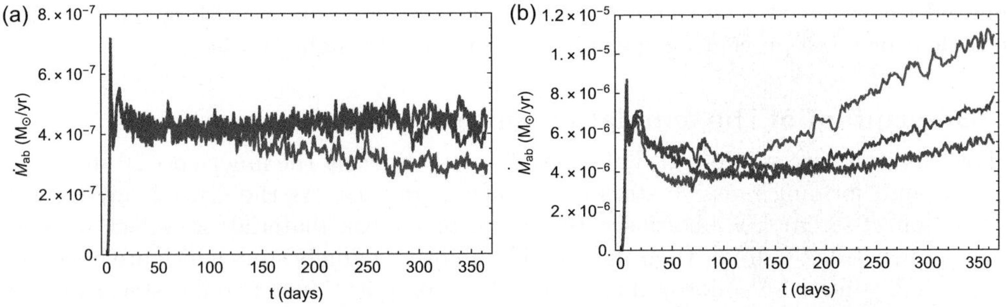

$M_* = 25M_\odot$ $M_* = 50M_\odot$

$M_* = 75M_\odot$ $M_* = 100M_\odot$

Figure 1. Disk destruction rate in solar masses per year for 25 (top, left), 50 (top, right), 75 (bottom, left), and 100 (bottom, right) solar mass stars. In each panel the three curves denote simulations with 10 (red), 20 (purple), and 30 (blue) $M_\odot$ disks. In the early simulation (the full simulation for the 25 $M_\odot$ star), the curves with the lowest, middle, and highest disk destruction rate are those with 10, 20, and 30 $M_\odot$ disks respectively. With the exception of the simulation with the 25 $M_\odot$ star, this ordering reverses as the simulation progresses.

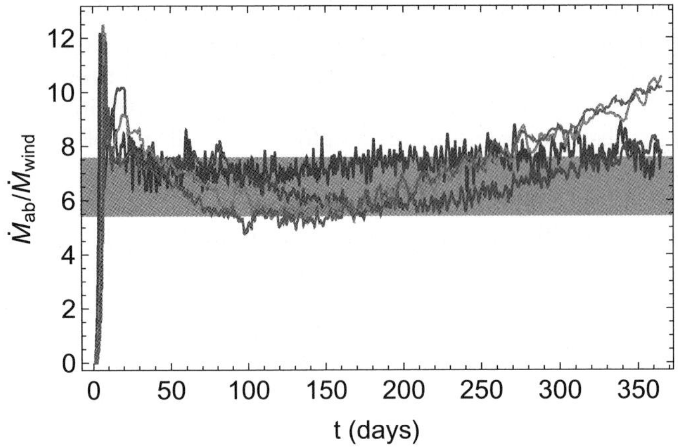

Figure 2. Disk destruction rate over spherically symmetric wind mass-loss rate for 25 (purple), 50 (red), 75 (orange), and 100 (green) solar mass stars. The grey bar at $\dot{M}_{\rm ab} = 6.5 \pm 1\ \dot{M}_{\rm wind}$ is put in by eye to draw attention to the overall uniformity of behavior between the simulations.

presence of accretion, we take the simulations with 30 $M_\odot$ disks, perform this rescaling, and plot the results in Figure 2. Immediately, it is evident that all simulations show disk evacuation at a rate $6 \sim 7$ times the wind mass-loss rate. Going forward this uniformity of behaviour, highlighted by the grey shading in Figure 2, allows us to analytically predict

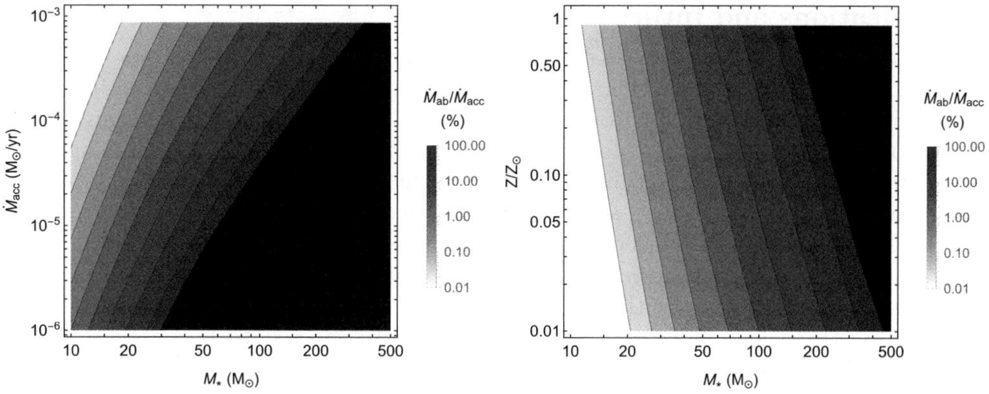

Figure 3. Ablation efficiency $\dot{M}_{\rm ab}/\dot{M}_{\rm acc}$ at Solar metallicity (left) and as a function of metallicity for an accretion rate of 10^{-4} $M_\odot$/yr (right). The black region on the right where $\dot{M}_{\rm ab} > \dot{M}_{\rm acc}$ denotes a forbidden region where stars can not form, such that the left edge of this region is the stellar upper mass limit as a function of accretion rate.

the rate at which forming massive stars can evacuate their disks simply by scaling the wind mass-loss rate up by a constant factor.

4. The stellar upper mass limit

With these results, we can now predict a stellar upper mass limit. To do so, we take the fully analytic Kudritzki et al. (1989) mass-loss rates computed for zero-age main-sequence stars using the mass-radius-luminosity relations taken from the Geneva stellar evolution code (Ekström et al. 2012, Yusof et al. 2013), boosted by the factor 6.5 found in the prior section. We then compare to a range of mass accretion rates $\dot{M}_{\rm acc}$ appropriate to forming massive stars, and compute the fraction of the accretion that would be removed from the disk by ablation. We plot a map of this 'ablation efficiency' at Solar metallicity in the left panel of Figure 3. The black region at the right of this figure is a forbidden region where $\dot{M}_{\rm ab} > \dot{M}_{\rm acc}$ and as such stars can not form. Therefore, for a chosen accretion rate the maximum mass of star that can form can be read off directly as the left edge of this forbidden region. Guided by observations and prior simulations, accretion rates for forming massive stars can reach $10^{-4} - 10^{-3} M_\odot$/yr, yielding an upper mass limit of a few hundred $M_\odot$, in good agreement with the most massive stars observed. Below this hard limit, it is important to note that order tens of percent of the accretion flow is already diverted for a star approximately half the maximum allowed mass, suggesting that ablation plays an important role in stellar feedback on the accretion disk even before the point where it alone can set the stellar upper mass limit.

Taking this one step further, we note that wind mass-loss rate, and as such ablation rate, is a function of metallicity Z, usually taken to be a power-law $\dot{M}_{\rm wind} \propto (Z/Z_\odot)^\alpha$. This means that ablation predicts a maximum allowed stellar mass also as a function of metallicity. To illustrate this, we take $\alpha = 0.5$ in the scaling relation above, and plot $\dot{M}_{\rm ab}/\dot{M}_{\rm acc}$ as a function of metallicity for a fixed accretion rate 10^{-4} $M_\odot$/yr (Figure 3, right). As would be predicted, a decrease of metallicity leads to decreases in both the available opacity for UV line-driving and the efficiency of ablation, and as such implies an increase in stellar upper mass-limit. This is again in good agreement with observations of star formation in the lower metallicity environments of the Magellanic Clouds.

5. Implications and future directions

Given that ablation operates at a notably smaller scale than other forms of feedback from forming massive stars on their environments (e.g. radiation pressure on dust grains, MHD driven outflows, and photoionization), it is likely that there is minimal direct interaction of these larger scale feedback mechanisms with ablation beyond modifying the rate of accretion that ablation must contend with. As such, the stellar upper mass limits predicted here are robust, as they are provided as a function of the accretion rate that reaches the near-star ablation zone. Looking back outwards, the disk surface flows launched by ablation likely do impact feedback on larger scales as larger scale feedback can be expected to interact with the additional kinematic component of the disk wind. Additionally, these low-angle disk winds can also be expected to be observed as they proceed to these larger scales (see Maud et al. 2018; for a recent potential observation of ablation at these larger scales). Therefore, future work will focus on directly including ablation in larger scale simulations to provide observational predictions, to examine this interaction of larger scale feedback with the flows launched by ablation, and to better constrain the accretion rates that reach the ablation zone and therefore the single value of the stellar upper mass limit as a function of the star formation environment.

6. Appendix A. Questions

D. Baade: The ablation rate appears to be rising near the end of each simulation. Could one have learned more by running these simulations for a longer period of time?
N. D. Kee: This is an artifact of the simulation set up. Since these simulations did not include accretion from larger radii, this rise is the accelerating ejection of the disk as the disk is detached from the star. Therefore, while it would be desirable to run the simulations longer to determine the asymptotic ablation rate, this will have to wait for the inclusion of the associated disk replenishment physics.

D. Hendriks: What is the relation between the outflow per unit surface area and the location where the outflow happens? Specifically, does outflow per unit surface area increase with distance from the star or is it mostly in the inner disk?
N. D. Kee: The ablation rate per unit surface area increases most strongly in the inner disk, namely most of the ablation flow is directly accelerated at the nearest point in the disk to the star, and minimal additional disk material is entrained further away.

A. de Koter: Does photoevaporation also play a role in the disk dispersal?
N. D. Kee: Moving from very massive stars to later spectral types does lead to a transition as ablation becomes less strong and photoevaporation increases in efficiency. Similarly, Tanaka et al. (2018) showed that photoevaporation is a dominant effect at low metallicity, where once again ablation is weak. The details of where both of these transitions occur, and how the two transitions interplay with one another at sub-Solar metallicity is a fertile topic for future research.

J. Regan: Lionel Haemmerlé in his talk discussed accretion rates of $> 1 M_{\odot}$/yr. Could you comment on how your findings might influence the formation of supermassive stars?
N. D. Kee: Since supermassive stars are formed at low metallicity, this is likely a part of the parameter space where ablation plays a more minor role. Future work could confirm this by computing UV line-driven wind mass loss rates for supermassive stars.

S. Oey: Is ablation accounted for in models of other circumstellar disks, e.g., excretion disks of Oe/Be stars? Presumably that sets the life expectancy of such disks?
N. D. Kee: Yes, earlier work in the series of papers on which this talk was based did consider ablation in classical Oe/Be star disks (Kee et al. 2016). A major finding of this work was that ablation alone is compatible with the rate of disk destruction in Be stars, and the dearth of Oe stars at solar metallicity.

References

Beltrán, M. T. & de Wit, W. J. 2016, Accretion disks in luminous young stellar objects. *A&AR*, 24, 6.

Castor, J. I., Abbott, D. C., & Klein, R. I. 1975, Radiation-driven winds in Of stars. *ApJ*, 195, 157–174.

Ekström, S., Georgy, C., Eggenberger, P., Meynet, G., Mowlavi, N., Wyttenbach, A., Granada, A., Decressin, T., Hirschi, R., Frischknecht, U., Charbonnel, C., & Maeder, A. 2012, Grids of stellar models with rotation. I. Models from 0.8 to 120 $M_\odot$ at solar metallicity ($Z = 0.014$). *A&A*, 537, A146.

Hosokawa, T., Yorke, H. W., & Omukai, K. 2010, Evolution of Massive Protostars Via Disk Accretion. *ApJ*, 721(1), 478–492.

Kee, N. D. & Kuiper, R. 2019, Line-driven ablation of circumstellar discs: IV. The role of disc ablation in massive star formation and its contribution to the stellar upper mass limit. *MNRAS*, 483(4), 4893–4900.

Kee, N. D., Owocki, S., & Kuiper, R. 2018, Line-driven ablation of circumstellar discs - III. Accounting for and analysing the effects of continuum optical depth. *MNRAS*, 479(4), 4633–4641.

Kee, N. D., Owocki, S., & Sundqvist, J. O. 2016, Line-driven ablation of circumstellar discs - I. Optically thin decretion discs of classical Oe/Be stars. *MNRAS*, 458(3), 2323–2335.

Klassen, M., Pudritz, R. E., Kuiper, R., Peters, T., & Banerjee, R. 2016, Simulating the Formation of Massive Protostars. I. Radiative Feedback and Accretion Disks. *ApJ*, 823(1), 28.

Kudritzki, R. P., Pauldrach, A., Puls, J., & Abbott, D. C. 1989, Radiation-driven winds of hot stars. VI. Analytical solutions for wind models including the finite cone angle effect. *A&A*, 219, 205–218.

Kuiper, R. & Hosokawa, T. 2018, First hydrodynamics simulations of radiation forces and photoionization feedback in massive star formation. *A&A*, 616, A101.

Maud, L. T., Cesaroni, R., Kumar, M. S. N., van der Tak, F. F. S., Allen, V., Hoare, M. G., Klaassen, P. D., Harsono, D., Hogerheijde, M. R., Sánchez-Monge, Á., Schilke, P., Ahmadi, A., Beltrán, M. T., Beuther, H., Csengeri, T., Etoka, S., Fuller, G., Galván-Madrid, R., Goddi, C., Henning, T., Johnston, K. G., Kuiper, R., Lumsden, S., Moscadelli, L., Mottram, J. C., Peters, T., Rivilla, V. M., Testi, L., Vig, S., de Wit, W. J., & Zinnecker, H. 2018, Chasing discs around O-type (proto)stars. ALMA evidence for an SiO disc and disc wind from G17.64+0.16. *A&A*, 620, A31.

Mignon-Risse, R., González, M., & Commerçon, B. 2021, Collapse of turbulent massive cores with ambipolar diffusion and hybrid radiative transfer. II. Outflows. *A&A*, 656, A85.

Oliva, G. A. & Kuiper, R. 2020, Modeling disk fragmentation and multiplicity in massive star formation. *A&A*, 644, A41.

Rosen, A. L. 2022, A Massive Star is Born: How Feedback from Stellar Winds, Radiation Pressure, and Collimated Outflows Limits Accretion onto Massive Stars. *arXiv e-prints*,, arXiv:2204.09700.

Tanaka, K. E. I., Tan, J. C., Zhang, Y., & Hosokawa, T. 2018, The Impact of Feedback in Massive Star Formation. II. Lower Star Formation Efficiency at Lower Metallicity. *ApJ*, 861(1), 68.

Yusof, N., Hirschi, R., Meynet, G., Crowther, P. A., Ekström, S., Frischknecht, U., Georgy, C., Abu Kassim, H., & Schnurr, O. 2013, Evolution and fate of very massive stars. *MNRAS*, 433(2), 1114–1132.

Massive Stars Near and Far
Proceedings IAU Symposium No. 361, 2024
J. Mackey, J. S. Vink, & N. St-Louis, eds.
doi:10.1017/S1743921322001909

Binarity in Massive Young Stellar Objects

Robert G. Shenton, René D. Oudmaijer and Stuart L. Lumsden

School of Physics & Astronomy, University of Leeds, Woodhouse Lane, LS2 9JT Leeds, UK
email: py15rgs@leeds.ac.uk

Abstract. Most massive stars (up to 100%) are thought to be in binary systems. The multiplicity of massive stars seems to be intrinsically linked to their formation and evolution, and so Massive Young Stellar Objects are key in observing this early stage of star formation. We have surveyed hundreds of MYSOs across the Galaxy from the RMS catalogue, using UKIDSS and VVV point source data. Preliminary results show binary fractions of $44 \pm 3\%$ for the UKIDSS sample and $32 \pm 3\%$ for the VVV sample. In addition we use the K-band magnitudes as a proxy for the companion mass, and find a significant fraction of the detected companions have estimated mass ratios greater than 0.5, which suggests a deviation from the capture formation scenario.

Keywords. binaries: general, stars: formation, stars: massive, stars: pre-main-sequence.

1. Introduction

Massive Young Stellar Objects (MYSOs) represent a crucial stage of star formation, where mass accretion and stellar multiplicity can be investigated through observations. Up to 100% of massive stars ($> 8M_{\odot}$) are thought to form in binary or higher order systems (Duchêne & Kraus 2013, Chini et al. 2012). To understand the origin of binarity, massive stars must be studied during their formation stage. However there have been very few studies which have looked into MYSO binarity (Pomohaci et al. 2019, Koumpia et al. 2019; 2021). We are investigating a sample of MYSOs and their companions, to determine their binary statistics and mass ratios, using data from the UKIDSS & VVV infrared surveys. This work considers hundreds of objects from the RMS catalogue of MYSOs (Lumsden et al. 2013) and as such concerns the largest study into MYSO multiplicity to date.

2. Methods

The UKIDSS (Lucas et al. 2008) and VVV (Saito et al. 2012) point source catalogues were used to search for potential companions around MYSOs. We employed statistical methods to determine whether a nearby object was physically associated with the MYSO, based on its separation from the primary and the density of background stars. The further an object is from the primary, or the denser the background source distribution, the less likely the object is to be classified as a physical companion (e.g. Pomohaci et al. 2019). Objects above a certain threshold are judged to be chance projections and are disregarded. Also, multi-colour information from UKIDSS/VVV was used to determine the mass ratios of the companions.

3. Results

In the separation range of 0-4 arcsec (corresponding to typical separations between 1000-20000 au), we find binary fractions of $44 \pm 3\%$ for the UKIDSS sample and $32 \pm 3\%$

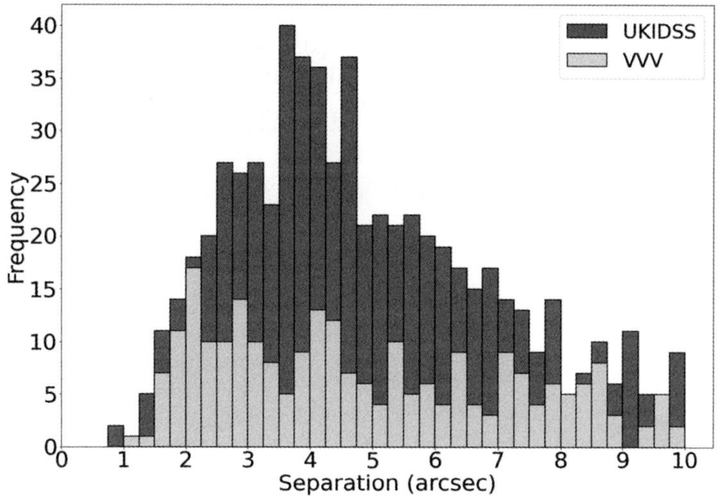

Figure 1. Histogram of object separation from its corresponding primary MYSO up to 10 arcsec. Only objects below the $P_{chance} < 20\%$ threshold (i.e. probable companions) are shown.

for the VVV sample. The larger UKIDSS fraction is due to variations in the galactic background density - lower densities increase companion likelihood, and so when excluding the low density region of the 'outer' galaxy (probed by UKIDSS but not VVV) the UKIDSS fraction is consistent with that of VVV. The separation distribution of the detected objects (up to 10 arcsec) can be seen in Figure 1. A large fraction of the mass ratios for the UKIDSS and VVV samples appear to be above 0.5.

4. Conclusions

The binary fractions found are larger than that of the MYSO binarity survey of Pomohaci et al. (2019), however when using the same separation distance and limiting magnitude, the statistics agree. The large fraction of mass ratios greater than 0.5 suggests a disagreement with the binary capture formation scenario, in which lower mass ratios are predicted (Salpeter 1955). As the binary fraction of massive stars is thought to be close to 100%, this study shows that 32% of them (in the VVV case) have observable companions at large separations. So far this work has focused on data from K-band imaging observations. Future work will involve using X-Shooter spectroscopy to identify close-in companions that cannot be resolved through imaging and help fill the gaps left by this work. Additionally, this spectra will be used to characterise the detected companions and study their environments.

Supplementary material

To view supplementary material for this article, please visit http://dx.doi.org/10.1017/S1743921322001909.

References

Chini, R., Hoffmeister, V. H., Nasseri, A., Stahl, O., & Zinnecker, H. 2012, MNRAS, 424, 1925
Duchêne, G., & Kraus, A. 2013, ARA&A, 51, 269
Koumpia, E., et al. 2019, A&A, 623, L5
—. 2021, A&A, 654, A109
Lucas, P. W., et al. 2008, MNRAS, 391, 136

Lumsden, S. L., Hoare, M. G., Urquhart, J. S., Oudmaijer, R. D., Davies, B., Mottram, J. C., Cooper, H. D. B., & Moore, T. J. T. 2013, ApJS, 208, 11
Pomohaci, R., Oudmaijer, R. D., & Goodwin, S. P. 2019, MNRAS, 484, 226
Saito, R. K., et al. 2012, A&A, 537, A107
Salpeter, E. E. 1955, ApJ, 121, 161

Massive Stars Near and Far
Proceedings IAU Symposium No. 361, 2024
J. Mackey, J. S. Vink, & N. St-Louis, eds.
doi:10.1017/S1743921322002691

In-Situ Field Massive Star Formation in the Small Magellanic Cloud

I. Vargas-Salazar[1], M. S. Oey[1], J. R. Barnes[1], X. Chen[1,2], N. Castro[3], K. M. Kratter[4] and T. A. Faerber[1,5]

[1]University of Michigan, Department of Astronomy, Ann Arbor, MI, USA, 48109
email: ivargasa@umich.edu

[2]Current address: Yale University, Department of Physics, New Haven, CT, USA, 06520

[3]Leibniz-Institut fur Astrophysik An der Sternwarte, 16 D-14482, Potsdam, Germany

[4]University of Arizona, Department of Astronomy, Tucson, AZ, USA, 85721

[5]Current address: West Virginia University, Department of Physics and Astronomy, Morgantown, WV, USA, 26506

Abstract. A fundamental question for theories of massive star formation is whether OB stars can form in isolation. We assess the contribution of any in-situ OB star formation by using 210 field OB stars in the Small Magellanic Cloud (SMC) from the Runaways and Isolated O-Type Star Spectroscopic Survey of the SMC (RIOTS4). We search for tiny, sparse clusters around our target OB stars using cluster-finding algorithms. Employing statistical tests, we compare these observations with random-field data sets. We find that $\sim 5\%$ of our target fields do show evidence of higher central stellar densities, implying the presence of small clusters. This frequency of small clusters is low and within errors, it is also consistent with the field OB population being composed entirely of runaway and walkaway stars. Assuming this small cluster fraction is real, it implies that some OB stars may form in highly isolated conditions. The low frequency could be caused by these clusters evaporating on a short timescale. However, another interpretation is that the low fraction of small clusters is observed because these form rarely, or not at all, implying a higher cluster lower-mass limit and generally consistent with a relationship between maximum stellar mass (m_{max}) and the cluster mass (M_{cl}).

Keywords. massive stars — field stars — Small Magellanic Cloud — open star clusters — star formation — runaway stars — multiple star evolution — OB associations — stellar populations

1. Introduction

Field stars account for $25-30\%$ of normal OB star populations (Oey et al. 2004). Field stars have two different possible origins: runaways, i.e., stars that are ejected from their parent cluster; or stars that form in situ in apparent isolation. The latter could happen, for example, if the stellar initial mass function (IMF) and cluster mass function (CMF) behave as probability density functions, allowing smaller, sparser clusters with a single massive star to form. We call the massive star in such a cluster a Tip-of-the-Iceberg Star (TIB). Simulations by Lamb et al. (2010) are consistent with the existence of these TIB clusters, and observations of field stars (Oey et al. 2013) have found evidence of potential in-situ formation. However, the frequency of in situ formed field OB stars has not been well established. Therefore, we used The Runaways and Isolated O-Type Star Spectroscopic Survey of the SMC (Lamb et al. 2016; RIOTS4) and OGLE III photometry (Udalski et al. 2008) to identify a 210-target sample to search for TIB clusters using cluster-finding algorithms to assess the contribution of any in-situ OB star formation in a complete sample of field stars.

2. Methods & Results

Our two cluster finding algorithms are Friends-of-Friends (FOF) and Nearest Neighbors (NN). FOF identifies associated stars ("friends") as those that are within a given clustering length (l_c) of another member. Each l_c is specific to each target field ($l_{c,field}$). Clusters are more likely to be those with a greater number (N_*) of stars identified or those with a higher M value, where $M = \frac{l_{c,SMC}}{l_{c,field}} N_*$ and $l_{c,SMC}$ is the average l_c for all our target-star fields. A higher M-value corresponds to a larger number of stars closely spaced together. The NN algorithm measures the stellar surface densities Σ_j associated with a given target star, calculated by counting the number of stars within radius r_j defined by its jth nearest neighbors: $\Sigma_j = \frac{j-1}{\pi {r_j}^2}$. We compare the resulting Σ_j with the background densities Σ_{bg} calculated for each target. The larger the difference, the more likely the target belongs to a cluster.

We compare the observed data with randomly generated fields for each target that serve as controls, and are therefore not independent of the observed data. We use two statistical tests for these comparisons: the Wilcoxon test (W), which looks for differences between two non-parametric, non-independent datasets; and the Rosenbaum test (R), which evaluates differences in the tails of two non-independent distributions, and are thus more sensitive to the presence of TIB clusters. We also compare known runaway stars (Oey et al. 2018), which should not show TIB clusters, with the rest of our targets, the non-runaways, using the K-S test and the Anderson-Darling (AD) test, which are valid when comparing independent samples. The AD again gives more weight to the tails of the distributions.

Both our FOF and NN results show a very small presence of positive detections. However, the statistical significance of the FOF results for the non-runaway sub-sample against the random fields is lower (W p-values: 0.00 – 0.1), relative to the NN results (W p-values: 0.00 – 0.03). The NN algorithm also shows a greater significant difference between the non-runaways and the runaways (KS and AD p-values: 0.02 – 0.07), which is expected of real detections. However, some statistical tests of the non-runaways against the random fields do not confirm a detection (R p-values: –0.50 – 0.50), thus these ambiguous results imply that TIB clusters are only marginally detected. Overall, we estimate a percentage of TIB clusters of $\sim 4\%$. For a complete discussion, please see Vargas-Salazar et al. (2020).

In summary, there is evidence that $\sim 4\%$ of field stars do form in situ in TIB clusters. The rest of the population is overwhelmingly composed of runaways. There are three possible scenarios for this result: 1) TIB clusters usually do not form with OB stars. 2) The cluster lower-mass limit is high, preventing the formation of TIB clusters. 3) Clusters evaporate quickly, thus our detections are clusters that have not yet evaporated. Our results place stringent constraints on the formation of OB stars in isolation.

Supplementary material

To view supplementary material for this article, please visit http://dx.doi.org/10.1017/S1743921322002691.

References

Lamb, J. B., Oey, M. S., Segura-Cox, D. M., Graus, A. S., Kiminki, D. C., Golden-Marx, J. B., & Parker, J. W. 2016, ApJ, 817, 113

Lamb, J. B., Oey, M. S., Werk, J. K., & Ingleby, L. D. 2010, ApJ, 725, 1886

Oey, M. S., King, N. L., & Parker, J. W. 2004, AJ, 127, 1632
Oey, M. S., Lamb, J. B., Kushner, C. T., Pellegrini, E. W., & Graus, A. S. 2013, ApJ, 768, 66
Oey, M. S., et al. 2018, ApJL, 867, L8
Udalski, A., et al. 2008, Acta Astronomica, 58, 89
Vargas-Salazar, I., Oey, M. S., Barnes, J. R., Chen, X., Castro, N., Kratter, K. M., & Faerber, T. A. 2020, ApJ, 903, 42

Chapter X:
Stellar endpoints and gravitational waves

Massive Stars Near and Far
Proceedings IAU Symposium No. 361, 2024
J. Mackey, J. S. Vink, & N. St-Louis, eds.
doi:10.1017/S1743921322002824

The Diversity of Massive Stellar Transients Found in Sky-surveys

Joe Lyman

University of Warwick, Department of Physics, Gibbet Hill Road, CV4 7AL
email: j.d.lyman@warwick.ac.uk

Abstract. The recent generation of dedicated wide-field, high-cadence sky-surveys have overwhelmed discovery statistics for all manner of extra-galactic transients, and uncovered new phenomena seemingly linked to the demise of massive stars. For the more established classes of transients, such as core-collapse supernovae, surges in discoveries are allowing true population studies to provide quantitative constraints not only on the explosion properties, but also on the progenitor populations. Crucially, such population insights are benefiting from creation of samples of transients constructed with largely unbiased methods for discovery and characterisation. Surrounding these discoveries are increasing samples of extreme transients that do not fit the standard core-collapse paradigm - requiring the invocation of exotic progenitor stars and placing demands on the stellar evolution of such systems. Here I will provide a high-level observationally-driven overview of recent results related to massive stellar transients.

Keywords. supernovae: general surveys, stars: evolution, techniques: photometric, circumstellar matter

1. Introduction

The relative dearth of massive stars, owing to stellar initial mass function (IMF) considerations, coupled with their short lives, makes the study of their final stages of stellar evolution extremely tricky – chiefly due to the corresponding rarity of suitable targets. At the ultimate end of their nuclear-burning lives massive stars are capable of undergoing spectacular explosive events, vastly increasing their luminosity, albeit for only a brief moment on astronomical time-scales. These explosive astrophysical transients can offer complementary insights into the nature of the final stages of stellar evolution for massive stars, and give us glimpses of the progenitor stars upon explosion. Although linking observables from the transients to the progenitor stars and their final evolution is a more indirect route, the great distances at which transients can be observed allows us to overcome the scarcity problem. Thanks to recent projects dedicated to the discovery of such transients, their study can now provide constraints on, and pose questions of, massive stellar evolution, not just with individual transients of interest, but with populations and statistical studies.

2. Transient sky-surveys

The discovery of transients is a time-domain problem in astrophysics, and requires continual snap-shots of the night-sky in order to pin-point the location of new (or missing) objects. Two axes of performance are ideally maximised for these sky-surveys:

• *Cadence:* A measure of the revisit time between a given location on the sky. A high cadence survey is critical such that observations for a given location recur faster than the time-scale of the transients to be discovered.

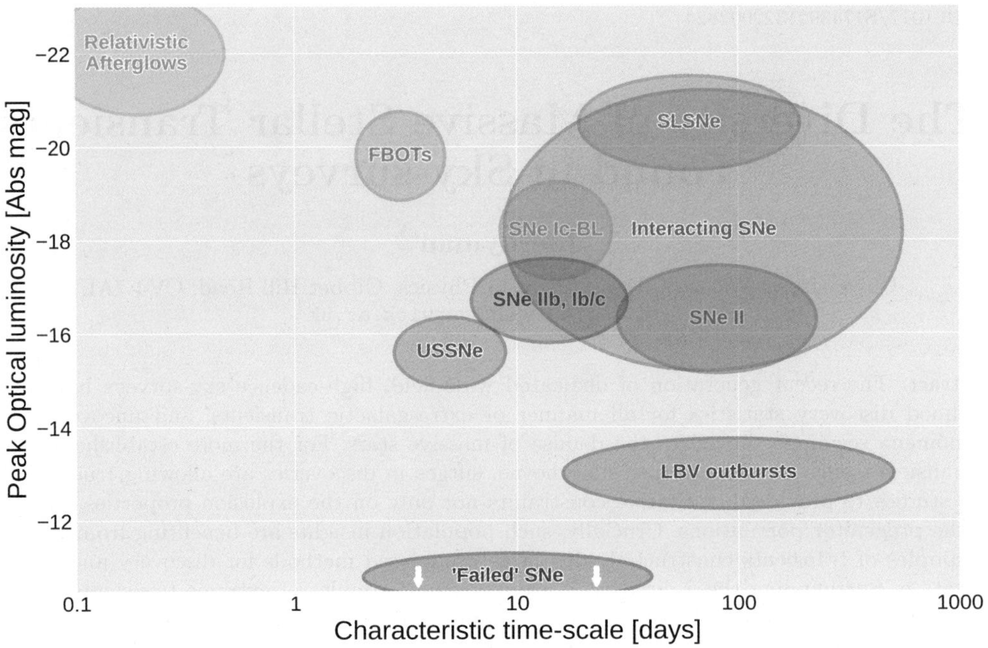

Figure 1. A visualisation of the luminosity-timescale parameter space of transients associated with massive stars. Characteristic time-scale is a loosely defined term, given the varied nature of light curves, but broadly encompasses a full-width-half-maximum-esque parameter – in any case the location of the ellipses for various sub-types are indicative only, and there are notable outliers in various sub-types. Data were collated from Smith et al. (2011b), Kasliwal (2012), Adams et al. (2017), Perley et al. (2020), Yao et al. (2020), Ho et al. (2022), Byrne & Fraser (2022).

• *Volume:* A measure of the volume of Universe probed, in which the observations would be sensitive to a given transient-type. This measure is a product of the area of the night-sky covered by the survey, as well as the flux-depth of the observations.

Each of these is crucial and of course comes with a cost to improve. New technical advances, particularly in the availability of large-format sensors, have had a major impact. Historically, transient surveys were comparatively focussed, typically optimised relatively narrow field-of-view instruments for the discovery of 'normal' supernovae. Supernovae evolve on timescales of weeks, and at the time displayed no obviously strong preference where they explode and largely tracing star-formation and/or stellar mass in the local universe (e.g. see review of Anderson et al. 2015). As such, the strategy employed was a sensible one – target massive, nearby galaxies with a recurrence time-scale of a few-days to a week. This intrinsically imposed a strong selection bias away from transients that: a) *do* have a strong preference in host-galaxy or environment (particularly any that favour low-mass/metallicity environments), b) evolve on a timescale of a few days, and/or c) are intrinsically very rare (since the volume probed with is comparatively small).

The typical sky-survey throughout the latter 2010s to current have been of a more synoptic, often multiplexed flavour. With wide fields of view in a single pointing – typically tens of square degree – a galaxy-targeted approach is no longer beneficial, and large swathes of the night sky can be covered in a given night. Prominent current and upcoming such surveys include (alphabetically) ASAS-SN (Shappee et al. 2014), ATLAS (Tonry et al. 2018), BlackGEM (Bloemen et al. 2015), GOTO (Steeghs et al. 2022), PSST (Huber et al. 2015), LSST (LSST Science Collaboration et al. 2009), and ZTF

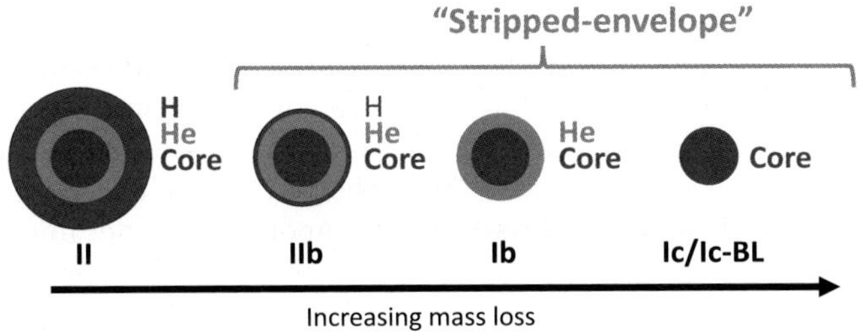

Figure 2. A schematic representation (not to scale) of the progenitor star structure for various CCSN sub-types. Moving from the hydrogen-rich Type II SNe, through to the H- and He-poor Type Ic is a manifestation of additional effects of mass-loss incurred by the progenitor stars prior to undergoing SN. The collectively name "stripped-envelope" CCSNe are indicated.

(Bellm et al. 2019). Each project fulfils a slightly different niche in the cadence and volume search space. Collectively the result is that most of the visible night sky is being observed to an optical depth of $\sim 19 - 21$ mag each night. Future surveys, particularly LSST, will provide additional depth across smaller regions of sky.

To discover transients automatically in the huge data streams from these surveys, *difference imaging* (e.g. Alard & Lupton 1998, Becker 2015, Zackay et al. 2016) is appealed to, where a historical image of the same patch of sky acts as a reference point, with which to compare new observations. Even using difference imaging, the task of finding the astrophysically-interesting objects from the many imaging artefacts, is a huge undertaking and typically requires the use of novel machine- or deep-learning techniques to sift the candidate discoveries in difference images (e.g. Brink et al. 2013, Duev et al. 2019, Killestein et al. 2021). Such measures are essential for the rapid dissemination of a new discovery to human vetters, who can announce to the community and trigger additional follow-up with other facilities. Studying systematically the intrigues of early, short-lived phenomena in transients as well as intrinsically fast-evolving transients, has only therefore only recently started to be possible with such streamlined processes taking us from observation to discovery to follow-up with minimal (and ideally no) human interaction.

The recent increase in statistics for common transient types, and the uncovering of rare, exotic explosions from these surveys is pushing forward and questioning our understanding of massive stars at the ends of their lives. A visual representation of luminosity-timescale parameter space for transients that is being evermore populated is shown in Figure 1. A few highlights for sub-types will be picked out in the following sections – more in-depth discussion of specific subtypes can be found elsewhere in this issue.

3. Core-collapse Supernovae

Core-collapse supernovae (CCSNe) are the most well-established transient associated with massive stars. An enduring, observationally driven classification scheme (Filippenko 1997, Modjaz et al. 2019) is used to discriminate sub-types largely based on the structure of the progenitor star upon explosion – specifically the degree of stripping the outer hydrogen- and helium-rich layers have undergone prior to SN. This is visually represented in Figure 2, where it is shown the distinction of Type II SNe (SNe II) from "stripped-envelope" SNe (SESNe). From single stellar evolution, an increase in mass loss would be expect to be driven primarily by an increase in zero-age main sequence (ZAMS) mass, or metallicity of the star. Following this, we may expect a mass and/or metallicity dependence of the progenitor star on the resultant CCSN type observed. However, it was noted already from earlier sky-surveys (e.g. LOSS (Filippenko et al. 2001)) that $\mathcal{R}\frac{\text{SESNe}}{\text{SNe II}}$ – i.e. the relative rate of SESNe – is larger than one might expect

(Li et al. 2011). Using straight-forward IMF arguments, and assuming a mass-segregation of the SNe II vs SESNe progenitors, the high relative abundance of SESNe requires a lower mass-limit of progenitors much lower than the observed and theoretical mass range of Wolf-Rayet (WR) stars (Smith et al. 2011a). WR stars possess sufficiently strong winds and mass-loss that they can strip themselves of their hydrogen and helium layers prior to SN and so appear as SESNe, but at lower masses another channel must be contributing to the creation of SESNe. This high relative rate of SESNe has survived the creation of larger and more homogeneously-selected samples of CCSNe (e.g. Perley et al. 2020), and remains significantly above that which is comfortable for single stellar evolution at around one quarter to one third.

The mechanism appealed to, in order to relieve the tension in relative rates of CCSN sub-types, has been the effect of binary stellar evolution on the progenitor stars (e.g. Yoon et al. 2010, Eldridge et al. 2013, Yoon et al. 2017) – an expected influential process given the prevalence of binary interactions in massive stars (e.g. Sana et al. 2012). The presence of a (close) binary companion to a SN progenitor strongly influences its evolution, and causes the binary parameters (i.e. mass and separation of companion) to be the dominating factor influencing its mass-loss rate. This will act towards washing out a strong ZAMS mass-dependence between CCSN sub-types. SESNe, therefore, are expected to have a large contribution of binary progenitors, and producing them in single star systems may actually be quite difficult, except for the most massive progenitors (Beasor & Smith 2022). It should be noted that other factors, such as top-heavy IMFs or a strong metallicity dependence on the observed CCSN sub-type may also explain the high SESN rate, but these factors would consequently come with a strong environmental and/or host-galaxy dependence on sub-type. The presence (or not) of such a dependence has been varyingly reported in the literature (e.g. Graur et al. 2017, Schulze et al. 2021), suggesting at least that, if present, they are not likely to be dominant factors.

In addition to volumetric rates, properties of the luminous SN signature also infer constraints on nature of the progenitor systems. In particular, following analytical relations describing the diffusion time-scale for SESNe (e.g. Arnett 1982, Khatami & Kasen 2019) an estimate for the ejecta mass can be obtained from the width of the bolometric light curve and photospheric velocity – the analytical estimate naturally coming with significant assumptions on the nature of the explosion. This prescription, although simplfied (see caveats in, e.g., Afsariardchi et al. 2021), does allow for population studies of large quantities of SESNe. A common theme in such studies is a distribution of low ejecta masses, typically clustered around 2–4 $\mathrm{M}_{\odot}$ (e.g. Drout et al. 2011, Lyman et al. 2016, Taddia et al. 2018, Prentice et al. 2019). This is at odds with expectations of single, massive WR progenitors, which may be expected to expel $\sim$8–10 $\mathrm{M}_{\odot}$, and points to the prevalence of more moderate mass progenitors producing the bulk of SESNe. New results from systematic sky-surveys however are revealing a significant population of spectroscopically-normal, broad-light curve SESNe that *are* consistent with explosions of very massive progenitors (Karamehmetoglu et al. in prep).

4. Superluminous Supernovae

The emergence of discoveries in the mid-2000s of extremely luminous supernovae (e.g. Quimby et al. 2007, Ofek et al. 2007) called into question the limits of the core-collapse scenario, and promoted more niche explosion mechanisms in order to explain these events. Since then, the recent and current generations of sky-surveys have figuratively exploded the detection rate of such luminous events – dubbed super-luminous SNe (SLSNe). Initially defined based on an peak luminosity cut (Gal-Yam 2012), the burgeoning SLSN population, inevitably, has since grown in observational diversity and a classification now includes a revised consideration of the peak luminosity (Gal-Yam 2019) as well as a search for spectroscopic signatures not typically seen in CCSNe (Quimby et al. 2011).

SLSNe are found as both H-rich and H-poor, designated Type I and II analogously to CCSNe. Succinctly, SLSNe-I display peak optical luminosities $M \lesssim 20$ mag, have longer rise times than CCSNe (in some cases much longer) and display hot photospheric spectra through peak, evolving at later times to resemble SESNe Type Ic. Another unusual feature are the presence of "bumps" in the light curves, both at early- (Nicholl & Smartt 2016) and late-times (Nicholl et al. 2016). SLSNe-II appear similar to SNe Type IIn (Section 6), albeit at higher luminosities and consequently different inferred circumstellar medium properties. Notable examples exist of broad H lines in SLSNe-II (e.g. SN 2008es Miller et al. 2009, Gezari et al. 2009), although as a whole SLSNe are the much poorer-studied of the type SLSN types.

Quickly it was established that the traditional energy deposition from radioactive decay – primarily ^{56}Ni – was unable to explain most SLSNe. For all but the most slowly evolving SLSNe (rise-times of months), the timescales of radioactive deposition and diffusion are too long when attempting to power their peak luminosities, and require implausibly high mass ratios of synthesised radioactive material to ejecta. In the case of the most slowly evolving SLSNe, a very high $M(^{56}\mathrm{Ni}) \simeq$ several $\mathrm{M}_\odot$ (cf. $\lesssim$0.1–0.2 $\mathrm{M}_\odot$ for CCSNe Lyman et al. 2016, Anderson 2019, Afsariardchi et al. 2021) is required, invoking $\mathrm{M}_{\mathrm{ej}} \simeq 100\,\mathrm{M}_\odot$. These properties are plausible in the pair-instability supernova (PISN) explosion of very massive stars ($M_{\mathrm{ZAMS}} > 100\,\mathrm{M}_\odot$). A very small handful of objects have been proposed as consistent with the PISN scenario (e.g. Kozyreva et al. 2018). The observational favourability of detecting these events in the current and future generation of sky-surveys, owing to their huge and long-lived luminosities, is tempered by the expected rate of such events, even from IMF arguments alone.

For SLSNe-II in particular, the dominance of narrow balmer emission lines in the spectra indicates an additional powering component from interaction of the ejecta with a slow-moving, dense pre-existing medium. Powerful, potentially episodic, mass-loss prior to explosion can furnish massive stars with a dense circumstellar-medium (CSM; Smith 2014), and it is likely similar phenomena seen in other interaction-driven SNe is at play, albeit with more extreme CSM properties driven by exteme mass-loss from more massive progenitors (Woosley et al. 2007) to power the high luminosities. The presence of hydrogen-free CSM can also allow interaction to contribute towards the energy-budget of SLSNe-I, and indeed may provide a means of explaining late time bumps in the light curves Inserra et al. (2017).

Another popular model that has emerged to explain SLSNe, particularly SLSNe-I, is through the presence of a nascent magnetar, formed during the collapse of the progenitor, which is able to inject an additional energy source into the SN ejecta via spin-down of the magnetar (e.g. Woosley 2010). With $P \sim$milli-seconds, and $B \sim 10^{14}$, satisfactory magnetar-powered models can reproduced the light curves (Inserra et al. 2013) and spectra (Dessart et al. 2012) of many SLSNe-I. The energy injection may also provide a means of powering the early light curve bumps (Kasen et al. 2016), although can have difficulty reproducing observed photospheric velocity evolution (Liu et al. 2017).

The nature of current and future sky-surveys being agnostic to any pre-existing "targets" to search for transients (i.e. nearby, massive galaxies) has contributed immensely to the discovery of large numbers of SLSNe, with detections now in their hundreds. When we assess the locations of SLSNe, they are dominated by a population of low-mass, and consequently low-metallicity, host galaxies (Angus et al. 2016, Chen et al. 2017). Identifying SLSNe early in their evolution is not trivial, yet is crucial to properly characterise fleeting phenomena such as their early bumps. Future progress will need to begin to take automated advantage of environmental information, particularly taking

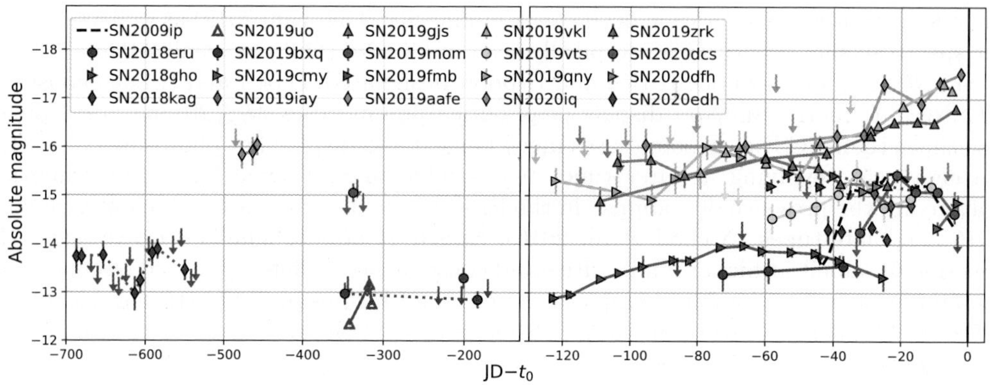

Figure 3. Precursor light curves for a sample of interaction-driven CCSNe from ZTF reproduced from Strotjohann et al. (2021). t_0 indicates the detection time of supernova. There is a prevalence of outbursts in the final few months of the stars' lives (5-69% of the analysed sample showed this behaviour), with indications of some comparatively luminous outbursts years prior to supernova. © AAS. Reproduced with permission.

advantage of wide-field multi-colour and multi-object-spectroscopic surveys to allow for rapid identification of luminous outbursts in faint galaxies.

5. Ultra-Stripped Supernovae

Evolving on comparatively short timescales, ultra-stripped SNe (USSNe) are a new classification of extra-galactic transient that bear many hall-marks of SESNe. Observationally, they bear many signatures of SESNe albeit on a compressed timescale: rising in a few days and transitioning to the nebular phase in a matter of weeks. Although detailed follow-up is sparse, their spectra and later-time luminosity evolution, which is described well by a radioactive power source, conform closely to what is seen in SNe Ic. SN 2005ek, one of the first, and most extreme examples of this empirical class exemplifies these properties (Drout et al. 2013). The proposed model for these events involves the core-collapse of an almost bare core, marginally above the Chandrasekhar limit, which expels only a few tenths of a solar mass of ejecta (Tauris et al. 2013; 2015). This scenario is extremely unlikely for single stellar evolution, and invokes exotic binary pathways where the neutron star remnant of a primary supernova in a binary acts to strongly strip its companion, prior to its own supernova. It is currently unclear how much USSNe can be said to arise from a separate stellar evolution channel to that of regular SESNe, or if there is a continuum of core-collapse events down to very low ejecta-masses. USSNe, more than most transient discussed here, demand high-cadence sky-survey data, and rapid identification procedures in order to grow the population of known events.

The interest in USSNe is compounded to by their potential role as progenitors of neutron star mergers – now being detected by the current generation of gravitational wave detectors (Abbott et al. 2017) – since the binary evolution pathways required for the system to undergo a high degree of He star stripping naturally leads to the formation of close binary neutron star (BNS) system (Tauris et al. 2017, Jiang et al. 2021). Indeed, the USSNe themselves are not too dissimilar (in the grand scheme of transient diversity) from the electromagnetic kilonovae arising from BNS mergers, WHICH CAN lead to their discovery and follow-up as interlopers within the electromagnetic follow-up campaigns of gravitational wave triggers (ENGRAVE collaboration in prep).

6. Interaction-driven Supernovae

Transients borne of massive stellar progenitors are more than likelyy to be conceived in complex environments. They may be still embedded within the comparatively dense interstellar medium of their natal star-formation complex, and they will have the history of their mass-loss (and that of their companion) imprinted in the immediate circumstellar surroundings – into both their freshly exploded ejecta will propagate. Such interaction between the fast-moving SN ejecta ($10^4 - 4.5\,\mathrm{km\ s^{-1}}$) and the surrounding media ($10^1 - 3\,\mathrm{km\ s^{-1}}$) will produce shocks and sources of luminosity that will imprint on the multi-wavelength photometric and spectroscopic evolution of the supernova (Chevalier 1977; 1982). Concurrent analysis of these multi-wavelength components can reveal the nature of the ambient medium (e.g. Tsuna et al. 2021).

Of relevance for extra-galactic transient timescales is the interaction with a circumstellar medium (CSM). The CSM of massive stars are dense, and can show varying profiles depending on the state of mass-loss history – primarily characterised as steady-state and/or eruptive. One of the most distinctive features of interaction-driven SNe is the presence of narrow emission lines, characterised by the velocity of the ambient medium. As such, the 'n' designation for the SNe type is used to indicate 'narrow' lines. This has been most prominently displayed by SNe IIn – those with narrow Balmer lines (Schlegel 1990) – but more recently, signatures of interaction with H-poor CSM have been found. The Ibn subclass are instead characterised by narrow He lines (Pastorello et al. 2008), and the even more recent Icn discoveries (Gal-Yam et al. 2022, Perley et al. 2022) display narrow C and O lines, and a conspicuous absence of H or He lines. Such H-poor CSM would be most naturally explained by WR progenitor stars, and indeed the inferred high density and velocity ($10^3\,\mathrm{km\ s^{-1}}$) of the media surrounding these SNe also agrees with this picture. The fast nature of these events, however, contradicts expectations of purported very massive WR progenitors. These latter sub-classes are therefore currently provoking resurgence in models of massive stellar collapse with significant fallback onto a nascent black-hole (Perley et al. 2022) and compact object accretion from WR stars (Metzger 2022), among others.

As well as impacting the SN itself, extensive mass-loss throughout a massive star's life may be responsible for the production of 'precursor' outbursts. The study of Strotjohann et al. (2021) found a high-occurrence of outbursts in the months preceding an interaction-driven SN (Figure 3). The cause of these outbursts is likely to be linked to eruptive mass-loss events, similar to that seen to occur in local massive stars, particularly luminous blue variable and η Car-like stars. Collisions between an euptive mass-loss event and the ambient medium could themselves also produce a weaker cousins of an interaction-powered SNe, manifesting as outbursts. The cause of such eruptive mass-loss is less clear. Late stage, unstable nuclear burning can drive strongly super-Eddington winds from stars (Fuller 2017, Fuller & Ro 2018), however in some cases the timescales for the precursors pre-date these late nuclear burning stages, and additionally appear much more luminous than may be expected. Constraints on pre-explosion outbursts, and even the variability of the progenitor star (e.g. Kilpatrick et al. 2021) will become more common place particularly with the commencement of the Vera Rubin Observatory and could give us the best means of studying the rare, final few years of massive stellar evolution.

Acknowledgements

JDL acknowledges support from a UK Research and Innovation Fellowship(MR/T020784/1)

References

Abbott, B. P., Abbott, R., Abbott, T. D., Acernese, F., & Ackley, K. a. 2017, ApJL, 848, L12
Adams, S. M., Kochanek, C. S., Gerke, J. R., Stanek, K. Z., & Dai, X. 2017, MNRAS, 468, 4968
Afsariardchi, N., Drout, M. R., Khatami, D. K., Matzner, C. D., Moon, D.-S., & Ni, Y. Q. 2021, ApJ, 918, 89
Alard, C., & Lupton, R. H. 1998, ApJ, 503, 325
Anderson, J. P. 2019, A&A, 628, A7
Anderson, J. P., James, P. A., Habergham, S. M., Galbany, L., & Kuncarayakti, H. 2015, PASA, 32, e019
Angus, C. R., Levan, A. J., Perley, D. A., Tanvir, N. R., Lyman, J. D., Stanway, E. R., & Fruchter, A. S. 2016, MNRAS, 458, 84
Arnett, W. D. 1982, ApJ, 253, 785
Beasor, E. R., & Smith, N. 2022, arXiv e-prints, arXiv:2205.02207
Becker, A. 2015, HOTPANTS: High Order Transform of PSF ANd Template Subtraction, Astrophysics Source Code Library, record ascl:1504.004
Bellm, E. C., et al. 2019, PASP, 131, 018002
Bloemen, S., Groot, P., Nelemans, G., & Klein-Wolt, M. 2015, in Astronomical Society of the Pacific Conference Series, Vol. 496, Living Together: Planets, Host Stars and Binaries, ed. S. M. Rucinski, G. Torres, & M. Zejda, 254
Brink, H., Richards, J. W., Poznanski, D., Bloom, J. S., Rice, J., Negahban, S., & Wainwright, M. 2013, MNRAS, 435, 1047
Byrne, R. A., & Fraser, M. 2022, MNRAS, 514, 1188
Chen, T.-W., Smartt, S. J., Yates, R. M., Nicholl, M., Krühler, T., Schady, P., Dennefeld, M., & Inserra, C. 2017, MNRAS, 470, 3566
Chevalier, R. A. 1977, ARA&A, 15, 175
—. 1982, ApJ, 258, 790
Dessart, L., Hillier, D. J., Waldman, R., Livne, E., & Blondin, S. 2012, MNRAS, 426, L76
Drout, M. R., et al. 2011, ApJ, 741, 97
—. 2013, ApJ, 774, 58
Duev, D. A., et al. 2019, MNRAS, 489, 3582
Eldridge, J. J., Fraser, M., Smartt, S. J., Maund, J. R., & Crockett, R. M. 2013, MNRAS, 436, 774
Filippenko, A. V. 1997, ARA&A, 35, 309
Filippenko, A. V., Li, W. D., Treffers, R. R., & Modjaz, M. 2001, in Astronomical Society of the Pacific Conference Series, Vol. 246, IAU Colloq. 183: Small Telescope Astronomy on Global Scales, ed. B. Paczynski, W.-P. Chen, & C. Lemme, 121
Fuller, J. 2017, MNRAS, 470, 1642
Fuller, J., & Ro, S. 2018, MNRAS, 476, 1853
Gal-Yam, A. 2012, Science, 337, 927
—. 2019, ARA&A, 57, 305
Gal-Yam, A., et al. 2022, Nature, 601, 201
Gezari, S., et al. 2009, ApJ, 690, 1313
Graur, O., Bianco, F. B., Modjaz, M., Shivvers, I., Filippenko, A. V., Li, W., & Smith, N. 2017, ApJ, 837, 121
Ho, A. Y. Q., et al. 2022, arXiv e-prints, arXiv:2201.12366
Huber, M., et al. 2015, The Astronomer's Telegram, 7153, 1
Inserra, C., et al. 2013, ApJ, 770, 128
—. 2017, MNRAS, 468, 4642
Jiang, L., Tauris, T. M., Chen, W.-C., & Fuller, J. 2021, ApJL, 920, L36
Kasen, D., Metzger, B. D., & Bildsten, L. 2016, ApJ, 821, 36
Kasliwal, M. M. 2012, PASA, 29, 482
Khatami, D. K., & Kasen, D. N. 2019, ApJ, 878, 56
Killestein, T. L., et al. 2021, MNRAS, 503, 4838
Kilpatrick, C. D., et al. 2021, MNRAS, 504, 2073

Kozyreva, A., Kromer, M., Noebauer, U. M., & Hirschi, R. 2018, MNRAS, 479, 3106
Li, W., et al. 2011, MNRAS, 412, 1441
Liu, Y.-Q., Modjaz, M., & Bianco, F. B. 2017, ApJ, 845, 85
LSST Science Collaboration et al. 2009, arXiv e-prints, arXiv:0912.0201
Lyman, J. D., Bersier, D., James, P. A., Mazzali, P. A., Eldridge, J. J., Fraser, M., & Pian, E. 2016, MNRAS, 457, 328
Metzger, B. D. 2022, ApJ, 932, 84
Miller, A. A., et al. 2009, ApJ, 690, 1303
Modjaz, M., Gutiérrez, C. P., & Arcavi, I. 2019, Nature Astronomy, 3, 717
Nicholl, M., & Smartt, S. J. 2016, MNRAS, 457, L79
Nicholl, M., et al. 2016, ApJ, 826, 39
Ofek, E. O., et al. 2007, ApJL, 659, L13
Pastorello, A., et al. 2008, MNRAS, 389, 113
Perley, D. A., et al. 2020, ApJ, 904, 35
—. 2022, ApJ, 927, 180
Prentice, S. J., et al. 2019, MNRAS, 485, 1559
Quimby, R. M., Aldering, G., Wheeler, J. C., Höflich, P., Akerlof, C. W., & Rykoff, E. S. 2007, ApJL, 668, L99
Quimby, R. M., et al. 2011, Nature, 474, 487
Sana, H., et al. 2012, Science, 337, 444
Schlegel, E. M. 1990, MNRAS, 244, 269
Schulze, S., et al. 2021, ApJS, 255, 29
Shappee, B. J., et al. 2014, ApJ, 788, 48
Smith, N. 2014, ARA&A, 52, 487
Smith, N., Li, W., Filippenko, A. V., & Chornock, R. 2011a, MNRAS, 412, 1522
Smith, N., Li, W., Silverman, J. M., Ganeshalingam, M., & Filippenko, A. V. 2011b, MNRAS, 415, 773
Steeghs, D., et al. 2022, MNRAS, 511, 2405
Strotjohann, N. L., et al. 2021, ApJ, 907, 99
Taddia, F., et al. 2018, A&A, 609, A136
Tauris, T. M., Langer, N., Moriya, T. J., Podsiadlowski, P., Yoon, S. C., & Blinnikov, S. I. 2013, ApJL, 778, L23
Tauris, T. M., Langer, N., & Podsiadlowski, P. 2015, MNRAS, 451, 2123
Tauris, T. M., et al. 2017, ApJ, 846, 170
Tonry, J. L., et al. 2018, PASP, 130, 064505
Tsuna, D., Kashiyama, K., & Shigeyama, T. 2021, ApJ, 914, 64
Woosley, S. E. 2010, ApJL, 719, L204
Woosley, S. E., Blinnikov, S., & Heger, A. 2007, Nature, 450, 390
Yao, Y., et al. 2020, ApJ, 900, 46
Yoon, S.-C., Dessart, L., & Clocchiatti, A. 2017, ApJ, 840, 10
Yoon, S. C., Woosley, S. E., & Langer, N. 2010, ApJ, 725, 940
Zackay, B., Ofek, E. O., & Gal-Yam, A. 2016, ApJ, 830, 27

Discussion

HIRSCHI: What are the environments of interacting supernovae?

LYMAN: For the H-poor events we do not have populations to draw significant conclusions as yet. The Type IIn SNe interesting display an extremely wide range of environments, and we see a much larger proportion of them, compared to SE SNe, for example, that occur in the outskirts of their host galaxies, or away from regions of ongoing star formation.

FRASER: There is the issue that what we label a Type IIn SN is probably an extremely heterogeneous collection of objects, ranging from faint SN 2008S-like events up to SLSNe.

BRENNAN: What are the causes of the pre-outburst bumps that have been seen?

LYMAN: Likely related to mass-loss events, but the underlying cause of those events is less clear. Certain luminosity and timescale constraints are expected from, e.g., late-stage unstable burning, and having much longer time-scale pre-SN variability constraints is needed to address this.

Massive Stars Near and Far
Proceedings IAU Symposium No. 361, 2024
J. Mackey, J. S. Vink, & N. St-Louis, eds.
doi:10.1017/S1743921322001910

Weak Mass Loss from the Red Supergiant Progenitor of SN 2021yja

Griffin Hosseinzadeh

Steward Observatory, University of Arizona, 933 North Cherry Avenue,
Tucson, AZ, 85721, USA
email: griffin0@arizona.edu

Abstract. Recent observations of Type II supernovae have revealed that their red-supergiant progenitors lose a significant amount of mass during the last years of their evolution. However, because it is difficult to discover supernovae within days of explosion, the diversity of mass loss in red supergiants has not yet been fully mapped. This talk presented the case of SN 2021yja, which was serendipitously imaged within hours of explosion and observed with a sub-day cadence during its rise to peak. From the exceptionally long plateau period and the high nickel mass, we infer a relatively massive red-supergiant progenitor star. However, archival imaging from the Hubble Space Telescope places a stringent upper limit of $\lesssim 9\ M_{\odot}$ on its progenitor mass. We discuss these conflicting constraints in the context of the larger sample of exploding red supergiants. Our analysis helps illuminate the poorly understood mechanism(s) behind red-supergiant mass loss.

Keywords. stars: mass loss, supergiants, supernovae: general, supernovae: individual (SN 2021yja)

Type II (hydrogen-rich) supernovae are the core-collapse explosions of red supergiants (see Smartt 2009 for a review). As such, they can serve as probes of red-supergiant mass loss, reaching greater sample sizes than are available in the Milky Way and Magellanic clouds, and recording the latest mass-loss stages, i.e., years to months before explosion. The difficulty lies in connecting the supernova observables to the properties of the circumstellar material. Gal-Yam et al. (2014) and Yaron et al. (2017) showed that short-lived narrow emission lines in the very early spectra of core-collapse supernovae indicate interaction with a confined shell of dense material ejected before explosion. Morozova et al. (2017; 2018) also showed that circumstellar material produces a faster rise and sharper peak in light curves of Type II supernovae. By modeling these emission lines and light curves, one can constrain the density profile of the circumstellar material. However, in both cases, very early observations are required. After only a few days, the ejecta overrun the circumstellar material and these observables disappear.

SN 2021yja was discovered on the outskirts of the nearby ($d_L = 23.4^{+5.4}_{-4.4}$ Mpc) galaxy NGC 1325 by the Asteroid Terrestrial-impact Last Alert System (ATLAS; Tonry et al. 2018), with prediscovery detections by the Distance Less Than 40 Mpc (DLT40) survey (Tartaglia et al. 2018) and an education and public outreach program on Las Cumbres Observatory (Kilpatrick 2021). These earliest images suggest that the progenitor exploded only about 5.4 ± 1.4 hours earlier. Starting immediately after discovery, we

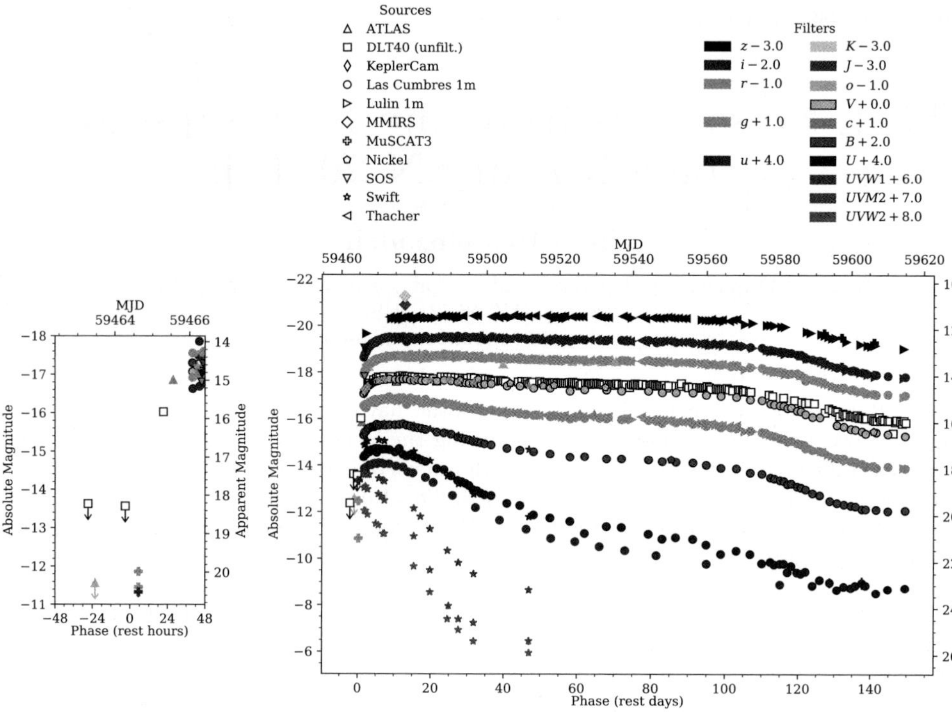

Figure 1. The light curve of SN 2021yja shows a fast rise, possibly indicating interaction with circumstellar material. The unusually long plateau, followed by a short fall onto the radioactive-decay-powered tail, indicate a large mass of ^{56}Ni, and possibly a massive progenitor star. The left panel shows the first two days of observations with no filter offsets. The supernova was serendipitously detected ≈5.4 hours after explosion by the MuSCAT3 imager at Las Cumbres Observatory. (From Hosseinzadeh et al. 2022; reproduced by permission of the AAS.)

obtained a densely sampled light curve (Figure 1) and spectral series. We also analyzed an archival image of NGC 1325 from the Hubble Space Telescope to place a limit on the luminosity, and therefore mass, of the progenitor star of SN 2021yja. I summarize the results of our analysis below; see Hosseinzadeh et al. (2022) for more detail.

In the absence of circumstellar material, the early light curves of core-collapse supernovae are powered by shock-cooling emission (see Waxman & Katz 2017 for a review). These models can be used to determine the radius of the progenitor star. We fit the shock-cooling model of Sapir & Waxman (2017) to the first 10 days of our light curve, and it appears to provide a good fit. However, it yields a best-fit radius of $\approx 2000\ R_\odot$, larger than expected for a red supergiant progenitor. This may an indication that the models are not a good description of the data, despite the reasonable match.

We also measured the mass of radioactive ^{56}Ni produced in the explosion by comparing the luminosity on the light-curve tail to the luminosity of SN 1987A at the same phases. This gives a nickel mass of $M_{\rm Ni} = 0.141^{+0.074}_{-0.049}\ M_\odot$, which is among the highest nickel masses for a Type II supernova (Anderson 2019). Eldridge et al. (2019) find a correlation between nickel mass and progenitor mass, suggesting a very massive ($\gtrsim 20\ M_\odot$) progenitor for SN 2021yja.

Despite the appearance of a good fit to the shock cooling models, SN 2021yja does show some evidence for circumstellar interaction. Its colors are very blue, and it is among

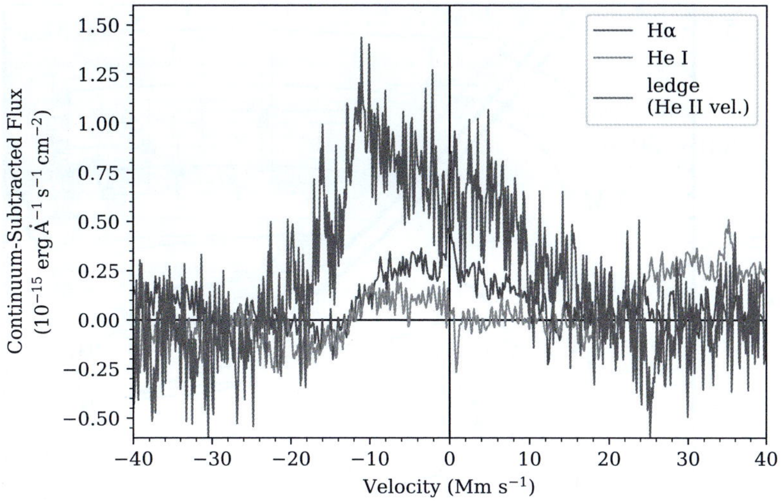

Figure 2. The ledge-shaped feature (green) in the earliest spectrum of SN 2021yja, compared to the P Cygni profiles of Hα (blue) and He I (orange) in the same spectrum. This feature, a blend of lines possibly including He II, has been interpreted as evidence for weak circumstellar interaction. It also resembles features in models of exploding red supergiants with extended atmospheres by Dessart et al. (2017). (From Hosseinzadeh et al. 2022; reproduced by permission of the AAS.)

most luminous Type II supernovae observed in the ultraviolet by the Neil Gehrels Swift Observatory (Brown et al. 2014). In addition, its earliest spectra (2–4 days after explosion) show an unusual broad "ledge-shaped" feature (Figure 2), which has previously been interpreted as evidence for weak circumstellar interaction (Bullivant et al. 2018, Andrews et al. 2019, Soumagnac et al. 2020, Bruch et al. 2021). This feature also appears in models of red supergiant explosions by Dessart et al. (2017) in which the photosphere lines within an extended atmosphere.

Despite the evidence above for a very massive red supergiant progenitor, the image of NGC 1325 taken 25 years before the explosion shows no point source at the position of SN 2021yja, down to very deep limits. By comparing to single stellar evolutionary tracks from Choi et al. (2016), this image allows us to place a limit of $\lesssim 9\ M_\odot$ on the initial mass of the progenitor of SN 2021yja (Figure 3). The conflict between this limit and our conclusions from the supernova itself can be lessened by assuming that the progenitor was variable; extreme dimming events in red supergiants are rare (Conroy et al. 2018) but not unheard of (Levesque & Massey 2020, Jencson et al. 2022). Variability may also be enhanced in the years to decades before explosion (e.g., Jacobson-Galán et al. 2022).

In summary, while our results regarding the mass of the progenitor of SN 2021yja are in conflict with each other, we do infer weak circumstellar interaction for the first several days of its evolution, despite the absence of the narrow emission lines described by Gal-Yam et al. (2014) and Yaron et al. (2017). This adds to the growing sample of red supergiant progenitors with preexisting circumstellar material, suggesting that some level of mass loss may be ubiquitous. Future analyses must consider that the supernova observables that indicate interaction are highly dependent on the details of the density profiles of both the progenitor atmosphere and its circumstellar material, in order not to miss cases of weak mass loss like we see here.

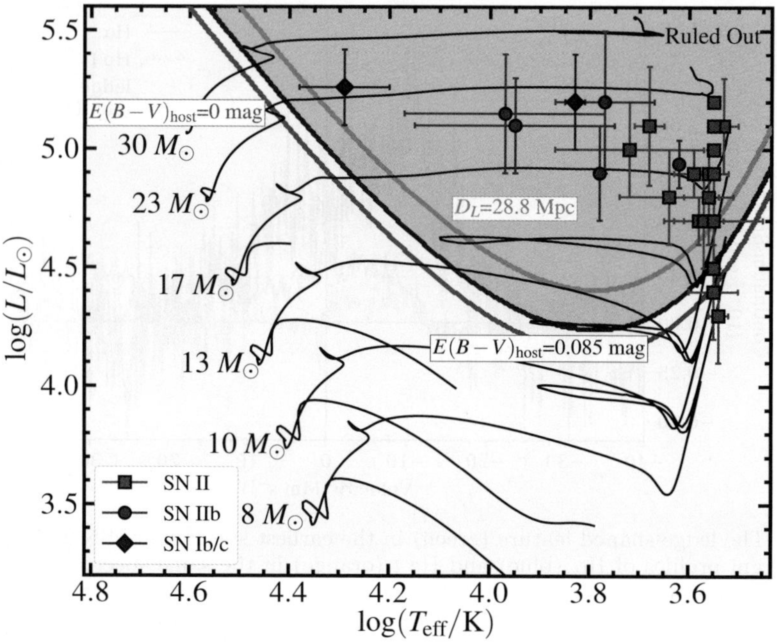

Figure 3. Hertzsprung-Russell diagram showing the single stellar evolutionary tracks of Choi et al. (2016), compared to core-collapse supernova progenitors observed by Cao et al. (2013), Smartt (2015), and Kilpatrick et al. (2021). The archival Hubble Space Telescope image constrains the initial mass of the progenitor of SN 2021yja to be $\gtrsim 9\ M_\odot$ (gray region, blue line), assuming a single-star origin. A larger distance estimate can weaken our limit (gold line), or a smaller estimate of the host galaxy extinction could strengthen it (red line). (From Hosseinzadeh et al. 2022; reproduced by permission of the AAS.)

References

Anderson, J. P. 2019, A&A, 628, A7
Andrews, J. E., Sand, D. J., Valenti, S., Smith, N., & Dastidar, R. 2019, ApJ, 855, 43
Brown, P. J., Breeveld, A. A., Holland, S., Kuin, P., & Pritchard, T. 2014, Ap&SS, 354, 89
Bruch, R. J., Gal-Yam, A., Schulze, S., et al. 2021, ApJ, 912, 46
Bullivant, C., Smith, N., Williams, G. G., et al. 2018, MNRAS, 476, 1497
Cao, Y., Kasliwal, M. M., Arcavi, I., et al. 2013, ApJ, 775, L7
Choi, J., Dotter, A., Conroy, C., et al. 2016, ApJ, 823, 102
Conroy, C., Strader, J., van Dokkum, P., et al. 2018, 864, 111
Dessart, L., Hillier, D. J., & Audit, E. 2017, A&A, 605, A83
Eldridge, J. J., Guo, N.-Y., Rodrigues, N., Stanway, E. R., & Xiao, L. 2019, PASA, 36, e041
Gal-Yam, A., Arcavi, I., Ofek, E. O., et al. 2014, Natur, 509, 471
Hosseinzadeh, G., Kilpatrick, C. D., Dong, Y., et al. 2022, arXiv:2203.08155
Jacobson-Galán, W. V., Dessart, L., Jones, D. O., et al. 2022, ApJ, 924, 15
Jencson, J. E., Sand, D. J., Andrews, J. E., et al. 2022, ApJ, 930, 81
Kilpatrick, C. D. 2021, TNSAN, 236, 1
Kilpatrick, C. D., Drout, M. R., Auchettl, K., et al. 2021, MNRAS, 504, 2073
Levesque, E. M., & Massey, P. 2020, ApJL, 891, L37
Morozova, V., Piro, A. L., & Valenti, S. 2017, ApJ, 838, 28
Morozova, V., Piro, A. L., & Valenti, S. 2018, ApJ, 858, 15
Sapir, N., & Waxman, E. 2017, ApJ, 838, 130
Smartt, S. J. 2009, ARA&A, 47, 63
Smartt, S. J. 2015, PASA, 32, e016
Soumagnac, M. T., Ganot, N., Irani, I., et al. 2020, ApJ, 902, 6

Tartaglia, L., Sand, D. J., Valenti, S., et al. 2018, ApJ, 853, 62
Tonry, J., Stalder, B., Denneau, L., et al. 2018, TNSTR, 707, 1
Waxman, E., & Katz, B. 2017, in Handbook of Supernovae, ed. A. W. Alsabti & P. Murdin (Cham: Springer)
Yaron, O., Perley, D. A., Gal-Yam, A., et al. 2017, NatPh, 13, 510

Massive Stars Near and Far
Proceedings IAU Symposium No. 361, 2024
J. Mackey, J. S. Vink, & N. St-Louis, eds.
doi:10.1017/S1743921322002472

Constraining massive star mass loss through supernova radio properties

Takashi Moriya[1,2]

[1]National Astronomical Observatory of Japan, National Institutes of Natural Sciences, 2-21-1 Osawa, Mitaka, Tokyo 181-8588, Japan
email: takashi.moriya@nao.ac.jp

[2]School of Physics and Astronomy, Faculty of Science, Monash University, Clayton, Victoria 3800, Australia

Abstract. Supernova properties in radio strongly depend on their circumstellar environment and they are an important probe to investigate the mass loss of supernova progenitors. Recently, core-collapse supernova observations in radio have been assembled and the rise time and peak luminosity distribution of core-collapse supernovae in radio has been obtained. In this talk, we will discuss the constraints on the mass-loss prescriptions of red supergiants obtained from the assembled radio properties of Type II supernovae. We take a couple of mass-loss prescriptions for red supergiants, calculate the rise time and peak luminosity distribution based on them, and compare the results with the observed distribution. We found that the widely spread radio rise time and peak luminosity distribution of Type II supernovae can only be explained by mass-loss prescriptions having strong dependence on the luminosity. Red supergiant mass-loss prescriptions should have steep luminosity dependence in the supernova progenitor range.

Keywords. circumstellar matter, stars: mass-loss, supergiants, supernovae: general

---

1. Introduction

Mass loss is a major uncertainty in massive star evolution. Massive stars keep losing their mass from birth to death. Massive stars that are about to explode as supernovae (SNe) are surrounded by the circumstellar matter (CSM) that is formed by their mass loss throughout their lives (Dwarkadas 2005. The subsequent supernova (SN) explosions occur within the CSM.

After a SN explosion, a forward shock travels through the CSM formed by its progenitor. Electrons are accelerated at the forward shock, resulting in synchrotron emission. The synchrotron emission makes SNe bright in radio (Weiler et al. 2002). The rise time and peak luminosity of SNe in radio are determined by the synchrotron emission and absorption processes. The major absorption processes are synchrotron self-absorption at the forward shock when the CSM density is low and free-free absorpbtion at the unshocked CSM when the CSM density is high. Thus, the rise time and peak luminosity of SNe in radio strongly depend on the CSM density and we can estimate it based on the SN radio properties. Because the CSM density is determined by mass-loss rates of the SN progenitors, we can constrain the mass-loss prescriptions of massive stars through SN radio observations (e.g., Chevalier et al. 2006). In this talk, we focus on mass loss of red supergiants (RSGs) that are progenitors of Type II SNe. We refer to Moriya (2021) for full details.

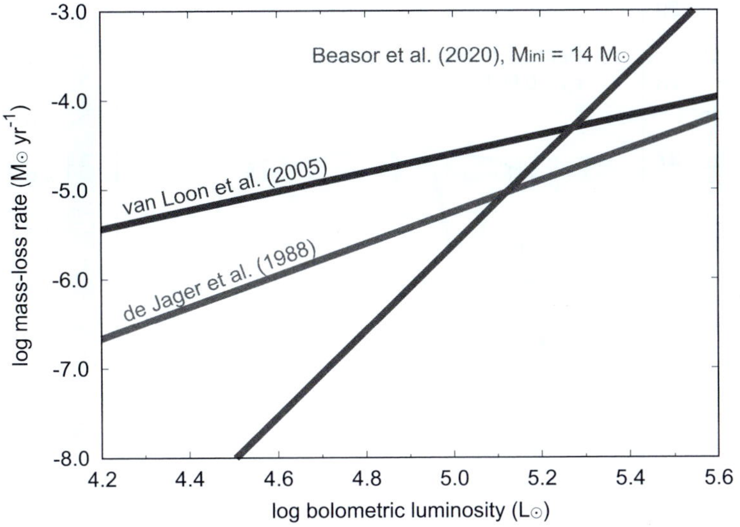

Figure 1. Representative RSG mass-loss prescriptions.

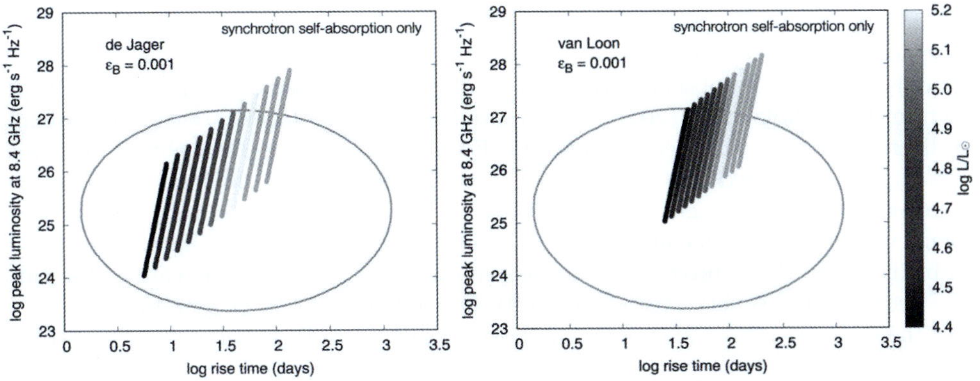

Figure 2. Rise time and peak luminosity distribution of Type II SNe and their comparison to the theoretical expectation from the de Jager prescription (left) and van Loon prescription (right).

2. RSG mass-loss prescriptions

Mass-loss mechanisms of RSGs are not well understood. Many different RSG mass-loss prescriptions have been used in stellar evolution calculations. Fig. 1 shows some examples of RSG mass-loss prescriptions. The RSG mass-loss rates by de Jager et al. (1988) and van Loon et al. (2005) have been often adopted. Recently, Beasor et al. (2020) proposed a RSG mass-loss prescription that depends on the initial mass ($M_{\rm ini}$). In Fig. 1, we show the case of $M_{\rm ini} = 14\ M_\odot$. Based on the three RSG mass-loss prescriptions, we estimate the expected rise time and peak luminosity distribution of Type II SNe in radio and compare it with the observed distribution provided by Bietenholz et al. (2021).

3. Comparison

Fig. 2 compares the expected rise time and peak luminosity distribution from the de Jager and van Loon rates. Each line shows the expected rise time and peak luminosity

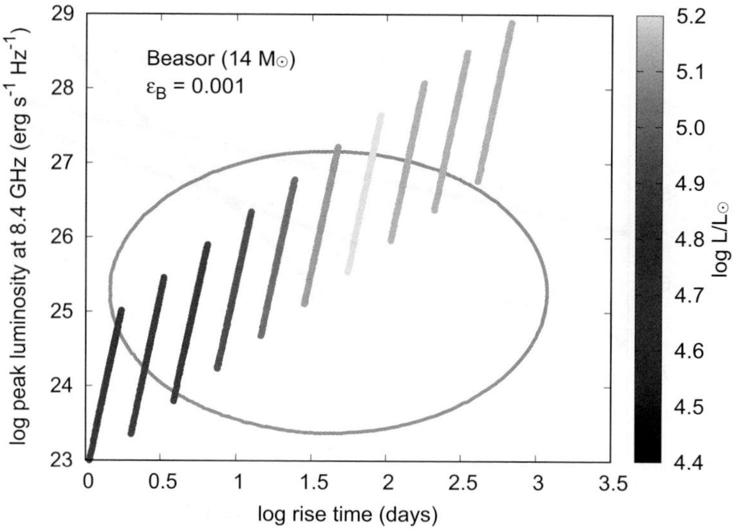

Figure 3. Same as Fig. 2, but for the Beasor mass-loss prescription with $M_{\rm ini} = 14\ M_\odot$.

range for the given progenitor luminosity. The gray lines are for log $L/L_\odot = 5.3, 5.4$, and 5.5, which may not be the luminosity range of RSG SN progenitors (Smartt 2015). The blue oblate shows the observed rise time and peak luminosity distribution of Type II SNe from Bietenholz et al. (2021). We can find that the de Jager rate can better explain the diverse rise time and peak luminosity distribution in radio light curves observed in Type II SNe compared to the van Loon rate. Especially, the de Jager rate can cover the lowest luminosity range. This is because the de Jager mass-loss rate has the steeper dependence on the luminosity and thus it can cover from low to high mass-loss rate within the Type II SN progenitor luminosity range. Fig. 3 shows the case of the Beasor rate. Because of the very steep luminosity dependence of the Beasor rate, it also can cover the entire rise time and peak luminosity range in radio observed in Type II SNe.

4. Conclusions

The widely-spread distribution in the rise time and peak luminosity in radio observed in Type II SNe indicates that mass-loss rates of RSG SN progenitors have steep dependence on their luminosity. This fact prefers RSG mass-loss prescriptions by de Jager et al. (1988) and Beasor et al. (2020). While RSG SN progenitors may often have mass-loss enhancement shortly before their explosion (Yaron et al. 2017; Förster et al. 2018), the radio observations in Type II SNe are mostly sensitive to the mass loss before such mass-loss enhancement. A similar study has been conducted for massive helium star SN progenitors by Moriya & Yoon (2022).

References

Beasor, E.R., Davies, B., Smith, N., van Loon, J.Th., Gehrz, R.D., & Figer, D.F. 2020, *MNRAS*, 492, 5994

Bietenholz, M.F., Bartel, N., Argo, M., Dua, R., Ryder, S., & Soderberg, A. 2021, *ApJ*, 908, 75

Chevalier, R.A., Fransson, C. & Nymark, T.K. 2006, *ApJ*, 641, 1029

Dwarkadas, V.V. 2005, *ApJ*, 630, 892

de Jager, C., Nieuwenhuijzen, H., & van der Hucht K.A. 1988, *A&AS*, 72, 259

Förster, F., et al. 2018, *Nature Astronomy*, 2, 808

Moriya, T.J. 2021, *MNRAS* (Letters), 503, L28

Moriya, T.J. & Yoon, S.-C. 2022, *MNRAS*, 513, 5606
Smartt, S.J. 2015, *PASA*, 32, 16
van Loon, J.T., Cioni, M.R.L., Zijlstra, A.A., & Loup, C. 2005, *A&A*, 438, 273
Weiler, K.W., Panagia, N., Montes, M.J., & Sramek R.A. 2002, *ARAA*, 40, 387
Yaron, O., et al. 2017, *Nature Physics*, 13, 510

Discussion

SOMEONE: Is it possible to classify SNe only with radio properites?

MORIYA: We can probably distinguish Type IIn SNe from others because of their large radio luminosity. Other SN types are difficult to distinguish just by radio observations.

HOSSEINZADEH: Now that I think about it, SN 2021yja does have all the constraints you need. There was a radio detection that is not published yet.

MORIYA: Very nice! We may be able to get a constraint on the progenitor mass loss and get some information on the progenitor.

SMARTT: Very good - go for it! Think the missing link in the combination of progenitor + very early detection and follow-up + radio + nebular spectra has been having consistent, very early detections, but that now is looking routine with ATLAS+ZTF combinations (and others). There's an opportunity to do this more consistently I think. Requires a lot of effort, there is significant observing resources to coordinate.

Massive Stars Near and Far
Proceedings IAU Symposium No. 361, 2024
J. Mackey, J. S. Vink, & N. St-Louis, eds.
doi:10.1017/S1743921322003209

Exploring a direct observational method to measure high-redshift cloud collapse timescales and GRB progenitor lifetimes

Jeff Cooke[1,2,3], Nandita Khetan[4], Sandra Savaglio[5], Jielai Zhang[1,3] and Mark Suhr[1,2]

[1]Centre for Astrophysics and Supercomputing, Swinburne University of Technology, Hawthorn, Victoria 3122, Australia
email: jcooke@astro.swin.edu.au

[2]ARC of Excellence for All Sky Astrophysics in 3 Dimensions *(ASTRO 3D)*

[3]ARC Centre of Excellence for Gravitational Wave Discovery *(OzGrav)*

[4]DARK Cosmology Centre, Niels Bohr Institute, University of Copenhagen, Jagtvej 128, 2200 Copenhagen, Denmark

[5]Department of Physics, University of Calabria, I-87036 Arcavacata di Rende, Italy

Abstract. Long gamma-ray bursts (LGRBs) and superluminous supernovae (SLSNe) are expected to result from massive star deaths. However to date, there has been no direct observational measurement of their cloud collapse timescales nor progenitor lifetimes to help constrain their mass. Our analyses of $z \gtrsim 2$ LGRB afterglow spectra and *Hubble Space Telescope* images find a higher fraction of host galaxies that are interacting, have a close companion, and/or may have experienced a recent galaxy 'fly by' as compared to the general $z \gtrsim 2$ galaxy population. A smaller set of $z \gtrsim 2$ SLSNe suggests a similar result. Under the hypothesis that galaxy interactions induce cloud collapse and star formation near their closest approach, we explore measurements of the host and companion galaxy velocities and separations at the time of the LRGB/SLSN event as a direct physical means to measure the timescale of cloud collapse plus progenitor star lifetime.

Keywords. gamma rays: bursts — galaxies: high-redshift — galaxies: interactions — supernovae: general — stars: evolution — stars: formation

---

1. Introduction

Galaxy interactions induce star formation, with direct evidence of massive star formation found, for example, by increased emission line strengths from HII regions as a function of separation and observed OB stars in-between interacting galaxies and within tidal tails (e.g., Patton et al. 1997, Barton et al. 2000, Barton Gillespie et al. 2003, de Mello et al. 2008). Under the assumption that long gamma-ray bursts (LGRBs) and superluminous supernovae (SLSNe) result from the deaths of short-lived massive stars, these events would be observed shortly after galaxy interactions. Here, we summarise our analyses to date of high-redshift ($z \gtrsim 2$) LGRB afterglow spectra and *Hubble Space Telescope (HST)* imaging that indicate a large fraction of LGRB host galaxies experiencing recent galaxy interactions. A subset of LGRBs with interacting hosts occur on the outskirts of their host galaxies in-between the host and the companion/interacting galaxy and/or in their tidal tails. We find a similar result for $z > 2$ SLSNe. We hypothesise that these recent galaxy interactions induced star formation that formed the LGRB/SLSN progenitor stars and that the physical separations of the galaxy pairs, combined with their relative velocities, are a direct physical measure of the timescales for progenitor cloud collapse and lifetimes. Finally, many of such interactions are expected to induce star formation in the outer regions of the host galaxies, which are typically low metallicity environments. This approach, applied to larger datasets, along with deep host and companion galaxy imaging and spectra, may provide an independent means to test for a metallicity preference for LGRB/SLSN progenitor star formation.

2. Data

2.1. *LGRB afterglow spectra*

Our LGRB afterglow spectra sample includes 191 spectra from 245 LGRBs with measured redshifts detected from 1997–2015 compiled by GHostS† (Savaglio et al. 2007). We focus on those at z $\sim$ 2–6 (94 LGRBs) to enable host and companion galaxy UV absorption line detection, including the Lyα feature. Assessment of additional spectra is underway. Of the 94 afterglow spectra, an initial sample of 40 were assessed that have sufficient signal-to-noise ratio (S/N), resolution, and spectral quality for the work discussed here.

2.2. *Host galaxy imaging and spectra*

Morphological assessment of z > 2 galaxies requires space-based imaging resolution. Our initial *HST* imaging sample consists of 54 z $\gtrsim$ 2 LGRB host galaxies with varying quality and depths acquired by various programs and extracted from the Mikulski Archive for Space Telescopes‡. Our z > 2 SLSN *HST* imaging sample is pulled from the Subaru HSC SHIZUKA survey (Moriya et al. 2019) in the COSMOS legacy field and ground-based imaging for SLSN SN2213-1745 (Cooke et al. 2012) from the Canada-France-Hawaii Telescope Legacy Survey (CFHTLS). Spectroscopy of SN2213-1745 and its companion galaxy were obtained 03 September 2010 with the Low-Resolution Imaging Spectrometer (LRIS; Oke et al. 1995, Steidel et al. 2004) mounted on the Keck-I telescope.

3. Analysis

3.1. *Pair absorbers*

A 'pair absorber' is defined here as an LGRB afterglow spectrum exhibiting two sets of galaxy absorption-line features assumed to be the LGRB host galaxy and a galaxy in the line of sight. We find 5/40 afterglow spectra (13%) with pair absorption having velocity separations < 1000 km s^{-1} (Figure 1). We do not include systems with $\lesssim$ 200 km s^{-1} velocity separation, e.g., those with small relative velocities or systems moving roughly tangential to the line of sight, as they are difficult to identify as separate systems in low-resolution spectra or discern from certain host galaxy rotation profile signatures. Thus, the true fraction of systems with line-of-sight pair absorption is > 13%.

3.2. *Geometry*

The geometry of LGRB host galaxies and the occurrence of pair absorption is illustrated in Figure 2. The host galaxy is located in the centre of the sphere and the LGRB is detected by the observer to the left. The radius, r, indicates the radius of interest. A companion galaxy with r < 30 kpc is defined as an imminent merger and a close pair is defined having r < 50 kpc. A companion galaxy behind the LGRB host will not be detected in absorption by the observer. Considering only this aspect, the observed line-of-sight > 13% pair absorption fraction becomes > 26% for randomly oriented systems.

† `http://www.grbhosts.org`
‡ `https://archive.stsci.edu/`

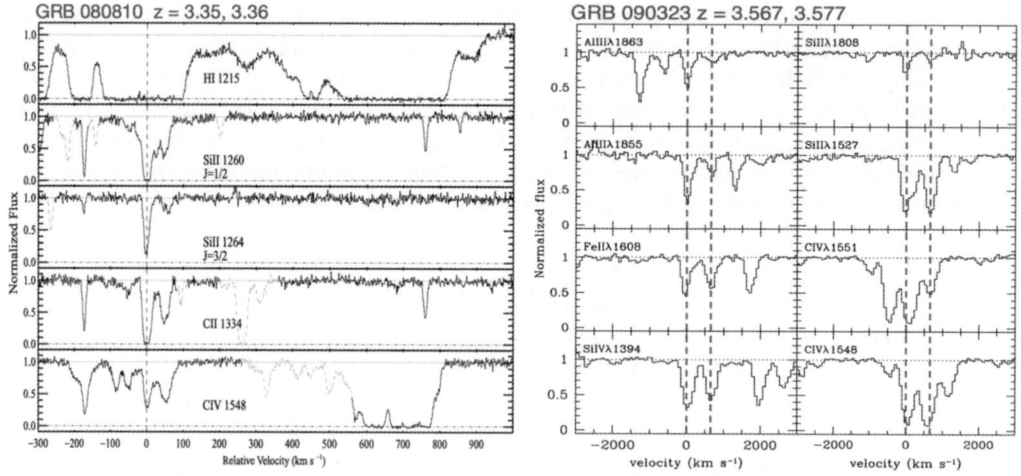

Figure 1. Two example z > 2 LGRB afterglow spectra (Page et al. 2009, Savaglio et al. 2012). The spectra show two sets of galaxy absorption lines, termed pair absorbers, with one set from the host galaxy and the other from a putative companion galaxy.

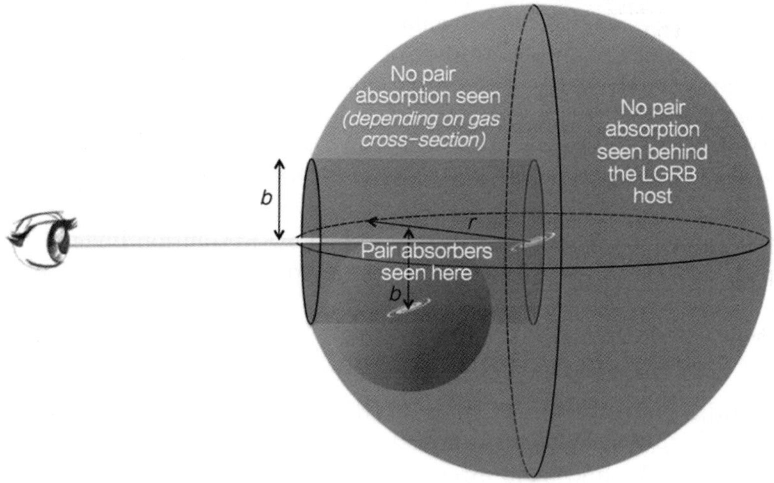

Figure 2. Geometry of an LGRB host galaxy and the ability to observe absorption by a companion galaxy. The LGRB host galaxy is in the centre, the LGRB is seen by the observer to the left, and 'r' indicates the radius of interest for the companion galaxy separation. Companion/interacting galaxies behind the host with respect to the observer are not seen in absorption. Thus, considering random orientations and nothing else, the fraction of pair absorbers observed in LGRB afterglow spectra is > 26% (see § 3.2). The separation between the companion galaxy centroid and the observer's line of sight is indicated by the impact parameter 'b'. The blue region illustrates that only companion galaxies within a cylinder with radius b in LRGB sightlines can be observed as pair absorbers *(see text)*. Using impact parameters from DLA sightlines to QSOs and background galaxies, the fraction of companion galaxies is estimated to be $\gtrsim 50\%$.

The impact parameter, b, is the separation between the LGRB line of sight and the foreground galaxy centroid in which the foreground galaxy absorption is seen in the afterglow spectrum consistent with observations. Thus, a companion galaxy needs to reside within the cylindrical-shaped region (blue in Fig. 2) to appear as a pair absorber.

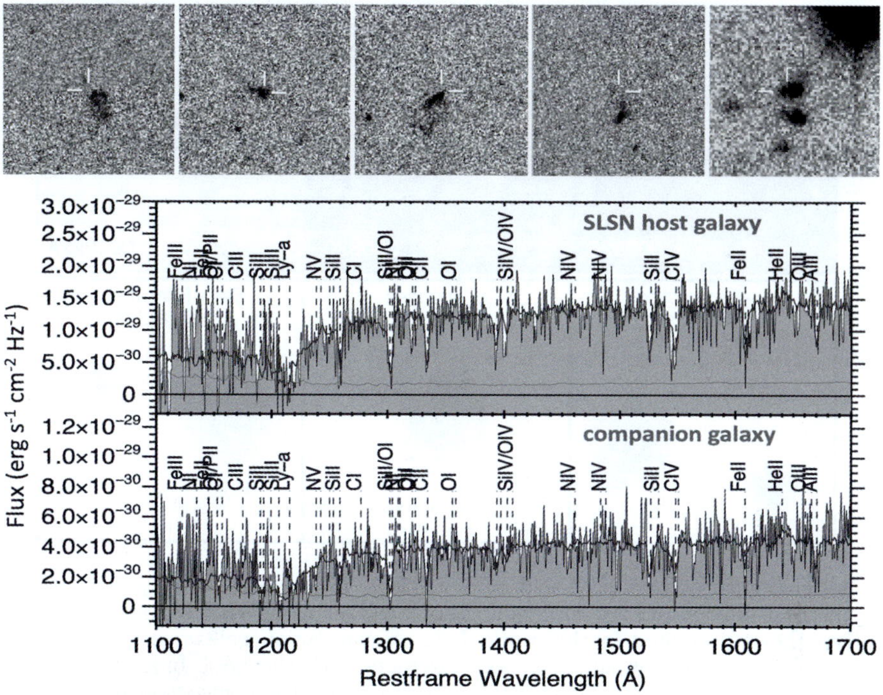

Figure 3. Upper panels: *HST* images of z > 2 SLSN host galaxies with the locations of the SLSNe indicated. Left to right: The first four images are *HST* images (~7.5 arcsec a side) of HSC17dbpf, $z_{spec} = 1.851$; HSC16adga, $z_{spec} = 2.399$; HSC17auzg, $z_{spec} = 1.965$; HSC16bvv, $z_{phot} = 2.5$, and the rightmost image is a CFHT Megacam image (15 arcsec on a side) of SN2213-1745, $z_{spec} = 2.046$ (Curtin et al. 2019, Moriya et al. 2019, Cooke et al. 2012) Lower panels: Keck 1D LRIS spectroscopy of SN2213-1745 host galaxy and companion reveals a ~150 km s^{-1} velocity difference. The spectrum (black with grey fill) has the positions of high-redshift star forming galaxy far-UV stellar and ISM lines marked by the vertical dashed lines and labelled. A Lyman break galaxy composite spectrum is overlaid (blue curve). The galaxy centroid separation in the image is 21.8 kpc.

Impact parameters for damped Lyman α systems (DLAs) seen in the sightlines to background QSOs and sightlines to background galaxies range from ~0–25 kpc, as a function of HI column density, i.e., higher column densities have smaller impact parameters (e.g., Krogager et al. 2012, Christensen et al. 2014, Cooke & O'Meara 2015). Using 13% as the observed line-of-sight pair absorber fraction in LGRB afterglow spectra, an impact parameter of ~21 kpc would correspond to 13% observed coverage (i.e., the fraction residing in the sightline cylinder) if 100% of LGRB host galaxies have interacting or companion galaxies within 50 kpc. Similarly, for host-companion galaxy separations of 30 kpc an impact parameter of ~13 kpc corresponds to 100% companion fraction. More distant companions or larger impact parameters would correspond to < 100% overall fraction. From the *HST* images, we find that the LGRB host galaxy interactions and close companions have ~10–20 kpc impact parameters (see Figure 4).

3.3. *Imaging*

An initial assessment of the LGRB *HST* host galaxy images finds 35% displaying interaction features i.e., disturbed morphology, tidal tails and/or mergers (e.g., Figure 4) and an additional 28% host galaxies having close pairs (in projection). A more rigorous assessment of these systems is underway. We note that the *HST* images were not acquired

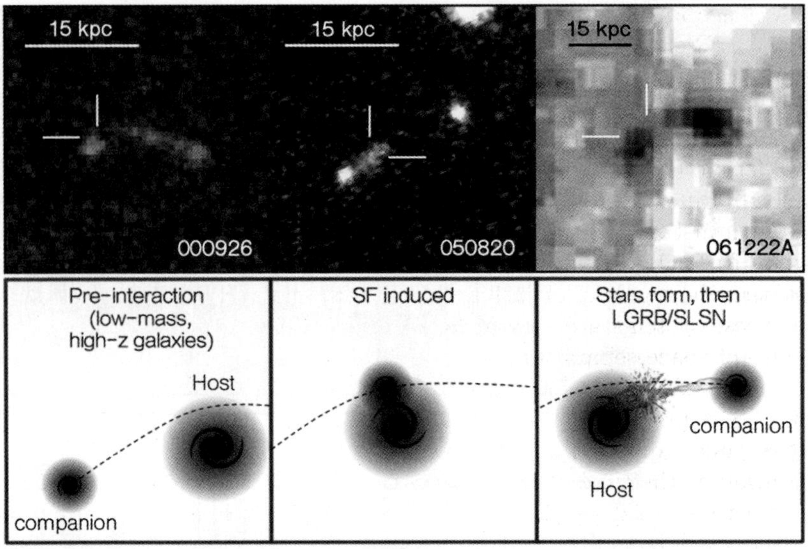

Figure 4. Upper three panels: A subset of LGRBs are found on host galaxy outskirts, in-between the host galaxy and a companion galaxy, and/or along visible tidal tails. Lower three panels: Hypothesis presented in this work. Left: A companion galaxy approaches the LGRB host galaxy. Centre: the companion galaxy interacts with the host, inducing star formation (SF). Right: After some time, the LGRB is observed and the companion has travelled some distance (typically ∼10–20 kpc on the sky). The right panel can be compared to the upper panels of interacting LGRB galaxies. Measuring the time that the companion galaxy would take to travel the physical separation between the two galaxies at the time of the observed LGRB at their velocities, yields the timescale of the induced cloud collapse plus lifetime of the progenitor star.

with the intent to reach sufficient depths to perform a thorough search for close companions, disturbed morphology, or tidal features at these high redshifts. Thus, the number of systems with a faint close companions or evidence of interaction may be higher.

This assessment can be compared to the ∼10–20% merger and close pair fractions for field galaxies at $z > 2$ from observation and simulations (e.g., Conselice et al. 2008, Bertone & Conselice 2009, Berrier & Cooke 2012, Ventou et al. 2019). The ∼10–20% fraction is dependent on host galaxy mass and corresponds to a mass range ($8 < \log M_*$ < 10) consistent with those for LGRB host galaxies.

Finally, we include a similar assessment of $z \gtrsim 2$ SLSNe. Figure 3 presents five SLSNe; four with *HST* imaging and one with ground-based CFHTLS imaging. The SLSNe appear to occur on the edge or outskirts of their host galaxies in the *HST* images (without adding projection) and some have signs of interaction or disturbed morphology. The SN2213-1745 host galaxy has a nearby companion and both galaxies appear to show tidal features. The two galaxies are relatively bright and help accommodate 8m-class spectroscopy. We observed the two galaxies 626 days (rest-frame) after the supernova initial detection and the 1D spectra are shown in Figure 3. Both galaxies have the same redshift to within ∼150 km s^{-1} and are separated on the sky by 21.8 kpc. As such, they exhibit a similar behaviour as LGRB galaxies with identified companions or interactions.

4. Discussion

Our initial analyses of LGRB afterglow pair absorbers and *HST* imaging, along with the geometry of the events, indicate that $z > 2$ LGRB host galaxies have a higher interaction and close pair fraction than ∼10–20% found for $z > 2$ galaxies. We find tentative evidence for a similar result for SLSNe. One reason for the higher fraction is that LGRBs/SLSNe may preferentially point to sites of galaxy interaction with recently induced star formation. Under the assumption that LGRB/SLSN progenitors are massive stars with short lifetimes, the two galaxies involved in the interaction will have only travelled a short distance during the time required for the expected cloud collapse plus progenitor lifetime (i.e., from birth to GRB/SLSN death). For example, galaxies with a relative velocity difference of 500 km s^{-1} would travel ∼10–20 kpc in ∼20–40 Myr.

Interestingly, an LGRB host galaxy subset shows imaging evidence of the above scenario, with ∼10–20 kpc host–companion separations and the LGRB location on the outskirts of their host galaxy or in-between the two galaxies in the tidal tails (Figure 4). Given that these systems are viewed at all orientations in projection, we can estimate that typical real space separations are roughly ∼10–40 kpc, for 0–60° pair inclinations (i.e., to include ∼88% of the ∼10–20 kpc separation cases). If our hypothesis that the LGRB (or SLSN) progenitor star formation is induced by the interaction near closest approach, the time required for each system of two interacting galaxies to reach the measured separation at the time of the observed LGRB (or SLSN) event, travelling at their measured velocities, is a direct measure of the cloud collapse and progenitor star lifetime.

Future deep space-based host galaxy imaging with *HST*, *Euclid*, *Roman*, and *JWST* and deep spectroscopic analysis of their velocities and metallicities with *JWST* and 30m-class telescopes will fully test this hypothesis and, if confirmed, will provide direct measurements of cloud collapse timescales to inform simulations and progenitor mass measurements of LGRBs, and potentially SLSNe, lending deeper insight into their nature.

References

Barton, E. J., Geller, M. J., & Kenyon, S. J. 2000, ApJ, 530, 660
Barton Gillespie, E., Geller, M. J., & Kenyon, S. J. 2003, ApJ, 582, 668
Berrier, J. C. & Cooke, J. 2012, MNRAS, 426, 1647
Bertone, S. & Conselice, C. J. 2009, MNRAS, 396, 2345
Christensen, L., Møller, P., Fynbo, J. P. U., et al. 2014, MNRAS, 445, 225
Conselice, C. J., Rajgor, S., & Myers, R. 2008, MNRAS, 386, 909
Cooke, J., Sullivan, M., Gal-Yam, A., et al. 2012, Nature, 491, 228
Cooke, J. & O'Meara, J. M. 2015, ApJL, 812, L27
Curtin, C., Cooke, J., Moriya, T. J., et al. 2019, ApJS, 241, 17
de Mello, D. F., Smith, L. J., Sabbi, E., et al. 2008, AJ, 135, 548
Krogager, J.-K., Fynbo, J. P. U., Møller, P., et al. 2012, MNRAS, 424, L1
Moriya, T. J., Tanaka, M., Yasuda, N., et al. 2019, ApJS, 241, 16
Oke, J. B., Cohen, J. G., Carr, M., et al. 1995, PASP, 107, 375
Page, K. L., Willingale, R., Bissaldi, E., et al. 2009, MNRAS, 400, 134
Patton, D. R., Pritchet, C. J., Yee, H. K. C., et al. 1997, ApJ, 475, 29
Prochaska, J. X., Neeleman, M., Kanekar, N., et al. 2019, ApJL, 886, L35
Savaglio, S., Budavári, T., Glazebrook, K., et al. 2007, The Messenger, 128, 47
Savaglio, S., Rau, A., Greiner, J., et al. 2012, MNRAS, 420, 627
Steidel, C. C., Shapley, A. E., Pettini, M., et al. 2004, ApJ, 604, 534
Ventou, E., Contini, T., Bouché, N., et al. 2019, A&A, 631, A87

Massive Stars Near and Far
Proceedings IAU Symposium No. 361, 2024
J. Mackey, J. S. Vink, & N. St-Louis, eds.
doi:10.1017/S1743921322002964

Metallicity distributions of core-collapse supernovae within 30Mpc: Evidence for a lack of single massive Ib progenitors at low metallicities.

Joanne L. Pledger[1], Rudi Ganss[1], Anne E. Sansom[1], Philip A. James[2], Joachim Puls[3] and Stacey M. Habergham-Mawson[2]

[1]Jeremiah Horrocks Institute, University of Central Lancashire, Preston, PR1 2HE, UK
email: jpledger@uclan.ac.uk

[2]Astrophysics Research Institute, Liverpool John Moores University, 146 Brownlow Hill, Liverpool L3 5RF, UK

[3]LMU München, Universitäts-Sternwarte, Scheinerstr, 1, D-81679 München, Germany

Abstract. Local HII environment metallicities of 65 supernovae (SNe), obtained with INT/IDS, have been determined using the N2 and O3N2 strong emission line methods. Resulting cumulative distribution functions reveal a narrower distribution for Ib SNe (standard deviation $\sigma \sim 0.06$ dex) compared to Ic and IIP distributions ($\sigma \sim 0.15$ dex). This narrow distribution of Ib SNe is confirmed with an extended dataset using data from Galbany et al. (2018). Statistical tests confirm a statistically significant difference between the Ib and II-P metallicity distributions with $< 5\%$ probability that they result from the same progenitors. This narrow distribution suggests a lack of Type Ib SNe in low metallicity environments and points towards single star progenitors for these Type Ib SNe, rather than binaries. It also suggests that single massive stars at low metallicity are not commonly able to produce helium-rich Type Ib supernovae.

Keywords. supernovae: general, ISM: HII regions, galaxies: abundances.

1. Introduction

Supernova subtypes are classified by the different chemical elements detected in their spectra and this is thought to represent the chemical composition of the progenitor star. Stripping of the outer envelopes from a star result in a H-poor Type Ib or a H+He-poor Type Ic core collapse supernova (ccSN), rather than a H-rich Type II. These observations lead to the question“is this stripping a result of stellar winds, binarity, or a combination of both?”

The progenitors of Type IIP supernova have been confirmed through direct observations to be Red Supergiants (RSG), for example SN2008bk with post-SN images confirming the disappearance of the progenitor star (Mattila et al. 2010). However, direct identification of Type Ibc progenitors has been challenging. For example, Cao et al. (2013) used archival HST imaging to identify the progenitor star of Type Ib SN iPTF13bvn and concluded that if it was a single star, the colours and luminosity were consistent with a Wolf-Rayet (WR) star, as was the mass-loss rate derived from radio emission. Groh et al. (2013) supported this conclusion, with Geneva models suggesting a single star mass of 31-35$M_\odot$. However the ejecta mass derived in the models is too high and the stellar radius too large compared to SN observational results.

Bersten et al. (2014) produced binary models of iPTFbv13 and concluded that a $19{+}20M_\odot$ binary progenitor is also consistent with the photometric measurements. Even post-SN HST imaging is unable to distinguish whether the flux is remaining from the SN itself or from a companion star (Folatelli et al. 2016). Similar binary versus single questions also exist for other SN e.g. SN2013ge (Fox et al. 2022) and SN2017ein (Van Dyk et al. 2018).

Other progenitor detections have raised many questions. For example, the progenitor of Type Ib SN2019yvr is identified in pre-SN imaging as a cool, inflated star much larger than expected for a single star with little to no hydrogen however it is consistent with a Type IIb SN with a small amount of hydrogen left (Kilpatrick et al. 2021). Maybe the most convincing argument for binary progenitors is Type Ic SN2002ap which has deep pre-SN imaging placing stringent constraints on the progenitor (Crockett et al. 2007).

2. Environments of Ibc Supernovae

In lieu of direct confirmation of progenitors we can look at and compare the distribution of different supernovae subtypes with their predicted progenitors and we can compare properties such as brightness, or metallicity. Metallicity is particularly of interest due to the dependence of stellar wind strength on metallicity. We use measurements of gas metallicity at the local environment to avoid issues with galaxy gradients. Early work by multiple groups (e.g. Anderson & James 2009) mainly found no statistical difference between subtypes, however more recent work by Galbany et al. (2018) using PISCO IFU data found that Ic supernovae occur in regions of high metallicity within their host galaxy suggesting the most massive progenitors compared with other ccSNe. But they also found that, on average, Ib SNe have higher metallicities than other subtypes.

To investigate SN environments we obtained new Isaac Newton Telescope/Intermediate Dispersion Spectrograph (INT/IDS) longslit observations of 65 spectra of Type Ib, Ic and IIP supernovae environments over 3 observing runs. Observations were taken at least 12 months after the supernovae had occurred to avoid contamination of the environment with the light from the SN itself. We placed the slit across the galaxy centre through the SN site using the SN position from the Open Supernova Catalogue. This allows us to identify the correct part of the slit to extract and we can also measure metallicity gradients if multiple HII regions are located along the slit position.

Data reduction was semi-automated and all standard steps included, with flux calibration achieved through observations of spectrophotometic standards on the same night as the observations. Figure 1 shows a typical spectrum achieved in $\sim$1hour for the environment of SN2012au with all the necessary diagnostic lines. We dereddened the spectra using the ratio of the Hβ and Hα lines and then used the updated calibrations from Marino et al. (2013) which uses the N2 and O3N2 indicators to determine the metallicity of the HII regions. The O3N2 indicator is preferred but [OIII] is often weak and not detected. The benefit of N2 is that the nitrogen line and Hα line are close so reddening is less of an issue.

With multiple observations of the same site over our 3 runs we could check our method consistency and found good agreement between final values. However, when we compared the N2 and O3N2 results for each source (Figure 2) we found an increasing scatter with increasing metallicity. This could be simply we dont have enough targets at high metallicity or if it could be a result of the [NII] line saturating at metallicities above solar in which case the O3N2 indicator would be preferable. Another explanation for the scatter could be the large calibration uncertainty, which is shown on Figure 2 as the black cross.

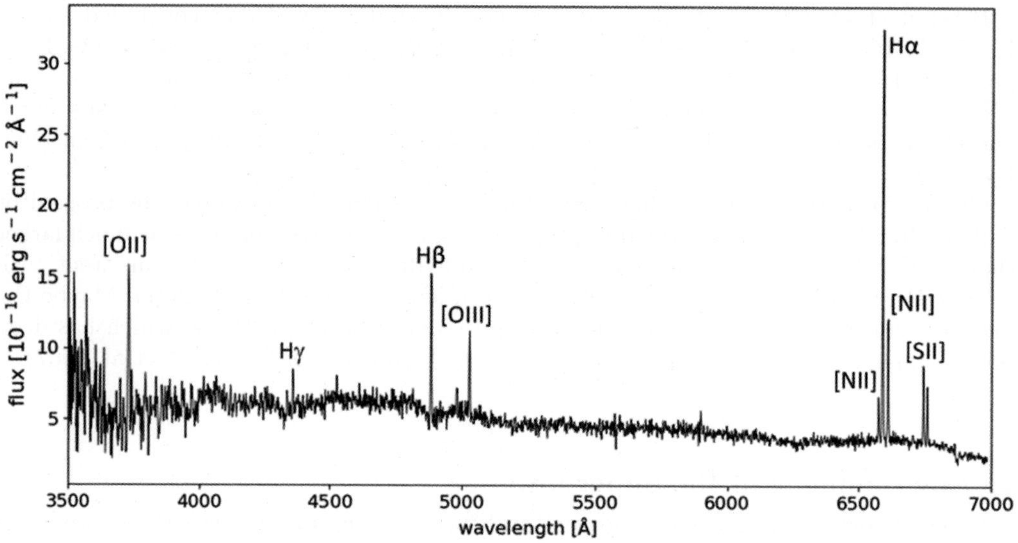

Figure 1. INT/IDS spectra of the host environment of Type Ib SN2012au showing all the diagnostic lines used for metallicity determination.

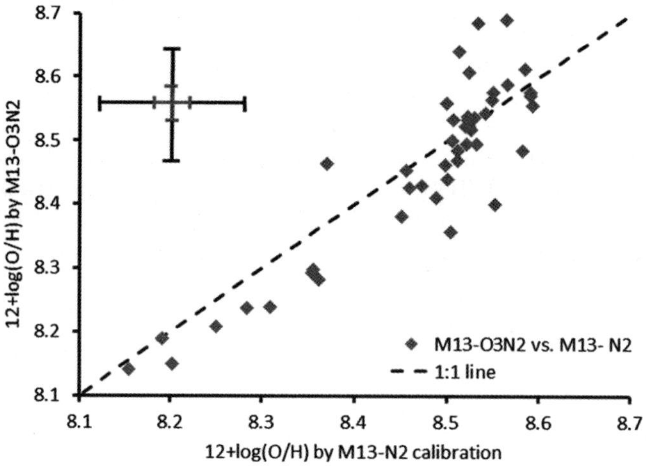

Figure 2. Comparison of metallicities determined using the N2 calibration from Marino et al. (2013) with those determined from the O3N2 calibration. The black error bar shows the calibration uncertainty and the blue error bar shows the observational uncertainty.

3. Results

From our analysis of the SN host environment metallicities we find that the differences in mean metallicity for each subtype are small with the largest difference being between Ib and IIP for the O3N2 calibration, but this not statistically significant and not found using the N2 calibration (Table 1). However both calibrations reveal a large difference between the standard deviation of Ib SN compared with Ic and IIP, at least a factor of 2. This difference is seen as the narrower distribution in the associated cumulative distribution functions (CDFs) shown in Figure 3 for the O3N2 calibration, and we see the same trend in the N2 plot.

Table 1. Comparison of the mean and standard deviation for Ib, Ic and IIP environment metallicities based on the N2 and O3N2 method

SN type	N2		O3N2	
	mean	σ	mean	σ
Ib	8.52	0.07	8.50	0.06
Ic	8.49	0.14	8.49	0.16
IIP	8.52	0.14	8.42	0.15

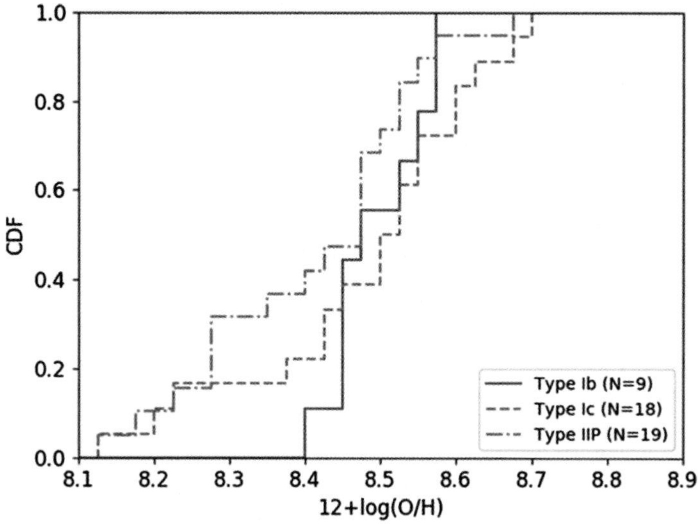

Figure 3. Cumulative Frequency Distribution of the SN environment metallicities measured with the O3N2 calibration from Marino et al. (2013).

To investigate our result further and see if the difference found was a real feature, and not just an effect of small sample sizes, we combined our sample with additional SNe environments from Galbany et al. (2018) which doubled our sample size. To do this we took the flux values presented in their Table 3 and determined the metallicity based on the Marino et al. (2013) calibration. The CDF shown in Figure 4 and subsequent analysis confirms our initial result, again finding the narrower distribution of Ib SN but again no difference in the average metallicity for each SN subtype. We performed both KS and AD tests on the combined sample which indicate a statistically significant difference of the O3N2 results between Type Ib versus IIP and Type Ib at a significance level of 5% (Table 2).

4. Discussion & Conclusions

So what is the explanation for the narrow Ib distribution? One explanation, is that single massive stars have fewer parameters determining their mass loss so this would indicate that at low metallicity single stars arent able to produce Ib SN. However, Ib should be able to be produced through binary evolution which we would expect to extend the distribution to lower metallicities. The fact we don't see this could suggest a lower binary contribution to Ib SN. Indeed, Massey et al. (2021) used BPASS models to investigate the RSG to WR ratio and found that increasing the binarity at low metallicity increased the number of WR stars, flattening out the RSG/WR ratio but if the binary fraction is low then these lower mass helium stars wont be produced nor will the resulting Ib SN.

Table 2. Probabilities (P-values) of the two sample KS- and AD-tests comparing Ib, Ic and IIP SN using metallicities determined from the O3N2 calibration from Marino et al. (2013). Combining our data with that of Galbany et al. (2018) indicate a statistically significant difference between the Type Ib versus IIP at the 5 per cent significance level.

SN type	KS-test	AD-test
Ib – Ic	0.311	0.086
Ib – IIP	0.027	0.020
Ic – IIP	0.189	0.165

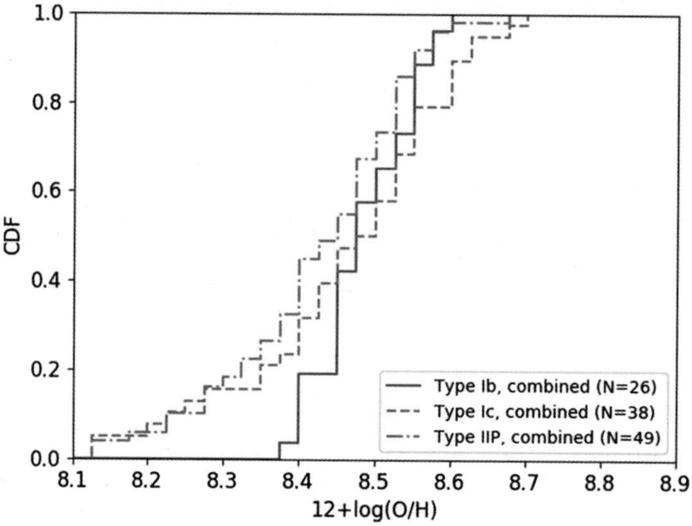

Figure 4. Cumulative Frequency Distribution of the SN environment metallicities combining our work in Ganss et al. (2022) and the Ib/Ic/IIP SN in Galbany et al. (2018).

Conversely, we do see Ic at low metallicity suggesting a significant fraction of binary progenitors for Ic, which would be consistent with observations of SN2002ap with a metallicity of around 8.35 dex. There has been discussion recently in the literature about IIb SN and how there might be contamination of the Ib sample if early-time spectra are not available, so we need to ensure rapid follow up for SN classifications wherever possible. The robustness of the KS/AD-test results will increase significantly with larger sample sizes therefore it is important to obtain additional observations of SN sites.

References

Anderson, J.P., & James, P.A. 2009, *MNRAS*, 399, 559
Bersten, M.C., et al. 2014, *AJ*, 148, 68
Cao, Y., et al. 2013, *ApJ*(Letters), 775, L7
Crockett, R.M. et al. 2007, *MNRAS*, 381, 835
Folatelli, G., et al. 2016, *ApJ*, 299, 219
Fox, O.D., et al. 2022, *ApJ*(Letters), 929, L15
Galbany, L., et al. 2018, *ApJ*, 855, 107
Ganss, R., et al. 2022, *MNRAS*, 512, 1541
Groh, J.H., Georgy, C., & Ekstrom, S. 2013, *A&A*, 558, L1
Kilpatrick, C.D. et al. 2021, *MNRAS*, 504, 2073
Marino, R.A., et al. 2013, *A&A*, 559, A114
Massey, P., Neugent, K.F., Dorn-Wallenstein, T.Z., Eldridge, J.J., Stanway, E.R., & Levesque, E.M 2021, *ApJ*, 922, 177

Mattila, S., Smartt, S., Maund, J., Benetti, S., & Mattias, E. 2010, *arXiv:1011.5494*
Mazzali, P.A., et al. 2002, *ApJ*(Letters), 572, L61
Van Dyk, S.D., et al. 2018, *ApJ*, 860, 90

Discussion

VINK: You mention the observed WR population in relation to the non-detection of the progenitor of SN2002ap, but the point is that the WR stars expected to explode are WO stars which are optically faint.

PLEDGER: Of course SN2002ap could still have had a single progenitor, although I believe it's observed ejecta mass suggests suggests a maximum initial mass of $\sim 20 M_{\odot}$ (Mazzali et al. 2021) which we wouldn't expect to evolve to a WO star or Ic SN. I think it is important that we continue to push to get deep pre-SN imaging in not just the optical but the UV too. It still doesn't explain why we aren't seeing Type Ib SN at low metallicity though.

FRASER: How much are mis-classifications of SN affecting your results?

PLEDGER: We have tried to use databases and literature to ensure only SN with accurate classifications are used. However, classifying supernovae is not something we have done ourselves so where poor or minimal spectra have been used to classify a supernova it is possible that SN have been mis-classified. For example, a relatively poor S/N spectrum taken a reasonable length of time after the SN event could suggest a Type Ic SN when a higher S/N spectrum close in time to the same SN explosion would have detected the helium. Indeed this could affect our CDFs and potentially explain our "missing" Ib SN at low metallicity so is something we will continue to look carefully at.

Massive Stars Near and Far
Proceedings IAU Symposium No. 361, 2024
J. Mackey, J. S. Vink, & N. St-Louis, eds.
doi:10.1017/S1743921322003192

The evolution of binary-stripped stars: consequences for supernovae and black hole formation

Eva Laplace[1,2], Fabian Schneider[1,3], Philipp Podsiadlowski[4], Selma de Mink[5,6,1], Stephen Justham[2,7], Mathieu Renzo[8,9], Ylva Götberg[10], Rob Farmer[5,2] and David Vartanyan[11]

[1]Heidelberger Institut für Theoretische Studien, Schloss-Wolfsbrunnenweg 35, 69118 Heidelberg, Germany

[2]Anton Pannekoek Institute of Astronomy and GRAPPA, Science Park 904, University of Amsterdam, 1098XH Amsterdam, The Netherlands
email:eva.laplace@h-its.org

[3]Zentrum für Astronomie der Universität Heidelberg, Heidelberg, Germany

[4]University of Oxford, St Edmund Hall, Oxford, OX1 4AR, United Kingdom

[5]Max Planck Institute for Astrophysics, Karl-Schwarzschild-Str. 1, 85748 Garching, Germany

[6]Center for Astrophysics, Harvard-Smithsonian, 60 Garden Street, Cambridge, MA 02138, USA

[7]School of Astronomy and Space Science, University of the Chinese Academy of Sciences, Beijing 100012, China

[8]Center for Computational Astrophysics, Flatiron Institute, New York, NY 10010, USA

[9]Department of Physics, Columbia University, New York, NY 10027, USA

[10]The Observatories of the Carnegie Institution for Science, 813 Santa Barbara Street, Pasadena, CA 91101, USA

[11]Department of Physics and Astronomy, University of California, Berkeley, CA 94720

Abstract. Gravitational-wave (GW) observations are revealing the population of compact objects from a new angle. Yet their stellar progenitors remain uncertain because few observational clues on their progenitors exist. Theoretical models typically assume that the progenitor evolution can be approximated with single-star models. We explore how binary evolution affects the pre-supernova (SN) structure of stars, and the resulting distribution of compact object remnants. We focus on the differences in the core properties of single stars and of donor stars that transfer their outer layers in binary systems and become binary-stripped. We show that the final structures of binary-stripped stars that lose their outer layers before the end of core helium burning are systematically different compared to single stars. As a result, we find that binary-stripped stars tend to explode more easily than single stars and preferentially produce neutron stars and fewer black holes, with consequences for GW progenitors.

Keywords. stars: evolution, stars: binaries, stars: interiors, supernovae: general black hole physics

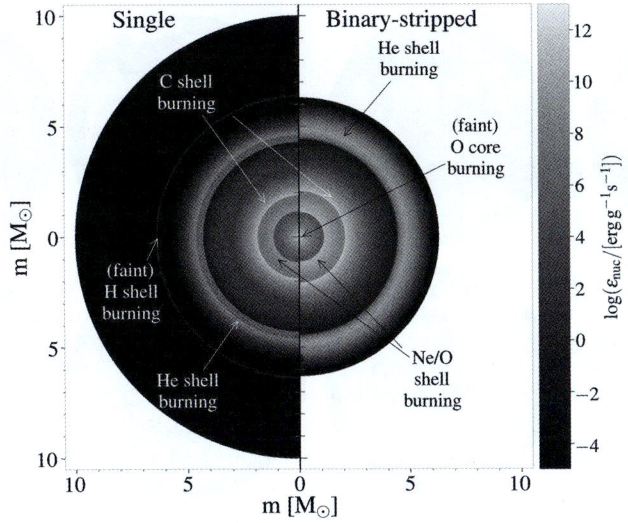

Figure 1. Core structure and nuclear energy generation rate of a single (left half-circle) and binary-stripped (right half-circle) star with the same reference helium core mass, shown at the end of core carbon burning (see also Laplace et al. 2021). Even for the same core mass, the location and strength of the nuclear burning shells is different, ultimately leading to changes in the chemical composition and final fate. The diagram was created with the TULIPS software (Laplace 2022).

1. Introduction

Understanding the final fate of massive stars is a pressing question in the era of GW sources. GW detectors have now detected almost a hundred compact objects, and yet their stellar progenitors remain uncertain (LIGO Scientific Collaboration and Virgo Collaboration et al. 2019). It is now established that most massive stars live in close multiple systems, with short enough separations to interact during their lifetime (Sana et al. 2012). Isolated binary systems are promising candidates for the progenitors of GW sources. Here, we investigate the impact of binary mass transfer on the final fate of massive stars and on the resulting population of compact objects.

2. Methods

We compute one-dimensional single and binary evolutionary models at solar metallicity using the MESA stellar evolution code (Paxton et al. 2011; 2013; 2015; 2018; 2019). In the calculations presented in Laplace et al. (2021) we focus on post main sequence mass transfer (case B) from primary stars with initial masses between 11 and 21 $M_{\odot}$, following a similar approach as Laplace et al. (2020). In (Vartanyan et al. 2021), the supernova explosion of these models is simulated self-consistently with the two-dimensional FORNAX code. To understand the effects of binary stripping on a stellar population, Schneider et al. (2021) explores a larger range of periods and initial masses and simulates the outcome of neutrino-driven supernova explosions of these progenitors based on an improved version of the parametric supernova code by Müller et al. (2016).

3. Results

As illustrated in Fig. 1, we find that binary-stripped stars that lose their outer layers before the end of core helium burning develop systematically different core structures compared to single stars, even for similar core masses (Laplace et al. 2021,

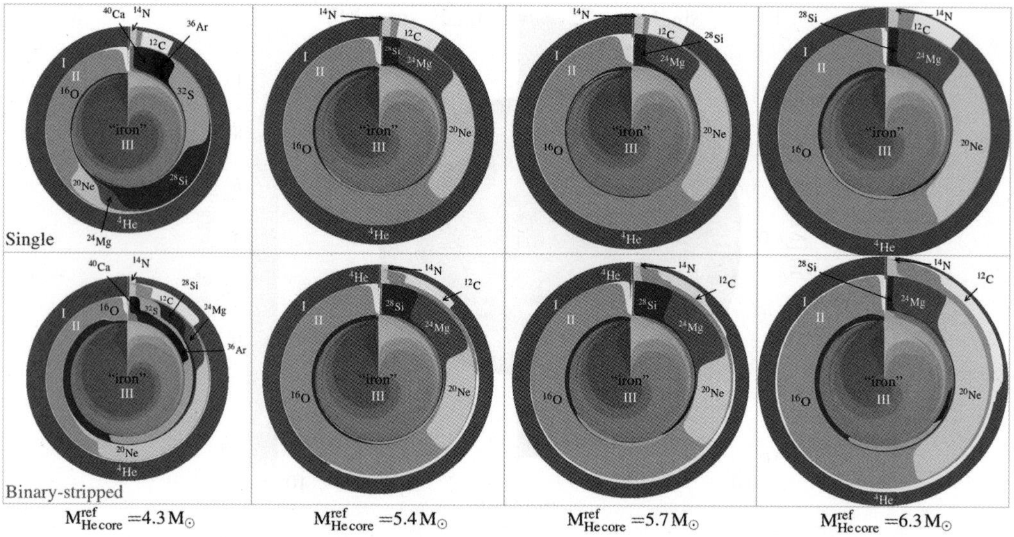

Figure 2. Chemical composition of the inner helium core of single and binary-stripped stars with the same reference helium core mass at the onset of core collapse (Laplace et al. 2021). Every isotope is represented by a different color. The radial direction is proportional to the square-root of the mass coordinate. Consequently, the surface area covered by each color is proportional to the mass of the corresponding isotope. Binary-stripped stars contain a gradient of carbon, oxygen, and neon around the helium-depleted core (region II), that is not present in single stars. The diagrams were created with the TULIPS software (Laplace 2022).

Schneider et al. 2021), in agreement with previous studies. Due to mass loss, the helium core shrinks during core helium burning, leading to a different chemical composition and density structure. From self-consistent simulations of the supernova explosions, we find that, for the same initial mass, binary-stripped stars tend to explode preferentially compared to single stars (Vartanyan et al. 2021). We also show that binary-stripped stars can populate the "mass gap" that may exist between the highest-mass neutron star (NS) and the lowest-mass black hole (BH) through fallback accretion. This could explain the recent report of a GW source containing a compact object within this gap (The LIGO Scientific Collaboration et al. 2020).

As shown in Fig. 2, the chemical composition of single stars and binary stripped stars is different, even for the same core mass. The most prominent difference is the presence of a layer above the helium-depleted core that contains a gradient of carbon, oxygen, and neon, and is not present in single stars. In Farmer et al. (2021), we show that this difference in chemical structure has also large consequences for the nucleosynthesis expected from single and binary-stripped stars. In particular, the binary-stripped stars we model are twice as efficient at producing ^{12}C compared to single stars.

In Fig. 3, we present the distribution of compact objects expected from a population of binary-stripped and single stars (Schneider et al. 2021). Binary-stripped stars create neutron stars for a larger range of initial masses than single stars, and for lower initial masses. In addition, they only create BHs for initial mass larger than 70 $M_\odot$ and for an "island of BH formation" at an approximate initial mass of 32 $M_\odot$. As a result, the number of gravitational-wave mergers from isolated binary evolution is expected to be significantly smaller than typically assumed in population synthesis calculations (Schneider et al. 2021). Based on a 1D parametric supernova code, we find that a number of models may experience fallback accretion and populate the NS-BH mass gap, in agreement with our expectations from 2D supernova simulations (Vartanyan et al. 2021).

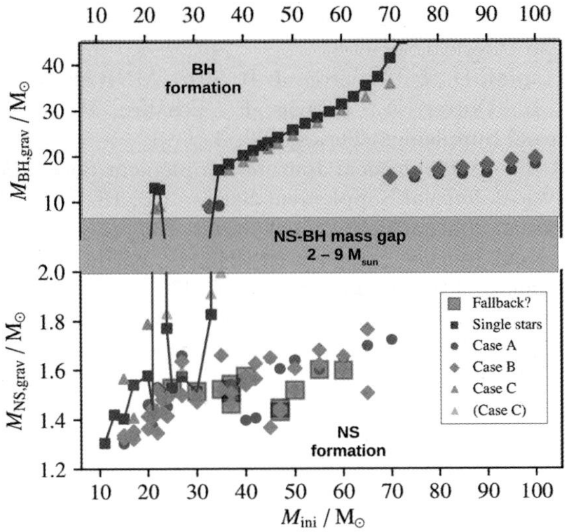

Figure 3. Compact remnants from single and binary-stripped stars (adapted from Schneider et al. 2021). Binary-stripped stars produce neutron stars for a much larger range of initial masses (up to 70 $M_\odot$) and black holes for a much smaller range (mainly for initial masses above 70 $M_\odot$).

4. Conclusions

We conclude that taking binary evolution into account is crucial to understand the formation of compact objects, and for interpreting the population of GW sources. Binary stripping is particularly important because it typically affects the deep interiors of stars, as well as their outer layers. This shifts the distribution of the compact objects to lower masses, leads to preferential explosion compared to single stars, and to differences in the nucleosynthesis. To obtain a full picture of the effect of binary interactions on the final fate of stars, other binary processes, such as accretion and merging, need to be investigated in the future.

Acknowledgements

This work was funded by the European Research Council (ERC) under the European Union's Horizon 2020 research and innovation program (Grant agreements No. 945806 and 715063), by the Netherlands Organisation for Scientific Research (NWO) with project number 639.042.728 and 614.001.501 and supported by the Klaus Tschira Stiftung. YG acknowledges the funding from the Alvin E. Nashman fellowship for Theoretical Astrophysics. DV acknowledges funding from the U.S. Department of Energy Office of Science and the Office of Advanced Scientific Computing Research via the Scientific Discovery through Advanced Computing (SciDAC4) program and Grant DE-SC0018297 (subaward 00009650).

References

Farmer, R., Laplace, E., de Mink, S. E., & Justham, S. 2021, ApJ, 923, 214

Laplace, E. 2022, Astronomy and Computing, 38, 100516

Laplace, E., Gtberg, Y., de Mink, S. E., Justham, S., & Farmer, R. 2020, Astronomy and Astrophysics, 637, A6

Laplace, E., Justham, S., Renzo, M., Götberg, Y., Farmer, R., Vartanyan, D., & de Mink, S. E. 2021, A&A, 656, A58

LIGO Scientific Collaboration and Virgo Collaboration et al. 2019, Physical Review X, 9, 031040, publisher: American Physical Society
Müller, B., Heger, A., Liptai, D., & Cameron, J. B. 2016, MNRAS, 460, 742
Paxton, B., Bildsten, L., Dotter, A., Herwig, F., Lesaffre, P., & Timmes, F. 2011, The Astrophysical Journal Supplement Series, 192, 3
Paxton, B., et al. 2013, The Astrophysical Journal Supplement Series, 208, 4
—. 2015, The Astrophysical Journal Supplement Series, 220, 15
—. 2018, The Astrophysical Journal Supplement Series, 234, 34
—. 2019, The Astrophysical Journal, Supplement, 243, 10, eprint: 1903.01426
Sana, H., et al. 2012, Science, 337, 444, eprint: 1207.6397
Schneider, F. R. N., Podsiadlowski, P., & Mller, B. 2021, Astronomy and Astrophysics, 645, A5
The LIGO Scientific Collaboration et al. 2020, The Astrophysical Journal, 896, L44, eprint: 2006.12611
Vartanyan, D., Laplace, E., Renzo, M., Gtberg, Y., Burrows, A., & de Mink, S. E. 2021, The Astrophysical Journal, 916, L5

Discussion

BERSTEN: I wonder how how we can use your results to analyze the SNe observable (LC and spectra). Because, you find differences on the chemical patron and density structure between single and binary but you are using some specific path of binary calculation. Is this path representative for what we expected for stripped star?

LAPLACE: Very good point. We focused on the most common channel that creates binary-stripped stars due to stable mass transfer, case B mass transfer. This means we expect most stripped stars (at solar metallicity) to come from this channel and our models to be representative of a large fraction of stripped SNe progenitors.

SANDER: Since you mentioned that you use a large network, are you distinguishing between ^{20}Ne and ^{22}Ne? If so, could you make any predictions about Ne surface abundances that could help to identify the different scenarios for H-depleted stars?

LAPLACE: Yes we can distinguish between ^{20}Ne and ^{22}Ne. The issue is that to save computational cost we switched to the larger network at the end of core oxygen burning. Before, we used a network of 21 isotopes. This means the core chemical structure is very well resolved, but we may miss differences in the chemical composition of the outer layers from earlier phases of the evolution.

NEGUERUELA: When you consider the difference between a stripped star and a single star, do you mean a fully stripped star (i.e. without any mantle, as your diagrams seem to imply)? Do your models produce stars that are partially stripped?

LAPLACE: These models are computed at solar metallicity. Due to wind mass loss, the binary-stripped stars lose all their hydrogen-rich envelope at the end of core helium burning. This means they may be partially stripped right after mass transfer, but at the end of their lives, they do not have a hydrogen-rich envelope any longer.

ROBERTI: I have a question about the explodability: do you evaluate it by using only the compactness or do you also consider other parameters?

LAPLACE: We compute the collapse self-consistently in 2D with the FORNAX code and identify explosion as the shock reaching runaway expansion at the end of the simulation (1s post bounce). Interestingly, we found that the compactness is not a sufficient diagnostic for successful explosion. For some models, we also need to look at the critical curve, so the interplay between the competing but coupled effects of accretion and neutrino luminosity in driving the explosion.

MORIYA: Do you find significant differences in explosion energies and ^{56}Ni masses from the explosions of single and binary progenitors? I especially wonder if you find significant differences in the progenitors with the similar core mass or compactness.

LAPLACE: For progenitors with the same initial mass, we found that for the duration of the simulation (up to one second post bounce) binary-stripped star models have higher net explosion energies than their single-star counterpart. However, the energy growth of the single stars models is higher than for the binary-stripped models (probably because of a larger accreting mass), suggesting that when integrated over the full time of the explosion, the single star models should have larger explosion energies. We would need 3D models to be certain about this. Unfortunately, this trend is less clear for progenitors with the same helium core mass. We did not compute the ^{56}Ni masses but we are planning to compute supernova nucleosynthesis abundances, including ^{56}Ni, in future work led by Rob Farmer.

Massive Stars Near and Far
Proceedings IAU Symposium No. 361, 2024
J. Mackey, J. S. Vink, & N. St-Louis, eds.
doi:10.1017/S1743921322002058

Low-frequency radio view of a fast-blue optical transient - AT 2018cow

A. J. Nayana[1] **and Poonam Chandra**[2,3]

[1]Indian Institute of Astrophysics, II Block, Koramangala, Bangalore 560034, India
email: nayana.aj@iiap.res.in

[2]National Radio Astronomy Observatory, 520 Edgemont Road, Charlottesville VA 22903, USA

[3]National Centre for Radio Astrophysics, Tata Institute of Fundamental Research, Pune, India

Abstract. We present low-frequency (0.40 − 1.25 GHz) radio observations and modeling of a Fast Blue Optical Transient (FBOT), AT2018cow (Nayana & Chandra 2021). Our data are best modeled as an inhomogeneous synchrotron emitting region expanding into an ionized circumstellar medium. We estimate the mass-loss rate of the progenitor star and shock parameters at multiple epochs post-explosion and find that the progenitor has gone through an enhanced phase of mass-loss close to its end-of-life.

Keywords. supernovae: individual: AT2018cow, radio continuum: general, stars: mass loss

1. Introduction

Fast Blue Optical transients (FBOTs) are characterised by high optical luminosities ($\gtrsim 10^{43}$ erg s^{-1}), rapid rise times to optical maxima (t <10 days), and blue colors (Drout et al. 2014). The short rise time and high peak luminosities can be explained in a scenario where the optical light curves are hydrodynamically powered (not radioactively) with small ejecta masses; $M_{\rm ej} \lesssim 10^{-1}$ (Lyutikov 2022).

AT 2018cow was discovered on 2018 June 16.44 UT in the galaxy CGCG 137−068 at a distance of 66 Mpc (Smartt et al. 2018). The event was classified as an FBOT based on the characteristics of the optical light curve and spectra (Prentice et al. 2018). Radio emission was detected from AT 2018cow (Nayana & Chandra 2018, Margutti et al. 2019, Ho et al. 2019) and was found to be synchrotron radiation from a shock ($v \sim 0.1$ c) created due to the interaction of the ejecta with the dense surrounding medium (Ho et al. 2019, Margutti et al. 2019). Modelling the early time ($t \sim 5 - 81$ days) radio observations, Ho et al. (2019) estimated the mass-loss rate of the progenitor to be $\dot{M} \sim 4 \times 10^{-4}$ $M_{\odot}$ yr^{-1}.

Late-time low-frequency radio observations of FBOTs will probe a longer period of mass-loss in the progenitors' life prior explosion. We present upgraded Giant Metrewave Radio telescope (uGMRT) observations of AT 2018cow up to $t \sim 570$ days at frequencies 0.40−1.25 GHz, and study the properties of its environment. Our findings are published in (Nayana & Chandra 2021) and here we summarize the main results presented in Nayana & Chandra (2021).

2. uGMRT observations and modeling

We observed AT 2018cow with the uGMRT during $t = 11 - 570$ days at frequencies 0.40, 0.75, and 1.25 GHz in the standard continuum mode. The radio light curves are shown in Fig. 1. We model the radio data as synchrotron radiation from a shock created

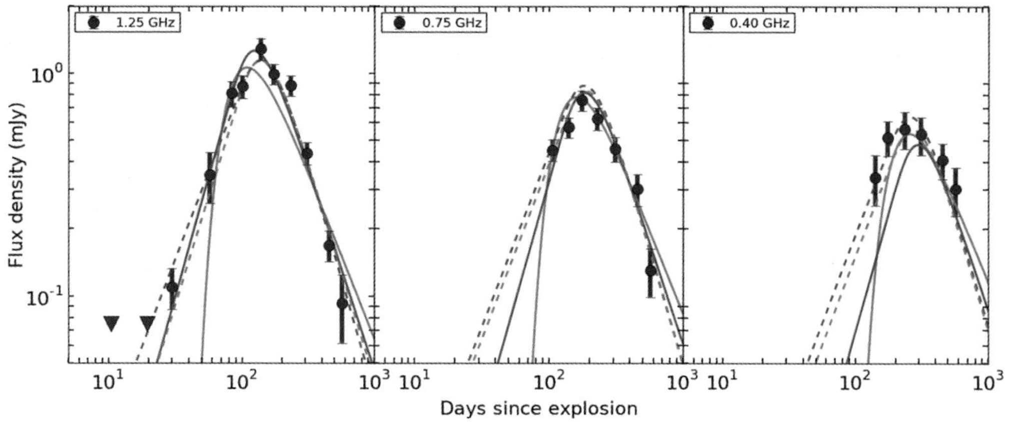

Figure 1. The best-fit homogeneous FFA (red solid lines), SSA (green solid lines), inhomogeneous FFA (red dotted lines), and inhomogeneous SSA models (green solid lines) are shown. The blue filled circles denote the observed flux densities at 0.40, 0.75 and 1.25 GHz. Figure reproduced from (Nayana & Chandra 2021).

due to the interaction of the ejecta from the explosion with the circumstellar medium, initially suppressed by free-free absorption (FFA; Chevalier 1982) or synchrotron self-absorption (SSA; Chevalier 1998). We model the data using standard homogeneous FFA and SSA models as well as inhomogeneous FFA and SSA models and find that the data is best represented by an inhomogeneous SSA model (see Fig. 1). The best-fit 1.25, 0.75 and 0.40 GHz light curves peak with flux densities $F_{\rm peak}$ = 1.1, 0.9 and 0.6 mJy at $t_{\rm peak}$ = 138, 182 and 257 days, respectively.

3. Results

We estimate shock radius (R) and magnetic field (B) at $t_{\rm peak}$ using equations 13 and 14 of Chevalier (1998). We also derive the mass-loss rates at multiple epochs using magnetic field scaling relation; $\frac{B^2}{8\,\pi} = \epsilon_{\rm B} \frac{\dot{M}}{4\pi R^2 v_{\rm w}} v^2$ (Chevalier 1998). We assume the fraction of shock energy distributed to magnetic fields to be $\epsilon_{\rm B} = 0.33$ and a wind velocity $v_{\rm w} \sim 1000$ km s^{-1}. The derived physical parameters are $R \sim$ (6.1, 9.3, 14.4) $\times$ 10^{16} cm, $B \sim$ 0.11, 0.07, 0.04 G, and $\dot{M} \sim$ (0.4, 0.3, 0.2) $\times$ 10^{-5} $M_\odot$ yr^{-1}, at $t_{\rm peak}$ = 138, 182 and 257 days (19 46 years prior explosion), respectively. These $\dot{M}$ values are $\sim$ 100 times smaller than the $\dot{M}$ values derived at $t_{\rm peak}$ = 22 days ($\sim$ 2 years prior explosion; Ho et al. 2019), indicating the presence of a dense shell in the viscinity of AT 2018cow.

Acknowledgements

Nayana A.J. acknowledges DST- INSPIRE Faculty Fellowship (IFA20-PH-259) for supporting this research. P.C. acknowledges the support of the Department of Atomic Energy, Government of India, under project no. 12-R&D-TFR-5.02-0700.

References

Chevalier, R. A. 1982, ApJ, 259, 302
—. 1998, ApJ, 499, 810
Drout, M. R., et al. 2014, ApJ, 794, 23
Ho, A. Y. Q., et al. 2019, ApJ, 871, 73
Lyutikov, M. 2022, arXiv e-prints, arXiv:2204.08366
Margutti, R., et al. 2019, ApJ, 872, 18
Nayana, A. J., & Chandra, P. 2018, The Astronomer's Telegram, 11950, 1

—. 2021, ApJL, 912, L9
Prentice, S. J., et al. 2018, ApJL, 865, L3
Smartt, S. J., et al. 2018, The Astronomer's Telegram, 11727, 1

Massive Stars Near and Far
Proceedings IAU Symposium No. 361, 2024
J. Mackey, J. S. Vink, & N. St-Louis, eds.
doi:10.1017/S1743921322002459

How Much Hydrogen is in Type Ib and IIb SN Progenitors?

Avishai Gilkis[1] and **Iair Arcavi**[1,2]

[1]The School of Physics and Astronomy, Tel Aviv University, Tel Aviv 6997801, Israel
email: agilkis@tauex.tau.ac.il

[2]CIFAR Azrieli Global Scholars program, CIFAR, Toronto, M5G 1M1, Canada

Abstract. We compare pre-supernova observations with synthetic photometry from stellar evolution models to infer the progenitor properties of the seven known progenitors of Type Ib and IIb supernovae. Our results are roughly consistent with a hydrogen mass threshold of $\approx 0.033\,\mathrm{M}_\odot$ for a Type II appearance.

Keywords. stars: evolution, stars: massive, supernovae: general

1. Introduction

Type Ib and IIb supernovae (SNe) are sub-types of core-collapse SNe which differ by the absence of hydrogen lines in the former, in contrast to the early appearance of broad hydrogen lines in the latter. Radiative-transfer simulations yield disparate values of the amount of hydrogen which will give rise to a type II appearance (Dessart et al. 2011, Hachinger et al. 2012). Here we address the question of the hydrogen mass that can be 'hidden' in SNe with a different approach, by finding the best-fitting models for all known type Ib and IIb SN progenitors.

2. Inferring progenitor properties from pre-explosion photometry

Binary evolution is computed with the MESA code (version r10398; Paxton et al. 2018) in a similar manner to Gilkis et al. (2019). The endpoints (core carbon depletion) of 494 stellar evolution tracks are assigned synthetic photometry values by convolution of empirical or computed spectra with dust extinction (Cardelli et al. 1989) and with the relevant filters. Spectra are taken from Pickles (1998) for $T_{\mathrm{eff}} < 15000\,\mathrm{K}$, from Lanz & Hubeny (2007) for $15000\,\mathrm{K} \leq T_{\mathrm{eff}} \leq 30000\,\mathrm{K}$, from Lanz & Hubeny (2003) for $30000\,\mathrm{K} < T_{\mathrm{eff}} \leq 55000\,\mathrm{K}$ and from Todt et al. (2015) for $55000\,\mathrm{K} < T_{\mathrm{eff}}$. We take the pre-SN photometry from Aldering et al. (1994) for SN 1993J, Folatelli et al. (2015) for SN 2008ax, Maund et al. (2011) for SN 2011dh, Van Dyk et al. (2014) for SN 2013df, Eldridge et al. (2015) for iPTF13bvn, Kilpatrick et al. (2021a) for SN 2016gkg and Kilpatrick et al. (2021b) for SN 2019yvr. For each SN, we generate 100 realizations of the photometry assuming a normal distribution with the error as the standard deviation. For each realization the best-fitting model is found, providing a distribution of progenitor properties. This is done for four assumptions on dust extinction: (*i*) the dust extinction parameter R_V and reddening $E\,(B-V)$ are fixed; (*ii*) R_V fixed and $E\,(B-V)$ variable; (*iii*) R_V variable and $E\,(B-V)$ fixed; (*iv*) R_V and $E\,(B-V)$ variable. We present the results in Fig. 1 and show that within the uncertainties a hydrogen mass threshold of $\approx 0.033\,\mathrm{M}_\odot$ to distinguish between Ib and IIb SNe is plausible.

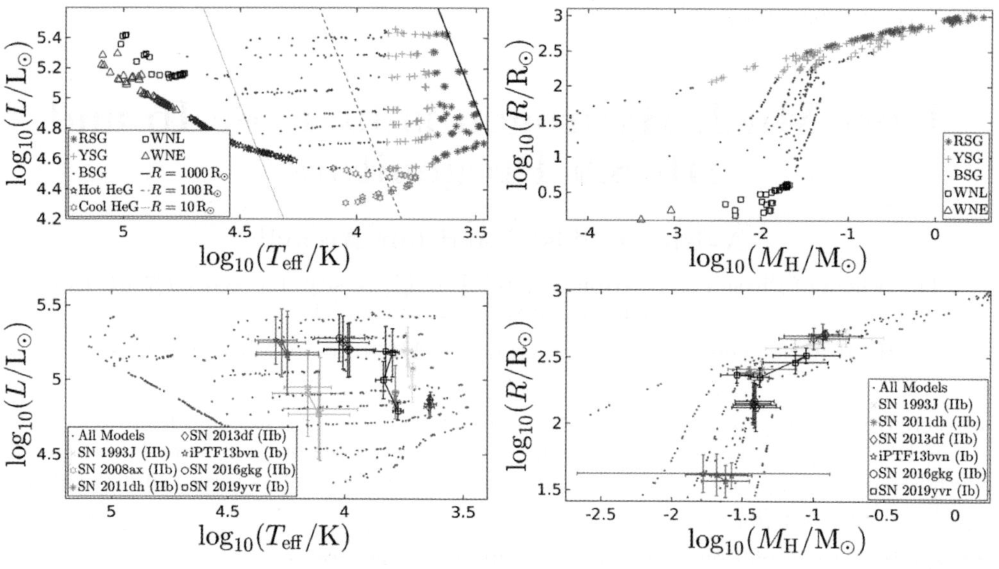

Figure 1. *Top-left:* Evolutionary endpoints marked by stellar types defined by the effective surface temperature $T_{\rm eff}$ and surface hydrogen mass fraction $X_{\rm s}$ as follows: red supergiant (RSG) - $T_{\rm eff} \leq 4.8\,{\rm kK}$, $0.01 \leq X_{\rm s}$; yellow supergiant (YSG) – $4.8\,{\rm kK} < T_{\rm eff} < 7.5\,{\rm kK}$, $0.01 \leq X_{\rm s}$; blue supergiant (BSG) – $7.5\,{\rm kK} \leq T_{\rm eff} \leq 55\,{\rm kK}$, $0.01 \leq X_{\rm s}$; hot helium giant (HeG) – $15\,{\rm kK} \leq T_{\rm eff} \leq 55\,{\rm kK}$, $X_{\rm s} < 0.01$; cool HeG – $T_{\rm eff} < 15\,{\rm kK}$, $X_{\rm s} < 0.01$; early nitrogen-sequence WR (WNE) – $55\,{\rm kK} < T_{\rm eff}$, $X_{\rm s} < 0.05$; late nitrogen-sequence WR (WNL) – $55\,{\rm kK} < T_{\rm eff}$, $0.05 \leq X_{\rm s}$. *Top-right:* Radii and total hydrogen masses for evolutionary endpoints marked by stellar types. *Bottom-left:* Inferred temperature and luminosity for Type Ib and IIb SN progenitors. *Bottom-right:* Inferred total hydrogen mass and radius for Type Ib and IIb SN progenitors.

References

Aldering, G., Humphreys, R. M., & Richmond, M. 1994, AJ, 107, 662

Cardelli, J. A., Clayton, G. C., & Mathis, J. S. 1989, ApJ, 345, 245

Dessart, L., Hillier, D. J., Livne, E., Yoon, S.-C., Woosley, S., Waldman, R., & Langer, N. 2011, MNRAS, 414, 2985

Eldridge, J. J., Fraser, M., Maund, J. R., & Smartt, S. J. 2015, MNRAS, 446, 2689

Folatelli, G., Bersten, M. C., Kuncarayakti, H., Benvenuto, O. G., Maeda, K., & Nomoto, K. 2015, ApJ, 811, 147

Gilkis, A., Vink, J. S., Eldridge, J. J., & Tout, C. A. 2019, MNRAS, 486, 4451

Hachinger, S., Mazzali, P. A., Taubenberger, S., Hillebrandt, W., Nomoto, K., & Sauer, D. N. 2012, MNRAS, 422, 70

Kilpatrick, C. D., Coulter, D. A., Foley, R. J., Piro, A. L., Rest, A., Rojas-Bravo, C., & Siebert, M. R. 2021a, arXiv e-prints, arXiv:2112.03308

Kilpatrick, C. D., et al. 2021b, MNRAS, 504, 2073

Lanz, T., & Hubeny, I. 2003, ApJS, 146, 417

—. 2007, ApJS, 169, 83

Maund, J. R., et al. 2011, ApJL, 739, L37

Paxton, B., et al. 2018, ApJS, 234, 34

Pickles, A. J. 1998, PASP, 110, 863

Todt, H., Sander, A., Hainich, R., Hamann, W. R., Quade, M., & Shenar, T. 2015, A&A, 579, A75

Van Dyk, S. D., et al. 2014, AJ, 147, 37

Massive Stars Near and Far
Proceedings IAU Symposium No. 361, 2024
J. Mackey, J. S. Vink, & N. St-Louis, eds.
doi:10.1017/S1743921322002575

Supplying angular momentum to the jittering jets explosion mechanism using convection in inner star layers

Dmitry Shishkin

Department of Physics, Technion, Haifa, 3200003, Israel
email: s.dmitry@campus.technion.ac.il

Abstract. The jittering jets explosion mechanism of core collapse supernovae proposes that stochastic convective motion, amplified by instabilities, results in intermittent accretion disk that launches jittering jets that explode the star. We conduct one-dimensional simulations of a wide range of stellar masses and show that the convective motion in the pre-collapse stellar core, when scaled to corresponding three-dimensional simulations, supply sufficient angular momentum fluctuation to form the intermittent accretion disk the launches the jittering jets. The resulting neutron star masses are consistent with observations.

Keywords. supernovae: general

1. Introduction

In the jittering jets explosion mechanism of core collapse supernovae (CCSNe), perturbations in the pre-collapsing core serve as the source of stochastic angular momentum fluctuations (Soker 2019) that are then further amplified by instabilities. The main relevant instability is the spiral standing accretion shock instability (spiral SASI). According to the jittering jets explosion mechanism, the amplified perturbations lead to the formation of intermittent accretion disks (or belts, Schreier & Soker 2016), which in turn launch the jittering jets (e.g., Papish & Soker 2011, Gilkis & Soker 2014). Supplying large enough perturbation seeds to the spiral SASI such that the final angular momentum fluctuations form the intermittent accretion disks is the key challenge of the jittering jets explosion mechanism.

In (Shishkin & Soker 2021) we followed the evolution of stars to core collapse velocities of $v_{\rm fall} > 1000$ km s^{-1}. We argued that the convective specific angular momentum fluctuations in the core are sufficiently large seed-perturbations to SASI for the formation of intermittent accretion disks around the newly born neutron star (NS). In (Shishkin & Soker 2022) we connect the properties of the pre-collapse core with the expected final remnant mass in the frame of the jittering jets explosion mechanism.

2. Results

We simulated stellar evolution up to core-collapse using MESA $r10398$. In Figure 1 we show a $M_{\rm ZAMS} = 15M_{\odot}$ model and the appearance of a vigorous inner convective zone at the edge of the iron core at the onset of collapse, with a convective velocity of $v_{conv} \sim 350$ km s^{-1}. Defining the minimal specific angular momentum fluctuations to explode the star as $j_{\rm jje}$ and the average value of specific angular momentum over a shell of at least $0.1M_{\odot}$ as $\bar{j}_{\rm conv,0.1}$, we extract the inner most convective zone to satisfy $M_{\rm rem,B} = m(\bar{j}_{\rm conv,0.1} > j_{\rm jje})$ which is our condition to create an accretion disk/belt. In

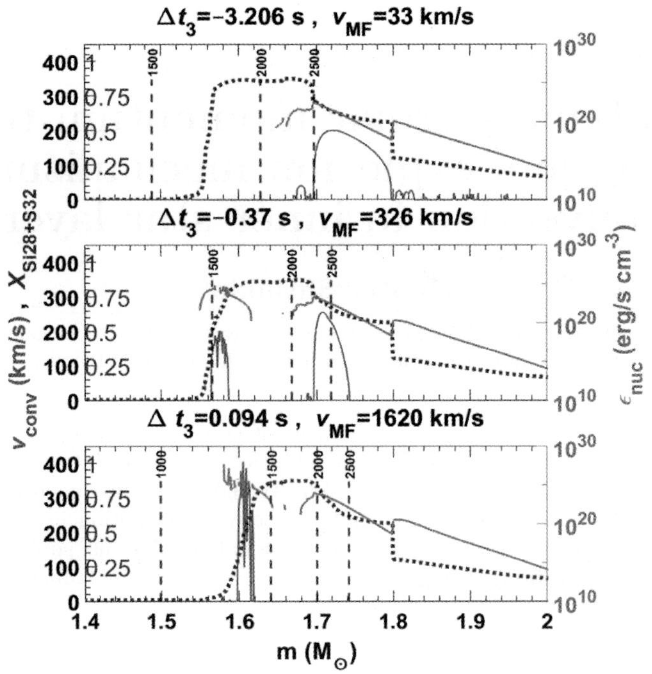

Figure 1. Evolution of the mass layer $m = 1.4 - 2M_\odot$ of a stellar model with $M_{\rm ZAMS} = 15M_\odot$ at three times from early to late core collapse. The time Δt_3 is measured relative to the time when the maximum infall velocity is $v_{\rm MF} = 1000$ km s^{-1}. We present the convective velocity (red solid line; scale on the left outside), the relative mass fraction of the ^{28}Si and ^{32}S combined (dotted gray line; scale on the left inside), and the nuclear power in erg s^{-1}cm^{-3} (solid orange line; scaled on the right). Vertical dashed lines indicate 4 radii as we mark in km. The $r = 1000$km line appears only in the last panel. Note the disappearance of the outer convective zone where oxygen burns and the appearance of a vigorous convective zone at the inner layer where silicon burns. Figure 1 from (Shishkin & Soker 2021).

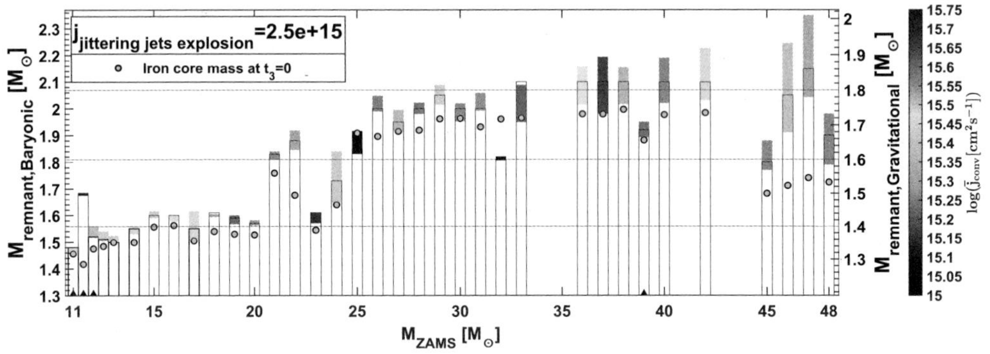

Figure 2. The expected remnant masses in the frame of the jittering jets explosion mechanism for a minimal specific angular momentum fluctuations to cause explosion value of $j_{\rm jje} = 2.5 \times 10^{15}$cm^{2}s^{-1}. The final remnant mass given by the height of the each column. The colored region depicts the $\bar{j}_{\rm conv}$ of the inner most convective zone that obeys our conditions to seed explosion. Figure 1 from (Shishkin & Soker 2022).

Figure 2 we display the expected remnant masses and the respective convective regions with sufficient angular momentum fluctuations to explode the star according to the jittering jets explosion mechanism, for many masses in the range of $M_{\rm ZAMS} = 11 - 48M_\odot$.

3. Conclusion

We find vigorous convective regions at core collapse to have sufficient angular momentum fluctuations to explode the star according to the jittering jets explosion mechanism.

References

Gilkis, A., & Soker, N. 2014, MNRAS, 439, 4011
Papish, O., & Soker, N. 2011, MNRAS, 416, 1697
Schreier, R., & Soker, N. 2016, Research in Astronomy and Astrophysics, 16, 001
Shishkin, D., & Soker, N. 2021, MNRAS: Letters, 508, L43
Shishkin, D., & Soker, N. 2022, MNRAS, 513, 4224
Soker, N. 2019, Research in Astronomy and Astrophysics, 19, 095

Massive Stars Near and Far
Proceedings IAU Symposium No. 361, 2024
J. Mackey, J. S. Vink, & N. St-Louis, eds.
doi:10.1017/S1743921322002034

Observations and modeling of two Type IIP supernovae in M61: similar yet so different

Rishabh Singh Teja[1,2], G. C. Anupama[1] and D. K. Sahu[1]

[1]Indian Institute of Astrophysics, II Block, Koramangala, Bengaluru-560034, Karnataka, India
email: rishabh.teja@iiap.res.in

[2]Pondicherry University, R.V. Nagar, Kalapet, Pondicherry-605014, UT of Puducherry, India

Abstract. We present photometric and spectroscopic studies of two type IIP SNe (SN 2008in and SN 2020jfo), to infer their physical parameters. These SNe exploded in the same host galaxy, M 61 (NGC 4303), at different epochs. SN 2008in was a normal Type IIP event with plateau length of $\sim$ 100 d and $M_V = -16.4 \pm 0.4$ mag, whereas SN 2020jfo was a short plateau ($\sim$ 60 d) event with $M_V = -17.4 \pm 0.4$ mag. Hydrodynamical modeling on these events using the MESA+STELLA framework suggests that the progenitors of both the objects were Red Super Giant (RSG) stars with mass $< 15 M_\odot$, but with different evolutionary history and environment.

Keywords. hydrodynamics, (stars:) supernovae: individual (SN 2008in, SN 2020jfo)

1. Introduction

Massive stars with zero age main sequence (ZAMS) mass of $\gtrsim 8\ M_\odot$ terminate their lives as Core-Collapse Supernovae (CCSNe), leaving neutron stars or black holes as remnants. Depending on their evolutionary history, environment and various physical processes such as mass loss and rotation, they result in various subclasses of CCSNe. In this work, we study two SNe in the same host, M 61 (Fig 1a), falling under the IIP sub-class, yet with different observational properties. SN 2008in had a plateau of $\sim$100 d (Roy et al. 2011), while SN 2020jfo (Teja et al. 2022) was a short plateau (<65 d) event with a fast plateau decline. Based on SDSS archival spectra and the study by Martin & Roy (1992) we find the environment around SN 2008in to be majorly sub-solar whereas for SN 2020jfo it is above solar.

2. Results

We modeled the light-curve and velocity evolution for both SNe using 1-D MESA and STELLA open source codes (Fig 1b, 1c). We estimated and ejecta mass of $\sim$9 $M_\odot$ and $\leq$ 5 $M_\odot$ for SN 2008in and SN 2020jfo respectively. Best fit models indicate both these SNe had RSG progenitor with ZAMS mass lying between $11 - 13\ M_\odot$. SN 2020jfos progenitor went through an extensive mass loss and shredded most of its hydrogen layer whereas, mass loss in the case of SN 2008in was fairly normal. Both the SNe were low energy explosions ($0.2 - 0.4\ foe$) along with a lower velocity evolution.

3. Conclusion

Early part of the pseudo-bolometric light curve for both SNe required an additional CSM component. Although, we could see hint of CSM signatures in the form of ionized He lines for SN 2020jfo (Fig 1f), no such evidence was seen in SN 2008in. But, we find increased flux in individual light curves for SN 2008in (Fig 1e). Models indicate an

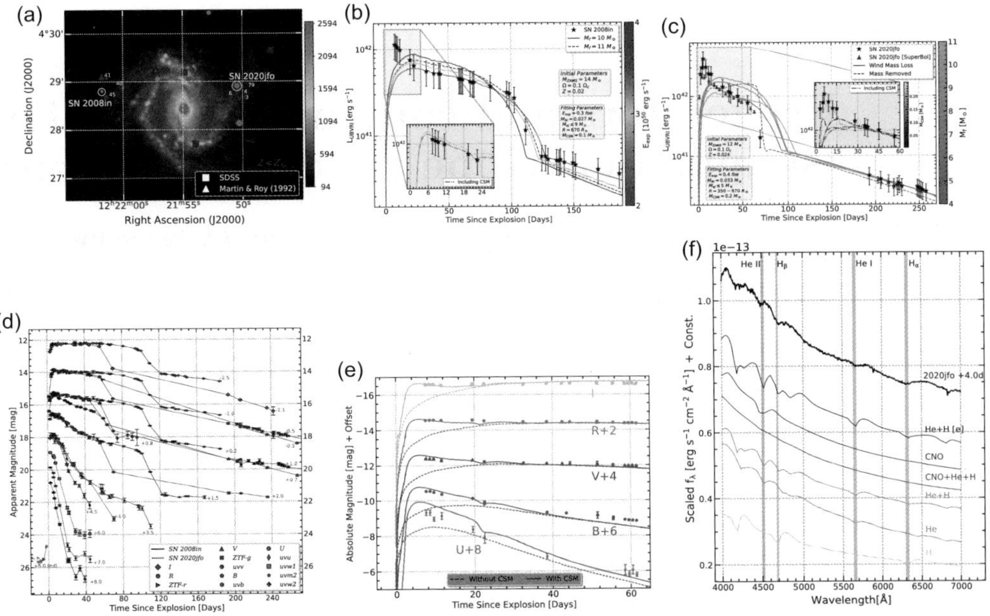

Figure 1. **a)** M 61 galaxy shown with SN 2008in and SN 2020jfo. **b)** MESA+STELLA fits for the quasi-bolometric light curve for SN 2008in. Inset shows the models including CSM. **c)** MESA+STELLA fits for the quasi-bolometric light curves for SN 2020jfo. Inset shows the models including different amount of CSM. **d)** Panchromatic light curves of SN 2008in and SN 2020jfo. **e)** Multiband light curves obtained from STELLA showing effect of CSM for initial 20 days post explosion in SN 2008in. **f)** TARDIS synthetic spectral models for SN 2020jfo to confirm ionized He lines due to nearby CSM.

extensive mass loss in the case of SN 2020jfo that could be due to either high metallicity environment or the presence of a binary companion, or both. Lack of clear spectral signatures indicate either a non-spherical CSM geometry, or that the initial bump is intrinsic to Type IIP SNe and requires further understanding. Detailed modeling and observations are required to discern signatures of a binary companion on the observables. Revisiting SNe sites, detailed 3-D modeling, constraining progenitor properties from observations, etc. would provide better understanding of such events.

Acknowledgements

We thank Avinash Singh, Brajesh Kumar and Nayana AJ, the co-authors of Teja et al. (2022) paper on SN2020jfo. Part of the data on SN 2008in is taken from the thesis of Uday Kumar Gurugubelli (2010)†.

References

Martin, P., & Roy, J.-R. 1992, ApJ, 397, 463
Roy, R., et al. 2011, ApJ, 736, 76
Teja, R. S., Singh, A., Sahu, D. K., Anupama, G. C., Kumar, B., & A. J., N. 2022, ApJ, 930, 34

† http://prints.iiap.res.in/handle/2248/6664

Massive Stars Near and Far
Proceedings IAU Symposium No. 361, 2024
J. Mackey, J. S. Vink, & N. St-Louis, eds.
doi:10.1017/S1743921322003362

Progenitors and companions of stripped-envelope supernovae

E. Zapartas[1], O. Fox[2], T. Fragos[3,4], M. Renzo[5], J. Andrews[6], S. S. Bavera[3,4], A. Dotter[6], M. Drout[7,8], S. van Dyk[9], D. Milisavljevic[10,11], K. Kovlakas[3], D. Misra[3], K. Rocha[5], N. Smith[12], B. Williams[13] and Z. Xing[3,4]

[1]IAASARS, National Observatory of Athens, Penteli, Greece.
email: ezapartas@gmail.com

[2]Space Telescope Science Institute, Baltimore, USA.

[3]Department of Astronomy, University of Geneva, Versoix, Switzerland.

[4]Gravitational Wave Science Center (GWSC), Universite de Geneve, Switzerland

[5]Center for Computational Astrophysics, Flatiron Institute, NY, USA.

[6]Center for Interdisciplinary Exploration and Research in Astrophysics (CIERA) and Department of Physics and Astronomy, Northwestern University, Evanston, IL, USA.

[7]David A. Dunlap Department of Astronomy and Astrophysics, University of Toronto, Canada.

[8]Observatories of the Carnegie Institution for Science, Pasadena, CA, USA.

[9]Caltech/IPAC, Pasadena, CA, USA.

[10]Department of Physics and Astronomy, Purdue University, West Lafayette, IN,USA.

[11]Integrative Data Science Initiative, West Lafayette, IN, USA.

[12]Steward Observatory, University of Arizona, Tucson, AZ, USA.

[13]University of Washington, Seattle, WA, USA

Abstract. In this poster, using the POSYDON code, we present results on binary progenitors of stripped-envelope SNe and their companions. We find that most progenitors are expected to explode, according to typical SN prescriptions (in contrast to single star progenitors). We also show the expected masses and position in the HR diagram of the companions of these SNe at the moment of explosion, allowing us to do a first statistical comparison with the compiled sample of observational detections (or upper limits) on these companions.

Keywords. stripped-envelope supernovae, massive stars, binary stars

---

Stripped-envelope supernovae (SNe Ib, Ic and IIb mainly) have stellar progenitors with (almost) no hydrogen-rich envelope. In Zapartas et al. (2021) we estimate the explodability of massive, single-star MESA models (e.g., Paxton et al. 2011) of solar metallicity that reach core carbon exhaustion, simulated with POSYDON (Fragos et al. 2022)that get stripped due to their winds, according to various core-collapse SN prescriptions. Overall, we find that most Wolf-Rayet progenitors of stripped-envelope SNe are not expected to explode, with higher rates of successful Type IIb SNe. Increased mass loss rates are found in principle to work in favor of successful hydrogen-poor SNe, but they are not supported by observations (e.g., Smith 2014).

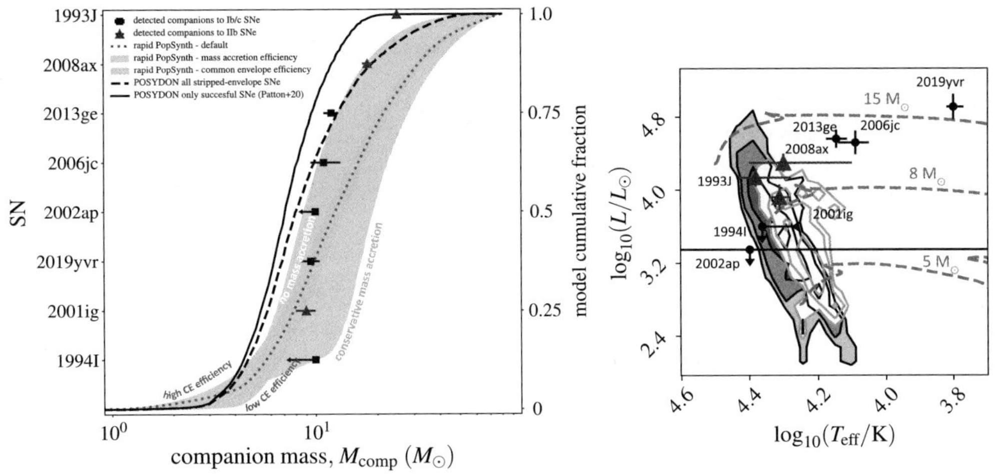

Figure 1. Left: Observed cumulative distribution of the detected mass of binary companions to stripped envelope SNe (squares for Ib/c, triangles for IIb), compared with model predictions from binary_c (dotted line and shadowed regions) and POSYDON (black solid and dashed line). Right: Hertzsprung Russel diagram of companions detections (points) compared to POSYDON predictions (blue contours for Ib/c, orange for IIb).

Stripping due to a binary companion is an alternative scenario for these SNe. In a binary population synthesis run with POSYDON (of only solar metallicity for now), we find an increased rate of successful Ib/c SNe (e.g. Laplace et al. 2021) of 81% (vs 1% for single stars) and 95% (vs 21% for single stars) according to Patton & Sukhbold (2020), which is the most pessimistic SN prescription in Zapartas et al. (2021). According to our models, all binary Ib/c SNe are in practice Ib, not Ic, having high surface helium abundance (e.g., Aguilera-Dena et al. 2022).

Observations of possible binary companions that stripped the SN progenitor provide extra information about the progenitor systems. The observed sample (e.g., Maund et al. 2004, Van Dyk et al. 2016, Fox et al. 2022) has reached a point where we can start looking at it from a population perspective. In the left panel of Fig. 1 we show the cumulative distribution of the companion mass at the moment of the stripped-envelope explosion and compare them with predictions from rapid population synthesis (Zapartas et al. 2017) and detailed binary populations with POSYDON. POSYDON's assumed mass-accretion cap due to spin-up of the accretor close to critical rotation, leads to low mass expected companions, consistent with those from rapid codes when no mass accretion is assumed. Observed distribution (including 2019yvr Ibn YHG companion and 2002ap Ic-BL's companion) shows instead a very sharp rise at around 10 $M_\odot$, with no broad tails on either side.

According to POSYDON, binary companions are expected to predominantly reside at their main sequence (MS, contours in the right panel of Fig 1), consistent also with Zapartas et al. 2017. Three detections of companions to Ib/c SNe are cooler than expected for MS stars. These could be stars evolved off-MS (which is a rare model prediction) or, in case of post-explosion detections, they can be companions that interacted with the SN ejecta (Hirai et al. 2018). Type IIb SNe companions are among the most massive observed (red triangles) and reside in the high luminosity regime of the predicted bulk of MS companions.

References

Aguilera-Dena, D. R., et al. 2022, arXiv e-prints, arXiv:2204.00025
Fox, O. D., et al. 2022, ApJL, 929, L15
Fragos, T., et al. 2022, arXiv e-prints, arXiv:2202.05892
Hirai, R., Podsiadlowski, P., & Yamada, S. 2018, ApJ, 864, 119
Laplace, E., et al. 2021, A&A, 656, A58
Maund, J. R., Smartt, S. J., Kudritzki, R. P., Podsiadlowski, P., & Gilmore, G. F. 2004, Nature, 427, 129
Patton, R. A., & Sukhbold, T. 2020, MNRAS, 499, 2803
Paxton, B., et al. 2011, ApJS, 192, 3
Smith, N. 2014, ARA&A, 52, 487
Van Dyk, S. D., de Mink, S. E., & Zapartas, E. 2016, ApJ, 818, 75
Zapartas, E., et al. 2017, ApJ, 842, 125
—. 2021, A&A, 656, L19

Massive Stars Near and Far
Proceedings IAU Symposium No. 361, 2024
J. Mackey, J. S. Vink, & N. St-Louis, eds.
doi:10.1017/S1743921322002897

Numerical modeling of IR SEDs of dusty CCSN within a Bayesian framework

Szanna Zsíros[1], Ilse De Looze[2,3] and Tamás Szalai[1,4,5]

[1]Dept. of Experimental Physics, Institute of Physics, University of Szeged, H-6720 Szeged, Dóm tér 9, Hungary
email: szannazsiros@titan.physx.u-szeged.hu

[2]Sterrenkundig Observatorium, Ghent University, Krijgslaan 281-S9, 9000 Gent, Belgium

[3]Dept. of Physics & Astronomy, University College London, Gower Street, London WC1E 6BT, United Kingdom

[4]Konkoly Observatory, Research Centre for Astronomy and Earth Sciences, H-1121 Budapest, Konkoly Thege Miklós út 15-17, Hungary

[5]MTA-ELTE Lendület Milky Way Research Group, Hungary

Abstract. We investigate the physical properties of dust in the environment of three core-collapse supernovae (CCSNe) through mid-infrared (mid-IR) spectral energy distribution (SED) modeling (both analytical and numerical methods) and interpret our results within a Bayesian framework. We provide evidence that the observed late-time mid-IR excess of the SNe can be described by dust models. We conclude that in case of various types of SNe, numerical dust models with a shell-like geometry can be reconciled with analytical models, regarding the essential properties of dust grains.

Keywords. supernovae: individual: SN 1980K, SN1993J, SN1996cr – infrared: stars

1. Introduction

Cosmic dust is one of the essential building blocks of the Universe. It originates from various sources, moreover, both theoretical expectations and observational results advocate that CCSNe are one of the main stellar sources of cosmic dust (for a review see e.g. Sarangi *et al.* 2018). Dust grains typically re-emit in the mid-IR and/or in the submilimeter regime. Thus, the mid-IR analysis of CCSNe provides an exceptional opportunity to reveal precious information about SN dust (e.g. Szalai *et al.* 2019).

We investigated the late-time mid-IR Spitzer data of SN 1980K (Sugerman *et al.* 2012), SN 1993J (Zsíros *et al.* 2022) and SN 1996cr (unpublished data). In order to have a fiducial understanding of the assumed dust properties, we fit a widely used analytical dust model adopted from Hildebrand (1983). The model describes the thermal emission of dust grains assuming a modified blackbody radiation from optically thin dust. We assumed a single dust composition – either amorphous carbon or silicate-type grains – in the models to fit the dust mass and temperatures (Zsíros *et al.* 2022).

We computed numerical models using the MOCASSIN (Monte Carlo SimulationS of Ionized Nebulae) radiative transfer code (Ercolano *et al.* 2003; 2005; 2007). The code follows the possible light-matter interactions and describes the re-emission by dust grains. We used a geometrical set-up with a central point source or a diffuse radiation field and a dust shell with a smooth density distribution. We applied a mixture of amorphous carbon and silicate dust.

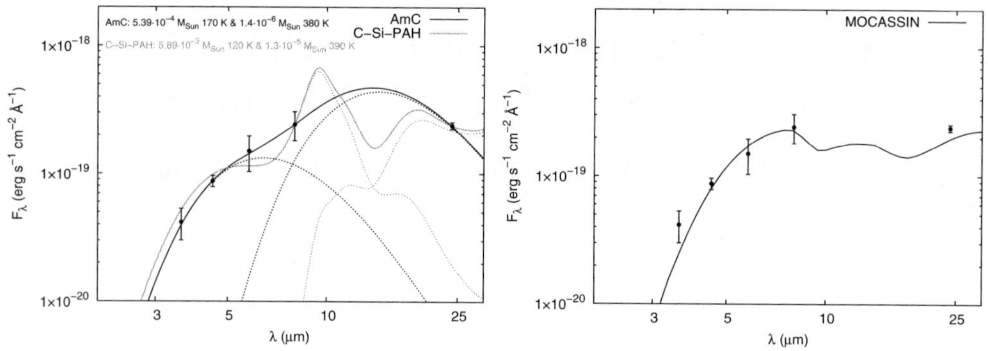

Figure 1. Our best-fit analytical dust models (left) with a numerical dust model (right) on mid-IR SED of SN 1980K (Sugerman *et al.* 2012).

Nevertheless, while numerical models provide more relevant representation of the possible physical background, they require a larger number of free parameters. Hence, we focused on the crucial parameters of the models. To have a deeper insight of the parameters, we interpreted our numerical models with a Bayesian inference method linked to a Markov Chain Monte Carlo (MCMC) algorithm (e.g. Bevan 2018, De Looze *et al.* 2019). For sampling, the method applies an affine invariant ensemble sampler (Goodman & Weare (2010)) implemented by an MCMC algorithm through the "emcee" package in *Python* (Foreman-Mackey *et al.* 2013). The sampling is executed by random walks of a selection of walkers, the posterior probabilities are drawn from the positions of the walkers, while the parameter ranges of the priors were determined from the respective numerical models.

We found that the observed late-time mid-IR emission can be described with dust models (see Fig. 1), thus supporting the presence of significant amounts of dust (in case of silicate dust: $\sim (1.4 - 5.9) \cdot 10^{-3} M_{\odot}$) in the vicinity of all three SNe in accordance with previous studies (e.g. Niculescu-Duvaz *et al.* (2022)).

Acknowledgements

This project has been supported by NKFIH/OTKA FK-134432 grant. S.Z. is supported by the National Talent Programme under NTP-NFTÖ-21-B-0262 Grant. I.D.L. acknowledge support from ERC starting grant #851622 DustOrigin. T.S. is supported by the János Bolyai Research Scholarship of the Hungarian Academy of Sciences and by the New National Excellence Program (UNKP-21-3) of the Ministry for Innovation and Technology from the source of the National Research, Development and Innovation Fund.

References

Bevan, A. 2018, *MNRAS*, 480, 4659

De Looze, I., Barlow, M. J., Bandiera, R., et al. 2019, *MNRAS*, 488, 164

Ercolano, B., Barlow, M. J., Storey, P. J., et al. 2003, *MNRAS*, 340, 1136

Ercolano, B., Barlow, M. J., & Storey, P. J. 2005, *MNRAS*, 362, 1038

Ercolano, B., Barlow, M. J., & Sugerman, B. E. K. 2007, *MNRAS*, 375, 753

Foreman-Mackey, D., Hogg, D. W., Lang, D., et al. 2013, *PASP*, 125, 306

Goodman, J. & Weare, J. 2010, Communications in Applied Mathematics and Computational Science, 5, 65

Hildebrand, R. H. 1983, *QJRAS*, 24, 267
Niculescu-Duvaz, M., Barlow, M. J., Bevan, A., et al. 2022, arXiv:2204.14179
Sarangi, A., Matsuura, M., & Micelotta, E. R. 2018, *Space Sci. Revs*, 214, 63
Sugerman, B. E. K., Andrews, J. E., Barlow, M. J., et al. 2012, *ApJ*, 749, 170
Szalai, T., Zsíros, S., Fox, O. D., et al. 2019, *ApJ*, 241, 38
Zsíros, S., Nagy, A. P., & Szalai, T. 2022, *MNRAS*, 509, 3235

Chapter XI:
Future instrumentation and facilities

Massive Stars Near and Far
Proceedings IAU Symposium No. 361, 2024
J. Mackey, J. S. Vink, & N. St-Louis, eds.
doi:10.1017/S1743921322003416

Ultraviolet observations of massive stars and the instrumentation to come

Ana I. Gómez de Castro[1,2]

[1]Joint Center for Ultraviolet Astronomy, Universidad Complutense de Madrid, Avda. Puerta de Hierro s/n, 28040 Madrid
email: aig@ucm.es

[2]S.D. Física de la Tierra y Astrofísica, Fac. CC Matemáticas, Plaza de Ciencias 3, 28040 Madrid

Abstract. The wealth of information obtained about massive stars from observations at ultraviolet (UV) wavelengths has been fundamental to understand their structure and the mechanisms regulating their evolution. There are however, important aspects that have not yet been addressed due to the lack of data and the relevant instrumentation such as the role of binarity and magnetic fields or the impact of low metallicity in the evolution of massive stars. There are plans to develop UV spectropolarimeters, UV monitors and very efficient telescopes with high collecting surfaces that will revolutionize the field. In this contribution, a short update on the current and foreseen UV instrumentation is provided.

Keywords. Massive stars, ultraviolet observations, ultraviolet instrumentation.

1. Introduction

The peak of the spectral energy distribution of the OBA stars is at ultraviolet (UV) wavelengths thus, massive stars have been the preferred targets and the most extensively observed sources since the very beginning of UV astronomy. Early UV telescopes such as the Carruther's lunar telescope (Carruthers & Page 1972) or the Copernicus (OAO-3, Rogerson et al. 1973) observatory had massive stars among their favourite targets and opened a fundamental window to study radiatively driven winds (Abbott 1978) as well as to evaluate their impact in the evolution of the interstellar medium (ISM).

The full understanding of the physics involved in the acceleration of the winds came later, with the systematic, high dispersion spectroscopic observations obtained with the International Ultraviolet Explorer (IUE, Boggess et al. 1978), the Hubble Space Telescope (HST, `www.stsci.edu/hst`) and the Far UV Spectroscopic Explorer (FUSE, Moos et al. 2000). Main results were the complete spectral characterization of massive stars, the understanding of line-driven winds, the determination of the relevance of the metallicity in the wind properties and terminal velocity, or the detection of clumps in the wind. The main diagnostic tools used for these works were the P-Cygni profiles produced by the resonance transitions of CIV, NV, OVI and the lines of the iron group which cover a broad range of ionization stages and enable the determination of the metallicity and other stellar properties such as the effective temperature (Najarro et al. 2006 and references therein). Later on, the GALaxy Evolution eXplorer (GALEX, Martin et al. 2005) mapped the extended star forming disks of spiral galaxies (Thilker et al. 2007) and provided a unique view of massive star formation triggering and the role of large scale galactic stresses on it. HST's high sensitivity allowed the study of massive stars in nearby galaxies and a deeper understanding of the role of metallicity in OBA stars physics and evolution (see i.e.

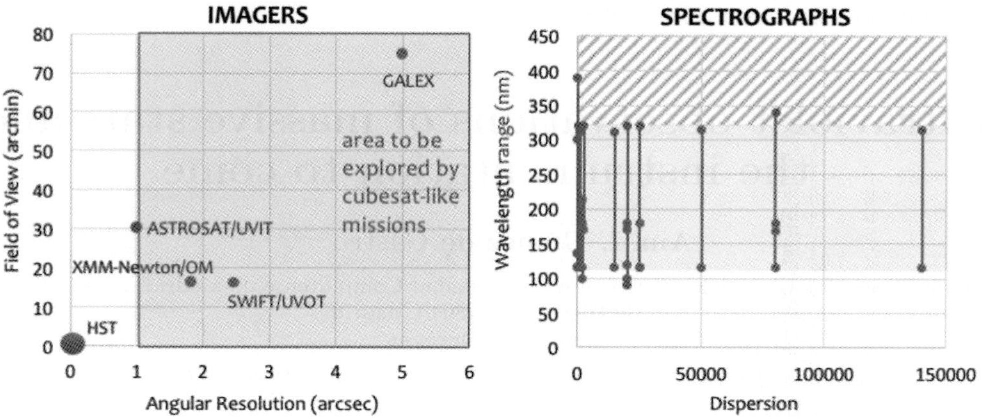

Figure 1. Main characteristics of the UV instrumentation already flown. *Left)* Imagers: the wide field of view and small angular resolution area, suitable to be explored for cubesat-like missions in the future is shadowed in yellow. *Right)* Spectrographs

Garcia et al. 2017, 2019). The impact of UV observations in massive star studies is well addressed in this volume, including interacting binaries and cataclysmic variables (see, also the publications of the network of ultraviolet astronomy, nuva.eu/nuva-publications). These studies are also fundamental to understand the star formation history of the Universe and characterize starbursts in galaxies.

The space missions being defined in the 2020's, seek raising this knowledge through the development of large space telescopes, such as the Large UV Optical Infrared Surveyor (LUVOIR, `asd.gsfc.nasa.gov/luvoir/`) or the European UV Visible Observatory (Gómez de Castro et al. 2022a). There is also, a significant on-going effort to carry out spectropolarimetric measurements to measure OBA stars magnetic fields and to investigate the circumstellar environment. In this contribution, I will describe the current status of UV facilities and the plans being outlined for future missions.

2. Characteristics of the UV instrumentation already flown

The properties of the UV instrumentation already flown are summarized in Figure 1. Very few imaging instruments have been developed and two of them are small telescopes flown together with high energy missions such as UVOT in SWIFT or OM in XMM-Newton to detect the UV counterparts to high energy sources. Both SWIFT and XMM-Newton have a field of view ≤ 20 arcmin and an angular resolutions of ~ 2". There has always been an UV imager on board HST, providing very high angular resolution and a broad range of UV filters (see Figure 1 in Gómez de Castro et al. 2022a). On the contrary, GALEX has provided a very wide field of view and low angular resolution.

All the imagers to date have been flown into Low Earth Orbit (LEO) and thus, observations at some specific wavelengths are hampered by the Earth geocoronal emission. As shown in Figure 2, the geocoronal emission is especially strong in the Lyman-α (Lyα) transition of Hydrogen and in several Oxygen transitions (resonance transition of OI at 130.4 nm, semiforbidden trransition of OI] at 135.6 nm and forbidden transition of [OII] at 247.1 nm). To avoid the impact of the airglow, some of the UV surveyors, such as GALEX, have avoided imaging the UV sky below 135 nm.

The effect of geocoronal emission can be removed in spectroscopic observations and, as shown in Figure 1, flown spectrographs have operated in the full UV range (from 90 to 350 nm). All sorts of dispersions have been utilized; from the very low dispersions achieved with prisms (*e.g.* the prism PR110L in the ACS instrument on board HST

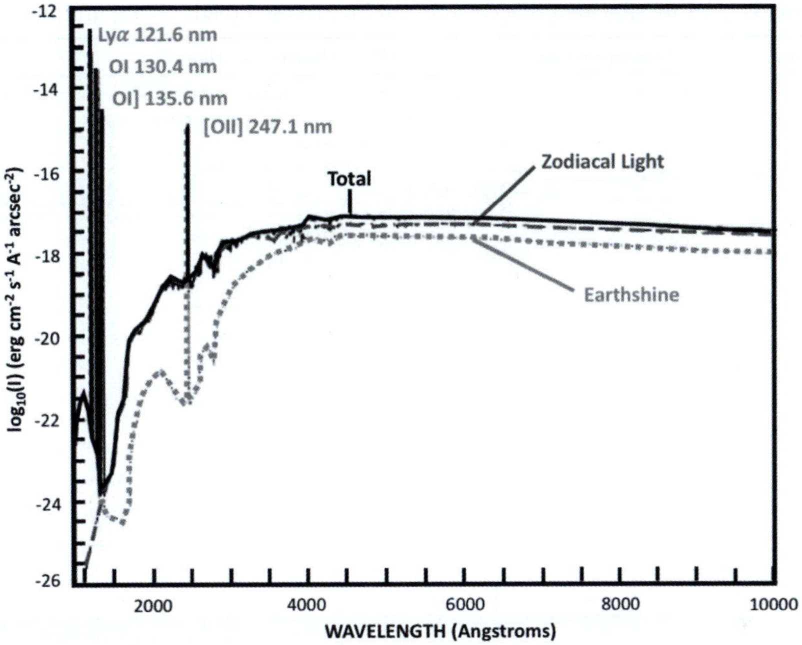

Figure 2. Sky background intensity. The zodiacal contribution corresponds to $m_V = 22.1$ arcsec^{-2}. The earthshine is for a target which is 38° from the limb of the sunlit Earth. The geocoronal airglow line intensities are plotted at "average" intensities (`hst-docs.stsci.edu/stisihb/chapter-6-exposure-time-calculations/6-5-detector-and-sky-backgrounds`).

provides dispersions as low as 80 at 150 nm) to the very high dispersion provided by the HST/STIS, echelle mode (Kimble et al. 1998).

Unfortunately, the only information available on UV spectropolarimetry was achieved with the Faint Objects Spectrograph on HST that was decommissioned early in the project to allocate the COSTAR. Also, efficient mid-dispersion (R∼ 20,000 - 30,000) spectrographs with simultaneous coverage of the 115 nm - 315 nm range have not been flown yet.

3. UV missions planning

In recent years, many proposals to build and operate UV telescopes have been submitted to the space agencies world-wide. Adding to the scientific interest of this range, is the perception that the end of the 32 years old HST mission is nearing. A summary-sketch with the space missions and the proposals being discussed at this time is outlined in Figure 3. The main characteristics of the missions to be flown in the coming 5 years are summarized in Table 1. Looking further ahead in the future, there are many projects and plans; they go from dedicated spectropolarimetric missions (PolStars, Arago, URIEL) to UV surveyors (CASTOR, INSIST, MESSIER) and a plethora of small, proposals (many cubesat like) designed as technology demonstrators (CASSTOR) or devoted to single-science cases (ESCAPE), see Gómez de Castro et al. 2021 for a description of these projects. All this activity is also fundamental to develop the scientific and technological knowledge that will pave the way to future, very ambitious missions such as LUVOIR, a 10-m class UV-optical-infrared telescope, or the European Ultraviolet-Visible Observatory (EUVO).

Among the coming missions, Spektr-UF/WSO-UV stands out by its versatility and its scientific program open to the world-wide community.

Table 1. List of approved missions (excluding those already flying in June 2022).

Project Name	Main operator	Main characteristics	Status
Xuntian	CNSA	UV-optical surveyor with FoV=1.1^{o}x1.1^o and Ang. Res.= 0.15?. NUV filter covers the 255nm-320nm range. Diameter of the primary: 200 cm Orbit: LEO	Launch end 2022
SPARCS	NASA	Dedicated cubesat mission or exoplanetary research. Imaging with filters in the 115nm-320nm spectral range. Diameter of the primary: 9 cm Orbit: LEO/Sun synchronous	Launch end 2023
Spektr-UF/WSO-UV	ROSCOSMOS	UV observatory operating in the with instrumentation for imaging and spectroscopy. Diameter of the primary: 170 cm Orbit: HEO/Geosynchronous	Launch 2025/26
ULTRASAT	ISA	Wide field (204 deg^2) UV transients monitor operating in the 230nm-290 nm range. Diameter of the primary: 33 cm Orbit: HEO/Geosynchronous	Launch 2025/26

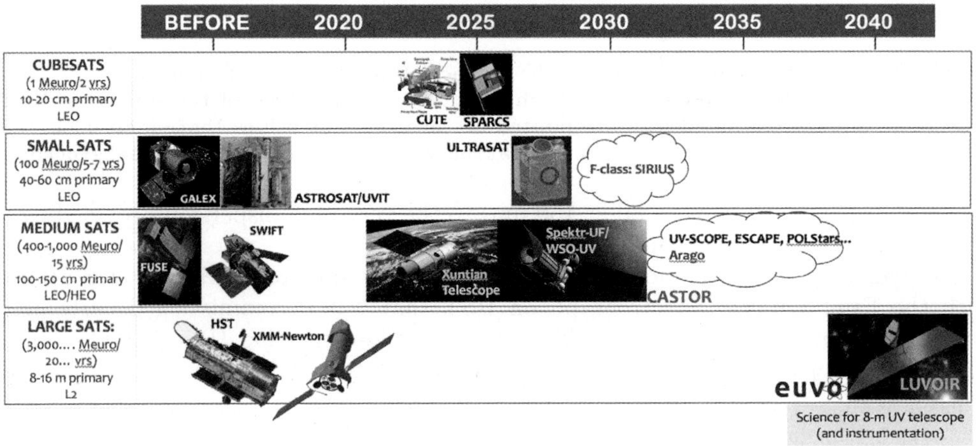

Figure 3. UV missions: on-going, programmed and proposed. The list is not exhaustive, especially at the low mass end.

3.1. *Spektr-UF/WSO-UV*

The World Space Observatory - Ultraviolet mission (Spektr-UF/WSO-UV) is a multipurpose space observatory to study the Universe in the 115 nm - 310 nm range using high- and low- resolution spectroscopy, high sensitivity imaging and slitless spectroscopy (Shustov et al. 2022). The project is included in the Russian Federal Space Program 2016-2025; it is the third mission of the Spektr series to be flown after Spektr-R (2011-2019) and Spektr-RG (launched in 2019, in operation). WSO-UV will operate from a geosynchronous orbit allowing for long (up to 10 h), uninterrupted observations and enjoying a dark UV sky. The scientific operation of WSO-UV is managed through three fundamental programs: the core program, the guaranteed time to the funding bodies of the project and the international program. With the launch coming near 2025, the calls for the various programs will be issued in upcoming years.

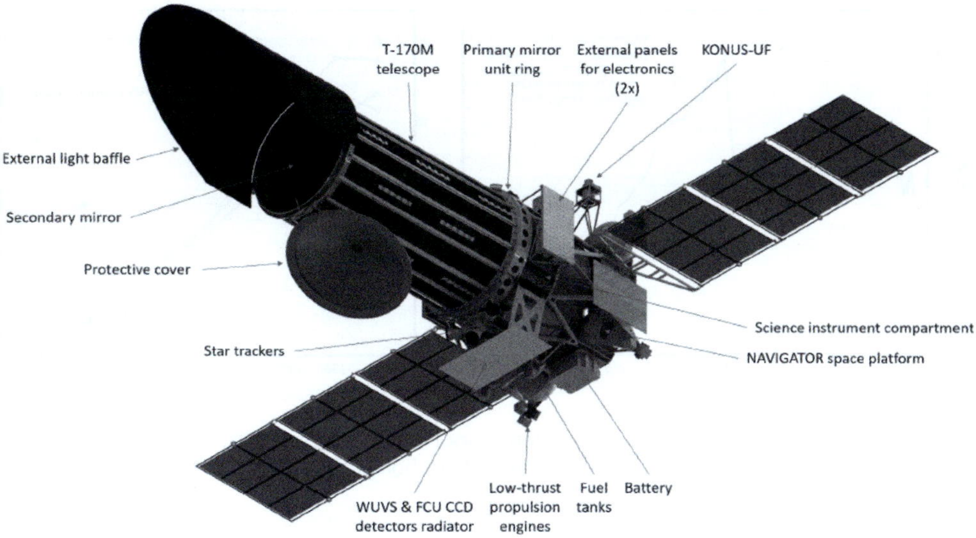

Figure 4. Spektr-UF/WSO-UV spacecraft.

The WSO-UV scientific payload includes the T-170M telescope, a 170 cm aperture telescope with a Ritchey-Chretien mounting, and the instrumentation: WUVS (WSO-UV Spectrograph), FCU (Field Camera Unit) and UVSPEX (see Fig. 4).

The WUVS includes three channels:

• VUVES channel providing high resolution (R $\sim$ 50,000) echelle spectra in the 115-176 nm spectral range.

• UVES channel providing high resolution (R $\sim$ 50,000) echelle spectra in the 174-305 nm spectral range.

• LSS (long-slit spectrograph) channel providing low resolution (R $\sim$ 1,000) spectra in the 115-310 nm spectral range.

Each of the three WUVS channels has an almost identical custom detector consisting of a custom CCD272-64 inside a vacuum enclosure. The main challenges of the WUVS detectors are to achieve high quantum efficiency in the FUV-NUV range, to provide low readout noise (≤ 3 e$^-$ at 50 kHz) and low dark current (≤ 12 e$^-$/pixel/hour), to operate with integral exposures of up to 10 hours and provide good photometric accuracy. Teledyne e2v designed three variants of a custom CCD272-64 sensor with different UV AR coatings, optimized for each WUVS channel. WUVS is being developed under the responsability of the Institute of Astronomy of the Russian Academy of Sciences (INASAN).

The FCU consists of two channels:

• The Far UV channel: provides an angular resolution of 0.1 arcsec over a field of view of 2$\times$ arcmin^2. The channel is equipped with an MCP detector using CsI as photosensitive substrate, *i.e.*, the detector is solar blind and sensitive to the 115 nm - 176nm spectral range. The channel is equipped with a set of step filters (similar to those available in the SBC channel of the HST/ACS instrument) for spectral band selection and, two prisms to carry out slit-less spectroscopy. PR122 has been designed to provide resolution, R$\sim$ 600, in Lyα and R$\sim$ 500, in the nearby OI resonance transition (130.4 nm) to detect transiting hot Jupiters and super-Earths. PR155 provides a resolution, R$\sim$ 600, in the

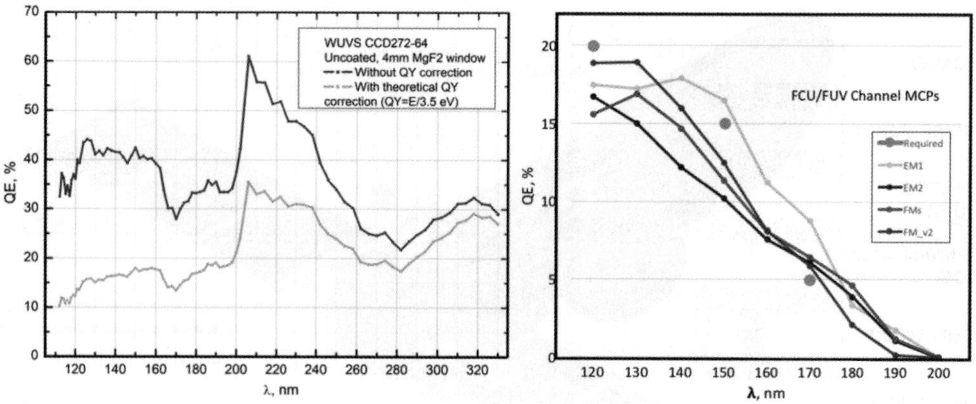

Figure 5. Quantum efficiencies of the detectors (engineering models, and also MCP flight models) used in the FCU instrument on board Spektr-UF/WSO-UV.

CIV resonance transition (155 nm) to measure terminal velocities of massive stars in nearby galaxies.

• Near UV channel: provides an angular resolution of 0.15 arcmin over a field of view of 7.5× arcmin^2. The detector is a Teledyne e2v CCD and the target spectral coverage is 174-305 nm. The CCD sensitivity goes up to 1000 nm thus, the channel is equipped with filters for UV, but also optical observations.

The quantum efficiency of the detectors measured in the engineering and flight models is shown in Figure 5. The FCU is a joint instrument developed by INASAN in collaboration with the Universidad Complutense de Madrid (UCM).

In 2020-2021, an additional instrument UVSPEX was incorporated to the project. UVSPEX is a UV spectrograph optimized to detect Oxygen in the exosphere of Earth-like exoplanets. UVSPEX will operate only in the 115-135 nm range with a spectral resolution of $\sim$ 10,000 to resolve the OI blend at 130.4 nm. UVSPEX is being developed by the Rikkyo University (Tokyo) and the Space Research Institute of the Russian Academy of Sciences (IKI).

Official, updated information on the project status, instrumentation and scientific calls can be found on the servers of the project (www.inasan.ru/en/sw/wso-uv/ and jcuva.ucm.es).

4. Seeking for a common framework for UV photometry

The advent of widely spread cubesat technology during the next decade is going to change space astronomy. Instead of a few large missions, many small missions dedicated to specific scientific purposes, surveys and time-domain astrophysics will be flown. It is in the best interest of the scientific community to define some basic standards to enable easy data comparison and test the reproducibility of the observations. The International Astronomical Union (IAU) recommends the implementation of some well defined bands which are summarized in Figure 6 and Table 2.

The main considerations for the selection of the bands in the system have been: a smooth integration of the dataset already available, such as the GALEX survey, a good coverage of the spectral range that avoids the Lyα geocoronal emission (since most of the cubesats will fly in low Earth orbits which is the most affected by geocoronal emission) and, to enable the study of the UV bump at 225 nm in the extinction curve. The UV bump is produced by small carbonaceous particles and depends on the line of sight.

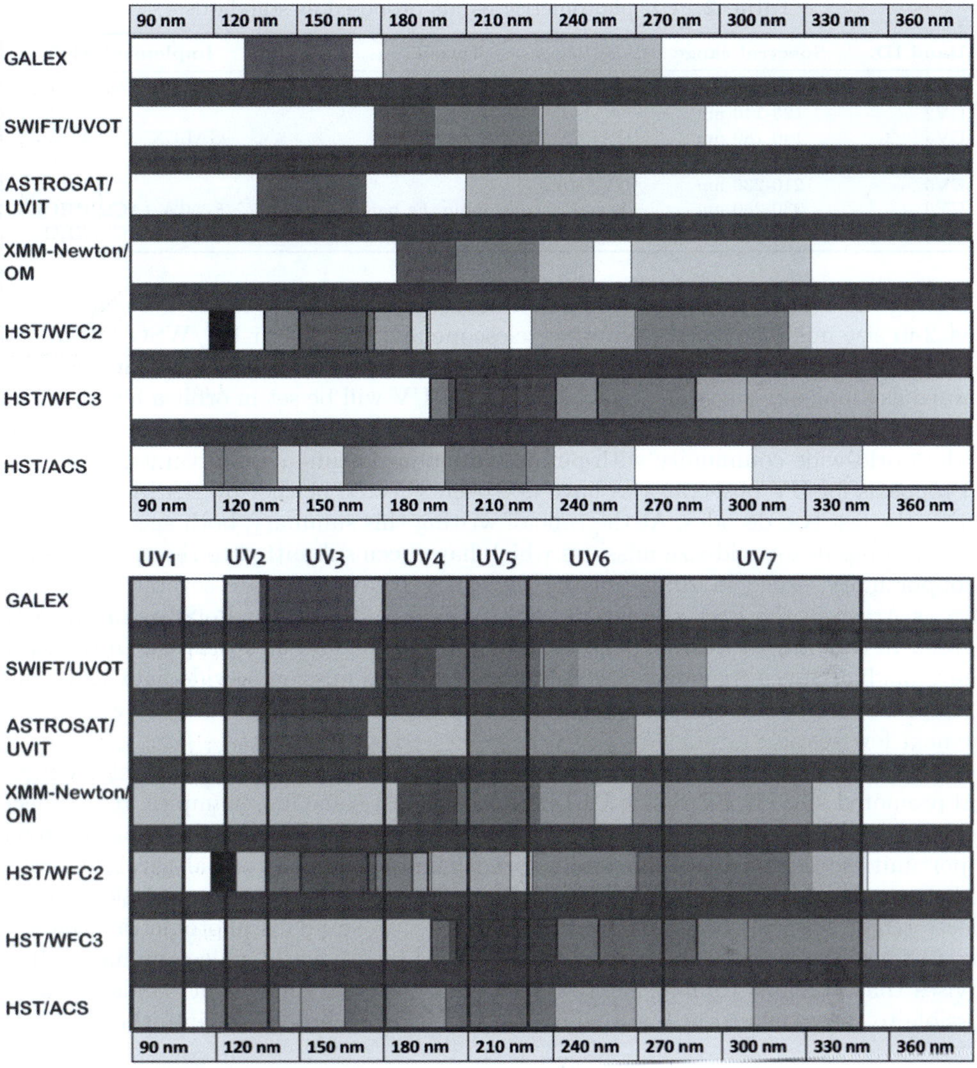

Figure 6. *Top:* Sketch outlining the spectral coverage of the UV filters (or bands) used for imaging purposes in astronomy (GALEX, HST/ACS, HST/STIS, XMM-Newton/OM, SWIFT/UVOT) from Gómez de Castro et al. 2022b. Note that the final transmittance also depends on the response of the detector. This is particularly relevant for the MCP type detectors often used in UV Astronomy since, they use photocathodes only sensitive to specific spectral ranges. We refer the reader to the Instruments manuals and handbooks for more details on the filters (precise transmittance curves, effective wavelengths, etc.). Note that, for instance, several ACS filters are step filters that cannot be adequately represented in this plot. Also, filters are many and overlap high in the HST instruments. *Bottom:* the windows of the recommended IAU filters (see Table 2), over-plotted on the filters sketch (grey boxes).

5. Summary

There are fundamental aspects of astrophysics requiring a deep understanding of massive stars, the relevance of binarity in their evolution and the role that magnetic fields and metallicity play in it. For this purpose, new instrumentation is required to measure magnetic fields (spectropolarimetric instruments) and study massive stars in low metallicity environments (*i.e.* increasing the effective area of the space telescopes).

Table 2. UV photometric bands proposed as standards.

Band ID.	Spectral range	Target	Implementation
UV1	90-110 nm		
UV2	125-140 nm		
UV3	140-180 nm	GALEX FUV	GALEX
UV4	180-210 nm	Continuum shortward of the UV bump	
UV5	210-230 nm	UV bump	
UV6	230-280 nm	Near UV continuum, Fe bands	F250W (ACS/HRC)
UV7	280-350 nm	Ozone cut-off window	F330W (ACS/HRC)

During the coming years, the lead in UV astronomy will move to eastern countries; the next 2-m size missions are the Xuntian telescope and the Spektr-UF/WSO-UV observatory. Xuntian will be launched in 2022-23 and operated by the Chinese Space Agency mainly, as an all-sky surveyor. Spektr-UF/WSO-UV will be set in orbit a few years later (2025-26) and will be operated by the Russian Space Agency as a space observatory open to the world-wide community with periodic announcements of opportunity.

Proposals for UV observatories have not been selected neither by NASA, nor by ESA during the last two decades. At the time of writing this summary, both agencies are evaluating proposals for mid-size missions which have been submitted to the announcements of opportunity issued in 2021. On the table, for study, there are several UV missions such as Arago in the ESA system and ESCAPE, UV SCOPE or PolStars in the NASA system. In case any of them is selected, it will fly in the mid 30's thus, far in the future. The Canadian Space Agency has selected a UV-optical mission, a wide-field surveyor, to be implemented in the coming 15 years; the consortium to run the project will be set in the next few years.

In the meantime, many small missions, including some cubesat like, are being designed and promoted: CUTE, SPARCS, ULTRASAT are just some few examples. It is expected that this number grows significantly in the coming years due to the rapidly growing opportunities to fly small satellites into space. The IAU is concerned about data curation and sharing in this context. For this reason, it recommends that those satellites running spectroscopic observations post-process the data into standard photometric bands (see Fig. 6) to feed a basic database that can be shared by the world-wide community. It also advises that missions running photometric surveys try to implement bands as close as possible to those defined in Figure 6 or transform their photometry into these bands.

Acknowledgements

This work has been funded by the Spanish Ministry of Research and Innovation through grant PID2020-116726RB-I00.

References

Abbott, D.C., 1978, *ApJ*, 225, 893

Boggess, A., Carr, F. A., Evans, D. C., et al. 1978, *Nature*, 275, 372

Carruthers, G. R., Page, T. 1972, *Science*, 177, 788

Garcia M., Herrero, A., Najarro, F., et al. 2017, *Proc IAU Symp*, 329, 313

Garcia M. Herrero A., Najarro F., Camacho I., et al. 2019, *MNRAS*, 484. 422

Gómez de Castro, A.I., Barstow, M.A., Brosch, N., et al. 2021, *UV Astronomy and the quest for the origin of life*, Elsevier, ISBN:978-0-12-819170-5

Gómez de Castro, A.I., Barstow, M.A., Brosch, N., et al. 2022a, *ExpA*, 44G, DOI: 10.1007/s10686-022-09854-9

Gómez de Castro, A.I., Brosch, N., Bettoni, D. et al. 2022a, *ExpA*, submitted

Kimble, R.A., Woodgate, B.W., Bowers, C.W., et al., 1978, *ApJ*, 492, L83

Martin, D. C., Fanson, J., Schiminovich, D., et al. 2005, *ApJ*, 619, L1

Moos, H. W., Cash, W. C., Cowie, L. L., et al. 2000, *ApJ*, 538, L1
Najarro, F., Herrero, A., Verdugo, E., 2006, *ApSS*, 303, 153
Rogerson, J. B., Spitzer, L., Drake, J. F., et al. 1973, *ApJ*, 181, L97
Shustov, B., Gómez de Castro, A.I., Sachkov, M., 2022, *Proc. of the SPIE*, 12181-104
Thilker, D. A., Bianchi, L Meurer, G Gil de Paz, A Boissier, S. et al. 2007, *ApJS*, 173, 538

Massive Stars Near and Far
Proceedings IAU Symposium No. 361, 2024
J. Mackey, J. S. Vink, & N. St-Louis, eds.
doi:10.1017/S174392132200326X

The CubeSpec space mission: Asteroseismology of massive stars from time-series optical spectroscopy

D. M. Bowman[1], B. Vandenbussche[1], H. Sana[1], A. Tkachenko[1], G. Raskin[1], T. Delabie[2], B. Vandoren[2], P. Royer[1], S. Garcia[1], T. Van Reeth[1] and the CubeSpec Collaboration

[1]Institute of Astronomy, KU Leuven, Celestijnenlaan 200D, 3001 Leuven, Belgium
email: dominic.bowman@kuleuven.be

[2]Arcsec NV, Blijde Inkomststraat 22, 3000 Leuven, Belgium

Abstract. The ESA/KU Leuven CubeSpec mission is specifically designed to provide low-cost space-based high-resolution optical spectroscopy. Here we highlight the science requirements and capabilities of CubeSpec. The primary science goal is to perform pulsation mode identification from spectroscopic line profile variability and empower asteroseismology of massive stars.

Keywords. instrumentation: spectrographs, stars: early-type, stars: oscillations

1. Scientific Context

There is a need to constrain the interior rotation, mixing and angular momentum transport mechanisms of massive stars, as they strongly influence stellar evolution. Asteroseismology – the study of stellar structure from pulsations – is an excellent method for probing the interiors of massive stars Bowman (2020). The ongoing NASA TESS mission is providing high-precision light curves of thousands of massive stars Ricker et al. (2015). Complementary time-series spectroscopy is highly advantageous in fully characterising pulsations: spectral line profile variability reveals the geometry of pulsation modes in terms of spherical harmonics (e.g. Aerts et al. 1994, Briquet et al. 2009). The combination of TESS photometry and CubeSpec spectroscopy will unlock the physics of massive star interiors through forward asteroseismic modelling.

2. CubeSpec Specifications

CubeSpec is a 6U cubesat on schedule to be launched in 2024. The platform includes a Cassegrain telescope with a rectangular primary mirror of $9 \times 19\ \ \mathrm{cm}^2$, and a compact high-resolution echelle optical spectrograph with a resolving power of $R = 55000$ Raskin et al. (2018). Schematics of the platform and optical design are shown in Fig. 1, and on the CubeSpec project website: https://fys.kuleuven.be/ster/instruments/cubespec.

To achieve the primary science goal to assemble time-series spectroscopy for massive-star asteroseismology, CubeSpec has the following science requirements: (i) High resolving power ($R > 50000$) to resolve perturbations to narrow spectral lines in slowly rotating stars; (ii) High signal-to-noise ratio (SNR > 200) spectra to detect small perturbations relative to the median line profile; (iii) Observing cadence of order 90 min to effectively sample the pulsation periods of order several hours; (iv) Short exposure times of order

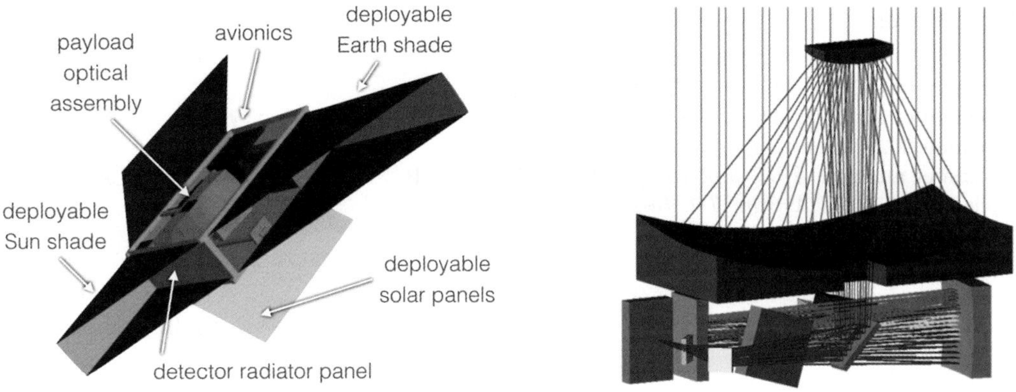

Figure 1. Platform layout (left) and optical design (right) of the CubeSpec space mission.

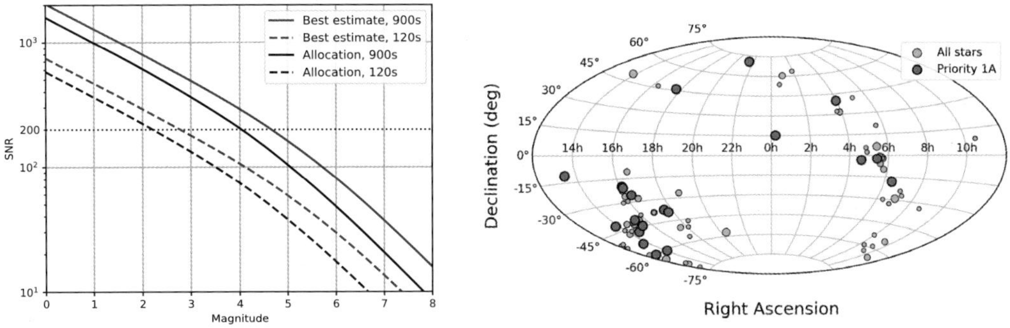

Figure 2. Left: SNR curves for simulated B0 V stars of different V mag brightnesses for conservative instrument performance allocation and best case performance scenario using exposure times of 120 and 900 s. The SNR is estimated at a wavelength of 5000 Å assuming $R = 55000$. Right: Sky distribution of high-priority CubeSpec targets from Bowman et al. (2022).

minutes to avoid pulsational smearing of spectral lines; (v) Mission duration exceeding three months to provide a high frequency precision; and (vi) Optical wavelength range to include the silicon triplet at λλ 4552, 4567, and 4574 Å, which is sensitive to pulsations.

3. Prioritised Target List

The compact design and science requirements of CubeSpec mean that bright (i.e. $V < 4$ mag) pulsating early-type stars are optimal targets to achieve its asteroseismic science case (see Fig. 2). CubeSpec targets β Cep stars, which have pulsation periods between $2 < P < 8$ hr and masses in the approximate range of $6 < M < 30$ $M_{\odot}$ Bowman (2020). A search for pulsating stars with $V < 4$ mag and known spectral types between O8 and B3 (including dwarfs, giants and supergiants) was performed a priori using previous literature studies of photometric and spectroscopic variability and new TESS light curves Bowman et al. (2022). In total, 23 β Cep stars were identified from a sample of 90 stars, with the highest priority targets being slow rotators with large amplitude and long period pulsations. Their sky location is shown in Fig. 2. Therefore, ample pulsating massive stars exist to be targeted by CubeSpec and empower asteroseismology of massive stars.

Supplementary material

To view supplementary material for this article, please visit http://dx.doi.org/10.1017/S174392132200326X.

References

Aerts et al. 1994, *A&A*, 286, 109
Briquet et al. 2009, *A&A*, 506, 269–280
Bowman et al. 2022, *A&A*, 658, A96
Bowman 2020, *Frontiers Astron. Space Sci.*, 7, 70
Raskin et al. 2018, *SPIE proceedings*, 10698, id. 106985R
Ricker at al. 2015, *Astron. Tel. Inst. Sys.*, 1, 014001

Massive Stars Near and Far
Proceedings IAU Symposium No. 361, 2024
J. Mackey, J. S. Vink, & N. St-Louis, eds.
doi:10.1017/S1743921322003167

Polstar : a FUV Spectropolarimetry Mission

Nicole St-Louis[1] and the Polstar consortium

[1]Département de physique, Université de Montréal,
Campus MIL, 1375 ave. Thérèse-Lavoie-Roux, Montréal (Qc), H2V 0B3, Canada
email: nicole.st-louis@umontreal.ca

Abstract. In this paper, we give a brief description of the Polstar instrument, an ultraviolet (UV) spectropolarimetric, Midex-class mission proposed to NASA to study the winds of massive stars as well as interstellar medium and protoplanetary discs topics.

Keywords. polarization, ultraviolet: stars, ultraviolet: ISM, Intrumentation: Polarimeters, techniques: polarimetric, techniques: spectroscopic

---

1. Introduction

Polstar (Scowen et al. 2021) is a Midex-class mission proposed to NASA that will offer the benefit of combining spectroscopy with polarimetry in a wavelength range critical for the study of hot, massive stars, the ultraviolet (UV). This will enable us to probe the intricacies of massive-star winds and in particular their large and small-scale structure, either from magnetic activity, instabilities or fast rotation. Its innovative combination of effective area and time coverage will allow us to study the great diversity of targets necessary to transform our understanding of many science areas in stellar astronomy as well as interstellar medium and proto-planetary disks studies.

Polstar will map stellar wind and magnetospheric structures by uniting time domain, polarimetry and spectroscopy capability in the near- and far-UV (NUV and FUV), which are densely populated with high-opacity resonance lines encoding a rich array of diagnostic information. The science program will be made possible thanks to an instrument that combines advances in high reflectivity UV coatings and delta-doped CCDs with high quantum efficiencies to provide dedicated FUV spectropolarimetry for the first time in 25 years. The planned 3 year mission is 100 times longer that of the *WUPPE* satellite, corresponding to orders of magnitude gains in stellar and interstellar observations.

2. The Instrument

The proposed satellite (Fig. 1) is a combination of the spacecraft, a classic 60-cm all reflective Cassegrain telescope and a 2-channel UV spectropolarimeter that will provide all 4 Stokes parameters. The modulator consisting of a stack of MgF_2 crytals, optimized for the 122–320 nm waveband and placed in a rotary mechanism, as well as the analyzer (a Wollaston prism) (Fig. 2) are placed in line with the output of the telescope near its Cassegrain focus. The modulator serves to isolate the entrance light's polarization and to pass it through the optical system. It is the multiple positions of the rotating modulator that provide the means to fully characterize the input light beam.

The spectropolarimeter will be equipped with 2 distinct channels. The FUV channel, covers 122-200 nm at a resolution of R=30k. This is more than 30 times better than that of WUPPE, with 10 times better effective area, while reaching shorter wavelengths. This will be crucial to gain the ability to access strong lines of species such as N IV and

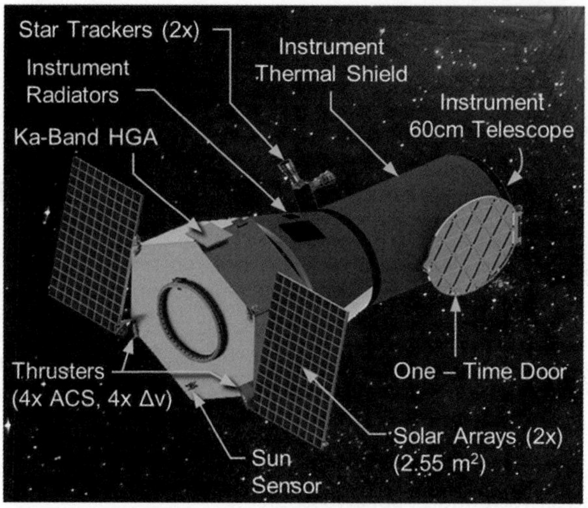

Figure 1. Polstar including the spacecraft, 60-cm Cassegrain telescope and the spectropolarimeter.

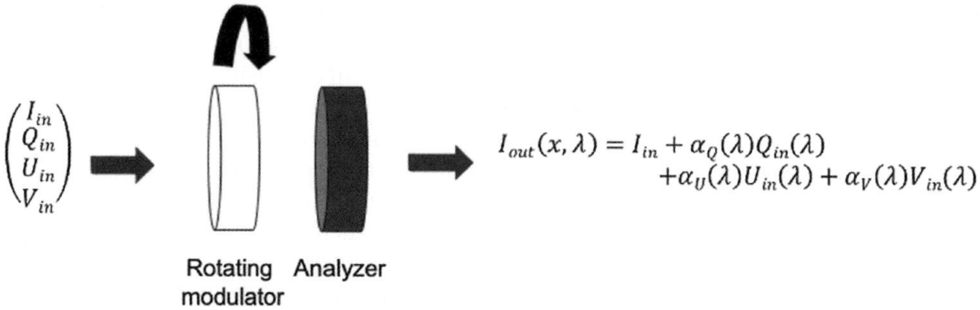

Figure 2. Simplified representation of the principles of the Polstar polarimeter showing the measured intensity (I_{out}) encoding of the input polarization in the 4 Stokes parameters of the incoming beam. Adapted from Figure 4 of Scowen et al. (2021) with permission.

SiIV. The NUV channel covers 122-320 nm at R 140–4k. It will be crucial to enable us to monitor fainter targets at a modest spectral resolution but at a high cadence.

The instrumental polarization stability in both channels is designed to provide signal-to-noise ratios (SNR) corresponding to UV polarimetry precision of $1\text{x}10^{-3}$ per exposure, per resolution element. The precision can be further improved with spectral binning and/or stacking multiple exposures.

3. Main Science Objectives

The science enabled by Polstar is wide ranging and covers 3 broad objectives. The first is *Massive Star Winds* and aims to study how magnetic fields, structured winds and rapid rotation are capable of significantly affecting their evolution. Strong fields will impart magnetic braking on the star and plasma in the magnetosphere will be trapped (ud Doula et al. 2022), altering its mass loss and spin-down rate (Schultz et al. 2022). Clumping and non-spherical winds will alter mass and angular momentum loss (Gayley et al. 2022). The structure of these winds can be probed using selective wind eclipses and colliding winds in binaries (St-Louis et al. 2022). Finally, rapidly rotating stars in massive binaries (Jones et al. 2022) can help understand dynamics at birth and evolution due to conservative or non conservative mass transfer (Peters et al. 2022). Polstar will also address topics in *Interstellar Medium* research, particularly the nature of UV interstellar extinction (Andersson et al. 2022), and *Protoplanetary Disks*, namely the processes governing the assembly and evolution of star and planetary systems (Wisniewski et al. 2022).

Supplementary material

To view supplementary material for this article, please visit http://dx.doi.org/10.1017/S1743921322003167.

References

Andersson, B.G., Clayton, G.C., Doney, K.D. et al. 2022, *Ap&SS*, Topical Collection, in press
Gayley, K.G., Vink, J.S., ud-Doula, A. et al. 2022, *Ap&SS*, Topical Collection, in press
Jones, C.E., Labadie-Bartz, J., Cotton, D.V. et al. 2022, *Ap&SS*, Topical Collection, in press
Peters, G.J.,Gayley, K.G., Ignace, R. et al. 2022, *Ap&SS*, Topical Collection, in press
Schultz, M.E., Casini, R., Cheung, M.C.M. et al. 2022, *Ap&SS*, Topical Collection, in press
Scowen, P.A., Gayley, K., Neiner, C. et al. 2021, *SPIE Conf. Series*, Vol. 11819
Scowen, P.A., Gayley, K., Neiner, C. et al. 2021, *SPIE Conf. Series*, Vol. 11819
St-Louis, N., Gayley, K.G., Hillier, D.J. et al. 2022, *Ap&SS*, Topical Collection, in press
ud Doula, A., Cheung, M.C.M., David-Uraz, A. et al. 2022, *Ap&SS*, Topical Collection, in press
Wisniewski, J.P., Berdyugin, A., Berdyugina, S. et al. 2022, *Ap&SS*, Topical Collection, in press

Author index

Abdellaoui, S. – 490
Abdul-Masih, M. – 157, 168, 184, 230, 267, 307
Acciari, V. A. – 105
Aerts, C. – 238
Alfaro, E. J. – 76
Alonso-Santiago, J. – 410
Andrews, J. – 612
Antoniadis, K. – 447, 454
Anupama, G. C. – 610
Apellániz, J. M. – 236
Araya, I. – 174
Arcavi, I. – 605
Arcos, C. – 174
Arellano-Córdova, K. – 46
Arias, L. – 177, 179
Arnett, W. D. – 353
Arora, B. – 323
Arrieta, A. – 177, 179
Arthur, S. J. – 499

Baade, D. – 307, 332
Balbus, S. A. – 212
Ballone, A. – 302
Banyard, G. – 267, 299, 307
Barbá, R. H. – 76, 292
Barnes, J. R. – 559
Barron, J. A. – 233
Bartlett, E. S. – 97
Bavera, S. S. – 612
Beasor, E. R. – 422
Berdnikov, L. – 460
Berg, D. A. – 26, 46
Berlanas, S. R. – 95, 122
Bernini-Peron, M. – 182
Bestenlehner, J. M. – 145, 163, 184
Bianchi, L. – 279
Bikmaev, I. – 321
Blomme, R. – 95
Bodensteiner, J. – 267, 286, 299, 307
Bonanos, A. Z. – 429, 432, 447, 454
Bonfini, P. – 454
Bordier, E. – 70, 307
Bouret, J. C. – 168
Bouret, J.-C. – 485
Boutsia, K. – 429
Bowman, D. M. – 218, 267, 376, 630
Brands, S. A. – 168, 184
Bressan, A. – 302
Brinchmann, J. – 485
Britavskiy, N. – 236, 429
Brose, R. – 510
Bruzual, G. – 58
Bugnet, L. – 238
Burssens, S. – 236
Butler, K. – 200

Caballero-Nieves, S. M. – 184
Cabezas, M. – 307
Calzoletti, L. – 194
Castellanos, R. – 97, 114
Castro, N. – 82, 151, 286, 559
Celeste, M. – 510
Chandra, P. – 602
Charlot, S. – 58
Chen, X. – 559
Chen, Z. – 26
Chené, A.-N. – 493
Chisholm, J. – 26
Chojnowski, S. D. – 332
Christodoulou, E. – 447, 454
Clark, I. – 46
Clark, J. S. – 97, 114
Colombo, E. – 105
Cooke, J. – 584
Cortina, J. – 105
Costa, G. – 302
Crespo Bellido, P. – 76
Crowther, P. A. – 15, 184
Curé, M. – 174

Daflon, S. – 95
Dallas, M. M. – 82
Davies, B. – 422
de Burgos, A. – 100
de Koter, A. – 168, 184, 286
de la Fuente, D. – 103, 111, 114
De Looze, I. – 615
de Mink, S. – 596
de Mink, S. E. – 286
De Wit, S. – 429, 432, 447, 454
Delabie, T. – 630
Delgado, C. – 105
Dhouib, H. – 238
Díaz, C. – 105
Dorda, R. – 279, 410
Dorigo Jones, J. – 82
Dotter, A. – 612
Dreizler, S. – 151
Driessen, F. A. – 168, 184
Drissen, L. – 515
Drout, M. – 612

Dsilva, K. – 267, 307, 496, 505
Duarte Puertas, S. – 88
Dumontier, C. – 515
Dupret, M.-A. – 241

Eggenberger, P. – 343
Egorov, O. – 513
Eikenberry, S. S. – 103, 111
Ekström, S. – 259, 343, 369
Eldridge, C. – 267
Eldridge, J. J. – 236
Evans, C. J. – 97

Fabry, M. – 267, 305
Faerber, T. A. – 559
Faigler, S. – 313
Farmer, R. – 596
Farnir, M. – 241
Fernández Aranda, R. – 76
Fierro-Santillán, C. – 179
Fierro-Santillán, C. R. – 177
Figer, D. – 194
Fink, D. – 105
Fiori, M. – 105
Folsom, C. P. – 233, 336
Fox, O. – 612
Fragos, T. – 612
Frost, A. J. – 70, 267, 305, 307
Fullard, A. G. – 502

Gagnier, D. – 359
Ganss, R. – 590
García, M. – 114, 184
Garcia, M. – 36, 52, 63, 97
Garcia, S. – 630
Gayley, K. – 493
Gebran, M. – 95
Geen, S. – 184
Georgy, C. – 259, 343, 353, 369
Gies, D. R. – 382
Gilkis, A. – 605
Gómez de Castro, A. I. – 621
Gonneau, A. – 108
González, M. P. – 236
González-Torà, G. – 151
Gordon, K. D. – 382
Gormaz-Matamala, A. C. – 190
Gosset, E. – 95, 248, 273
Götberg, Y. – 596
Gräfener, G. – 184
Green, S. – 510
Grunhut, J. H. – 336
Guberman, D. – 105
Guerrero, M. A. – 88
Guha, S. – 251
Gvaramadze, V. – 460

Habergham-Mawson, S. M. – 590
Hadrava, P. – 307, 332
Haemmerlé, L. – 369, 388
Hamann, W.-R. – 139
Hassan, T. – 105
Hastings, B. – 63, 286
Hawcroft, C. – 168, 184, 267, 305, 307
Haworth, T. J. – 510
Heap, S. R. – 485
Heckman, T. – 26
Heida, M. – 307
Hekker, S. – 311
Henneco, J. – 311, 339
Herrero, A. – 95, 103, 111, 122, 197, 292
Higgins, E. – 391
Higgins, E. R. – 224, 416
Hillier, D. J. – 127
Hirschi, R. – 224, 343, 353, 369
Hoffman, J. L. – 502
Holgado, G. – 236, 292, 388
Hosseinzadeh, G. – 575
Hubeny, I. – 485
Hunt, L. K. – 88
Hutsemékers, D. – 273

Iglesias-Páramo, J. – 88

James, B. L. – 26
James, P. A. – 590
Janssens, S. – 267, 313
Jiménez Martínez, I. – 105
Jiménez-Hernández, P. – 499
Johnson, R. A. – 502
Justham, S. – 596

Kaiser, E. – 343
Kamann, S. – 151
Kaper, L. – 184
Kavanagh, R. D. – 510
Kee, N. D. – 404, 550
Kehrig, C. – 88
Keszthelyi, Z. – 184
Khetan, N. – 584
King, T. – 26
Klapp, J. – 177, 179
Klement, R. – 307, 332
Kluska, J. – 273
Kniazev, A. – 460
Kochukhov, O. – 233, 336
Koumpia, E. – 397
Kovlakas, K. – 612
Kratter, K. M. – 559
Kraus, M. – 251
Krtička, J. – 490
Kuiper, R. – 528, 550
Kurfürst, P. – 490

Langer, N. – 1, 63, 184, 253, 267, 279, 286, 313
Lanthermann, C. – 273, 316
Lanz, T. – 485
Laplace, E. – 311, 339, 596
Larkin, C. J. K. – 108
Le Bouquin, J.-B. – 305
Lecoanet, D. – 218
Lefever, R. R. – 505
Leitherer, C. – 46
Lennon, D. – 279
Lennon, D. J. – 63, 286
Lenoir-Craig, G. – 493, 507
Lobel, A. – 95, 273
Lohr, M. – 114
Lomax, J. R. – 502
Lorenzo, M. – 52
Lumsden, S. L. – 556
Lyard, E. – 105
Lyman, J. – 565

Machuca, N. – 174
Mackey, J. – 510
MacLeod, M. – 302
Maeder, A. – 343
Mahy, L. – 95, 168, 267, 273, 299, 305, 307
Maíz Apellániz, J. – 76, 95
Mangano, S. – 105
Mapelli, M. – 302
Maravelias, G. – 429, 432, 447, 454
Marchant, P. – 267, 286, 305, 313, 496
Marco, A. – 103, 111, 197
Marcolino, W. L. F. – 182
Mariotti, M. – 105
Marston, A. P. – 499
Martin, C. L. – 26
Martinet, S. – 369
Martínez, G. – 105
Martins, F. – 95
Maryeva, O. – 460, 513
Massa, D. – 205
Massa, D. L. – 192
Mathis, S. – 238
Matsukoba, R. – 534
Mazeh, T. – 313
McDonald, M. – 26
McDonald, S. L. E. – 422
Meakin, C. – 353
Mehner, A. – 437
Mérand, A. – 70
Merand, A. – 307
Meynet, G. – 259, 343, 369
Milisavljevic, D. – 612
Milone, A. – 286
Mirzoyan, R. – 105
Misra, D. – 612
Moe, M. – 82
Moffat, A. F. J. – 493, 507
Moiseev, A. – 513
Molina Lera, J. A. – 76
Morel, T. – 95
Moriya, T. – 580
Motch, C. – 319
Moutzouri, M. – 510
Muñoz-Sánchez, G. – 114
Muñoz-Sanchez, G. – 454
Munoz-Sanchez, G. – 432, 447
Murphy, A. S. – 353

Najarro, F. – 52, 97, 114, 194
Naletto, G. – 105
Nandal, D. – 369
Nayana, A. J. – 602
Nazé, Y. – 236, 248, 319, 502
Negueruela, I. – 103, 111, 197, 279, 410
Nikolaeva, E. – 321
Njoh Ekoume, T. – 105
Noels, A. – 241
Nordsieck, K. H. – 502

Oey, M. S. – 82, 559
Ohlmann, S. T. – 212
Oliva, G. A. – 528
Omukai, K. – 521
Oskinova, L. – 479
Oskinova, L. M. – 192
Oudmaijer, R. D. – 397, 556
Owocki, S. P. – 261

Pablo, H. – 507
Pakmor, R. – 212
Pandey, J. C. – 323
Pantaleoni González, M. – 76
Panzera, T. – 502
Parsons, T. N. – 192
Patrick, L. – 197
Patrick, L. R. – 97, 103, 111, 114, 279, 410
Pauli, D. – 116
Pauwels, T. – 326
Payen-Sandoval, A. – 177, 179
Peimbert, A. – 334
Phillips, G. D. – 82
Pledger, J. L. – 590
Podsiadlowski, P. – 212, 339, 596
Polo, M. – 105
Poniatowski, L. – 505
Prinja, R. – 205
Prinja, R. K. – 192
Produit, N. – 105

Przybilla, N. – 151, 200
Puls, J. – 168, 184, 194, 590

Quintana, A. L. – 119

Rainot, A. – 326
Ramachandran, V. – 329
Ramos-Larios, G. – 88
Raskin, G. – 630
Rauw, G. – 241, 248, 319
Reed, B. C. – 76
Regan, J. A. – 539
Reggiani, M. – 273, 307, 326
Renzo, M. – 596, 612
Rieutord, M. – 359
Rivinius, T. – 307, 332
Rizzuti, F. – 353
Rocha, K. – 612
Rodríguez, J. J. – 105
Röpke, F. K. – 212, 339
Rosu, S. – 241, 248
Roth, M. M. – 151
Royer, P. – 630
Rubio-Díez, M. M. – 194
Rübke, K. – 197
Ruest, M. – 515

Sabhahit, G. – 391
Sabhahit, G. N. – 416
Sahu, D. K. – 610
Sana, H. – 70, 168, 184, 267, 273, 286, 299, 305, 307, 313, 326, 496, 630
Sánchez Arias, J. P. – 251
Sander, A. – 224
Sander, A. A. C. – 182, 416, 473, 505
Sansom, A. E. – 590
Savaglio, S. – 584
Schaefer, G. H. – 382
Schneider, F. – 596
Schneider, F. R. N. – 184, 212, 311, 339
Schootemeijer, A. – 63, 279, 286
Schürmann, C. – 63, 253, 313
Schweizer, T. – 105
Selman, F. – 307
Sen, K. – 267
Shahaf, S. – 313
Shenar, T. – 184, 267, 299, 307, 313, 465, 496, 505
Shenton, R. G. – 556
Shepard, K. – 332
Shimonishi, T. – 534
Shishkin, D. – 607
Simón-Díaz, S. – 236, 292, 388
Simon-Díaz, S. – 100
Simon-Diaz, S. – 197
Sixtos, A. – 58, 334
Smith, L. – 58
Smith, M. A. – 319
Smith, N. – 612
Sota, A. – 76
Springel, V. – 212
St-Louis, N. – 493, 507, 515, 633
Stacey, E. – 336
Stark, D. P. – 26
Sterken, C. – 248
Suhr, M. – 584
Sundqvist, J. O. – 168, 184, 194
Szalai, T. – 615

Tabernero, H. – 197
Tanaka, K. E. I. – 534
Teja, R. S. – 610
Thilker, D. – 279
Tkachenko, A. – 630
Toalá, J. A. – 499
Todt, H. – 505
Traficante, A. – 194
Trager, S. C. – 108
Tramper, F. – 256, 429, 432, 447, 454
Tsiatsiou, S. – 259

ud-Doula, A. – 261
Urbaneja, M. A. – 100, 151

van Dyk, S. – 612
Vandenbussche, B. – 630
Vandoren, B. – 630
Van Reeth, T. – 218, 238, 267, 630
Vargas-Salazar, I. – 559
Vartanyan, D. – 596
Verro, K. – 108
Vetter, M. – 339
Vílchez, J. M. – 88
Vink, J. S. – 184, 224, 391, 416

Wade, G. A. – 233, 336
Walter, R. – 105
Wang, C. – 253, 286, 299
Weßmayer, D. – 200
Wei, D. – 339
Weiler, M. – 76
Williams, B. – 612
Winch, E. – 391
Wofford, A. – 58, 334
Wright, N. J. – 119
Wunderlich, C. – 105

Xing, Z. – 612
Xu, X.-T. – 253, 286

Yang, M. – 429, 432, 447, 454
Yarovova, A. – 513

Zampieri, L. – 105
Zapartas, E. – 454, 612
Zapartas, M. – 429, 447
Zargaryan, D. – 510
Zargó, J. – 177
Zhang, J. – 584
Zhang, Y. – 534
Zsargo, J. – 179
Zsíros, S. – 615